# Proceedings of the 52nd INDUSTRIAL WASTE CONFERENCE

## May 5–7, 1997

## Purdue University
## West Lafayette, Indiana

**The Purdue Industrial Waste Conference is under the direction of:**

**The School of Civil Engineering**
**Continuing Education, Purdue University**

**in cooperation with the**

**Indiana Department of Environmental Management**
**Indiana Department of Natural Resources**
**Indiana Environmental Management Board**
**Indiana Pollution Prevention & Safe Materials Institute of Purdue University**
**Indiana Section of American Society of Civil Engineers**
**Indiana Section of the American Water Works Association**
**Indiana State Board of Health**
**Indiana Stream Pollution Control Board**
**Indiana Water Pollution Control Association**

CRC Press
Taylor & Francis Group
Boca Raton London New York

CRC Press is an imprint of the
Taylor & Francis Group, an **informa** business

First published 1998 by Ann Arbor Press

Published 2019 by CRC Press
Taylor & Francis Group
6000 Broken Sound Parkway NW, Suite 300
Boca Raton, FL 33487-2742



ISBN 13: 978-1-57504-098-1 (hbk)



**Visit the Taylor & Francis Web site at**
**http://www.taylorandfrancis.com**

**and the CRC Press Web site at**
**http://www.crcpress.com**

ISSN 0073–7682

# PREFACE

The 52nd Purdue Industrial Waste Conference was co-sponsored by Purdue's School of Civil Engineering and the Indiana Clean Manufacturing Technology and Safe Materials Institute. This year's program included 18 sessions which, along with the Plenary Session, featured a total of 72 technical speakers. Sixty written manuscripts were, in turn, submitted for inclusion within the Proceedings publication. The following speakers provided featured presentations during the course of this 52nd annual event:

- Plenary #1: Tom Neltner, Assistant Commissioner, IDEM
  *Waste Management and Pollution Prevention for Indiana*
- Plenary #2: Bruce Carter, President, Carter and Associates, Inc.
  *Innovative Approaches to Environmental Management*
- Luncheon, Monday: F. Tom Turpin, Professor
  *Leaders Can't Be Cowards*
- Luncheon, Tuesday: Morgan J. Burke, Purdue Athletic Director
  *The True Value of the Intercollegiate Athletic Experience*

Commensurate with Purdue's cutting-edge, high-tech transition into the Twenty-First Century, our 52nd PIWC program also provided an innovative, Web-based access format. Interested 'browsers' should check out the following special URL - *http://ce.ecn.purdue.edu/-piwc*. Along with featuring digital copies of this year's entire paper inventory, this new PIWC home page also provides an extremely powerful search engine covering the full Table of Contents for this event's entire 52-year history. And as a courtesy service to our interested browsers, we would be happy to assist you with copying any of our historic papers!

Yet another unique feature of this 52nd program was our development of a special Web-based "Student Home Page" contest. This creative venture, sponsored graciously by CH2M-HILL Inc., triggered inquiries from participants and interested browsers which literally spanned the globe. In turn, the following students received special awards for their contribution to this effort:

$500 - 1st Place Winner: Mike McAdams
Civil and Environmental Engineering, University of Texas
*Optimization of LNAPL Free Product Recovery*

$300 - 2nd Place Winner: Kowei John Hou
School of Civil Engineering, Purdue
*Prototype of Hazardous Material Transport and Environment Monitoring System: An Application of the General Corridor Alignment Problem*

$200 - 3rd Place Winner: Charity M. Deyton
Dept. of Civil and Environmental Engineering, Tennessee Technological University
*Industrial Waste Minimization*

$100 - 4th Place Winner: James F. Russell
Bioresource Engineering Department, Rutgers University
*Walhalla Technologies Incorporated*

$50 each - Honorable Mention Winners:
Amit Gupta, Civil and Environmental Engineering, University of Cincinnati
Adjo Amekudzi, Dept. of Civil and Environmental Engineering, Carnegie Mellon University

Lastly, these proceedings have been subdivided into 14 major sections, as follows:

- Sections 1-8: Biological
- Section 9-10: Physical-Chemical
- Section 11: Oil & Petrochemical Wastes
- Section 12: Management and Reuse Strategies
- Section 13: International Activities
- Section 14: Pollution Prevention

A full keyword index covering the past 10 years is also appended! However, the new Web-based search engine mentioned earlier provides a considerably more robust means of checking the entire coverage of our past proceedings, starting from Professor Don Bloodgood's first program in 1944!

James E. Alleman, *Conference Chairman*
Rita Butz, *Conference Secretary*

# CONTENTS

## Section One. Biological: Aerobic Thermophilic Systems



## Section Two. Biological: Inhibition and Toxicity



## Section Three. Biological: Bioremediation and Biotechnology



## Section Four. Biological: Biodegradation and Treatability

**Section Five. Biological: High Strength Wastes**



**Section Six. Biological: Animal and Dairy Wastes**



**Section Seven. Biological: Phenolic Waste Treatment**



**Section Eight. Biological: Anaerobic Systems**



**Section Nine. Physical-Chemical: General**

# Proceedings of the 52nd
# INDUSTRIAL WASTE CONFERENCE
# May 5–7, 1997

Purdue University
West Lafayette, Indiana

# Section One
# BIOLOGICAL: AEROBIC THERMOPHILIC SYSTEMS

## 1 DESIGN AND OPERATIONAL CONSIDERATIONS FOR THERMOPHILIC AEROBIC REACTORS TREATING HIGH STRENGTH WASTES AND SLUDGES

**Alan F. Rozich**
Rozich Technologies, Inc.
Exton, Pennsylvania

**Richard J. Colvin**
ERM
Exton, Pennsylvania

### ABSTRACT

Thermophilic aerobic treatment systems offer unique advantages for treatment of high strength waste streams and for sludge reduction. These systems combine the best features of aerobic systems which have rapid biodegradation kinetics and anaerobic systems which have low cell yields. Application of these systems can result in substantial economic benefit and cost savings by reducing off-site disposal costs for many industrial facilities. This chapter will overview modified thermophilic aerobic systems, discuss various applications, and review case histories. The focus of the chapter will be on design and operational considerations for these systems. These considerations include reactor engineering, process kinetics, and essential ancillary support equipment considerations.

### INTRODUCTION

The dual requirements of environmental compliance and cost reduction are prompting environmental and operations managers to consider new approaches for waste treatment. Residuals which are generated as a result of waste treatment activities are often a major consideration when implementing treatment of target streams. For example, when treating organic waste streams, the ultimate disposal of sludges can frequently represent the largest operational expense. Future liability considerations and the current regulatory climate often make the selection of treatment processes "end-product driven." Owners and operators must carefully evaluate potential treatment processes in order to ensure that a residual by-product will not be produced that will have substantive cost or regulatory implications or both.

The purpose of this paper is to present the conceptual and operational application of an innovative biological treatment process for treating high strength hazardous organic wastes, sludges, and difficult-to-degrade waste streams. Thermophilic aerobic biological treatment technology offers an attractive, economical alternative to conventional aerobic and anaerobic treatment processes. These aerobic biological processes function at temperatures from 110°F to 160°F. Proprietary versions of this technology can minimize or virtually eliminate the production of organic residuals requiring off-site disposal. This offers an extremely attractive and cost-effective alternative to incinerating a number of organic waste streams. This paper will present an overview of waste treatment technology, an overview of engineering thermophilic aerobic biological treatment, a discussion of characteristics of treatment situations which would suggest the application of thermophilic aerobic treatment systems, and case histories involving full-scale applications of

---

*52nd Purdue Industrial Waste Conference Proceedings*, 1997, Ann Arbor Press, Chelsea, Michigan 48118. Printed in U.S.A.

the thermophilic aerobic biological treatment of difficult waste streams. The case histories will describe both technical and economic considerations.

## OVERVIEW OF WASTE TREATMENT TECHNOLOGY

Current approaches for waste treatment are segmented into three major classes of waste treatment technology:

- biological
- physical/chemical
- thermal (incineration)

When practically applicable, biological processes are often preferred because of economical considerations. Physical/chemical processes tend to be expensive and are prone to generating large quantities of sludge which can be difficult to handle and costly for off-site disposal. Incineration processes provide a total destruction technology but can be very costly and have substantive permitting issues, particularly under the new Clean Air Act. Incineration systems are also frequently plagued with siting problems because of public opposition and pressure. Biological processes are generally classified as either aerobic or anaerobic. The general thinking on these two different systems is that aerobic systems tend to be more resilient in being able to handle a larger variety of waste streams and more difficult-to-degrade waste streams, but that anaerobic systems produce much less sludge. Anaerobic systems also have lower energy requirements because they do not require aeration. Thermophilic aerobic systems combine the best features of both anaerobic and aerobic processes. Thermophilic aerobic systems still require aeration but have the kinetic resilience of conventional aerobic systems while having the low sludge production characteristics of anaerobic systems.

Experience has indicated that thermophilic aerobic systems can handle the same range of waste types that aerobic systems can, and more. The high temperatures employed in thermophilic systems tend to facilitate the biodegradation of waste streams that contain components which are more soluble at higher temperatures (e.g., fats, oils, and grease [FOG] wastes). It has also been observed that aerobic thermophilic processes are able to deal with waste streams with high salt concentrations which are problematic for anaerobic systems.

## ENGINEERING THERMOPHILIC AEROBIC BIOLOGICAL TREATMENT SYSTEMS

Thermophilic aerobic biological treatment systems are technological cousins of a more familiar technology, composting. Composting is usually an aerobic process where air is added to a pile of organic matter (e.g., yard wastes or other materials) in order to stimulate microbial degradation of the target organics. As organic degradation occurs, the heat from the exothermic reaction is allowed to accumulate in the reactor, the compost pile, and the pile heats up. Thermophilic microbes in the compost pile are responsible for the degradation of the target organics. With thermophilic aerobic treatment systems, the biomass and wastes are fluidized much like the activated sludge process. The aeration system and reactor are designed in a way to enable the reactor to accumulate heat and operate at thermophilic temperatures (110°F to 160°F). These systems were originally called "fluidized composters." One thermophilic system that has been used for sludge treatment and pasteurization is the Autothermal Aerobic Digestion (ATAD) process. Proprietary modifications of these systems which incorporate an integral chemical treatment system have demonstrated the ability to treat high strength organic sludge and waste streams and achieve over 90% reductions in organic sludge.

There are some key issues when utilizing thermophilic aerobic biological treatment systems. Since thermophilic systems have a tendency to be applied for the treatment of high strength wastes, careful attention must be given to the design of the oxygen transfer system. Many practitioners profess the mythology that it is difficult, if not impossible, to design oxygen transfer systems which are capable of meeting the high COD loading rates that thermophilic aerobic systems handle. This contention is based on the perception that it is difficult to transfer oxygen in high temperature fluids because the driving force for oxygen transfer is reduced due to the reduced saturation value for oxygen at higher temperatures. Field experience has, however, shown that oxygen transfer can be achieved at high temperatures with results similar to that achieved with lower temperature systems. The key is to focus on the oxygen transfer requirements and to provide a sufficient cushion in the design of the aeration system (more reactor depth or a more aggressive transfer system) to meet oxygen transfer needs.

Foaming can also be an issue with thermophilic aerobic biological treatment systems. A foaming tendency exists because of two combined factors. One factor is that thermophilic aerobic biological treatment systems will tend to operate at high biomass concentrations (10,000 to 30,000 mg/L of suspended solids). The maintenance of high solids concentrations also results in the accumulation of relatively high levels of microbial surfactants which are foaming agents. This situation along with the fact that relatively high levels of air are introduced into a thermophilic aerobic biological reactor creates the potential for foaming and associated operational problems (e.g., foam-overs, etc.). Foaming can be addressed by the use of antifoaming agents or the use of mechanical foam breakers. Another foam control strategy involves addressing one of the root causes of foaming by controlling surfactant accumulation in the thermophilic aerobic biological reactor. One proprietary modified thermophilic aerobic biological treatment system incorporates a chemical treatment step which has this process feature. The best overall approach for foam control is to employ an integrated control strategy.

## APPLICATION CONSIDERATIONS FOR THERMOPHILIC AEROBIC BIOLOGICAL TREATMENT SYSTEMS

Thermophilic aerobic treatment systems can fit a large niche for treatment applications. There are key criteria which can be utilized for determining whether it makes sense to consider the application of a thermophilic aerobic treatment system as part of an overall scheme to address a particular treatment situation. Everything ultimately comes to cost. It should be stressed that "cost" must include all cost items involved with a particular system including capital, operational, residuals disposal and expenses associated with ancillary support equipment. Process performance risk or the relative tendency for a biological process to endure or resist process upset is not as easy to quantify but is often a consideration. For example, if a waste is subject to variation in either qualitative or quantitative loading characteristics, the level of process performance risk can also represent a significant cost.

Key characteristics of potential treatment situations which would warrant the potential application of thermophilic aerobic treatment systems include:

- residuals disposal
- high strength/low flow streams
- high salt concentration streams
- toxic or hazardous organic waste streams
- resistance to shock loads

These characteristics are discussed below.

### Residuals Disposal

The high cost for the disposal of residuals is a strong driver for prompting the consideration of thermophilic aerobic technology. Off-site organics disposal can be due to a sludge stream or waste stream. Many manufacturers generate either high strength sludge or waste streams that can only be handled by expensive destruction technologies such as incineration. Modified thermophilic aerobic treatment systems can be implemented which can virtually destroy these types of organics. The key criteria is to determine what the current or projected off-site disposal costs are and what the implementation costs are for a modified thermophilic aerobic treatment system. If off-site disposal costs are $300,000 a year or more, it is appropriate to consider applying a modified thermophilic aerobic treatment system as a cost-savings measure. It should be noted that these systems can be implemented as either a "greenfield" system or using a retrofit approach or some combination of these approaches.

### High Strength Organic Waste Streams

These streams which can either be simply high strength organic streams or have onerous characteristics such as having high salt or high concentrations of toxic or hazardous organic compounds can be ideal candidates for modified thermophilic aerobic treatment systems. Alternate treatment strategies for "easily degraded" high strength streams usually consist of anaerobic treatment technologies. However, if other issues are represented such as turn-up and turn-down capability, the presence of potential inhibitory agents, or sulfates, then a modified thermophilic aerobic treatment system may warrant consideration.

For difficult organic streams where the only other option may be incineration, thermophilic aerobic treatment systems should be in serious consideration because they bring the low sludge production characteristics of anaerobic systems along with the kinetic resilience of an aerobic system. Source treatment applications using modified thermophilic aerobic treatment are appearing to warrant more serious economic consideration as a means to increase the capacity of existing systems while radically reducing off-site sludge disposal issues.

### Resistance to Shock Loads

One of the myths that seems to have been promulgated about thermophilic aerobic treatment systems is that they are unstable. It appears that much of this belief is rooted in the fact that thermophilic systems have a tendency toward foaming. This tendency is also associated with the fact that thermophilic reactors are often high biomass concentration systems which inherently carry large concentrations of biosurfactants which result in foaming. The tendency toward foaming and kinetic capability are two different issues. It has been determined with thermophilic respirometry that a thermophilic biomass has the same or superior biokinetic capability on target wastes as a mesophilic aerobic biomass. Field experience has shown that thermophilic units are extremely resilient to shock loads. For example, a modified thermophilic system which was treating a combined waste activated sludge/FOG (fats, oils, and grease) endured a shock load which lowered reactor pH to 5 for over 14 hours. The reactor foamed badly (as would any biological system which had a 2.5 pH unit shock) but, after adjusting the pH back to 7.5, the unit was able to be put back online within 12 hours at full loading. Field experience has shown that thermophilic aerobic systems are not delicate and as hardy or hardier than mesophilic aerobic treatment systems when dealing with shock loads.

## CASE HISTORIES

### Thermophilic Aerobic Treatment of High Strength Ground Water

An investigation of a hazardous waste site found groundwaters which contained high strength organic acids and benzene with chemical oxygen demand (COD) values ranging from 80,000 to 200,000 mg/L. Ancillary hydrogeological investigations determined that the site contained in excess of 5,000,000 gallons of this material. Estimated costs for off-site disposal were over $1 per gallon. Other geological data and plant production records indicated that the site may contain as much as 40,000,000 gallons. These facts suggested that on-site treatment of the material followed by discharge of the treated groundwater to a local publicly owned treatment works (POTW) might be the most cost-effective option.

A review of treatment and disposal technologies showed that on-site aerobic treatment would prove most cost-effective. Anaerobic technologies were shown not to be effective and to be too costly for this application primarily because of the presence of salts and sulfates. Because the waste has such a high COD and the off-site disposal of excess biological sludges would be so expensive, it was decided to implement a modified thermophilic aerobic treatment unit in order to treat the waste and minimize the off-site disposal of waste sludges.

System startup and operation went relatively smoothly from a kinetic perspective. After two months of operation, the system was receiving a feed chemical oxygen demand (COD) of 30,000 mg/L and producing an effluent COD of 200 mg/L which is a COD removal rate of 99%. Most of the operating and startup problems were related to mechanical issues as opposed to process issues. Foaming and foam-overs tended to be a persistent problem, particularly during startup. Other experience has indicated that the use of an integrated chemical treatment step for controlling sludge production in the system along with other proactive strategies will adequately address foaming issues. The high COD removals in combination with the prospect of little or no sludge wasting and off-site disposal of residuals indicates that this modified thermophilic system is performing very well from both environmental and economic perspectives.

### Thermophilic Aerobic Treatment of Dairy FOG Wastes and Waste Activated Sludge

A food products company operates a dairy which produces milk, milk products, and various fruit juices. Wastewater generated by the dairy operations is discharged to a treatment plant consisting of equalization, primary treatment to remove settleable and flotable solids, and activated sludge treatment. Solids from the treatment systems are sent to a thickener for storage and thickening prior to implementing off-site disposal. Flows to the treatment system average approximately 75,000 gallons per day with influent chemical oxygen demands (COD) averaging around 5,000 mg/L. The dairy wastewater treatment plant had a history of odor problems. Expected increases in production prompted the dairy owners to undertake a large expansion of the dairy. As part of the expansion, the dairy desired to address the chronic odor problems which had been plaguing the wastewater treatment facility.

The modifications to the treatment system resulted in implementing a dissolved air flotation (DAF) system to remove FOG in order to relieve the organic loading to the activated sludge system. Both FOG and waste activated sludge were removed in the DAF float. Since storage and hauling of the float produced occasional odor problems, the dairy desired to implement a modified thermophilic aerobic treatment unit to destroy potential odor-producing materials as they are generated and to reduce off-site sludge disposal requirements.

Treatability testing indicated that a modified thermophilic treatment system could successfully treat the combined FOG/waste activated sludge float stream. System implementation of a modified thermophilic aerobic process is being effected by retrofitting an existing 30,000 gallon thickener with aeration equipment, incorporating foam control measures, and using an ancillary storage tank for chemical treatment and for foam containment, if needed. The system has been operating for almost a year and off-site residuals disposal costs have been cut by 95%.

### Thermophilic Aerobic Treatment of High Strength Chemical Manufacturing Sludges

A specialty chemical manufacturer is spending over $1 million a year for the off-site disposal of organic sludges that are produced from manufacturing operations. Plant operations staff decided to evaluate the application of a thermophilic aerobic biological treatment system for reducing the levels of materials requiring off-site disposal. Bench-scale treatability testing, concept engineering, and economical analysis efforts were initiated. Treatability and pilot work showed extremely positive results. Additionally, the economic analysis showed that implementation of a modified thermophilic aerobic biological treatment system will realize an internal rate of return (IRR) of over 100% on this investment. Since the company's IRR hurdle rate is 30%, system implementation is heavily driven by the economics. (We have seen similar economic scenarios in many other situations regarding the implementation of modified thermophilic aerobic biological treatment systems.) The project is currently in the engineering phase with system installation and start-up slated for the summer and fall of 1997, respectively.

## SUMMARY

The purpose of this paper was to present the conceptual and operational application of an innovative biological treatment process for treating high strength organic wastes, sludges, and difficult-to-degrade waste streams. Thermophilic aerobic biological treatment technology offers an attractive, economical alternative to conventional aerobic and anaerobic treatment processes. These systems also have a number of advantages over other conventional technologies used for the treatment of organic wastes. These systems are highly flexible and applicable in situations where treatment and residuals minimization are a priority. Experience has indicated that this technology has significant economic advantages over other technologies used for the treatment of difficult-to-degrade wastes and sludges. Proprietary modifications of these systems which incorporate an integral chemical treatment system have demonstrated the ability to treat high strength organic sludge and waste streams and achieve over 90% reductions in organic sludge destruction.

# 2 EVALUATION OF THE OXYGEN TRANSFER CHARACTERISTICS OF AERATION SYSTEMS OPERATING IN THERMOPHILIC TEMPERATURE RANGES

**Brion F. Wynn**

**Walter V. Linden**, P.E.

**Robert C. Backman**

WATERLINK Mass Transfer Systems, Inc.
Fall River, Massachusetts 02723

## ABSTRACT

There has been a large volume of research on oxygen transfer which has helped to define factors affecting the design of aeration systems. There are, however, many misconceptions regarding the effects that high liquid temperatures have on the oxygen transfer process in both clean water and biological reactors. This chapter will discuss the effect of various mesophilic and thermophilic temperatures on dissolved oxygen saturation concentrations (beta factor), the overall mass transfer coefficient (alpha factor), and the temperature correction coefficient (theta factor). The operating performance of jet aerators over the past 12 years in both thermophilic activated sludge processes and autothermal thermophilic aerobic digesters will be discussed. In addition, the results of a performance evaluation of jet aerators and membrane diffusers will be presented. This comparison will show the oxygen transfer characteristics of the two devices operating in different temperature ranges.

Ongoing research by Mass Transfer Systems is being carried out to further define the effect of various components of the aerobic treatment on oxygen transfer performance. This is a continuation of previous work conducted in conjunction with Northeastern University[1,2] and now being continued in association with Purdue University.[6]

## INTRODUCTION

Over the past 12 years, various full scale oxygen transfer systems in industrial effluent biological processes or thermophilic aerobic digesters have exhibited excellent performance in terms of mass transfer per unit kW. These aeration systems are typically installed in activated sludge processes or autothermophilic aerobic digesters (ATADs) in the Pharmaceutical and Food Processing industries. In addition to exhibiting excellent oxygen transfer performance, these processes have other similar features, such as: high COD concentrations, thermophilic temperature ranges, high TDS concentrations, tendency to foam (i.e., low surface tension), and very low sludge production rates while operating at conventional F/M ratios.

It is believed that the relationship between thermophilic temperature, foaming, and low sludge production are related. It would appear that the increased biological activity in these temperature ranges promotes a high degree of endogenous respiration resulting in the breakdown of bacterial cells into simpler components. In particular, the breakdown of protein results in the release of extracellular enzymes which tend to cause foam. The effect that this reduction in surface tension, in conjunction with high TDS concentrations, has on oxygen transfer performance is the basis for this ongoing research.

*52nd Purdue Industrial Waste Conference Proceedings*, 1997, Ann Arbor Press, Chelsea, Michigan 48118. Printed in U.S.A.

## BACKGROUND

Prior to attempting full-scale testing, it was determined that some of the basic relationships that temperature, TDS concentration, and surface tension exhibit with respect to oxygen transfer parameters should be investigated at pilot scale. Most of the data were collected using a jet aerator. At the end of the study, some of the testing was duplicated using a fine/medium bubble diffuser. A portion of the testing with the jet aerator was conducted to compare to previous research performed by others.[1,5]

Lewis and Whitman[3] are credited with the advancement of the most widely used and accepted theory describing the absorption of gas by a liquid. Oxygen transfer in wastewater treatment involves the physical mass transport across a two-film layer consisting of a gas film and a liquid film. The oxygen transfer rate can be expressed by the following equation:

$$N = K_L a\,(C_s - C_l) \qquad (1)$$

where:
- $N$ = Mass of oxygen transferred per unit time
- $K_L$ = Liquid film mass transfer coefficient
- $a$ = Interfacial area per unit volume
- $C_s$ = Saturation concentration of oxygen at the gas/liquid interface
- $C_l$ = Concentration of oxygen in the bulk liquid

The value $K_L a$ is considered as the overall mass transfer coefficient and includes the liquid film coefficient and the interfacial area per unit volume. This "combined" coefficient is used because it is impractical to measure the liquid film coefficient or the interfacial area, and $K_L a$ can easily be determined for nonsteady state oxygenation tests by a computer using nonlinear regression.

The overall oxygen transfer process has been hypothesized to occur in three phases:

1. transfer of oxygen molecules to the liquid surface resulting in an equilibrium or saturation condition at the gas/liquid interface
2. the passing of oxygen molecules through the liquid film into bulk solution by molecular diffusion
3. oxygen is mixed in the bulk solution by diffusion and convection.

Aeration equipment suppliers in the United States rate the performance of their equipment by its oxygen transfer efficiency in clean water (SOTR), assuming 0.0 mg/L dissolved oxygen, 1.0 atmosphere pressure, and a temperature of 20°C. To translate from process conditions to standard conditions the following equation is used:

$$SOR = \frac{AOR}{\alpha * \dfrac{\left((\beta^* C_{walt}\ D_c) - C_L\right)}{(C_{20}{}^* D_c)} * \theta^{(T_w - 20)}} \qquad (2)$$

where:
- $SOR$ = Standard Oxygen Requirement, (lb/hr)
- $AOR$ = Actual (Process) Oxygen Requirement, (lb/hr)
- $\alpha$ (Alpha) = Ratio of oxygen transfer coefficient ($K_L a$) of the wastewater to that of tap water
- $\beta$ (Beta) = Ratio of oxygen saturation of the wastewater to that of tap water
- $C_{walt}$ = Surface saturation dissolved oxygen concentration in clean water at the wastewater temperature (Tw) and basin elevation, mg/L
- $D_C$ = Depth Correction Factor, $D_c = \dfrac{\text{Water Depth (ft.)}}{100} + 1$
- $C_L$ = Residual Dissolved Oxygen concentration
- $C_{20}$ = Surface saturation dissolved oxygen concentration of air in clean water at 20°C and 760 mm Hg (9.09 mg/L per ASCE)
- $\theta$ (Theta) = Temperature correction coefficient
- $T_w$ = Wastewater Temperature (°C)

Wastewater contaminants, temperature, dissolved oxygen concentration, type of aeration device, and turbulence all affect oxygen transfer rate. The three parameters that account for the impact of these influences on oxygen transfer are alpha ($\alpha$), beta ($\beta$), and theta ($\theta$).

Alpha ($\alpha$), the ratio of process water to clean water volumetric mass transfer coefficient, is one factor that has received a lot of research and is still widely misunderstood within the industry. It has been well documented that alpha is quite different for different aeration devices. With certain aeration devices, wastewater contaminants depress the oxygen transfer in relation to clean water performance and for some devices it significantly enhances oxygen transfer. The impact of wastewater contaminants can be attributed to dissolved salts and surface active agents. The calculation of alpha is represented by the following equation:

$$\alpha = \frac{K_L a_{waste}}{K_L a_{clean}} \tag{3}$$

where: $\alpha$ = Alpha
$K_L a_{waste}$ = Overall mass transfer coefficient of waste (process) water
$K_L a_{clean}$ = Overall mass transfer coefficient of clean (potable) water

Estimation of $K_l a$ is calculated by using the nonlinear regression method detailed in ASCE *Standard for the Measurement of Oxygen Transfer in Clean Water* (1984), which compares D.O. (mg/L) versus time (min). These results were used to calculate the alpha values reported in Table II.

Beta ($\beta$) is defined as the ratio of the saturation dissolved oxygen concentration of actual process water to that in clean water. Temperature, barometric pressure, and dissolved solids are the key variables that affect the appropriate beta value. Generally, the aeration device is not considered to have an impact on the beta factor. The calculation of beta is represented by the following equation:

$$\beta = \frac{C^*_{waste}}{C^*_{clean}} \tag{4}$$

where: $\beta$ = Beta
$C^*_{waste}$ = Saturation concentration of oxygen in waste (process) water, mg/L
$C^*_{clean}$ = Saturation concentration of oxygen in clean (potable) water, mg/L

The theta ($\theta$) factor is used to relate the overall mass transfer coefficient at a specific temperature (typically standard conditions, 20°C) to that at a different temperature. Temperature strongly affects aeration in a variety of ways. The greatest effect is on saturation dissolved oxygen concentration. The effect of saturation levels is not included in the theta factor, but handled in the conversion equation from AOR to SOR. Temperature correction of the mass transfer coefficient is achieved by applying the following equation:

$$K_L aT = K_L a_{20} \theta G^{(T-20)} \tag{5}$$

where: $K_L aT$ = Mass transfer coefficient at temperature, T
$K_L a_{20}$ = corrected to Standard Conditions (Barometric Pressure = 1 atm, water temperature = 20°C)
$\theta G$ = Geometric temperature correction coefficient, theta
$T$ = process water temperature, °C.

This geometric equation is the one most commonly used. The correction for the effect of temperature is empirical and attempts to lump all possible factors influencing the transfer coefficient, such as viscosity, surface tension, diffusivity of oxygen, etc. This approach has produced a wide range of correction factors (from 1.05 to 1.01), with 1.024 as the typical value used. This broad range in estimated values concludes that the current assumptions are inadequate.

An alternate method postulated tries to correlate the effect of temperature on $K_L a$ to the level of turbulence generated by the aeration device.[4] Since it has be shown that different aeration devices have different transfer characteristics in the presence of surface active agents, it is plausible that each type of aeration device has a different temperature correction based on temperature and turbulence.

## TEST EQUIPMENT

Testing was conducted on a portable trailer-mounted pilot unit located inside our manufacturing facility. Data collection was done via an automated computer program operating with input from dissolved oxygen probes.

### Pilot Unit

The MTS alpha pilot unit consists of a trailer-mounted pilot unit with two polyethylene tanks, aeration devices, recirculation pump, blower, air and liquid piping, valves and fittings mounted on a galvanized steel skid with FRP grating. Each tank is 4′ diameter (1200 mm) × 6′ high (1800 mm). Design water level is 5.1′ (1530 mm), yielding an aeration volume of approximately 480 gallons (1819 liters). The tanks are prepiped and prewired so that the entire testing sequence can be conducted from the two control panels. Figure 1 (Appendix A) shows a plan view of the test apparatus.

The aeration tanks are fitted with threaded suction and discharge ports for the recirculation pumps and a drain valve. A model MT0.3 jet nozzle is fitted on the inside of one tank using schedule 80 PVC fittings. The nozzle discharges horizontally at approximately 45° off the radius. The nozzle is also fitted with a 0.75-inch pipe for air supply (via aspirated air or an outside air source), or attached to a gas cylinder for deoxygenation with nitrogen gas (field supplied). Liquid recirculation is supplied by a 1.0 Hp dry pit pump. Low pressure air can be supplied by an oil-free, 1.5 HP centrifugal compressor. Air flow is measured using a rotometer which is located in the hydraulic control panel. The other tank is fitted with a membrane diffuser and air supply piping. The jet aerator was operated at a liquid flow of 32 gpm and an air flow of 80 cfh. The diffuser was operated at 180 cfh.

The pumps and blower are located inside a protective enclosure which can also be used as an observation stand. The electrical control panel is designed to accept either 20 amp/220 volt or 2 @ 20 amp/110 volt single-phase power. The hydraulic control panel is prepiped and includes pressure gauges, vacuum gauges, rotometers and all valves and fittings necessary to conduct the testing.

To facilitate the efficient heating and cooling required to test the same water samples at different temperatures, two hot water heaters and copper heating pipe were installed. The primary heater is an in-home electric water heater rated for 150°F and the secondary heater is an in-line instantaneous heater rated for 130°F. The required maximum test temperature was 50°C (122°F), so these heaters were able to increase the water temperature from 20°C to 50°C when left to heat overnight. Cooling of the test water was achieved by running tap water (approximately 15°C) through the heating pipe system. The heating pipes are designed to isolate each tank and has coils on the floor of the tanks. The hot or cold water is pushed through the pipe system via two ½ HP recirculation pumps. Keeping the test water well-mixed and continually moving across the pipes is necessary to raise and lower the temperature as quickly as possible.

### Data Collection Setup and Measurement Devices

A dissolved oxygen data collection system was used for testing, consisting of YSI Model 57 dissolved oxygen meter with Model 5739 dissolved oxygen probes, Remote Measuring Systems Model ADC-1 analog-to-digital converter, and Macintosh Powerbook 160 laptop computer. The analog-to-digital converter takes an analog 4–20 µamp signal from the YSI meters and sends a digital signal to the laptop. The computer then collects and stores data through a custom designed program within the application Dataview, which allows for multiple probe readings simultaneously and at preselected time intervals.

Temperature of the test waters was taken with a Cooper Instruments Model No. SH66A Electro-Therm electronic thermometer. Surface tension of the detergent samples was measured via a surface tension apparatus utilizing the method outlined in Findlay's *Practical Physical Chemistry*, 8$^{th}$ edition, pp. 94–97. The surface tension apparatus consists basically of a 250-mm borosilicate glass capillary tube, graduated from 0 to 10 cm in 1-mm increments; a large outer tube, open at one end and with a glass side-arm opening near the top; and a one-hole cork to hold

**Table I. Data Sets for Four Different Water Conditions**

| JET | | | | | DIFFUSER | | |
|---|---|---|---|---|---|---|---|
| Water Condition | Temp (°C) | | | | Water Condition | Temp (°C) | |
| Clean water | 20 | 30 | 40 | 50 | Clean water | 20 | 20 |
| 3% Salt | 20 | 30 | 40 | 50 | 3% Salt | 20 | 50 |
| 3% Salt w/detergent | 20 | 30 | 40 | 50 | 3%Salt w/detergent | 20 | 30 |
| Detergent | 20 | 30 | 40 | 50 | Detergent | 20 | 50 |

the capillary inside the outer tube. The distance measured between the liquid level in the capillary tube and the liquid level in the outer tube is inserted to a correction equation to solve for surface tension (in dynes/cm), which corrects for the liquid density, acceleration due to gravity, and altitude and latitude of test location.

## TESTING PROCEDURE

Two data sets were collected, one each for the jet aerator and the membrane diffuser. Each device was tested in four different water conditions, as shown above in Table I.

Testing in clean water was performed to establish the baseline as well as to determine the effect temperature has on the theta value. The two aeration devices were operated under conditions resulting in similar $K_La$ values in clean water at 20°C. Clean water $K_La$ determination was done in accordance with the ASCE Standard "Measurement of Oxygen Transfer in Clean Water" dated July 1984. All data collection procedures, reductions, modeling, and calculations were done by this standard. Nitrogen gas was used for deoxygenation in lieu of sodium sulfite in order to prevent any possible chemical interference.

Once clean water tests were complete, 125 lb of sodium sulfate ("salt," Table I) were added to raise the total dissolved solids concentration of the test water to 3%. Testing was conducted at all test points indicated in Table I for each device. Mixing of the water was continuous throughout the testing to ensure a homogenous sample. At the conclusion of the "salt" tests, one dry cup of detergent was added to the sample to investigate the combined effects of salts and surfactants.

Upon completion of the "salt" and "salt w/detergent" tests, the test tanks were emptied and cleaned out prior to being refilled with clean water. The detergent tests were then conducted on samples that had a surface tension of 50 dynes/cm or less, requiring a minimum of six dry cups of detergent added to the water. Surface tension measurements were taken between each test run to ensure that excessive surfactant was not lost through foam overflow. The sample temperature was recorded at the beginning and end of each test run for all test conditions.

## RESULTS

The main focus of this series of experiments was to look at the effects of temperature in conjunction with TDS concentration and surface tension with respect to oxygen transfer parameters. It has been previously reported[2] that alpha is enhanced to varying levels in high-strength industrial waste waters when jet aerators are used. By looking at individual components of TDS and surfactants, their individual impacts can be shown on different aeration devices.

As indicated by the results in Table II, Theta (θ) remains relatively unchanged for the jet aerator at different temperatures in clean water, calculated to be 1.024. This corresponds with published figures for jet aerators.[4]

Apparent alpha of the jet aerator becomes less enhanced as the temperature of the wastewater is increased under all test conditions. This value is calculated based on two factors, $K_L$ and $a$, which are difficult to measure independently. Since this experiment was not geared toward the effects of temperature on either $K_L$ or $a$, but to observe what the resulting alpha value would be at each temperature, further testing is recommended which would isolate the effects on both or either parameter and attempt to define the cause(s) of the falling alpha values. However, the comparison of the jet aerator to the diffuser alpha is even more significant than the decreasing jet alpha with temperature.

Apparent alpha is enhanced under high TDS conditions for both the jet aerator and the diffuser, as shown in Tables II and III. The alpha of the jet is significantly higher than that of clean

**Table II. Results Using Jet Aerator at Different Temperatures**

| Water Condition | Temp. (°C) | Clean Water Theta | Beta |
|---|---|---|---|
| Clean water | 21.4 | 1.024 | 1.00 |
| | 31.3 | 1.044 | 1.00 |
| | 42.8 | 1.028 | 1.00 |
| | 49.3 | 1.024 | 1.00 |
| **Water Condition** | **Temp. (°C)** | **Apparent Alpha** | **Beta** |
| 3% Salt | 20.7 | 5.20 | 0.76 |
| | 30.0 | 5.04 | 0.78 |
| | 40.0 | 4.37 | 0.79 |
| | 51.0 | 3.97 | 0.80 |
| Detergent (<a50 dynes/c,) | 20.5 | 2.49 | 0.99 |
| | 30.3 | 2.21 | 1.00 |
| 3% Salt w/detergent | 20.3 | 4.18 | 0.76 |
| | 30.3 | 3.91 | 0.78 |
| | 39.8 | 3.46 | 0.79 |
| | 50.3 | 3.16 | 0.80 |

water, even at high temperatures. The alpha of the diffuser is also enhanced well above that of clean water under high TDS conditions, but not nearly as significantly. Both devices experience a reduction in alpha when surface active agents are added to the high TDS wastewater. The jet aerator alpha, however, was still enhanced above clean water while the diffuser alpha dropped below that of clean water.

The presence of surfactants producing a surface tension of 50 dynes/cm (or less) yields an enhanced alpha for the jet aerator. This apparent alpha is about half that of the high TDS results, but is still significantly higher than that of clean water. There are no data on the effect of surfactants on the performance of the diffuser.

Beta ($\beta$) has been calculated based on dissolved oxygen saturation levels detailed by Hunter,[7] attached in Appendix B (1978, Table IA). Calculation of beta based on test result $C^*_{waste}$ and

**Table III. Test Results—Diffuser**

| Water Condition | Temp. (°C) | Clean Water Theta | Beta |
|---|---|---|---|
| Clean water | 49.7 | 1.013 | 1.00 |
| **Water Condition** | **Temp. (°C)** | **Aparent Alpha** | **Beta** |
| 3% Salt | 20.2 | 1.62 | 0.76 |
| | 49.1 | 1.46 | 0.80 |
| Detergent (<a50 dynes/cm) | 48.9 | n/a | 1.00 |
| 3% Salt w/detergent | 19.7 | 0.84 | 0.76 |
| | 51.0 | 0.92 | 0.80 |

$C^*_{clean}$ values is not appropriate because they rely on the accuracy of the D.O. recording equipment and estimation of the nonlinear regression program.

Please refer to Appendix C for graphs comparing the results.

## CONCLUSIONS AND OBSERVATIONS

Due to logistical problems, the data set is not as complete as desired, but still yielded very real and usable results. Several trends were noticed that confirmed documented results and warrant further investigation with respect to jet aerators and higher temperatures (+40°C).

- Apparent alpha of the jet aerator is greatly enhanced under high TDS.
- Alpha enhancement of the jet aerator under high TDS conditions decreases as temperature increases.
- Apparent alpha of the jet aerator is enhanced in the presence of surfactants when surface tension is less than 50 dynes/cm.
- Theta for the jet aerator remains fairly constant at temperatures from 20°C to 50°C.
- Under high TDS conditions, jet aerator alpha enhancement is greater than that of a diffuser.

## REFERENCES


1. Backman, R.C., et al. "Evaluation of Oxygen Transfer Parameters for Industrial Wastewaters," 41st Annual Industrial Waste Conference, Purdue University, West Lafayette, IN, 1986.
2. Backman, R.C., et al. "Chemical Enhancement and Depression of Oxygen Transfer Parameters for Industrial Wastewaters," 42nd Annual Industrial Waste Conference, Purdue University, West Lafayette, IN, 1987.
3. Lewis, W.K. and W.G. Whitman, "Principles of Gas Adsorption," *Ind. Engr. Chem.* 16, pp. 1215, 1924.
4. Stenstrom, M.K. and R.G. Gilbert, "Effects of Alpha, Beta and Theta Factor upon the Design, Specification and Operation of Aeration Systems," *Water Res.* 15, pp. 643-654, 1981.
5. Zlokarnik, M. "Tower Shaped Reactors for Aerobic Biological Waste Water Treatment," *Biotechnol.*, 2, pp. 538–569, 1985.
6. Alleman, J.E. and LaPara, T.M. "Thermophilic Aerobic Biological Treatment of High-Strength Organic Wastewaters," proposal to National Science Foundation, 1997.
7. Hunter, J.S. "Accounting for the Effects of Water Temperature in Aerator Test Procedures," *Proceedings: Workshop Toward and Oxygen Transfer Standard*, U.S. EPA, 1978.


## APPENDIX A

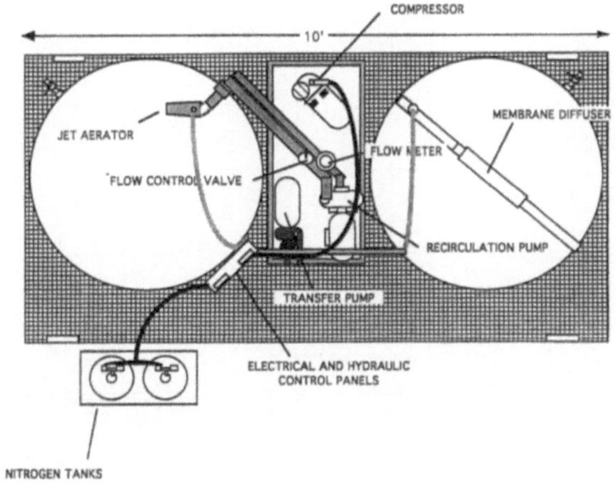


**Figure 1. A plan view of the test equipment.**

## APPENDIX B

**Table IA. Solubility of Oxygen in Water Exposed to Water Saturated Air (mg/L)[a,b]**

| Temp (°C) | Chloride Concentration in Water (mg/L) 0 | 5,000 | 10,000 | 15,000 | 20,000 | 30,000* |
|---|---|---|---|---|---|---|
| 0 | 14.60 | 13.72 | 12.90 | 12.13 | 11.41 | 10.64 |
| 5 | 12.75 | 12.02 | 11.32 | 10.67 | 10.05 | 9.40 |
| 10 | 11.27 | 10.65 | 10.05 | 9.49 | 8.96 | 8.40 |
| 15 | 10.07 | 9.53 | 9.01 | 8.53 | 8.07 | 7.59 |
| 20 | 9.07 | 8.60 | 8.16 | 7.73 | 7.33 | 6.91 |
| 25 | 8.24 | 7.83 | 7.44 | 7.06 | 6.71 | 6.34 |
| 30 | 7.54 | 7.17 | 6.83 | 6.49 | 6.18 | 5.85 |
| 35 | 6.93 | 6.61 | 6.30 | 6.01 | 5.72 | 5.43 |
| 40 | 6.41 | 6.12 | 5.84 | 5.58 | 5.33 | 5.07 |
| 45 | 5.95 | 5.69 | 5.44 | 5.20 | 4.98 | 4.74 |
| 50 | 5.54 | 5.31 | 5.08 | 4.87 | 4.66 | 4.45 |

**Beta Factor[C]**

| Temp (°C) | Chloride Concentration in Water (mg/L) 0 | 5,000 | 10,000 | 15,000 | 20,000 | 30,000* |
|---|---|---|---|---|---|---|
| 0 | 1.00 | 0.94 | 0.88 | 0.83 | 0.78 | 0.73 |
| 5 | 1.00 | 0.94 | 0.89 | 0.84 | 0.79 | 0.74 |
| 10 | 1.00 | 0.94 | 0.89 | 0.84 | 0.80 | 0.75 |
| 15 | 1.00 | 0.95 | 0.89 | 0.85 | 0.80 | 0.75 |
| 20 | 1.00 | 0.95 | 0.90 | 0.85 | 0.81 | 0.76 |
| 25 | 1.00 | 0.95 | 0.90 | 0.86 | 0.81 | 0.77 |
| 30 | 1.00 | 0.95 | 0.91 | 0.86 | 0.82 | 0.78 |
| 35 | 1.00 | 0.95 | 0.91 | 0.87 | 0.83 | 0.78 |
| 40 | 1.00 | 0.95 | 0.91 | 0.87 | 0.83 | 0.79 |
| 45 | 1.00 | 0.96 | 0.91 | 0.87 | 0.84 | 0.80 |
| 50 | 1.00 | 0.96 | 0.92 | 0.88 | 0.84 | 0.80 |

[a]From J.S. Hunter, "Accounting for the Effects of Water Temperature in Aerator Test Procedures."[7]
[b]Values for 30,000 mg/L (3%) are interpolated.
[c]Beta is calcualted based on Equation 4.

# APPENDIX C

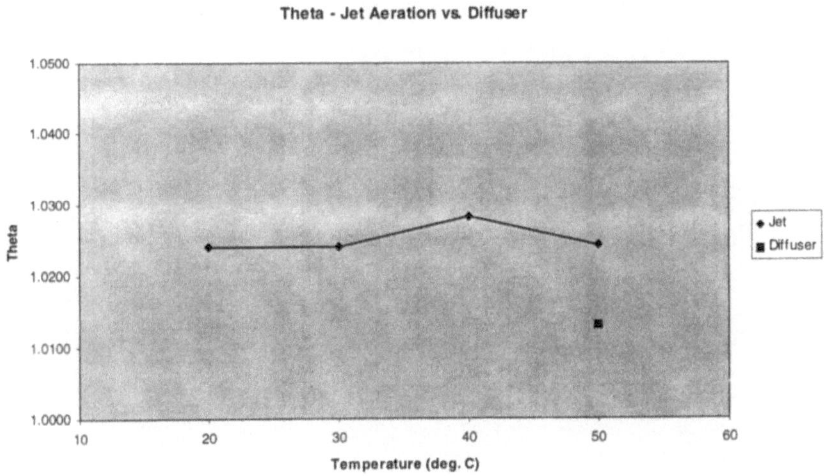
Theta - Jet Aeration vs. Diffuser
1.0500
1.0400
1.0300
1.0200
1.0100
1.0000
Theta
10
20
30
40
50
60
Temperature (deg. C)
Jet
Diffuser

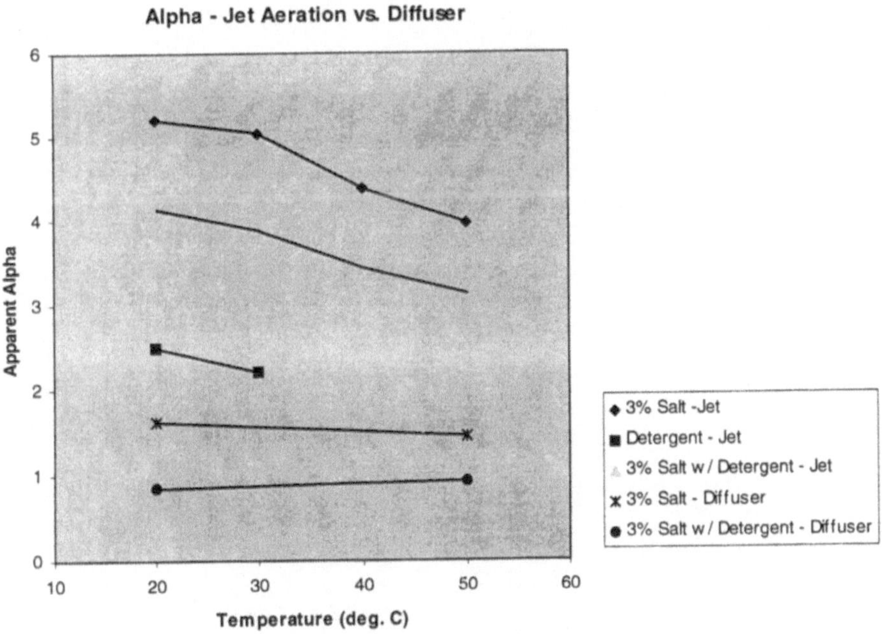
Alpha - Jet Aeration vs. Diffuser
6
5
4
3
2
1
0
Apparent Alpha
10
20
30
40
50
60
Temperature (deg. C)
3% Salt -Jet
Detergent - Jet
3% Salt w / Detergent - Jet
3% Salt - Diffuser
3% Salt w / Detergent - Diffuser

# 3 HIGH RATE THERMOPHILIC PRETREATMENT OF HIGH STRENGTH INDUSTRIAL WASTEWATERS

**Enos L. Stover**, President
**G. Joshua Samuel**, Project Engineer
THE STOVER GROUP
Stillwater, Oklahoma 74076

## ABSTRACT

Bench-scale treatability studies were performed with both synthetic and actual high strength industrial wastewaters to evaluate the concept of employing high-rate biological thermophilic activated sludge technology with conventional gravity solids/liquid separation as a pretreatment alternative. The thermophilic activated sludge processes were operated at 55°C and 60°C. The major advantage noted compared to lower temperature mesophilic processes was significantly lower sludge production at comparable levels of treatment. The bench-scale studies show promise for conventional activated sludge technology operating under thermophilic conditions for treating high temperature, high strength, or combinations of high temperature high strength industrial wastewaters as a pretreatment alternative under high loading rate (F/M) conditions.

## INTRODUCTION

The autoheated thermophilic aerobic digestion process has been developed for stabilization of waste biological sludges. The primary objective of this process is to operate under thermophilic conditions (40° to 80°C), preferably 50° to 60°C range, without supplemental heat. In order to accomplish this objective, adequate carbon source (volatile suspended solids or chemical oxygen demand) must be supplied to the process. The organic carbon loading can be supplied in either particulate (volatile solids) or soluble COD form. One of the advantages of the thermophilic process is minimal biomass production or yield coefficient which equates to maximum heat generation. Biological heat production as high as 6,300 BTU/lb oxygen utilized has been reported.[1-3] In terms of biological volatile suspended solids with an oxygen equivalent of 1.42 lb $O_2$/lb VSS, this equates to about 9,000 BTU/lb VSS destroyed. Therefore, if the VSS or soluble COD loading rates are high enough and proper precautions are designed into the treatment system to manage heat loss, the aerobic thermophilic treatment process can be equally effective for treatment/stabilization of biological waste sludges or high strength industrial wastewaters. The general requirements to maintain appropriate thermophilic temperatures includes sufficient biodegradable organics (volatile solids or soluble COD) to provide heat of oxidation up to 25 to 30 Kcal/liter, an insulated reactor, and adequate mixing and oxygen transfer efficiency to minimize excessive heat loss. The purpose of this paper is to present the results of bench-scale thermophilic activated sludge studies with soluble high-strength wastewaters.

## MATERIALS AND METHODS

A schematic of the bench-scale biological thermophilic activated sludge system used in this study is presented in Figure 1. The aeration tank was a 5.2 liter (liquid volume) Plexiglas reactor. The aeration tank reactor was mixed and aerated with two 4-inch aquarium diffuser stones. The clarifier was a glass clarifier with a liquid volume of 2.6 liters. The wastewater feed was pumped into the aeration tank reactor, and the effluent flowed by gravity to the clarifier. The clarifier effluent flowed by gravity into an effluent collection bottle, and the settled solids were recycled

*52nd Purdue Industrial Waste Conference Proceedings*, 1997, Ann Arbor Press, Chelsea, Michigan 48118. Printed in U.S.A.

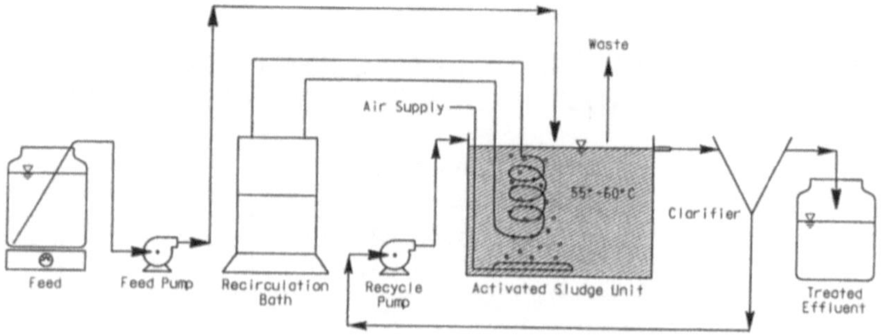


**Figure 1. Schematic of bench-scale thermophilic aerobic treatment system.**

(pumped) back to the aeration tank. Waste sludge consisted of the solids lost in the clarifier effluent and intentional sludge wastage from the aeration tank. The aeration tank reactor temperature was maintained by a copper heating coil inserted within the tank with a hot water recirculation system from a temperature controlled water bath.

The activated sludge system was initially seeded with sludge from a previously completed high-strength industrial wastewater thermophilic treatability study plus 0.5 liters of sludge from an autoheated thermophilic aerobic sludge digester. The wastewater feed used during this study was a synthetic sugar wastewater with an average COD and BOD content of 5,500 mg/L and 3,100 mg/L, respectively. Adequate ammonia-nitrogen and phosphorus were added to the wastewater to Ensure proper biological treatment with the carbon source being the growth-limiting nutrient. The aeration reactor contents were routinely monitored for pH, temperature, dissolved oxygen, and mixed liquor suspended solids. The clarifier effluent was routinely monitored for COD and BOD, along with TSS, nitrogen, and phosphorus contents.

The system was operated under steady state conditions at two different temperatures (55° and 60°C). A temperature shock study was also conducted to evaluate the impacts on the system of operating in and out of the thermophilic temperature range. At the end of the 60°C steady-state data collection period the temperature of the reactor was lowered into the mesophilic range (22°C) for 12 days and then heated back to 60°C to monitor the performance of the system from thermophilic conditions to mesophilic conditions and then back to thermophilic conditions again.

## TEST RESULTS

The average conditions and treatment performance of the thermophilic activated sludge system treating the synthetic wastewater are summarized in Table I. The influent feed conditions, reactor operating conditions, effluent quality, and total system performance are all presented in Table I. Throughout the entire study period (over three months) the influent feed was maintained between 5.2 to 5.3 liters/day with average COD and BOD concentrations of 5,500 mg/L and 3,100 mg/L, respectively. The aeration reactor temperatures were carefully controlled, and the mixed liquor pH was controlled between pH 7.0 to 8.0.

A trend plot of COD treatment performance at the operating temperatures of 55°C and 60°C is presented in Figure 2. A trend plot of BOD treatment performance at 60°C is presented in Figure 3. Effluent BOD data collection was initiated during the 60°C study period. Table I presents the influent and effluent COD and BOD values in terms of both concentration and mass loading conditions. However, the treatment performance figures are presented in mass loading rates only, in order to project the true performance of the systems due to the high water evaporation rates at 55°C and 60°C. At 55°C and 60°C the average treated effluent volumes were 3.49 and 3.08 liters/day, respectively, which correlates to evaporation rates of 34% and 42%, respectively.

Effluent total and soluble COD and BOD values are both presented in Table I and Figures 2 and 3. System treatment performance in terms of total influent COD and effluent-soluble COD was virtually identical at both temperatures of 55°C and 60°C. The COD treatment performance averaged 88% removal, and the BOD treatment performance averaged 91% removal at 60°C. The primary differences in system performance at 55°C and 60°C were noted to be mixed liquor solids settling characteristics and effluent-suspended solids concentrations. Within 24 hours after

**Table I. Summary of Thermophilic System Operating Conditions and Treatment Performance**

| | | Steady State | | Temperature Shock | |
|---|---|---|---|---|---|
| Parameter | Units | 55°C | 60°C | 22°C | 60°C |
| Influent | | | | | |
| Flow rate | (L/day) | 5.29 | 5.32 | 5.27 | 5.19 |
| COD: Conc. | (mg/L) | 5,500 | 5,500 | 5,500 | 5,500 |
| Mass | (mg/day) | 29,100 | 29,250 | 28,985 | 28,545 |
| BOD: Conc. | (mg/L) | 3,100 | 3,107 | 3,100 | 3,100 |
| Mass: | (mg/day) | 16,402 | 16,526 | 16,337 | 16,089 |
| Reactor | | | | | |
| pH (range) | (s.u.) | 7,2–7.9 | 7.2–8.1 | 7.5–8.0 | 7.0–7.8 |
| SRT | (days) | 16.9 | 23.9 | 18.7 | 16.1 |
| MLSS | (mg/L) | 5,067 | 7,200 | 7,030 | 5,710 |
| MLVSS | (mg/L) | 4,529 | 5,838 | 5.700 | 4,630 |
| Wasting Rate | (ml/day) | 267 | 86 | 25 | 25 |
| RAS | (mg/L) | | 12,000 | | |
| Effluent | | | | | |
| Flow rate | (L/day) | 3.49 | 3.08 | 4.78 | 2.90 |
| tCOD: Conc. | (mg/L) | | 2,111 | 1,454 | 3,268 |
| Mass | (mg/day) | | 6,495 | 6,949 | 9,477 |
| sCOD: Conc. | (mg/L) | 1,016 | 1,160 | 962 | 2,472 |
| Mass | (mg/day) | 3,542 | 3,571 | 4,596 | 7,168 |
| tBOD: Conc. | (mg/L) | | 892 | 236 | 1,551 |
| Mass | (mg/day) | | 2,745 | 1,130 | 4,499 |
| sBOD: Conc. | (mg/L) | | 504 | 95 | 1,181 |
| Mass | (mg/day) | | 1,550 | 452 | 3,425 |
| TSS | (mg/L) | 892 | 441 | 626 | 653 |
| VSS | (mg/L) | 794 | 391 | | |
| Performance | | | | | |
| F/M(tCOD/MLVSS) | (mg/mg-day) | 1.32 | 1.00 | 0.99 | 1.21 |
| U(COD/MLVSS) | (mg/mg-day) | 1.16 | 0.87 | 0.83 | 0.90 |
| Removal(COD) | (%) | 87.78 | 8.78 | 84.2 | 74.9 |
| Sludge Yield (COD) | (mg/mg) | 0.13 | 0.07 | | |
| F/M(tBOD/MLVSS) | (mg/mg-day) | 0.74 | 0.57 | 0.56 | 0.68 |
| U(BOD/MLVSS) | (mg/mg-day) | | 0.50 | 0.50 | 0.54 |
| Removal (BOD) | (%) | | 90.6 | 97.2 | 78.6 |
| Sludge Yield (BOD) | (mg/mg) | | 0.13 | | |

increasing the aeration tank temperature to 60°C, the mixed liquor settling characteristics improved and the effluent suspended solids decreased significantly. The average effluent suspended solids concentrations were 892 mg/L and 441 mg/L at 55°C and 60°C, respectively. The best effluent suspended solids concentrations achieved throughout the entire study period was around 150 mg/L.

The average mixed liquor suspended solids concentrations were 5,067 mg/L and 7,200 mg/L at 55°C and 60°C, respectively, corresponding to sludge retention times of 17 days and 24 days, respectively. The corresponding volatile solids fractions were 89% and 81%, respectively, at 55°C and 60°C. The average food to microorganism (F/M) ratios were 1.32 and 1.00 with corresponding specific substrate utilization rates (U) of 1.16 and 0.87 in terms of COD at 55°C and 60°C, respectively. The average F/M ratios in terms of BOD were 0.74 and 0.57 at 55°C and 60°C, respectively. The U value at 60°C in terms of BOD was 0.50. The average sludge production rates or sludge yield values in terms of COD removed were 0.13 and 0.07 at 55°C and 60°C, respectively. The average sludge yield value in terms of BOD removed at 60°C was 0.13.

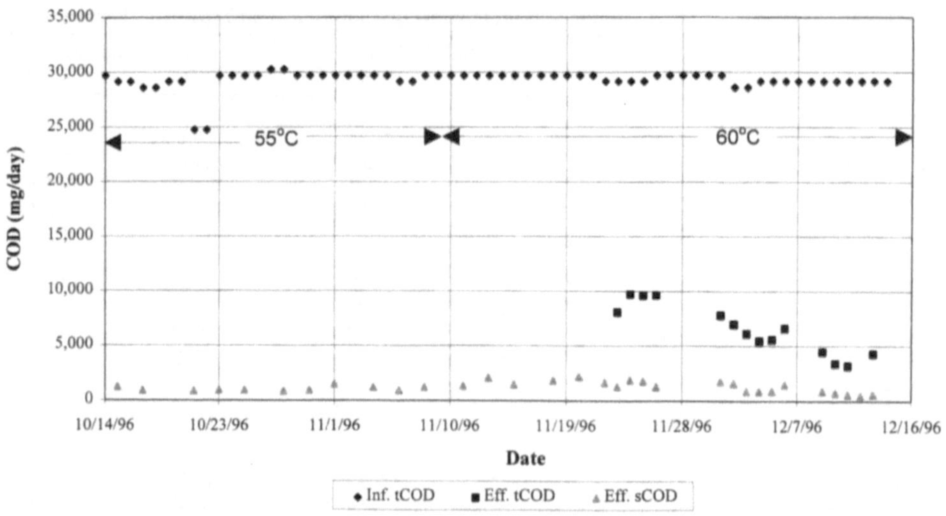


**Figure 2. COD treatment performance at 55°C and 60°C.**

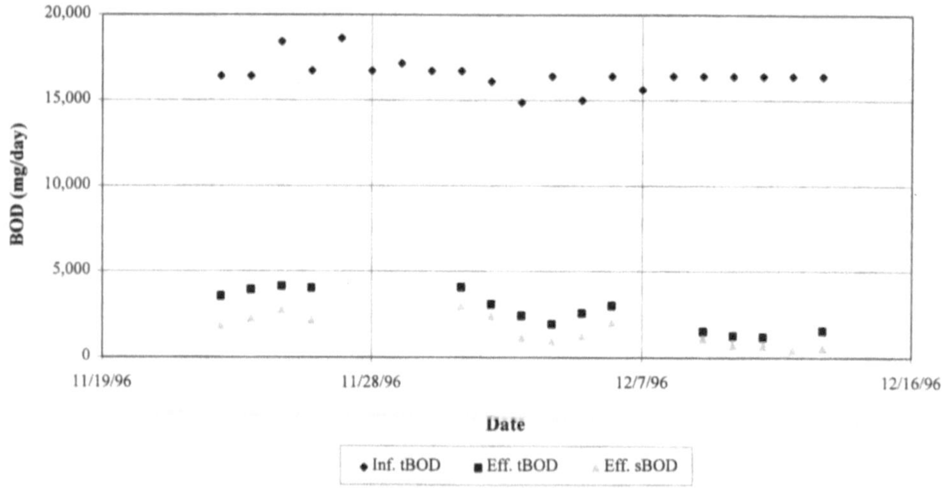


**Figure 3. BOD treatment performance at 60°C.**

The COD and BOD treatment performance of the thermophilic system under mesophilic shock load conditions is presented in Figures 4 and 5, respectively, and the total system operating conditions and performance before, during, and after the shock temperature study are summarized in Table I. The temperature shock study was performed at the end of the data collection period with the system operating at 60°C. The reactor temperature was lowered to 22°C for 12 days and then increased again to 60°C. The system performance did not deteriorate during the mesophilic operating period. In fact, the COD removed was comparable to the thermophilic conditions while the BOD removal increased from 91% to 97%. Going back into the thermophilic temperature range the system performance appeared to initially deteriorate.

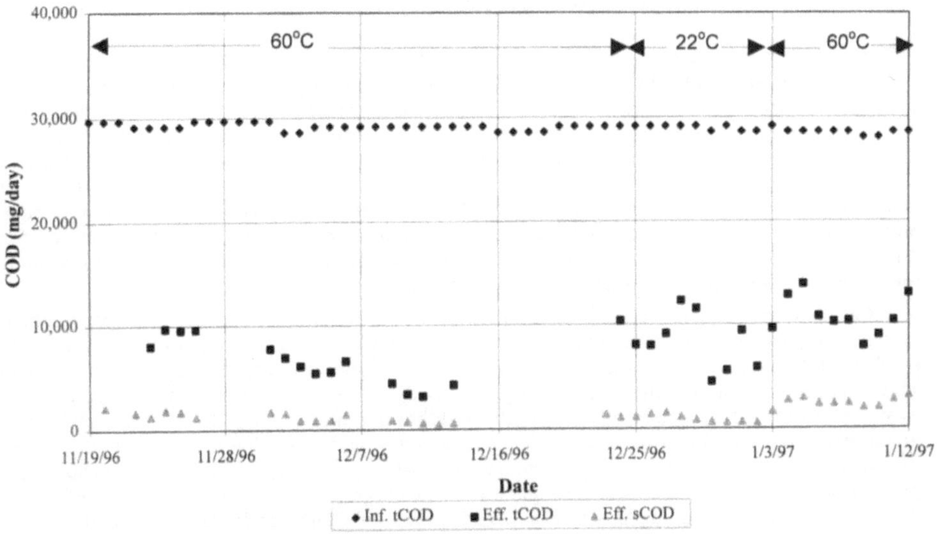


**Figure 4. COD treatment performance during temperature shock study.**

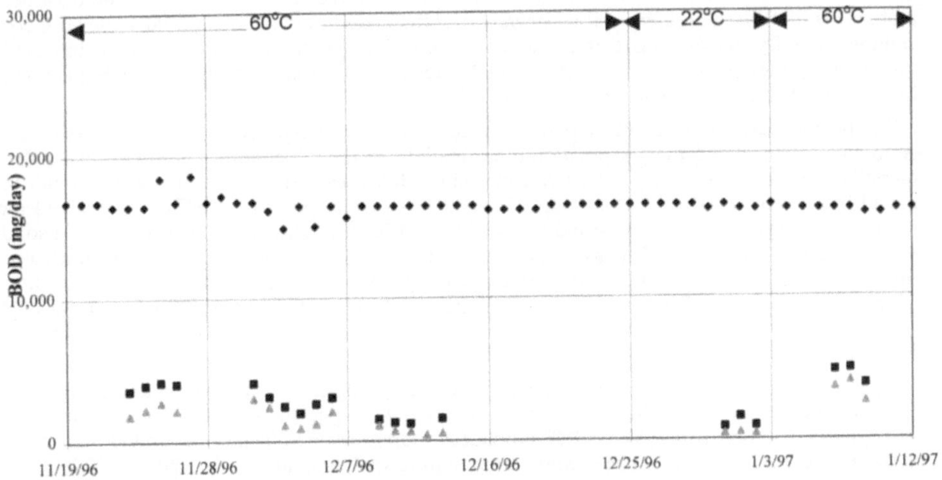


**Figure 5. BOD treatment performance during temperature shock study.**

A COD material balance was performed around the system in order to estimate the oxygen utilization rates or dissolved oxygen uptake rates in the system, as follows:

$$\text{Influent COD} - \text{Effluent COD} = \text{Oxygen Utilized} + \text{COD Waste Sludge}$$

or

$$\text{Oxygen Utilized} = \Delta\text{COD} - \text{COD Waste Sludge}$$

In performing the COD balance around the thermophilic system at 60°C, the dissolved oxygen uptake rates were calculated to be between 150 to 200 mg/L/hr.

## DISCUSSION AND CONCLUSIONS

The primary objective of the bench-scale thermophilic activated sludge study discussed herein was to evaluate the applicability of using this technology as a pretreatment alternative for high strength industrial wastewaters. The primary issues of the investigation centered around system operations, treatment performance, and biological sludge generation rates. Due to the nature and inherent limitations of the bench-scale system and the influent COD concentration there were no attempts to operate the system as an autothermal system. As discussed previously, it is known that biological heat production as high as 6,300 BTU/lb oxygen utilized can be used to autoheat biological systems. With sufficient organic COD content to produce heat of oxidation up to 25 to 30 Kcal/liter, an insulated reactor, and adequate mixing and oxygen transfer efficiency to minimize heat loss, appropriate thermophilic temperatures can be maintained.

Initial concerns centered around operability of a thermophilic activated sludge pretreatment system relative to mixed liquor solids settling characteristics and effluent quality, especially suspended solids concentrations. Prior industrial wastewater high rate once-through activated sludge studies under both mesophilic and thermophilic conditions by the authors indicated comparable treatment performance under both mesophilic and thermophilic conditions. The major difference noted between the two high rate once-through systems was the fact that the thermophilic system exhibited significantly higher oxygen uptake rates with corresponding lower biological sludge production rates. During the once-through studies (average SRT of 1.0 day) the observed sludge production rates or yields were 0.43 and 0.75 in terms of COD and BOD, respectively, for the mesophilic system while the yields were 0.15 and 0.20 in terms of COD and BOD, respectively, for the thermophilic system. The thermophilic sludge yields averaged in the 25% to 30% range of the mesophilic sludge yields. With this inherent advantage, the decision was made to perform a research study with a conventional activated sludge system approach to determine if the thermophilic process could be used in the conventional sludge recycle mode as a pretreatment process for high strength industrial wastewaters. If the mixed liquor solids would settle with conventional gravity clarification, then thermophilic activated sludge technology could be applicable for either high temperature wastewaters or high strength industrial wastewaters with high enough COD concentrations to provide autoheating.

The bench-scale thermophilic activated sludge studies discussed herein have confirmed that it is possible to achieve gravity separation (clarification) of the mixed liquor produced under thermophilic operating conditions. Clarifier underflow solids concentrations as high as two times the mixed liquor solids concentrations were observed. However, the clarifier effluent remained turbid throughout the entire study period with the best effluent quality in terms of suspended solids as low as around 150 mg/L. The average effluent suspended solids during the 60°C operating period was 441 mg/L. During the best operating time periods the effluent suspended solids varied from around 150 to 300 mg/L (both at 55°C and 60°C). Operating the aeration reactor at 60°C produced significantly better sludge settling characteristics and effluent suspended solids compared to operating at 55°C.

Operating the thermophilic activated sludge system with gravity clarification and sludge recycle at around 15 to 20 day SRTs compared to earlier high-rate thermophilic once-through activated sludge studies at a 1.0 day SRT produced sludge production rates of around 50% of the high-rate (low SRT) studies. The aerobic thermophilic sludge yields at 15 to 20 day SRTs during this study were 0.07 and 0.13 in terms of COD and BOD, respectively, compared to aerobic thermophilic sludge yields at a 1.0 day SRT (earlier studies) of 0.15 and 0.20 in terms of COD and BOD, respectively. Since the thermophilic sludge yields are lower than mesophilic sludge yields, the corresponding F/M ratios are higher under the thermophilic operating conditions. At the 15- to 20-day SRT operating conditions of this study, the COD and BOD F/M ratios were typically greater than 1.0 and 0.5, respectively.

The results of the research study reported herein indicates that aerobic thermophilic activated sludge technology with gravity solids/liquid clarification should be applicable for pretreatment of high temperature and/or high strength industrial wastewaters. Final clarifier effluent quality suitable for direct discharge applications was not achieved during this study. Additional solids/liquid separation following gravity clarification or additional treatment would be required prior to direct discharge. The major advantage of aerobic thermophilic activated sludge pretreatment over mesophilic pretreatment is significantly reduced sludge production rates. The sludge production rates observed during this study were comparable to anaerobic sludge production rates.

## REFERENCES


1. Andrews, J.F. and Kambhu, K., Thermophilic Aerobic Digestion of Organic Solid Wastes, EPA-670/2-73-061, NTIS PB 222 396, 1973.
2. Jewell, W.J., et. al., Autoheated Aerobic Thermophilic Digestion with Air Aeration, EPA Project No. R804636, MERL Report, NTIS PB 82-196908.
3. Cooney, C.L., Wang, D.I.C., and Mateles, R.I., "Measurement of heat evolution and correlation with oxygen consumption during microbial growth," *Biotechnol. Bioeng.*, 11, p. P269, 1968.

# 4 AUTOTHERMAL THERMOPHILIC AEROBIC WASTE TREATMENT SYSTEMS: A STATE-OF-THE-ART REVIEW

**Timothy M. LaPara**, Graduate Research Assistant

**James E. Alleman**, Professor
Purdue University
West Lafayette, Indiana 47907-1284

## ABSTRACT

Thermophilic aerobic biological treatment systems have many advantages compared to conventional techniques for high-strength wastewaters, including faster biodegradation rates, greater overall process efficiencies, and low rates of residual biosolids production. High reactor temperatures, however, alter the physical, chemical, and biological characteristics of the treatment process, so that the enormous knowledge-base for conventional activated sludge operations no longer directly applies. Several of these unique operating conditions are discussed. Process kinetics are also examined, with a particular focus on the high rate of endogenous decay typically observed with thermophilic systems.

## INTRODUCTION

Microbial life at temperatures exceeding 45(C has long been a peculiar phenomenon to scientists and engineers. Recent advances in biotechnology, however, have propelled the study of thermophilic bacteria from mere curiosity to the search for potential solutions to real problems. The most well-known beneficial use of thermophilic bacteria is the Taq polymerase isolated from *Thermus aquaticus*,[1] used in DNA amplification during the polymerase chain reaction (PCR). This DNA polymerase, stable at temperatures well above those of more common enzymes, allows for repeated cycles of strand separation (95°C), primer hybridization (54°C), and DNA synthesis (72°C) without supplying any additional enzyme (for a review of PCR, see Reference 2).

Along with the demonstrated utility of thermostable enzymes with PCR, engineers and scientists have searched for other uses of thermophilic microorganisms and their enzymes for the production of antibiotics, chemical feedstocks, fuels, etc.[3] Advantages of thermophilic biotechnology include faster reaction rates (production) and a significantly decreased potential for process contamination. Extending beyond the use of thermophiles for production purposes, this chapter reviews thermophilic treatment processes for the destruction and remediation of environmental contaminants and wastewaters. As with biotechnological applications, thermophilic treatment processes benefit from faster reaction rates (degradation) and, in this case, a significantly improved ability to destroy pathogens.

Clearly the limiting factor preventing the widespread use of thermophilic biotechnology in environmental engineering is the cost of raising reactor temperatures. Two situations exist in which this cost can be avoided: (1) when wastewaters are produced hot; and (2) when the wastewaters are highly concentrated such that the heat released during contaminant biodegradation is sufficient for autothermal operation. The former condition exists at a select few industries (e.g., pulp and paper facilities), while the latter circumstance is much more common. Indeed, in addition to the multitude of industries which produce high-strength wastewaters, virtually every publicly-owned treatment works (POTW) produces residual biosolids which can be readily concentrated to allow for autothermal thermophilic aerobic digestion (ATAD).

*52nd Purdue Industrial Waste Conference Proceedings*, 1997, Ann Arbor Press, Chelsea, Michigan 48118. Printed in U.S.A.

Although thermophilic aerobic wastewater treatment certainly represents a promising technology for high-strength wastewaters, a surprisingly lean knowledge-base makes design and operation an unpredictable proposition; even the most innovative engineers are reluctant to implement thermophilic aerobic treatment without a sufficient background of laboratory, pilot, and full-scale results. Eckenfelder and Musterman[4] have reported that as few as five thermophilic aerobic wastewater treatment facilities exist at industrial facilities in the United States. In addition, the information contained within the technical literature generally indicates that poor sludge flocculation occurs at high temperatures, preventing solids separation via conventional clarifiers.

## THERMOPHILIC AEROBIC DIGESTION

Despite the potential advantages of thermophilic aerobic digestion, relatively few thermophilic aerobic facilities exist in the United States. As of 1990, no ATADs were in operation at POTWs in the United States.[5] ATAD is particularly popular in Germany where strict requirements for pathogen destruction are enforced. Under the present conditions in the United States, ATAD probably cannot compete with conventional anaerobic digestion for cost-effectiveness in meeting current pathogen and vector attraction limits for disposal on agricultural lands; should further restrictions on sludge disposal be enacted, ATAD could become an increasingly popular technology, as its ability to produce a pathogen-free residue is unmatched.

Kambhu and Andrews[6] performed the first detailed analysis on thermophilic aerobic sludge digestion, demonstrating through computer simulation that autothermal operation was possible with high efficiency aeration systems and an influent waste at 4–6% solids. Within the next decade or so, autothermal operation was empirically demonstrated using high-purity oxygen aeration[7] and self-aspirating aeration units using air.[8] The overwhelming majority of ATADs in operation today are in Europe (particularly Germany), where fairly stringent requirements for pathogen destruction make ATAD an attractive sludge treatment alternative. The interested reader is referred to a recent document published by the United States Environmental Protection Agency[5] for a more detailed discussion of ATAD technology.

## THERMOPHILIC AEROBIC WASTEWATER TREATMENT

The technical literature contains several reports of laboratory and pilot-scale studies performed on the thermophilic aerobic biodegradation of wastewaters. Many of these papers have been written on the treatment of wastewaters from the pulp and paper industry, a process which tends to produce a high-temperature, relatively high-strength wastewater conducive to thermophilic treatment (either anaerobic or aerobic). Other researchers have investigated the treatment of a variety of wastewaters, including: municipal, livestock, slaughterhouse, beer brewery, synthetic, and citrus fruit processing.

Thermophilic aerobic wastewater treatment appears to be very similar to conventional aerobic systems in process stability with regard to variable influent loading and temperature changes.[9] A primary concern for thermophilic aerobic wastewater treatment is that extremely high microbial activity may lead to an oxygen uptake rate (OUR) that cannot be matched by commercially available aeration equipment. Indeed, existing full-scale thermophilic aerobic treatment facilities have reported that maintaining reactor dissolved oxygen (DO) concentration above 2.0 mg/L is often difficult, but possible.

### Pulp and Paper Wastewater

Pulp and paper industries commonly discharge a waste stream that is quite hot (~50°C) compared to other industrial wastewaters. Conservative treatment process operations cool these hot wastewaters to a temperature typical for conventional anaerobic and/or aerobic treatment. Periodically, different investigators have attempted to treat pulp and paper wastewaters at the thermophilic temperatures at which they are discharged in an attempt to reduce the energy costs for treatment; these papers are reviewed below.

Rudolfs and Amberg[10] investigated the biological treatment of white water discharged from board mills at temperatures up to 50°C. BOD removal rates were highest at thermophilic temperatures, but dispersed microbial growth limited overall removal efficiencies. BOD removal in excess of 90% was possible only with the addition of chemical coagulant (alum).

Gehm[11] successfully treated kraft mill effluent for three months at temperatures from 50° to 53°C and a pH of 9.5 to 9.8. Compared with treatment at lower temperatures (30° to 38°C), similar BOD removals were observed. Dissolved oxygen concentrations were nondetectable, but this apparently had no adverse effect on process performance. Settling characteristics were "excellent," with a sustainable mixed liquor solids concentration up to 3,000 mg/L.

Carpenter et al.[12] compared the aerobic treatment of pulp and paper wastewater at temperatures of 26°, 37°, 42°, 47° and 52°C. Effluent BOD increased at higher temperatures from a maximum removal percentage of 80.5% (37°C) to a low of 37% (52°C). A low sludge volume index (SVI) was measured at thermophilic temperatures (47° and 52°C), indicative of dispersed microbial growth.

Pertulla et al.[13] treated sulfite mill condensates containing acetate and ammonia at 65°C in packed-bed bioreactors, attempting to save energy by decreasing cooling requirements before treatment and reheating prior to recycling the treated effluent. Extensive surface growth was observed, even to the point of clogging an activated carbon packing media. Overall, acetate removal was similar to that of mesophilic processes.

Rintala and Lepistö[14] treated bleached kraft mill effluent by both anaerobic-aerobic and aerobic bench-scale processes at 55°C. The second stage of the anaerobic-aerobic configuration removed virtually no COD when the anaerobic stage operated properly, but provided excellent treatment during periods of anaerobic reactor upsets. Overall, both systems studied provided levels of COD and adsorbable organic halogen (AOX) removal similar to analogous mesophilic systems.

Barr et al.[15] investigated the effects of hydraulic residence time (HRT), solids residence time (SRT), and temperature on the aerobic treatment of kraft pulping effluent. Similar BOD removal was observed at 41–50°C as at mesophilic temperatures, but a higher COD removal was reported. Settling characteristics improved with higher temperatures. The authors attributed this unique result to the use of an external clarification system which extended the SRT to approximately 15 days.

## Livestock Wastes

Pöpel and Ohnmacht[16] presented theoretical considerations and experimental evidence for the autothermal thermophilic aerobic treatment of wastewater solids, industrial waste, and liquid cow/pig manure at temperatures of 45–50°C. Ginnivan et al.[17–19] subsequently treated liquid pig manure at 55°C, successfully reducing TSS, COD, and $BOD_5$ concentrations. Experiments demonstrated rapid destruction of a pathogenic *Salmonella* sp., and the feasibility of reutilizing dried slurry effluent as a supplementary foodstuff for pigs. Beaudet et al.[20] also treated swine waste at 55°C, with excellent removals of COD and ammonia (via air stripping).

## Municipal Wastewater

Thermophilic treatment of municipal wastewater has been studied by Tischer et al.,[21] Hunter et al.[22] and Shindala and Parker.[23] Due to the enormous energy requirement to heat municipal wastewater, this process is almost certainly an economically infeasible treatment process. Tischer et al.[21] isolated a *Bacillus* sp. obligate for thermophily from oxidation ponds treating municipal wastewater. Experiments on synthetic and domestic sewage demonstrated the technical feasibility of COD removal at 55°C. Respirometry analysis also indicated that organisms had a growth requirement for carbon dioxide, but these results were inconclusive, as the effect of $CO_2$ concentration versus pH could not be distinguished. A turbid reactor supernatant was reported for all experiments. Shindala and Parker[23] treated municipal wastewater at 55°C utilizing biosolids recycle to improve the process efficiency. Filtered BOD removal frequently exceeded 90%, but high supernatant turbidity generally limited total BOD removal to less than 70%. Hunter et al.[22] compared process performance at multiple temperatures (4–55°C), and observed an optimum treatment efficiency at 45°C.

## High-Strength Wastewaters

One of the major problems of wastewater treatment is that very large reactors are often needed to provide sufficient capacity. Thermophilic aerobic systems present an elegant solution for the treatment of high-strength wastewaters. High biodegradation rates at thermophilic temperatures clearly represent a potential advantage compared to conventional mesophilic systems, but providing the heat to raise reactor temperatures limits the cost-effectiveness of this approach. When treating high-strength wastewaters, this heat energy can be derived from the waste itself—released during microbial metabolism—to raise reactor temperatures to 45°C and higher. The technical literature regarding thermophilic aerobic treatment of high-strength wastewater is reviewed below.

McNary et al.[24] discussed the treatment of citrus wastewaters at pilot-scale in which the heat released as a result of biological activity raised reactor temperatures from 32° to 41°C. Dougherty and McNary[25] subsequently performed laboratory studies at temperatures ranging from 21° to 46°C. The temperature was raised slowly in a single laboratory reactor. Treatment efficiency remained high up to 38°C, but further temperature increases adversely affected performance. At temperatures greater than 43°C, a poor settling mixed liquor was observed and microscopic analysis indicated that no protozoa were present.

Sürücü et al.[26,27] treated a high-strength synthetic wastewater (glucose plus other nutrients) to evaluate the feasibility of thermophilic treatment at 58°C with subsequent recovery of single-cell protein. Compared with conventional activated sludge systems, extremely fast biodegradation rates and surprisingly low growth yields were observed. Kinetic analysis in chemostats indicated that low growth yields were the result of a high bacterial maintenance requirement, which exceeded commonly reported activated sludge values by an order of magnitude. Single-cell protein production as an ancillary benefit to thermophilic treatment was determined to be economically feasible for wastewaters with a COD and flow rate of at least 12,500 mg/L and 0.1 MGD, respectively.

Jackson[28] treated a high-strength ($BOD_5$~2,000 mg/L) corrugated cardboard wastewater at 53°C. Kinetic analysis in chemostats again indicated extremely low growth yields resulting from a high cell maintenance requirement. Pilot-scale studies indicated that moderate liquid-solids separation could be achieved by gravitational settling, however the clarifier overflow contained TSS concentrations up to 300 mg/L. Polymer addition drastically improved clarification efficiency.

Pig slaughterhouse wastewater was treated at 45°, 52° and 58°C by Couillard et al.,[29] Gariépy et al.[30] and Couillard and Zhu,[31] respectively. Extremely high substrate utilization rates were observed, exceeding that of mesophilic treatment systems by a factor of 10. Again, low growth yields resulted from a high bacterial maintenance requirement. COD removals of 85% or higher were consistently achieved at all temperatures. A biomass containing 64–78% crude protein was harvested; amino acid analysis indicated that levels met or exceeded the nutritional requirements of many different animals.

Zvauya et al.[32] studied the aerobic treatment of Zimbabwean traditional opaque-beer brewery wastewater at 50°C. This wastewater had high $BOD_5$ (6,500 mg/L) and TSS (4,800 mg/L) concentrations; the thermophilic aerobic treatment system removed 67% and 75% of these water quality parameters, respectively.

Rozich et al.[33] performed a bench-scale treatability study to determine the feasibility of treating high-strength landfill leachate (COD~80,000–200,000 mg/L) via an autothermal aerobic treatment system. The thermophilic reactor (50°C) demonstrated a COD removal (~95%) similar to a mesophilic reactor (20°C); projected residual biosolids disposal costs, however, were significantly lower for the thermophilic system, because the cell yield was less than 0.10 mg TSS/mg COD. The results of this bench-scale study were subsequently used to design, construct, and operate a full-scale facility operated at 48–50°C.[34] Full-scale cell yields were approximately 0.05 mg VSS/mg COD for a system maintaining more than 99% COD removal.

## RELATED THERMOPHILIC TREATMENT ISSUES

Although thermophilic aerobic wastewater treatment systems appear suitable for a broad spectrum of different wastewaters and utilize similar unit operations as conventional treatment processes (e.g., aeration systems), the physical, chemical, and biological characteristics of the system are altered in such a way that the enormous knowledge-base regarding the design and operation of conventional digestion and activated sludge no longer directly applies. Particularly pertinent aspects of thermophilic aerobic process operation are oxygen transfer, reactor temperatures, biochemical kinetics, and bacterial settling. These issues are discussed below.

**Table I. Physical-Chemical Parameters of Water Affected by High Temperatures**

| Water Characteristic | Expected Change | Impact on Thermophilic Aerobic Treatment |
|---|---|---|
| Viscosity | Decrease | • Improved gas transfer efficiency<br>• Improved mixing efficiency |
| Surface tension | Decrease | • Improved gas transfer efficiency<br>• Potentially increased foaming |
| Diffusivity | Increase | • Improved gas transfer efficiency<br>• Improved mixing efficiency |
| Gas-liquid solubility | Decrease | • Reduced oxygen transfer efficiency<br>• Improved off-gas stripping efficiency (e.g., $CO_2$, $NH_3$, etc.) |
| Solid-liquid solubility | Increase | • Higher possible concentrations of most organics and inorganics<br>• Exceptions include the solubility of carbonate salts |

## Physical-Chemical Aspects

Changes in the physical-chemical properties of water at high temperatures alter the operational characteristics of wastewater treatment. The more significant effects are reviewed in Table I. Most importantly, it appears that a decrease in the solubility of dissolved oxygen would adversely affect oxygen transfer, but the concomitant changes in viscosity and diffusivity correspondingly improve the overall mass transfer coefficient of aeration ($k_La$). As a result, the overall oxygen transfer rate is not affected greatly, and can slightly increase with temperature.[35]

Although the acid-base chemistry of thermophilic treatment has not been studied, the increased stripping of $CO_2$ and $NH_3$ no doubt plays an important role in the high pH (~8) typically observed with ATAD systems.[5] The basic equilibrium chemistry of these compounds can undergo significant changes at high temperatures which may affect aerobic wastewater treatment operation characteristics. For example, ammonia stripping is much more efficient at thermophilic temperatures because partitioning of free ammonia to the gas phase is favored, and the fraction of free ammonia increases with temperature (Figure 1). Such an enhancement of ammonia stripping may, in fact, represent the sole opportunity to remove, as biological nitrification is inhibited at temperatures greater than 43°C or so. Another important example of the unique equilibrium

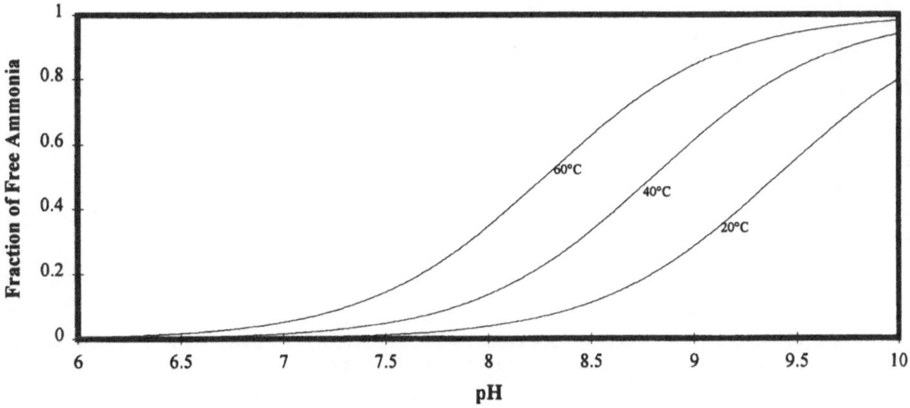


**Figure 1. Fraction of free ammonia.**

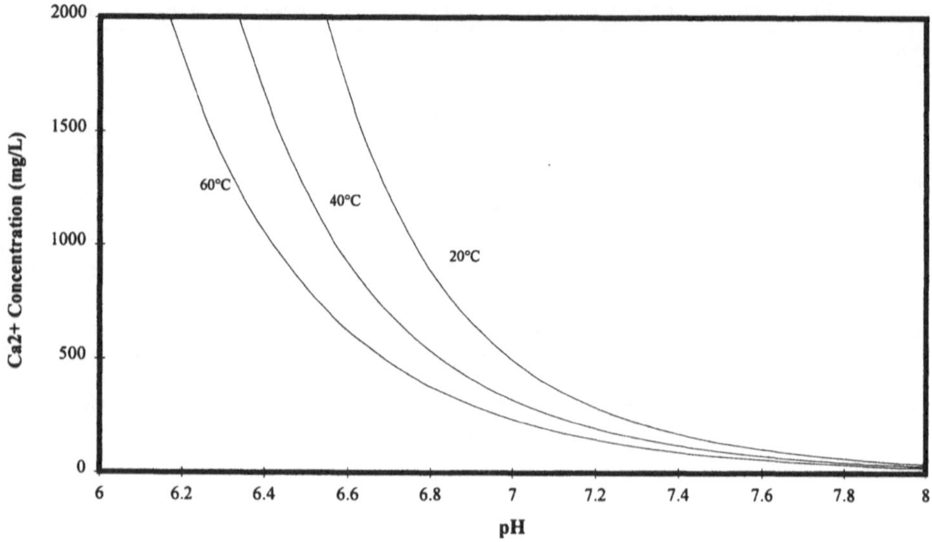


**Figure 2. Calcium solubility (total carbonates = 1 mM; pH 6–8).**

chemistry at thermophilic temperatures is the solubility of calcium (Figures 2 and 3): calcium carbonate solubility decreases with temperature in addition to a shift in the carbonate system toward a higher carbonate concentration (i.e., lower carbonic acid and bicarbonate concentrations).

## Microbial Growth and Substrate Utilization Kinetics

Modeling the kinetics of microbial growth and substrate utilization provides a powerful tool during the design of biological wastewater treatment facilities. In practice, kinetic models, combined with reactor design principles, are used to predict: (1) effluent concentrations of surrogate

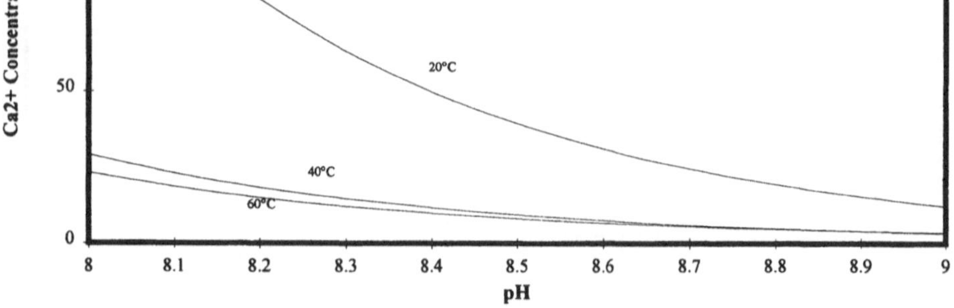


**Figure 3. Calcium solubility (total carbonates = 1 mM; pH 8–9).**

pollutant parameters such as COD, $BOD_5$, etc.; (2) effluent concentrations of specific compounds designated as priority pollutants; (3) residual biomass production rates; and (4) oxygen requirements for aerobic treatment systems. The rates of microbial growth and substrate utilization, respectively, in biological treatment systems are often described by the following equations originally presented by Herbert:[36]

$$\left(\frac{dX}{dt}\right)_g = \frac{\mu_m S}{K_s + S} X - bX \tag{1}$$

$$\left(\frac{dS}{dt}\right)_u = -\frac{1}{Y_T} \frac{\mu_m S}{K_s + S} X \tag{2}$$

where $\mu_m$ is the maximum specific growth rate (time$^{-1}$); $X$ is the cell mass concentration (mass volume$^{-1}$); $K_s$ is the Monod constant (mass volume$^{-1}$); $S$ is the substrate concentration (mass volume$^{-1}$); $Y_T$ is the maximum yield of growth (mass mass$^{-1}$); and $b$ is the endogenous decay coefficient (time$^{-1}$). These equations were later applied by Lawrence and McCarty[37] to produce the familiar equation predicting sludge age ($\theta_c$) at steady-state in a completely-mixed aeration basin with biomass recycle:

$$\frac{1}{\theta_c} \frac{\mu_m S}{K_s + S} - b \tag{3}$$

The maximum specific rate of substrate utilization per unit mass of microorganisms ($k_{max}$) has been reported from only a few laboratory or pilot studies on thermophilic aerobic wastewater treatment. As can be observed from Figure 4, there is a clear increase in $k_{max}$ with temperature. Rather obviously, higher substrate utilization rates increase the amount of COD removed in a given reactor. This advantage becomes particularly pertinent for the treatment of high-strength wastewaters in which very large and expensive aeration basins are often required.

Another unique characteristic of growth at high temperature is a high maintenance requirement which reduces the observed growth yield ($Y_{obs}$) at higher mean cell residence times, as described by the following equation:

$$Y_{obs} = \frac{Y_T}{1 + b\theta_c} \tag{4}$$

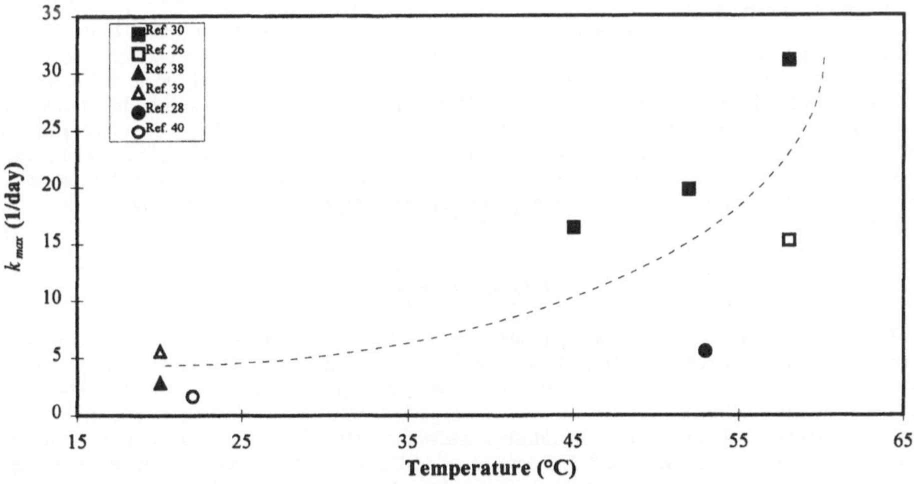


**Figure 4. Effect of temperature on substrate utilization rate.**

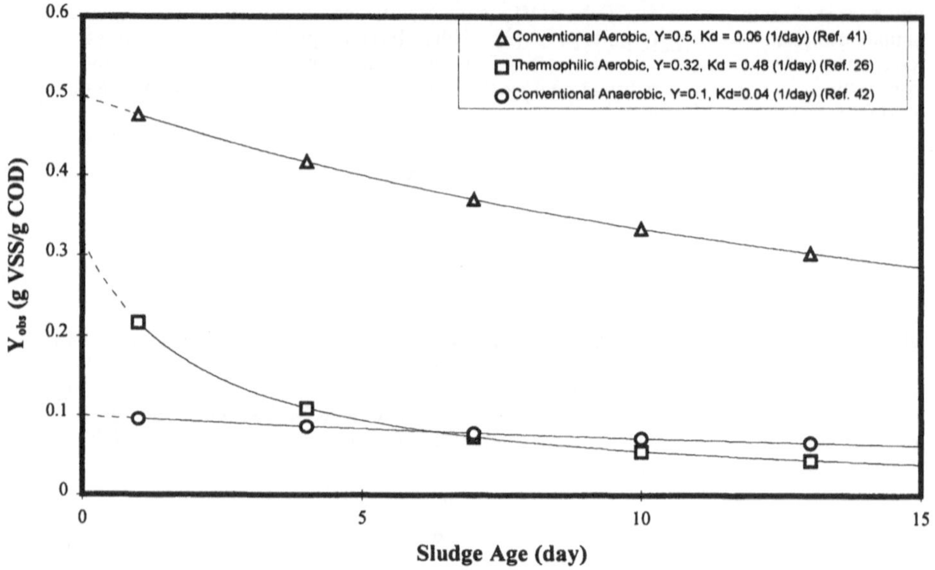


**Figure 5. Observed growth yields.**

The high maintenance requirement observed has been attributed to an additional energy requirement to repair enzymes and other cell macromolecules which are less stable at higher temperatures. In most wastewater treatment systems at ambient temperatures, *b* is often on the order of 0.05 $day^{-1}$; in thermophilic aerobic treatment systems, *b* is closer to 0.5 $day^{-1}$. This extremely high maintenance requirement reduces the amount of residual biomass requiring disposal during biological treatment at thermophilic temperatures, thereby decreasing the overall economic cost of wastewater treatment. At sludge ages typically used in wastewater treatment, thermophilic aerobic systems produce sludge at a rate similar to anaerobic treatment (Figure 5).

## Microbial Flocculation

Poor microbial settling characteristics are often observed with thermophilic treatment systems, hindering solids separation with conventional clarifiers. Most researchers report a highly turbid effluent, which is generally attributed to dispersed growth. Only Gehm[11] and Barr et al.,[15] however, have reported good settling. The operating strategies in these studies on wastewaters from the pulp and paper industry were high reactor pH (~9.5–9.8) and extended mean cell residence times, respectively. The inability to develop a flocculant microbial consortia may be the leading hindrance to thermophilic aerobic systems becoming a feasible technology for high-strength wastewater treatment. Two options currently exist: (1) operate without solids recycle; or (2) separate biological solids with membrane filters as used by Colvin et al.[34] The former alternative severely reduces process efficiency and requires further processing downstream, while the latter may be too costly for widespread implementation.

## CONCLUSIONS

Thermophilic aerobic biological treatment systems have many advantages compared to conventional techniques for high-strength wastewaters, including: faster degradation rates, greater overall process efficiencies, and low rates of sludge production. Despite these apparent benefits, however, thermophilic aerobic systems remain largely unexplored as a viable treatment technology for high-strength and high-temperature wastewaters. High bioreactor temperatures alter the physical, chemical, and biological characteristics of the treatment process, such that the enormous knowledge-base for conventional activated sludge operations no longer directly applies. Many of these process-related differences were discussed herein. Thermophilic biodegradation kinetics appear to offer greatly accelerated degradation rates well beyond that of conventional aerobic treatment, while producing residual biosolids at rates similar to anaerobic systems. These systems also frequently exhibit poor settling characteristics which hinder biosolids separation by conventional gravitational settling techniques.

## REFERENCES

1. Brock, T.D. and H. Freeze, "*Thermus aquaticus* gen. n. and sp. n., a non-sporulating extreme thermophile," *J. Bacteriol.*, 98, p. 289, 1969.
2. Arnheim, N. and H. Erlich, "Polymerase chain reaction strategy," *Annu. Rev. Biochem.*, 61, p. 131, 1992.
3. Wiegel, J. and L.G. Lungdahl, "The importance of thermophilic bacteria in biotechnology," *CRC Crit. Rev. Biotechnol.*, 3, p. 39, 1983.
4. Eckenfelder, W.W. and J. L. Musterman, *Activated Sludge Treatment of Industrial Wastewater*, Technomic Publishing, Inc., Lancaster, PA, 1995.
5. United States Environmental Protection Agency, *Environmental Regulations and Technology: Autothermal Thermophilic Aerobic Digestion of Municipal Wastewater Sludge*, EPA/625/10-90/007, 1990.
6. Kambhu, K. and J. F. Andrews, "Aerobic thermophilic process for the biological treatment of wastes—simulation studies," *J. Water Pollut. Control Fed.*, 41, p. R127, 1969.
7. Matsch, L. C. and R. F. Drnevich, "Autothermal aerobic digestion," *J. Water Pollut. Control Fed.*, 49, p. 296, 1977.
8. Jewell, W.J. and R. M. Kabrick, "Autoheated aerobic thermophilic digestion with aeration," *J. Water Pollut. Control Fed.*, 52, p. 512, 1980.
9. Carter, J.L. and W.F. Barry, "Effects of shock temperature on biological systems," *J. Sanit. Eng. Div., ASCE*, 101, p. 757, 1975.
10. Rudolfs, W. and H. R. Amberg, "White water treatment v. aeration with nonflocculent growths," *Sewage Indust. Wastes*, 25, p. 70, 1953.
11. Gehm, H.W., Activated Sludge at High Temperatures and High pH Values, In: *Biological Treatment of Sewage and Industrial Wastes, Volume I: Aerobic Oxidation,* McCabe, J., and W. W. Eckenfelder, eds., Reinhold Publishing Corporation, New York, 1956, p. 352.
12. Carpenter, W.L., J.G. Vamvakias, and I. Gellman, "Temperature relationships in aerobic treatment and disposal of pulp and paper wastes," *J. Water Pollut. Control Fed.*, 40, p. 733, 1968.
13. Perttula, M., M. Konrádsdóttir, J. Pere, J.K. Kristjánsson, and L. Viikari, "Removal of acetate from NSSC sulphite pulp mill condensates using thermophilic bacteria," *Water Res.*, 25, p. 599, 1991.
14. Rintala, J. and R. Lepistö, "Thermophilic anaerobic-aerobic and aerobic treatment of kraft bleaching effluents," *Water Sci. Techn.*, 28(2), p. 11, 1993.
15. Barr, T.A., J.M. Taylor, and S.J.B. Duff, "Effect of HRT, SRT and temperature on the performance of activated sludge reactors treating bleached kraft mill effluent," *Water Res.*, 30, p. 799, 1996.
16. Pöpel, F. and C. Ohnmacht, "Thermophilic bacterial oxidation of highly concentrated substrates," *Water Res.*, 6, p. 807, 1972.
17. Ginnivan, M.J., J.L. Woods, and J.R. O'Callaghan, "Thermophilic aerobic treatment of pig slurry," *J. Agric. Eng. Res.*, 26, p. 455, 1981.
18. Ginnivan, M.J., J.L. Woods, and J.R. O'Callaghan, "Survival of *Salmonella dublin* in pig slurry during aerobic thermophilic treatment in batch, cyclic and continuous systems," *J. Appl. Bacteriol.*, 49, p. 13, 1980.
19. Ginnivan, M.J., J.L. Woods, and J.R. O'Callaghan, "Evaluation of thermophilically-treated pig slurry as a source of nutrients in pig diets," *Agric. Environ.*, 5, p. 227, 1980.
20. Beaudet, R., C. Gagnon, J. G. Bisaillon, and M. Ishaque, "Microbiological aspects of aerobic thermophilic treatment of swine waste," *Appl. Environ. Microbiol.*, 56, p. 971, 1990.
21. Tischer, R.G., L.R. Brown, and D.W. Cook, "Decomposition of wastewater by thermophilic microorganisms," *J. Water Pollut Control Fed.*, 34, p. 1244, 1962.
22. Hunter, J.V., E.J. Genetelli, and M.E. Gilwood, Temperature and Retention Time Relationships in the Activated Sludge Process, In: *Proceedings of the 21st Industrial Waste Conference*, Purdue University, Lafayette, IN, 1966, p. 953.
23. Shindala, A. and J.E. Parker, "Thermophilic activated sludge process," *Water Wastes Eng.*, 7, p. 47, 1970.

24. McNary, R.R., R.W. Wolford, and M.H. Dougherty, "Pilot plant treatment of citrus waste water by activated sludge," *Sewage Indust. Wastes*, 28, p. 894, 1956.
25. Dougherty, M.H. and R. R. McNary, "Elevated temperature effect on citrus waste activated sludge," *Sewage Indust. Wastes*, 30, p. 1263, 1958.
26. Sürücü, G.A., E.S.K. Chian, and R.S. Engelbrecht, "Thermophilic microbiological treatment of high strength wastewaters with simultaneous recovery of single cell protein," *Biotechnol. Bioeng.*, 17, p. 1639, 1975.
27. Sürücü, G.A., E.S.K. Chian, and R.S. Engelbrecht, "Aerobic thermophilic treatment of high strength wastewaters," *J. Water Pollut. Control Fed.*, 48, p. 669, 1976.
28. Jackson, M.L., Thermophilic Treatment of a High-Biochemical Oxygen Demand Wastewater: Laboratory, Pilot-Plant and Design, In: *Proceedings of the 37th Purdue Industrial Waste Conference*, West Lafayette, IN, Ann Arbor Science Publishers, Ann Arbor, MI: 1983, p. 753.
29. Couillard, D., S. Gariépy, and F.T. Tran, "Slaughterhouse effluent treatment by thermophilic aerobic process," *Water Res.*, 23, p. 573, 1989.
30. Gariépy, S., R.D. Tyagi, D. Couillard, and F. Tran, "Thermophilic process for protein recovery as an alternative to slaughterhouse wastewater treatment," *Biological Wastes*, 29, p. 93, 1989.
31. Couillard, D. and S. Zhu, "Thermophilic aerobic process for the treatment of slaughterhouse effluents with protein recovery," *Environ. Pollut.*, 79, p. 121, 1993.
32. Zvauya, R., W. Parawira, and C. Mawadza, "Aspects of aerobic thermophilic treatment of Zimbabwean traditional opaque-beer brewery wastewater," *Bioresource Technol.*, 48, p. 273, 1994.
33. Rozich, A.F., S.G. Clay, and R.J. Colvin, Thermophilic Aerobic Treatment of High-Strength Ground Water: Bench-Scale Results, In: *Proceedings of the 47th Purdue Industrial Waste Conference*, West Lafayette, IN, Ann Arbor Science Publishers, Ann Arbor, MI: 1992, p. 261.
34. Colvin, R.J., A.F. Rozich, V. D'Aco, G.H. Hollerbach, and D. Mitchell, "Thermophilic Aerobic Treatment of High Strength Groundwater: Full-scale System Operations," Presented at the 69th Annual Conference & Exposition, Dallas, Texas, paper 9620001, 1996.
35. Boogerd, F.C., P. Bos, J.G. Kuenen, J.J. Heijnen, and R.G.J.M. van der Lans, "Oxygen and carbon dioxide mass transfer and the aerobic, autotrophic cultivation of moderate and extreme thermophiles: A case study related to the microbial desulfurization of coal," *Biotechnol. Bioeng.*, 35, p. 1111, 1990.
36. Herbert, D., Some Principles of Continuous Culture, In: *Recent Progress in Microbiology*, G. Tunevall, ed., Springfield, IL: 1958, p. 381.
37. Lawrence, A.W., and P.L. McCarty, "Unified basis for biological treatment design and operation," *J. Sanit. Eng. Div., ASCE*, 96, p. 757, 1970.
38. Stack, V.T. and R.A. Conway, "Design data for completely-mixed activated sludge treatment," *J. Water Pollut. Control Fed.*, 31, p. 1181, 1959.
39. Benedek, P. and I. Horvath, "A practical approach to activated sludge kinetics," *Water Res.*, 1, p. 663, 1967.
40. Lovett, D.A., S.M. Travers, and K.R. Davey, "Activated Sludge Treatment of Abattoir Wastewater—I: Influence of Sludge Age and Feeding Pattern," *Water Res.*, 18, p. 429, 1984.
41. Tchobanoglous, G. and F.L. Burton, *Wastewater Engineering: Treatment, Disposal, and Reuse*, Third Edition, Metcalf & Eddy, Inc., McGraw-Hill, Inc., 1991.
42. Speece, R.E., *Anaerobic Biotechnology for Industrial Wastewaters*, Archae Press, Nashville, TN, 1996.

# Section Two
# BIOLOGICAL: INHIBITION AND TOXICITY

## 5 WHOLE EFFLUENT TOXICITY IN STORMWATER: COMMONLY IDENTIFIED TOXICANTS AND COMPOUNDING FACTORS

**Micheal B. Copenhaver**, Project Manager
**Douglas J. Fort**, Vice President
**Enos L. Stover**, President
**John K. Delphon**, Laboratory Manager
**Thomas W. Gorman**, Laboratory Supervisor
THE STOVER GROUP
P.O. Box 2056
Stillwater, Oklahoma 74076

### INTRODUCTION

Approaches used today to regulate the discharge of contaminants in stormwater to the environment are similar to those used by National Pollutant Discharge Elimination System (NPDES) permits for wastewater effluent discharge. Stormwater regulations require the development and implementation of a specific management plan for controlling contaminant release, as well as specific water quality monitoring which includes Whole Effluent Toxicity testing. Unlike wastewater effluents, many facilities discharging stormwater do not provide pretreatment prior to release. Thus, the number of Whole Effluent Toxicity problems emanating from stormwater are escalating, particularly in the midwestern United States.

Stormwaters from two industrial facilities were studied to determine the cause(s) and source(s) of aquatic toxicity, and provide recommendations on practical remedial actions needed to abate Whole Effluent Toxicity. The Case Study 1 facility is a mirror manufacturing company, and the Case Study 2 facility manufactures wire. Results from both facilities' stormwater toxicity tests indicated that the runoff from the respective collection points induced acute toxicity to both *Daphnia pulex* and *Pimephales promelas*. Both cases were superficially straightforward in terms of identifying the causes of toxicity. The Toxicity Identification Evaluation (TIE) studies, however, were complicated by the physical/chemical characteristics typical of the stormwater encountered in the midwest region. First, virgin rainwater is extremely low in ionic strength containing virtually no hardness ($CaCO_3$) and second, the pH of the regional stormwater is typically acidic. Thus, the nature of the stormwater itself poses problems in terms of Whole Effluent Toxicity. These stormwater chemical characteristics were the principal issues in the two studies. The results from the associated investigations from Case Study 1 and Case Study 2 are presented herein.

### INVESTIGATIVE PROGRAM

#### Approach

Stormwater in the midwestern United States may be ionically imbalanced or deficient in critical ions necessary to support a homeostatic environment for aquatic organisms, particularly the

*52nd Purdue Industrial Waste Conference Proceedings*, 1997, Ann Arbor Press, Chelsea, Michigan 48118. Printed in U.S.A.

macroinvertebrate species. The low concentrations of conductive dissolved solids can create a stressful environment by disrupting osmoregulation. The deficiency in hardness also reduces the capacity of the stormwater to buffer the aquatic toxicity of many cationic metals. The release of even small quantities of these heavy metals can easily result in Whole Effluent Toxicity, and because the regional stormwater is acidic, it tends to be corrosive, causing the release and solubilization of corrosion-based metals such as copper, lead, nickel, and zinc. Preliminary evaluation of both case study facilities indicated a high probability of metals-induced toxicity and stormwater matrix issues. Investigative procedures in Case Study 1 included a Phase I: Acute Toxicity Characterization/Identification Evaluation of discharge from the stormwater collection point (Outfall 001), toxicity screening tests on ion exchange-treated roof discharge (Roof 001), and toxicity screening tests with pH and conductive dissolved salts adjustments on discharge from both Outfall 001 and Roof 001 collection points. Case Study 2 included a Phase I: Acute Toxicity Characterization/Identification Evaluation of discharge from the stormwater collection point (Outfall 001), a hardness (as $CaCO_3$) spiking study, and an anion-cation screen and balance. Acute toxicity screening tests of the virgin rainwater were also conducted.

## Methods

### *Phase I: Acute Toxicity Characterization/Identification*

The Phase I: Acute Toxicity Characterization/Identification studies were conducted in accordance with the protocol described in Methods for Aquatic Toxicity Identification Evaluations: Phase I Toxicity Characterization Procedures (EPA/600/6-91/003).[1] The protocol is outlined in Figure 1. Because the initial stormwater toxicity tests had indicated greater toxicity to *P. promelas* than to *D. pulex*, the Case Study 1 characterization was performed on Outfall 001 discharge using *P. promelas* only. However, the study included a partial characterization conducted on Roof 001 discharge utilizing the $C_{18}$-Solid Phase Extraction (SPE) and mixed-bed (MB) ion exchange treatments using both *D. pulex* and *P. promelas*. Case Study 2 initially indicated equal toxicity to *D. pulex* as well as *P. promelas*; thus, both species were used in the characterization of Outfall 001 discharge.

### *Phase I Toxicity Testing*

Definitive acute toxicity tests used to evaluate the characterization treatments were conducted as described in Methods for Measuring the Acute Toxicity of Effluents to Freshwater and Marine

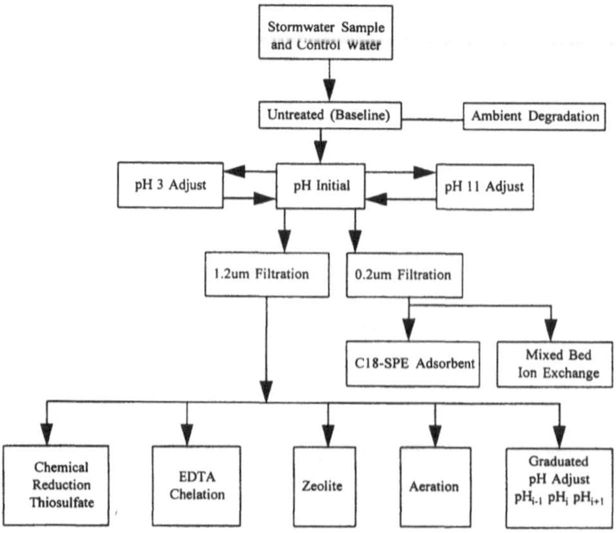


**Figure 1. Schematic of Phase I: Toxicity characterization/identification evaluation.**

Organisms: Fourth Edition (EPA/600/4-90/027F).[2] For each treatment, a sample aliquot was diluted with moderately hard reconstituted water to the following concentrations: 100%, 75%, 50%, 25%, and 0% (control). Five 24-hour-old organisms, *D. pulex* or *P. promelas*, were placed in a test vessel of each sample concentration for each given treatment. Median lethal concentrations (LC50) were determined by the Trimmed Spearman-Karber method when possible.

Toxicity tests were conducted on the following treatments and reconstituted water control fractions: unfiltered aliquot of the initial pH fraction (baseline); unfiltered pH-adjusted aliquots (pH 3.0 and pH 11.0); unfiltered aerated aliquot of the initial pH fraction; 1.2 μm filtered aliquot of the initial pH fraction; 0.2 μm filtered $C_{18}$-Solid Phase Extraction (SPE) treated aliquot of the initial pH fraction; 0.2 μm filtered mixed-bed (MB) ion exchange treated aliquot of the initial pH fraction; 1.2 μm filtered zeolite (ammonium ion exchange resin) treated aliquot of the initial pH fraction; 1.2 μm filtered EDTA-chelated aliquot of the initial pH fraction; and 1.2 μm filtered sodium thiosulfate reduced aliquot of the initial pH fraction.

The $C_{18}$-SPE and MB ion exchange treatments were conducted using automated Ion Chromatography methodology. Use of this method required the stormwater to be micro-filtered (0.2 μm) prior to injection into the instrument. Therefore, as a standard QA/QC procedure, a 0.2 μm filtered aliquot of the initial pH fraction was also tested. Additionally, an unfiltered fraction of Case Study 2 stormwater was set aside under ambient laboratory conditions for 24 hours to test for possible degradation of toxicity. Identical test procedures were then used to test for ambient toxicity degradation.

### *Acute Toxicity Screening Tests*

The acute toxicity screening tests were conducted as described in Methods for Measuring the Acute Toxicity of Effluent to Freshwater and Marine Organisms: Fourth Edition (EPA/600/4-90/027F).[2] The tests in both case studies were conducted in 30 mL plastic soufflé cups with 20 mL of test solution (0% and 100%) per cup. The Case Study 1 tests were conducted on three treated aliquots of stormwater discharge from Outfall 001 and Roof 001 using *D. pulex* and *P. promelas*. The chemical parameters of the aliquots from each discharge point were manipulated by: (1) adjusting the pH to 7.0 ; (2) addition of salts to increase hardness; and (3) combining pH adjustment to 7.0 with the addition of conductive salts. Magnesium sulfate, sodium bicarbonate, potassium chloride, and calcium sulfate salts were added at concentrations of 0.25 g/L, 0.2 g/L, 0.008 g/L, and 0.12 g/L, respectively. The Case Study 2 tests were conducted on aliquots of virgin rainwater using *D. pulex* and *P. promelas*. No manipulation of the virgin rainwater was performed prior to toxicity testing.

### *Supplemental Procedures*

Because preliminary evaluations had suggested possible metals-induced toxicity, additional investigative procedures were performed in both case studies. Case Study 1 included analyses for total cadmium, chromium, copper, lead, nickel, silver, and zinc on the untreated stormwater using atomic absorption spectroscopy. Case Study 2 included analyses for total cadmium, chromium, cobalt, copper, iron, lead, manganese, mercury, nickel, silver, and zinc. Additionally, in order to investigate the potential for toxicity resulting from an ionic imbalance and to ascertain if unknown ionic constituents were present, an anion-cation screen was conducted on the Case Study 2 discharge utilizing IC/HPLC methodology. The resulting data were evaluated and balanced with the computer model, GRI-FW STR™ (version 1.02). This computer model was developed by Gas Research Institute, Inc. to predict the toxicity of ionically complex wastewaters and effluents to freshwater organisms and can predict ionic imbalances in the effluent.

A hardness spiking study was conducted on the Case Study 2 discharge in conjunction with the Phase I characterization. The spiking study was designed to help confirm the suspected metals-induced toxicity, as well as to indicate on a preliminary basis the hardness required to reduce/remove such toxicity. The hardness standard solution was prepared by mixing 10 grams of calcium chloride into one liter of deionized water. Hardness of the standard solution was determined to be 9,000 mg/L (as $CaCO_3$). Six graduated hardness values were produced by spiking the standard solution into six different aliquots of the stormwater sample. Actual hardness values were then measured in each of the aliquots in order to ascertain a precise range in which toxicity may be reduced/removed. The resulting hardness values were 140, 188, 238, 342, 440, and 538 mg/L (as

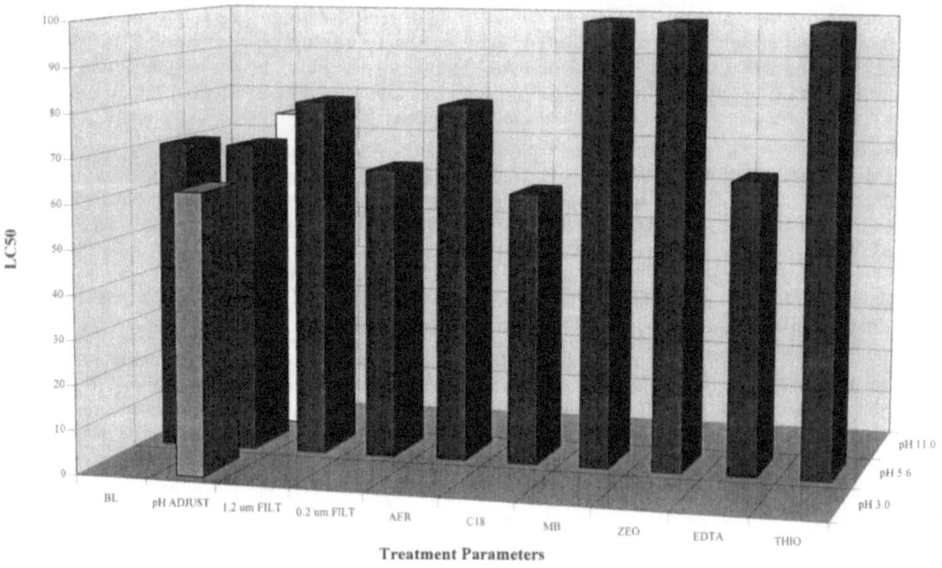


**Figure 2. Case Study 1 - Stormwater toxicity characterization (*P. promelas*).**

$CaCO_3$). Five 24-hour-old organisms, *D. pulex* or *P. promelas*, were placed in a test vessel of each hardness concentration. Percent mortality was determined at 24-hour intervals.

## RESULTS

### Case Study 1

*Phase I: Acute Toxicity Characterization/Identification*

The results of the Phase I: Acute Toxicity Characterization/Identification study performed on the Outfall 001 sample are presented in Figure 2. In Figure 2, bar height is inversely related to the level of toxicity, such that increasing bar height corresponds to decreasing sample toxicity. Full-scale deflection (LC50 100% sample) represents complete toxicity removal.

The MB ion exchange, zeolite, and sodium thiosulfate treatments completely removed all toxicity to *P. promelas*. Toxicity removal by the MB ion exchange treatment indicated that the toxicant(s) possessed ionic properties and, specifically, characteristics of divalent cations. Likewise, successful results from the zeolite treatment suggested the possible presence of a cationic toxicant(s). Although zeolite treatment is principally used to target ammonia-nitrogen-induced toxicity, the level detected in the sample (1.58 mg/L as NH3-N) did not support ammonia-nitrogen as a primary toxicant. Because zeolite acts as a cation ion exchange resin, toxicity removal may have been the result of nonspecific toxicant(s) binding, such as heavy metals and/or organic chelates.

The sodium thiosulfate treatment also removed all toxicity. Thiosulfate determines the extent to which the toxicant(s) can be reduced to nontoxic forms. The treatment is normally most effective in the presence of oxidants such as chlorine. However, in this instance, reduction in toxicity may have resulted from a chemical alteration of the stormwater matrix [oxidation-reduction (redox) potential] or chemical chelation of the metals which reduced the toxicity of the suspected toxicant(s) by effectively decreasing bioavailability of the metals.

**Table I. Case Study 1 - Acute Toxicity Screening Results from Roof 001 and Outfall 001 Stormwater Discharge**

| Collection Point | Treatment | Percent Mortality (24-hr) | |
|---|---|---|---|
| | | *D. pulex* | *P. promelas* |
| Roof 001 | pH adjustment (7.0) | 100% | 100% |
| | Salts added | 100% | 100% |
| | Salts added w/ pH adjustment | 100% | 100% |
| | C18-SPE | 100% | 100% |
| | MB ion exchange | 6% | 0% |
| Outfall 001 | pH adjustment (7.0) | 100% | 100% |
| | Salts added | 0% | 0% |
| | Salts added w/ pH adjustment | 0% | 0% |

*Acute Toxicity Screening Tests*

The results of the acute toxicity screening tests are presented in Table I. The initial pH values of the samples from Outfall 001 and Roof 001 were 5.6 and 4.9, respectively, and the hardness of both samples was determined to be 16 mg/L (as CaCO3). The toxicity tests with pH and salts manipulation were conducted because samples with an acidic pH and/or weak ionic strength can directly induce toxicity, as well as affect the bioavailability of some toxicants, especially heavy metals. The pH adjustment (7.0) to both Roof 001 and Outfall 001 discharge did not reduce toxicity to either *D. pulex* or *P. promelas*. Likewise, the addition of salts to the Roof 001 sample had no effect on toxicity. However, the addition of the salts to the Outfall 001 sample effectively eliminated toxicity to both species. The sample from Outfall 001 appeared to contain a high concentration of suspended organic solids, while the roof sample appeared colorless and contained little suspended material. The removal of the toxicity by the addition of salts to the Outfall 001 sample, therefore, may have been a result of the binding of the toxicant(s) by the organic solids. The increase in matrix ionic strength caused by adding the salts may have further increased sorption of the metals to the solids in the Outfall 001 sample, effectively decreasing metals bioavailability and thus, toxicity.

Of the ion exchange treatments conducted on the Roof 001 sample, only the MB ion exchange treatment removed the toxicity. These results again suggested that the toxicant(s) possessed ionic characteristics specific to divalent cations (EDTA-chelatable heavy metals). The results added credibility to the theory that the toxicity in Outfall 001 was metals-induced and possibly could be removed by sorption to suspended organic solids as a function of increasing the ionic strength of the matrix.

*Supplemental Findings*

The physical/chemical water analytical data, including metals analyses, are presented in Table II. Elevated levels of copper, silver, and zinc were detected in both the Outfall 001 and Roof 001 samples. Total concentrations of copper and silver in the Outfall 001 discharge were determined to be 110 µg/L each. Total concentration of zinc was 290,000 µg/L. Total concentrations of copper, silver, and zinc in the Roof 001 sample were determined to be 150 µg/L, 27 µg/L, and 390,000 µg/L, respectively. Although total metals did not necessarily represent the entire bioavailable fraction (i.e., the soluble fraction), the concentrations of copper, silver, and zinc appeared to be above levels which would be expected to cause acute toxicity based on the pH and hardness of the two samples.

The metals analyses further suggested that the copper and zinc were originating from the roof, while the silver was probably originating from an internal source prior to Outfall 001. Copper and zinc concentrations were lower in the Outfall 001 sample, indicating some dilution had occurred between the two collection points. The higher concentrations of silver, and to a less extent the introduction of nickel and lead, in the Outfall 001 sample suggested the possibility of contamination from a process waste stream. Although lead and nickel concentrations were detectable,

**Table II. Stormwater Discharge Physical/Chemical Water Quality Characteristics[a]**

| | Case Study 1 | | Case Study 2 | |
|---|---|---|---|---|
| Parameter | Outfall 001 | Roof 001 | Rainwater | Outfall 001 |
| Temperature (°C) | 25 | 25 | 25 | 25 |
| Dissolved Oxygen (mg/L) | 9.5 | 9.6 | — | 9.1 |
| pH (s.u.) | 5.6 | 4.9 | 3.8 | 7.0 |
| Conductivity (µmhos/cm$^2$) | 43 | 11 | 34 | 126 |
| Hardness (mg/L as $CaCO_3$) | 16 | 16 | 12 | 42 |
| Alkalinity (mg/L as $CaCO_3$) | ND[b] | ND | ND | 22 |
| Ammonia-Nitrogen (mg/L) | 1.58 | 0.42 | 0.48 | 1.1 |
| Total Residual Chlorine (mg/L) | <0.1 | <0.1 | <0.1 | <0.1 |
| Total Metals | | | | |
| Cadmium (µg/L) | <1.0 | <1.0 | | <1.0 |
| Chromium (µg/L) | <2.0 | <2.0 | | <0.05 |
| Cobalt (mg/L) | — | — | | <0.05 |
| Copper (µg/L) | 110.0 | 150.0 | | 58.6 |
| Iron (µg/L) | — | — | | 276.5 |
| Lead (µg/L) | 10.0 | 2.2 | | 100.0 |
| Manganese (µg/L) | — | — | | 14.0 |
| Mercury (µg/L) | — | — | | <0.2 |
| Nickel (µg/L) | 17.0 | <2.0 | | 248.5 |
| Silver (µg/L) | 110.0 | 27.0 | | <10.0 |
| Zinc (µg/L) | 290,000 | 390,000 | | 1,946 |
| Anions | | | | |
| Bromide (mg/L) | | | | <0.01 |
| Fluoride (mg/L) | | | | 0.76 |
| Phosphate-P (mg/L) | | | | 0.13 |
| Nitrate-N (mg/L) | | | | 0.83 |
| Nitrite-N (mg/L) | | | | <0.01 |
| Sulfate (mg/L) | | | | 9.78 |
| Chloride (mg/L) | | | | 12.89 |
| Cations | | | | |
| Lithium (mg/L) | | | | <0.001 |
| Sodium (mg/L) | | | | 3.19 |
| Ammonium (mg/L) | | | | 1.23 |
| Potassium (mg/L) | | | | 2.37 |
| Magnesium (mg/L) | | | | 0.77 |
| Calcium (mg/L) | | | | 15.26 |

[a]Case Study 1 samples collected March 28, 1996. Case Study 2 samples collected November 4, 1996.
[b]Not detected.

copper, silver, and zinc concentrations were several times that which might be expected to induce acute toxicity with respect to the water quality characteristics of the samples.

## Case Study 2

*Phase I: Acute Toxicity Characterization/Identification*

The results of the Phase I: Acute Toxicity Characterization/Identification study performed on the Outfall 001 sample using *P. promelas* and *D. pulex* are presented in Figures 3 and 4, respectively. As with Case Study 1, the physical/chemical water analytical data, including metals and anion-cation analyses, are presented in Table II. In the figures, bar height is inversely related to the level of toxicity, so that increasing bar height corresponds to decreasing sample toxicity. Full-scale deflection (LC50 ± 100% sample) represents complete toxicity removal.

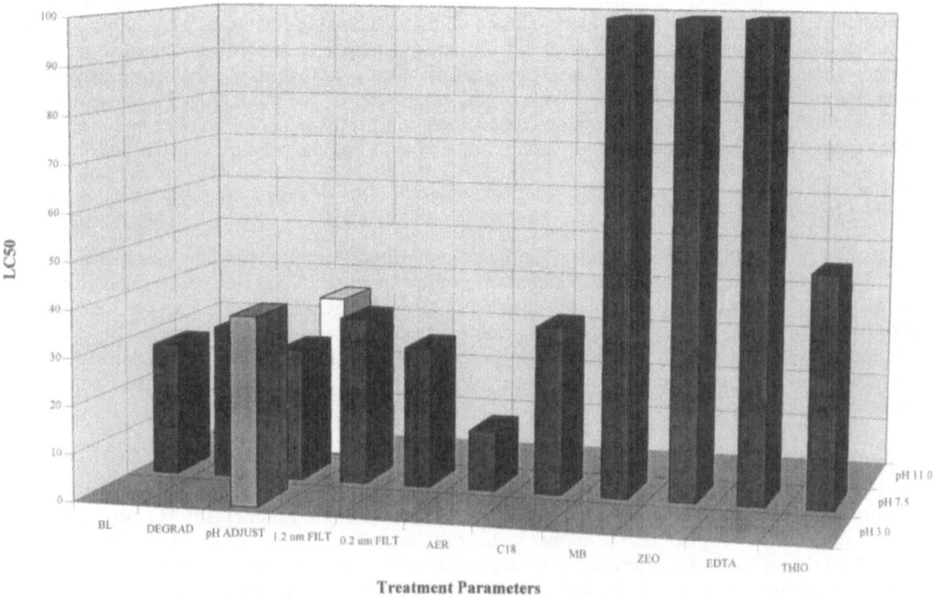


**Figure 3. Case Study 2—Stormwater toxicity characterization (*P. promelas*).**

As illustrated by Figures 3 and 4, results of the major treatments on toxicity to both test organisms were notably similar. Moreover, the results from Case Study 2 were comparable to the results from the Case Study 1. The MB ion exchange, zeolite, and EDTA chelation treatments removed all toxicity to both *D. pulex* and *P. promelas*. As previously discussed, the successful removal of

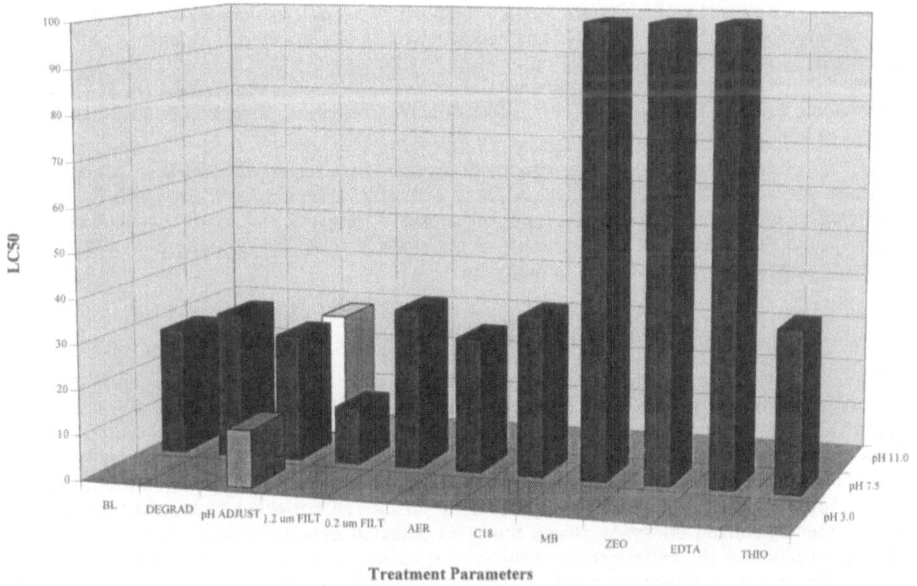


**Figure 4. Case Study 2—Stormwater toxicity characterization (*D. pulex*).**

toxicity following these treatments indicated that the toxicant(s) possessed ionic properties and, specifically, characteristics of divalent cations (EDTA-chelatable heavy metals). Although no effect was observed in the *D. pulex* test, the sodium thiosulfate treatment may have somewhat reduced toxicity to *P. promelas*. As observed in Case Study 1, it is likely that any thiosulfate-induced toxicity reduction was the result of the alteration of the toxicant(s) themselves or the stormwater matrix [oxidation-reduction (redox) potential]. This phenomenon has been recorded in other cases in which divalent cationic metals were found to be the primary cause of toxicity.

*Acute Toxicity Screening Tests*

Because weak ionic strength and/or acidic pH can directly induce toxicity, acute toxicity tests and water quality analyses were conducted on virgin rainwater. No toxicity was induced to either *D. pulex* or *P. promelas*; however, the pH of the rainwater was determined to be 3.8 (Table II). Rainwater typically has a pH in the range of 4.5 to 6.5; thus a pH of 3.8 would be considered acidic and would certainly be corrosive to exposed surface areas. Under these conditions, metal surfaces would be particularly subject to corrosion and the soluble metals would be readily transported with the runoff.

*Supplemental Findings*

Because ionic constituents were suggested as causing toxicity, analyses for total cadmium, chromium, cobalt, copper, lead, manganese, iron, mercury, nickel, silver, and zinc were conducted on the unfiltered stormwater discharge (Table II). Total concentrations of nickel and zinc were determined to be 249 µg/L and 1,946 µg/L, respectively. Based on the physical/chemical characteristics of the stormwater matrix (hardness of 42 mg/L as $CaCO_3$), the individual concentrations of nickel and zinc in the discharge were well beyond the ranges believed capable of inducing toxicity to either test species. Considering the weak ionic strength of the matrix, the concentration of copper (58.6 µg/L), and possibly lead (100 µg/L), were also within the toxicity range, and the additive effect of all the metals detected could have further exacerbated the toxicity. The anion-cation balance indicated no ionic imbalance and no toxicity was predicted by the GRI-FW STR™ (version 1.02) computer model; thus, the ionic balance of the stormwater studied did not appear to problematic.

The hardness spiking study was conducted in conjunction with the Phase I characterization study to help confirm the suspected metals-induced toxicity, as well as indicate on a preliminary basis the hardness required to reduce/remove such toxicity. Results from the *P. promelas* and *D. pulex* tests are presented in Figures 5 and 6, respectively. The *D. pulex* test results indicated that a hardness of 440 mg/L (as $CaCO_3$) was necessary to reduce toxicity to *D. pulex* for the first 48-hours. At 72-hours, however, 100% mortality was observed in the test concentration with the greatest ionic strength (i.e., 538 mg/L as $CaCO_3$). These results suggested that, in order to remove acute toxicity to *D. pulex*, the hardness of the stormwater discharge would have to be in excess of 538 mg/L (as $CaCO_3$).

The *P. promelas* test results were more definitive. Test results determined that a minimum hardness of 342 mg/L (as $CaCO_3$) was necessary to remove toxicity to *P. promelas* for the full 4-day testing period. Vertebrate species are less sensitive to heavy metals toxicity than are invertebrate species. Therefore, the total removal of toxicity at a lower ionic strength than that affecting the more sensitive *D. pulex* was not unexpected.

## DISCUSSION

The Phase I characterization studies performed in Case Study 1 and Case Study 2 indicated that the toxicants possessed ionic properties and, specifically, characteristics of divalent cations (EDTA-chelatable heavy metals). Because stormwater is deficient in hardness which buffers the aquatic toxicity of many cationic metals, release of small quantities of these heavy metals can easily cause Whole Effluent Toxicity problems. This was the circumstance with both midwestern facilities studied. Analytical analyses of the discharge collected from two locations at the mirror manufacturing company (Case Study 1) detected elevated levels of copper, silver, and zinc. Evaluation of the metals concentrations further indicated two possible sources of contamination. Because of the higher concentrations in the Roof 001 sample, copper and zinc probably originated from the roof. As the samples were collected at the same time, the lower concentra-

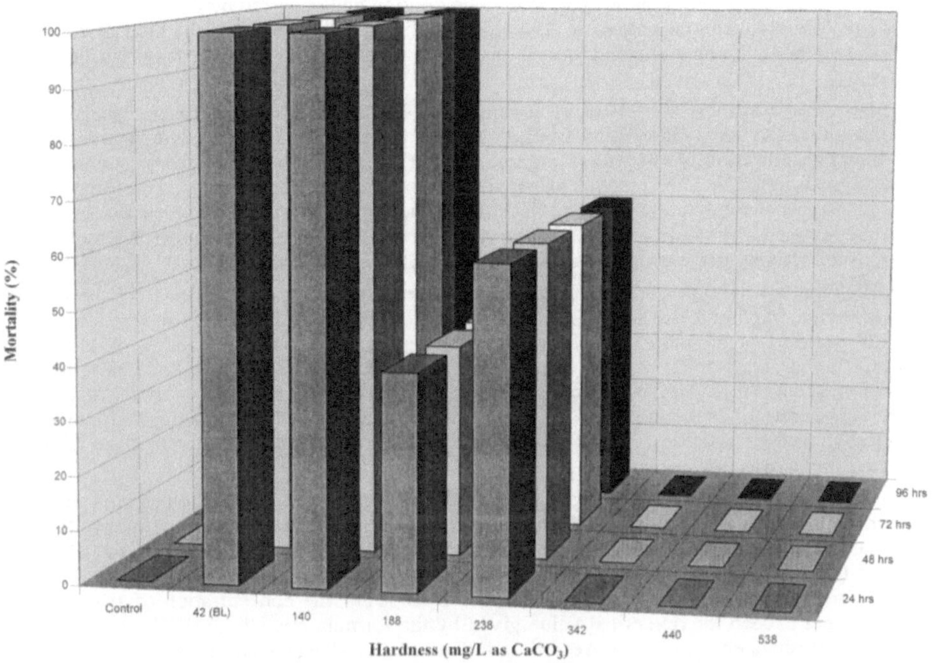


**Figure 5. Case Study 2 - Hardness spiking study (*P. promelas*).**

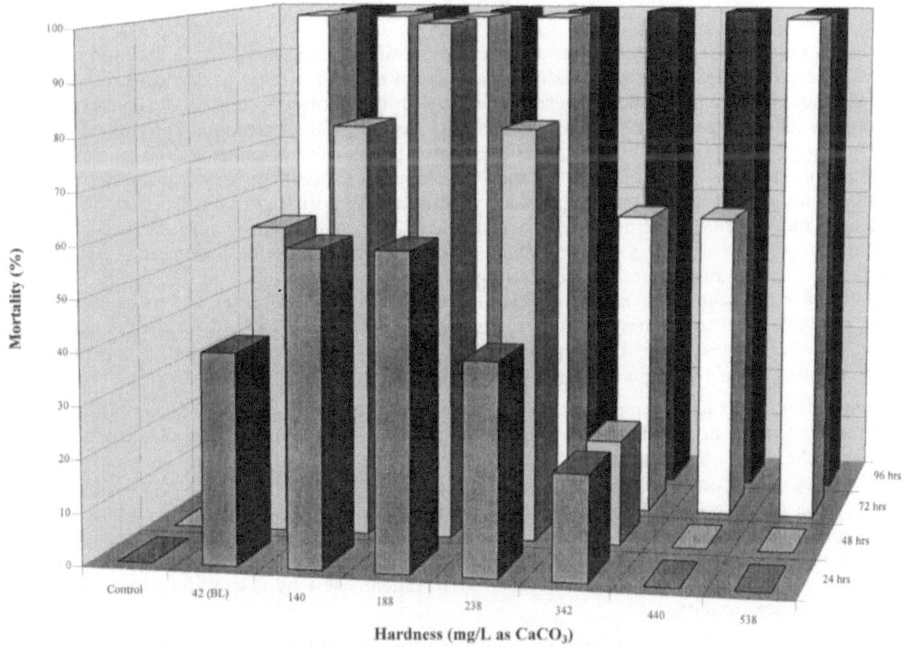


**Figure 6. Case Study 2 - Hardness spiking study (*D. pulex*).**

tions in the Outfall 001 sample indicated some dilution had occurred between the two collection points. Silver concentrations, as well as lead and nickel, were higher in the Outfall 001 sample; thus, originating from a source closer in proximity to the outfall, possibly from within the facility.

Although increasing the pH of each sample to 7.0 had no effect on toxicity, the addition of conductive salts removed toxicity from the Outfall 001 sample. This sample appeared to contain a high concentration of suspended organic solids (TSS), while the roof sample contained little suspended material. The removal of the toxicity by the addition of salts, therefore, may have been a result of the binding of the toxicant(s) by the organic solids. The increase in matrix ionic strength caused by adding the salts may have further increased sorption of the metals to the solids, effectively decreasing metals bioavailability and thus, toxicity.

The hardness spiking study conducted on the stormwater sample collected from the wire manufacturing company (Case Study 2) further illustrates the importance of the matrix ionic strength with regard to metals-induced toxicity. The test results indicated that a hardness of 440 mg/L (as $CaCO_3$) was necessary to remove toxicity to *D. pulex* for the first 48-hours and a minimum hardness of 342 mg/L (as $CaCO_3$) was necessary to remove toxicity to *P. promelas* for the full 4-day testing period. Since the minnows are less sensitive to metals-induced toxicity than are macroinvertebrates, the reduction of toxicity at a lower ionic strength than that affecting the more sensitive *D. pulex* was not surprising.

Metals toxicity is influenced by a number of chemical factors including calcium and magnesium salts, hardness, alkalinity, pH, and matrix ionic strength. These factors affect the toxicity of metals by either influencing the proportion of a particular metal ion available to the organism or inhibiting the sorption of the available metal ions by biological tissue. For example, as described in Quality Criteria for Water, 1986 (EPA 440/5-86/001),[3] the concentration of silver (in µg/L) should not exceed the numerical value given by the formula $e^{(1.72\,[\ln(\text{hardness})]\,-\,6.52)}$ for the survival of freshwater aquatic life. According to the equation, the concentration of silver in the Case Study 1 stormwater greatly exceeded the concentration believed to be capable of inducing acute toxicity relative to the hardness of the sample. Although formulas such as the one given should not be used to definitively predict toxicity, they do serve as general guidelines to predict potential toxicity problems, as well as illustrate the dependence of metals-induced toxicity on the matrix hardness.

## CONCLUSIONS

The two classes of toxicants most commonly identified in stormwater are cationic metals such as, copper, lead, zinc, nickel, cadmium, chromium, and silver; and pesticides (primarily organophosphates). Surfactants, miscellaneous semipolar organic toxicants, and cyanide have also been identified in specific case studies, but are much less common. The physical/chemical properties of stormwater which include extreme softness (low calcium and/or magnesium), low overall ion strength, ionic imbalance, and low pH; may exacerbate aquatic toxicity, particularly when involving cationic metals. Rainwater in the midwestern U.S. typically has an acidic pH, and until the natural buffering capacity is increased, the runoff can be very corrosive to exposed surface areas. Corrosiveness of metal surfaces is high under these environmental conditions and solubilized metals in elevated concentrations are readily transported with the runoff. Since heavy metals are more soluble at acidic pH ranges, decreasing pH increases the proportion of metal ions available, and soft water contains fewer salts to compete with the increased available metal ions for uptake sites in the test organisms. Thus, even the release of small quantities of these toxic metals can easily cause Whole Effluent Toxicity problems.

In addition to exacerbating potential metals-induced toxicity, in many cases, virgin rainwater will not support aquatic life due to the osmotic stress induced by the aforementioned physical/chemical characteristics. As previously stated, virgin stormwater contains virtually no hardness ($CaCO_3$) and is extremely low in ionic strength. The pH of the stormwater is typically acidic with pH ranging from 4.5 to 6.5. Stormwater in the midwestern U.S. region is ionically imbalanced or deficient in critical ions necessary to support a homeostatic environment for the aquatic organisms, particularly the macroinvertebrate species. The low concentrations of conductive dissolved solids can create a stressful environment itself by disrupting osmoregulation and possibly induce toxicity.

Based on the results of these studies, control of toxicity in the stormwater discharge from both case study facilities was determined to be dependent on the control of solubilized metals in the runoff. The acidic pH and softness of stormwater discharge provided the ideal environment to induce toxicity with only trace amounts of heavy metals. Once the suspect metals were identified through analyses, sources of the particular metals were easily defined in both case studies. Sources of the metals included corrosion from a metal roof, stormwater-induced cooling tower corrosion, and metals originating from specific operations at the facilities. Remedies for the Whole Effluent Toxicity problems included minimizing stormwater contact with metal surfaces, changes in stormwater collection and discharge, increased housekeeping and enhanced operational control, and increased stormwater pretreatment.

## REFERENCES

1. Methods for Aquatic Toxicity Identification Evaluations: Phase I Toxicity Characterization Procedures, EPA/600/6-91/003, U.S. Environmental Protection Agency, Cincinnati, OH, 1991.
2. Methods for Measuring the Acute Toxicity of Effluents to Freshwater and Marine Organisms: Fourth Edition, EPA/600/4-90/027F, U.S. Environmental Protection Agency, Cincinnati, OH, 1993.
3. Quality Criteria for Water, 1986, EPA 440/5-86/001, U.S. Environmental Protection Agency, Cincinnati, OH, 1986.

# 6 DETERMINATION OF INHIBITORY CONCENTRATIONS OF THE SURFACTANT AFFF ON BIOLOGICAL NITRIFICATION PROCESS AND PASS-THROUGH TOXICITY

**M. Erten-Unal**, Assistant Professor

**G. C. Schafran**, Associate Professor

**Sudhanva Paranjape, Edwin Garcia-Cardona, and Han Yan**, Graduate Students
Old Dominion University Department of Civil and Environmental Engineering, KDH 135, Norfolk, VA, 23529-0241

**Fred Williams**
Naval Research Laboratory

**David Cotnoir and Ken Clark**
NAVFAC LANTDIV

**Douglas Kirk**
NAS Oceana

## ABSTRACT

This chapter will evaluate the impact of the surfactant AFFF to a biological nutrient removal process and determine whether pass-through toxicity occurs in the effluent of a biological process receiving AFFF wastewater. A bench-scale study has been conducted to investigate the potential inhibitory effects of untreated AFFF wastewater to a BNR process.

## INTRODUCTION

The U.S. Navy utilizes a surfactant in firefighting water that improves the ability to control petroleum-based fires. The surfactant, which is widely used by the Navy including facilities in the Hampton Roads region, is currently manufactured by up to five companies and is commonly referred to as AFFF (aqueous film-forming foam) conforming to military specifications mil-F-24385F. The AFFF chemical makeup is proprietary and likely varies among manufacturers. The U.S. Navy is exploring a number of options that include disposal of the firefighting water to wastewater collections systems where the components of AFFF wastewater would be transported to a wastewater treatment facility and removed biologically.

Current disposal of firefighting water that includes AFFF wastewater has been limited by concerns for the environmental/toxic effects associated with AFFF. Disposal of the firefighting foam to sanitary sewers has been considered as an option; however, concern for the potential toxic or inhibitory effects associated with AFFF wastewater have generally led to a ban from introduction of AFFF to wastewater collection systems.

Present concerns over inhibitory effects of AFFF wastewater have resulted in the prohibition of its disposal to the Hampton Roads Sanitation District (HRSD) collection system where it would eventually enter one of the biological wastewater treatment plants operated by the District. HRSD is particularly concerned with how AFFF wastewater might interfere with biological nutrient removal (BNR) processes at its Virginia Initiative Plant (VIP). The VIP plant incorporates biological nitrogen, phosphorous, and organic matter (BOD/COD) removal through a sequential series of anaerobic, anoxic, and oxic reactors. Nitrogen removal occurs through microbially-mediated nitrification and denitrification and phosphorous removal occurs through enhanced uptake by poly P bacteria. It is well known that the nitrification and denitrification processes can be inhibited in the

*52nd Purdue Industrial Waste Conference Proceedings*, 1997, Ann Arbor Press, Chelsea, Michigan 48118. Printed in U.S.A.

presence of various chemicals and Poly P bacteria have been observed to be inhibited by $H_2S$ and 2,4-dinitrophenol. Because of the sensitivity of these microbial processes to inhibition, it is important to characterize the relationship between concentrations of various chemicals and the rates of conversion of nitrogen, phosphorous, and organic substrate.

Several studies have been performed on the disposal and treatment of AFFF surrogate wastewater. Bench-scale anaerobic and aerobic reactors were used to investigate the potential inhibition of the AFFF surrogates to nitrification, denitrification, and phosphorus release and uptake in a BNR process.[1,2] These effects were investigated for both untreated and pretreated AFFF wastewater. The results showed that pretreated wastewater allowed for complete nitrification and denitrification and untreated or pretreated wastewater did not have any adverse effect on denitrification and phosphorus release. The use of coagulants such as alum, ferric chloride, calcium chloride, and cationic polymers also have been observed to be capable of reducing the organic content of AFFF.[1-4]

Treatability studies have also been conducted with a high-purity oxygen activated sludge system. The results showed that acceptable untreated firefighting wastewater effluent could be obtained with a dilution factor of 100. The use of dissolved air flotation further reduced the dilution ratio to lower levels for acceptable effluent quality.[5,6]

The use of chemical pretreatment with dissolved air flotation (DAF) provided consistent removal of COD, BOD, TSS, and firefighting surfactants.[7] Overall, the use of coagulation, flocculation, and clarification aided in the reduction of organics prior to discharge to a BNR process. Upon chemical pretreatment and using aerobic and anaerobic sequencing batch reactors, it was found that an acceptable effluent that is amenable to an aquatic stream could be processed.[8-10] Additional studies were performed to determine the biodegradability of AFFF wastewater. Some of the additional work included the use of wastewater containing actual firefighting water and AFFF. However, detailed testing on the effects of actual AFFF wastewater on biological nutrient removal was not performed in these studies.[11,12] The biodegradability of commonly used AFFF surrogates which have been used in training activities were evaluated on bench-scale, continuous-feed activated sludge processes.[13-15] The AFFF dose that was fed to the reactor increased gradually from 100 ppm to 250 ppm. The results exhibited very good BOD and COD removal rates; however, nitrification was inhibited with increasing AFFF concentrations.

The overall objective of this study was to evaluate the impact of AFFF wastewater to a biological nutrient removal process and determine whether pass-through toxicity occurs in the effluent of a biological process receiving AFFF wastewater. A bench-scale study has been conducted to evaluate the potential inhibitory effects of untreated AFFF wastewater to the VIP BNR process. Under this testing, bench-scale reactors simulating the BNR process have been loaded at various AFFF concentrations and the influence on the process performance has been evaluated. The purpose of this effort was to determine the level of AFFF that could be incorporated into the influent of biological nutrient removal process without causing process upsets. Toxicity pass-through testing has also been performed to determine maximum loadings of the untreated AFFF wastewater without increasing effluent toxicity of the effluent from a BNR process. Acute toxicity of the effluent to *Mysidopsis bahia* (mysid shrimp) and *Cyprinodon variegatus* (sheepshead minnow) has been examined in toxicity testing of both control and AFFF-loaded inhibition reactors.

## METHODS AND MATERIALS

In order to maintain a continuous supply of uniform nitrifying microorganisms, a 100-liter capacity fill-and-draw type batch reactor was used at the Civil and Environmental Engineering laboratory of Old Dominion University. The reactor operation was under cyclical anaerobic and aerobic conditions to establish and maintain a nitrifying, phosphorus-accumulating biomass that would show similar nutrient removal performance as in the VIP plant. The reactor was seeded with a target mixed-liquor suspended-solids concentration using sludge from the VIP plant to obtain simultaneous nitrogen and phosphorus removal. Initially, an appropriate period of time was allowed to stabilize the reference reactor bacterial population. Reactor operation was sequenced with a programmable timer to activate mechanical mixing, aeration, solution feed, and mixed liquor and supernatant withdrawal. The reactor was operated on a 4-hour cycle of feed, anoxic react, aerated react, settle, and decant. The wastewater feed was prepared in the laboratory and included bactopeptone, beef extract, and other inorganic compounds and minerals. It was delivered to the reactor over the periods specified each day from a feed tank that is kept in the refrigerator. Aeration was provided through in-house air supply source and a dissolved oxygen concentration of approximately 4 mg/L was targeted during the feed cycle and react cycle in the reactor. Additional mixing was supplied by a mechanical mixer. After a specified settling period, supernatant (effluent) from the reactor was withdrawn by a solenoid valve and collected in a sample bottle for analysis.

The analytical methods employed in this study for evaluating the effects of AFFF wastewater inputs on biological treatment performance consisted of procedures as prescribed by the USEPA or in *Standard Methods*. All chemicals used were reagent grade or better and all quality assurance/quality control procedures were followed as closely as possible.

Measurements of organic strength were determined through $CBOD_5$, COD, and TOC measurements. $CBOD_5$ (determined with a nitrification inhibitor added to BOD bottles) were measured to eliminate potential interferences that nitrification could have on the evaluation of organics removal with the BOD test. CBOD, COD, and TOC analyses were determined using filtered samples on reactor effluent and filtered and unfiltered samples in the influent. Samples were filtered through a glass fiber filter to eliminate microorganisms and other particulate materials that are not related to the organic components of the AFFF or the dissolved organic compounds that are in the wastewater before AFFF introduction. Since the AFFF components are water soluble and will be dissolved in solution, filtration should not directly interfere with their accurate detection. Measurements of total suspended and volatile suspended solids (TSS and VSS, respectively) were used to determine organic solids loading, reactor MLSS concentrations, and nonsettleable TSS concentrations in reactor effluent. In order to reduce variability of TSS and VSS data, the tests were performed on the same days that solids concentrations feeding into the reactor. The nitrogen series were determined by three different analytical techniques. Persulfate digestion followed by ammonia analysis by ion selective electrode was utilized to determine total Kjeldahl nitrogen concentrations. Ammonia concentrations were measured by ion selective electrode without sample digestion. Total and dissolved phosphorous concentrations were determined on unfiltered and filtered samples by the ascorbic acid method following persulfate digestion.[16]

As part of this study, butyl carbitol, a major component of AFFF, is also being analyzed by ion chromatography utilizing electrochemical detection. Decomposition of fluoro-organic compounds are being evaluated by ion chromatography through determination of inorganic fluoride directly and inorganic fluoride following persulfate digestion. The change in fluoride concentration between pre- and postdigestion will give an indication of the amount of fluoride that is tied up in organic compounds. In addition to this indirect determination, the investigators are working with Dionex Corporation to develop a method that will directly measure fluoro-organic compounds. If method development is successful, this procedure will also be used to characterize the influence of treatment on the conversion of these compounds. These results will be made available after the completion of the analyses.

## BNR Inhibition Batch Assays

Nitrification inhibition was assessed in series for untreated AFFF wastewater using a batch assay procedure. Inhibition tests were performed with different concentrations of AFFF and controls using six, 6-liter batch reactors. Uniform seed biomass of approximately 4,000 mg/L was obtained from the reference reactor for each batch reactor. Approximately 2,000 mL of appropriate organic nutrients, inorganic nutrients and AFFF concentrate were added from the stock nutrient tank. Each batch reactor was equipped with an air supply source, an air stone, and a mixer. After the uniform seed biomass was added to each reactor, the air was turned on and the feed stock solution was introduced manually at four intervals during the two-hour fill cycle. The reactors were mixed and aerated during the feed cycle. At the end of this cycle, samples were withdrawn and the reactors were covered with lids to maintain anaerobic conditions. Mixing was continued to maintain the biomass in suspension.

At the end of two hours another sample was withdrawn from each reactor, air was turned on, and the lids were removed. Aeration and mixing were continued for another two hours, and additional samples were taken at the end of the aerobic cycle. Finally, the reactor contents were settled out and samples were taken from the supernatant after the decant cycle. Each sample from the reactors was analyzed for ammonia, $NO_3^-$, $NO_2^-$, phosphorus, COD, and TKN concentrations. Comparisons were made to the control which did not contain any AFFF.

All reactors were standardized for phosphorus, $NO_3^-$, $NO_2^-$, or ammonia concentrations if the AFFF addition significantly changed (i.e., mean concentrations varied by more than 20% between treatments and control) the total concentration of each respective parameter in the reactors. The degree of ammonia oxidation in beakers receiving a loading of AFFF wastewater was compared to the degree of oxidation in control reactors receiving similar volumes of tap water. All samples were held for less than 48 hours prior to use in testing. During the inhibition batch assay experiments, oxygen uptake rates were measured and the respiration rates were determined by SOUR (OUR/MLVSS) measurements. This procedure provided an indication of the effects of the untreated AFFF wastewater on the microorganisms.

### Toxicity Pass-Through Testing

Toxicity pass-through testing was performed on the inhibition reactors to estimate what the maximum loading of the untreated AFFF wastewater to the VIP plant would be without increasing effluent toxicity at the VIP plant. The acute toxicity pass-through tests were performed using the procedures outlined by the EPA.[17] At the end of the BNR inhibition batch aeration period, the mixed liquor was allowed to settle and clarified supernatant was decanted from each reactor and filtered through a coarse glass fiber filter. This filter is of the same type that is used for suspended solids analysis with 10 micrometer nominal size and without organic binder. Prior to use, the glass fiber filters were rinsed thoroughly by filtering high-purity water through the filter. The filtration apparatus was rinsed between each sample aliquot using 10% $HNO_3$, acetone, and high purity water. The filter toxicity was also checked by testing filtered dilution water.

Toxicity samples were submitted to a qualified bioassay laboratory for acute toxicity testing using *M. bahia* and *C. variegatus* following the current EPA procedures. It was ensured that the laboratory would perform a standard reference toxicant test on a regular basis and develop accompanying quality control charts. All samples were held for less than 48 hours prior to use in testing unless an approved method of preservation was utilized.

## RESULTS AND DISCUSSION

Initially, a range of AFFF wastewater concentrations were used for inhibition studies. This range aided in narrowing the span of concentrations to be tested in the further biological nutrient removal inhibition evaluation tests. The concentrations of AFFF used were 1,050 ppm, 105 ppm, 60 ppm, 10.5 ppm, 1.05 ppm and a control. The results indicated that nitrification inhibition occurred at AFFF concentrations of 60 ppm, 105 ppm, and 1,050 ppm in the feed wastewater. The ammonia nitrogen concentrations in the supernatant were 0.1 mg/L for the control, 1.2 mg/L for 1.05 ppm, and 0.2 mg/L for 10.5 ppm indicating little or no inhibition as seen in Figure 1. Nitrate

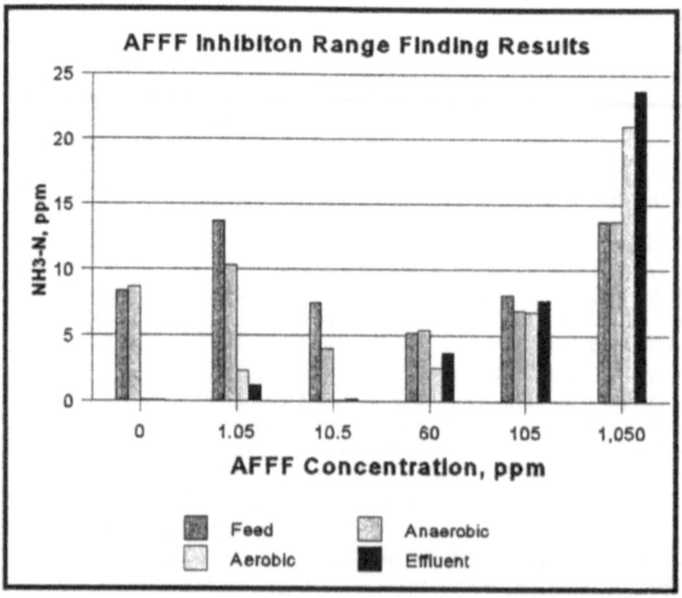


**Figure 1. AFFF inhibition study range finding results.**

production rates were also in accordance with the ammonia removal rates. An excellent mass balance on the nitrogen species was observed overall.

The COD removal rates decreased with increasing AFFF concentrations. The COD removal in the control reactor was 87% and decreased to 27% at the highest AFFF concentration. The COD of the 1% AFFF by volume (300 ppm concentrate) was measured to be 8,200 mg/L. Therefore, for all AFFF concentrations, additional COD was contributed to each reactor from the AFFF itself.

After determining a specific range of AFFF wastewater that exhibited inhibitory effects to the biological nutrient removal process, four concentrations of AFFF were tested in addition to paired controls. During each inhibition testing, one set of triplicate reactors (6-liter volume) were used as control which did not include any AFFF wastewater exposure. The remaining three reactors were used for one specific AFFF concentration. The inhibition concentrations that were tested include 10 ppm, 30 ppm, 50 ppm, and 60 ppm of AFFF in the feed wastewater and mixed liquor from the reference reactor.

The results showed no significant nitrification inhibition for any of the AFFF concentrations tested as compared to the control reactors as shown in Figures 2a and 2b. It was observed that nitrification started to occur during the initial two hours of the aerated feed stage for all of the reactors. At the end of the anaerobic cycle, some of the ammonia nitrogen was released in all tests, possibly due to bacterial reduction of nitrates and nitrites and due to incomplete denitrification. The elevated nitrite concentrations at the end of the anaerobic cycle for all tests supported these data as well. The ammonia nitrogen concentrations decreased significantly at the end of the aerobic cycle and in the effluent for each reactor exhibiting no nitrification inhibition. The nitrate data for each inhibition test also supported the occurrence of nitrification in the reactors. The nitrification occurring in each reactor can also be seen in Figure 3, which shows the ammonia nitrogen removal rate during different stages for each AFFF concentration tested. The effluent from each reactor exhibited between 98% to greater than 99% ammonia nitrogen removal rates.

There were significant COD reduction rates observed for each AFFF concentration tested as well. However, the COD removal rates in the inhibition reactors were less than that of the control reactors and the COD reduction decreased with increasing AFFF concentrations in the inhibition reactors. These results are shown in Figures 4a and 4b. During this study, there was an increase in the foaming in the inhibition reactors with increased AFFF concentrations. This foaming was specifically heavy during the aerated feed stage of the inhibition testing. The major influence on the reactor performance was the loss of solids (MLSS) at higher concentrations. However, the foaming did not have an inhibitory effect on the nitrification potential.

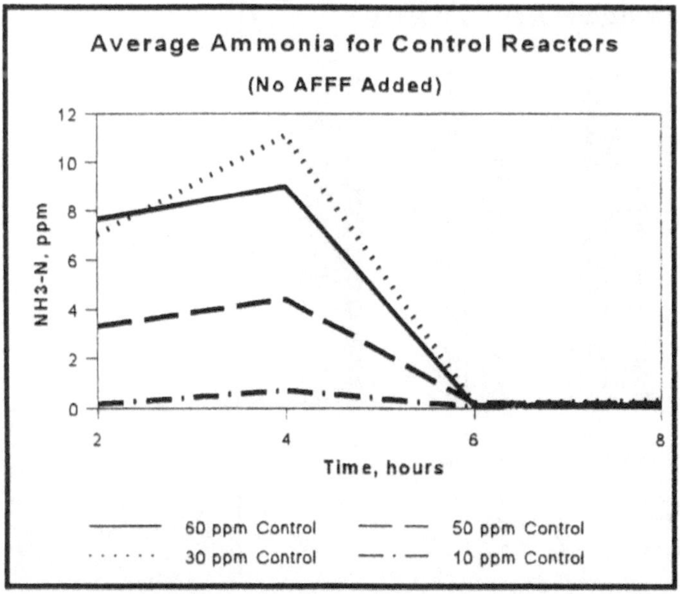


**Figure 2a.** Average ammonia concentrations for control reactors.

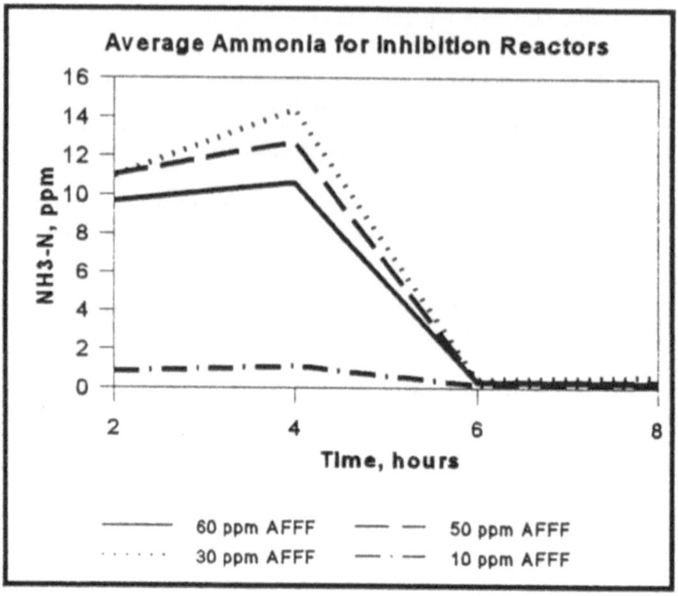


Figure 2b. Average ammonia concentrations for inhibition reactors.

The results of the plant toxicity pass-through tests conducted with the mysid shrimp and minnows did not exhibit any pass-through toxicity. The response measured during the acute toxicity tests was survival over the exposure period. The toxicity test results for the samples collected at the end of each inhibition testing from the reactors with and without AFFF had $LC_{50}$ values greater than 100% for both test organisms. The only sample that exhibited consistent toxicity was the influent feed to the reactors which was attributed to the high ammonia concentrations present in the feed mix which ranged from 30 to 35 ppm of $NH_3$-N.

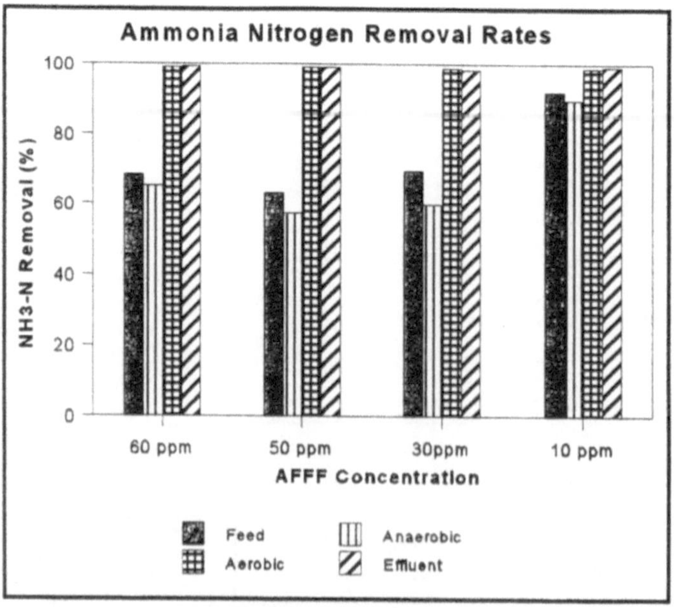


Figure 3. Ammonia nitrogen removal rates.

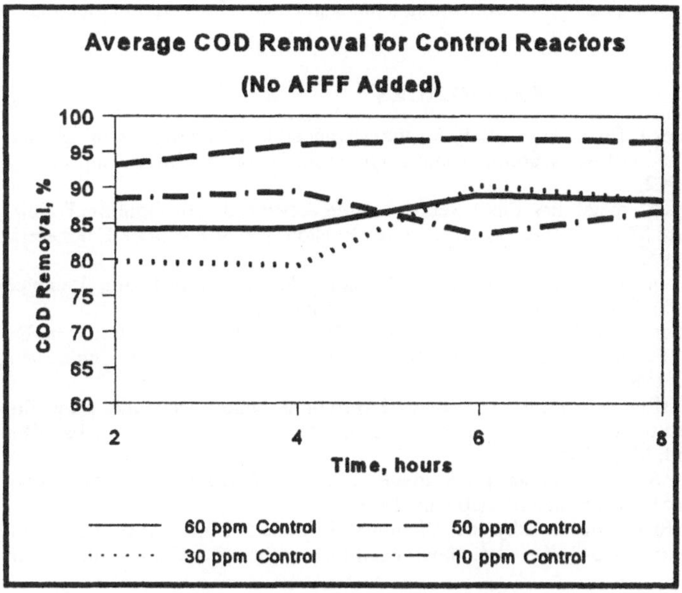


**Figure 4a.** Average COD removal rates for the control reactors.

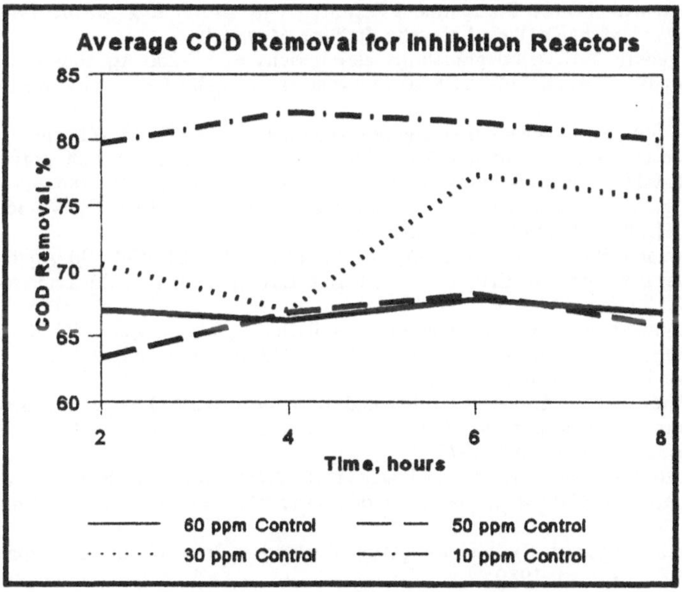


**Figure 4b.** Average COD removal rates for the inhibition reactors.

## SUMMARY

The results of the nitrification inhibition study showed that the AFFF concentrations tested in the range between 10 ppm to 60 ppm did not show any inhibition to biological nitrification. The effluent from each reactor did not exhibit any pass-through toxicity as well. The intensity of foaming increased with the increasing AFFF concentrations as well as loss of solids. Uninhibited nitrification was also observed among the reactors that had excessive foaming. Some reductions in the COD removal rates were observed as the AFFF concentrations increased. Additional testing is being performed by using AFFF and defoaming agents and their effects to nitrification inhibition.

## REFERENCES


1. $CH_2M$ Hill; Wastewater Effluent Pilot Study for the Advanced Fire Fighting Training Facility, Naval contract N62470-91-R-6650, Atlantic Division, Naval Facilities Engineering Command, October 1992.
2. $CH_2M$ Hill; Wastewater Treatability Final Report for the Advanced Fire Fighting Training Facility, Navy Contract N62470-91-C-6650, Atlantic Division, Naval Facilities Engineering Command, January 1995.
3. Chan, D.B., Disposal of Wastewater Containing Aqueous Film Forming Foam, Technical Memorandum No. M-54-78-06, Civil Engineering Laboratory, April 1978.
4. Chan, D.B., and Pan Bingham, AFFF-laden Wastewater Treatment Technology Initiation Decision Report (IDR), Technical Memorandum No. TM-71-88-11, Naval Civil Engineering Laboratory, December 1988.
5. Toxicity of Selected Effluents from the U.S. Navy Firefighting School, Norfolk, VA to Embryos of Eastern Oysters. Technical Report by EG&G, Atlantic Division, Naval Facilities Engineering Command, May 1978.
6. Union Carbide, Unox System Treatability Study Report, U.S. Navy Firefighting School, Naval Facilities Engineering Command, February 1978.
7. Engineering-Science Incorporated, Physical-Chemical Treatment from Navy Firefighting Schools, Contract No. N00025-74-C-0004, Naval Facilities Engineering Command, November 1986.
8. Saam, R., and Rakowski, P., Firefighting School Wastewater Study, Technical Memorandum No. 54-79-14, Civil Engineering Laboratory, June 1979.
9. Saam, R., Rakowski, P, and Aydlett, G., Treatability of Firefighting School Wastewaters: U.S. Navy Compliance with POTW Pretreatment Requirements, *Proceedings of the 34th Purdue Industrial Waste Conference*, West Lafayette, Indiana, May 1979.
10. Thomas, J.F., and LeFebvre, E.E., Biodegradability and Toxicity of FC-200 Aqueous Film Forming Foam, Report No. EHL(K) 74-3, USAF Environmental Health Laboratory, Kelly AFB, Texas, 1973.
11. Grace and Associates Inc., Engineering Investigation of Impact of AFFF on Wastewater Treatment Performance at Naval Air Station, Memphis, Millington, Tennessee, U.S. Navy Contract N62467-86-C-0351, Naval Facilities Engineering Command, November 1986.
12. Lefebvre, E.E., Biodegradability and Toxicity of Light Water. Report No. EHL (K) 71-36, USAF Environmental Health Laboratory, November 1971.
13. Lefebvre, E.E., and Inmand, R.C., Biodegradability and Toxicity of ANSUL K74-100, Aqueous Film Forming Foam, Report No. EHL(K) 75-3, USAF Environmental Health Laboratory, Kelly AFB, Texas, January 1975.
14. Lefebvre, E.E., and Inmand, R.C., Biodegradability and Toxicity of Light water FC-206,, Aqueous Film Forming Foam, Report No. EHL(K) 74 26, USAF Environmental Health Laboratory, Kelly AFB, Texas, November 1974.
15. Lefebvre, E.E., and Thomas, J.F., Biodegradability and Toxicity of AER-O-Water 3 and 6 Aqueous Film Forming Foam, Report No. EHL(K) 73-22, USAF Environmental Health Laboratory, Kelly AFB, Texas, December 1973.
16. American Public Health Association, American Water Works Association, and Water Environment Federation, *Standard Methods for the Examination of Water and Wastewater*, 19th ed., Washington, D.C., 1995.
17. Environmental Protection Agency, Toxicity Reduction Evaluation Protocol for Municipal Wastewater Treatment Plants, April 1989.

# 7 INHIBITION OF MESOPHILIC SOLID SUBSTRATE ANAEROBIC DIGESTION (DASS) BY AMMONIA-RICH WASTES

**Héctor M. Poggi-Varaldo**
Ecatepec Institute of Technology-TESE and P3 Consulting Engineers, P.O. Box 75-202, Mexico D.F., 07300, MEXICO

**E. Arce-Medina**
ESIQIE del IPN, Graduate School of Chem. Engng., Mexico City, MEXICO

**G. Fernández-Villagómez**
Centre for Disaster and Hazards Prevention (CENAPRED), Mexico City, MEXICO

**S. Caffarel-Méndez**
Ecatepec Institute of Technology-TESE and P3 Consulting Engineers, P.O. Box 75-202, Dexico D.F., 07300, MEXICO

## ABSTRACT

It is known from the anaerobic digestion practice with liquid wastes that ammonia could be toxic and inhibitory to methanogenic bacteria and the process. However, no similar information is available for the solid substrate anaerobic digestion of municipal and industrial wastes (DASS). Therefore, this work focused on determining the effects of ammonia-nitrogen concentration (TAN) on the specific methanogenic activity of microbial consortia from DASS and the stability of DASS continous process.

First, specific methanogenic activity tests (SMA) were carried out with DASS mesophilic and thermophilic inocula at several doses of ammonia nitrogen (range 500 to 7000 mgTAN/kg). TAN inhibition concentration 50% (IC50) was determined based on SAM. Second, bench-scale, semi-continuous, mesophilic reactors were operated at 21 days mass retention time and dosed with $NH_4Cl$ such that the corresponding COD/N ratios in their feeds were 90, 80, 65, and 50 (reactors R1 or control, R2, R3, and R4, respectively).

Based on the IC59 from the SMA tests, mesophilic DASS consortia seemed to be more resistant to ammonia-nitrogen than thermophilic cultures. Also, results indicated that aceticlastic cultures were affected less than the thermophilic ones. Regarding the mesophilic continuous DASS, the process deteriorated with increasing dosages of ammonia-N, with practically process cessation at COD/N=50 (R4). Inhibition was characterized by a gradual decline of efficiency and biogas productivity and a more sudden drop of methane content in biogas and pH. A significant increase of propionic, butyric, and valeric acid was found in reactors receiving the highest doses of ammonia-N (R3 and R4), suggesting the inhibition of syntrophic bacteria. Luong and Pearson inhibition models adequately represented the acute effects of N-supplementation on DASS; however, the Luong model could also account for the process cessation at a critical ammonia-N concentration of 2800 mg/kg mixed solids.

## INTRODUCTION

Solid substrate anaerobic digestion (DASS, digestion at high total solids content), has become an attractive and demonstrated technology for stabilizing agricultural, food, yard and municipal solid wastes [1-11]. However, little is known regarding the effects of potential inhibitors or toxi-

---

*52nd Purdue Industrial Waste Conference Proceedings*, 1997, Ann Arbor Press, Chelsea, Michigan 48118. Printed in U.S.A.

cants on this process.[5] This is an important problem in developing countries, where mixing of municipal solid waste with other types of wastes is likely to occur due to the difficulty of implementing appropriate solid waste handling schemes. Ammonia nitrogen-containing waste or its precursors (such as a slaughterhouse, fertilizer, animal wastes) are of concern, due to the potential inhibitory effects of ammonia on the liquid or slurry mode anaerobic digestion process.[12-17] Unfortunately the information regarding ammonia effects on DASS is scarce in the refereed literature.

Thus, the objectives of this work were:

- to study the effects of ammonia-Nitrogen on the activity of aceticlastic and hydrogenotrophic mesophilic and thermophilic DASS consortia, and
- to determine the impact of ammonia-nitrogen supplementation on the mesophilic, continuous DASS process performance.

## METHODOLOGY

### Experimental Design

*Effects of Ammonia-Nitrogen on the SMA of Meso- and Thermophilic DASS Consortia*

Wide mouth, 200 mL Gerber juice bottles were used as batch reactors. They were inoculated in an anaerobic glove box with 40 g of meso and thermophilic inocula from DASS reactors. Ammonia nitrogen was added as $NH_4Cl$ (to give a range of 500 to 7000 mg TAN/kg, each dose by duplicate), and $NaHCO_3$ was added for alkalinity compensation. Bottles were flushed with three cycles of anoxic nitrogen and vacuum, and finally vented at lab pressure under water seal. One meso and one thermophilic series of duplicate bottles received acetic acid (HAc) as substrate (target 1800 mg HAc/kg in the bottle; MA and TA consortia, respectively), while the second meso and the second thermophilic series were injected with a $H_2/CO_2$ mixture (4:1) at 2 atm (MH and TH consortia, respectively). The bottles were statically incubated at 35°C or 55°C for 16 to 20 hr. Biogas from bottles was sampled for methane analysis several times during the incubation period and the specific methanogenic activity (SMA) for each consortium at each dose was determined[18]. Passive controls (only inoculum, no ammonia nitrogen nor substrate) and active controls (inoculum plus substrate, no ammonia nitrogen) were also run in parallel. The results were compared in terms of the TAN inhibition concentration 50% (IC50); that is, the ammonia-nitrogen concentration at which the SMA is half of the SMA of the active control bottle.

*Impact of Ammonia-Nitrogen on Mesophilic, Continuous DASS*

Semicontinuous reactors, 3 L each, were employed in this experiment. All the reactors were operated in a draw-and-fill mode at 39°C and the MRT of 21 days. Ammonia nitrogen was supplemented as $NH_4Cl$ to the base feedstock to give COD/N ratios of 90, 80, 65 and 50 (COD, Chemical Oxygen Demand; N, Total Kjeldahl Nitrogen). Reactors were labeled R1, R2, R3, and R4, respectively. Besides, $NaHCO_3$ was also dosed in stoichiometric proportion to the alkalinity destroyed by the acid hydrolisis of $NH_4Cl$. Reactors 2, 3, and 4 received an initial or slug dose of $NH_4Cl$ at start, with the purpose of bringing each reactor to the corresponding COD/N ratio. This aimed at minimizing the transient state and avoiding any acclimation to the ammonia nitrogen inhibitory effects. The reactor standard deviations reported in this work reflect the performance variation with time.

### Feedstock

The base feedstock was prepared by mixing newspaper, food waste, and biosolids from a conventional activated sludge plant. The newspaper and the food waste simulated the organic fraction of municipal solid waste (OFMSW). Biosolids addition was calculated to give a ratio biosolids:OFMSW of 1:1 total mass basis or equivalently 0.090 TS biosolids/g TS OFMSW.

## Methods and Analysis

Biogas volumes in continuous reactors were measured once or twice a day, depending on the production, by the acidified brine displacement method. Preparation of samples to measure pH, alkalinity, and volatile organic acids (VOA) was based on Kasali & Senior's technique[19] and further modified by Poggi-Varaldo and Rinderknecht.[20] The pH and alkalinity were determined in the resulting slurry according to the *Standard Methods*.[21] Dry matter, volatile solids, and ash contents were determined as outlined elsewhere.[6] Total Kjeldahl Nitrogen (TKN), ammonia-nitrogen, and Chemical Oxygen Demand (COD) were determined according to the *Standard Methods*[21] in homogenized 3g, 3g, and 1g samples, respectively. The biogas composition was determined by gas chromatography with a Gow Mac-350 equipped with a thermal conductivity detector and a 1.8 m length x 6 mm diameter stainless steel column packed with Porapak Q using Helium as the carrier gas. The volatile organic acids (VOA) were analyzed by gas chromatography in a Tracor 570 equipped with a flame ionization detector, and a Carbowax capillary column 30 m × 0.32 mm × 0.25 µm film was used.

# RESULTS AND DISCUSSION

## Effects of Ammonia-Nitrogen on the Activity of Meso- and Thermophilic DASS Consortia

Figures 1 and 2 show the overall pattern of the SMA of DASS consortia. Mesophilic consortia exhibited a montonic decline of SMA, while thermophilic consortia seemed to have some type of stimulation at the lowest dose of $NH_4Cl$, after which the SMA started to decrease.

The SMA of both aceticlastic (MA and TA) active controls were of the same order, around 22 µmol/h.gVS, while the SMA of hydrogenotrophic DASS cultures were 3 to 4 times higher (80 and 65 µmol/hr.gVS for MH and TH, respectively) than those of the aceticlastic, see Table I. In general, the SMA of DASS consortia were 5 to 6 times lower than the SMA of aceticlastic and hydrogenotrophic consortia from slurry anaerobic digestion of manures (Table I).

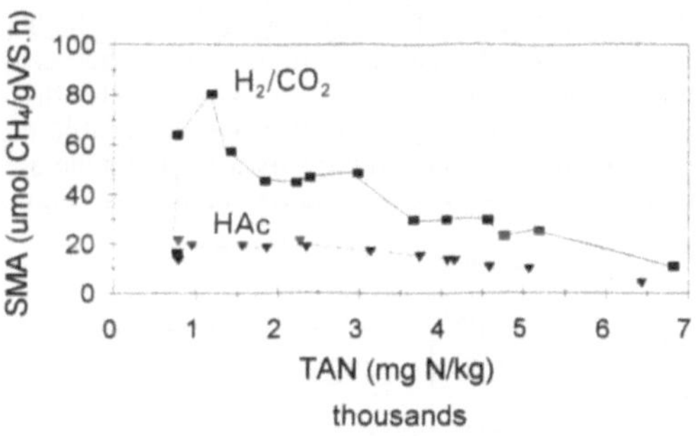


**Figure 1. Effect of ammonia nitrogen on the SMA of aceticlastic DASS consortia.**

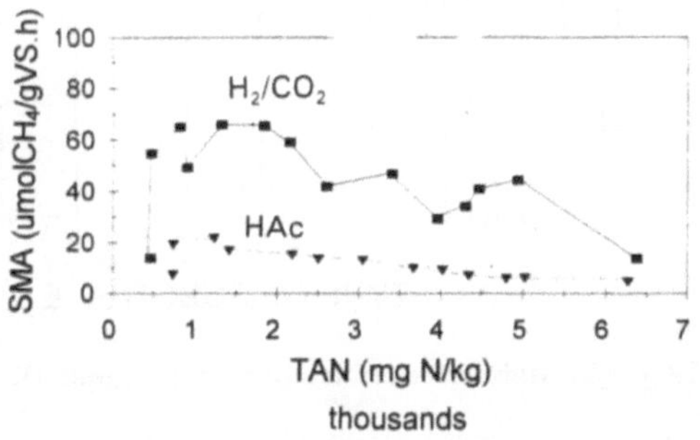


**Figure 2. Effect of ammonia nitrogen on the SMA of hydrogenothrophic DASS consortia.**

**Table I. Specific Methanogenic Activity of Consortia from Several Anaerobic Digestion Processes**

| Substrate | Inoculum | Temperature (°C) | SMA ($\mu$molCH4/gVS.hr) | Reference |
|---|---|---|---|---|
| Acetic acid 30mM | DASS, fed MSW & ISW | 35 | 21.2 | this work |
| Acetic acid 30mM | " | 55 | 21.7 | this work |
| $H_2/CO_2$ 4:1@ 2 atm | " | 35 | 80.4 | this work |
| $H_2/CO_2$ 4:1@ 2 atm | " | 55 | 64.9 | this work |
| Acetic acid 30mM | slurry DAN[a] fed cattle manure | 55 | 59.0–61.5 | 18 |
| $H_2/CO_2$ 4:1@ 2 atm | slurry DAN fed cattle manure | 55 | 326–424 | 18 |

[a]DAN: anaerobic digestion, at 4.5% TS.

When considering the pooled data, the IC50 for DASS mesophilic consortia was 10% higher than that of thermophilic cultures, Figure 3A. It is worth emphasizing that the lower the IC50, the higher sensitivity (or less resistant) to the toxicant and vice versa. The IC50 for aceticlastic consortia was 8% higher than that of the hydrogenotrophic, Figure 3B. However, there was a strong interaction between substrate type and temperature regime. In effect, the MA displayed the highest IC50 of all (4560 mg TAN/kg) and seemed to be significantly more resistant than both TA and MH (IC50 around 3250 mgTAN/kg and 3280 mgTAN/kg, respectively). The TH consortium was second with an IC50 of 4080 mgTAN/kg. In the mesophilic regime, the aceticlastic culture was less sensitive than the hydrogenotrophic consortium, while in the thermophilic mode it was the opposite. Since 70% of the methanogenesis in waste anaerobic digesters is thought to be effected via the aceticlastic pathway,[22] our findings indicate an advantage of the mesophilic regime over thermophilic regarding the resistance to ammonia Nitrogen inhibition.

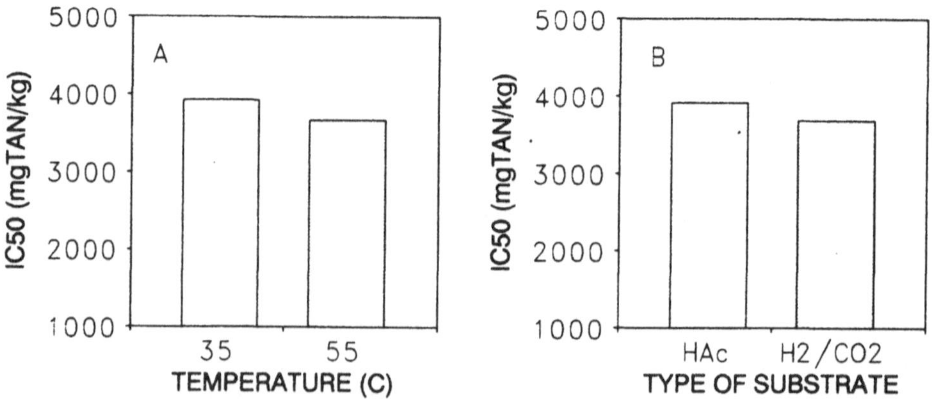


**Figure 3. Inhibition concentration 50% of TAN: (A) Main effect of temperature regime, (B) Main effect of substrate type.**

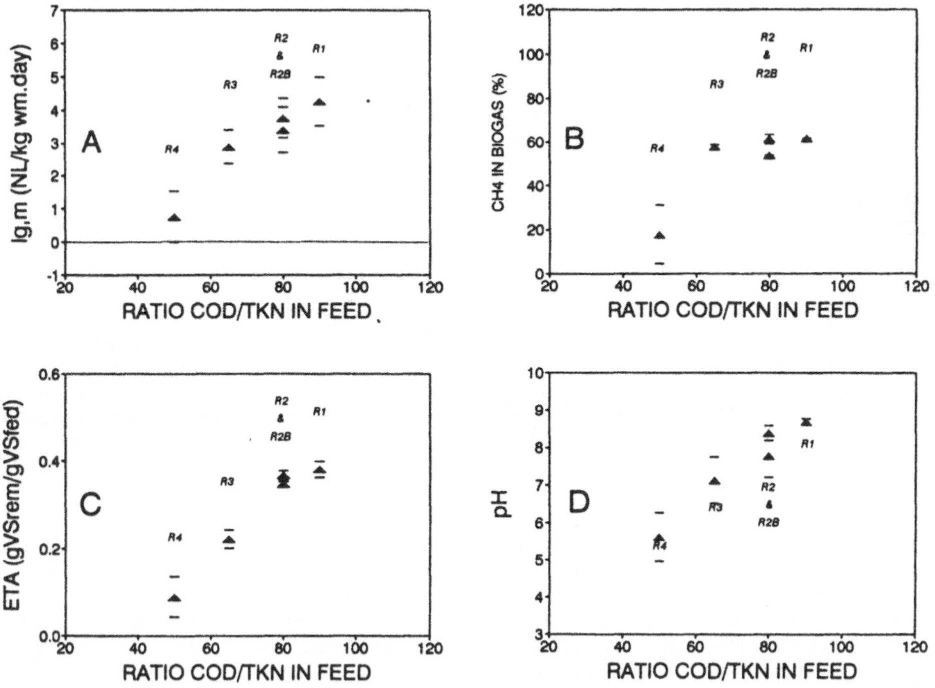


**Figure 4. Effect of feedstock nitrogen supplementation on DASS process performance:**
**(A) biogas productivity ($I_{g,m}$)**
**(B) methane content in biogas**
**(C) volatile solids removal efficiency (_)**
**(D) pH in reactors.**

## Impact of Ammonia-Nitrogen on Mesophilic, Continuous DASS

In the following discussion, it should be noted that increasing levels of nitrogen supplementation resulted in decreasing values of the COD/N ratio. All the process performance responses in Figure 4 (biogas productivity, $I_{g,m}$; biogas methane content; organic matter removal efficiency, $\eta$ VS basis; and pH in the reactors) showed declining patterns with increasing ammonia nitrogen supplementation, which were consistent with inhibitory conditions. Reactor R4—which received the highest dose of $NH_4Cl$ (or feedstock with the lowest COD/N ratio)—showed significantly lower $I_{g,m}$, methane content, $\eta$ and pH than reactor R1 (control, no supplementation). Reactors R2 and R2B, which received the smallest $NH_4Cl$ dose and a stimulatory nutrient effect would be expected, presented an $I_{g,m}$ and $\eta$ slightly lower than those of reactor R1.

As shown in Figure 4B, the methane content in reactor R4 biogas was significantly lower (17% v/v) than that in control reactor R1 (60% v/v). This drop in biogas methane concentration suggested an inhibitory effect of ammonia nitrogen on the methanogenesis. The biogas methane content was a somewhat less sensitive deterioration indicator than biogas productivity since control reactor R1, R2, and R3 displayed similar methane percentage (in the range 50-60% v/v). Control reactor R1 reached 40% total VS removal (or 62% on biodegradable VS basis) similar to those exhibited by mesophilic DASS reactors operated at 21 days MRT in a screening experiment[6]. The smallest addition of ammonia nitrogen (reactors R2 & R2B) did not seem to have any stimulatory effect on the process since its efficiency was only 6% lower than the corresponding to R1 (see Figure 4C).

The low pH value of R4 (approximately 5.7 average, Figure 4D) and the relatively low value of 7.3 in R3 were consistent with the process deterioration suggested by their low $I_{g,m}$ and $\eta$ re-

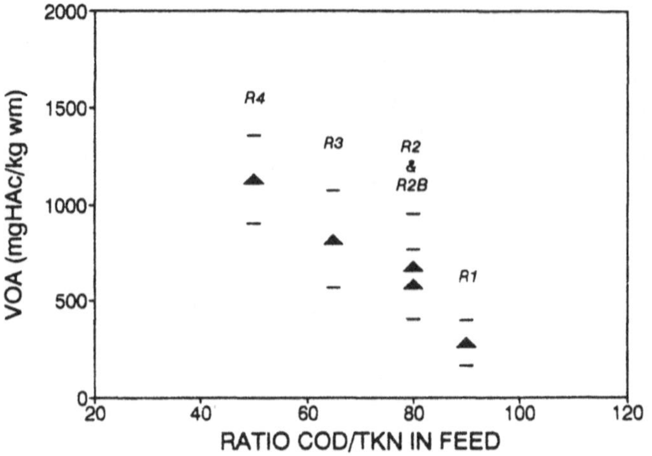


**Figure 5. Effect of feedstock nitrogen supplementation on reactor volatile organic acids concentration.**

sults. The pH drop was associated to an organic volatile acids buildup in reactor R4 (Figure 5). Increase percentages of total VOA in reactors R3 and R4 (based on the corresponding values in control reactor R1) were 190% and 300%, respectively. The highest VOA concentrations were found in reactor R4, where propionic acid predominated (590 mg/kg) and both butyric and valeric acid concentrations also reached important concentrations. This suggested that syntrophic bacteria could also be inhibited by ammonia nitrogen.

Luong and Pearson models were chosen for representing the process inhibition (responses $I_{g,m}$ and $\eta_b$, efficiency based on biodegradable VS, see Table II), as a function of TAN or unionized ammonia nitrogen.[12,15,23,24] The latter is recognized as the most important inhibitory species to methanogenic bacteria.[25-28] In this work, however, the independent variable selected as the inhibitor concentration was the TAN, since it was determined experimentally.

Model fitting results are shown in Table II and Figure 6. The nonlinear least square algorithm was used (SigmaPlot v.5.0). The smallest norm indicates the best fitting. In general, the Luong model represented the results somewhat better than the Pearson model. Both models predicted

**Table II. Equations and Parameter Values of Luong and Pearson Inhibition Models for Mesophilic DASS**

| Model | Response | $N_c$ (mg TAN/kg) | a or b | $R_o$ or $A_o$ | $N_{50}$ (mg TAN/kg) | Norm |
|---|---|---|---|---|---|---|
| Luong[a] | $I_{g,m}$ | 2799 | 4.70 | 4.03 | (2380)[c] | 0.4640 |
| | $\eta_b$ | 2871 | 3.31 | 0.651 | (2375)[c] | 0.0236 |
| Pearson[b] | $I_{g,m}$ | (+infin.)[c] | 13.08 | 3.82 | 2406 | 0.5939 |
| | $\eta_b$ | (+infin.)[c] | 7.92 | 0.6095 | 2313 | 0.0149 |

[a]$Resp = R_o\,[1 - (N/N_c)^a]$
[b]$Resp = A_o\,[1 + (N/N_{50})^b]^{(-1)}$
[c]Values between parentheses are not model parameters but estimated values by interpolating or extrapolating using the model equation.

Symbols:
*a*, exponent in Luong model; $A_o$, maximum activity in Pearson model; *b*, sensitivity exponent in Pearson model; $I_{g,m}$ biogas productivity; N, ammonia nitrogen concentration; $N_c$, ammonia nitrogen critical concentration at which the process ceases; $N_{50}$, ammonia nitrogen concentration at which the process response is reduced to 50% or the control value; $R_o$, maximum response value in Luong model; TAN, total ammonia nitrogen.

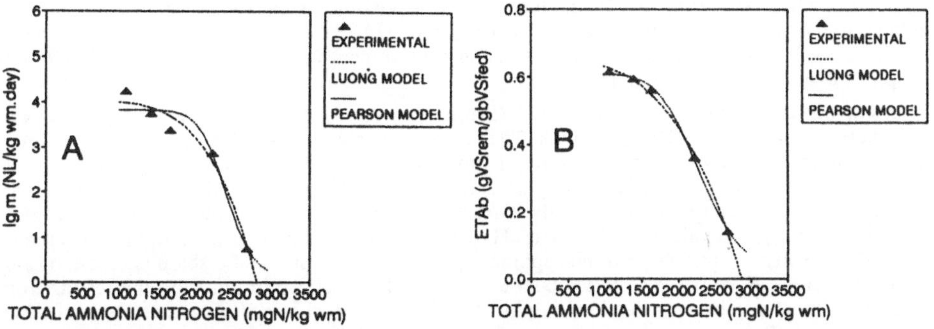


**Figure 6. Fitting DASS process performance as a function of the total ammonia nitrogen concentration in reactors with Luong and Pearson models:**
**(A) biogas productivity ($I_{g,m}$)**
**(B) biodegradable volatile solids removal efficiency ($\_{b}$).**
**The thick line represents the Luong model fitting, the thin line is the Pearson model fitting.**

similar values for the 50% concentration effect $N_{50}$ (i.e., the TAN level which reduces the process response to half of that in the control) in the range 2300–2400 mg/kg. However, only the Luong model could represent the process cessation at a critical value of the inhibitor ($N_c$) which was approximately 2800 mg/kg. Interestingly, the $N_{50}$ and $N_c$ values were quite close; this feature indicated that the inhibitory effect was very acute.

A comparison of Pearson and Luong model parameter values for mesophilic DASS in this work (Table III and Table IV) with those obtained for batch anaerobic digestion (liquid or slurry mode, DAN) results suggested that the inhibitory effect of the toxicant TAN was more drastic in the DASS process than in DAN[12,15]. The DASS $N_{50}$ was nearly 50% lower than that in DAN, while the Pearson sensitivy coefficient *b* for DASS was almost four times higher than that in batch DAN (Table III). Similar trends were observed for $N_c$ and Luong sensitivity coefficient *a*

**Table III. Pearson Parameter Values for DASS and Liquid Batch Anaerobic Digestion**

| Process | Variable | $N_{50}$ (mg TAN/kg) | b | Reference |
|---|---|---|---|---|
| DASS | $I_{g,m}$ | 2406 | 13.06 | this work |
| Liq. DAN[a] | methane production | 3920 | 3.60 | 12 |

[a] Liquid phase anaerobic digestion.

**Table IV. Luong Parameter Values for DASS and Liquid Batch Anaerobic Digestion**

| Process | Variable | $N_c$ (mg TAN/kg) | $N_{50}$ (mg TAN/kg) | a | Reference |
|---|---|---|---|---|---|
| DASS | $I_{g,m}$ | 2800 | (2380)[a] | 4.67 | this work |
| liq. DAN[b] | growth rate (based on $CH_4$ production) | 5825 | (—) | 1.62 | 15 |

[a]Numbers between parentheses are values interpolated using the model.
[b]Liquid phase anaerobic digestion.

(Table IV) when comparing DASS and batch DAN. A possible explanation of this result could be related to the washout of active bacteria in continuous DASS (while no washout occurs in batch DAN), plus the disadvantage of a more stressing environment for the microbes in DASS (for example, a lower water activity in DASS[29]).

In experiments with a thermophilic DASS pilot reactor[30] it was found that the process was completely inhibited at 2500 mgTAN/kg. The reactor was operated at 30 days MRT and acclimation to increasing TAN occurred. In spite of this, the $N_c$ level was lower than that found in our work with mesophilic DASS (2800 mg/kg). This indicated that mesophilic continuous DASS could be more resistant than thermophilic DASS. The basis of this resistance could be related to the fact that for given values of TAN and pH, the UAN concentration in thermophilic reactors is several times higher than those in mesophilic digesters. This is due to the sharp decrease of ammonium cation $pK_a$ with a rise in temperature.[31] The above is also in line with the results from our SMA studies, since aceticlastic mesophilic DASS consortia seemed to be more resistant than thermophilic cultures to ammonia nitrogen (Figure 3).

A comparison of data gathered from several works on ammonia nitrogen effects on anaerobic digestion (Table V) also confirms the trends found in our work: batch is less sensitive than con-

**Table V. Inhibition by Ammonia Nitrogen in Several Anaerobic Digestion Processes**

| Critical TAN (mg/kg) | Process | Substrate | Reactor | Temperature (°C) | Acclimation | Reference |
|---|---|---|---|---|---|---|
| 5800 | batch DAN[a] liq.[b] | acetate | bottles | 35 | — | 15 |
| > 5000 | cont. DAN[c] liq. | sludge | lab reactors | 35 | yes | 31 |
| 3830[d] | cont. DAN liq. | cattle | lab CMR[e] manure @4.5% TS | 55 | yes | 18 |
| 3000 | cont. DAN liq. | cattle | lab reactors manure @10% TS | 50 | yes | 28 |
| 6000 | batch DAN | cattle | lab reactors manure @10% TS | 35 | — | 11 |
| 5000 | cont. DAN semi-liq. | agro-industrial wastes | lab reactors | 35 | yes | 31 |
| 3500 | cont. DAN semi-liq. @ 9% ST | food wastes @20% ST | lab reactors | 55 | yes | 31 |
| 4100 | batch DASS | HAc | bottles | 35 | no | this work |
| 2500 | cont. DASS | MSW | pilot @35% TS | 55 IMR[g] | yes | 30 |
| 2800 | cont. DASS | MSW/sludge @30% TS | lab IMR | 39 | no | this work |

[a] DAN = anaerobic digestion
[b] liq.= liquid or slurry mode
[c] cont. = continuous
[g] IMR = intermittently mixed reactor
[d] estimated from $pK_a$ = 8.35 and pH = 7.7
[e] CMR = complete mix reactorç
[f] based on *net* SMA of aceticlastic DASS consortia

tinuous anerobic digestion, liquid phase anaerobic digestion seems to be more resistant than DASS, and finally, mesophilic regime is less sensitive than thermophilic.

The engineering significance of our results are related to the operation of continuous DASS in the case of transient or chronic toxicity by ammonia nitrogen. In order to minimize its effects, we recommend:[31] (1) select mesophilic DASS; (2) if the process is in thermophilic mode and a change to mesophilic regime could be inconvenient, then set the temperature to the lowest side of the thermophilic range; (3) increase operating mass retention time; and (4) increase process resistance to TAN and/or reduce TAN concentration by diluting the reactor content with water and/or adding substrates poor in TAN precursors.

## CONCLUSIONS

From the tests with the specific methanogenic activity of DASS consortia, using the IC50 as criterion, it can be concluded that:

- Mesophilic DASS consortia seemed to be more resistant to ammonia nitrogen than thermophilic DASS cultures;
- Aceticlastic consortia were more resistant than hydrogenotrophic consortia;
- There was a strong interaction between temperature regime and substrate type, which deserves further research.

Regarding the impact of ammonia Nitrogen on mesophilic, continuous DASS performance, it can be stated that:

- The DASS process deteriorated with increasing dosages of ammonia-N, with practically process cessation at COD/N=50 in the feedstock. Inhibition was characterized by a significant fall of efficiency and biogas productivity and a more sudden drop of methane content in biogas and pH.
- C3 to C5 organic acid concentration increases occurred in reactors receiving the highest doses of ammonia-N (R3 and R4), suggesting the inhibition of the syntrophic bacteria present in the methanogenic consortia.
- Luong and Pearson inhibition models represented very well the acute effects of Nitrogen supplementation on DASS, although the Luong model could also show the process cessation at a critical ammonia-N concentration of 2800 mg/kg.
- Comparison of our results with literature data suggests that the mesophilic DASS was more sensitive to ammonia nitrogen inhibition than liquid or slurry mode anaerobic digestion. By contrast, mesophilic DASS seemed to be more resistant to the toxicant than thermophilic DASS.
- Our results and those from other researchers suggest the following operational strategies for DASS in case of transient or chronic inhibitory effects of ammonia Nitrogen: first, if possible, select mesophilic DASS; if the process is in thermophilic mode and a change to mesophilic regime would be inconvenient, then set the operating temperature to the lowest side of the thermophilic range; and/or increase operating mass retention time; and/or increase process resistance to TAN; and/or reduce TAN concentration by diluting the reactor content with water and/or adding substrates poor in TAN or TAN precursors.

## ACKNOWLEDGMENTS

The authors wish to thank Mr. Máximo Vázquez-Cruz for his help with reactor operation and SMA tests, and Prof. Elvira Ríos-Leal for her assistance with chromatographic analysis. Training in anaerobic glove box techniques provided by Prof. Richard Sparling (University of Manitoba, Dept. of Microbiology, Canada) to one of the authors (HMP-V) is gratefully acknowledged. Professor Noemi Rinderknecht-Seijas (ESIQIE del IPN and *P3* Consulting Engineers) critically read the manuscript.

## REFERENCES


1. Cecchi, F., Traverso, P.G., Mata-Alvarez, J., Clancy, J., and Zaror, C. State of the art of R&D in the anaerobic digestion process of municipal solid waste in Europe. *Biomass* 16, pp. 257–284, 1988.
2. de Jong, H.B.A., Koopmans, W.F., van der Knijff, A., and Jumelet, H.J. Conversion Techniques for Vegetable, Garden and Fruit Waste. Haskoning Report, Nijmegen, The Netherlands, 1993.
3. Kayhanian, M., and Tchobanoglous, G. Innovative two-stage process for the recovery of energy and compost from the organic fraction of solid wastes. *Water Sci. Technol.* 27(2), pp. 133–140, 1993.
4. Poggi-Varaldo, H.M., Valdés, L., Fernández, G., and Esparza, F. Solid substrate anaerobic co-digestion of municipal and paper-mill sludge. *Water Sci. Technol.* In press, 1996.
5. Poggi-Varaldo, H.M., Yu, A., and Rinderknecht, N. Dry anaerobic digestion: a review and feasibility study *R'95 Recovery, Recycling, Re-integration.* Geneve, Switzerland, February 1–4, 1995.
6. Poggi-Varaldo, H.M., and Oleszkiewicz, J. Anaerobic co-composting of municipal solid waste and waste sludge at high total solids levels. *Environ. Technol.* 13(3), pp. 409–421, 1992.
7. Poggi-Varaldo, H.M., and Valdéz-Ledezma, L. Optimization of Dry Anaerobic Digestion Start-Up. *17th Canadian Waste Manage. Conf.* Sept. 11–14, 1995, Quebec, QC, Canada, 1995.
8. Poggi-Varaldo, H.M., Rodríguez-Vázquez, R., Fernández-Villagómez, G., Rinderknecht-Seijas, N., Oleszkiewicz, J., and Sparling, R. Research and Development on Solid Susbtrate Anaerobic Digestion of Municipal and Indsutrial Waste in Mexico. *Proc. 2nd Europ. Mtg. Chem. Industry Environ. EMCHIE 96,* Sept. 11–13, 1996, Alghero, Italy, 1996.
9. Rivard, C.J., Rodríguez, J.B., Nagle, N.J., Self, J.R., Kay, B.D., Soltanpour, P.N., and Nieves, R.A. Anaerobic digestion of municipal solid wastes: utility of process residues as a soil amendment. *Appl. Biochem. Biotechnol.* 51/52, pp. 125–135, 1995.
10. Rivard, C.J., Himmel, M.E., Vinzant, T.B., Adney, W.S., Wyman, C.E., and Grohmann, K. Anaerobic digestion of processed municipal solid waste using a novel high solids reactor: maximum solids levels and mixing requirements. *Biotechnol. Lett.* 12(3), pp. 185–190, 1990.
11. Wujcik, W.J., and Jewell, W.J. Dry anaerobic fermentation *Biotechnol. Bioeng. Symp.* 10, pp. 43–65, 1980.
12. Heinrichs, D., Poggi-Varaldo, H.M., and Oleszkiewicz, J. Effects of ammonia on the anaerobic digestion of simple organic substrates. *J. Environ. Eng. (ASCE),* 116(4), pp. 698–711, 1990.
13. Hendriksen, H.V., and Ahring, B.K. Effects of ammonia on growth and morphology of thermophilic hydrogen-oxidizing methanogenic bacteria. *FEMS Microbiol. Ecol.* 85, pp. 241–246, 1991.
14. Koster, I.W., and Lettinga, G. The influence of ammonia nitrogen on the specific activity of pelletized methanogenic sludge. *Agric. Wastes,* 9, pp. 205–216, 1984.
15. Poggi-Varaldo, H.M., Tingley, J., and Oleszkiewicz, J. Inhibition of growth and acetate uptake by ammonia in batch anaerobic digestion. *J. Chem. Technol. Biotechnol.* 52(1), pp. 135–142, 1991.
16. Poggi-Varaldo, H.M. and Estrada-Vázquez, C. Agricultural Wastes. *Water Environ. Res.* In press, 1997.
17. Robbins, J.E., Gerhardt, S.A., and Kappel, T.J. Effects of total ammonia on the digestion of and an example of digester performance from cattle manure protein mixtures. *Biol. Wastes,* 27 (1), pp. 1–14, 1989.
18. Angelidaki, I., and Ahring, B.K. Anaerobic thermophilic digestion of manure at different ammonia loads: Effect of temperature, *Water Res.* 28(3), pp. 727–731, 1994.
19. Kasali, G.B., and Senior, E. Effects of temperature and moisture on the anaerobic digestion of refuse. *J. Chem. Technol. Biotechnol.* 44(1), pp. 31–41, 1989.
20. Poggi-Varaldo, H.M., and Rinderknecht, N. *Mini-Handbook of Solid Waste Analysis,* CINVESTAV del IPN, México DF, México, 1990. In English, available upon request.
21. *Standard Methods for the Examination of Water and Wastewaters*, 16th ed., APHA-WPCF-AWWA, Washington, D.C., 1985.

22. Tilche, A. and Shan, Y.X. Light and Scanning Electron Microscope Observations on the Granular Biomass of Experimental SBAF and HABR Reactors. *International Workshop Granular Anaerobic Sludge-Microbiology and Technology.* GASMAT, Lunteren, Netherlands, 25–27 October, 1987.
23. Mulchandani, A., and Luong, J.H.T. Microbial inhibition kinetics revisited. *Enzyme Microb. Technol.* 11, pp. 66–73, 1989.
24. Pearson, F., Chang, S.C., and Gautier, M. Toxic inhibition of anaerobic biodegradation. *J. Water Pollut. Control Feder.* 52(3), pp. 472–482, 1980.
25. Sprott, G.D., and Patel, G.B. Ammonia toxicity in pure cultures of methanogenic bacteria. *System Appl. Microbiol.* 7, pp. 358–363, 1986.
26. Sprott, G.D., Shaw, K.M., and Jarrel, K.F. Ammonia/potassium exchange in methanogenic bacteria. *J. Biol. Chem.* 259, pp. 12602–12608, 1984.
27. Wiegant, W.M., and Zeeman, G. The mechanism of ammonia inhibition in the thermophilic digestion of livestock wastes. *Agric. Wastes*, 16, pp. 243–253, 1986.
28. Zeeman, G., Wiegant, W.M., Koster-Treffers, M.E., and Lettinga, G. The influence of the total ammonia concentration on thermophilic digestion of cow manure. *Agric. Wastes,* 14, pp. 19–35, 1985.
29. Hahn-Hägerdal, B. Water activity: a possible external regulator in biotechnical processes. *Enzyme Microb. Technol.,* 8(6), pp. 322–327, 1986.
30. Kayhanian, M. Performance of a high-solids anaerobic digestion process under various ammonia concentrations, *J. Chem. Technol. Biotechnol.,* 59, pp. 349–352, 1994.
31. Poggi-Varaldo, H.M., Rodríguez-Vázquez, R., Fernández-Villagómez, G., and Esparza-García, F. Inhibition of solid substrate anaerobic digestion by ammonia nitrogen. *Appl. Microbiol. Biotecthnol.* In press, 1997.

## LIST OF ABBREVIATIONS

| | |
|---|---|
| a | exponent in Luong model |
| $A_o$ | control value of response in Luong model |
| b | sensitivity (exponent) in Pearson model |
| bVS | biodegradable volatile solids |
| COD | chemical oxygen demand |
| DAN | anaerobic digestion at low solids contents (liquid phase or slurry mode) |
| DASS | solid substrate anaerobic digestion; |
| HAc | acetic acid |
| IC50 | inhibition concentration 50%, ammonia-nitrogen concentration at which the specific methanogenic activity is half of that in the control |
| MA | mesophilic aceticlastic DASS consortium |
| MH | mesophilic hydrogenotrophic DASS consortium |
| MRT | mass retention time |
| MSW | municipal solid waste |
| N | nitrogen or total Kjeldahl nitrogen |
| $N_c$ | ammonia nitrogen critical concentration (at which the process ceases) |
| $N_{50}$ | ammonia nitrogen concentration at which the DASS process response is reduced to 50% of the control value (for Pearson and Luong models) |
| OFMSW | organic fraction of municipal solid waste |
| $R_j$ | reactor "j" (j = 1, 2, 2B, 3 and 4) |
| $R_o$ | control value of response in the Pearson model |
| SMA | specific methanogenic activity |
| TA | thermophilic aceticlastic DASS consortium |
| TAN | total ammonia nitrogen concentration |
| TH | thermophilic hydrogenotrophic DASS consortium |
| TKN | total Kjeldahl nitrogen concentration |
| TS | total solids |
| UAD | ultimate anaerobic degradability |
| UAN | un-ionized ammonia nitrogen concentration |
| VOA | volatile organic acids concentration |
| VS | total volatile solids |

Greek letters:

| | |
|---|---|
| $\eta$ | organic matter removal efficiency, based on total volatile solids |
| $\eta_b$ | organic matter removal efficiency, based on biodegradable volatile solids |
| $\theta$ | mass retention time |

# 8 EFFECTS OF COPPER AND ZINC ON SEQUENCING BATCH REACTORS

**Christine Goldman Robinson**, Project Engineer
Montgomery Watson Americas, Inc.
Metairie, Louisiana 70002

**Dennis D. Truax**, Professor

**Adnan Shindala**, Professor
Dept. of Civil Engineering
Mississippi State University
Mississippi State, Mississippi 39762

## ABSTRACT

This chapter will focus on the effects of copper and zinc, acting individually, in bench-scale sequencing batch reactors which were studied at different mean cell retention times (MCRTs). Results indicate that MCRT and the metal mass to biological mass ratio present are the major factors in determining the impacts of the copper or zinc on the treatment process. Organic removal efficiency and suspended solids levels were observed to be indicators of potential inhibition.

## INTRODUCTION

Municipal solid waste landfills (MSWLFs) and hazardous waste landfills (HWLFs) produce leachate that must be treated in accordance with Subtitle D of the Resource Conservation and Recovery Act or the Hazardous and Solid Waste Amendments of 1984. Leachate composition is highly variable and may contain significant heavy metal constituents as well as the more common high concentrations of biodegradable organics. Because of the high organic content, biological treatment systems, such as those found at publicly-owned treatment works (POTWs) and used for the treatment of municipal sewage, have been demonstrated effective for the treatment of leachate from HWLFs[1,2] and MSWLFs.[3,4] Copper and zinc are two heavy metals that are found consistently in MSWLF and HWLF leachates, due to the high occurrence of their use in many industries. This study was conducted using bench-scale Sequencing Batch Reactor (SBR) activated sludge systems to research the independent effects of copper and zinc on the treatment systems and trace the fate of each metal through the system.[5] Emphasis was placed on relating the effects of each metal to the MCRT of the biomass. To meet the above goals, the following objectives were pursued in this research:

1. Determine the concentration of copper and zinc at which each can be present in the feed to the treatment process without significantly inhibiting biomass synthesis.
2. Evaluate the effects of varying levels of copper and zinc on the organic removal efficiency of the biomass, noting differences over the range of MCRTs studied.
3. Evaluate the relationship between metals removal and MCRT for each metal and note any differences between the two.

## PRIOR RESEARCH

Hannah et al.[6] conducted a pilot plant study comparing the capabilities of five less-than-secondary treatment processes to the conventional activated sludge system for removal of priority

*52nd Purdue Industrial Waste Conference Proceedings*, 1997, Ann Arbor Press, Chelsea, Michigan 48118. Printed in U.S.A.

pollutants. The activated sludge process was the most effective in removal of conventional pollutants and toxic priority pollutants when compared to the facultative lagoon, aerated lagoon, high-rate trickling filter system, chemical clarification system, and primary clarification and filtration system studied. The typical activated sludge process is operated either on a continuous-flow basis, as in a typical aeration basin/clarifier style wastewater treatment plant, or as a batch system where flows are introduced intermittently, as in a Sequencing Batch Reactor (SBR) treatment system.

SBRs preceded the use of continuous-flow activated sludge technology, but fell from common use due to the high level of operator attention required to sequence the reactor through its various phases to achieve the desired performance. As new automation tools were developed, a revival of interest in and use of the SBR technology has occurred. As a result, SBRs have become increasingly popular in the United States for small systems and industrial applications. Glenn et al.[7] state, "Batch processing of industrial wastewaters is an alternative technology being explored due to more stringent discharge requirements and lower maintenance costs. The process is also easily adaptable to new demands because the equipment can be operated in several different modes." Others claim that because an SBR tank serves as an equalization basin during fill, SBRs can tolerate peak flows and shock loads of BOD or COD without degradation in effluent quality.[8,9]

The resiliency of the SBR to difficult waste streams makes it ideal for treatment of landfill leachate, whether on-site or at a POTW. Additionally, the relative ease of operation further supports the use of SBRs as an on-site treatment process, when required. Because SBR construction is relatively simple and much yard piping is eliminated, the capitol costs for SBRs are much less that a conventional activated sludge treatment plant. It is for these reasons that a bench-scale SBR system was chosen for study in this project.

Neufeld and Hermann[10] reported data on the kinetics of heavy metal interactions in the activated sludge process with an acclimated biomass for mercury, cadmium, and zinc. The researchers shock loaded up to 1,000 mg/L of each heavy metal in a soluble chloride form into a batch-type laboratory reactor. They observed a rapid uptake of the metal by the sludge that tends to level off after three hours of contact. Uptake is apparently not a function of the metal concentration or the organism viability such that the initial heavy metals removal is not a biological phenomenon, but is related to the physical, chemical, and probably surface properties of the biological mass. The researchers recommended varying sludge ages to achieve a satisfactory level of effluent water quality. A high level of solids under aeration was recommended as the best way to increase sludge age without increasing aeration time. This type of acclimated culture can then be used to treat a waste with high food value (COD) which also contains high concentrations of heavy metals.

A study by Cheng et al.[11] was conducted similar to that of Neufeld and Hermann using batch-operated activated sludge units; however, the reactors were dosed with metal concentrations of approximately 26 mg/L or less. The purpose of this study was to define the effects of nickel, copper, lead, cadmium, and zinc on the biokinetic process. The researchers found that activated sludge has the ability to remove and accumulate metal ions from solution in a rapid initial phase of three to ten minutes followed by a second and slower phase of uptake. Removal of the metal in the slow phase was relatively insignificant when compared with that of the first phase. For example, the association of copper with the biomass was essentially complete after ten minutes of contact time.

## METHODOLOGY

Return activated sludge from the Ernest Jones Wastewater Treatment Facility, Starkville, Mississippi, was used for all of the research reported herein. It was acclimated to the mean cell retention time (MCRT) used in the research initially in 22 liter carboys maintained in a batch mode. The biomass was then transferred into the bench-scale reactors, split equally among five reactors (with one reactor acting as a control) so as to receive different levels of metal while maintaining the same MCRT in each. MCRTs of 6, 10, 15, and 20 days were studied at soluble copper doses of 1, 2, 4, and 6 mg/L. For zinc, the extreme MCRTs of 6 and 20 days were examined for dosages of 1, 2, 4, and 6 mg/L as soluble zinc. The chloride form of each metal was used during this research. The activated sludge was further acclimated in the bench-scale reactors until acclimation was confirmed by achieving steady-state levels for Mixed Liquor Suspended Solids (MLSS), Mixed Liquor Volatile Suspended Solids (MLVSS), Chemical Oxygen Demand

(COD) removal, and Oxygen Uptake Rate (OUR). A more detailed discussion of the methodology used is presented in the thesis by Robinson.[5]

The organic needs of the biomass were supplied by a mixture of glucose and sodium acetate at an initial theoretical COD level of 250 mg/L in the entire eight liter volume of each reactor (2000 mg/day). In addition, Solution A (see Table I), containing potassium phosphate (monobasic and dibasic, $K_2HPO_4$ and $KH_2PO_4$) and ammonium chloride ($NH_4Cl$), was added to the synthetic substrate as a supplement to meet the nutrient needs of the biomass as well as provide a pH buffer. Solution B, containing trace metals and calcium, was also part of the synthetic substrate. Concentrated stock solutions of the glucose, sodium acetate, Solution A, and Solution B were prepared using distilled, deionized water every week and preserved in the 4(C cooler. Synthetic substrate was prepared every day by diluting the required amount of stock solutions with tap water to obtain a total feed solution of 250 mg/L in the reactor. Periodic COD tests were performed as quality control to confirm the synthetic substrate strength. Table I gives the synthetic substrate composition.

The 10 bench-scale reactors used in this research each had a capacity of nine liters. The liquid level was maintained at eight liters to allow adequate freeboard. Oxygen was provided by sparging compressed air at a flow rate of five standard cubic feet per hour (SCFH) through each reactor. At this flow rate, mixing was provided and the dissolved oxygen level was maintained above 5 mg/L. The average reactor mixed liquor temperature during the copper loadings was 20°C. The average reactor mixed liquor temperature during the zinc loadings was 23°C. The temperatures in each reactor were measured daily.

The reactors were maintained as Sequencing Batch Reactors (SBRs) with a 24-hour cycle. The SBR cycle is given in Table II. Once a day, the air supply to the units was turned off and the activated sludge allowed to settle. The clear supernatant was then decanted down to a level of four liters of combined biomass and supernatant in a way as not to disturb the biomass and to keep the supernatant removal constant. After decanting, the units were refilled with synthetic substrate.

The analyses were performed on samples collected from the influent, reactors, and the effluents. Periodic testing of the stock solutions occurred to verify the COD and metals concentrations.

Analyses for COD, suspended solids, oxygen uptake rate, and pH conducted during this research were performed in accordance with *Standard Methods* (1992). The procedures prescribed

**Table I. Synthetic Substrate Components**

| Constituent | | Final Concentration |
|---|---|---|
| Glucose and Sodium Acetate | | 250 mg/L as COD |
| Solution A | $K_2HPO_4$ | 142 mg/L as P |
| | $KH_2PO_4$ | 91 mg/L as P |
| | $NH_4Cl$ | 13 mg/L as N |
| Solution B | $MgSO_4$ $7H_2O$ | 3.7 mg/L as Mg |
| | $FeSO_4$ $7H_2O$ | 0.25 mg/L as Fe |
| | $ZnSO_4$ $7H_2O$ | 0.28 mg/L as Zn |
| | $MnSO_4$ $H_2O$ | 0.33 mg/L as Mn |
| | $CaCl_2$ | 1.8 mg/L as Ca |

**Table II. Sequencing Batch Reactor Cycle**

| Cycle Period | Time of Day | Duration |
|---|---|---|
| React | 1:00 p.m. – 12:00 p.m. | 23 hours |
| Settling | 12:00 p.m. – 12:30 p.m. | 30 minutes |
| Decant | 12:30 p.m. – 12:45 p.m. | 15 minutes |
| Fill | 12:45 p.m. – 1:00 p.m. | 15 minutes |

in the *Spectrophotometer Manual, AA/AE,* (1976) were followed for analysis of the suspended and soluble forms of copper and zinc. Samples were filtered using a 0.45 micron cellulose membrane filter paper, as required for separation of metals samples by *Standard Methods, Part 3030 B* (1992), "Preliminary Filtration." The soluble copper or zinc concentration in the samples was obtained by measuring the copper or zinc concentration in the liquid collected that passed through the filter. For the determination of the suspended form of the metal in the mixed liquor or in the effluent (the metal retained on the filter), the "Nitric Acid Digestion" technique was used for preparation of the samples for analysis in accordance with *Standard Methods, Part 3030 E* (1992).

Mean cell retention time (MCRT), commonly called sludge age, was adopted as the main operational control parameter during the study. The amount of daily wasting of the sludge from the reactors was calculated by substituting the suspended solids data collected from the previous 24-hour cycle period. This method allows for accommodation of daily reactor fluctuations, including changes in solids lost in the effluent.

During the FILL period, the substrate and the metal were added and the air turned on. Initial samples were taken from each reactor immediately after complete mixing was achieved (five to twenty minutes after air supply was returned). These samples were analyzed for soluble COD, MLSS, MLVSS, soluble metals and metals adsorbed on or within the biomass (suspended metals). Dissolved Oxygen (DO) uptake, pH, and reactor temperature were monitored once per cycle, typically during the eighth to tenth hours of the contact time. Three times during the contact period of the SBR cycle, samples were pulled for monitoring the phase of the metals. Soluble COD, MLSS and MLVSS were performed on the supernatant (effluent) as well as metals phase sampling. The sampling times are shown in Table III. After each experiment was performed, the reactors were emptied and acid rinsed to prevent any contamination from the previous experiment.

The temperature and pH of the reactors were observed to be the same for each reactor studied within a given MCRT. The reactor temperatures were observed to be 20°C during the copper studies and 23°C during the zinc experiment. The pH was measured as approximately 7.2 to 7.4 during all experiments.

## RESULTS AND CONCLUSIONS

A statistical analysis was determined not to be viable, given the amount of data collected. Therefore, various figures were generated for trend analysis and the conclusions are based on the trends observed. The conclusions are grouped together for each metal studied.

### Copper Research Conclusions

- The system biomass population (measured as MLVSS) generally decreased with an increase in copper loading. In this study, all reactors were maintained at the same volume; therefore, the MLVSS increased with the increase in MCRT. As shown in Figure 1, the higher MCRT reactors, which had a higher MLVSS concentration than the lower MCRT reactors, showed a significant drop in reactor MLVSS at the high end of copper loading. In addition, the effluent MLVSS of the highest MCRT reactor increased with the copper dosage, indicating an inhibitory effect of reduced settleability. A possible explanation is that the older biomass populations are more sus-

**Table III. Sampling Times for Reactors**

| Cycle Time | Actual Time | Samples Taken |
|---|---|---|
| 0 | 1:05–1:25 p.m. | soluble COD, MLSS, MLVSS, soluble metals, suspended metals |
| 8 hours | 9:00 p.m. | soluble metals, suspended metals |
| 8–10 hours | 9:00–11:00 p.m. | DO uptake, pH, and temperature |
| 20 hours | 9:00 a.m. | soluble metals, suspended metals |
| 22 hours | 11:00 a.m. | soluble metals, suspended metals |
| 23.5 hours | 12:30 p.m. | EFFLUENT: soluble COD, MLSS, MLVSS, soluble metals, suspended metals |

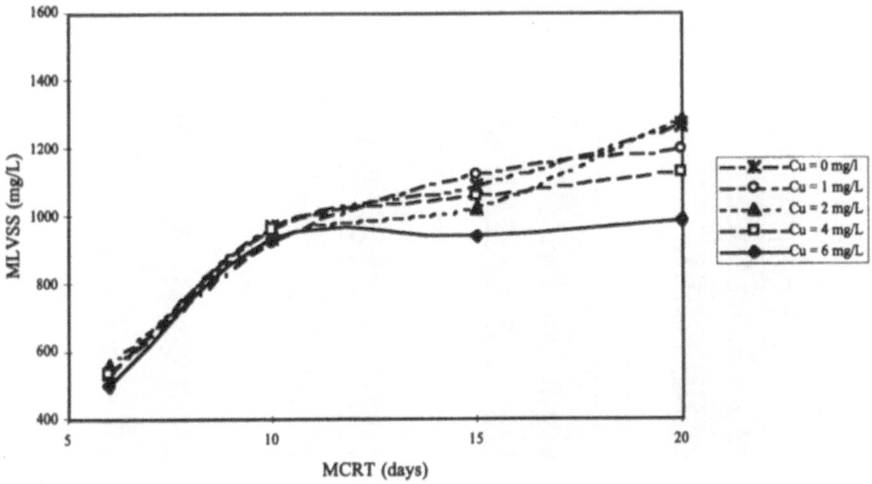


**Figure 1. MLVSS vs. MCRT for varying influent copper.**

ceptible to the toxic effects of the higher copper doses than the younger biomass populations.

- Figure 2 depicts the specific OUR present in the reactors which increased as the dosage of copper increased for each MCRT studied. Hence, an increase in the specific OUR may be an indicator of pending toxicity problems for wastewaters containing soluble copper.
- As shown in Figure 3, COD removal efficiency decreased with the increase in the dosage of copper, particularly at the lower MCRTs of 6 and 10 days. At the higher MCRTs of 15 and 20 days, a small change in the removal efficiency and effluent soluble COD was observed. At the lower MCRTs, the biomass under aeration was lower than at the higher MCRTs, which may account for the greater variability in organic removal.

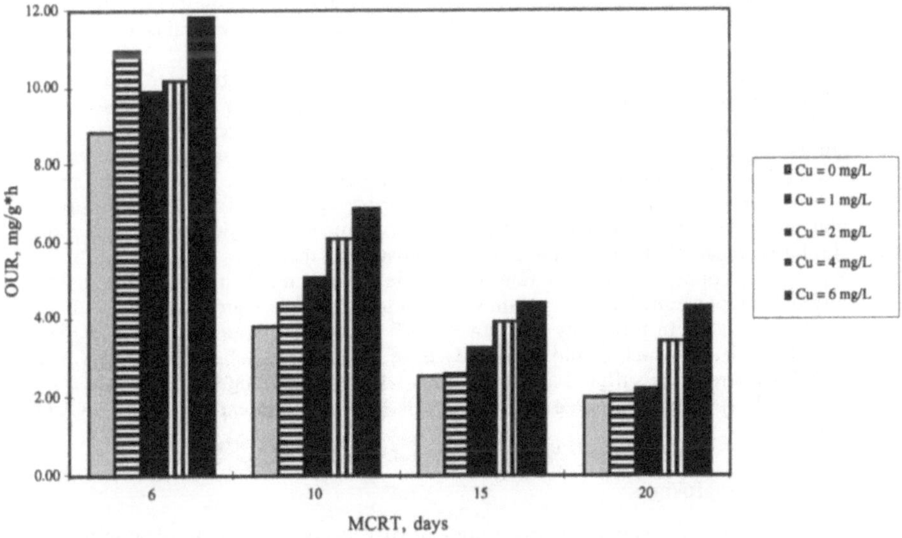


**Figure 2. OUR vs. MCRT for varying influent copper.**

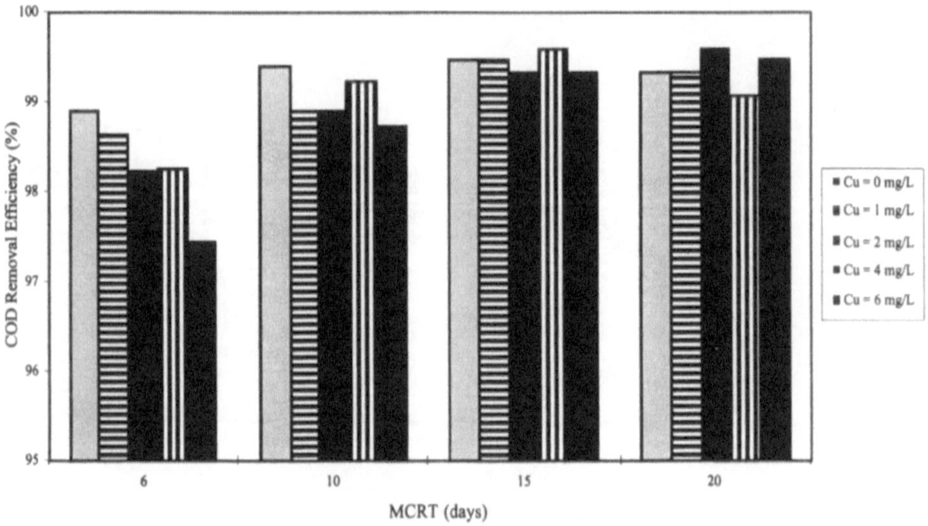


**Figure 3. COD removal efficiency vs. MCRT for various influent copper.**

- The effluent copper levels increased with the increase in influent copper dosages for all MCRTs studied. However, the higher MCRTs showed a much lower increase in effluent copper (therefore a higher removal efficiency) than the reactors maintained at the lower MCRTs. This suggests that the level of biomass present in the reactors plays a key role in the removal of the copper for each concentration studied.
- While the MLVSS levels of the higher MCRT reactors showed more sensitivity to increases in copper loadings than the lower MCRTs, the effluent COD and copper levels were higher in the lower MCRTs, where the metal mass to biomass ratio is lower. An explanation for these phenomena is that the higher MCRT reactors contained a higher concentration of MLVSS than the lower MCRT reactors. Therefore, there were either more adsorption sites available for the copper at the higher MCRT or the greater biomass populations are more capable of assimilating a certain mass of copper dosage than lesser populated reactors, or some combination thereof (see Figure 4). In other words, the mass of copper escaping in the effluent (that is, the copper which was not somehow suspended in the biomass) was close for each MCRT studied at the lowest copper dosage. Hence, this mass of copper introduced was somewhat similarly assimilated by all levels of biomass studied (low to high). However, at the remaining levels of copper introduced to each reactor, there was a greater variability in the mass of effluent copper observed, with the lower biomass reactors retaining less of the influent copper mass. This suggests that there may be some saturation point for copper (and possibly organic) removal tied to the biomass populations, either due to the complete occupation of all adsorption sites or the assimilative capacity of the biomass, or a combination of the two.
- Toxicity of copper and reactor removal efficiency are not so much factors of the influent copper concentration as they are factors of MCRT and the copper mass to biomass ratio. In this study, the low MCRT reactors suffered a loss in COD removal efficiency which would eventually have an affect on solids. Conversely, the high MCRT reactors suffered a loss of solids while maintaining COD removal efficiency. However, a continued trend of solids loss would eventually impact COD removal.

## Zinc Research Conclusions

- The system biomass population (measured as MLVSS) observed showed no trend for decreasing or increasing with an increase in zinc loading for the two MCRTs (6-day

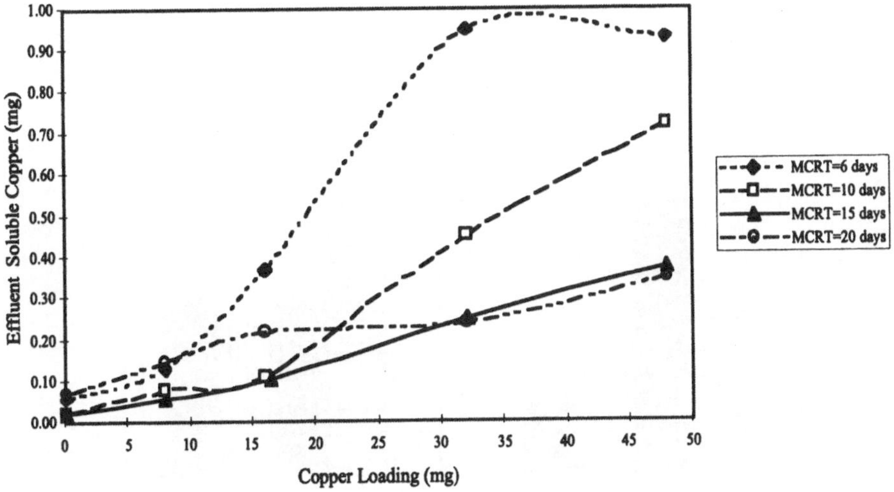


**Figure 4. Effluent soluble copper vs. copper loading for various MCRT.**

and 20-day) studied. In this study, as with the copper study, all reactors were maintained at the same volume and the MLVSS increased with the increase in MCRT. The higher MCRT reactors showed more variability in MLVSS concentration over the range of zinc loading than the lower MCRT reactors.

- Specific OUR present in the reactors revealed no identifiable trend as the dosage of zinc increased for each MCRT studied, though the lower MCRT reactors did have higher OUR than the higher MCRT reactors, as with the copper reactors. However, OUR was not found to be an effective indicator of zinc toxicity in this study.
- COD removal efficiency decreased with the increase in the dosage of zinc as shown on Figure 5, particularly at the lower MCRT of 6 days, where the metal mass to biomass ratio is lower. At the higher MCRT of 20 days, a small change in the removal efficiency and effluent soluble COD was observed.
- Figure 6 illustrates how the effluent zinc levels increased with the increase in influent zinc dosages for both MCRTs studied. However, the higher MCRT showed a lower increase in effluent zinc (hence a higher removal efficiency), as expected, than the reactors maintained at the lower MCRT until the highest zinc loading, where both MCRTs studied showed a significant increase over the reactor with the next highest loading rate.
- The effluent COD and zinc levels were higher in the 6-day MCRT reactors than in the 20-day MCRT reactors. One explanation for this is that the higher MCRT reactors contained a higher concentration of MLVSS than the lower MCRT reactors. Therefore, there were either more adsorption sites available for the zinc at the higher MCRT or the greater biomass populations are more capable of assimilating a certain mass of zinc dosage than lesser populated reactors, or through an unknown combination of the two.
- As with the copper study, the inhibitory effects of the zinc and the reactor removal efficiencies appear to be a function of the MCRT and the metal mass to biomass ratio rather than solely the concentration level of the zinc.

## General Conclusions

As a result of this study, it can be concluded that the SBR process is an appropriate treatment technology for removal of copper and zinc at the mass loadings studied. In particular, the SBR designed for a 15-day MCRT with a high level of biomass appears to be a viable process which can handle both metals studied with small treatment efficiency interrupts. Of course, there may

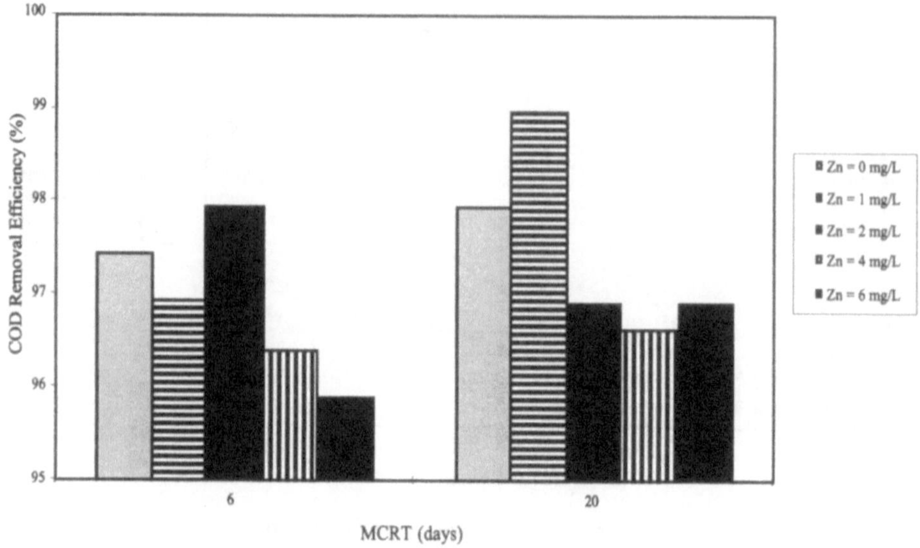


**Figure 5. COD removal efficiency vs. MCRT for varying influent zinc.**

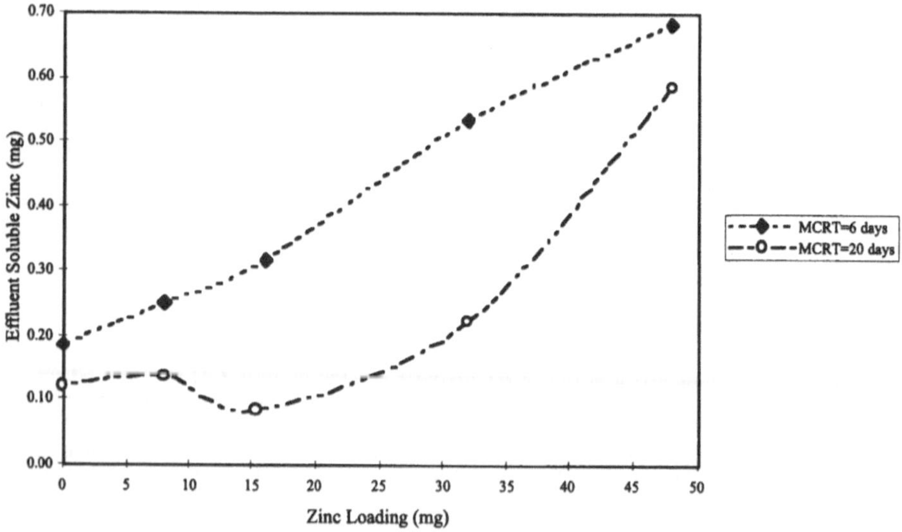


**Figure 6. Effluent soluble zinc vs. zinc loading for various MCRT.**

be a total mass toxicity which could occur after continued dosage of metals to the system which could not be observed in the experiments conducted.

The SBR process is simple and very resilient to varying wastewater conditions. The higher MCRT reactors studied exhibited a stable metal removal and effluent quality did not appear vulnerable to increasing metal loads. The highest MCRT did exhibit increased suspended solids in the effluent, which may indicate poor settleability of the biomass. It is for this reason that the highest MCRT is not recommended as the ideal design.

## RECOMMENDATIONS

Further investigation of the metals observed in this study is needed for more thorough understanding of the processes involved in removal of the metals and its immediate or long-term effects on the removal of organics. Topics recommended for further research are given below.

- At each of the metal loadings studied, there was not a significant reduction in organic treatment capacity of the reactors, which leads to the conclusion that either the level of metals loading was not high enough to cause a breakdown in the kinetic processes of the activated sludge or that the SBR is truly a resilient activated sludge process suitable for treatment of problem wastewaters. While the latter is likely true, further research should examine higher loading rates of copper and zinc to determine where the toxic threshold lies for each, as determined by reduction in organic removal efficiency or washout of the biomass.
- A study should be performed to evaluate the removal efficiencies of SBR systems maintained at the same MCRT with varying biomass populations, ranging from sparse to dense. MCRTs studied should range from low to high. The reason for this recommended study comes from the 20-day MCRT reactor observed in the copper portion of this research. This reactor showed a high variability in MLVSS concentration indicative of a toxic response at higher copper loadings, yet performed well when measured by organic and copper in the effluent. It is possible that the biomass in this reactor was nearing the endogenous respiration phase and was sensitive to the copper, but performed well due to the relatively high population of biomass present. The further research recommended would define the true optimum parameters where metal removal is both achieved satisfactorily and represents the least toxicity to the biomass. This would enable the engineer to design a treatment system to maintain the required ratio of expected influent metal concentration to biomass population at the best growth phase.
- Further, similar studies should be conducted on other metals acting individually and in combination to determine which metals should be of greatest concern when treating landfill leachates. Determination of synergistic or antagonistic relationships between metals would be especially useful.
- Studies on the effect of metals on nitrification, denitrification, and biological nutrient removal processes are important due to the increasing presence of tertiary limits in POTW discharge permits. The MCRTs equal to or greater than 10 days evaluated in this study will give some indication of results that may be expected in these recommended studies. Since nitrification is a sensitive biological process, a study which examines nitrite and nitrate levels in relation to metal removal, COD removal and suspended solids may provide a good indicator of inhibition and pending toxicity to an activated sludge process.
- Investigations of metals removal by attached growth treatment processes and continuous flow suspended growth treatment processes which review similar parameters as discussed above would be useful as well, since many POTWs in place utilize one of these processes.
- Based on the results of this study, the SBR process merits additional study for metals removal. SBRs would be a viable alternative for landfill leachate management, given their relatively low capital and operating costs and the low level of mechanical equipment required for a full-scale system which translates into reduced operator attention. If metals are expected in the wastewater, the SBR should be designed with a longer MCRT (greater than 10 days) and a high level of biomass to achieve higher total removals of the metals and stability with continued loading.

## REFERENCES


1. Brochi, P.A. *Biological Treatment of Hazardous Waste Landfill Leachate Using an Anaerobic/Aerobic Process.* M.S. Thesis, Lamar University, Beaumont, TX, 1991.
2. Hayes, K.W. Aerobic Ammonia-Nitrogen Oxidation During the Treatment of Hazardous Waste Leachate. M.S. Thesis, Lamar University, Beaumont, TX, 1991.
3. Schuk, W.W. and James, S.C. Treatment of landfill leachate at publicly owned treatment works. *Waste Manage. Res.*, 4, p. 265, 1986.

4. Thirmurthi, D., Austin, T.P., Ramalingaiah, and Khakhria, S. Anaerobic/aerobic treatment of municipal landfill leachate. *Water Pollut. Res. J. Can.*, 21, p. 8, 1986.
5. Robinson, C.G. Effects of Copper and Zinc on Sequencing Batch Reactors. M.S. Thesis, Mississippi State University, Mississippi State, MS, 1996.
6. Hannah, S.A., Austern, B.A., and Eralp, A.E. Comparative removal of toxic pollutants by six wastewater treatment processes, *J. Water Pollut. Control Fed.*, 58, p. 27, 1986.
7. Glenn, S.L., Norris, R.T., Jr., and Sommerfeld, J.T. Discrete-event simulation in wastewater treatment. *J. Envir. Sci. and Health, Pt. A Envir. Sci. and Engr.*, 25, p. 407, 1990.
8. Kreissl, J.F. Sequencing Batch Reactors: U.S. Overview. U.S. EPA Water Engr. Res. Laboratory, Office of Res. and Devel., Cincinnati, OH, EPA/600/D-87/024, 1987.
9. Tseng, D-H. and Yeh, J-H. Effects of various operating strategies on the performance of sequencing batch biological reactors. *Proc. Natl. Sci. Council Repub. China, Part A: Phys. Sci. Engr.*, 12, p. 196, 1988.
10. Neufeld, R.D. and Hermann, E.R. Heavy metal removal by acclimated activated sludge. *J. Water Pollut. Control Fed.*, 47, p. 310, 1975.
11. Cheng, M.H., Patterson, J.W., and Minear, R.A. Heavy metal uptake by activated sludge. *J. Water Pollut. Control Fed.*, 47, p. 362, 1975.

# Section Three
# BIOLOGICAL: BIOREMEDIATION AND BIOTECHNOLOGY

## 9 BIOLUMINESCENCE AND COMETABOLIC DEGRADATION OF TOLUENE AND TRICHLOROETHYLENE BY *PSEUDOMONAS PUTIDA* TVA8

**Jan G. Pieters**
Department of Civil & Environmental Engineering
The University of Tennessee, Knoxville, Tennessee 37996
Department of Agricultural Engineering
The University of Ghent, Belgium

**John Sanseverino**
Center for Environmental Biotechnology
The University of Tennessee, Knoxville, Tennessee 37996

**Charles L. Wright, Kevin G. Robinson, Catherine L. Cheng, Ke Liu, and Chris D. Cox**
Department of Civil & Environmental Engineering
The University of Tennessee, Knoxville, Tennessee 37996

**Staci R. Kehrmeyer, Bruce M. Applegate, and Gary S. Sayler**
Center for Environmental Biotechnology
The University of Tennessee, Knoxville, Tennessee 37996

### ABSTRACT

In this chapter, the toluene degradation kinetics of two *Pseudomonas* strains, *P. putida* F1 and *P. putida* TVA8, in liquid batch are compared. Furthermore, the TCE degradation kinetics for TVA8 are assessed. The relationship between toluene/TCE degradation and bioluminescence production in *P. putida* TVA8 was explored.

### INTRODUCTION

Trichloroethylene (TCE) is a widespread solvent used in a large variety of industrial processes. Since it is also readily mobile in the environment, contamination of soils and groundwater by TCE is a ubiquitous problem. Remediation is mostly based on removing the TCE and other volatile organic compounds (VOCs) by soil vapor extraction, air stripping, or air sparging, resulting in the release of high volumes of air contaminated with low concentrations of VOCs. Trichloroethylene, however, is considered a dangerous (carcinogenic) air pollutant. Hence, subsequent treatment of these air streams is required.

Treatment of these air streams is mainly based on adsorption of the contaminant to granular activated carbon, on absorption in liquids, or on incineration. Although these techniques have very high efficiencies (typically larger than 99%), they have some obvious disadvantages. Absorption and adsorption only turn a gas waste problem into a solid or liquid waste problem. Incineration has the disadvantage of releasing not only carbon dioxide ($CO_2$) and water ($H_2O$), but also (very) small amounts of oxides of nitrogen, carbon monoxide, smoke, odors, and other undesirable compounds. Furthermore, since high inputs of fuel are required, this technique cannot be considered sustainable.

*52nd Purdue Industrial Waste Conference Proceedings*, 1997, Ann Arbor Press, Chelsea, Michigan 48118. Printed in U.S.A.

Most of these disadvantages can be overcome by bioremediation, which is based on the capability of microorganisms to degrade a large variety of organic compounds to $CO_2$ and $H_2O$ and biomass, at least when placed in a favorable environment. The most popular and least expensive bioremediation technique is biofiltration, in which the contaminant is degraded by a biofilm supported on a porous medium, such as compost, peat, soil, ceramic or plastic packing material. This system is mechanically simple since it only requires a biofilter bed, a fan, a moisturizing system and some means to provide the necessary nutrients. Generally, no hazardous by-products are created during these processes at ambient temperatures. Except for small amounts of nutrients, no additional chemicals or fuel are required. As a consequence, biooxidation is an inexpensive means for the removal of VOCs, especially for low concentration, high volume air streams.[1,2]

In spite of the major advantages bioremediation has, it is not yet applied on a large scale for control of chlorinated organics. This is mainly due to its low efficiency for the removal of highly halogenated compounds, such as TCE. Recently, lab-scale experiments dealing with the cometabolic degradation of TCE have been described reporting removal efficiencies ranging from 20.7 to more than 99% but mainly between 50 and 60%.[3-8]

Another disadvantage of biooxidation for air pollution control is the difficulty of accurately controlling the oxidation process. One way to increase the reliability of bioremediation is to describe the processes that take place in biofilters mathematically, allowing feed-forward control strategies to be applied. Since the pioneering work of Ottengraf and Van de Oever,[9] several models describing the kinetics of biodegradation have been developed and tested for the description of the biodegradation of different kinds of pollutants. Bioluminescent reporter organisms may represent a more promising control variable for biofiltration process control. Successful applications of the use of bioluminescent bacteria for online monitoring of naphthalene bioavailability have been described by Heitzer and coworkers.[10] Recently, Applegate and coworkers[11] inserted a *tod-lux* gene fusion into the chromosome of *Pseudomonas putida* F1, resulting in *Pseudomonas putida* TVA8. This organism metabolizes toluene, cometabolizes TCE, and produces light as an indicator of toluene dioxygenase production. Since gene expression of bioluminescence is under the same genetic control as toluene dioxygenase, light production may be used to monitor and provide feedback control for the degradation process.

The overall goal of this research was to determine if bioluminescence may be used as a process control mechanism in the degradation of toluene and/or TCE. The specific objectives were to assess toluene and TCE degradation kinetics for *P. putida* F1 and *P. putida* TVA8 and to determine if a qualitative and/or quantitative relationship exists between toluene/TCE degradation and bioluminescence production in *P. putida* TVA8.

## MATERIALS AND METHODS

### Microorganisms

The bacterial strains used in this study are listed in Table I. *P. putida* strains were grown on a modified mineral salts buffer (MSB) (Hunters medium[12]). The phosphate buffer of Hunters medium was replaced with a buffer solution containing $KH_2PO_4$ (2.15 g/L) and $K_2HPO_4$ (5.3 g/L). Final pH was 7.0. Toluene (200 µL) was supplied in the vapor phase in a suspended glass bulb. *Escherichia coli* strains were grown in LB broth. Kanamycin (final concentration 25 µg/mL) was added to *P. putida* TVA8 cultures. Ampicillin (final concentration 100 µg/mL) was added to *E. coli* ES2 (pLJS-*lux*).

Once a week, cells were taken from a frozen stock (–80°C) and grown in a 250 mL Erlenmeyer flask containing 50 mL of MSB and toluene supplied as discussed below. Each day, a new

**Table I. Bacterial Strains Used in This Study**

| Strain | Characteristics | Reference |
|---|---|---|
| *P. putida* F1 | Contains a chromosomally-encoded *tod* operon for toluene degradation | 13 |
| *P. putida* TVA8 | *P. putida* F1 containing the *tod* promoter fused with the *lux* genes; kanamycin resistant | 11 |
| *E. coli* ES2 | *E. coli* DH5α containing plasmid pLJS-*lux*; ampicillin resistant | 14 |

subculture was prepared by transferring 1 mL of overnight culture ($OD_{546}$ ~ 1.3) to 50 mL of MSB and toluene. All cultures were incubated at 28°C at 200 rpm. All optical densities were measured by a Beckman DU®-70 spectrophotometer (La Jolla, CA) at a wavelength of 546 nm.

### Cell Preparation

For each experiment, overnight cultures were centrifuged at 8000 rpm for 10 min, washed once in MSB and diluted to a final $OD_{546}$ of 1.0. Five mL of the diluted culture was added to a series of 20 mL vials. The vials were closed with Mininert® valves (Dynatech) for the toluene experiments and with Teflon-faced silicone septa and open-top polypropylene caps for the TCE experiments. Subsequently, the vials were put in an ice bath for at least half an hour. The use of the ice bath was required since most experiments had to be carried out sequentially to obtain accurate results because the degradation was rather fast compared to the time interval between two GC measurements.

### Toluene Experiments

All time readings for one experiment were expressed with respect to the moment at which the vial was taken from the ice bath and brought in contact with the ambient air, having a constant temperature of 20°C. For *P. putida* TVA8, bioluminescence was measured at $t = 0$, 5, and 10 min. Toluene (0 to 100 ppm) was added to the cells as toluene-saturated MSB (saturated concentration about 515 ppm) at $t = 14$ min, so that the first headspace sample for the determination of the toluene concentration was taken at $t = 15$ min, all other samples being injected after the previous GC cycle had been completed; i.e., at approximately 2.5 min intervals. Bioluminescence readings were performed immediately following each GC injection. Each experiment also included a vial containing pure MSB, a vial containing 5 mL of the culture without toluene (the blank), and finally a killed control vial containing dead cells (cells were killed by adding 0.5 mL of 2 N $H_2SO_4$). The same amount of toluene-saturated MSB was injected in both the MSB and the killed control vials, allowing possible absorption of toluene by the cells or loss of toluene from the vials to be determined. The blank was used as a reference for the bioluminescence measurements. Experiments were run in triplicate. For the bioluminescence experiments in which toluene concentrations were not monitored, the same procedure was followed except that experiments were run simultaneously instead of sequentially.

### Toluene plus TCE Experiments

The toluene plus TCE experiments were conducted in basically the same way as the toluene experiments. However, TCE (0 to 10 μg) was injected as TCE-saturated MSB (saturated concentration of 1100 ppm) immediately following the injection of the toluene-saturated MSB. Bioluminescence was measured every 2 to 3 minutes. In addition to the controls described above, a treatment with toluene only was included.

### Analytical Procedures

Headspace analysis of toluene and TCE was performed using a Hewlett Packard 5890 Series II gas chromatograph, equipped with a HP 624 capillary column (0.53 mm inner diameter, 3.0 μm film thickness, 30 m length). Gas sample injections (25 μL) were made manually. Toluene was detected by a flame ionization detector (FID), while TCE was detected by an electron capture detector (ECD). Helium was the carrier gas (10 mL/min), and nitrogen was the make-up gas (25 mL/min). Injection and detector temperatures were 150°C and 250°C, respectively. Injector split ratio was 4:1. The capillary column contained a splitter which directed equal carrier gas flow to the FID and the ECD simultaneously.

For experiments where both toluene and TCE concentrations where measured, temperature programming was applied: the oven was operated at 40°C for 2 min, after which the temperature was increased by 25°C/min until a final temperature of 140°C was reached. One measurement took about 10 min. For experiments where only the toluene concentration was to be determined, the total time needed for one measurement was reduced to about 2.5 min by operating the oven isothermally at 102°C.

Bioluminescence was measured with the use of an Oriel digital display (Model 7070) with a photomultiplier tube (Model 77340) (Stanford, CT) connected to a flexible liquid light pipe and collimating beam probe. Since the photon flux was converted in an electric current, bioluminescence was read in nanoampere [nA].

## Data Treatment

Degradation rates for the toluene experiments were expressed as the slopes of the linear parts of the degradation curves (i.e., the curves giving the toluene concentration in the headspace of the vials as a function of time). These slopes were determined by visually selecting the linear part of the curve and by applying linear regression analysis to the corresponding data. For the TCE experiments, the preliminary results showed no linearity, but rather an exponential decrease of the TCE concentration as a function of time. Since data were only obtained at 10 min intervals and the low number of experiments did not allow the development of a mathematical description, TCE degradation kinetics were described by means of the absolute and relative removal of TCE after 1 hour. Removal efficiencies were obtained as the ratio of the final (3 hours after injection of the TCE) and the initial TCE concentrations in the headspace.

Bioluminescence was measured in nanoamps [nA] and plotted versus time together with headspace concentrations of toluene and TCE. For each run, total bioluminescence was determined by numerically integrating the bioluminescence over time for both the sample and the blank; subtraction of the latter from the former gave the total net bioluminescence. If at the end of an experiment the bioluminescence of the sample was still higher than the bioluminescence of the blank, corrections for the integral were applied by assuming that the bioluminescence for the sample decreased linearly while the bioluminescence for the blank remained constant. The total bioluminescence was obtained as the integral of the corresponding electric current over time and is expressed in Coulomb [C].

## RESULTS AND DISCUSSION

## Toluene Degradation Assays

Both *P. putida* F1 and *P. putida* TVA8 were efficient in degrading toluene; both strains removed all the added toluene from the headspace, although at different rates. Figure 1 shows the toluene degradation rates for both F1 and TVA8 as a function of the initial concentrations in the liquid phase over a range of 20 to 85 ppm, corresponding to injected masses of toluene between 0.1 and 0.5 mg. Rates are expressed with respect to disappearance of toluene from the headspace in the vials and not from the liquid phase.

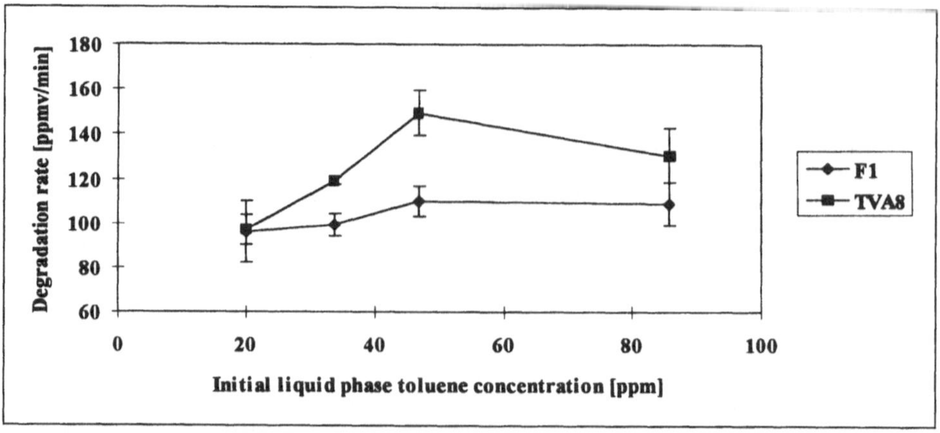


**Figure 1. Toluene degradation rates for F1 and TVA8 as a function of the initial amount of toluene.**

Under these experimental conditions, toluene degradation rates for *P. putida* TVA8 increased from 100 ppmv/min for 20 ppm initial toluene concentration to approximately 130-140 ppmv/min for an initial toluene concentration of 50 ppm. Degradation rates remained almost constant for higher toluene concentrations up to 85 ppm. For *P. putida* F1, the increase of the degradation rate as a function of toluene concentration is statistically insignificant. The maximum degradation rates were less than 100ppmv/min. From the control vials, no or statistically insignificant absorbance or loss from the vial occurred, therefore the disappearance of toluene from the headspace was due to (bio)degradation.

The only difference between *P. putida* F1 and *P. putida* TVA8 was the addition of the *lux* genes under the control of the *tod* promoter in *P. putida* TVA8. It was hypothesized that the luciferase enzyme (the enzyme responsible for light production) may be oxidizing toluene, giving an apparently higher toluene degradation rate. To test this hypothesis, *E. coli* ES2 (pLJS-*lux*) was used. This strain naturally does not metabolize toluene. The *lux* genes are constitutively expressed in this strain. If luciferase was oxidizing toluene, then a loss of toluene should be observed in this *E. coli* strain. A toluene degradation assay was performed as described above. No toluene removal was observed compared to control assays (*E. coli* without pLJS-*lux*; data not shown). Therefore, luciferase does not appear to be oxidizing toluene and some other mechanism was responsible for the increased toluene degradation rate in *P. putida* TVA8.

## Bioluminescence

Figure 2 shows the bioluminescence response of TVA8 to the addition of three different amounts of toluene. As a reference, the toluene concentration in the headspace is plotted on the same graph. In most experiments, the injection of toluene at $t$ = 14 min resulted in an almost immediate and relatively high increase in bioluminescence. It was also observed that during toluene degradation, the bioluminescence was approximately 40 nA. Only after most of the toluene had been removed did the bioluminescence increase by a factor of two or more, as can be seen in Figures 2b and 2c. It was also observed that for a given amount of toluene injected, the peak level of bioluminescence occurred at the same time interval from the toluene injection at 14 min. For example, in Figure 2b, peak bioluminescence occurred at approximately 34 min after toluene was injected. This interval was reproducible. Lower levels of toluene gave rise to shorter intervals before peak light; higher levels gave rise to longer intervals. The time elapsed between toluene injection and peak bioluminescence increased from about 10 min for 0.10 mg of toluene injected to about 30 min, 50 min, and more than 80 min for 0.18, 0.26 and 0.52 mg of toluene, respectively.

In contrast to the high reproducibility of the elapsed time of maximum bioluminescence, bioluminescence levels and the total bioluminescence produced for each experiment showed more scatter. This was clearly demonstrated by the results of Figure 2. For these experiments, which were all conducted (in triplicate) on different days, no relationship between these different levels and the initial toluene concentration could be found. Throughout several days of experiments, however, it was observed that the bioluminescence at the start of the experiment; i.e., before toluene was injected, varied from day to day, while much less variance was found during one day. Based on this observation, results were normalized using the following expression:

$$B_n = E[b_{10}] \cdot \frac{B}{E_d[b_{10}]} \tag{1}$$

with
- $B_n$ = normalized total bioluminescence [C]
- $B$ = total bioluminescence calculated as described in section 3.4. [C]
- $b_{10}$ = bioluminescence determined at $t$ = 10 min; i.e., at the last measurement point before injection of the toluene [A]
- $E[b_{10}]$ = expected value of $b_{10}$, determined as the average of all $b_{10}$ values measured over all days [A]
- $E_d[b_{10}]$ = expected value of $b_{10}$, determined as the average of all $b_{10}$ values measured on day $d$ [A]

Multiplying by the constant $E[b_{10}]$ is not strictly necessary, but it was introduced in order to obtain the same average values for the normalized and for the raw total bioluminescence data, and in order to obtain the same dimensions for both quantities.

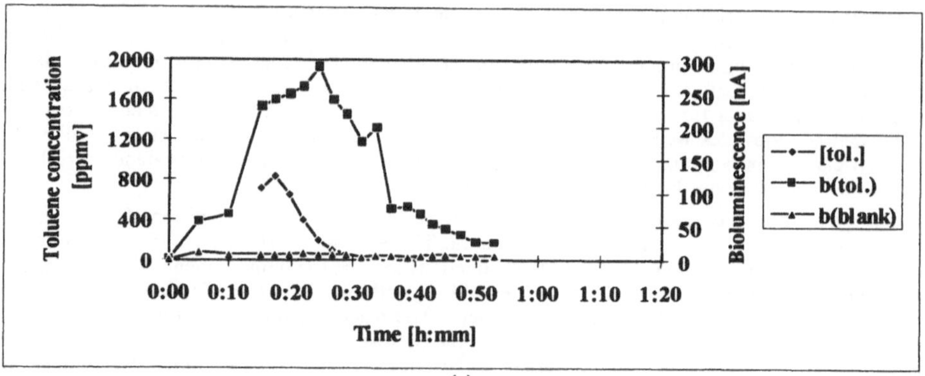


(a)

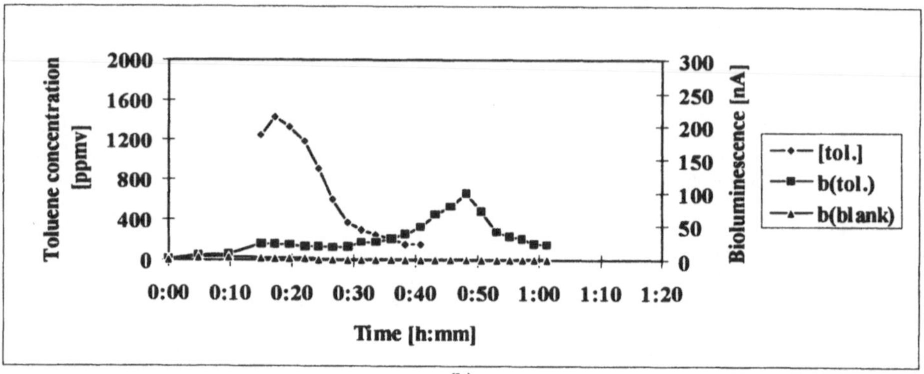


(b)

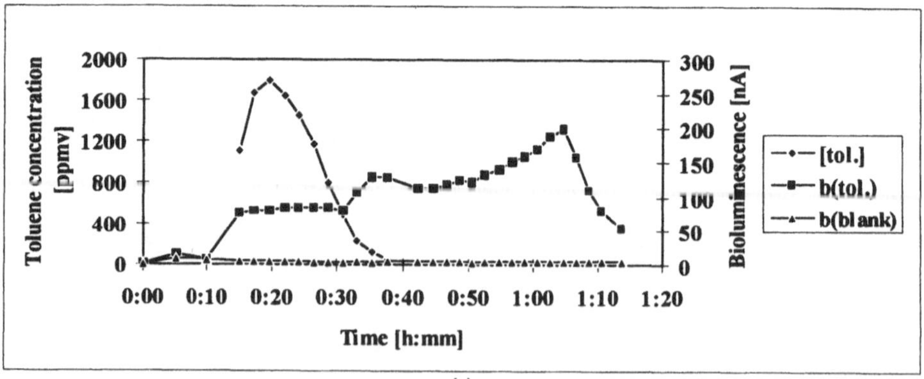


(c)

**Figure 2. Headspace toluene concentration [tol.] and bioluminescence responses [b] of TVA8 to injections of (a) 0.10 mg, (b) 0.18 mg, and (c) 0.26 mg of toluene.**

This normalization method was applied to the experiments described above. By means of linear regression, prediction of the untreated ($B$) and the normalized ($B_n$) values of the total bioluminescence for an injection of 0.52 mg (100 ppm) of toluene was attempted. The predicted values were compared to the results of two series of 0.52 mg injections obtained on two different days, with different $E_d[b_{10}]$ values. The results of this experiment are shown in Figure 3. From

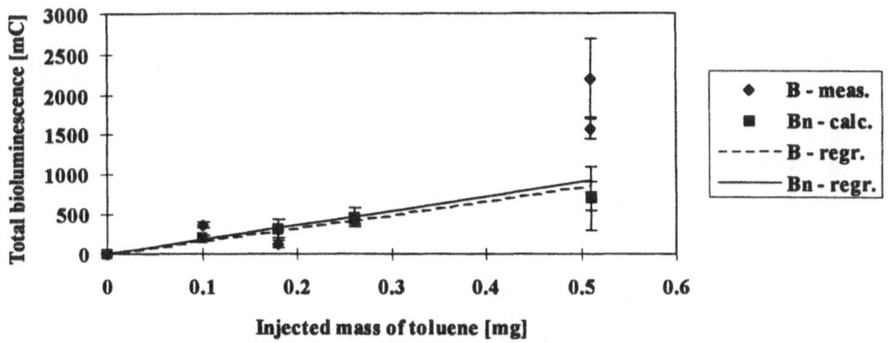


**Figure 3. Absolute (B) and normalized (Bn) total bioluminescence data as measured or calculated and regression predictions for TVA8 as a function of injected mass of toluene. Data points represent the average of three replicate values with error bars representing the standard deviations. The regression lines, based on the measurements for toluene injections between 0 and 0.26 mg, are given by the lines. The data point for 0.52 mg toluene is not included in the regression analysis.**

this experiment it was concluded that the data for absolute bioluminescence showed no regular pattern, while the normalized data are situated on a straight line. The coefficients of determination for the two regression lines were 0.502 and 0.995, respectively. Furthermore, the untreated data obtained for the two series in which 0.52 mg of toluene was injected are found to be far from one another, while their error bars do not overlap. In both cases, the data points were situated far above the corresponding regression line. After normalization of the total bioluminescence, however, the results are almost coincidental and situated slightly below the regression lines.

Based on the previous findings, it was hypothesized that the total bioluminescence of *P. putida* TVA8 as a function of the injected mass of toluene measured on any day would be described by a straight line through the origin, while the slope of the line would be proportional to the $E_d[b_{10}]$ value. This hypothesis was verified by means of simultaneous "bioluminescence only" experiments; i.e., experiments in which the toluene degradation was not investigated. In these experiments, the bioluminescence response of *P. putida* TVA8 to seven different concentrations of toluene was measured in triplicate. An example of the results from such an experiment are given in Figure 4. In this experiment a linear relationship was found ($r^2 = 0.958$) between bioluminescence produced vs. toluene concentration. Applegate and coworkers[11] also observed this relationship. Bioluminescent experiments performed on different days did give different light levels and different slopes. These differences may be due to minor variations in time of cell harvest, temperature at which the experiment is conducted (20 ± 2°C), etc. The biochemistry of bioluminescence is complicated and may be affected by a number of subtle factors. Normalization, as described above, will remove any variability. For all experiments, application of the normalization method gave rise to a slope that was statistically the same as for all other days, even though the maximum slope calculated for the untreated (B) data was more than 10 times higher than the minimum slope.

## Toluene plus TCE Degradation Assays

The main purpose of these assays was to establish a method of investigation and to describe the behavior of *P. putida* TVA8 when exposed to toluene and TCE. These results are only preliminary. TCE degradation data for *P. putida* TVA8 are summarized in Table II. Toluene degradation data were not included since the GC cycle of 10 minutes did not allow for detailed results, most toluene being degraded after 10 or 20 minutes for toluene injections of 0.10 and 0.26 mg, respectively. In control vials, headspace TCE concentrations above dead cells were not lower than TCE

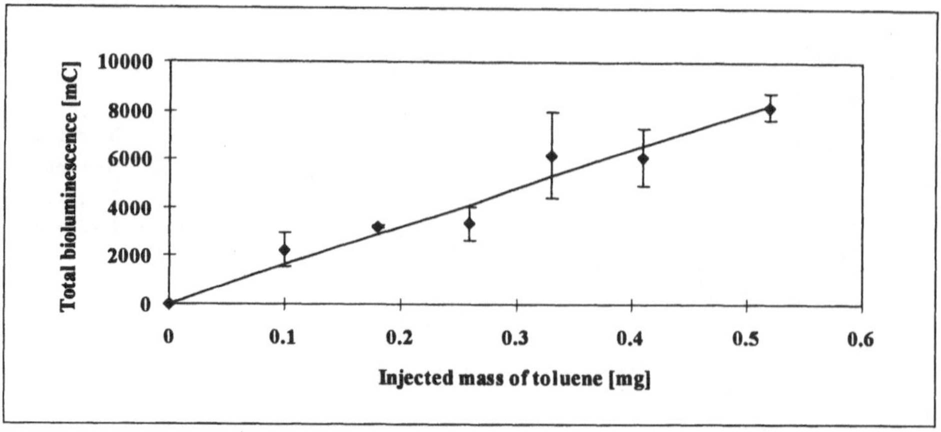


**Figure 4. Measured and regressed total bioluminescence versus injected mass of toluene obtained from simultaneous measurements. Each data point represents the average of triplicate values. Error bars represent the standard deviation.**

concentrations above MSB. The disappearance of TCE from the headspace of the vials was concluded to be the consequence of biodegradation only.

Table II shows that the majority of TCE degraded occurred in the first hour. For 100 (g of toluene, TCE removal efficiencies decreased with increasing TCE concentrations. This was expected since the amount of toluene dioxygenase produced will be proportional to the amount of toluene injected. Since the enzyme has only a limited half-life and assuming that the TCE degradation is not purely substrate limited, higher amounts of TCE will give rise to lower final degradation efficiencies. It was also observed that TCE degradation was not detectable during the first 10 minutes after TCE injection, corresponding to the time needed to degrade all the toluene.

At a higher toluene concentration (260 (g), higher amounts of TCE were degraded; 62% after 1 hr and 74% after 3 hr. Although it took more time to degrade the larger amount of toluene and thus to start TCE degradation, the relative TCE removal was much higher than for the corresponding experiment at a lower toluene concentration. This is probably due to higher amounts of toluene dioxygenase being produced by the increased amount of toluene present. Absolute amounts of toluene dioxygenase in the cells were not measured. It is important to note that TCE is not degraded until the majority of toluene has been removed. This is due to competitive inhibition. Toluene will outcompete TCE for the active site of the toluene dioxygenase. Once toluene is removed, gene expression ceases. Any TCE removed is dependent on any toluene dioxygenase remaining in the cell.

**Table II. Toluene/TCE Degradation Data for P. putida TVA8**

| Amount of Toluene [µg] | Amount of TCE [µg] | Initial TCE Conc. [ppmv] | TCE Conc. after 1 hr [ppmv] | TCE Removed after 1 hr [ppmv] | TCE Removal after 1 hr [%] | TCE Removal after 3 hr [%] |
|---|---|---|---|---|---|---|
| | 3 | 24 | 1 | 23 | 97 | 100 |
| | | 17 | (40 min)a | (25)b | 100 | 100 |
| | | 17 | (50 min)[a] | (20)[b] | 100 | 100 |
| 100 | | 22 | 1 | 21 | 94 | 100 |
| | 5 | 40 | 12 | 28 | 70 | 87 |
| | 10 | 90 | 52 | 38 | 42 | 52 |
| | | 82 | 49 | 33 | 40 | 58 |
| 260 | 10 | 90 | 34 | 56 | 62 | 74 |

[a]Time after which all TCE has disappeared.
[b] Linearly extrapolated value.

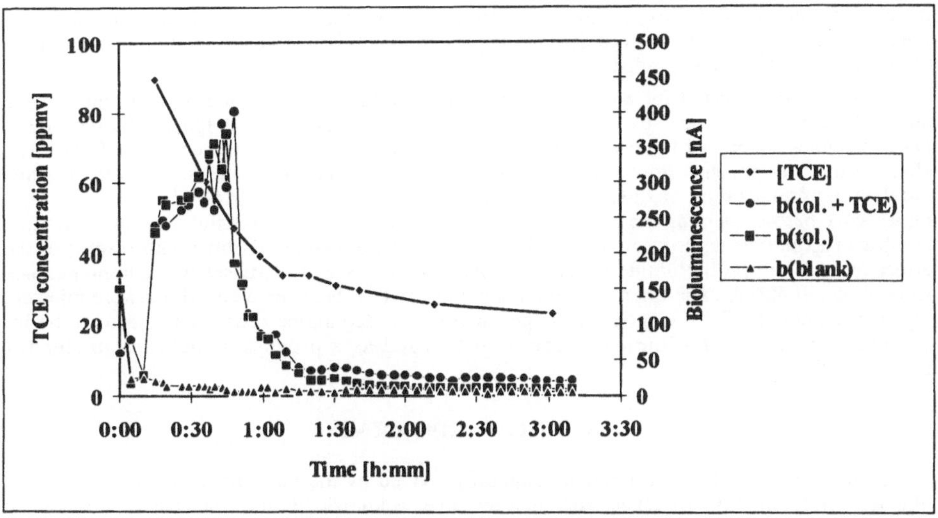


**Figure 5. TCE concentration [TCE] in the headspace, and bioluminescence response [b] of TVA8 to the injection of ("tol. + TCE") 100 µg toluene plus 10 µg of TCE, ("tol.") 100 mg of toluene, compared to ("blank") the bioluminescence of living cells without toluene and TCE.**

### Bioluminescence

Since accurate bioluminescence data were only obtained from a few experiments, no decisive answer concerning the influence of TCE on bioluminescence could be given. Figure 5 shows the bioluminescence response of *P. putida* TVA8 to 100 µg of toluene and to 100 µg of toluene plus 10 mg of TCE. The addition of TCE did not significantly affect the bioluminescence.

Decreases or increases of bioluminescence were found from other experiments in which amounts of TCE between 3 and 10 (g were injected (higher amounts of bioluminescence being observed for higher TCE concentrations). Future experiments will try to find out whether this might be explained by two opposite effects of TCE on the cells: inhibitory versus solving effects, the latter giving rise to a continuous destruction of the cell membrane and consequently stimulating the metabolism for membrane repairing. Heitzer and coworkers[10] showed that perturbing cells with organic solvents will generate an increase in bioluminescence. This is due to increased cell cycling of fatty acids (a required substrate for bacterial luciferase) and not due to increased gene expression.

## SUMMARY AND CONCLUSIONS

A genetically engineered organism, *P. putida* TVA8 was derived from *P. putida* F1 by inserting the *lux* operon behind the *tod* promoter.[11] *P. putida* TVA8 was tested in liquid batch assays to determine its toluene and TCE degradation capabilities. For *P. putida* TVA8, toluene degradation rates increased from about 100 ppmv/min for a toluene injection of 0.10 mg to about 130-140 ppmv/min for a toluene injection of 0.26 mg, while they remained almost constant for higher toluene injections of up to 0.52 mg. For the same experimental conditions, toluene degradation rates for *P. putida* F1 were shown to be lower. Experiments using a nontoluene degrading *E. coli* strain in which the *lux* operon was constitutively expressed revealed that luciferase did not degrade toluene or TCE, implying that the difference in degradation kinetics between TVA8 and F1 was not directly related to luciferase production.

As far as the degradation of TCE in the presence of toluene was concerned, for a given mass of toluene, the absolute TCE degradation was slower for lower concentrations of TCE. It was also found that lower values of the toluene/TCE ratio gave rise to lower degradation efficiencies after three hours. This was explained by the fact that relatively high toluene concentrations lead to relatively large amounts of dioxygenase being produced, resulting in quicker TCE removal during the limited lifetime of the enzymes. Results suggested that TCE degradation took place only after all the toluene had been degraded.

The data were analyzed for a qualitative and/or quantitative relationship between the toluene and TCE degradation processes and light production by *P. putida* TVA8. A linear relationship between the total amount of toluene being degraded and the total bioluminescence was found. The slope of this straight line, however, was found to vary between experiments. To overcome this problem, a normalization method in which the total bioluminescence was divided by the bioluminescence before injecting the toluene, was proposed and proved successful. Due to the limited number of experiments, no decisive conclusions could be drawn at this time with respect to the effect of TCE on the bioluminescence of TVA8. Decreases and increases of bioluminescence were observed during experiments in which amounts of TCE between 3 and 10 µg were injected, higher amounts of TCE giving rise to larger increases in bioluminescence. This needs to be investigated further if bioluminescence is to be used as a process control parameter for toluene/TCE degradation.

## ACKNOWLEDGMENT

J. Pieters acknowledges the financial support provided by the Fund for Scientific Research - Flanders (Belgium). B. Applegate and S. Kehrmeyer acknowledge the financial support of the Tennessee Valley Authority (Grant TV-94002V) and the United States Air Force (Grant (F49620-89-C-0023). All the authors acknowledge the financial support for this research provided by the University of Tennessee's Waste Management Research and Education Institute.

## REFERENCES

1. Bohn, H. "Consider biofiltration for decontaminating gases," *Chem. Eng. Prog.*, 88(4), pp. 34–40, 1992.
2. Yudelson, J.M. and P.D. Tinari, Economics of Biofiltration for Remediation Projects, *In: Bioremediation 3* (R.S. Skeen) Battelle Press, OH, 1995, pp. 205–209.
3. Speitel, G.E. and D.S. McLay, "Biofilm reactors for treatment of gas streams containing chlorinated solvents," *J. Environ. Eng.*, 119, 1993, pp. 658–678.
4. Devinny, J.S., T.S. Webster, E. Torres, and S. Basrai, "Biofiltration of PCE and TCE vapors from contaminated air," *Haz. Waste Haz. Mater.*, 12, 1995, pp. 283–293.
5. Enomoto, K. and K. Nakamura, Degradation of Trichloroethylene by Vapor-Phase Bioreactors Using Toluene Degrading Bacteria, Poster Abstracts of In-Situ and On-Site Bioreclamation/Third International Symposium, Group I(c), 1995.
6. Watwood, M.E., and S. Sukesan, "Biodegradation of trichloroethylene in finished compost materials," *Compost Sci. Util.*, 3(4), 6–19, 1995.
7. Anon. "Co-metabolic biofiltration destroys TCE," *Air Pollut. Consultant*, 6(3), 1.9–1.11, 1996.
8. Woo, H.-J., C. Cox, and K.G. Robinson, Biofiltration of TCE and Toluene Vapors Using *Pseudomonas putida* F1, Extended abstract presented at the I&EC Special Symposium, American Chemical Society, Birmingham, AL, September 9–12, 1996.
9. Ottengraf, S.P.P. and A.H.C. Van de Oever, "Kinetics of organic compound removal from waste gases with a biological filter," *Biotechnol. Bioeng.*, 25, 1983, pp. 3089–3102.
10. Heitzer, A., K. Malachowsky, J.E. Thonnard, P.R. Bienkowski, D.C. White, and G.S. Sayler, "Optical biosensor for environmental on-line monitoring of naphthalene and salicylate bioavailability with an immobilized bioluminescent catabolic reporter bacterium," *Appl. Environ. Microbiol.*, 60, 1994, pp. 1487–1494.
11. Applegate, B.M., S.R. Kehrmeyer, and G.S. Sayler, "Generation of a chromosomally-based tod-lux reporter for detection and quantification of BTEX compounds in aqueous solutions," Submitted to *J. Bacteriol.*, 1997.
12. Stanier, R.Y., N.J. Palleroni, and M. Doudoroff, "The aerobic Pseudomonads: a taxonomic study," *J. Gen. Microbiol.*, 41, 1966, pp. 159–271.
13. Wackett, L.P. and D.T. Gibson, "Degradation of trichloroethylene by toluene dioxygenase in whole-cell studies with *Pseudomonas putida* F1," *Appl. Environ. Microbiol.*, 54, 1988, pp. 1703–1708.
14. Applegate, B. M., personal communication, 1997.

# 10 REDUCTIVE DECHLORINATION OF PERCHLOROETHYLENE IN AN AEROBIC SEQUENCING BATCH REACTOR

**Enos C. Inniss**, Doctoral Student

**Lloyd Ketchum Jr.**, Associate Professor

Department of Civil Engineering and Geological Sciences

University of Notre Dame, Notre Dame, Indiana 46556

## ABSTRACT

Microorganisms mediate the sequential dechlorination of perchloroethylene (PCE) to trichloroethylene, dichloroethylene, vinyl chloride, and, ultimately, ethylene, a nontoxic compound. Researchers have focused on PCE dechlorination in strict anaerobic environments while little or no data have been published on the effects of oxygen concentration on this process.

In this experiment, sequencing batch reactors contained aerobic/facultative anaerobic mixed cultures which could dechlorinate PCE during anaerobic periods, with glucose as a carbon and energy source, despite experiencing an aerobic period during the cycle. The acclimated enrichment cultures were able to dechlorinate 90 to 95% of 10 mg/L of PCE to cis-1,2-dichloroethylene (cDCE). The cultures were exposed to 8–9 mg/L of oxygen for 1.0 hour every 24 hours. This report represents a major step toward the creation of a single biological reactor than can promote both PCE dechlorination and oxidation of the dechlorination products.

## INTRODUCTION

Tetrachloroethylene (i.e., perchloroethylene or PCE), and its dechlorination products, trichloroethylene (TCE), dichloroethylene (1,1-, cis-1,2-, and trans-1,2-DCE), and vinyl chloride (VC), are common groundwater contaminants regulated under the Safe Drinking Water Act Amendments of 1986.[1] DiStephano et al., Vogel and McCarty, and others have presented evidence of transformation of PCE, a four chlorine substituted ethylene, to its dechlorination products under strict anaerobic or reducing conditions.[2–4] But PCE does not appear to be transformable under aerobic conditions. Comparatively, its transformation products, namely TCE and DCE, are more labile to anoxic and aerobic transformation pathways due to the lesser chlorine substitutions.[5–8]

This chapter addresses the question of whether a process can be established which is efficient at utilizing both anaerobic and aerobic processes; i.e., can the two processes occur within the same reactor? If an environment may be engineered which allows for PCE to be the initial preferential electron acceptor while still allowing some members of the population to maintain their ability to use other electron acceptors such as oxygen, then it may be possible for PCE to be dechlorinated and for its products to undergo subsequent transformation, possibly through oxidation and mineralization.

Previous studies at the University of Notre Dame have shown success at reductively dechlorinating PCE to cis-DCE as the predominant end product under anaerobic conditions within a 24-hour period. The results of these experiments have been consistent and reproducible.[3] Our study tests whether consistent reductive dechlorination may also be achieved in a reactor system which is not strictly anaerobic. The ability to subject the reactor to both aerobic conditions and anaerobic conditions will allow for the transition from an environment favorable for transforming PCE

*52nd Purdue Industrial Waste Conference Proceedings*, 1997, Ann Arbor Press, Chelsea, Michigan 48118. Printed in U.S.A.

to one which may efficiently transform the dechlorination products to a nonchlorinated end product.

The reactor system employed is the Sequencing Batch Reactor (SBR). The SBR is a periodically operated system and thus allows for temporal variations depending on the operating strategy.[9] Thus the reactor used is particularly useful in time rather than in space. The reactor may be operated to control oxygen concentrations and, to a certain extent, the oxidation-reduction potential and thus create a reactor environment with suitable reaction conditions for PCE transformation and subsequent transformation of the dechlorination products.

## MATERIALS AND METHODS

### Reactor Setup

As illustrated in Figure 1, a 4.0-L Pyrex® Kettle reaction flask (Corning® 6947-4L, Fisher Scientific) was used as the laboratory-scale SBR. Glucose was separated from the rest of the feed solution. It was introduced to the reactor through a needle placed through the stopper which housed the air flow tube. The needle was connected to a syringe by flexible Tygon® tubing and the syringe was connected to an infusion/withdrawal pump (Model 230, kd-Scientific). The reactor feed was taken from the feed container and entered the reactor through glass tubes. Movement of the solution was accomplished by connecting the two glass tubes with Masterflex® Viton® flexible tubing which was threaded through the pump head of a Masterflex® pump controller (Model 7553-50, Cole Parmer Instrument Co.). Similarly, the draw tube out of the reactor was connected to a waste container. Air was supplied through an air stone connected by flexible Tygon® tubing to the compressed air spout of the fume hood where the reactor was located. Air flow was controlled by a solenoid valve (Model 8260A54, Automatic Switch Co.). Mixing was accomplished with a magnetic stirrer placed in the bottom of the reactor. The kettle sat on top of a stir table which was used to control the magnetic stirrer. The system was automated by plugging the syringe pump, the pump controllers, the stir table, and the solenoid valve to ChronTrol® XT Series timers. Samples were manually taken from a fourth port in the lid of the kettle.

The reactor was seeded with a mix of 4500 mg of anaerobic digester sludge and 4500 mg of secondary clarifier waste-activated sludge from the South Bend Wastewater Treatment Facility placed in 3.0-L liquid volume. Organisms which could transform PCE were preferentially selected for within the reactor because it was operated with PCE as the predominant electron acceptor.

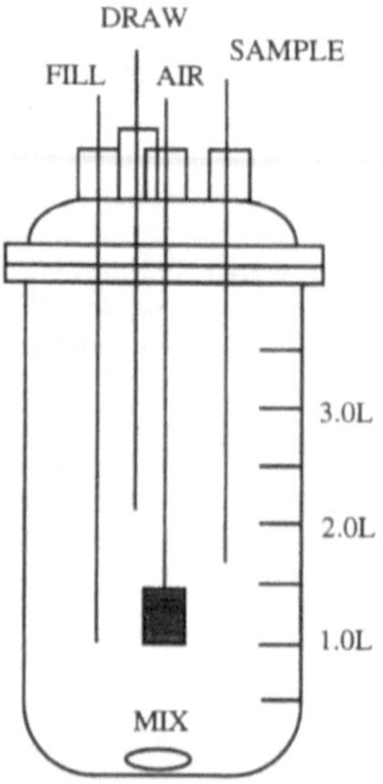


**Figure 1. Physical reactor setup.**

## Operating Strategy

At the start of each 24-hour cycle, 10 mg/L (60.3 μmol/L) of PCE and 600 mg/L of glucose were added with the mineral media to produce a 1.0-L working volume. This was added to the 2.0-liter volume already present in the reactor to yield a 3.0-liter mixed liquor volume and a hydraulic residence time of 3 days. The periods of the cycle are as follows:

| | Length of Time Period | Time After Start of Cycle |
|---|---|---|
| MIXED FILL ($F_m$) | 0.7 hour | 0.70 hour |
| REACT (R) | 20.22 hours | 20.92 hours |
| SETTLE (S) | 1.83 hours | 22.75 hours |
| DRAW (D) | 0.12 hour | 22.87 hours |
| STATIC IDLE ($I_s$) | 0.05 hour | 22.92 hours |
| AERATED IDLE ($I_a$) | 1 hour | 23.92 hours |
| MIXED IDLE ($I_m$) | 0.08 hour | 24.00 hours |

*MIXED FILL*

The fill period, which is considered the start of the cycle, introduces the PCE and glucose to the system. The purpose for having this period mixed was to create a completely mixed environment which allows reaction to begin during fill as the organisms are exposed to the substrate, PCE, and the electron donor, glucose. The proximity allows the organisms to immediately begin to take up the compounds and undergo biological reaction.

*REACT*

The react period was also mixed to continue the exposure of the organisms to the PCE and glucose. The mixing allows for slight air transfer such that the reactor tends to remain near anoxic conditions during the cycle.

*SETTLE*

The stir bar is turned off so that the reactor liquid may become quiescent. This period allows for the sedimentation of the biomass in solution. The resulting treated effluent may then be drawn off while the organisms remain within the reactor. Unlike with a continuous flow system, this period separates the hydraulic from the organism retention times as reaction and settling take place in the same reactor.

*DRAW*

Removal of the treated effluent toward the end of each cycle allows for the addition of more PCE and glucose at the beginning of the next cycle.

*STATIC IDLE*

The purpose for incorporating a STATIC IDLE in the reaction cycle is to ensure that DRAW period does not overlap with AERATED IDLE period. If the two were to overlap then some of the biomass would be washed out.

*AERATED IDLE*

An aeration period was part of the research goals. This period allows for oxidative processes to occur during the cycle.

*MIXED IDLE*

Air bubbles are removed from the system during this cycle. The reactor begins a transition from oxidative processes to reductive processes as it moves from aerobic conditions toward anoxic/anaerobic conditions. This is in preparation for the reception of PCE in the next cycle which cannot be transformed if oxygen is present as a competing electron acceptor.[3]

*PCE-Saturated Water:* The experiment introduced the PCE to the reactor in the form of PCE-saturated water. Therefore, the mass of PCE added was the solubility of PCE, 150 mg/L (904.5 μmol/L), times the volume of PCE-saturated water. This promotes the ability of the organisms present in the reactor to see exposure to all of the PCE in solution.[3] Preparation of the PCE-saturated water involved placing double-distilled water in a glass tower. Pure PCE, a dense non aqueous-phase liquid (DNAPL), was also placed in the tower along with a stirrer. The DNAPL-water combination was then mechanically mixed using a stir table. The resulting solution has the PCE in equilibrium between the aqueous and DNAPL phases.

*Mineral Media:* Ingredients per liter—51.1 g of $NaH_2PO_4 \cdot H_2O$, 36.8 g of $K_2HPO_4$, 29.8 g of $Na_2HPO_4$, 2.3 g of $NH_4H_2PO_4$, 1.2 g of $MgSO_4 \cdot 7H_2O$, 0.408 g of $CaHPO_4$, and 0.012 g of $Fe(NO_3)_3 \cdot 9H_2O$. The mineral media was added to double distilled water in a 1 to 10 ratio to yield a 0.1M stock solution which was used as the growth media in the SBR. The recipe was designed to keep the background chloride counts low so that the transformations taking place in the reactor, which may also be detected by monitoring the change in chloride ions present in the solution, may be more easily quantified.[3] Also chloride ions present in solution are expected to be in equilibrium with HCl, therefore the buffering capacity of the stock solution is designed to offset the production of the acid.

*Dissolved Oxygen:* The dissolved oxygen (DO) profile was taken by following Method #4500-O G, the Membrane Electrode Method (Standard Methods). The dissolved oxygen measurements were used to determine whether the reactor was being exposed to oxygen saturated conditions during the aerobic portion of the cycle. The DO reading taken under well-mixed conditions is expected to be representative of the reaction environment. Oxygen saturated conditions at room temperature are defined as approximately 9 mg/L.[10]

*Oxidation-Reduction Potential:* The oxidation reduction potential (ORP) for the reactor was determined using Method #2580 (Standard Methods). Calibration of the ORP probe was done using ZoBell's solution®. The electrode was cleaned using aqua regia. The ORP measurements were used to give a general indication to what phase the reactor environment was in. The convention is that ORP values above +200 mV are typical of aerobic environments and that values below -200 mV are typical of anaerobic environments. Therefore the values between +200 mV and -200 mV are considered to be in the anoxic region.[11]

*Mixed Liquor Suspended Solids:* Mixed liquor suspended solids (MLSS) were measured to provide some idea of the growth rate of the organisms in the reactor. Determination of MLSS was done following Method #2540 (Standard Methods).

*Chlorinated Ethylene Analysis:* The chlorinated ethylenes from the liquid phase were quantified using a Tekmar® LSC 2000 Purge and Trap™ Concentrator connected to a Varian® 3400 Gas CHROMATOGRAPH (GC) WITH A FLAME IONIZATION DETECTOR (FID). THE PURGE-AND-TRAP UNIT WAS USED to concentrate the ethylene derivatives from a auto-liquid sampler (Tekmar® ALS 2016) so that they would more easily be introduced to the GC. The GC was then used to separate the compounds for detection. The method used was modified from EPA Methods TO-2 and TO-3.[3] The separation column contained 60/80 Carbopack B/1% SP-1000 (8′ × 1/8″ OD stainless steel; Supelco, Inc.). The oven temperature was held at 40°C for 4 minutes. Then the temperature was increased to 220°C at 10°C/min. Desorption was 3 min at 220°C. The carrier gas was helium. The retention times for PCE, TCE, and cDCE were 22.4 minutes, 17.8 minutes, and 13.2 minutes, respectively. (Note: If VC and ETH were products coming out of the reactor then they could also be detected at 5.8 minutes and 1.1 minutes, respectively. But these compounds were not detected during the experiment.)

*Chlorinated Ethylene Standards:* The standards used to make calibration curves for quantification of the chlorinated ethylenes were taken from 100 mg/L stock solutions of the compound in methanol. The expected detection range was between 0.1 μg and 0.9 μg. This amount of PCE, TCE, and cDCE was added to 5 mL of double distilled water and placed in the purge-and-trap unit for analysis. Statistical reliability of the curves was promoted by running triplicates of each amount. the $R^2$ values were typically 0.98 and higher.

## Results and Discussion

### *Reactor History*

Consistent operation of the reactor was defined as showing the ability to transform the 180.9 μmol of PCE added to the 3.0-L reactor to TCE and then to predominantly,[12] the cis-isomer of DCE (cDCE) within the 24-hour cycle. After 83 days of operation, consistent operation was achieved. After 138 days of operation, the reactor was physically moved and in the process spent 24 hours idle. After the move the reactor was able to achieve consistent operation almost immediately.

## Environmental Conditions

In establishing what the environmental conditions were for reactor operation, the DO concentration during a cycle was profiled as well as the ORP. These profiles were compared to the distribution of chlorinated ethylenes throughout the cycle to get some idea of how the various conditions correlated to reactor performance.

## Track Study

The distribution of the chlorinated ethylenes was tracked during the cycle 151 days after startup. The profile (Figure 2) shows that the total amount of chlorinated ethylenes introduced to the reactor during FILL is in the form of PCE. The amount present in solution is quickly transformed from PCE to TCE and then from TCE to cDCE. The decrease in PCE concentration corresponds to the increase in cDCE concentrations. At the end of REACT the total concentration of chlorinated ethylenes exceeds the amount originally seen in the bulk solution. This discrepancy can be explained by the idea that the PCE was taken up by the organisms during the FILL period of the cycle to undergo chemistry and therefore was not present in the bulk fluid.

## DO Profile

The dissolved oxygen (DO) profile taken during a cycle shows that during aeration the DO approached 8 mg/L (Figure 3). Obligate anaerobes are not expected to tolerate such an exposure to oxygen. Therefore the organisms present are more likely a mixed culture of both aerobes and, predominantly, facultative anaerobes.[13] The flat line trend of the profile which initiates during the FILL period indicates that all of the oxygen introduced to the system is quickly used up by the conorganisms and continues until the AERATED IDLE of the next cycle begins.

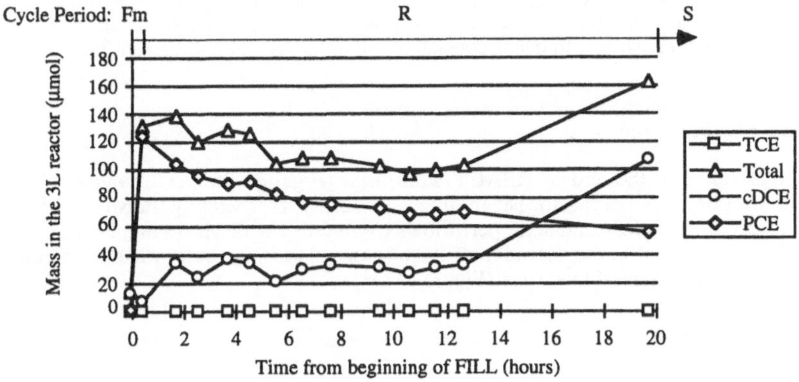


**Figure 2. Chlorinated ethylenes versus time.**

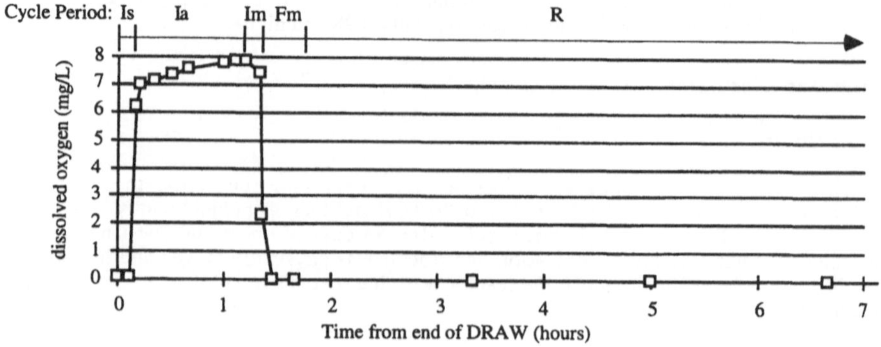


**Figure 3. Dissolved oxygen profile during a cycle.**

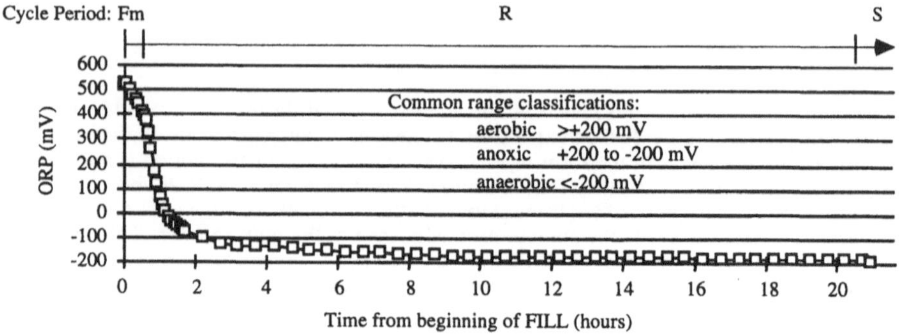


**Figure 4. Oxidation-reduction potential profile.**

## ORP Profile

Similarly the oxidation-reduction potential (ORP) showed a trend from aerobic conditions during aeration toward anoxic conditions during fill and then anoxic/anaerobic conditions during the REACT period of the cycle (Figure 4). It is predominantly during this reaction period that the PCE is taken up and the two successive transformations to TCE then to cDCE took place. These transformations were shown to occur consistently over several months given the operation of the reactor (Figure 5).

## Reactor Performance

Samples taken at the end of REACT were used to gauge the performance of the reactor. Around the 140th day the reactor began to lose its ability to efficiently transform the PCE to cis-DCE (Figure 5). This loss in reactor performance coincided with MLSS numbers exceeding 2000 mg/L. Up to this point in the operation of the reactor no wasting strategy had been employed and MLSS gradually increased over time (Figure 6).

## Growth Hypothesis

This observation led to the hypothesis that the ratio of electron donor (i.e., glucose) to biomass could no longer sustain both growth and PCE transformation. An experiment was run in which

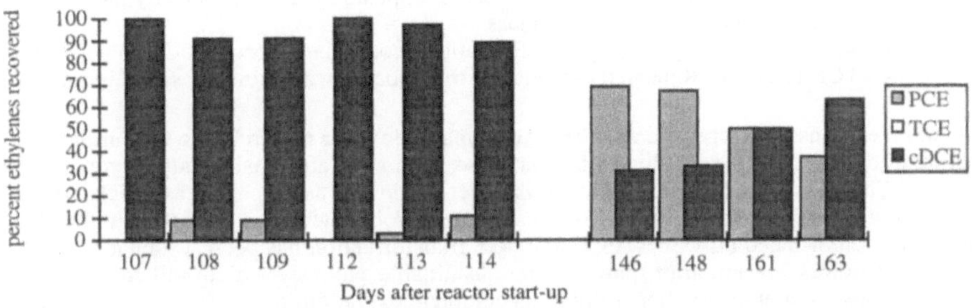


**Figure 5. Ethylenes sampled at the end of REACT.**

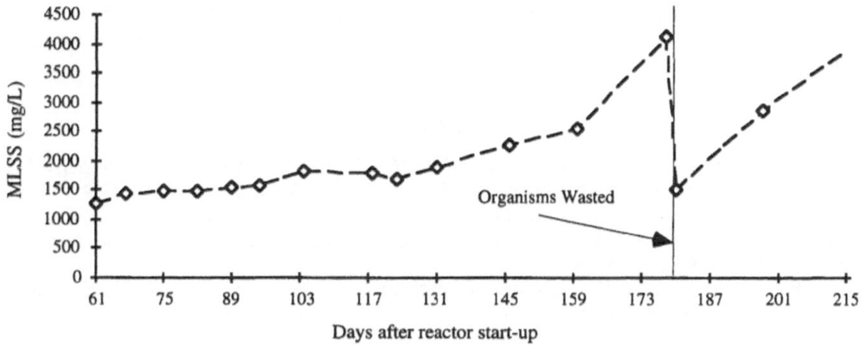


**Figure 6. Distribution of mixed liquor suspended solids.**

the glucose concentration was increased to regain the original proportion of glucose to biomass. The result of the glucose concentration increase was recovery of transformation ability after day 160 (Figure 5). This operating strategy of increasing the glucose concentration proportional to the increase in MLSS did not solve the problem completely. Glucose introduced to the system continued to promote the growth of the organisms such that growth began to become exponential. So the best approach was not to continue to increase the glucose concentration as the biomass concentration increased, but rather to begin a wasting strategy of the organisms.

## Wasting Strategy

The first wasting of the organisms came on the 180th day. After this wasting the organisms quickly recovered their previous numbers (Figure 6). Therefore the wasting strategy employed would have to be more regimented (i.e., part of the normal operation during a cycle such that the organisms would be wasted often enough to maintain a constant MLSS).

## CONCLUSIONS

- The dissolved oxygen profile demonstrates the degree to which the system is aerated, supporting the notion that the system selects predominantly for organisms which are facultative anaerobes.
- The oxidation-reduction potential does show a shift from an aerobic phase of the reaction to an anoxic and subsequently a borderline anaerobic phase of the reaction.
- Transformation of PCE takes place predominantly during the anaerobic phase, but the system is not required to remain anaerobic throughout the entire cycle.

- Effective transformation of the PCE to cDCE appears to be influenced by the ratio of electron donor (i.e., glucose) to biomass.
- During the 24-hour cycle two transformations consistently occur, PCE to TCE and then TCE to cDCE. Repeated evidence of this process was found for several months.

Continued consistent operation of the reactor may allow for research into the further transformation of the cDCE using aerobic pathways. Research in this area has already been done by Folsom and Chapman[5] in a bioreactor and Hopkins, et al.[6] *in situ* using phenol as the electron donor. The goal of this direction in research is to establish a process which is consistent in transforming the various chlorinated ethylenes. PCE to cDCE transformation has been sustainable. But cDCE presence requires investigation into another sustainable pathway which will transform it to a nonhazardous compound and hopefully to its complete oxidation.

## ACKNOWLEDGMENTS

The research was supported in part by the McGuire Fellowship, the General Electric Fund Graduate Fellowship, and the Graduate Research Traineeship Program (NSF-GER93 54889). Analysis was done with the help of the Center for Bioengineering and Pollution Control at the University of Notre Dame. Reactor operation was maintained with the assistance of Kristine Mitchell. Much of the motivation and help with the laboratory portion of the project came from Patrick J. Hirl.

## REFERENCES

1. Safe Drinking Water Act Amendments of 1986: Conference report (to accompany S.124). United States Congress (99th House, 2nd Session: 1986), Washington, D.C.: U.S. G.P.O.
2. DiStephano, T.D., Gossett, J.M., and Zinder, S.H. "Hydrogen as an electron donor for dechlorination of tetrachloroethene by an anaerobic mixed culture." *Appl. Environ. Microbiol. 1992, pp.* 3622–3629.
3. Hirl, P.J. The Biological Degradation of Perchloroethylene Using Periodically Operated Reactors, Doctor of Philosophy Dissertation, University of Notre Dame, Notre Dame, IN, 1996.
4. Vogel, T.M. and McCarty, P.L. "Biotransformation of tetrachloroethylene to trichloroethylene, dichloroethylene, vinyl chloride, and carbon dioxide under methanogenic conditions." *Appl. Environ. Microbiol.*, 1985, pp. 1080–1083.
5. Folsom, B.R. and Chapman, P.J. "Performance characterization of a model bioreactor for the biodegradation of trichloroethylene by *Pseudomonas cepacia* G4." *Appl. Environ. Microbiol.* 1991, pp. 1602–1608.
6. Hopkins, G.D., Semprini, L., and McCarty, P.L. "Microcosm and in situ field studies of enhanced biotransformation of trichloroethylene by phenol-utilizing microorganisms." *Appl. Environ. Microbiol.* 1993, pp. 2277–2285.
7. Nelson, M.J.K., Montgomery, S.O., Mahaffey, W.R., and Pritchard, P.H. "Biodegradation of trichloroethylene and involvement of an aromatic biodegradative pathway." *Appl. Environ. Microbiol.* 1987, pp. 949–954.
8. Wackett, L.P. and Gibson, D.T. "Degradation of trichloroethylene by toluene dioxygenase in whole-cell studies with *Pseudomonas putida* F1." *Appl. Environ. Microbiol.*, 1988, pp. 1703–1708.
9. Irvine, R.L. and Ketchum, L.H., Jr. "Sequencing batch reactors for biological wastewater treatment." *CRC Crit. Rev. Environ. Control, 18(4)*, 1989, pp. 255–294.
10. *Standard Methods for Examination of Water and Wastewater: 18th ed.*, American Public Health Association, Washington, D.C., 1992.
11. Sheker, R.E. Biological Nitrogen Removal Using Biofilm, Suspended Growth and Hybridized Systems, Master of Science Thesis, University of Notre Dame, Notre Dame, IN, 1992.
12. Parsons.
13. Garrett, R.H. and Grisham, C.M. *Biochemistry*. Saunders College Publishing, Fort Worth, TX, 1995.

# 11 ACCELERATED IN-SITU BIOREMEDIATION OF CUTTING OIL CONTAMINATED SEDIMENTS USING OXYGENATED NUTRIENT WATER

**Glenn R. Gustafson**, Graduate Student

**Walter J. Maier**, Professor

Dept. of Civil Engineering

University of Minnesota, Minneapolis, Minnesota 55455

## ABSTRACT

The Superfund site near Morgan Park, Minnesota has been severely contaminated by cutting oils (6.47 wt %) and small concentrations of polynuclear aromatic hydrocarbons (PAHs). In-situ bioremediation studies were carried out on a composite sediment sample using a series of continuous flowthrough columns to study the treatability at high and low oxygen additions (i.e., air or pure oxygen sparging). It was shown that the masses of oil and PAHs were removed at a significantly larger rate using high oxygen addition. The removal of oil and PAHs using high oxygen addition was 30.6% and 41.6%, respectively.

## INTRODUCTION

Sediments contaminated with petroleum products are a continuing source of pollution for the waters in the Great Lakes, in general, and harbor areas specifically. Polynuclear aromatic hydrocarbons (PAH's) are of particular concern because they are toxic to benthic communities, cause tumor epizootics in fish, and public health concerns. Relatively little progress has been made in cleaning up these sources of water pollution and new technologies are needed that can be applied at relatively low cost.

Researchers have shown that in-situ bioremediation of petroleum contaminated soils and sediments is a promising and cost-effective technology. Hildebrandt and Wilson[1] performed one case study where aerated-nutrient water was cycled through a berm of petroleum contaminated soil. The soil was determined to be contaminated with oil and grease at a concentration of 1000 mg/kg dry soil and was successfully bioremediated below the available detection limits of 50 mg/kg dry soil after an eight week period. This is an overall approximate removal of 95% of the contaminants added in a relatively short period of time.

In contrast to the claim that PAHs are recalcitrant to microbial populations, research has shown that PAH removal does occur. Aprill et al.[2] showed that PAH removal was as high as 71%. Aprill showed that even the higher molecular weight PAHs (i.e., fluoranthene, pyrene, benzo(a)anthracene, and chrysene), which are much more recalcitrant than low molecular weight compounds, show a degradation as high as 53%.

This research concentrates on a Superfund site contaminated with cutting oils which contain low concentrations of PAHs. The site was contaminated from the USX steel mill operations located at Morgan Park, MN on the St. Louis River. The site is approximately 230 acres in area and the contamination extends to levels of 2 to 20 ft into the sediments. Most of the contaminated sediments are in a shallow area of the estuary with water levels ranging from 0 to 2 ft in depth.

Preliminary site investigations showed that the dissolved oxygen concentrations were below the limits of detection for the groundwater in the estuary. Exploratory laboratory tests also showed that addition of oxygen resulted in rapid aerobic biodegradation. This indicates that engi-

*52nd Purdue Industrial Waste Conference Proceedings*, 1997, Ann Arbor Press, Chelsea, Michigan 48118. Printed in U.S.A.

neered systems to stimulate in-situ bioremediation may be a possible approach for the cleanup of this site.

The overall objective of the research was to carry out a combination of bench-scale and field studies. These studies would demonstrate the engineering feasibility and practicality of applying accelerated in-situ bioremediation technology for treating PAH and oil contaminated sediments. In-situ biological treatment is recognized as a low-cost approach for cleanup of soils. The technology is based on findings from previous studies in our lab of PAH and oil contaminated soils showing that in-situ biodegradation can be greatly accelerated by providing adequate supplies of molecular oxygen and nutrients to stimulate indigenous microbial populations. Furthermore, these previous studies showed that all the identified PAHs as well as the carrier oils are completely mineralized into carbon dioxide and water. This chapter presents results from the bench-scale column studies only.

Bench-scale continuous flow column studies were carried out to determine the treatability of the sediments by the indigenous microbial population. The batch sample which was taken from the proposed field study treatment site was composited into a large uniform sample. This sample was used to determine the overall bioremediation potential. The findings are discussed in terms of their practical implications.

Several questions were answered from the results, relating to the use of pure oxygen and air-sparged nutrient water. Questions like: Does oxygen sparged water enhance the biodegradation of oil and PAHs with no addition of extraneous microbial cultures, and Does nitrate reduction account for any removal of the organic pollutants? The answers to these and other questions are explained in this paper or subsequent papers concerning this research.

## MATERIALS AND METHODS

The bench-scale continuous flow through columns studies are represented in Figure 1. Column dimensions were 5.5 cm in diameter and 12 cm in length. The columns were packed with a contaminated composite sample prepared from a 2-foot section of sediments obtained near the shore line in the vicinity of the outflow from the wire mill pond. The sample was well mixed to form a representative composite. Two columns were flushed with air-sparged water containing excess nutrients and two columns were flushed with pure oxygen-sparged water containing excess nutrients. The composite was analyzed for grain size distribution, porosity, hydraulic conductivity, moisture content, polynuclear aromatic hydrocarbon (PAHs), oils, and total organic carbon concentration. Bacterial concentrations were measured by plate counts. At the end of the 312-day column study, sediments were subdivided into 2-cm sections and analyzed for the above con-

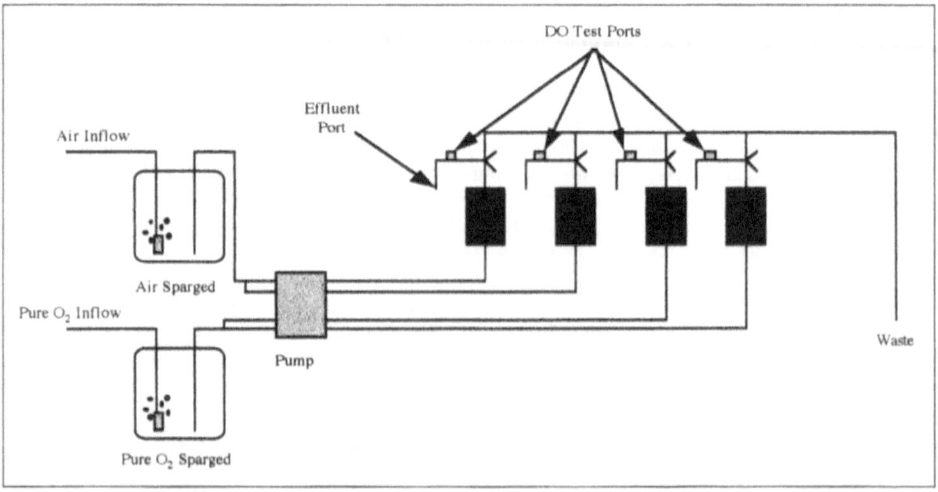


**Figure 1. Bench-scale continuous-flow column setup.**

stituents. Water samples were collected from the influent and effluent sampling port and were analyzed for dissolved oxygen (DO), nitrate ($NO_3$), carbon dioxide ($CO_2$), and total organic carbon (TOC) concentration. The measurements were used to calculate total oxygen consumption, carbon dioxide production as evidence of biodegradation, and to estimate total organic carbon removed.

Sediment samples were analyzed using Soxhlet extraction with a 1:1 acetone/hexane solvent mix to extract polar and nonpolar hydrocarbons. Concentrating the extract was performed using a standard Kuderna-Danish concentrator. The concentrate was measured for oil by evaporation of the excess solvent and weighing the residue. The PAHs were measured on a Hewlett Packard 5890 Gas Chromatograph with auto-sampler using EPA Methods 8000 and 8100. Biological experiments were performed to determine biomass concentrations using standard plate counts with a 1:10 LB agar media and a tryptone surfactant solution to separate the bacteria from the sediments.

Influent and effluent water samples were analyzed for dissolved oxygen (DO), nitrate, carbon dioxide ($CO_2$), total organic carbon (TOC), and biomass. DO was measured using an Orion high oxygen test probe. Nutrient utilization was measured using a Dionex DX-100 ion chromatograph. The carbon dioxide and TOC were measured using a Skalar total carbon analyzer. Biomass analysis was performed with the same plating procedure used for the sediments.

## RESULTS AND DISCUSSION

### Site Characterization Studies

As part of the initial survey of the site, a series of nine cores were taken at selected locations of 1 to 30 ft from the shore line and at 50, 75, and 100 ft south of the outlet of the wire mill pond. The sediment cores were used to determine the average porosity and hydraulic conductivity of the site and to provide preliminary information about the spatial distribution of organic pollutants. The porosity was determined using the measured volume of a saturated sample and determining the moisture content assuming that all of the voids were filled with water. The porosity was found to be 0.375. The hydraulic conductivity was calculated using a method outlined by Fletcher Driscoll[3] which is based on grain size distribution. The core used for the hydraulic conductivity was divided into sections and analyzed for grain size distribution. The vertical range of hydraulic conductivity was .0012 to .0125 cm/s which indicates a fairly homogeneous sample. The weighted average hydraulic conductivity for the sections was found to be 0.0048 cm/s.

The sediments used in these column studies came from the composite sediment sample described above. The % dry weight of oil and the concentration of 8 PAHs in mg/kg of dry sediment are presented in Table I and will be referred to as the initial chemical composition.

The sum of the eight PAHs measured make up only 1.1% of the extractable oil. Additional analysis were made for eight higher molecular weight compounds which were not detectable because they made up such a small percentage of the oil.

**Table I. Initial Chemical Composition of Sediment from Extraction Well Installation Site, Located 100 feet South of the Wire Mill Pond Outlet**

| Contaminant | Concentration |
|---|---|
| Naphthalene, mg/kg dry sediment | 7.8 |
| Acenaphthylene, mg/kg dry sediment | 14.9 |
| Acenaphthene, mg/kg dry sediment | 4.7 |
| Fluorene, mg/kg dry sediment | 16.9 |
| Phenanthrene, mg/kg dry sediment | 97 |
| Anthracene, mg/kg dry sediment | 69.8 |
| Fluoranthene, mg/kg dry sediment | 284.5 |
| Pyrene, mg/kg dry sediment | 194.7 |
| | |
| Total of 8 PAH's, mg/kg dry sediment | 690.3 |
| Total oil, dry weight % of sediment | 6.47 |

## Bench-Scale Column Studies

The composite was analyzed for treatability using a laboratory continuous flow column study. The columns were packed with a calculated quantity of sediment to obtain the physical properties described previously, in order to replicate field conditions. As shown in Figure 1, two columns were fed with air-sparged nutrient water and two columns were fed with pure oxygen-sparged nutrient water. The flow was initiated on February 29, 1996 and ran until January 6, 1997. A base case flow rate of 1 L/d which gave a pore velocity of 1 m/d. This rate was used throughout most of the study except for short periods of variable flow testing to obtain additional information on the effects of changing flow rate. This velocity was chosen based on calculations using a groundwater program to estimate the horizontal flow that could be achieved in the field using a coupled system of extraction and recharge wells. Throughout the experiment, the influent and effluent waters were analyzed for DO utilization, nitrate utilization, $CO_2$ production, and TOC flushing in order to determine a mass balance for the utilization of the oil. The total volume of water passed through each column was 454 liters over a period of 312 days.

Data from each of the two sets of replicate columns were averaged because the concentrations were similar.

## Influent and Effluent Water

Influent water had an average DO of 8.8 mg/L and 41.5 mg/L for the air and pure oxygen columns, respectively. Over the course of the experiment, the DO was used completely within the columns except near the end of the experiment where the pure oxygen columns saw oxygen breakthrough. Over the course of the experiment, the total mass of oxygen utilized was calculated to be 3.83 g for the air columns and 17.89 g for the pure oxygen columns.

Nitrate utilization was found by subtracting the effluent concentration from the influent concentration. Nitrate utilization in the air columns ranged from 12.6 mg/L near the start of the experiment to 9.8 mg/L at the end of the experiment. The nitrate utilization in the pure oxygen columns ranged from 19.2 mg/L at the beginning of the experiment to 0 mg/L near the end when oxygen breakthrough occurred. The fact that nitrate utilization went to zero after oxygen breakthrough indicates that the nitrate was utilized as an alternate electron acceptor and not for microbial growth. The latter was achieved from the ammonia addition. Nitrate measurements were performed on only two occasions, once near the beginning and once near the end of the experiment. For mass balance calculations, an average concentration was multiplied by the total volume of water passed through the columns in order to obtain an estimate of the mass of nitrate utilized. The masses of nitrate utilized were 5.08 g and 4.36 g for the air and pure oxygen columns, respectively.

Measurements of TOC were used to calculate the total mass of organic chemicals flushed from the columns. The concentration of TOC flushed remained relatively constant over the course of the entire experiment for all of the columns ranging from 1.68 to 1.87 mg/L in the air columns and 3.88 to 4.18 mg/L in the pure oxygen columns. Effluent concentrations of the TOC flushed was 1.78 mg/L and 4.18 mg/L as carbon for the air and pure oxygen columns, respectively. This gives an overall mass of 0.808 g and 1.90 g of organic carbon flushed from the air and pure oxygen columns, respectively. The mass of organic chemicals flushed was calculated to be 0.943 g and 2.21 g for the air and pure oxygen columns, respectively, assuming the pollutant's composition is $((CH_2)_n)$ which is representative of oils. It should be noted that the TOC concentration flushed from the air columns was considerably less than in the pure oxygen columns. One possible explanation is that the higher TOC values are related to the release of metabolites that have surfactant like properties (biosurfactants). It is hypothesized that when a higher concentration of oxygen is available, there will be more microbial growth. Hence, a greater biosurfactant concentration results in an increase in TOC concentrations due to increased solublization of oils.

Masses of carbon dioxide production, oxygen and nitrate utilization were analyzed to determine ratios of $CO_2$ production to oxygen and nitrate utilization.

Table II shows the data obtained after oxygen breakthrough in the pure oxygen columns. It shows that there was no nitrate removed after oxygen breakthrough. This implies that all of the carbon dioxide then exiting the pure oxygen columns is a result of respiration using oxygen only. Therefore, when oxygen breakthrough occurred in the pure oxygen columns the ratio of carbon dioxide produced to dissolved oxygen used (i.e., $[(CO_2)n]/[O_2]$) is 0.868 mg $(CO_2)$/mg $O_2$. The following stoichiometric equation and the values from Table II were used to calculate the aerobic

**Table II. Parameters Used for Mass Balance Calculations for the Air and Pure Oxygen Columns**

| Measurement | Air Columns | Pure Oxygen Columns |
|---|---|---|
| DO utilized, mg/L | 8.05 | 35.15 |
| $CO_2$ produced, mg/L | 12.74 | 30.49 |
| $NO_3^-$ utilized, mg/L as ($NO_3^-$) | 9.77 | 0.0 |

**Table III. Aerobic and Anaerobic Coefficients for Mass Balance Calculations**

| Ratio | Aerobic | Anaerobic |
|---|---|---|
| Y, yield coefficient: mg cell mass / mg $(CH_2)_n$ used | 0.235 | 0.011 |
| $(CH_2)_n$ used/$O_2$ used, mg/mg | 0.323 | NA[a] |
| $(CH_2)_n$ used/($NO_3^-$) used, mg/mg | NA* | 0.189 |

[a]Not applicable.

respiration ratios. It was assumed that all nitrogen for bacteria ($C_5H_7NO_2$) assimilation came from ammonia. Effects of oxygen utilization associated with endogenous metabolism were neglected.

$$0.81^*CH_2 + 1.11^*O_2 + 0.024^*NH_3 \rightarrow 0.024^*C_5H_7NO_2 + 0.69^*CO_2 + 0.76^*H_2O \qquad (1)$$

The ratio of organic chemicals removal to oxygen utilization was then used to calculate the yield of cell mass produced (Y, yield coefficient), using the values of DO utilized and $CO_2$ produced in the pure oxygen columns. The calculated ratios for aerobic biodegradation are shown in Table III.

A similar mass balance calculation of oxygen, $CO_2$, nitrate, and cell mass was used to analyze the anaerobic test results. The ratio of organic chemicals utilized by nitrate reduction was determined from the $CO_2$ measurements that were adjusted for the amount of $CO_2$ produced from dissolved oxygen utilization using the ratio from the pure oxygen columns. The remaining carbon dioxide was then assumed to be produced by nitrate reduction. Since ammonia was present, it was assumed that it was used as a source of nitrogen for assimilation and that all the nitrate utilized was as an alternate terminal electron acceptor and ultimately reduced to nitrogen gas. The calculated ratio of the carbon dioxide produced to the nitrate utilized (i.e., $[CO_2]/[NO_3^-]$) was determined to be 0.590 mg $CO_2$/mg $NO_3^-$. The following stoichiometric equation and the values for the air columns in Table II were used to determine the parameters for growth and organic chemical removal by anaerobic respiration.

$$0.132^*CH_2 + 0.157^*HNO_3 + 0.0002^*NH_3 \rightarrow 0.0002^*C_5H_7NO_2 + 0.131^*CO_2 + 0.209^*H_2O + 0.079^*N_2 \qquad (2)$$

The calculated ratios for organic chemical removal to oxygen utilization and cell mass produced to organic chemicals removed (Y, yield coefficient) are shown in Table III. Due to the uncharacteristically low yield coefficient, it was determined that this value is very sensitive to the accuracy of the carbon dioxide and nitrate measurements. This was determined by manipulating the $CO_2/NO_3$ ratio by as little as a 10% difference, which brought the yield within range of expected values which are similar to the value for yield in the aerobic condition. This shows that these stoichiometric calculations do not provide a good estimate for the yield coefficient.

Each column was packed with 481 g (dry weight) of sediment which contained a calculated mass of 31.1 g of oil. The total mass of oil removed was found using the above ratios and the masses of oxygen utilized, nitrate utilized, and TOC flushed (converted to $(CH_2)_n$). The calculated removal of oil was 10.1% and 28.2% for the air and pure oxygen columns, respectively. It should be noted that the sensitivity effect seen for the yield coefficient was not seen for the ratio of oil removed. This says that these values would not change significantly due to small measurement error. Hence, this method of calculation provides a good estimate for oil removal. From these calculations it can be seen that the addition of pure oxygen increased the removal of the oil by a factor of 2.8 over the use of air sparging.

## Final Sediment Analysis

The columns were shut down on January 6, 1997 and the sediment was removed in 2-cm sections for analysis of remaining oil and PAHs. The data on oil and total PAHs are described below. A more detailed analysis of individual PAHs will be presented in a subsequent publication.

The remaining oil concentration ratios expressed as the concentration remaining at the end of the experiment divided by the initial concentration (C/Co) for the air and pure oxygen columns are shown in Figure 2 as a function of depth. It can be seen that all ratios are less than 1.0. The air columns only showed significant oil removal in the initial part of the section because of the limited availability of oxygen and limited utilization of nitrate. Oil removed in the latter part of the columns was due to anaerobic biodegradation and flushing as evident from the TOC concentrations. In contrast to the air columns, the pure oxygen columns had more removal across the entire column due to deeper oxygen penetration into the bed. Nitrate utilization was also observed in the pure oxygen columns before oxygen breakthrough was detected. This contributed to the oil removal. Flushing of the TOC out of the column also contributed to the oil removal. Hence, in both the air and pure oxygen columns there were three mechanisms of removal. They included aerobic biodegradation, anaerobic biodegradation, and dissolution flushing which was also biologically enhanced.

After reviewing all the concentration data, including that of the individual PAHs, it has come to light that the measured concentration of the first 2-cm segment of the bed, may not be representative of a column operating with uniform flow distribution. It appears that the flow pattern at the inlet short-circuited some of the sediments in a cone-shaped volume as illustrated in Figure 3, even though a thin layer of clean Jordan sand had been place at the inlet to ensure uniform distribution. The clean sand probably minimized the maldistribution but did not eliminate the problem. As shown in Figure 3, it appears that the sediments in the cone-shaped volume were not subjected to the same throughflow of water as the central core. As a result, there was less aerobic biodegradation as well as less flushing activity. The entire column was saturated, even near the inlet, but the water in the pore spaces of the sediment was coming from dispersion or a backflow circulation of water. Hence, the sediments in the first 2 cm of the column have not seen as high a mass of oxygen as at the 3 cm depth of the column. This maldistribution effect is believed to be the cause of the lower rate of removal of oil as well as PAHs near the inlet. The overall average oil removal for the entire mass of sediments in the columns was 12.0% and 30.6% for the air and pure oxygen columns. These values are in reasonably good agreement with the calculated values

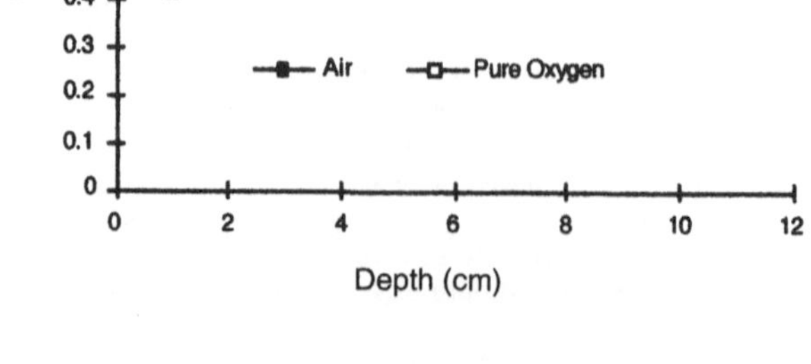


**Figure 2. Oil concentration ratios remaining (C/Co) vs. depth (air and pure oxygen columns).**

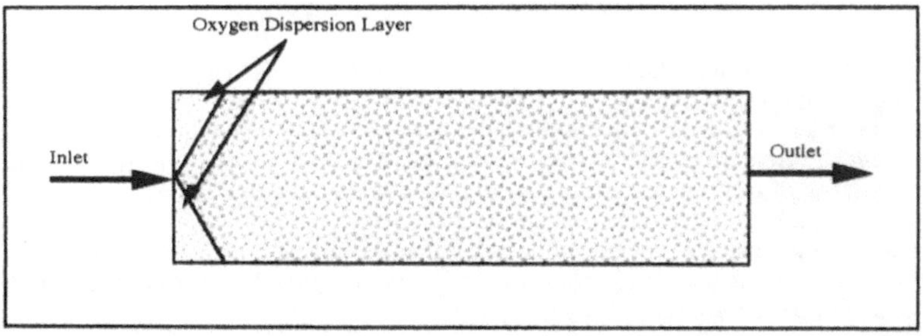


**Figure 3. Oxygen/back flow dispersion effect.**

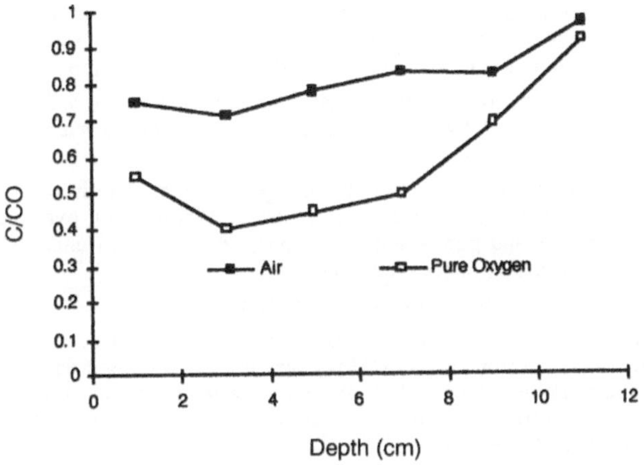


**Figure 4. Total PAH concentration ratios remaining (C/Co) vs. depth (air and pure oxygen columns).**

from the water analysis, suggesting that the assumption that organic chemicals were in the form of $(CH_2)_n$ was valid.

The C/Co concentration ratios of the total PAHs versus column depth for the air and pure oxygen columns are shown in Figure 4. The ratios for both the air and pure oxygen columns across the depth are significantly lower than 1.0 indicating a significant removal of PAHs. However, removal of PAHs in the pure oxygen columns was much greater, similar to the case for the removal of oil. In contrast to this, the air columns still saw some removal which is based on studies which have shown that the lower molecular weight PAHs can be removed in the presence of nitrate reduction.[4] The maldistribution effect explained above also explains the apparent lower removal of PAHs in the first 2 cm of the columns. The effect is more pronounced in the pure oxygen columns than in the air columns.

Comparison of the overall removals of PAHs and oil shows that the PAH C/Co ratios are consistently lower than the oil C/Co ratios, indicating that proportionately a larger fraction of the PAHs was removed. The overall removal of PAHs for the entire sediment columns was 18.9% and 41.6% for the air and pure oxygen columns, respectively. Therefore, the PAH removal was 1.58 times and 1.36 times greater than the oil in the air and pure oxygen columns, respectively.

Studies aimed at simulating the removal of oils and PAHs by aerobic biodegradation and flushing are in progress. This phase is aimed at obtaining quantitative descriptions for rates of dissolution and biodegradation using an available computer program which couples the effect of dissolution of chemicals, biodegradation, oxygen availability, and biomass accumulation.

## PRACTICAL IMPLICATIONS

As indicated in the introduction, this continuous flow bench-scale column study was carried out to provide basic information about the potential merits for providing dissolved oxygen and nutrients in order to stimulate in-situ biodegradation of organic contaminants. It was designed to complement the field study that was set up on site. The field test was designed to induce lateral subsurface flow of oxygenated and nutrient spiked water to stimulate in-situ biodegradation. In essence the same approach was used in the field study; namely, to monitor the availability of oxygen and to measure the pretest and post-test distribution of organic pollutants (oils and PAHs) in the affected sediments. The field work has been completed and results are being analyzed. These data will be reported in subsequent papers.

Several basic questions were answered in the continuous flowthrough column studies performed on the composite sediment sample; they are:

1. *Does the introduction of oxygen-sparged water with nutrients stimulate the biodegradation of the cutting oils which account for 99% of the organic chemicals in the sediments without the addition of extraneous microbial cultures?*

The answer to this question is yes, the addition of oxygen and nutrients accelerates in-situ biodegradation substantially and does not require addition of extraneous organisms. Comparison of the air and pure oxygen column data clearly shows that oxygen availability is the limiting factor. Removal of the oil was 30.6% in the pure oxygen columns and 12% in the air columns.

2. *Does the introduction of oxygen-sparged water with nutrients stimulate the biodegradation of the polynuclear aromatic hydrocarbons (PAHs) without the addition of extraneous microbial cultures?*

The data clearly show that PAH removal is higher in the pure oxygen columns than in the air columns and does not require addition of extraneous organisms. Average PAH removal over the columns was 18.9% and 41.6% for the air and pure oxygen columns, respectively. Hence, increasing the available concentration of oxygen enhances the removal of PAHs.

3. *Does the addition of nitrate stimulate biodegradation of oils and PAHs?*

Based on the observation that nitrate was utilized under anaerobic conditions, it can be concluded that nitrate served as an alternate electron acceptor. However, molecular oxygen is more effective as proven by the observation that less than 22.6 % of the nitrate fed to the air columns was utilized, whereas 100% of the oxygen in the air columns was utilized. The data do not provide any definitive proof as to whether PAHs were biodegraded under anaerobic conditions.

4. *Does the mass balance on oxygen and nutrients added and carbon dioxide produced give any clear indication of whether oxygen is being used in the oxidation of inorganic chemical?*

Results of mass balance calculations on oxygen, nitrate, carbon dioxide, and organic carbon in the effluent and the overall measured removal of organic chemicals from the sediments show that essentially all the oxygen is being utilized for biodegradation. This is an important finding because the sediments contain large amounts of magnetic iron and there has been concern that these inorganic materials could exert a large oxygen demand. The results of this study show that this is not a likely problem.

5. *Does the introduction of flowing water with or without oxygen result in the solublization of organic chemicals?*

Although the data were not present in the main text of the paper, the tests of the effluent water showed that all PAHs were below detection limits. However, there was evidence of transport of other organic chemicals as indicated by TOC measurements. Higher TOC concentrations were found in the pure oxygen columns. Effluent TOC concentrations decreased with time after oxygen breakthrough had occurred. This trend suggests that downgradient transport of TOC is due to column dynamics in which there is insufficient time for the solublized organic chemicals to undergo biodegradation before they are flushed out of the columns. However, the rate of biodegradation of this TOC material was not measured directly. As noted above, higher average TOC concentrations were measured in the pure oxygen column effluent. Review of the literature indicates that this may be caused by a large production/release of metabolic intermediates that are surface-active and therefore enhance the solubility of slightly soluble chemicals. Use of higher concentrations of oxygen results in increased concentrations of biomass, in turn increasing the concentration of biosurfactants. Thus, the addition of pure oxygen in place of air affects solubility as well as biodegradation.

6. *In view of the fact that the PAHs in the sediments are seen as the chemicals of greatest concern for this research but represent only 1% of the total organic pollutants, is there a clear indication that PAH removal is limited by the removal of oil which accounts for 99% of the organic pollutants?*

This question is partly answered by comparing the fraction of PAHs removed versus the fraction of oil removed, which shows that PAH removal is proportionately higher than oil removal. These results show that PAH removal is not limited by the presence of high oil concentrations but is in fact favored over oil removal, particularly in the first half of the column.

However, it is unlikely that complete PAH removal could be obtained without concurrent removal of a substantial fraction of the oils. The quantitative relationship of PAH to oil removal is currently being studied by a detailed mathematical model analysis of the removal of each individual PAH species. The results of this phase of the work will be reported in the near future.

## REFERENCES

1. Hildebrandt, W.W. and Wilson, S.B. "On-site bioremediation systems reduce crude oil contamination." *J. Pet. Technol.* 43(1), pp. 18–22, 1991.
2. Aprill, W., Sims, R.C., Sims, J. L., and Matthews, J. E. "Assessing detoxification and degradation of wood preserving and petroleum wastes in contaminated soil." *Waste Manage. Res.* 8(1), pp. 45–65, 1990.
3. Driscoll, F. G. In: *Groundwater and Wells;* 2nd ed., F. G. Driscoll, Ed., Johnson Filtration Systems, Inc., St. Paul, MN, pp. 737–738.
4. Fetter, C. W. In: *Contaminant Hydrogeology*, R. A. McConnin, Ed., Prentice-Hall, Inc., Upper Saddle River, NJ, p. 319.

# 12 METABOLISM DECHLORINATION OF TRICHLOROETHYLENE BY NITRIFYING BACTERIA IN AN ENRICHED BATCH CULTURE

**Lei Yang**, Associate Professor
Department of Marine Environment
National Sun Yat-sen University
Kaohsiung, Taiwan

## INTRODUCTION

Trichloroethylene (TCE) is used widely in industry and laundry as solvent for degreasing and dry cleaning. TCE migrates quickly and exhibits slow breakdown rates in subsurface to make it become the most widely distributed organic groundwater and soil pollutant in the United States.[1,2] According to the report from the USEPA, this chlorinated hydrocarbon is the most prevalent hazardous organic compound at 246 out of 1,035 Superfund sites.[3] TCE has been found to be highly potentially carcinogenic and extremely toxic.[4] Due to its suspected carcinogenicity and toxicity, an extensive study of its fate in the environment is necessary to reduce the risk to human health and to apply remediation techniques successfully.

Biological processes, instead of chemical or physical methods, have received much attention to remove organic compounds from contaminated soil and groundwater, principally because the costs of bioremediation are potentially low. Thus, a variety of biological treatment technologies for the destruction of chlorinated hydrocarbons, such as TCE, are now under development. Although the microorganism which can grow on TCE as a sole carbon or energy source has not yet been isolated, it was found that through cometabolism, alternative approaches have been achieved by several physiological types of bacteria which can cometabolically dechlorinate and then partially or fully degrade TCE.[5–9] Vogel and McCarty[10] found that TCE could be cometabolically transformed as a nongrowth-supporting electron acceptor by methanogens in an anaerobic condition, in which the reduction of TCE resulted in dechlorination to generate products with progressively decreasing levels of chlorination. As to aerobic biodegradation of TCE, it was first reported by Wilson and Wilson[11] in their study of stimulating soil with natural gas. They found that methanotrophic (methan-oxidation) bacteria in soil were in charge of this biological reaction. The mechanism of this reaction was found to be triggered from cometabolic dechlorination of TCE catalyzed by the enzyme of methane monooxygenase (MMO) generated in the bacterial cells initially to catalyze the oxidation of methane as its primary substrate to sustain growth while gaining no benefit from TCE degradation.[12] Since then, research on microbial processes capable of degrading TCE and other chlorinated hydrocarbons has been focused largely on organisms that oxidize compounds by monooxygenase under aerobic conditions.[4,13] Up to the present, aerobic oxidation of TCE has been demonstrated for pure and mixed cultures, including bacteria grown on methane,[14] toluene,[15] phenol,[7] propane,[16] propylene,[6] cumene,[17] isoprene,[18] and ammonia.[9,19–22] Also due to broad substrate specificities of monooxygenase enzymes, the bacteria possessing the enzymes potentially could be used in a wide variety of bioremediation situations. However, because it goes on a cometabolic pathway, we have to make sure that there are sufficient primary substrates provided in the treatment systems to sustain the bacterial growth, which can stimulate more enzymes generated for cometabolism. This is usually the major limitation of using cometabolism to remove TCE or other chlorinated organic compounds out of the contaminated sites in in-situ bioremediation.

The applications of such bioremediation technologies are based on the study of pure and mixed cultures of aerobic bacteria that cometabolize chlorinated hydrocarbons. Methanotrophic bacteria, for both pure and mixed cultures, and their ability to cometabolize TCE have been

*52nd Purdue Industrial Waste Conference Proceedings*, 1997, Ann Arbor Press, Chelsea, Michigan 48118. Printed in U.S.A.

mostly studied with both suspended[23–25] and biofilm systems.[26,27] However, it is necessary to provide methane gas to the remediation sites using methanotrophic cometabolism to biotransform TCE. Due to the low solubility of methane gas, it is difficult to handle this primary substrate for methanotrophs during operation. Hence, using higher solubility of ammonia/ammonium instead of methane as primary substrate for nitrifying bacteria to cometabolize TCE by ammonia monooxygenase (AMO) might increase the operation efficiencies in in-situ bioremediation of TCE. The research of using nitrifying bacteria to cometabolize halogenated hydrocarbons has been studied by Rasche, Hyman, Vannelli, and their research fellows, respectively.[9,20–22,28] In some of their studies,[9,20] they have used a pure culture of *Nitrosomonas europaea to* test its cometabolic ability to TCE and found that these autotrophic bacteria were capable of degrading TCE with limited amounts of TCE initially applied to the system,[20] and otherwise TCE would show irreversible inhibitory effects to ammonia-oxidizing activity.[9] According to the cometabolic kinetics model developed by Ely and his research fellows,[4,29] it was shown that TCE was a competitive inhibitor of ammonia oxidation, with TCE affinity for AMO being about four times greater than that of ammonia for the enzyme. They also thought that inactivation of AMO occurred more gradually and was proportional to the rate and amount of TCE oxidized. However, no mixed culture of nitrifying bacteria has been further studied for TCE cometabolism by AMO to make this technology unable to apply in fields. Thus, the purpose of this study is to use an enriched nitrifying culture to investigate the degradation efficiencies of TCE and its effects of inhibition to ammonia-oxidizing activity in cometabolism of TCE by AMO generated in this enriched culture.

## MATERIALS AND METHODS

### Seeding and Culturing an Enriched Nitrification System

An enriched stock nitrifying culture was used during this research effort. This stock culture was supplied with a solution containing ammonia sulfate [$(NH_4)_2SO_4$], sodium bicarbonate ($NaHCO_3$), and other trace nutrients ($MgSO_4$, $FeCl_3$, $KH_2PO_4$, and $CuSO_4$) in a chemostat-culturing system. The original seeding of nitrifying bacteria was sampled from the wasted activated sludge in Kaohsiung City Human Body Waste Treatment Plant in Taiwan. The biomass concentrations (expressed as MLSS or MLVSS) and nitrificational oxygen uptake rate (NOUR, expressed in units of mg $O_2$/L-hr.) were measured weekly to check the growth rates and activities of the nitrifying bacteria cultured in the system until a high value of NOUR (about 80–100 mg $O_2$/L-hr.) was reached at a steady state. The culture was then ready to be used as stock nitrifying bacteria for TCE cometabolism dechlorination tests.

### Experimental Procedures

When the enriched culture of nitrifying bacteria was at steady state and the NOUR was reached up to the values that we expected, the batch tests of cometabolism dechlorination of TCE by AMO existing in the enriched nitrifying culture were begun. The mixed liquor of nitrifying bacteria taken from the stock chemostat system was first rinsed by deionized water (DI) water three to four times to remove the interfering materials. The rinsed nitrifying sludge was then mixed with the air or pure oxygen gas aerated nutritious DI water for dilution. The additives in the nutritious DI water included $MgSO_4$, $CaSO_4$, Fe $SO_4$, $KH_2PO_4$, $NaHCO_3$, and $CuSO_4$.

There were three different sets of batch tests run in this experiment. In the first set of batch tests, the rinsed and diluted nitrifying sludge sample was put into a 300-mL BOD bottle, and then different amounts of $NH_3$ and TCE were added to fill up the bottle and covered by water seal. The BOD bottle was then immediately put into a constant-temperature tank maintained at 20°C to let the cometabolism dechlorination occur in a quiet condition (i.e., without shaking). The samples would be taken from the bottle daily for analyzing the concentrations of pH, DO, TCE, $NH_3$, $NO_2^-$, and $NO_3^-$ changed with time. In the second set of batch tests, the rinsed and diluted nitrifying bacteria sample was put into a 150-mL brown bottle, and then it was filled up with different amounts of $NH_3$ and TCE. The bottle was then covered and shaken in a shaking table to keep the sample mixing. The analyzing parameters of the samples taken from the bottle daily were still the same as the ones in previous batch tests. In the final set of batch tests, the experimental procedures were as same as the second set of batch tests except for the final volume filled inside the sample. The final volume was controlled at 125 mL and thus 25 mL air space was left inside the bottle for supplementing the DO consumed by the nitrifying bacteria in the

**Table I. Controlling Conditions for Each Set of Batch Tests**

| Batch Test Run No. | Biomass (mg VSS/L) | $NH_3$-N (mg N/L) | TCE (μg/L) |
|---|---|---|---|
| The first set of batch tests (300 mL, full, without shaking): | | | |
| 1 | 6.3 | 4.0 | 75 |
| 2 | 4.1 | 4.0 | 100 |
| 3 | 4.8 | 4.0 | 200 |
| The second set of batch tests (150 mL, full, with shaking): | | | |
| 4 | 28.7 | 2.0 | 150 |
| 5 | 28.7 | 20.0 | 150 |
| 6 | 28.7 | 40.0 | 150 |
| The third set of batch tests (150 mL, 125 mL full and 25 mL void, with shaking): | | | |
| 7 | 36.8 | 16.0 | 125 |
| 8 | 36.8 | 16.0 | 200 |
| 9 | 36.8 | 16.0 | 300 |
| 10 | 65.4 | 12.0 | 150 |
| 11 | 105.5 | 12.0 | 150 |
| 12 | 141.2 | 12.0 | 150 |

batch system. In this test, the sample bottles were also mixed by a shaking table. A blank test was run as a control in each set of batch tests by adding allythiourea (ATU), which was a specific inhibitor of the enzyme AMO. The amounts of nitrifying biomass (as MLVSS), $NH_3$ and TCE controlled in each set of batch tests are shown in Table I.

### Methods of Analysis

Nitrificational oxygen uptake rate (NOUR) was analyzed by using the Gilson Model 5/6H Oxygraph respirometer. Nitrifying biomass was tested as mixed liquor volatile suspended solids (MLVSS) and measured by using the method shown in *Standard Methods* (209D).[30] $NO_2^-$ and $NO_3^-$ were determined by using a "TOA ICA-5000 System" Ion Chromatography (IC), while $NH_3$ was measured by using the Nesslerization colorimetric method[30] with a spectrophotometer (HACH, DR-2000). The measurements of TCE were made by using the equipment of GC-FID (Shimadzu, GC-14A) with the samples pretreated through a purge-and-trap system (Tekmar, Purge & Trap Concentrator-2000).

## RESULTS AND DISCUSSION

### The First Set of Batch Tests (300 mL, Full, Without Shaking)

In the first set of batch tests, small amounts of nitrifying bacteria (4.1–6.3 mg VSS/L) were added into the test bottle for the reason of limited dissolved oxygen supplied inside the bottle through aerating the dilution water with air. That was also the reason why a low concentration of $NH_3$ (4.0 mg N/L) was applied to this batch system. Three tests with different initial concentrations of TCE applied (75, 100 and 200 μg/L) were run in this experiment, and the experimental results are shown in Figures 1 to 4. As seen from Figure 1, the amounts of TCE removed were not significant for all of the three test runs, while about half the amounts of the removed TCE were likely owing to the loss of TCE vapor through evaporation, learned from the results of blank tests as shown in Figure 2. However, according to Figures 3 and 4, the effects of nitrification were still obvious for both runs 1 and 2 with initial TCE concentrations of 75 and 100 μg/L applied to the system, respectively, while no ammonia was oxidized to nitrite or nitrate in the third run with initially applied TCE concentration of 200 μg/L. This meant that in the level of 200 μg/L of TCE comparing 4 mg/L of NH3 applied, TCE might be a potent competitive inhibitor of ammonia oxidation by nitritifying bacteria due to TCE affinity for AMO being about four times greater than that of ammonia for the enzyme.[29] This inference has been verified in the previous studies using a pure culture of *Nitrosomonas europaea*.[9,20] In addition, it appeared that the level of TCE (about 1500 μg/L) showing inhibition to ammonia oxidation used in those studies9 was much higher than the one applied in this set of batch tests of this study. The reason

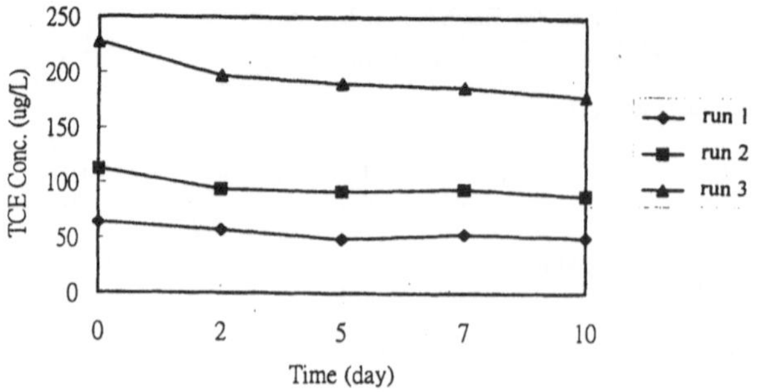


**Figure 1. The concentrations of TCE varied with time in the first set of batch tests with conditions of biomass of 4.1–6.3 mg VSS/L, $NH_3$-N of 4.0 mg N/L, and TCE of 75–200 µg/L.**

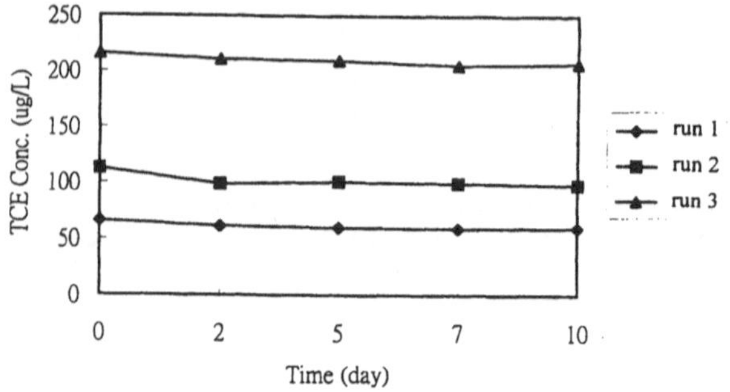


**Figure 2. The concentrations of TCE varied with time in the first set of batch tests with conditions of biomass of 4.1–6.3 mg VSS/L, $NH_3$-N of 4.0 mg N/L, TCE of 75–200 µg/L, and ATU of 0.5 mL/100 mL (blank test).**

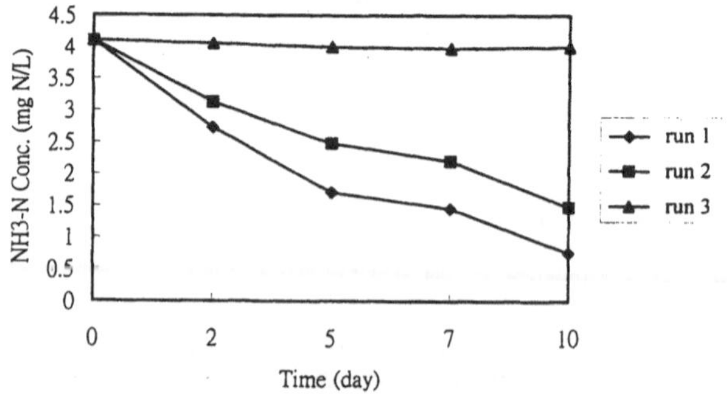


**Figure 3. The concentrations of $NH_3$-N varied with time in the first set of batch tests with conditions of biomass of 4.1–6.3 mg VSS/L, $NH_3$-N of 4.0 mg N/L, and TCE of 75–200 µg/L.**

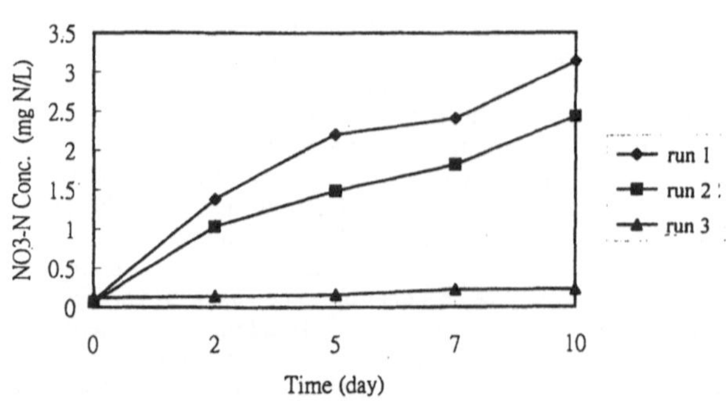


**Figure 4. The concentrations of $NO_3$—N varied with time in the first set of batch tests with conditions of biomass of 4.1–6.3 mg VSS/L, $NH_3$-N of 4.0 mg N/L, and TCE of 75–200 µg/L.**

might be that an enriched nitrifying culture with small amounts (4 mg VSS/L) was used in this test run.

## Second Set of Batch Tests (150 mL, Full, with Shaking)

In the second set of batch tests, the amounts of nitrifying sludge controlled in the test bottle were increased to 28.7 mg VSS/L, and thus the dilution water was aerated by pure oxygen gas instead of air before being added into the bottle in order to increase the initial DO level. This set of batch tests was designed to compare the efficiencies of cometabolism dechlorination achieved by the enriched nitrifying culture under the conditions of the same biomass (28.7 mg VSS/L) and initially applied TCE (150 µg/L) concentrations, which might not show inhibition to ammonia oxidation, but different $NH_3$ levels (2.0, 20.0, and 40.0 mg N/L, respectively) initially applied to this batch system. The experimental results are shown in Figures 5 to 7. According to Figures 5

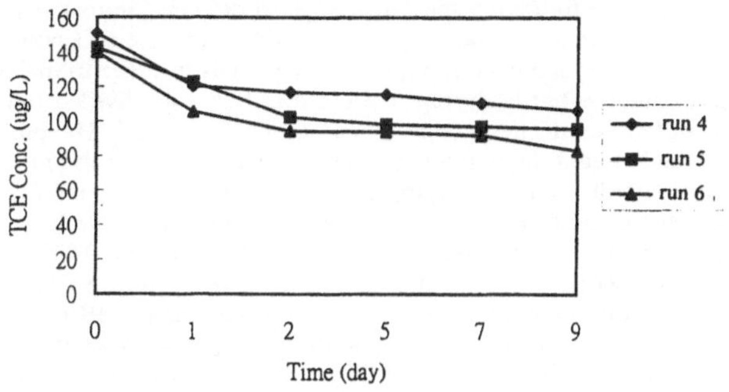


**Figure 5. The concentrations of TCE varied with time in the second set of batch tests (biomass: 28.7 mg VSS/L, $NH_3$-N: 2.0–40.0 mg N/L, TCE: 150 µg/L).**

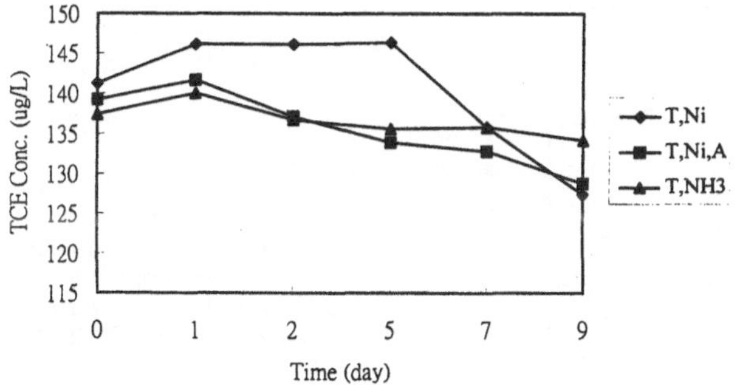


**Figure 6. The concentrations of TCE varied with time in the second set of batch blank tests (biomass: 28.7 mg VSS/L, $NH_3$-N: 2.0–40.0 mg N/L, TCE: 150 µg/L, or ATU of 0.5 mL/100mL).**

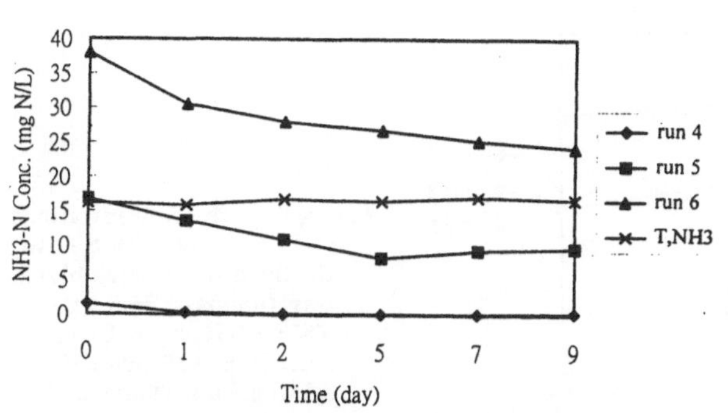


**Figure 7. The concentrations of $NH_3$-N varied with time in the second set of batch tests (biomass: 28.7 mg VSS/L, $NH_3$-N: 2.0–40.0 mg N/L, TCE: 150 µg/L).**

and 6, most of the TCE removed from the batch systems was likely to be through cometabolism dechlorination accompanied with lesser amounts of it through vapor loss. However, as seen in Figure 6, the cosubstrate of TCE could be cometabolized by nitritifying bacteria without the presence of the primary substrate of $NH_3$, while no significant removal of TCE was observed in both of the systems with TCE and $NH_3$ only, and with TCE, $NH_3$, and ATU, respectively. It was inferred that cometabolism of TCE required AMO activity, which could be inhibited by ATU. In this set of batch tests, although the removal efficiencies of TCE through cometabolism dechlorination were higher than those in the first set of batch tests, nitrification was still not completed even after 9-day's running in the systems with higher NH3 concentrations applied. There might be two reasons: one was TCE-mediated inactivation, and the other was deficiency of DO in the bottle. Due to the TCE level of 150 µg/L showing no inhibition to ammonia oxidation in the system with initially applied $NH_3$ concentration of 2 mg N/L, the reason might be the second one. Thus, the third set of batch test was designed to solve the problem of DO supplement.

## Third Set of Batch Tests (150 mL, Not Full, with Shaking)

In the third set of batch tests, in addition to following the same steps to prepare the test samples as in the second set of batch tests, the final volume of sample liquid inside the test bottle was kept at 125 mL, and this left 25 mL void space containing pure oxygen gas in order to make sure that enough DO would be kept inside the bottle during the test run. Six tests divided into two groups were run in this experiment with controlled conditions as shown in Table I. The purpose of the first group of tests was to study again the maximum level of TCE showing inhibitory effects to ammonia oxidation in conditions of higher nitrifying biomass concentrations and DO levels provided to the system. The experimental results of this test are shown in Figures 8 to 10. Deducting the interference of TCE loss from evaporation, we found that most of the TCE removed from the system was due to cometabolism dechlorination as shown in Figures 8 and 9. However, as seen from the same figures, the removal rates of TCE were similar for all of the three systems with initially applied TCE levels of 125, 200, and 300 µg/L, respectively, while the amounts of ammonia oxidation were increased with decreased TCE levels initially applied to the

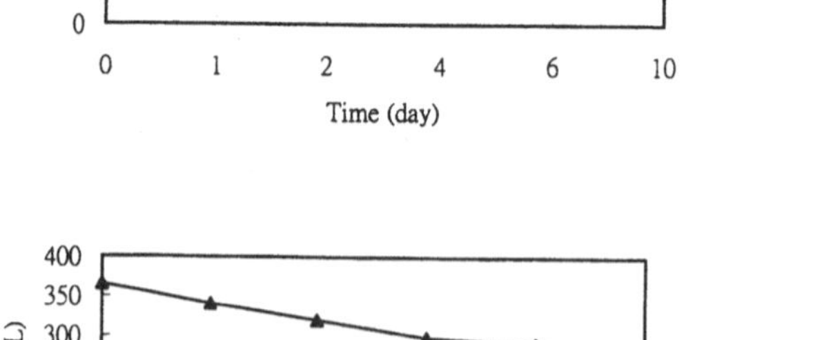


**Figure 8. The concentrations of TCE varied with time in the third set of batch tests (biomass: 36.8 mg VSS/L, $NH_3$-N: 16.0 mg N/L, TCE: 125–300 µg/L).**

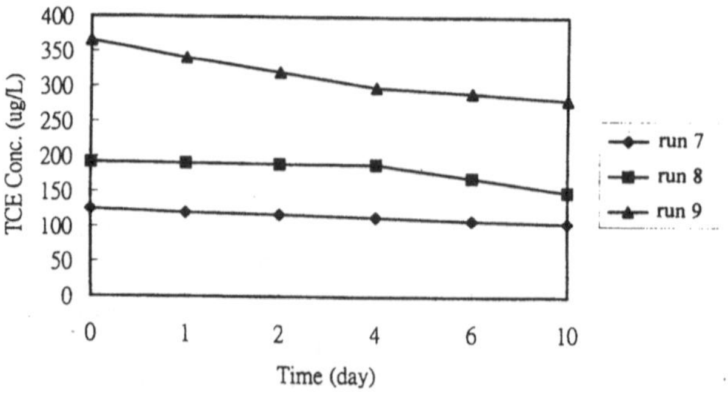


**Figure 9. The concentrations of TCE varied with time in the third set of batch blank tests (biomass: 36.8 mg VSS/L, $NH_3$-N: 16.0 mg N/L, TCE: 125–300 µg/L, ATU: 0.5 mL/100mL).**

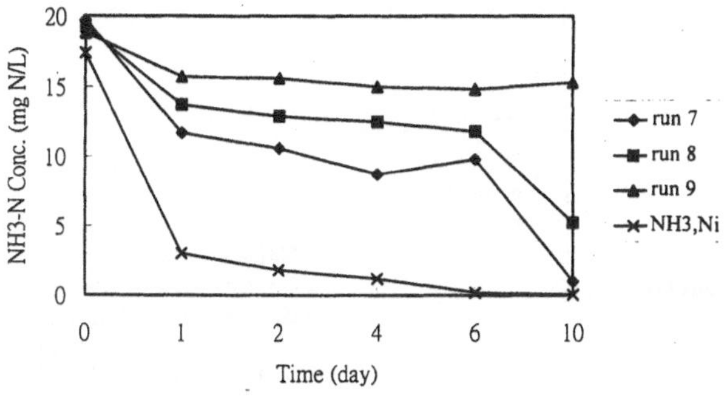


**Figure 10. The concentrations of $NH_3$-N varied with time in the third set of batch tests (biomass: 36.8 mg VSS/L, $NH_3$-N of 16.0 mg N/L, and TCE of 125–300 μg/L).**

system, as shown in Figure 10. Therefore, it was concluded that the maximum level of TCE showing inhibitory effect to nitrification was increased up to 300 μg/L in an enriched batch nitrifying system with biomass of 28.7 mg VSS/L and initially applied $NH_3$ levels of 16 mg N/L.

In the second group of tests of this experiment, the initially applied ammonia and TCE levels were maintained constant (12 mg N/L and 150 μg/L, respectively), while the concentrations of nitrifying biomass controlled in the systems were changed from 65.4 to 141.2 mg VSS/L in order to test the effects of biomass concentration on the efficiencies of cometabolism dechlorination of TCE. The experimental results are shown in Figures 11 to 13. According to Figure 11, the amounts of TCE removed by cometabolism dechlorination were the largest ones among the three sets of batch tests in this study after deducting the amounts of vapor loss through evaporation learned from the blank tests shown in Figure 12. The reasons might be that the three different biomass concentrations used in this test were the highest ones in this study. Also, it was found that the higher the concentrations of biomass applied to the systems, the more TCE was removed

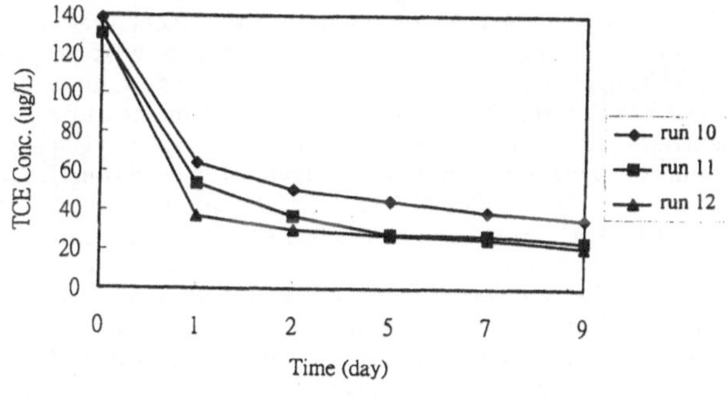


**Figure 11. The concentrations of TCE varied with time in the third set of batch tests (biomass: 65.4–141.2 mg VSS/L, $NH_3$-N: 12.0 mg N/L, TCE: 150 μg/L).**

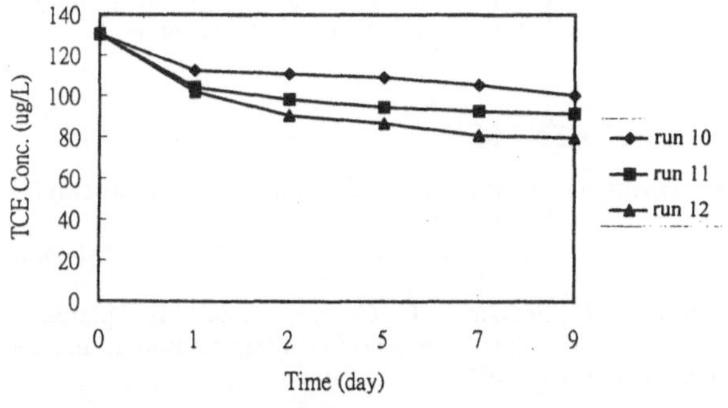


**Figure 12. The concentrations of TCE varied with time in the third set of batch blank tests (biomass: 65.4–141.2 mg VSS/L, $NH_3$-N: 12.0 mg N/L, TCE: 150μg/L, ATU: 0.5 mL/100mL).**

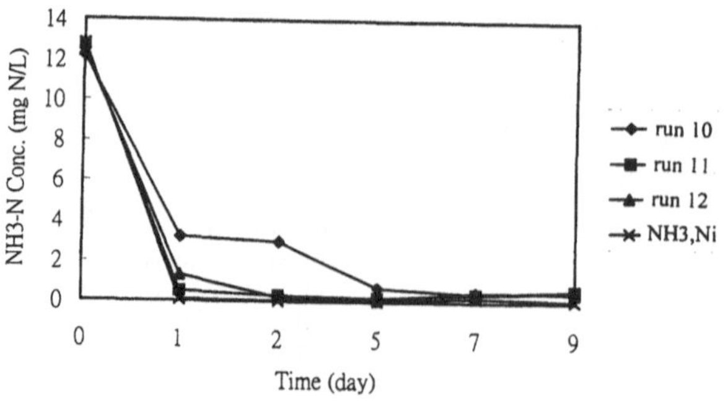


**Figure 13. The concentrations of $NH_3$-N varied with time in the third set of batch tests (biomass: 65.4–141.2 mg VSS/L, $NH_3$-N of 12.0 mg N/L, and TCE of 150 μg/L).**

from the systems through cometabolism dechlorination by an increasing amount of AMO generated by nitritifying bacteria. As seen in Figure 13, no inhibition to ammonia oxidation was exhibited in this test, while over 80% of ammonia was removed in the first day after the test was begun for all of the three systems. However, in test run 5, the TCE level of 150 μg/L showed inhibition to ammonia oxidation, which was because the biomass concentrations used in that test were smaller (28.7 mg VSS/L). The amounts of ammonia removed were also proportional to the biomass concentrations applied to the systems. Hence, it was inferred that in order to let more TCE be removed through cometabolism dechlorination by AMO, it required more nitrifying bacteria growing in the treatment systems for both in-situ and off-site bioremediation processes.

## CONCLUSION

TCE was found capable of being dechlorinated through cometabolism by the enzyme of ammonia monooxygenase (AMO) generated by *Nitrosomonas europaea* in a pure culture. However, no further work has been studied on this enzymatic cometabolism by using an enriched or mixed culture. In this study, the cometabolism chlorination of TCE was investigated by using an enriched nitrifying culture, initially sampled from a human body waste treatment plant, controlled under different environmental conditions in a batch test. According to the experimental results, it was found that the amounts of TCE removed through cometabolism dechlorination were proportional to the initial concentrations of TCE applied to the systems. Both of the high initially applied TCE concentrations and low DO levels maintained in the batch system might cause less ammonia oxidation by nitrifying bacteria. In this study, we found that under enough oxygen supplied, TCE showed inhibition to ammonia oxidation with an initial level of over 200 μg/L applied to the system, which is much lower than the one (about 1500 μg/L) by using a pure culture of nitrifying bacteria (*Nitrosomonas europaea*)[9] instead of an enriched culture. In addition, according to the results, the optimal initial applied concentrations of TCE to the system would be 150 ug/L for both ammonia oxidation and TCE removal through cometabolism by AMO generated by nitrifying bacteria under sufficient DO maintained in the system. Besides, the amounts of TCE removed were found to be proportional to the concentrations of nitrifying biomass, and thus the amounts of AMO, applied to the system, while it appeared that there was no relationship between the amounts of TCE removed and initially applied ammonia concentration to the systems. Hence, it was concluded that an enriched nitrifying culture might cometabolize TCE well as long as an environment was provided with sufficient oxygen, some ammonia, enriched nitrifying biomass, suitable pH level, and TCE concentrations below the inhibitory levels to ammonia oxidation.

## REFERENCES

1. Love, O.T., Jr., and R.G. Eilers, "Treatment of drinking water containing trichloroethylene and related industrial solvents," *J. Am. Water Works Assoc.*, 74, pp. 413–425, 1982.
2. Westrick, J.J., J.W. Mello, and R.F. Thomas, "The groundwater supply survey," *J. Am. Water Works Asso.*, 76, pp. 52–59, 1984.
3. Aziz, C.E., M.W. Fitch, L.K. Linquist, J.G. Pressman, G. Georgiou, and G.E. Speitel, Jr., "Methanotrophic Biodegradation of trichloroethylene in a hollow fiber membrane bioreactor," *Environ. Sci. Technol.*, 29, pp. 2574–2583, 1995.

4. Ely, R.L.,K.J. Williamson, R.B. Guenther, M.R. Hyman, and D.J. Arp, "A cometabolic kinetics model incorporation enzyme inhibition, inactivation, and recovery: I. Model development, analysis, and testing," *Biotech. Bioengr.*, 46, pp. 218–231, 1995.
5. Alvarez-Cohen, L., and P.L. McCarty, "Effects of toxicity, aeration, and reductant supply on trichloroethylene transformation by a mixed methanotrophic culture," *Appl. Environ. Microbiol.*, 57, pp. 228–235, 1991.
6. Ensign, S.A., M..R. Hyman, and D.J. Arp, "Cometabolic degradation of chlorinated alkenes by alkene monooxygenase in a propylene-grown *Xanthobacter* strain," *Appl. Environ. Microbiol.*, 58, pp. 3038–3046, 1992.
7. Folsom, B.R., P.J. Chapman, and P.H. Pritchard, "Phenol and trichloroethylene degradation by *Pseudomonas cepacia* G4: Kinetics and interactions between substrates," *Appl. Environ. Microbiol.*, 56, pp. 1279–1285, 1990.
8. Malachowsky, K.J., T.J. Phelps, A.B. Teboli, D.E. Minnikin, and D.C. White, "Aerobic mineralization of trichloroethylene, vinyl chloride, and aromatic compounds by *Rhodococcus* species," *Appl. Environ. Microbiol.*, 60, pp. 542–548, 1994.
9. Rasche, M.E., M.R. Hyman, and D.J. Arp, "Factors limiting aliphatic chlorocarbon degradation by *Nitrosomonas europea*: cometabolic inactivation of ammonia monoosygenase and substrate specificity," *Appl. Environ. Microbiol.*, 57, pp. 2986–2994, 1991.
10. Vogel, T.M., and P.L. McCarty, "Biotransformation of tetrachloroethylene to trichloroethylene, dichloroethylene, vinyl chloride, and carbon dioxide under methanogenic conditions," *Appl. Environ. Microbiol.*, 49, pp. 1080–1083, 1985.
11. Wilson, J.T., and B.H. Wilson, "Biodegradation of trichloroethylene in soil," *Appl. Environ. Microbiol.*, 49, pp. 242–243, 1985.
12. Chu, K.-H., L. Alvarez-Cohen, "Trichloroethylene degradation by methane-oxidizing cultures grown with various nitrogen sources," *Water Environ Res.*, 68, pp. 76–82, 1996.
13. Ensley, B.D., "Biochemical diversity of trichlorothylene metabolism," *Annu. Rev. Microbiol.*, 45, pp. 283–299.
14. Oldenhuis, R., J.Y. Oedzes, J.J. van der Waarde, and D.B. Janssen, "Kinetics of chlorinated hydrocarbon degradation by *Methylosinus trichosporium* OB3b and toxicity of trichloroethylene," *Appl. Environ. Microbiol.*, 57, pp. 7–14, 1991.
15. Shields, M.S., S.O. Montgomery, P.J. Chapman, S.M. Cuskey, and P.H. Pritchard, "Novel pathway of toluene catabolism in the trichloroethylene-degrading bacterium G4," *Appl. Environ. Microbiol.*, 55, pp. 1624–1629, 1989.
16. Wackett, L.P., G.A. Brusseau, S.R. Householder, and R.S. Hanson, "Survey of microbial oxygenase: trichloroethylene degradation by propane-oxidizing bacteria," *Appl. Environ. Microbiol.*, 55, pp. 2960–2964, 1989.
17. Dabrock, B.,J. Riedel, J. Bertram, and G. Gottschalk, "Isopropylbenzene (cumene)-a new substrate for the isolation of trichloroethylene-grading bacteria," *Arch.. Microbiol.*, 158, pp. 9–13, 1992.
18. Ewers, J., D. Frier-Schroder, and H.-J. Knackmuss, "Selection of trichloroethylene (TCE) degrading bacteria that resist inactivation by TCE," *Arch. Microbiol.*, 154, pp. 410–413, 1990.
19. Arciero, D., T. Vannelli, M. Logan, and A.B. Hooper, "Degradation of trichloroethylene by the ammonia-oxidizing bacterium *Nitrosomonas europaea*," *Biochem. Biophys. Res. Commun.*, 159, pp. 640–643, 1989.
20. Hyman, M., S.A. Russell, R.L. Ely, K.J. Willamson, and D.J. Arp., "Inhibition, inactivation, and recovery of ammonia-oxidizing activity in cometabolism of trichloroethylene by *Nitrosomonas europaea*," *Appl. Environ. Microbiol.*, 61, pp. 1480–1487, 1995.
21. Rasche, M.E., M.R. Hyman, and D.J. Arp, "Biodegradation of halogenated fumigants by nitrifying bacteria," *Appl. Environ. Microbiol.*, 56, pp. 2568–2571, 1990.
22. Rasche, M.E., R.E. Hicks, M.R. Hyman, and D.J. Arp, "Oxidation of monohalogenated ethanes and n-chlorinated alkanes by whole cells of *Nitrosomonas europeae*," *J. Bacteriol.*, 172, pp. 5368–5373, 1990.
23. Broholm, K., B.K. Jensen, T.H. Christensen, and L. Olsen, "Toxicity of 1,1,1-trichloroethane and trichloroethylene on a mixed culture of methane-oxidizing bacteria," *Appl. Environ. Microbiol.*, 56, pp. 2488–2493, 1990.
24. Little, C.D., A.V. Palumbo, S.E. Herbes, M.E. Lindstrom, R.L. Tyndall, and P.J., "Trichloroethylene biodegradation by a methane-oxidizing bacterium," *Appl. Environ. Microbiol.*, 54, pp. 951–956, 1988.

25. Tsien, H.C., G.A. Brusseau, R.S. Hanson, and L.P. Wackett, "Biodegradation of trichloroethylene by *Methylosinus trichosporium* OB3b," *Appl. Environ. Microbiol.*, 55, pp. 3155–3161, 1989.
26. Phelps, T.J., J.J. Niedzielski, R.M. Schram, S.E. Herbes, and D.C. White, "Biodegradation of trichloroethylene in continuously-recycled, expanded-bed bioreactor," *Appl. Environ. Microbiol.*, 56, pp. 1702–1709, 1990.
27. Strandberg, G.W., T.L. Donaldson, and L.L. Farr, "Degradation of trichloroethylene and trans-1,2-dichloroethylene by a methanotrophic consortium in a fixed-film, packed-bed bioreactor," *Environ. Sci. Technol.*, 23, pp. 1422–1425, 1989.
28. Vannelli, T., M. Logan, D.M. Arciero, and A.B. Hooper, "Degradation of halogenated aliphatic compounds by the ammonia-oxidizing bacterium *Nitrosomonas europaea*," *Appl. Environ. Microbiol.*, 56, pp. 1169–1171, 1990.
29. Ely, R.L., M.R. Hyman, D.J. Arp, R.B. Guenther, and K.J. Williamson, "A cometabolic kinetics model incorporation enzyme inhibition, inactivation, and recovery: II. Trichloroethylene degradation experiments," *Biotech. Bioengr.*, 46, pp. 232–245, 1995.
30. *Standard Methods for the Examination of Water and Wastewater*, 18th ed., APHA, AWWA and WEF, Washington, D.C., 1992.

# Section Four
# BIOLOGICAL:
# BIODEGRADATION AND TREATABILITY

## 13 IMPROVED RAPID ANAEROBIC TREATABILITY/TOXICITY SCREENING PROCEDURE

**Thomas R. Wynn** and **Enos L. Stover**
The Stover Group
P.O. Box 2056
Stillwater, Oklahoma 74076

**John N. Veenstra**, Professor
School of Civil and Environmental Engineering
Oklahoma State University, Stillwater, Oklahoma 74078

### ABSTRACT

An experimental procedure was developed by Brooks et al.[1] for anaerobic screening studies using 125 mL syringes with luer-lock valves. This procedure was performed under static conditions with intermittent mixing twice per day. The new modified procedure involves continuous mixing on a shaker table. A specific test protocol has been developed with standardized operating conditions, including organic loads and volatile suspended solids concentrations. Advantages of this procedure are reduction in test duration and minimal variability.

### INTRODUCTION

Anaerobic treatment of municipal and industrial waste has been used with great success in stabilizing organic waste. The popularity of anaerobic treatment has risen in recent years, mostly due to the high energy yielding by-product, methane. Other advantages include lower sludge yields and treatment of high strength organic waste streams more economically than aerobic systems. Anaerobic treatment involves the decomposition of waste in the absence of dissolved oxygen and consists of sequences of reactions in which one group of microorganisms produces substrate for another group of microorganisms. The synergistic relationship produces a delicately balanced environment which can diminish rapidly if environmental conditions are not satisfied. Anaerobic treatability/toxicity screening tests provide needed information to maintain favorable environmental conditions.

Anaerobic treatability/toxicity screening tests are bench-scale tests that simulate full-scale responses to waste streams. Information from these tests will determine the anaerobic treatability and waste characteristics of a particular waste stream and can assist in developing large-scale treatability studies. Types of treatability/screening tests include batch, continuous or semicontinuous feed assays. Each type of assay has advantages and disadvantages, with the ultimate selection depending on the desired information. Batch assays are quick, inexpensive, reproducible, and tend to produce a conservative value for a toxicity evaluation.[2] Inhibition and toxicity can be evaluated from total volume and rate of gas produced. Gas production relates to the health of the inoculum and a decrease in the rate or total gas produced indicates a negative effect from the waste sample.[3] Accurate gas measurements are essential when gas production is measured for indication of toxicity or inhibition.

A recent study by Brooks et al.[1] evaluated the use of 125 mL glass luer-lock syringes for developing an anaerobic screening procedure. The glass luer-lock syringe was the reactor vessel

*52nd Purdue Industrial Waste Conference Proceedings*, 1997, Ann Arbor Press, Chelsea, Michigan 48118. Printed in U.S.A.

and contained the biogas produced. Both liquid and gas samples can be removed without introducing oxygen. Gas production may be measured on a routine time sequence, to help in evaluating lag periods, substrate removal, kinetics and toxicity, or inhibition. The procedure was simple and testing periods lasted from 7 to 21 days.[1] The goal of this study was to improve the methodology using 125 mL luer-lock glass syringes as the reactor vessel. The improvements were focused on minimizing the duration of the test and reducing variability through continuous mixing on a shaker table along with developing a protocol with standard operating procedures, such as organic loading and environmental conditions. Constant mixing results in improved gas production rates, which is important relative to using the syringe in an anaerobic screening procedure. Improper test conditions can result in high gas production rates which can may expel the plunger from the syringe, damaging the plunger and terminating the test. Therefore, operating parameters must be defined under mixing conditions to prevent excessive labor requirements for supervision of the syringes.

There are three primary objectives of this study. The first objective was to determine specific operating conditions such as solids loading and food-to-microorganism ratio (F/M ratio) in terms of mg COD/mg VSS (VSS is volatile suspended solids and COD is chemical oxygen demand). Defining these operating conditions is necessary to prevent the expulsion of the plunger from the syringe. A second objective was to evaluate the syringes for variability and reproducibility under mixing conditions. The third objective was to compare the response of continuous mixing versus static testing.

## METHODS AND MATERIALS

The anaerobic seed source used in the study was from a hybrid anaerobic reactor treating high strength carbohydrate wastewater. The sludge was stored in an airtight container and kept at a temperature of approximately 35° Celsius. Gas was released once a day until gas pressure was minimal, which occurred after approximately seven days. At that time, pH, alkalinity, volatile fatty acids (VFA), and soluble COD analysis were conducted every two to three days to determine endogenous conditions. Experimental procedures were initiated when the seed source reached endogenous conditions, represented by insignificant changes in COD and VFA analysis.

The setup procedure was similar for all test runs. In the first experiment, a 120 mL stock solution of sucrose diluted with BOD dilution water and anaerobic seed was mixed thoroughly, and then a 60 mL subsample was withdrawn and placed into the syringe. The pH was adjusted with sodium hydroxide or sulfuric acid to 7.0 to 7.3 s.u. before the 60 mL subsample was withdrawn. The remaining 60 mL of solution was used for chemical analysis. The setup procedure was similar for experiments two and three. In experiments three and four one stock solution containing a mixture of sucrose, anaerobic seed, phosphate buffer and BOD dilution water was mixed thoroughly. All 60 mL syringe samples were withdrawn from the stock solution. Micronutrients were not added, since micronutrient additions occurred in the process from which the seed source was collected.

The first experiment was designed to determine the optimum range of volatile suspended solids and F/M ratio, for practical administration of the test. Syringes loaded with high VSS concentrations or high organic loads may produce gas too quickly, possibly separating the plunger from the syringe, and terminating the test. In addition, high VSS and F/M ratios may require intense supervision and multiple evacuations of the biogas, increasing test variability. Conversely, a low volatile suspended solids concentration or low F/M ratio may produce a low, inconclusive volume of biogas. An F/M ratio range must be determined to prevent reactor failure.

The objective of test A of the first experiment was to determine the optimum range of volatile suspended solids; syringe A was loaded with 1,000 mg/L VSS, syringe B with 2,000 mg/L, syringe C with 3,000 mg/L, syringe D with 4,000 mg/L, syringe E with 5,000 mg/L and syringe F with 6,000 mg/L VSS. All syringes contained a F/M ratio around 0.25. This value was selected to ensure that gas production rates were low enough to monitor the syringes over time and that a high organic load would not impair reactor performance. The objective of test B was to determine an acceptable range of F/M ratio. Two VSS concentrations were selected based on the results of test A. The criteria for VSS selection were the volume of gas produced, number of times gas was evacuated, and reactor performance. F/M ratios were set at 0.5, 1.0, and 1.5 for each VSS concentration selected. Criteria for selecting the F/M ratio were based on practical consideration of administering the test. The test procedure was developed so that the syringes may be examined twice daily without supervision between inspections.

Biogas production relates to performance of the anaerobic bacteria, thus syringe variability must be distinguished from inhibition produced from the test media. The objective of the second experiment was to examine the variability and reproducibility of the syringes. Six syringes were prepared at a VSS concentrations of 3,000 and 4,000 mg/L with a F/M ratio of 0.5. The test media for each syringe was withdrawn from a stock solution of anaerobic seed, sucrose, and phosphate buffer to ensure uniformity. Two test runs were used to assess variability of the method. Since comparisons were made between tests, the accuracy of each setup was important for maintaining uniformity; therefore, a reasonable expectation of the amount of error was needed.

Improvements in the test procedure focused on reducing the duration of the test and improving contact between the anaerobic seed and the substrate or toxicants. This was accomplished by utilizing a mixing device (shaker table). The objective of the third experiment was to evaluate the effects of mixing versus static conditions. Three syringes were placed on a shaker table and three syringes were mixed twice per day remaining under static conditions between mixing.

All samples for wet chemistry analysis were obtained from residual solutions made for each syringe or syringes. The contents were mixed well before a subsample was withdrawn. Samples which required filtration were filtered through 4.25 micron glass filters. The sCOD was determined colorimetrically using the reactor digestion method and HACH chemical reagents. The detection range used for all sCOD analysis was 0-1,500 mg/L. For this method, 5 mL subsamples were filtered. Total and volatile suspended solids, alkalinity, and volatile fatty acids (VFA) were determined in accordance with *Standard Method*'s 18th edition.[4] The final pH was measured by removing the syringe valve and pushing the liquid sample through the end of the syringe, directing the contents into a beaker with a pH probe in position. This procedure produced an accurate representation of the final pH. The pH increases when exposed to ambient conditions due to $CO_2$ release; therefore, the pH was determined first.

The gas production rate was measured twice a day by subtracting the previous volume gas produced from the total volume gas measured at that time. The volume of gas was measured from the tip of the plunger, or if the plunger extended past the measurable volume, the syringe was placed vertically and measured from the meniscus of the sample. Gas was evacuated, if it was believed gas production rates would cause the expulsion of the plunger between inspections. Carbon dioxide was determined by injecting a 0.7 mL sample of biogas into an Astro 2001 TOC/TIC/TC analyzer. A silicon tube was connected to the end of the luer-lock valve, the valve was then opened, flushing the silicon tube with biogas and then clamped to capture the biogas. This procedure required approximately 10 to 15 mL of gas. A sample was extracted from the tube using a 5 mL syringe. A 30% $CO_2$ and 70% helium standard was made using the same type of syringe used in this study. Three to four standards were analyzed and averaged for calibration during each test.

## RESULTS AND DISCUSSION

The results of this study are summarized in Tables II through IV. Cumulative gas curves for the syringes are presented in Figures 1 through 5. The results are discussed in three sections. The first section discusses results of the experiment which was developed to define the operating conditions, such as VSS concentrations and F/M ratio. The second section discusses the variability and reproducibility of the syringes, and the third section evaluates the effects of mixing.

The scope of this project was to improve an existing anaerobic treatability/toxicity screening procedure. The objectives of this study, as outlined in Table I, were to define operational conditions, such as volatile suspended solids concentrations and F/M ratios, while evaluating the variability of the test procedure and reproducibility. Another objective was to minimize the length of the test period while maintaining accurate, reproducible results. The procedure employed 125 mL glass luer-lock syringes as the reactor vessel under mixing conditions on a shaker table. The syringes are unique in that both the test media and biogas are contained in the same vessel, allowing measurement of biogas production.

The results of test A of experiment one are presented in Table II and Figure 1. In test A, syringes were loaded with varied VSS concentrations and an F/M ratio near 0.25 mg COD/mg VSS. Gas production increased with the increase in VSS concentrations and organic load. The syringes loaded with 5,750 and 4,600 mg/L VSS required numerous gas evacuations the first five days. Gas was evacuated three times for the syringe with 3,850 mg/L VSS, twice for the syringe with 2,800 mg/L VSS, and gas was not evacuated in the two lowest concentrations. The two

**Table I. Objectives of Each Experiment**

| Experiment Number | Objective |
|---|---|
| 1 | Determine Optimum Range of VSS<br>Determine Optimum F/M Ratio |
| 2 | Examine the Variability and<br>Reproducibility of the Syringes |
| 3 | Evaluate the Effects of Mixing<br>Versus Static Conditions |

highest concentrations required frequent inspections to prevent expulsion of the plunger. The time required for inspecting the two syringes made the VSS loads impractical. The gas production rate was slow for syringes with 900 and 1,600 mg/L VSS, requiring three days to produce enough gas for analysis, and subsequent gas production was not adequate for additional gas analysis. A larger total volume of gas would be more beneficial for distinguishing between responses to toxicity or treatability and variability among the syringes. The VSS loads of 2,800 and 3,600 mg/L VSS required a maximum of three evacuations and produced volumes of gas adequate for gas analysis. The optimum range of VSS for this test sludge was determined to be 3,000 to 4,000 mg/L.

**Table II. Initial Operating Conditions and Final Conditions for Experiment 1; Test A (VSS Determination) and Test B (F/M Ratio Determination)**

| | Syringe | | | | | |
|---|---|---|---|---|---|---|
| Parameter | A | B | C | D | E | F |
| Test A | | | | | | |
| Initial Conditions | | | | | | |
| VSS (mg/L) | 900 | 1,600 | 2,700 | 3,850 | 4,650 | 5,800 |
| Sucrose (mg/L) | 250 | 500 | 750 | 1,000 | 1,250 | 1,500 |
| COD (mg/L) | 350 | 460 | 750 | 900 | 1,250 | 1,450 |
| Alkalinity (mg/L) | 500 | 500 | 800 | 1,150 | 1,450 | 2,400 |
| pH (s.u.) | 7.2 | 7.2 | 7.2 | 7.2 | 7.2 | 7.2 |
| Final Conditions | | | | | | |
| Total Gas (mL) | 22 | 46 | 86 | 106 | 141 | 160 |
| CODr (%) | 74 | 39 | 46 | 90 | 92 | 92 |
| Alkalinity (mg/L) | 250 | 850 | 1,200 | 1,650 | 2,050 | 2,400 |
| pH (s.u.) | 6.5 | 6.5 | 6.8 | 7.0 | 7.0 | 7.1 |
| $CO_2$ (%) | 9.2 | 15 | 15 | 15 | 14 | 14 |
| Test B | | | | | | |
| Initial Conditions | | | | | | |
| VSS (mg/L) | 2,700 | 3,100 | a | 3,650 | 4,200 | a |
| Sucrose (mg/L) | 1,500 | 3,000 | 4,500 | 2,000 | 4,000 | 6,000 |
| COD (mg/L) | 1,025 | 2,415 | 3,000 | 1,305 | 3,000 | 5,000 |
| Alkalinity (mg/L) | 705 | 720 | 670 | 906 | 916 | 950 |
| pH (s.u.) | 7.1 | 7.7 | 7.7 | 7.2 | 7.2 | 7.3 |
| Final Conditions | | | | | | |
| Total Gas (mL) | 54 | 31 | – | 121 | 55 | – |
| CODr (%) | 38 | 0 | – | 90 | 0 | – |
| Alkalinity (mg/L) | 990 | 760 | – | 1322 | 980 | – |
| pH (s.u.) | 6.5 | 5.4 | – | 6.9 | 5.5 | – |
| VFA (mg/L) | 300 | 1,125 | – | 100 | 1,275 | – |
| $CO_2$ (%) | 32 | 48 | – | 33 | 59 | – |

[a]These syringes were shut down and VSS were measured at the end of the test

**Table III. Initial Operating Conditions and Final Conditions for Experiment 2; Variability and Reproducibility**

| | Syringe | | | | | |
|---|---|---|---|---|---|---|
| Parameter | A | B | C | D | E | F |
| Test A | | | | | | |
| Initial Conditions | | | | | | |
| VSS (mg/L) | 4,100 | 4,100 | 4,100 | 4,100 | 4,100 | 4,100 |
| Sucrose (mg/L) | 2,000 | 2,000 | 2,000 | 2,000 | 2,000 | 2,000 |
| COD (mg/L) | 2,125 | 2,125 | 2,125 | 2,125 | 2,125 | 2,125 |
| Alkalinity (mg/L) | 1,600 | 1,600 | 1,600 | 1,600 | 1,600 | 1,600 |
| pH (s.u.) | 7.2 | 7.2 | 7.2 | 7.2 | 7.2 | 7.2 |
| $KH_2PO_4$ (mg/L) | 20 | 20 | 20 | 20 | 20 | 20 |
| Final Conditions | | | | | | |
| Total Gas (mL) | 108 | 110 | 104 | 108 | 105 | 105 |
| CODr (%) | 95 | 95 | 95 | 96 | 86 | 96 |
| Alkalinity (mg/L) | 1,850 | 1,700 | 1,611 | 1,700 | 1,750 | 1,750 |
| pH (s.u.) | 6.9 | 7.0 | 7.0 | 7.0 | 7.0 | 7.0 |
| $CO_2$ (%) | 24 | 25 | 23 | 24 | 23 | 22 |
| Test B | | | | | | |
| Initial Conditions | | | | | | |
| VSS (mg/L) | 3,650 | 3,650 | 3,650 | 3,650 | 3,650 | 3,650 |
| Sucrose (mg/L) | 2,000 | 2,000 | 2,000 | 2,000 | 2,000 | 2,000 |
| COD (mg/L) | 1,900 | 1,900 | 1,900 | 1,900 | 1,900 | 1,900 |
| Alkalinity (mg/L) | 2,250 | 2,250 | 2,250 | 2,250 | 2,250 | 2,250 |
| pH (s.u.) | 7.2 | 7.2 | 7.2 | 7.2 | 7.2 | 7.2 |
| $KH_2PO_4$ (mg/L) | 20 | 20 | 20 | 20 | 20 | 20 |
| Final Conditions | | | | | | |
| Total Gas (mL) | 103 | 98 | 105 | 100 | 104 | 103 |
| CODr (%) | 94 | 95 | 95 | 95 | 95 | 96 |
| Alkalinity (mg/L) | 1,800 | 1,850 | 1,700 | 1,750 | 1,750 | 1,850 |
| pH (s.u.) | 7.0 | 6.9 | 6.9 | 7.0 | 7.0 | 7.0 |
| $CO_2$ (%) | 30 | 30 | 32 | 29 | 28 | 28 |

The results of test B are presented in Table II and Figure 2. Three F/M ratios near 0.5, 1.0, and 1.5 were evaluated with VSS concentrations near 3,000 to 4,000 mg/L. Gas production rates were too high for syringes with F/M ratios near 1.5, as seen in Figure 2. These syringes required constant monitoring and were terminated the first day. Results of the chemical analysis presented in Table II indicated that the syringes with F/M ratios near 1.0 failed. The effluent pH was below 6.0, VFAs were over 1,200 mg/L $CO_2$ comprised nearly 50% of the gas and no COD was removed. Reactor failure may be due to inadequate buffering. Initial alkalinities for syringes B and E were 760 and 980 mg/L as $CaCO_3$, respectively. Brooks et al.[3] did not encounter reactor failure at a F/M near 1.0 with an initial alkalinity near 1,400 mg/L as $CaCO_3$. However, percent COD removal (CODr) was lower under these conditions. Syringes with F/M ratios near 0.5 were easily monitored and produced adequate volumes of gas for analysis. The VFAs were 300 mg/L in syringe A, with an initial alkalinity of 990 mg/L, where syringe D had a VFA of 100 mg/L and an initial alkalinity of 1,322 mg/L as $CaCO_3$. Percent COD removed was much higher in syringe D than syringe A. Buffers were not initially added since the anaerobic seed source had an alkalinity of 2,570 mg/L. Phosphate buffer ($KH_2PO_4$) was, however, added in the subsequent experiments. The optimum range for the F/M ratio was determined to be 0.5 to 1.0.

Gas production is an indication of the health of anaerobic bacteria and is monitored to determine the effects of a chemical component or waste stream. One advantage of syringes is the variable time scale during which the gas production may be monitored, daily or hourly. However, differentiating between inhibition or toxicity effects and the variability of the syringes must be understood. If variability of the syringes is minimal, an accurate threshold inhibition concentra-

**Table IV. Statistical Results of the Variability Study**

| | Total Gas Production (mL) |
|---|---|
| Test A | 108 |
| | 110 |
| | 104 |
| | 108 |
| | 105 |
| | 105 |
| Test B | 103 |
| | 98 |
| | 105 |
| | 100 |
| | 104 |
| | 103 |
| Combined Test One and Two | |
| Sum | 1253 |
| Count | 12 |
| Statistical Results | |
| Mean. | 104.4 |
| Standard Error | 0.96 |
| Standard Dev. | 3.3 |

tion may be determined. This information can be beneficial for troubleshooting tests, process control and developing data for large-scale treatability studies or full-scale operation assistance. The objective of the third experiment was to evaluate the variability and reproducibility of the syringes used in this study. Two tests were duplicated to evaluate variability and reproducibility. Variability among the syringes was low in terms of total gas produced and gas production rates, as seen in Figures 3 and 4. Also, the results of the % $CO_2$ analysis in Table III show an insignificant amount of variability. Total gas production from the two tests were combined to calculate

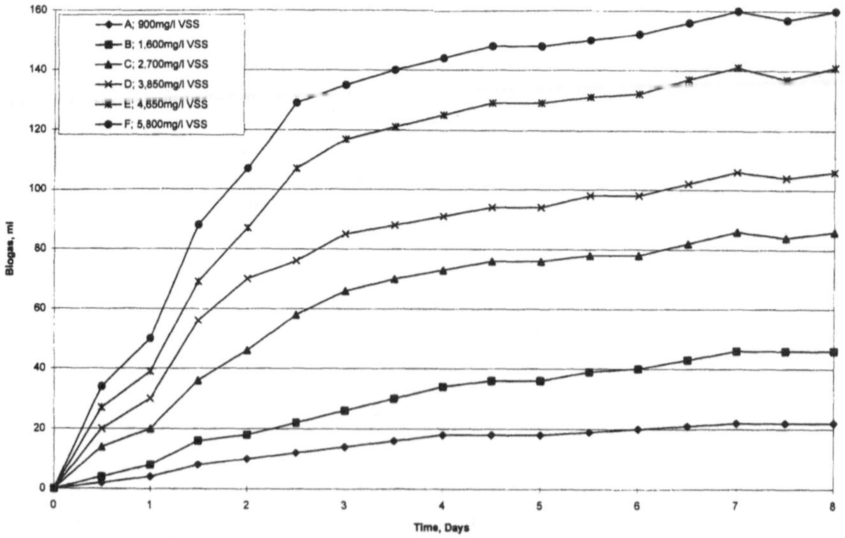


**Figure 1. Cumulative gas curves for experiment 1; test A (VSS determination) at a F/M OF 0.25.**

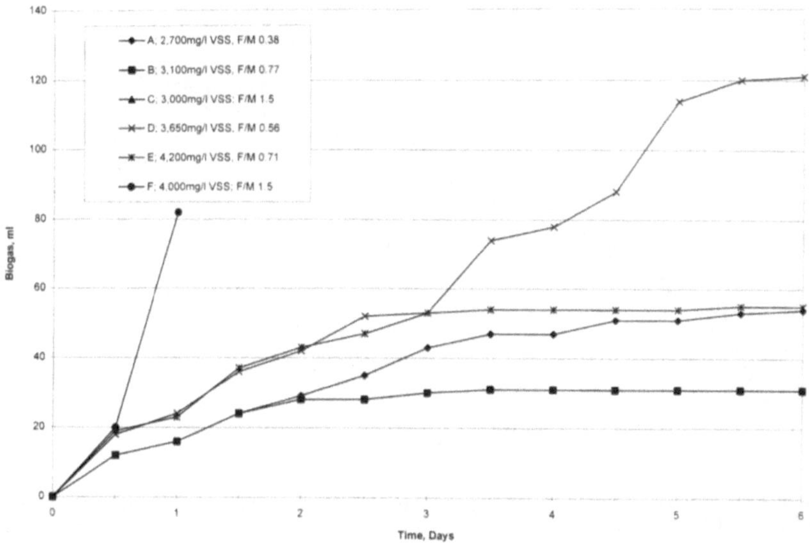


**Figure 2. Cumulative gas curves for experiment 1; test B (F/M determination).**

the mean, standard deviation, standard error, and 95% confidence interval. These results are summarized in Table IV. Additional tests should be conducted to further investigate and refine estimates of variability and reproducibility. An easy method for evaluating variability using a control chart is presented in Figure 6. This figure was developed from the statistical results presented in Table IV. The control chart illustrates the trends of previous duplicate tests by plotting the mean and one standard deviation with total gas production of the duplicate tests. The objec-

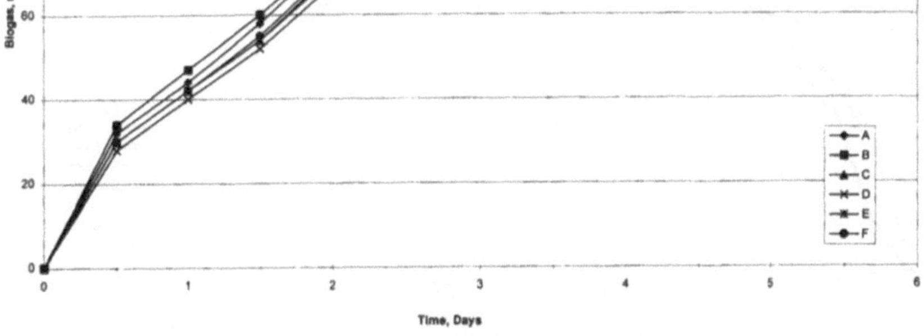


**Figure 3. Cumulative gas curves for experiment 2; test A (variability/reproducibility) at 4,100 mg/L VSS, F/M ratio of 0.52.**

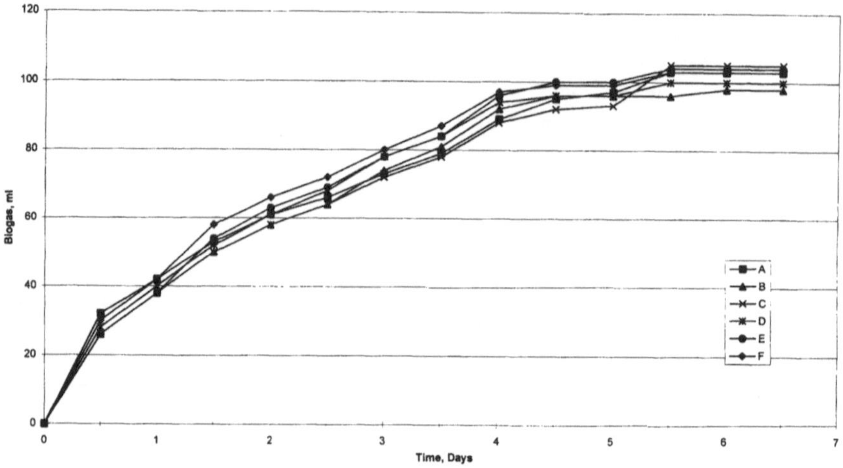


**Figure 4. Cumulative gas curves for experiment 2; test B (variability/reproducibility) at 1,900 mg/L VSS, F/M ratio of 0.52.**

tive of the control chart is to monitor the performance of the syringes. If the gas production of a syringe is outside the acceptable range of variability; i.e., one or two standard deviations, the syringe or the plunger may need to be replaced. Duplicate tests should be run periodically or before a treatability test is initiated, to develop an accurate control chart. Results of the duplicate tests in this study indicate minimal variability, with a standard deviation of 3.3 mL. In addition to monitoring syringes, the control chart may serve as a quality control chart and illustrate the treatability of a wastewater or the toxicity of a chemical. Once several syringe tests have been completed at a defined VSS and F/M ratio with an easily degradable substrate, such as sucrose, a well-defined standard deviation can be defined. Results may be compared to the standard devia-

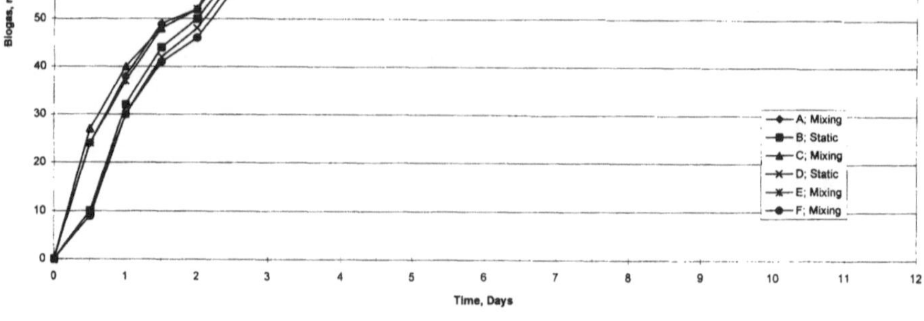


**Figure 5. Cumulative gas curves for the static versus mixing experiment 3; at 3,750 mg/L VSS, F/M ratio of 0.56.**

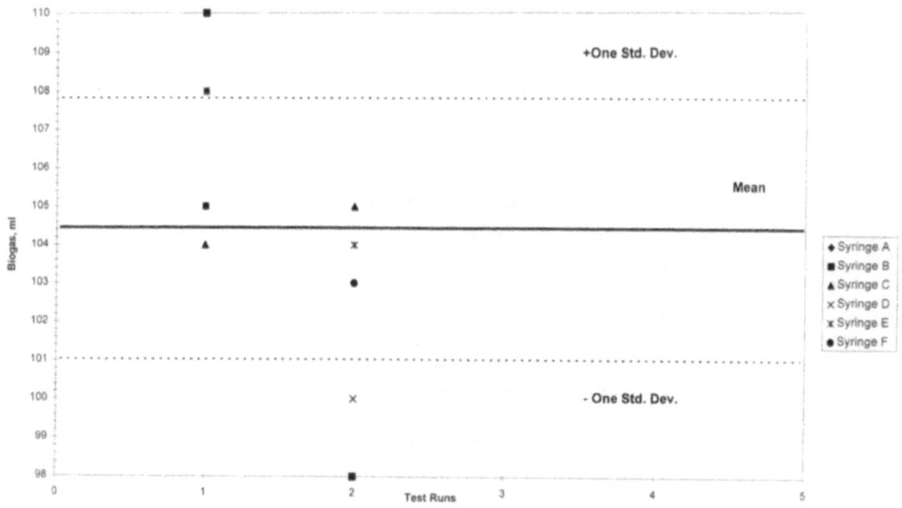


**Figure 6. Control chart for monitoring syringe variability and reproducibility.**

tion to determine the degree of toxicity or the biodegradability of a wastewater relative to sucrose. Also, gas production in liters of gas produced per gram of COD removed can be calculated for total gas volume for evaluating the treatability or toxicity of a wastewater or chemical compound. A quality control assessment may be administered using the control chart.

A previous study with this type of syringe utilizing intermittent mixing lasted up to two to three weeks.[1] An objective of this study was to minimize the test period and reduce variability by continuously mixing the syringes. A comparison study of three static syringes and three continuously mixed syringes showed significant reduction in the duration of the test. The static syringes exceeded the continuously mixed syringes by five days. The total volume of gas was less in the static syringes; however, this may be a function of greater variability. The cumulative gas curves in Figure 5 indicate a lag period of two days with the static syringes. The mixing conditions increased contact between the anaerobic seed source and the sucrose. A similar study should be performed with toxicants to further evaluate the effect of mixing.

## CONCLUSION AND SUMMARY

A recent study by Brooks et al.[1] evaluated the use of 125 mL glass luer-lock syringes for developing an anaerobic screening procedure. This procedure was performed under static conditions with intermittent mixing twice per day. A new modified procedure was developed using constant mixing on a shaker table. Significant advantages of this procedure were reducing the test duration and minimizing variability. Also, the test media can be removed quickly from the syringe to accurately measure pH. The syringes were convenient for extracting gas for analysis and measuring gas production. A specific test protocol has been developed with standardized operating parameters, including VSS loads and F/M ratios. These parameters must be defined for practical administration of the test. Constant mixing results in faster gas production, which may expel the plunger from the syringe between inspections. The results of this study indicate that the VSS load should range between 3,000 to 4,000 mg/L with this seed source used during these studies, and the F/M ratio should range between 0.5 to 1.0. Syringe variability must be defined to differentiate between inhibition and variability among the syringes. Variability was found to be minimal after evaluating duplicate tests. To ensure accuracy, the syringes should be tested under identical conditions periodically or before conducting a treatability study. An easy method for evaluating variability is with a control chart (Figure 6). The control chart illustrates the trends of previous tests with the mean and standard deviation plotted with total gas production results.

This procedure will ensure that the syringes are consistent and will identify syringes that may be faulty.

New studies should focus on the versatility of the syringe test procedure. In addition to convenient gas measurements, samples may be withdrawn for analysis during the test, without introducing oxygen, which allows for studies evaluating the kinetics of waste streams and other chemical compounds.

## REFERENCES

1. Brooks, S., Stover, E.L., and Clarkson W.W., Anaerobic Treatability/Toxicity Screening Study Procedure. Presented at the 1994 Food Industry Environmental Conference, Atlanta, Georgia, 1994.
2. Stucky, D.C., Owen, W.F., McCarty, P.L., and Parkin, G.F. Anaerobic toxicity evaluation by batch and semi-continuous assays. *J. Water Poll. Contr. Fed.* 52(4), 1980.
3. Stover, E.L., Brooks, S., and Munirathinam, K., Control of Bulk Liquid Sulfide Toxicity During Anaerobic Treatment. Proceedings of the Georgia Tech Industrial Food Conference, Atlanta, GA, 1992.
4. *Standard Methods for the Examination of Water and Wastewater*, 18th ed., American Public Health Association, Washington, D.C., 1992.

# 14 INFLUENCE OF So/Xo RATIO AND MEDIUM COMPOSITION ON ANAEROBIC BIODEGRADABILITY TEST

**Germán Buitrón**, Research Leader

**Gloria Moreno**, Graduate Student
Environmental Bioprocesses Department,
Institute of Engineering
National University of Mexico
C.U., Apartado Postal 70-472
04510 Mexico, D.F., MEXICO

## ABSTRACT

The effect of the composition of the culture medium and the variation of the substrate/microorganism ratio during the development of anaerobic biodegradability test was studied. An automatic monitoring system was developed for continuous and parallel monitoring of nine reactors measuring the pressure generated by the biogas produced during the anaerobic biodegradation of the substrate. With this system, the experimentation was carried out in serum bottles at 35°C with anaerobic sludge acclimated to ethanol. The influence of three different culture media as well as the effect of the initial substrate concentration to the initial biomass concentration ratio, So/Xo, on the extent of anaerobic biodegradability was studied. It was observed that the composition of the medium has an important influence on the results. The degradation rate of substrate increased when the So/Xo ratio increased (under agitation). Because of that, it is necessary to make an effort to unify criteria including media composition and So/Xo ratio to achieve a possible standardization of the test.

## INTRODUCTION

Biodegradability tests are conducted in order determine the susceptibility of an effluent to be treated under specific environmental conditions; i.e., aerobic or anaerobic environment. In the case of the anaerobic biodegradability tests, a common procedure utilized is the measure of the biogas produced by the microorganisms during the organic substrate degradation. The quantification of the biogas production can be achieved with manual or automatic methodologies. Pressure increase, due to the biogas generation in a closed bottle, is one of the techniques frequently used to monitor online the extent of the biodegradation process.[1,2]

It has been observed that the results of the biodegradation tests could vary upon the methodology followed. The most important variables which influence the results are: the initial activity of the microorganisms, the medium composition and the initial substrate concentration to the initial biomass concentrations ratio, So/Xo. Some authors have concluded that the So/Xo ratio is the most important factor in batch experiments.[3,4] Chudoba et al.[4] and Chang et al.[5] have observed that the substrate uptake kinetics depend on the initial So/Xo ratio and it has been demonstrated that this ratio determines whether multiplication or biomass increase is taking place.[6,7]

In this work, the influence on the composition medium and the So/Xo ratio on the anaerobic biodegradation extent was studied, using an inoculum with constant metabolic activity. An automated device measuring the pressure generated by the biogas during degradation was utilized.

*52nd Purdue Industrial Waste Conference Proceedings*, 1997, Ann Arbor Press, Chelsea, Michigan 48118. Printed in U.S.A.

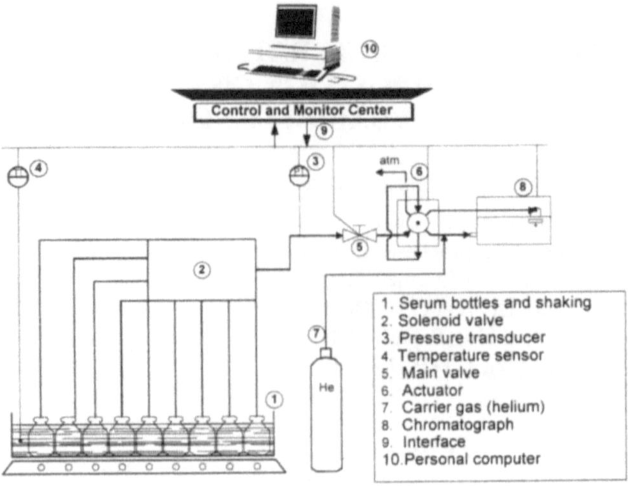


**Figure 1. Anaerobic biodegradability device.**

## EXPERIMENTAL PROCEDURE

### Anaerobic Biodegradability Test

The biodegradability tests were performed in a device constructed in the Institute of Engineering (Figure 1). The device is based on the monitoring (9 bottles in parallel) of the pressure generated by the biogas produced during the organic substrate degradation. The pressure generated during the biogas production is registered through a pressure transducer (Cole Parmer) and senT to a computer at constant periods of time (minimum, every 10 minutes). After the pressure is registered, the bottle is purged in order to reach the atmospheric pressure. Thus, the pressure increments are added and plotted as a function of time. A portion of the biogas purged is sent to a gas chromatograph to determine the methane concentration. This procedure is followed twice a day. The curves of pressure are transformed to moles of biogas with the ideal gas equation, and the moles of methane produced are evaluated from the quantity of biogas, and their composition determined with the gas chromatograph. A detailed description of the test is presented elsewhere.[8] The biodegradability tests were conducted in serum bottles of 160 mL; each particular condition was run by duplicate. For all the tests final volume was 80 mL. The biodegradability tests were conducted in a temperature controlled chamber (35 + 2°C). The substrate and sludge transfer were conducted in anaerobic conditions using an anaerobic chamber (McCoy).

### Acclimation of the Anaerobic Sludge

In order to have an inoculum with constant activity, the follow methodology was applied. The inoculum was harvested from an upflow anaerobic sludge blanket reactor, UASB, treating a malting industry effluent. Analysis of the inoculum was performed according to the *Standard Methods.*[9] The granular sludge had the next initial conditions: specific methanogenic activity = 0.5 mg COD-$CH_4$/mg VSS-d, volatile suspended solids concentration, VSS, = 45 g/L, sludge volumetric index, SVI, = 6.1 mL/g, settling velocity = 12.6 m/hr. Granulometric analysis showed that 85% of the sludge was composed of granules with a diameter larger than 0.6 mm. Granular sludge was acclimated to ethanol as the sole source of carbon and energy, and nutrients.[10] Ethanol was used as substrate because: (a) it is easily biodegradable by anaerobic bacteria and (b) from this compound acetate, one of the principal substrates of acetoclastic methanogenic bacteria, is produced. The reactor was operated in batch, with cycles of 24 hours. The total volume of the reactor was 12 l. Every day one liter of supernatant was removed and replaced with new substrate and nutrients, giving an organic load of 0.5 gCOD/gVSS-d. Temperature was controlled at 35°C and the mixture was stirred. Biogas produced by the acclimation reactor was measured through an electronic device constructed in the Institute of Engineering. Gas composition was

Table I. WV, LV and WOV Media Used in the Study[a]

| Compound, mg/L | WV | LV | WOV | Compound mg/L | WV | LV | WOV |
|---|---|---|---|---|---|---|---|
| $KH_2PO_4$ | 300 | – | 270 | p-aminobenzoic acid | 0.05 | 0.005 | – |
| $K_2HP0_4$ | 300 | – | 350 | lipoic acid | 0.05 | – | – |
| $NH_4Cl$ | 120 | 399 | 530 | nitrilotriacetic acid | 15 | – | – |
| $(NH_4)_2HPO_4$ | – | 80 | – | thioctic acid | – | 0.005 | – |
| NaCl | 610 | – | – | $FeSO_4{\cdot}7H_2O$ | 21 | | |
| | | | | $FeCl_2{\cdot}4H_2O$ | | 370 | 20 |
| KCl | – | 130 | – | $CoSO_4$ | 1 | | |
| | | | | $CoCl_2{\cdot}6H_2O$ | | 0.30 | 0.5 |
| $MgCl_2{\cdot}6H_2$ | 105 | 180 | 100 | $CuSO_4$ 5H2O | 0.1 | 2.7 | |
| | | | | $CuCl_2{\cdot}2H_2O$ | | | 0.03 |
| $MgSO_4{\cdot}7H_2O$ | 30 | – | – | $ZnSO_4$ | 1 | 2.1 | |
| | | | | $ZnCl_2$ | | | 0.05 |
| $MnSO_4{\cdot}H_2O$ | 5 | | | $NiCl_2$ | 0.5 | | |
| $MnCl_2{\cdot}4H_2O$ | | 20 | 0.5 | $NiCl_2$ $6H_2O$ | | | 0.05 |
| $CaCl_2{\cdot}2H_2O$ | 9 | 250 | 750 | $Al_2(SO_4)_3$ | 0.1 | – | – |
| Biotin | 0.02 | 0.002 | – | $H_3BO_3$ | 0.1 | 5.7 | 0.05 |
| Folic acid | 0.02 | 0.002 | – | $NaMoO_4.2H_2O$ | 0.1 | 2.55 | 0.01 |
| Pyridoxine HCl | 0.1 | 0.01 | – | yeast extract | 1000 | – | – |
| Thiamin HCl | 0.05 | 0.005 | – | casein peptone | 1000 | – | – |
| Riboflavin | 0.05 | 0.005 | – | $NaHCO_3$ | 3000 | 4700 | 1200 |
| Nicotinic acid | 0.05 | 0.005 | – | cysteine | 500 | – | – |
| D-L Calcium pantothenate | 0.05 | | – | $Na_2S$ $9H_2O$ | – | 500 | 500 |
| pantothenic acid | 0.05 | 0.005 | – | | | | |
| Vitamin $B_{12}$ | 0.001 | 0.0001 | – | $Na_2SeO_3$ | – | – | 0.05 |

[a]10 mL of each medium were added to the serum bottles and completed to 80 mL.

determined with a gas chromatograph (SRI model 8610) equipped with a thermal conductivity detector.

## Effect of the Medium Composition

To study the effect of the mineral medium composition on the biodegradability results, different culture media were tested. The biodegradability tests were conducted with three culture media: medium with vitamins,[11] WV, medium low in vitamins,[12] LV, and medium without vitamins,[13] WOV (Table I). In order to study a substrate easily biodegradable and an another with toxic effects, glucose and phenol were used as the sole source of organic carbon and two So/Xo ratios were studied (0.1 and 1.0). Each experimental condition was complemented with two blanks: the first one took into account the endogenous respiration of the sludge (and a possible biogas production due to the organic compounds present in the medium composition) and was prepared adding to the bottle the mineral medium of interest and acclimated sludge. The second blank was used to verify the activity of the sludge and was prepared by adding to the bottle acclimated sludge, the mineral medium of interest and ethanol (1 g ethanol/ gVSS).

## Effect of So/Xo Ratio

To study the influence of the So/Xo ratio on the extent of the substrate biodegradation, a set of experiments was carried out using acetate as the sole source of carbon and the WV medium. Three sets of factors with two levels were studied: initial substrate concentration (625 and 12,500 mg acetate/L), initial biomass concentration (2,500 and 12,500 mg/L) and shaking influence (with and without shaking). For the last parameter, an orbital shaker (Avos 160) was used. Shaking was programmed at 150 rpm with cycles of 2 hours shaking and 30 minutes rest. A $2^k$ experimental design was applied, where 2 represented the number of levels of each variable and k the number of factors to study. Thus, with the experimental design we obtained the next So/Xo ratios: 0.05, 0.025, 1.0 and 5, each one with and without shaking. For each condition, two blanks were run in parallel in order to determine the endogenous respiration and the sludge activity.

## RESULTS AND DISCUSSION

### Anaerobic Biodegradability Device

To verify the reproducibility of the lectures made by the device, a set of nine anaerobic biodegradation tests were conducted with glucose under identical conditions (Figure 2). We can observe that the maximal biodegradation rate was obtained in the first 15 hours. An excellent reproducibility within the test was obtained: a variation coefficient of 2.51 % for the first 30 hours, and 1.86 % for all the 181 hours of the experiment, was obtained.

### Acclimation of the Anaerobic Sludge

Initial methane production of the sludge to be acclimated was 0.15 $l$ $CH_4/l_{reactor}$-d. After acclimation this production was increased to 0.22 $l$ $CH_4/l_{reactor}$-d and was constant during the 32 days of operation of the reactor. Methane content in the biogas was 69%. As we can observe in Figure 3, acclimation of the sludge to ethanol was obtained in only 6 days. This can be explained because the sludge had a high initial methanogenic activity (0.5 g COD-$CH_4$/g VSS-d). After acclimation, the sludge characteristics were: specific methanogenic activity = 0.6 mg COD-$CH_4$/mg VSS-d, VSS, = 46 g/L, SVI, = 7 mL/g, settling velocity = 14 m/hr. Throughout all the biodegradation tests these characteristics were maintained. In days 9 and 27 a diminution in methane production was observed. This was attributed to the fact that the reactor was not fed these days. As we can see, it took up to 3 days to recover the constant gas production.

Different sources of inoculum could lead to different biodegradability test results as a consequence of the different level of microbial adaptation and population composition. Acclimating the sludge to a specific substrate can stabilize the microbial community, maintain the microbial activity and composition constant independently of the source of inoculum. The principal objection to this procedure is that not all the bacteria needed for the biodegradation of a specific substrate could be present, especially with industrial effluents. But in such cases, if a doubtful result is found, a treatability test must be conducted, acclimating bacteria to the specific effluent.

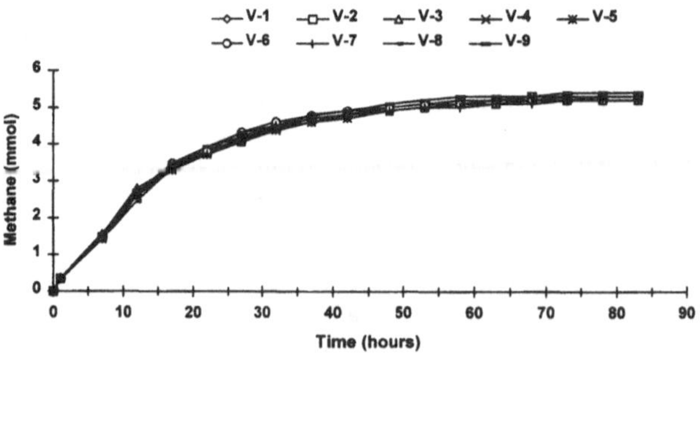


**Figure 2. Reproducibility of the Anaerobic Biodegradability Device. Glucose as sole source of carbon and energy (1.0 g DQO/ g VSS). Data were corrected with the endogenous blank.**

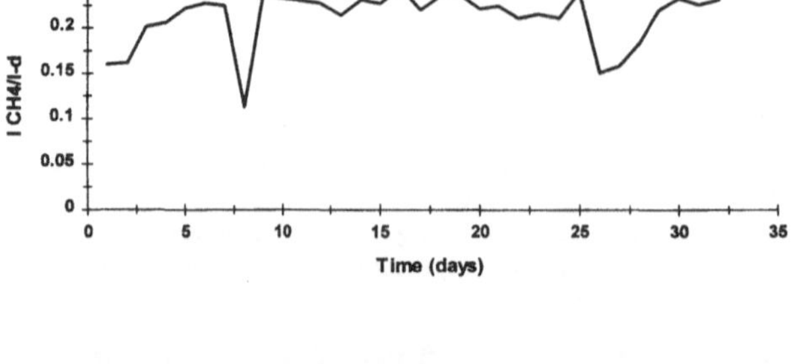


**Figure 3. Granular anaerobic sludge acclimation.**

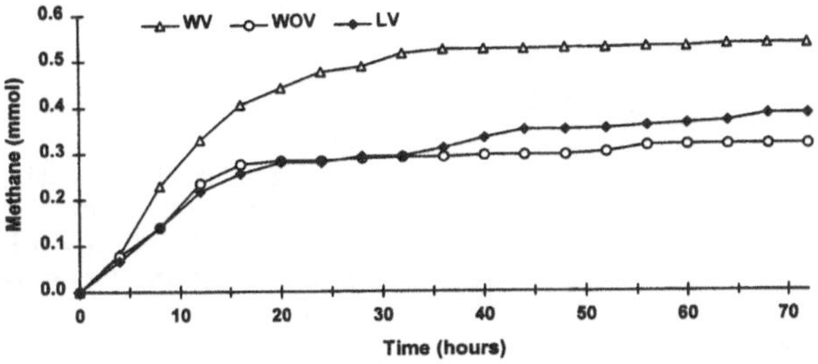


**Figure 4. Anaerobic biodegradation of glucose under three different culture media. So/Xo = 0.1 g glucose/ g VSS. Data were corrected with the endogenous blank.**

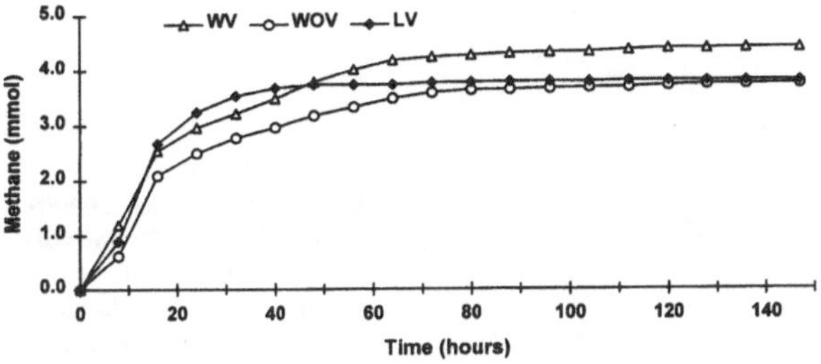


**Figure 5. Anaerobic biodegradation of glucose under three different culture media. So/Xo = 1.0 g glucose/ g VSS. Data were corrected with the endogenous blank.**

## Effect of the Medium Composition

In Figure 4, the results for the biodegradation of glucose with a So/Xo = 0.1 are presented. Most of the substrate was degraded within the first 15 hours of the test. A maximal methane production was obtained with the WV medium. Similar results were obtained with a So/Xo ratio of 1.0 (Figure 5), since the maximal methane production was obtained with the WV medium. Specific methane production rate was evaluated by computing the slope of the biodegradation curve (zero order model) and dividing it by the initial biomass concentration. We can observe in Table II that the specific methane production rate for the So/Xo ratio of 1.0 was higher than for the 0.1 ratio.

Table II presented the principal parameters obtained for the biodegradability test with glucose at 0.1 and 1.0 So/Xo ratios. It seems that the medium composition had no influence on the reaction rates and the extent of biodegradation of an easily biodegradable compound, since similar results were obtained for the three media and for the two So/Xo ratios.

In the case of phenol degradation (Figures 6 and 7) the maximal methane production was obtained with the WV medium for the 0.1 as well as for the 1.0 So/Xo ratio. For both So/Xo ratios a lag phase was observed (10 hr for So/Xo = 0.1 and 37 hr for So/Xo = 1.0). The lag phase was

**Table II. Results for the Anaerobic Biodegradability of Glucose Under Three Different Media**

| Parameter | So/Xo = 0.1 | | | So/Xo = 1.0 | | |
|---|---|---|---|---|---|---|
| Medium | WV | LV | WOV | WV | LV | WOV |
| Net production of $CH_4$, mmol | 0.53 | 0.38 | 0.32 | 4.40 | 3.80 | 3.70 |
| Specific reaction rate, mg $CH_4$/gVSS-h | 1.04 | 0.73 | 0.73 | 13.53 | 14.57 | 12.50 |
| % Biodegradability | 91 | 90 | 91 | 98 | 98 | 98 |

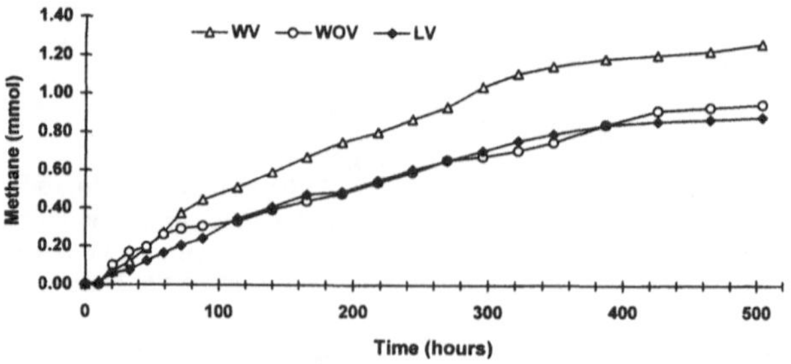


**Figure 6. Anaerobic biodegradation of phenol under three different culture media. So/Xo = 0.1 g phenol/ g VSS. Data were corrected with the endogenous blank.**

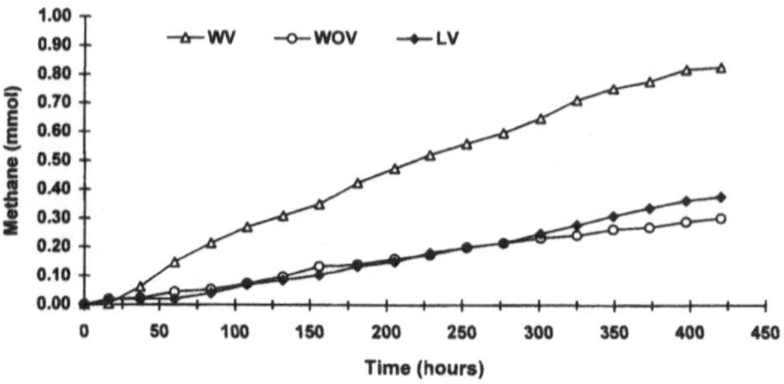


**Figure 7. Anaerobic biodegradation of phenol under three different culture media. So/Xo = 1.0 g phenol/ g VSS. Data were corrected with the endogenous blank.**

higher for the LV and WOV media and So/Xo = 1.0 than for the WV medium. It is important to mention that phenol concentration in the serum bottles was inferior to 100 mg/L in order to avoid toxicity effects.[13] The lag period showed for the phenol biodegradation could be considered as the time needed for the bacteria for their acclimation. The low lag period observed with the WV medium can be explained by looking at the medium composition (Table I). We can see that besides vitamins, yeast extract and casein peptone are found. These compounds can function as co-substrates, stimulating the phenol biodegradation.

Table III shows that not only the net production of methane was increased with the WV medium, but also the specific reaction rate and the extent of biodegradability were increased, regardless of the So/Xo ratio studied.

The objective for using the culture media in the biodegradability tests is to cover the physiological needs of the microorganisms and to avoid limitations. In this manner the biodegradation of a substrate is not affected. This point is specially important in the evaluation of industrial effluents which may contain toxic compounds. For a good stimulation of the anaerobic bacterial activity, the most important nutrients are, in a decreasing order: nitrogen, phosphorous, iron, cobalt, nickel, molybdenum, selenium, riboflavin and B12 vitamins.

**Table III. Results for the Anaerobic Biodegradability of Phenol Under Three Different Media**

| Parameter | So/Xo = 0.1 | | | So/Xo = 1.0 | | |
|---|---|---|---|---|---|---|
| Medium | WV | LV | WOV | WV | LV | WOV |
| Net production of $CH_4$, mmol | 1.28 | 0.89 | 0.95 | 0.83 | 0.38 | 0.31 |
| Specific reaction rate, mg $CH_4$/gVSS-h | 0.31 | 0.10 | 0.21 | 3.85 | 1.87 | 1.46 |
| % Biodegradability | 95 | 90 | 92 | 89 | 87 | 84 |

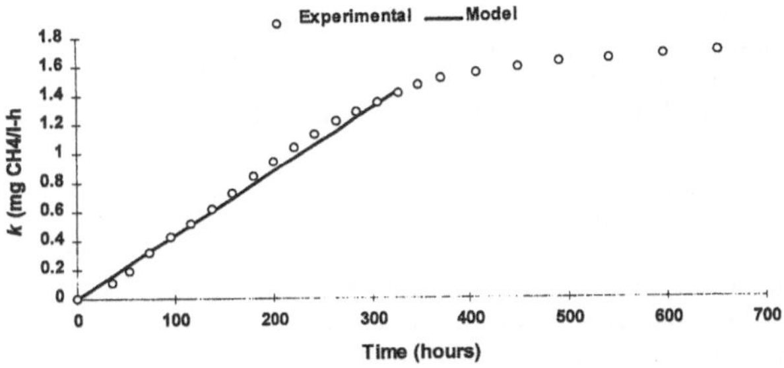


**Figure 8. Adjustment of Equation 1 to experimental data.**

The most important difference within the three media tested in this study are the concentration and presence of vitamins (see Table I). The WV medium has the highest quantity of vitamins, followed by the LV medium. The WOV did not contain any vitamin. Also, the LV and the WOV media did not contain yeast extract, casein peptone, lipoic acid, nitriloacetic acid, and cysteine. It is known that the vitamins and the yeast extract act as metabolic stimulants for the methanogenic bacteria. Yeast extract, cysteine, p-amino benzoic acid, and some amino acids are essential factors for growth.[14] Thus, methane production can be achieved in the absence of some elements such as the vitamins, but the methane rate is reduced as well as the extent of biodegradation, as was observed for the toxic substrate with the LV and the WOV media.

## Effect of the So/Xo Ratio

In order to analyze the results of the influence of the So/Xo ratio on the biodegradation test, a Modod-type model was adjusted to the methane production results (Equation 1).

$$\frac{d[CH_4]}{dt} = k\frac{[CH_4]}{([CD_4]+K_s)} \tag{1}$$

Where $CH_4$ = methane production concentration, mg $CH_4/L_{liquid}$
$k$ = reaction rate mg $CH_4/L_{liquid}$-h
$K_s$ = saturation constant, mg $CH_4/L_{liquid}$

The estimation of $k$ and $K_s$ was performed using a computer program written in Turbo Pascal IV, integrating first, by the 4$^{th}$ order Runge-Kutta procedure the Equation 1, and estimating next the kinetic parameters with the Hook and Jeev's method.[15] Figure 8 presents a typical result obtained and the theoretical model adjusted.

Table IV summarizes the biokinetic coefficients obtained with the model adjustment. Reaction time was computed by plotting $dCH_4/dt$ vs. t and taking the intersection with the ordinate. Specific methane production rate, $q$, was obtained by dividing $k$ by the initial biomass concentration, Xo. First of all, we can observe that shaking increases the reaction rate and decreases the reaction time, due to a better mass transfer. The maximal reaction rates were obtained for the highest

**Table IV. Biokinetic Coefficients Obtained**

| Initial Parameter | | | Without Shaking | | | With Shaking | | |
|---|---|---|---|---|---|---|---|---|
| So, mg/L | Xo, mg/L | So/Xo | $q$ mg $CH_4$ gVSS-h | $K_s$, mg/L | Reaction time, days | $q$ mg $CH_4$/ gVSS-h` | $k_s$, mg/L | Reaction time, days |
| 625 | 12500 | 0.05 | 0.26 | 1.5 | 9.6 | 0.32 | 1.76 | 5.0 |
| 625 | 2500 | 0.25 | 0.52 | 1.2 | 10.4 | 1.80 | 0.7 | 2.8 |
| 12500 | 12500 | 1.0 | 3.05 | 1.4 | 9.5 | 4.24 | 6.2 | 2.3 |
| 12500 | 2500 | 5.0 | 3.6 | 2.7 | 24.9 | 7.04 | 59.3 | 6.3 |

So/Xo ratio, regardless of whether shaking was switched. When the acetate concentration was increased and the biomass concentration decreased, as in the case of the So/Xo ratio of 5, the reaction time increased. It seems that a better affinity was obtained with the So/Xo ratio of 5 and shaking. In this case, the model presented in Equation 1 can be simplified with a zero order model.

It was found that the initial substrate concentration and the initial biomass concentration have a significant effect on the reaction rate. The biomass effect is negative; i.e., low biomass concentrations gave higher rates than high concentrations; on the contrary, substrates have a positive effect: a maximal reaction rate was obtained with the maximal substrate concentration. As we see in Table II, combining the substrate and biomass concentration effects, a relationship between specific methane production rate and the So/Xo ratio exists. This signifies the importance of taking into account this parameter for a suitable biodegradation test.

## CONCLUSIONS

Previous acclimation of the sludge to a substrate can stabilize the microbial community, maintaining the microbial activity and composition constants independently of the source of inoculum. In order to have a suitable biodegradability test, all the necessary components must be present. The presence of vitamins and oligoelements in the culture medium are very important factors that can maximize the quantity of biogas produced and increase the biodegradation rate, mainly if a toxic effluent is tested. The initial substrate concentration to initial biomass concentration ratio has a great influence on the results of the biodegradation test. Within the interval of So/Xo ratios studied in this work, a ratio of 5.0 gave the highest values, considered as the specific methane production rate. All these results indicate the importance of the choice of correct conditions under which the biodegradability tests are performed. It would be adequate to standardize the different methodologies used during the biodegradability tests in order to have better and reproducible results.

## REFERENCES


1,. Owen W.F., D.C. Stuckey, J.B. Healy, L.Y. Young and P.L. McCarty, "Bioassay for monitoring biochemical methane potential and anaerobic toxicity," *Water Res.*, 13, pp. 485–492, 1979.
2. Cohen A., "Effects of some industrial chemicals on anaerobic activity measurements by sequential automated methanometry (SAM)," *Water Sci. Technol.*, 25, pp. 11–20, 1992.
3. Pitter P. and J. Chudoba, *Biological Degradation of Organic Substances in the Aquatic Environment*, Chap. 2, CRC Press, Boca Raton, FL, 1989.
4. Chudoba P., J.J. Chevalier, J. Chang, and B. Capdeville, "Effect of anaerobic stabilization of activated sludge on its production under batch conditions at various So/Xo ratios," *Water Sci. Technol.*, 23, pp. 917–926, 1991.
5. Chang J., P. Chudoba, and B. Capdeville "Determination of the maintenance requirements of activated sludge," *Water Sci. Technol.*, 28(7), pp. 139–142, 1993.
6. Speece R.E., R.S. Engelbrecht, and D.R. Aukamp, "Cell replication and biomass in the activated sludge process," *Water Res.*, 7, pp. 361–374, 1973.
7. Chudoba P., B. Capdeville, and J. Chudoba, "Explanation of biological meaning of the So/Xo ratio on batch cultivation," *Water Sci. Technol.*, 26(3–4), pp. 743–751, 1991.
8. Moreno G., C. Sanchez, A. Noyola, L. Santiago, and G. Buitr6n, Automatizaci6n de una prueba de biodegradabilidad anaerobia," In: *Memorias del X Congreso Nacional de Ingeniería Sanitaria y Ambiental*, SMISAAC. Toluca, Mexico, 1996.
9. *Standard Methods for the Examination of Water and Wastewater*, 18th ed., APHA, AWWA, and WPCF, American Public Health Association, New York, 1992.
10. Young C.J. and N.K. Reza, Anaerobic Protocol for Assessing Industrial Waste Treatability, *Proceedings of the 50th Industrial Waste Conference*. Purdue University, West Lafayette, IN, 8–10 May, 1995.
11. Balch, W.E., G.E. Fox, L.J. Magrum, C.R. Woese, and R.S. Wolfe, "Methanogens: Reevaluation of a unique biological group," *Microbiol. Rev.*, 43, pp. 260–269, 1979.
12. Shelton D. and J.M. Tiedje, "General method for determining anaerobic biodegradation potential," *Appl. Environ. Microbiol.*, 47, pp. 850–857, 1984.
13. Fedorak P.M. and S.E. Hrudey, "Batch anaerobic methanogenesis of phenolic coal conversion wastewater," *Water Sci. Technol.*, 17, pp. 143–154, 1984.

14. Oremland R.S., Biogeochemistry of Methane Production, In: *Biology of Anaerobic Microorganisms*, A.J.B. Zehnder, Ed., John Wiley and Sons, New York, 1988, pp. 707–743.
15. Buitrón G., Biodegradation de composes xenobiotiques par un procédé discontinu de type SBR, Ph.D. thesis, No.254, Institut National des Sciences Appliquees de Tolouse, France, 1993.

## ACKNOWLEDGMENT

This research was partly supported by the Programa de Apoyo a las Divisiones de Estudios de Posgrado, Apoyo a Proyectos de Investigación, Tesis Doctorales through Grant No. 003328.

# 15 BIOLOGICAL TREATABILITY OF RDX-CONTAMINATED SOIL

**W. Craig Light**, Graduate Research Assistant
School of Civil and Environmental Engineering
Oklahoma State University

**Gregory G. Wilber**, Associate Professor

**William W. Clarkson**, Associate Professor
School of Civil and Environmental Engineering
Oklahoma State University
Stillwater, Oklahoma 74078

## ABSTRACT

Soils from the Department of Energy's Pantex plant in Amarillo, Texas were examined for biological removal of nitroaromatic contaminants, principally RDX. Bench-scale reactors were used to test treatability of soil slurries, water extracts of soil contaminants, and reagent nitroaromatic solutions. Results of these and numerous control reactors were analyzed to determine: relative differences in the rates of RDX desorption and transformation; extent and rate of biotransformation under aerobic, nitrate-reducing, sulfate-reducing, and methanogenic conditions; and effect of TNT on RDX bioconversion.

## INTRODUCTION

Hexahydro-1,3,5-trinitro-1,3,5-triazine (RDX) is used by the military in high-impact explosives. It is commonly found in wastewaters from munitions manufacturing plants and handling facilities. These process waters generally contain appreciable concentrations of 2,4,6-trinitrotoluene (TNT) as well. For example, RDX and TNT concentrations of 75 and 150 mg/L, respectively, have been found in process waters.[1] Improper disposal of these wastewaters can result in contamination of surface waters and soils. RDX primarily enters the soil by leaching from waste lagoons. Due to its low affinity for soil, RDX tends to migrate through the soil where seepage into groundwater can occur.[2,3]

Anaerobic biodegradation of RDX has been reported. McCormick et al.[4] reported essentially complete disappearance of RDX (50-100 mg/L) after 4 days using static batch reactors inoculated with anaerobic sewage sludge. Kitts et al.[5] observed RDX transformation by pure cultures isolated from explosives-contaminated soil. Funk et al.[6] reported successful removal of RDX (30 mg/L) from soil slurry batch reactors within 24 days with no identifiable intermediates.

Early studies by McCormick et al.[4] indicated that RDX was not amenable to microbial degradation under aerobic conditions. Similar results were also reported by Kitts et al.[5] and Smith-Simon and Goldhaber.[3] However, several recent studies have indicated aerobic transformation and/or mineralization of RDX under nitrogen-limited conditions.[7,8] RDX transformation and/or mineralization have also been reported in studies using aerobic composting techniques.[9–11]

RDX is one of several nitroaromatic compounds detected in soils at the Department of Energy (DOE) Pantex plant in Amarillo, Texas. Various areas of contaminated soils including playa lake, landfill, and ditches have been identified.[12] Concentrations of RDX range from 0.18 to 15,000 mg/kg in these contaminated soils. Although RDX levels are very high in portions of the Pantex soils, solubility in groundwater is limited to a range of about 38–42 mg/L,[3,13,14] limiting availability for biological conversion.

*52nd Purdue Industrial Waste Conference Proceedings*, 1997, Ann Arbor Press, Chelsea, Michigan 48118. Printed in U.S.A.

The proposed strategy to be employed in remediation of the soil at the DOE's Pantex facility is to excavate and treat the soil using ex-situ bioremediation. The method of interest is a prepared-bed or a bioslurry reactor process. Although biological degradation of RDX has been observed in laboratory studies, application of this approach to RDX-contaminated soil has not been fully developed. There also appear to be inconsistencies in the aforementioned studies, primarily in the ability of aerobic microorganisms to degrade RDX. Several other questions including the optimal conditions for RDX degradation and the fate of RDX and its intermediates are yet to be answered. Therefore, this study, consisting of a limited set of preliminary bench-scale experiments, was performed to define the effect of certain operating conditions on the efficiency of such a process for the bioconversion of RDX. The aim of this research was to study the biological treatability of RDX-contaminated soil under aerobic, nitrate-reducing, sulfate-reducing and methanogenic conditions.

The main objectives of this study were the following:

1. To study the ability of native soil bacteria to degrade RDX under various electron accepting conditions.
2. To study the use of a primary organic substrate to support cometabolic bioconversion of RDX.
3. To determine if competing substrates or sorption to soil particles limit the rate of RDX biotransformation.
4. To detect any possible inhibitory effects of RDX on sulfate-reduction and methanogenesis.

## MATERIALS

Soil samples from the DOE's Pantex plant in Amarillo, Texas were provided by the U.S. Army Corps of Engineers, Tulsa District. The soil samples (approximately 0.5 $ft^3$), collected from a landfill area at the Pantex site, were believed to contain RDX in the range of 10–100 mg/kg. However, analysis of this soil (as described below) indicated that the average initial concentrations of RDX, HMX, and TNT were considerably higher than expected. Average concentrations (and ranges of four samples) measured were approximately 5100 (4900–5300) ppm, 2100 (1800-2500) ppm, and 2700 (1025–5200) ppm for RDX, HMX, and TNT, respectively.

Reagent-grade RDX, HMX, and TNT (1000 μg/mL in acetonitrile or methanol) were purchased from either Chem Service (West Chester, PA) or Supelco (Bellefonte, PA). Other chemicals were of analytical grade and, unless stated otherwise, were obtained from Fisher Scientific, Chem Service, Sigma Chemical Company (St. Louis, MO), or Supelco.

## BIOLOGICAL REACTORS

Three primary series of experiments were performed, each including four sets of reactors under different electron-accepting conditions (aerobic respiration, nitrate reduction, sulfate reduction, and methanogenesis). The initial series of experiments used soil-water slurries of the Pantex soil. The purpose of these reactors was to simulate the actual operating conditions of the soil-slurry reactor with all its complexities. The second series of reactors contained no soil, but instead an aqueous extract from the Pantex soil. As such, these reactors contained all of the soluble constituents initially present in the first set of reactors, but did not contain a source of continuously desorbing or dissolving nitroaromatic contaminants. The third set of reactors was fed reagent-grade RDX as the only nitroaromatic constituent, the purpose being to further simplify the initial soil-water reactors by eliminating other nitroaromatics that might compete as substrate or otherwise inhibit RDX degradation.

All test reactors were set up using a concentrated enrichment medium and trace metal solution adapted from Manning et al.[15] and Shah.[16] The enrichment medium consisted of $NH_4Cl$ (0.40 g/L), NaCl (0.05 g/L), $CaCl_2$ (0.04 g/L), $MgCl_2$ (0.01 g/L), $NaHCO_3$ (0.50 g/L), $K_2HPO_4$ (1.40 g/L), and $KH_2PO_4$ (0.60 g/L). The trace metal solution consisted of $FeSO_4$-$7H_2O$ (200 mg/L), $ZnSO_4$-$7H_2O$ (10 mg/L), $MnCl_2$-$4H_2O$ (3 mg/L), $CoCl_2$-$6H_2O$ (20 mg/L), $CuCl_2$-$2H_2O$ (1 mg/L), $NiCl_2$-$6H_2O$ (2 mg/L), and $Na_2MoO_4$-$2H_2O$ (3 mg/L). Unless otherwise specified, each test reactor was also fed an initial dose of yeast extract (100 mg/L) to minimize the likelihood of nutrient limitations. Each reactor was also dosed with two to three drops of a concentrated resazurin solution to serve as a color indicator of the redox potential of the reactors. Note that all reactors were run in duplicate. The initial pH of each reactor was adjusted to 7.1 using either 10% HCl (hydrochloric acid) or NaOH (sodium hydroxide) as needed. Each reactor was incubated at ambient conditions (22°C) on a shaker table. Prior to being incubated, each anaerobic reactor (nitrate-reducing, sulfate-reducing, and methanogenic) was purged with either argon or nitrogen gas for 15–30 minutes, and the headspace in the reactor was also filled with the inert gas.

Within each set of reactors, some were fed an additional carbon substrate to investigate the usefulness of such an amendment. Lactate was added to the sulfate-reducing reactors and glucose to the other three types. The final concentration of the additional substrate was 1 g/L. All the reactors within a particular set were also amended with the appropriate electron acceptor (oxygen, nitrate, sulfate or bicarbonate). The aerobic reactors were continuously aerated with compressed air throughout the duration of the experiments, which required some volume adjustment periodically using distilled water. Nitrate and sulfate were added to the appropriate reactor sets to provide an initial concentration of 660 mg/L and 1000 mg/L, respectively. The concentrations of nitrate, sulfate, and the soluble substrate were monitored over time and reintroduced as necessary.

Each set of water extract and reagent-grade reactors was seeded with a 2% (v/v) inoculum from the soil slurry reactors with the corresponding redox condition. A sulfate-reducing culture could not be established from the Pantex soil. Therefore, the sulfate-reducing reactors were inoculated (2% v/v) with an active sulfate-reducing mixed culture established from digested sludge supernatant from the Stillwater, OK municipal wastewater treatment plant, and leachate and contaminated soil from an aquifer adjacent to the landfill at Norman, OK.

Abiotic control reactors were also set up for each set of reactors and monitored for comparison to the biologically active reactors. The control reactors were identical to the active reactors except that they were dosed with sodium azide to a final concentration of 4.6 mM to prevent biological activity.

### Soil Slurry Reactors

Aerobic reactors were prepared in 1-liter glass bottles with a reactor volume of 500 mL. The reactors were then dosed with 100 grams of RDX-contaminated soil from the Pantex site. All other soil-water reactors were prepared in 500-mL glass bottles with a reactor volume of 250 mL. These reactors were dosed with 50 grams RDX-contaminated soil from the Pantex site.

RDX in the aqueous-phase reached its solubility limit of about 40 mg/L within the first few days' incubation in all of the soil-water reactor sets. Aqueous-phase HMX and TNT were not monitored in these reactors, though their presence was noted. Initial solid-phase RDX, HMX and TNT concentrations were estimated to be the average concentrations determined for the Pantex soil.

### Water Extract Reactors

The water extract reactors were prepared using the aforementioned media and a water-extract from the Pantex soil. The extract was prepared by adding 600 grams of the RDX-contaminated soil to 3 liters of distilled water. The soil-water mixture was placed on a shaker table for 5 days before being allowed to settle. The supernatant was then collected and filtered through 0.45 μm glass fiber filters for use in these reactors.

The aerobic test reactors were set up in 1-liter bottles with a liquid volume of 500 mL. All other reactors were set up in 120-mL serum bottles with a liquid volume of 100 mL. The initial concentration of RDX in these reactors was close to its aqueous solubility of about 40 mg/L. The initial concentration of TNT in these reactors was approximately 70 mg/L. Aqueous-phase HMX was not monitored in these reactors.

### Reagent-Grade Reactors

Reagent-grade RDX reactors were prepared using the aforementioned media and spiked with reagent-grade RDX to a final concentration of 10 mg/L. The aerobic test reactors were set up in

1-liter bottles with a liquid volume of 500 mL. All other reactors were set up in 120-mL serum bottles with a liquid volume of 100 mL.

## ANALYTICAL METHODS

Before analysis, aqueous samples (4-5 mL) from each type of test reactor were filtered using Gelman 0.45 um Metricel (25 mm) membrane filters. Solid-phase extraction of munitions residues from the soils was conducted using the wrist-action shaker method as described by Jenkins and Grant.[17]

RDX and other nitroaromatic compounds were analyzed by high performance liquid chromatography (HPLC) using methods similar to those proposed by Bauer et al.[18] and Jenkins et al.[19] During this study, a Beckman liquid chromatograph equipped with two Model 127 solvent pumps, a Model 166 absorbance detector set at 254 nm and a System Gold controller was used. The mobile phase was methanol-water (45:55 v/v) at a flow rate of 1.5 mL/min. Samples (20 μL) were injected onto a Beckman C-18 reverse-phase column (Ultrasphere ODS, 5 um particle diameter, 4.6 mm x 25 cm column). The results were integrated on a Hewlett Packard (HP) 3396 Series II integrator. Analytical standards of reagent-grade RDX, HMX, and TNT were used for quantification of the reactor samples.

Methane production was monitored using a Gow-Mac Gas Chromatograph (GC) Series 350 with Thermal Conductivity Detector. Gas samples (250 μl) taken from the headspace of the reactors were injected onto a Gow-Mac Porapak column (80/100 mesh). Helium (He) at 60 mL/min was used as a carrier gas. The column temperature, detector temperature, and injector temperature were set to 55, 170, and 105°C, respectively. The bridge current was set to 70 mA. The results were integrated on a HP 3380A integrator. Quantification of the gas samples was made using pure methane gas.

Volatile Fatty Acids (VFA), particularly acetic and propionic acids, were analyzed by gas chromatography. Aqueous samples (1 μl) were injected on a packed glass column (Supelco 60/80 Carbopak C/0.3% Carbowax 20M/0.1% $H_3PO_4$, inner diameter 2 mm, length 24 in.) in a model 5890 HP Series II GC equipped with a flame ionization detector (FID). Helium (He) at 50 mL/min was used as a carrier gas. The column temperature, detector temperature, and the injector temperature were set to 120, 200, and 200°C, respectively. The results were integrated on a HP 3396 Series II integrator. Quantification was achieved using standard solutions of the VFA under consideration.

Nitrate and sulfate reduction were monitored using a Dionex ion chromatograph (IC), series 2000i/sp. The procedure used in this analysis is consistent with *Standard Methods*.[20] Diluted aqueous samples (30 μl) were injected onto a Dionex IonPac AS4A-SC 4 mm analytical column. The system was pressurized at 50 psi using compressed nitrogen gas. The eluent consisted of 1.8 mM $Na_2CO_3$ (sodium carbonate) and 1.7 mM $NaHCO_3$ (sodium bicarbonate) at a flow rate of 2.0 mL/min. The column regenerant was a 25 mN $H_2SO_4$ (sulfuric acid) solution. The results were integrated on a HP 3380A integrator. Quantification was made using standard solutions of known concentrations of the particular anion under consideration.

All other tests, unless otherwise stated, were performed in accordance to procedures recommended by *Standard Methods*,[20] with the exception of total dissolved organics, which were monitored as COD.[21]

## RESULTS AND DISCUSSION

Slurries of soil contaminated with approximately 5100 mg/kg RDX, 2100 mg/kg HMX, and 2700 mg/kg TNT were not entirely inhibitory to biological activity, as evidenced by the production of gas in those amended with additional carbon substrate and by the transformation of TNT. However, the rate of RDX biotransformation by native microorganisms is significantly slower than the rate of desorption/dissolution. Therefore, even in the presence of bioconversion, no measurable change in soluble RDX concentration occurred for some time. The net loss of aqueous RDX was negligible in all four soil slurry reactor sets, even after approximately one year of incubation. At the end of this study, the soils from the slurry reactors were extracted for solid-phase RDX, HMX, and TNT. Results from this extraction showed that significant loss of RDX from soils (almost to background equilibrium levels) occurred only in nitrate-reducing slurry reactors. These results are presented in a subsequent section.

## Aerobic Reactors

Although RDX did not appear to inhibit the growth of aerobic microorganisms (evident by the large amount of biomass produced in those reactors fed glucose), no significant loss of RDX was observed under aerobic conditions in any of the test reactors, despite their relatively long incubation periods (357 days for the soil-water reactors, 286 days for the water extract reactors, and 270 days for the reagent-grade RDX reactors). Biological activity was further indicated in some of these reactors by the transformation of other compounds, primarily TNT. TNT biotransformation was essentially complete after 50 days in both the aerobic soil slurry and water extract reactors that had been amended with glucose.

## Nitrate-Reducing Reactors

The nitrate-reducing reactors were the most consistent of any of the test reactors. Despite no net losses in aqueous RDX in the soil slurry reactors, the soil slurry reactors fed glucose maintained active nitrate-reduction throughout the duration of the experiment. Nitrate (660 mg/L) was consistently reduced in these reactors in approximately 7 to 10 days. However, no significant reduction of nitrate was observed in those reactors fed no additional carbon substrate during the 372 days these reactors were in operation.

Among the water extract reactors, RDX degraded quickest under nitrate reducing conditions. In one glucose-amended reactor, the RDX (42 mg/L) was completely transformed after about 70 days of incubation (Figure 1). The duplicate reactor started out similarly, but then RDX degradation and nitrate reduction stalled after 25 days. Subsequent addition of buffers to raise the pH, which had fallen to near 6, resulted in renewed degradation of RDX at a slower rate, but rapid nitrate-reduction was observed (Figure 2). This reactor was then split in half, and one of the new reactors (labeled Glucose: B2) was reinoculated (2% v/v) with an active nitrate-reducing inoculum. RDX quickly disappeared in the r-inoculated reactor (Figure 1).

Analysis of volatile fatty acids was performed on samples from the stalled reactor. Results indicated no significant buildup of acetic or propionic acid in this reactor. Additional experiments were conducted to determine if the reactors were poorly buffered. Three sets of reagent-grade reactors were set up and incubated as previously described except that the concentration of phosphate in each reactor was adjusted to 12, 25, or 50 mM. Each of the reactors (including the duplicates) was spiked with RDX (10 mg/L) and a 2% (v/v) active nitrate-reducing inoculum. In each of the test reactors, RDX was completely transformed in 25 days with relatively no difference in transformation rates. However, the pH of the reactors buffered with 12 mM phosphate increased

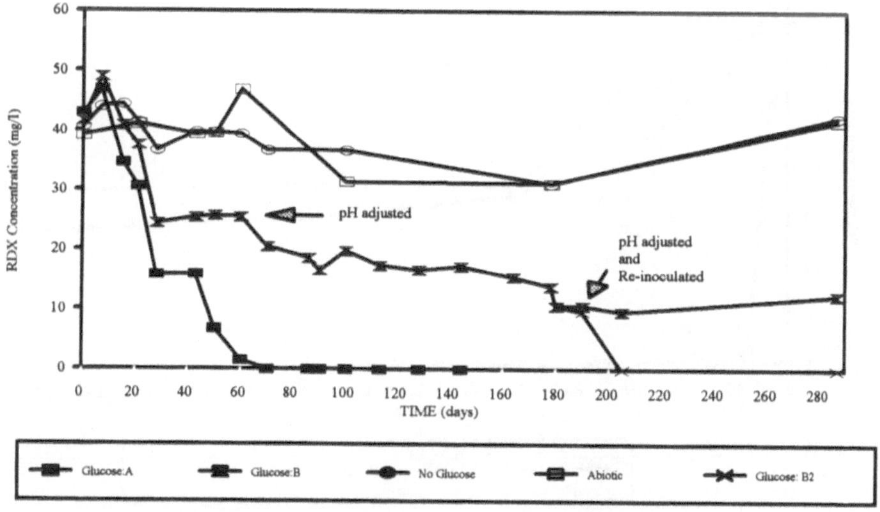


**Figure 1. RDX concentrations in nitrate-reducing 'water extract' reactors.**

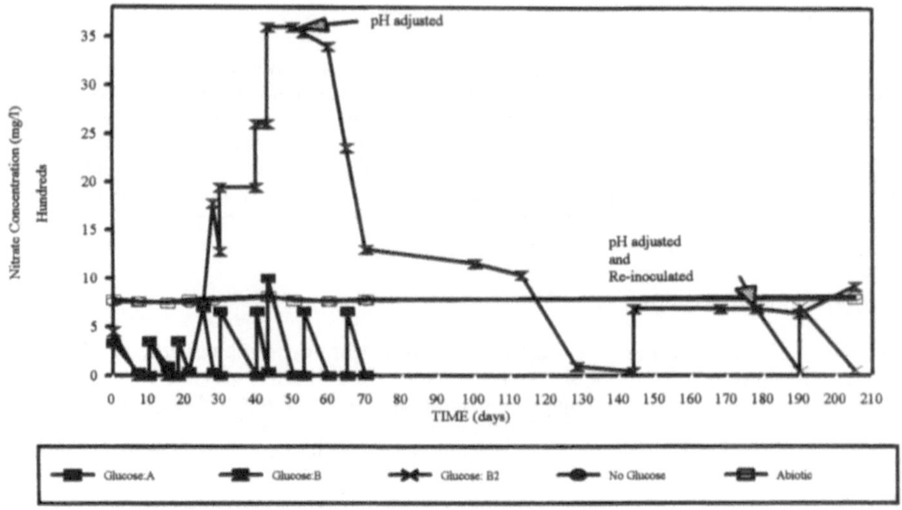


**Figure 2. Nitrate concentrations in nitrate-reducing 'water extract' reactors.**

from 7.1 to 7.6 in 25 days (corresponding to approximately three feeding cycles). The pH of the 25 mM and the 50 mM phosphate buffered reactors increased from 7.1 to 7.3 in the same time. Therefore, it is possible that the initial reactors (also set up with 12 mM phosphate) were poorly buffered, and it is recommended that future experiments be carried out using a minimum of 25 mM phosphate for buffering.

TNT transformation also occurred in the nitrate-reducing water extract reactors fed glucose. TNT transformation was essentially complete after 20 days of incubation (Figure 3) in both of the duplicate reactors. Those reactors which received no additional carbon substrate, as well as the abiotic control, showed some activity toward TNT. However, no significant reduction of nitrate was observed (Figure 2). The rates of TNT disappearance in the reactors fed no glucose and the control were similar, indicating TNT removal by some abiotic mechanisms. Possible abiotic

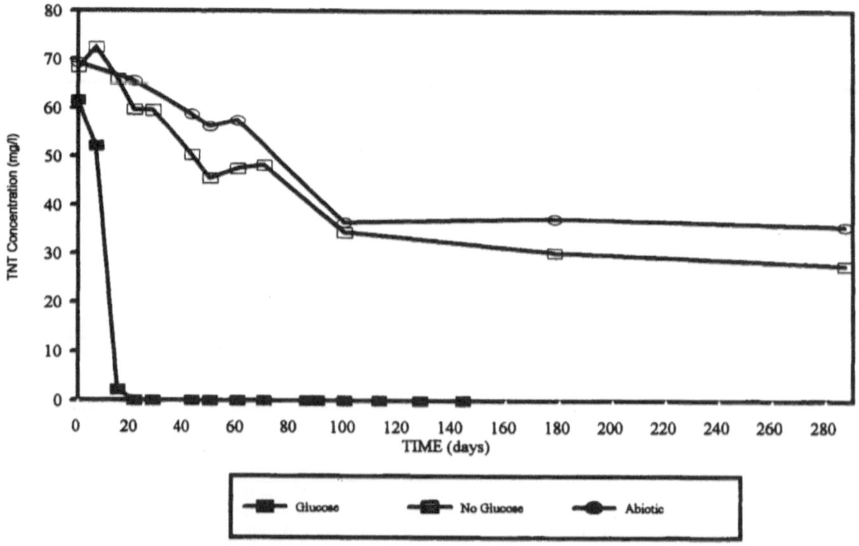


**Figure 3. TNT concentrations in nitrate-reducing 'water extract' reactors.**

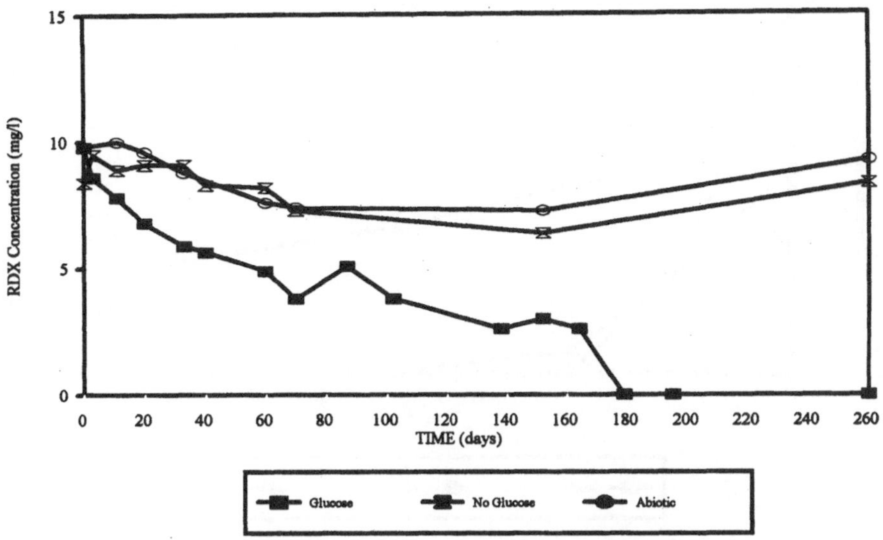


**Figure 4. RDX concentrations in nitrate-reducing 'reagent-grade RDX' reactors.**

mechanisms include reactions with the trace metal solution or UV (ultraviolet) photodegradation. Hydrolysis does not appear to be a viable explanation of RDX transformation in these reactors, based on studies by Semmens et al.[1] and Heilmann et al.,[22] which indicate that TNT, RDX, and HMX transformation by hydrolysis occurs only at elevated temperature and pH.

RDX (10 mg/L) was completely transformed in the nitrate-reducing reagent-grade reactors fed glucose in 120 days (Figure 4). The curves shown in Figure 4 are representative of both duplicate glucose-fed reactors. It is noteworthy that transformation of RDX in these reactors is slower than observed in the water extract reactors. The biomass concentration of the initial culture in the reactor was likely very small, which could explain some of the initial delay. A second possible explanation is the lack of complex substrate that was available in the water extract reactors, which in turn could limit the rate of biological growth in the reagent-grade reactors. However, based on the rate of gas production in these reactors, it was clear that nitrate-reducing organisms grew very rapidly. Very little change was observed in the reactors fed no additional glucose, as well as in the abiotic control (Figure 4).

Unlike the nitrate-reducing water extract reactors, both nitrate-reducing reagent-grade reactors fed glucose maintained active nitrate-reduction throughout the duration of the experiment. Nitrate (660 mg/L) was repeatedly reduced in these reactors in 7 to 10 days.

An experiment was conducted to determine if TNT degradation must precede that of RDX under nitrate-reducing conditions. Two sets of reactors were inoculated with nitrate-reducing bacteria from the active water extract reactor, spiked with reagent-grade RDX (10 mg/L), and incubated as previously described. RDX was completely transformed in both sets within 25 days. After this initial cycle, one set of reactors was spiked with RDX (11 mg/L) and the other with RDX (11 mg/L) and TNT (20 mg/L). The set spiked with only RDX showed continued activity toward RDX, while the other set of reactors showed little activity toward RDX until TNT was completely transformed. Typical results are depicted in Figure 5 and indicate that TNT is a preferred substrate in these systems under nitrate-reducing conditions. Therefore, any system designed to biologically treat this soil under nitrate-reducing conditions must incorporate the kinetics of TNT biotransformation.

## Sulfate-Reducing Reactors

Due to nitrate present in the Pantex soil samples, the sulfate-reducing soil slurry and water extract reactors operated at first under nitrate-reducing conditions. The initial concentration of

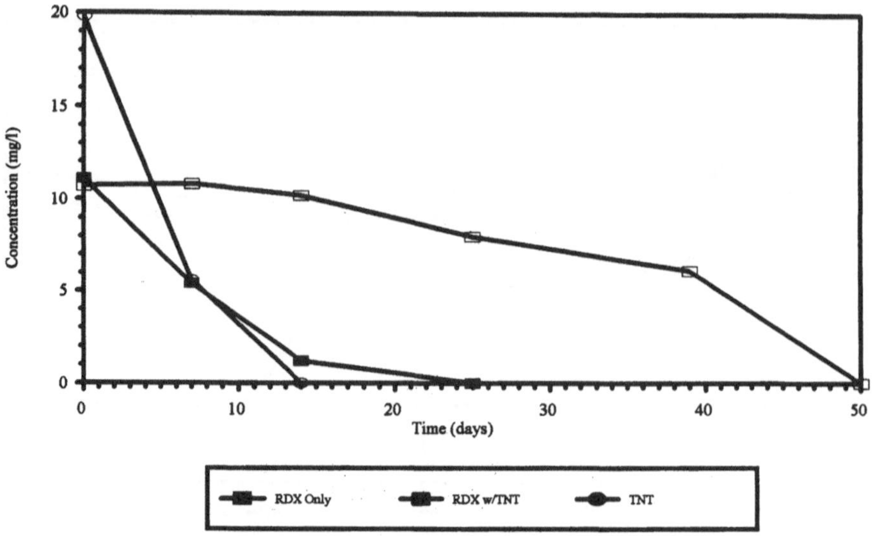


**Figure 5. Effect of TNT on RDX transformation under nitrate-reducing conditions.**

aqueous-phase nitrate in these reactors was approximately 120 mg/L. This concentration was depleted in approximately 10 days in both reactor sets where an additional substrate (lactate) had been added. No significant reduction of nitrate was observed in those reactors fed no additional substrate. This was consistent with the actual nitrate-reducing set of reactors fed no additional substrate. Although nitrate was reduced quickly in the sulfate-reducing soil slurry and water extract reactors where an additional carbon substrate was amended, those reactors failed to produce the desired electron-accepting conditions. No measurable reduction of sulfate was observed in either set of reactors. This was true even after being spiked several times with an active sulfate-reducing inoculum. The lactate-fed reactors continued to produce gas during the duration of the experiment, indicating some sort of biological activity. It is possible that something else, possibly iron or manganese, acted as a preferred electron acceptor in these reactors. No data on soil composition were available to confirm the presence of these metals in the Pantex soil, although it is likely that significant concentrations are present.

Although no loss of RDX was observed in either the soil slurry or the water extract reactors, TNT (70 mg/L) was completely transformed in 40 days in the water extract reactors fed lactate. Initial reduction of TNT occurred along with the reduction of nitrate that was present in the water extract (120 mg/L). Some reduction of TNT was also observed in the reactors fed no lactate. TNT (70 mg/L) was reduced to below 10 mg/L in 284 days. In the control reactor, TNT was also reduced from 70 mg/L to 38 mg/L in 284 days. Possible abiotic mechanisms for TNT loss in the control reactor were described in the previous section.

In the reagent-grade RDX reactors, RDX (10 mg/L) was completely transformed after 75 days in those reactors fed lactate (Figure 6). The curves shown in Figure 6 are representative of both duplicate reactors. Approximately 20% of the RDX was transformed in 165 days in both the reactors fed no lactate and the abiotic control. Possible reasons for the loss of RDX are as previously described for TNT. In this set of experiments, the concentration of yeast extract in those reactors fed no lactate was reduced to less than 10 mg/L from 100 mg/L used in previous sulfate-reducing unamended control reactors that showed significant reduction of RDX, presumably because these organisms were using the yeast extract as a primary substrate rather than as a source of minor nutrients as intended.

Although RDX was transformed in these reagent-grade reactors, no significant reduction of sulfate was observed (Figure 7). However, while the measured sulfate removal in the organic-amended reactors was small (relative to the amount of sulfate present), a definite sulfide odor was present, as was darkening of the reactors, indicating active sulfate reduction in those reactors fed lactate. A biologically active control reactor fed no RDX showed complete sulfate-reduc-

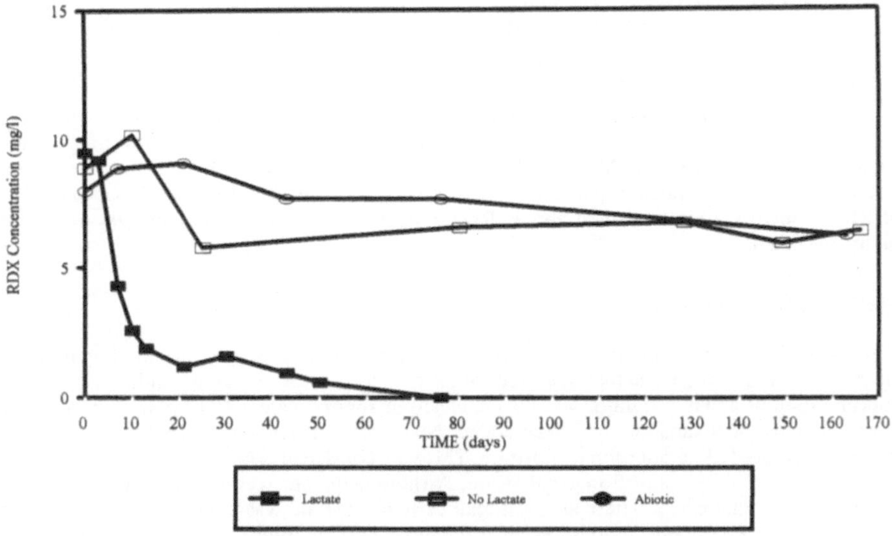


**Figure 6. RDX concentrations in sulfate-reducing 'reagent-grade RDX' reactors.**

tion (at similar concentration) during this time period (Figure 7). As indicated in Figure 7, measurable sulfate reduction did not occur until RDX had been completely transformed. This indicates some inhibition of sulfate-reducing microorganisms by the presence of RDX.

In subsequent experiments using reagent-grade reactors with sulfate concentrations of 50 mg/L and spiked with a 2% (v/v) inoculum from the previous reactors, RDX (10 mg/L) was completely transformed in 5 days with no measurable reduction of sulfate, although a definite sulfide odor was present. Using this same approach, RDX (40 mg/L) was reduced by 98% after 175 days of incubation. Again, while no measurable reduction of sulfate was observed, a definite sulfide

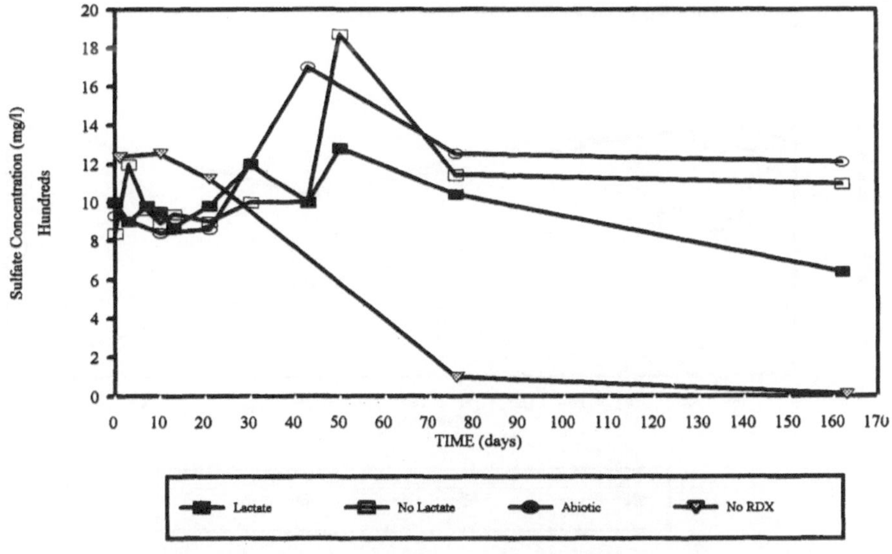


**Figure 7. Sulfate concentrations in sulfate-reducing 'reagent-grade RDX' reactors.**

odor was present. The fact that sulfate is being slightly reduced and the RDX is degrading significantly in these reactors (in which no other nitroaromatics are present) suggests that the sulfate-reducing organisms were further inhibited in those reactors which contained significant amounts of nitroaromatics other than RDX.

An additional experiment was conducted to investigate if RDX reacts directly with sulfide. Autoclaved reagent-grade reactors spiked with RDX (10 mg/L) and sulfide (100 mg/L) showed some activity toward RDX. After 30 days of incubation, less than 40% of RDX had been reduced. These results indicate that RDX reacts slowly with sulfide. However, based on the rates of RDX transformation presented above, RDX reduction in these reactors appears to be microbially enhanced.

## Methanogenic Reactors

As with the sulfate-reducing soil slurry and water extract reactors, the methanogenic soil slurry and water extract reactors operated initially under nitrate-reducing conditions. The initial concentration of about 120 mg/L was also depleted in approximately 10 days in both reactor sets where an additional substrate had been added. No significant reduction of nitrate was observed in those reactors fed no additional substrate. This was consistent with the actual nitrate-reducing set of reactors fed no additional substrate. Although nitrate was reduced quickly in these methanogenic reactor sets where an additional carbon substrate was amended, the soil slurry reactors failed in producing the desired electron-accepting conditions. No measurable concentration of methane was observed during the 337 days the reactors were incubated, although these reactors continued to produce gas during the duration of the experiment. Besides analysis for methane, aqueous samples from the methanogenic reactors fed an additional substrate were analyzed for volatile fatty acids. Results showed the presence of both acetic and propionic acids, indicating active fermentation in these reactors.

Despite no loss of RDX or production of methane in the soil slurry reactors, these were observed in the methanogenic water extract reactors fed glucose. RDX (42 mg/L) was reduced to approximately 2 mg/L in 284 days (Figure 8). Methane production was first confirmed in the glucose-fed reactors after 25 days of incubation. No significant reduction of RDX nor measurable production of methane was observed in those reactors fed no glucose or the abiotic control reactor.

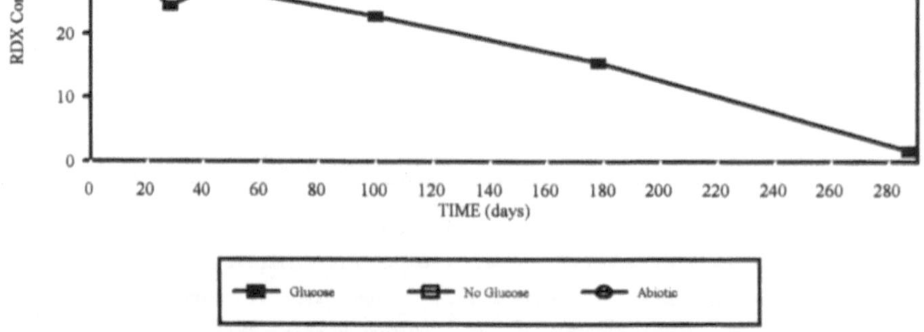


**Figure 8. RDX concentrations in methanogenic 'water extract' reactors.**

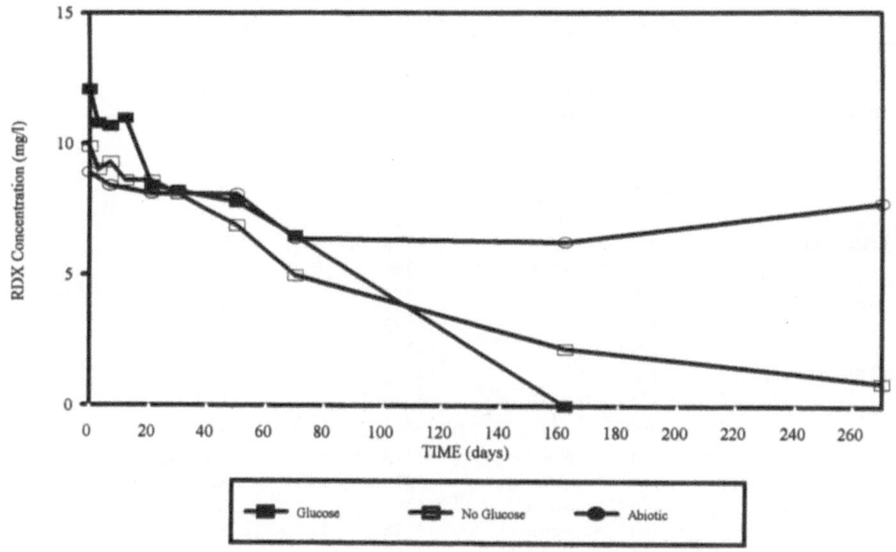


**Figure 9. RDX concentrations in methanogenic 'reagent-grade RDX' reactors.**

TNT (70 mg/L) was completely transformed in those reactors fed glucose in approximately 20 days. However, initial reduction of TNT corresponded with the reduction of nitrate that was present in the water extract. Some reduction of TNT was also observed in both the reactors fed no glucose and the abiotic control. The rate of TNT transformation in these reactors was consistent with that of the sulfate-reducing reactors discussed above.

In the reagent-grade reactors, complete transformation of RDX (12 mg/L) was also observed in those reactors fed additional glucose after approximately 160 days (Figure 9). Methane production was confirmed after 50 days. RDX (10 mg/L) was also reduced by 80% in those reactors fed no glucose in 265 days. However, no measurable amount of methane was produced. Since RDX was the only available organic substrate in these reactors and its concentration was small, measurable methane production was not expected. Some reduction of RDX (from 8.9 mg/L to 7.8 mg/L) was also observed in the control reactor. The rate of RDX transformation in the abiotic control reactor was consistent with that of sulfate-reducing reactors discussed above. As such, RDX appears to be reduced in these reactors due to similar, unidentified abiotic mechanisms.

Since a highly active methane producing culture was not established from the Pantex soil, additional experiments were conducted to determine if RDX inhibits methanogenesis. Reactors inoculated with anaerobic digester sludge and spiked with RDX (10 mg/L) produced no measurable amount of methane after one week of incubation. Control reactors produced approximately 15 mL of 12% methane gas in the same period. Note that the percent methane in the headspace in the reactors where RDX was not present is somewhat lower than would be expected. It is possible that some factor other than RDX also limited methane production in these reactors. Methane production appears to be not completely inhibited, but slowed in the presence of RDX, as indicated by the water extract and reagent grade reactors discussed previously.

## Soil Slurry Reactors Revisited

At the conclusion of the soil slurry experiments, a solid-phase extraction was performed on soils from each of the slurry reactors. Results from this extraction are summarized in Table I. The average initial concentrations of RDX, HMX, and TNT in these soils were 5100 ppm, 2100 ppm, and 2700 ppm, respectively. As mentioned previously, initial concentrations of solid-phase RDX, HMX, and TNT ranged from 4900–5300, 1800–2500, and 1025–5200 mg/kg, respectively.

Note that some loss of RDX and TNT present in the solid-phase occurred in each reactor due to the solubility of these compounds in water. Based on the maximum aqueous solubility of RDX

**Table I. Summary of Soil-Extraction Results**

| Set | Reactor Description | RDX (ppm) | HMX (ppm) | TNT (ppm) |
|---|---|---|---|---|
| | Glucose: A | 1776 | 1455 | nd |
| | Glucose: B | 4069 | 1470 | nd |
| Aerobic | No Glucose: A | 4663 | 1466 | nd |
| | No Glucose: B | 1722 | 1674 | nd |
| | Abiotic | 1800 | 1666 | nd |
| | Glucose: A | 66 | 1555 | nd |
| | Glucose: B | 14 | 1367 | nd |
| Nitrate-reducing | No Glucose: A | 3875 | 2074 | 46 |
| | No Glucose: B | 4364 | 2096 | 116 |
| | Abiotic | 3971 | 1635 | 187 |
| | Lactate: A | 3132 | 2209 | nd |
| | Lactate: B | 2775 | 1452 | nd |
| Sulfate-reducing | No Lactate: A | 4350 | 2110 | 42 |
| | No Lactate: B | 4420 | 1913 | 46 |
| | Abiotic | 4862 | 2560 | 266 |
| | Glucose: A | 3516 | 960 | nd |
| | Glucose: B | 3276 | 805 | nd |
| Methanogenic | No Glucose: A | 4394 | 1300 | 22 |
| | No Glucose: B | 4176 | 1661 | 10 |
| | Abiotic | 3742 | 1245 | 191 |

nd = not detectable (TNT detectable limit = 5 ppm

(42.3 mg/L),[3,14] the estimated loss of solid-phase RDX in these reactors due to dissolution of the compound into the aqueous media would be 212 ppm. Similarly, for soluble TNT (130 mg/L),[16] the estimated loss of solid-phase TNT would be 650 ppm.

Although it is possible that some solid-phase RDX was transformed in all of the soil slurry reactors including the abiotic controls, the only reactors that showed significant loss of solid-phase RDX were the nitrate-reducing reactors fed glucose (Table I). The loss of solid-phase RDX in these reactors based on the average initial concentration was 99%. Note that TNT was completely transformed in these nitrate-reducing reactors as well. By far, this appears to be the most promising mode of operation for soil slurry reactor treatment of soils contaminated with RDX.

Results from the solid-phase extractions listed in Table I also indicate that TNT was readily transformed in each of the reactors where additional carbon substrate was amended. TNT was no longer detectable in the aqueous-phase of these reactors as well. The incubation times required for the complete disappearance of TNT in the aqueous-phase of the substrate-amended reactors are summarized in Table II. As mentioned previously, the sulfate-reducing and methanogenic reactors operated initially under nitrate-reducing conditions due to nitrate in the Pantex soil.

No substantial reduction of solid-phase HMX was observed in these reactors even after approximately one year of operation. Due to the low solubility of HMX in water (5 mg/L at 25°C)[13] and its early HPLC retention time (2-3 minutes) based on the mobile-phase used in these experiments, the aqueous-phase HMX could not be monitored successfully.

**Table II. Incubation Times for Disappearance of Aqueous TNT**

| Reactor Description | Time (days) |
|---|---|
| Aerobic | 50 |
| Nitrate-reducing | 14 |
| Sulfate-reducing | 28 |
| Methanogenic | 28 |

## CONCLUSIONS

Based on the results of this study, the following conclusions can be drawn:

- Although TNT was successfully transformed in all soil-water reactors where additional carbon substrate was amended, significant reduction of solid-phase RDX was only observed under nitrate-reducing conditions, indicating that this is the most promising mode of operation for soil slurry reactor treatment of soils contaminated with RDX.
- In the nitrate-reducing soil slurry reactors, initial solid-phase RDX concentration of 5100 ppm (average of 4 samples) were reduced to the equivalent of aqueous concentration limits of 40 ppm (average of 2 samples) in the presence of other nitroaromatic contaminants after 372 days of incubation on a shaker table at 22°C.
- TNT acts as a preferred substrate under nitrate-reducing conditions. Significant reduction of RDX does not occur until TNT is completely transformed. Furthermore, based on results from the solid-phase extractions, RDX must likewise be transformed before HMX.
- Slurries of soils containing approximately 5100 ppm RDX, 2100 ppm HMX, and 2700 ppm TNT are not inhibitory to biological activity. However, sulfate- reducing and methanogenic bacteria do appear to be inhibited by the complex mixture of nitroaromatics in the Pantex soil.
- RDX transformation proceeds quickest under sulfate-reducing conditions when RDX is the only nitroaromatic present. However, reduction of sulfate appears to be somewhat inhibited by the presence of RDX. Significant reduction of sulfate does not occur until RDX is completely transformed.
- Methane production appears to be inhibited in the presence of RDX. Methanogensis is possible at aqueous RDX concentrations of 10 mg/L. However, the rate of methane production is slower than in control reactors where no RDX is present.

## REFERENCES


1. Semmens, M.J., Barnes, D., and O'Hara, M. "Treatment of an RDX-TNT waste from a munitions factory." *Proceedings of the 39th Purdue Ind. Waste Conf.,* West Lafayette, IN, May 8–10, 1984, pp. 837–842, 1985.
2. Haderlein, S.B., Weissmahr, K.W., and Schwarzenbach, R.P. "Specific adsorption of nitroaromatic explosives and pesticides to clay minerals." *Environ. Sci. Technol.* 30, pp. 612–622, 1996.
3. Smith-Simon. C., and Goldhaber, S. Toxicology Profile for RDX. Prepared by: Research Triangle Institute Under Contract No. 205-93-0606 for the U.S. Department of Health and Human Services Agency for Toxic Substances and Disease Registry, 1995.
4. McCormick, N.G., Cornell, J.H., and Kaplan, A.M. "Biodegradation of hexahydro-1,3,5-trinitro-1,3,5-triazine." *Appl. Environ. Microbiol.* 42(5), pp. 817–823, 1981.
5. Kitts, C.L., Cunningham, D.P., and Unkefer, P.J. "Isolation of three hexahydo-1,3,5-trinitro-1,3,5-triazine degrading species of the family *Enterobacteriaceae* from nitramine explosive-contaminated soil." *Appl. Environ. Microbiol.* 60(12), pp. 4608–4711, 1994.
6. Funk, S.B., Roberts, D.J., Crawford, D.L., and Crawford, R.L. "Initial-phase optimization for bioremediation of munition compound-contaminated soils." *Appl. Environ. Microbiol.* 59(7), pp. 2171–2177, 1993.
7. Binks, P.R., Nicklin, S., and Bruce, N.C. "Degradation of hexahydro-1,3,5-trinitro-1,3,5-triazine (RDX) by *Stenotrophomonas maltophilia* PB1." *Appl. Environ. Microbiol.* 61(4), pp. 1318–1322, 1995.
8. Jones, A.M., Greer, C.W., and Hawari, J. "Highly energetic organonitro pollutants: Assessment of the aerobic biodegradation potential of RDX, TNT, GAP, and NC." *J. Cell. Biochem.* S21A S21A, p. 48, 1995.
9. Williams, R.T., and Myler, C.A. "Bioremediation using composting." *Biocycle* Nov., pp. 78–82, 1990.
10. Bayman. P., Ritchey, S.D., and Bennett, J.W. "Fungal interactions with the explosive RDX (hexahydro-1,3,5-trinitro-1,3,5-triazine)." *J. Ind. Microbiol.* 15, pp. 418–423, 1995.
11. Doyle, C.D., Kitchens, J.F., and Erickson, M.D. "Composting of soils contaminated with explosives." *Proceedings of the 1992 Federal Environmental Restoration Conference and Exhibition,* pp. 373–376, 1992.

12. U.S. Army Corps of Engineers, Tulsa District. Jacobs Engineering Group, Inc., October 1994 Draft Preliminary Data Package for Construction Landfill 3, Pantex Plant, Amarillo, TX (SWMU 54), 1994.
13. Gorontzy, T., Drzyzga, O., Kahl, M.W., Bruns-Nagel, D., Breitung, J., von Loew, E., and Blotevogel, K-H. "Microbial degradation of explosives and related compounds." *Crit. Revs. Microbiol.* 20(4), pp. 265–284, 1994.
14. McLellan, W., Roberts, W.C., and Hartley, W.R. Health Advisory for Hexahydro-1,3,5-trinitro-1,3,5-triazine (RDX). In: *Drinking Water Health Advisory: Munitions.* Lewis Publishers, Chelsea, MI, 1992, pp. 133–180.
15. Manning, J.F., Jr., Boopathy, R., and Kulpa, C.F. A Laboratory Study in Support of the Pilot Demonstration of a Biological Soil Slurry Reactor. U.S. Army Environmental Center, Report No. SFIM-AEC-TS-CR-94038, 1995.
16. Shah, S.R. Anaerobic Degradation Intermediates of 2,4,6-trinitrotoluene (TNT) under Different Electron Accepting Conditions. M.S. Thesis, Oklahoma State University, 1995.
17. Jenkins, T.F., and Grant, C.L. "Comparison of extraction techniques for munitions residues in soil." *Anal. Chem.* 59, pp. 1326–1331, 1987.
18. Bauer, C.F., Grant C.L., and Jenkins, T.F. "Interlaboratory evaluation of high-performance liquid chromatographic determination of nitroorganics in munition plant wastewater. *Anal. Chem.* 58, pp. 176–182, 1986.
19. Jenkins, T.F., Leggett, D.C., Grant, C.L., and Bauer, C.F. "Reversed-phase high-performance liquid chromatographic determination of nitroorganics in munitions wastewater." *Anal. Chem.* 58, pp. 170–175, 1986.
20. *Standard Methods for the Examination of Water and Wastewater*, 17th ed,, American Public Health Association, Washington, D.C., 1989.
21. *HACH Water Analysis Handbook*, 2nd ed., HACH Company, Loveland, CO, 1996.
22. Heilmann, H.M., Wiesmann, U., and Stenstrom, M.K. "Kinetics of the alkaline hydrolysis of high explosives RDX and HMX in aqueous solutions and adsorbed to activated carbon." *Environ. Sci. Technol.* 30, pp. 1485–1492, 1996.

# 16 BIODEGRADATION POTENTIAL OF ALKYLBENZENES AND PHENOLS IN LANDFILLS

**Yu-Sheng Wang**, Doctoral Candidate

**Morton A. Barlaz**, Associate Professor
Department of Civil Engineering
North Carolina State University
P.O. Box 7908, Raleigh, North Carolina 27695

## ABSTRACT

This research evaluated the anaerobic biodegradability of benzene, toluene, yethylbenzene, xylenes (BTEX), phenol, and p-cresol using microbial consortia derived from decomposing refuse samples. The effects of environmental conditions, pertinent to landfills, on phenol and p-cresol degradation were also studied. This study demonstrated that landfills harbor anaerobic microorganisms capable of degrading toluene, phenol, and p-cresol. In one consortium, phenol biodegradation was stimulated by both cellulose and nutrient additions and degradation was optimal at pH 6.5 to 7.0. An accumulation of the fermentation intermediates, benzoate and acetate, inhibited phenol biodegradation. The results show that it is possible to design landfill management strategies to promote contaminant biodegradation.

## INTRODUCTION

Prior to 1980, there were few restrictions on the types and quantities of industrial waste buried with municipal solid waste (MSW) in landfills. As a result of past waste disposal practices, groundwater downgradient of uncontrolled landfills is frequently contaminated by leachate and poses environmental hazards. Thus, it is not surprising that approximately 25% of the sites on the EPA's National Priorities List of cleanup sites are landfills that accepted both hazardous and municipal solid wastes.[1] Benzene, toluene, ethylbenzene, xylenes (BTEX) and phenolic compounds are among the more common contaminants in landfill leachate.[2,3] Benzene, as well as several substituted phenols, including chlorophenols and cresols, are designated priority pollutants by the USEPA.

Landfills represent an active anaerobic ecosystem in which the conversion of cellulosic substrates to methane and carbon dioxide is well documented.[4] The long residence time of refuse buried in landfills enhances the opportunity for microbial degradation of recalcitrant compounds. However, there has been only limited research on the fate of trace organic pollutants in landfills. BTEX has been shown to biodegrade under denitrifying, sulfate-reducing, and methanogenic conditions in habitats that include soil, aquifer sediment, and wastewater treatment sludge.[5–7] However, to our knowledge the only research regarding BTEX in landfills was conducted by Cossu and Serra (1989)[8] who showed that the co-disposal of BTEX bearing sludge was not inhibitory to refuse methanogenesis. BTEX biodegradation was not documented.

Similarly, there has been only limited research on the biodegradation of phenols in the landfill ecosystem.[9] Reinhart and Pohland (1991)[10] reported on the attenuation of some priority pollutants, including two phenolic compounds, using a laboratory-scale refuse column. The importance of biological processes to compound attenuation was not specifically addressed in their study. Watson-Craik and Senior (1989)[11] have reported phenol biodegradation in refuse columns operated with leachate recycle. Anaerobic degradation of o-cresol was also measured by microbial consortia enriched from refuse.[12]

*52nd Purdue Industrial Waste Conference Proceedings*, 1997, Ann Arbor Press, Chelsea, Michigan 48118. Printed in U.S.A.

The objectives of this research were to investigate (1) whether landfills harbor anaerobic microorganisms capable of BTEX, phenol, and p-cresol degradation and (2) the effects of environmental conditions, pertinent to landfills, on phenol degradation.

## MATERIALS AND METHODS

### Experimental Design

In the first phase of this research, the degradative ability of decomposing refuse samples toward BTEX, phenol, and p-cresol was tested for six samples. Tests were conducted in serum bottles containing anaerobic media, a liquid extract of each refuse sample as the inoculum, target contaminants, and other additions in some treatments. Tests were conducted under several conditions including: (1) BTEX; (2) BTEX plus ground refuse; (3) BTEX plus phenol, p-cresol, and benzylalcohol; (4) phenol, p-cresol, and benzylalcohol; and (5) abiotic controls corresponding to treatments 1, 2, and 4.

Serum bottles in which a significant toluene loss was measured relative to abiotic controls were dosed with [ring-$^{14}C$] toluene (Sigma, St. Louis, MO). Fresh abiotic controls were initiated in parallel. The total of two $^{14}C$ spikes to each bottle provided about 1.894 $10^7$ dpm and the final toluene concentrations after spiking were about 1368 µg/L. Degradation of the $^{14}C$-toluene spike in the live bottles was monitored by gas chromatograph (GC) analysis of a liquid sample and comparison to the abiotic controls. At the end of the incubation, $^{14}C$ activity in both the liquid and headspace was analyzed for $^{14}CO_2$ and $^{14}CH_4$.

The results of initial screening studies showed broad degradative activity toward phenolics. Therefore, a refuse sample was selected for further study of phenol degradation. A phenol-degrading enrichment, MR-P, was developed and maintained on phenol (0.3 mM) as the sole substrate for more than 6 months. The effects of environmental conditions on phenol degradation were then tested using this enrichment culture as the inoculum. Conditions tested included the presence of alternate carbon sources, pH, and the accumulation of degradation intermediates.

As landfills contain a wide variety of cellulose-based carbon sources, experiments were conducted to evaluate the effects of added carbon on contaminant biodegradation. Carbon substrates tested initially included fresh refuse (0.3 g), decomposed refuse (0.5 g), and filter paper (Whatman No.1, 0.1 g). Based on the results of these treatments, additional tests were conducted with a refuse liquid extract, cellobiose (0.3 mM), acetate (30 mM as sodium acetate), and yeast extract (0.05 % w/v). A phenol only control was used in every test.

Refuse decomposition typically proceeds through an acid phase in which the pH is between 5 and 6.[4] Thus, the effect of pH on phenol biodegradation was tested over the range of pH 5.6 to 7.7. Because the pH adjustment resulted in elevated concentrations of $Na^+$ or $Cl^-$, controls containing similar $Na^+$ or $Cl^-$ concentrations were prepared with NaCl at pH 7.

During the acid phase of refuse decomposition, fermentation intermediates typically accumulate. To evaluate the importance of intermediates on contaminant degradation, phenol degradation was tested in the presence of benzoate and acetate. Treatments included: (1) phenol; (2) phenol, benzoate and acetate; and (3) phenol, benzoate, acetate, and bromoethanesulfonic acid (BESA). BESA is an inhibitor of methanogenic bacteria. Constituent concentrations were: phenol 0.32 mM, benzoate 1.3 mM, acetate 16.6 mM, and BESA 1mM. The benzoate concentration was selected based on preliminary work in which phenol degradation was inhibited at benzoate concentrations greater than 1 mM. The acetate concentration was selected on the basis of a previous study in which the addition of 20 mM sodium acetate resulted in inhibition of benzoate degradation.[13]

### Inocula Source

Microbial inocula tested in the first phase were formed from two laboratory refuse reactors and four landfill samples. One of the laboratory reactors contained a mixture of grass and branches while the other contained mixed residential refuse. The refuse in both reactors had been producing methane for more than one year and was well decomposed based on its minimal methane production rate. Four refuse samples were excavated from landfills in High Point and Greensboro, NC. Refuse samples were excavated by a backhoe from approximately 10 feet below the cover soil. In the field, a portion of the excavated sample was packed into a mason jar full of anaerobic water for later analysis. Leachate from the two refuse reactors as well as the liquid extracts of four excavated samples were analyzed for pH, total aromatics (mg/l as p-cresol), BTEX, phenol, and p-cresol. Within 24 hr of excavation, about 1.5 kg of each landfill sample was packed into a 2-liter plastic refuse reactor and sparged with 80:20 $N_2/CO_2$. Before inoculum formation, all reactors were operated with leachate recycle and gas collection until methane production was observed.

## Serum Bottle Assays

Refuse inocula were formed by blending approximately 400 gm of refuse in 1 liter of 24 mM anaerobic phosphate buffer as described previously.[4] After blending, 15 mL of this mixture was inoculated into 85 mL of defined anaerobic medium in a 125-mL serum bottle. The medium contained (mL solution per liter medium): $PO_4$ solution (100), M3 solution (100), trace mineral solution (10), vitamin solution (10), 0.1% resazurin (2), deionized water (768), sodium bicarbonate (3.5g), and 0.2M $Na_2S{\bullet}9H_2O$ (10).[4,14] After inoculation, the headspace of serum bottles was flushed with an 80:20 $N_2/CO_2$ gas mixture. Abiotic controls were autoclaved at 250°F for one hour on two consecutive days. Inoculated serum bottles, including abiotic controls, were then spiked with BTEX (20 to 50 μM each) or phenol, p-cresol, benzylalcohol (0.25 mM each) in an anaerobic glove box. Abiotic controls were also spiked with mercury chloride (2mM). Serum bottles were incubated in the dark at 38 C. Liquid samples were withdrawn for BTEX and phenols analyses and headspace samples were analyzed for methane and carbon dioxide.

Ground refuse was used in some serum bottle tests as described above. The fresh refuse was residential refuse collected in Raleigh, NC. The decomposed refuse was recovered from a laboratory refuse reactor.[15] Both refuse samples and filter paper (99% cellulose) were ground to a particle size of 0.5 mm, redried and stored at 4°C until use. Fresh refuse, filter paper, cellobiose, and acetate were all added in amount designed to have a methane potential of 40 mL. The methane potential of the fresh refuse and old refuse were measured in a biochemical methane potential (BMP) test,[16] while the methane potentials of filter paper and cellulose were calculated. The methane potential of decomposed refuse was minimal and the amount of decomposed refuse added in the test was arbitrary.

## Analytical Methods

BTEX was analyzed by a Perkin-Elmer gas chromatograph equipped with a DB-VRX capillary column (30 m by 0.45 mm, J & W Scientific) and a flame ionization detector. Phenols and benzoate were analyzed by high-performance liquid chromatography (HPLC) equipped with a $C_{18}$ reverse-phase column and a UV detector. Gas composition was analyzed by a gas chromatograph equipped with a thermal conductivity detector and a HayeSep Q 80/100 column. Gas volume was measured by a wetted syringe method.[16] To calculate methane production, total gas volume, including the overpressure gas and headspace volume, was multiplied by the methane concentration measured by GC. All gas data are presented at standard temperature and pressure.

### *$^{14}C$ Activity Measurements*

To measure dissolved $^{14}C$, 5 mL of culture fluid was withdrawn from a serum bottle and added to 1 mL of 2N NaOH. A 0.5-mL subsample was then transferred to 18 mL of scintillation fluid for measurement of total dissolved $^{14}C$ activity. A 2-mL subsample was added to 1 mL of 1N HCl, leaving an acidified and a basified sample. These samples were simultaneously purged with $N_2$ (500 mL/min) for 20 min, after which 1 mL of each sample was added to 18 mL of scintillation fluid. $^{14}CO_2$ was calculated by the difference in activity of the base-purged and acid-purged samples. The volatile fraction was calculated from the difference between the total activity and the base-purged sample. Activity present in the acid-purged sample was taken to represent nonvolatile intermediates.

Headspace $^{14}C$ was measured by sparging a serum bottle with approximately 3 liters of oxygen-free 80:20 $N_2/CO_2$. Gas was collected in a Tedlar gas bag. The gas was then pumped through a series of two ORBO-91 tubes (Supelco, Bellefonte, PA) into a second gas bag. The contents of the ORBO tubes were removed and extracted by 2 mL of $CS_2$. The extract (0.1 mL) was counted to determine the $^{14}C$-toluene present in the headspace. To measure the $^{14}CH_4$ and $^{14}CO_2$ content of the gas, a 200-mL sample from the second gas bag was injected into a system of two $CO_2$ traps (2 M NaOH, 15 mL), followed by a tube furnace in which the $CH_4$ was oxidized to $CO_2$ at

700°C, and then a second set of two $CO_2$ traps. After trapping, 2 mL of each NaOH trap were transferred to scintillation fluid, refrigerated for 24 hours to eliminate chemiluminescence, and then counted.

The volume of gas collected in the gas bag was measured in a specially fabricated stainless steel cylinder of known volume (3.461 L). The pressure in the cylinder was measured after evacuation and again after venting a gas bag to the steel cylinder. The amount of gas transferred from a gas bag to the steel cylinder was calculated by using the ideal gas law. Methane production data are reported as dry gas at standard temperature and pressure.

### Data Analysis

All experiments were conducted in triplicate. To determine the effects of the various treatments on phenol and p-cresol degradation, degradation rates (mM/day) were analyzed by analysis of variance, *F*-tests with Tukey's test to compare effects between treatment sets or with Dunnett's test to compare a treatment with a nontreatment control.

## RESULTS AND DISCUSSION

### Degradative Activity of Refuse Inocula

*BTEX Degradation*

Inocula derived from six refuse samples were tested in serum bottle assays for their ability to degrade BTEX, phenol, and p-cresol. Toluene was the only BTEX component for which biodegradation was measured. After 380 days, significant toluene loss relative to abiotic controls was only observed in assays with two excavated samples from the Greensboro landfill. However, subsequent respikes of regular and $^{14}C$-toluene to GB1 did not result in additional toluene biodegradation, and only the results of biodegradation with GB2 are presented (Table I). Toluene degradation (>95%) was initially measured after 389 days in three serum bottles from the treatment of BTEX plus refuse, but not in the other treatments. These three bottles were spiked with $^{14}C$-toluene on days 409 and 495. GC analyses showed that >95% of the first and second spikes of $^{14}C$-toluene, added at 717 and 651 μg/L, respectively, were degraded within 75 days of each addition, while 50% of the added $^{14}C$-toluene still remained in abiotic controls (data not shown). Ninety days after the second spike, recovery of $^{14}C$-toluene as $^{14}CH_4$ and $^{14}CO_2$ was 27.6% and 35.1%, respectively, thus documenting mineralization. Total $^{14}C$ recovery ranged from 63% to

**Table I. Toluene Concentration in Tests with GB 2 (μg/L)[a]**

| Treatment | Day 3 | Day 265 | Day 389 | Day 602 |
|---|---|---|---|---|
| BTEX | 3719<br>(223) | 2054<br>(315) | 2090<br>(150) | 2 of six replicates below 60 μg/L |
| BTEX (Abiotic) | 3787<br>(150) | 2124<br>(264) | 2117<br>(59) | 1414<br>(219) |
| BTEX + Phenols | 3875<br>(390) | 2013<br>(140) | 1921<br>(121) | 1 of six replicates below 60 μg/L |
| BTEX + Refuse | 3220<br>(489) | 1474<br>(82) | 3 of six replicates below 40 μg/L | All six replicates below detection[b] |
| BTEX + Refuse | 3467<br>(242) | 1808<br>(217) | 1718<br>(209) | 1570<br>(159) |

[a]Data are the average of triplicates. The number in parentheses is the standard deviation. Intermediate sampling dates are not shown.
[b]Detection limit for toluene is 20 μg/L.

85% in triplicate tests. Toluene that dissolved into the black butyl rubber stoppers used here was not measured, although this represents a toluene sink. Grbic-Galic and Vogel (1987)[5] reported that about 40% of $^{14}C$-labeled toluene dissolved into red rubber sleeve stoppers. The ratio of $^{14}CH_4$ to $^{14}CO_2$ was 0.5, 0.6, and 1.4 in the three tests, which is considerably lower than the theoretical ratio of 1.8.

While anaerobic toluene degradation has been reported in a number of habitats including marine sediment and aquifers, this is the first time to our knowledge that a landfill has been shown to harbor a toluene-degrading population. Toluene was the only BTEX component degraded and degradation was confirmed with only one inoculum. Under methanogenic conditions, toluene, among BTEX components, has been reported to be the most susceptible to anaerobic biodegradation.[17]

The extended lag period prior to toluene biodegradation (265 days) cannot be readily explained by the presence of other readily degradable substrates because tests with BTEX but no ground refuse did not exhibit as rapid toluene degradation as tests with ground refuse (Table I). It has been shown that acclimation to xenobiotics can be an important process for the indigenous population in a natural environment.[18] However, microbial acclimation rates vary and the precise reasons for this variation may be difficult to determine. There was no information regarding the history of previous exposure of sample GB2 to BTEX, and an extended acclimation period was required.

*Phenolic Degradation*

The inocula derived from the various decomposing refuse samples showed broad potential to biodegrade phenol, p-cresol, and benzylalcohol. Benzylalcohol (0.25 mM) was depleted rapidly in every test (data not shown). Phenol was depleted in tests with three of six inocula within 180 days (Figure 1A), while p-cresol was degraded in four tests within 100 days (Figure 1B). p-Cresol did not degrade after 200 days in assays inoculated with the High Point landfill samples. These two inocula also exhibited lower activity toward phenol. The phenol and p-cresol degradation rates were highest in tests with the yard waste inoculum.

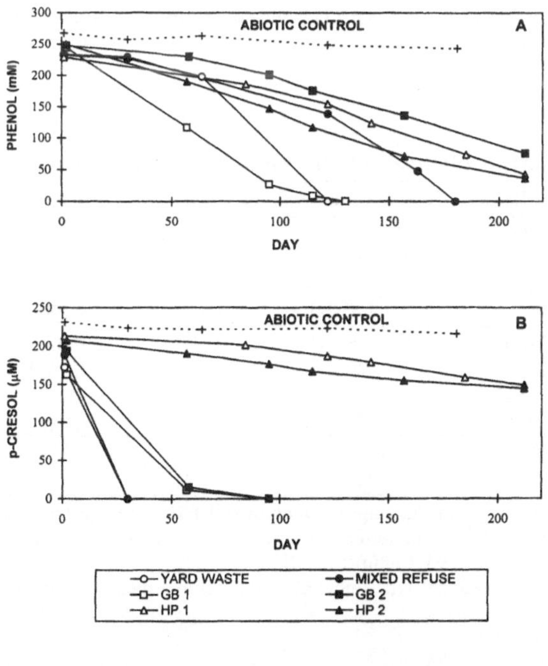


**Figure 1. Degradation of phenol and p-cresol by six refuse inocula.**

Methane was produced from each inoculum. The methane recovery was calculated as the measured methane production divided by the stoichiometric amount of methane that was expected from the total degraded phenolics. This fraction varied from 42% to 121%. The methane recovery in excess of 100% for the mixed refuse inoculum likely reflects a combination of small analytical errors as the total methane potential of the phenols added in serum bottle assays was relatively low. Methane production was close to the stoichiometric amount for the yard waste, mixed refuse, and HP2 inocula. However, low methane production did not correspond with the absence of phenol or p-cresol degradation activity. For example, GB1 had a low methane recovery (60%) but was active in both phenol and p-cresol degradation.

Based on both the methane recovery and degradation data, the yard waste inoculum, among six refuse inocula, was most active in phenolics degradation. One explanation for this is the long-term exposure of microorganisms in the yard waste reactor to aromatic compounds present in the leachate. The total aromatic concentration in leachate from the yard waste reactor (55 mg/L as p-cresol) was the highest among the tested refuse samples (2 to 12 mg/L as p-cresol). Thus, microbes from the yard waste reactor had substantial exposure to aromatic compounds. As wood and yard trimmings represent about 15% of the volume of MSW landfilled,[19] high aromatic concentrations would also be expected in landfills. Refuse consists of many lignocellulose substrates and the presence of fulvic-like material with a high aromatic hydroxyl group density in landfill leachate has been documented.[20] The presence of abundant phenolics in landfills provides an opportunity for the indigenous microbial populations to acclimate to phenolics degradation.

The inocula derived from both High Point landfill samples exhibited relatively low degradative activity toward phenol and p-cresol. One potential explanation for the low activity is that these two samples had an initial pH between 5.5 to 5.7, suggesting a lack of healthy methanogenic populations at the time of excavation.

## The Effects of Environmental Conditions

### *Presence of Other Carbon Substrates*

The effect of carbon addition on phenol degradation and on methane production is presented in Table II. Results of the first test suggested that fresh refuse and filter paper stimulated phenol degradation while decomposed refuse had no apparent effect (Figure 2A). This stimulation resulted from a biological mechanism as fresh refuse addition did not cause a loss of phenol in uninoculated abiotic controls. The stimulatory effect of fresh refuse was verified when two serum bottles from both the phenol-only control and the decomposed refuse treatment were respiked with both phenol and fresh refuse (Figure 2A).

**Table II. The Effects of Carbon Addition on Phenol Degradation Rate**

| 1st Test | Control | Decomposed Refuse | Fresh Refuse | Filter Paper | |
|---|---|---|---|---|---|
| Ratea | 2.89 0.15(a) | 2.73 0.12(a) | 4.73 0.02(b) | 4.13 0.15(c) | |
| CH4a, b | 1.3 0.4(a) | 0.7 0.2(a) | 14.5 4.9(b) | 10.1 0.5(b) | |
| **2nd Test** | **Control** | **Acetate** | **Refuse Extract** | **Fresh Refuse** | **Cellobiose** |
| Ratea | 2.18 0.05(a) | 2.22 0.09(a) | 4.51 0.08(b) | 4.19 0.33(b) | 2.90 0.16(c) |
| CH4a,b | 2.2 0.1(a) | 39.5 0.8(b) | 3.6 0.2(a) | 16.2.9(c) | 25.4 0.7(d) |
| **3rd Test** | **Control** | **Refuse Extract** | **Yeast Extract** | | |
| Rate[a] | 2.71 0.12(a) | 4.58 0.00(b) | 4.22 0.23(b) | | |
| CH4a, b | 1.8 0.1(a) | 3.3 0.1(b) | 13.4 0.5(c) | | |

[a]Phenol or p-cresol degradation rate (mg/L/day) and methane production (ml $CH_4$) are presented as average of triplicates +/- standard deviation. Different letters in parentheses indicate statistically significant differences as analyzed by Tukey's test at 95% confidence limit.

[b]Methane production was measured in every treatment when phenol in the nontreatment control was depleted.

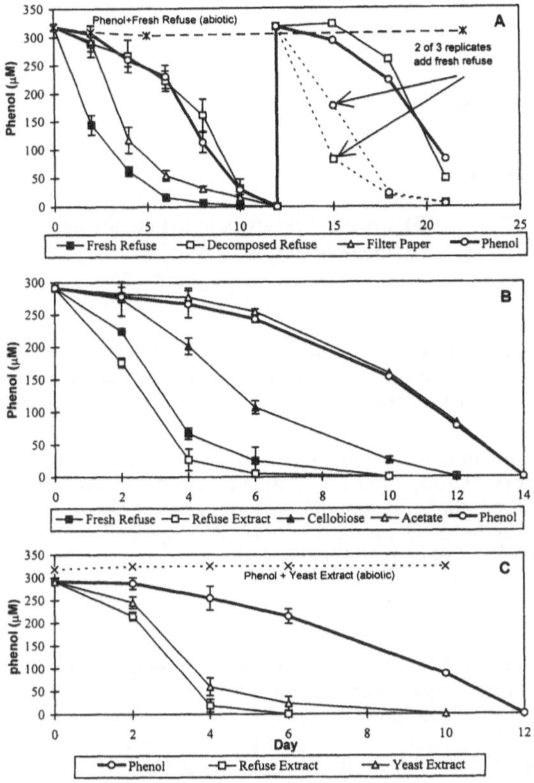


**Figure 2. The effects of carbon addition on phenol degradation by MR-P.**

To evaluate whether it was the carbon in fresh refuse, or a trace component that resulted in the stimulation of phenol degradation, additional tests were conducted with a refuse liquid extract, cellobiose and acetate (Figure 2B). Fresh refuse, the refuse extract, and cellobiose were all stimulatory, while the acetate addition had no effect on phenol degradation, even though it resulted in the highest methane production. The cellobiose treatment confirmed the observed enhancement of filter paper in the first test, indicating that polysaccharides and cellulose-stimulated phenol degradation by MR-P. In the first two tests, both filter paper and cellobiose treatments resulted in equal or more methane relative to that of fresh refuse and refuse extract treatments. However, the latter two treatments were more stimulatory with respect to phenol degradation (Table II). Also in the second test, the refuse liquid extract, without significant methane production, provided statistically comparable stimulation as the fresh refuse treatment. These results suggest that some trace component in fresh refuse other than carbon was stimulatory. In a subsequent test, both a refuse extract and yeast extract stimulated phenol degradation (Figure 2C), and the stimulation of both treatments was statistically similar. Thus, the data suggest that both excess carbon and nutrients stimulated phenol degradation by MR-P. Because ground refuse, refuse extract, and yeast extract all contained both carbon and nutrients, they were more stimulatory than carbon alone (cellobiose). However, stimulation of methanogens by acetate addition did not stimulate phenol degradation and higher methane production did not correspond with higher phenol degradation (Table II).

Substrate interactions occur when alternative carbon sources are provided to microbial populations. Additional substrate may enhance biodegradation by providing a growth factor or increasing cell populations, or inhibit biodegradation as a result of substrate competition. We found little literature focused on substrate interaction in anaerobic phenol degradation. Under aerobic conditions, results suggest that the type of substrate interaction may be both organism- and substrate-specific. For example, glucose enhanced the removal rate of 2,4-dinitrophenol by a micro-

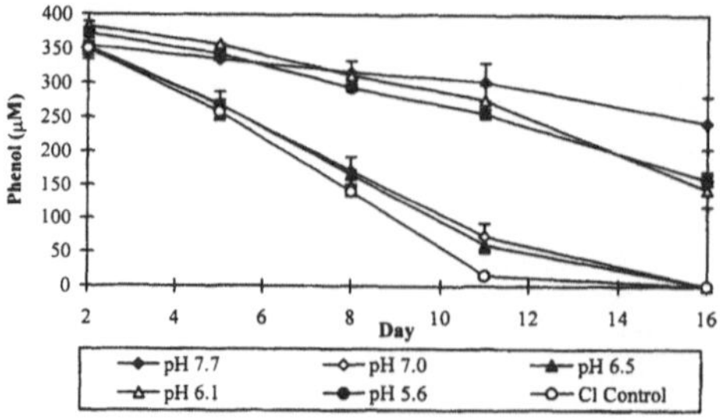


**Figure 3. The effect of pH on phenol degradation.**

bial consortium.[21] The author attributed the enhancement to the growth of 2,4-dinitrophenol degrading bacteria on glucose. In contrast, glucose inhibited phenol degradation by a phenol-acclimated enrichment culture.[22] It was suggested that glucose was a preferred substrate for the phenol degraders.

Other than providing carbon sources, the stimulatory effects of additional substrates may be due to the supply of certain growth factors. Working with an obligate syntrophic consortium, Knoll and Winter (1989)[23] suggested that the phenol-degrading bacterium may be cross-fed by the benzoate-degrading bacterium present in the same consortium. Proteose peptone has been reported to provide an essential growth factor for a phenol and p-cresol-degrading consortium.[24]

### *pH*

The effect of pH on phenol degradation by MR-P enrichment culture is presented in Figure 3. The degradation rates of controls prepared with sodium chloride at pH 7.0 were comparable to the rates of the pH 7.0 control, indicating that the $Na^+$ and $Cl^-$ additions did not affect the results. The optimum pH for MR-P was between pH 6.5 and 7.0. This pH optimum is slightly different from the previously reported pH optimum of 7.0 to 8.0 for phenol and o-cresol degradation.[12, 24] However, this pH is similar to the pH optimum for methane production in anaerobic ecosystems.[25]

### *Accumulation of Degradative Intermediates*

The effect of efficient removal of benzoate on phenol degradation with MR-P is presented in Figure 4. The addition of BESA stopped acetate utilization by methanogens and there was no measurable methane production. In this treatment, the benzoate concentration remained high and phenol degradation was inhibited. In the absence of BESA, phenol degradation was also inhib-

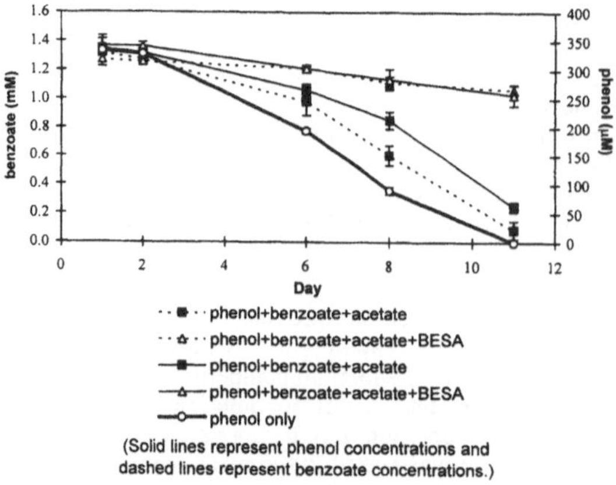


**Figure 4. The effect of benzoate and acetate accumulation on phenol degradation by MR-P.**

ited by a high initial benzoate and acetate concentration. However, phenol and benzoate did ultimately degrade. The phenol concentration decreased in parallel with the benzoate concentration. Methane production on Day 6 in this treatment was 15.1 mL. This represents 67% of the theoretical methane potential of the acetate spike. Inhibition was not caused by the addition of sodium since phenol degradation in a sodium control was not significantly different from the nontreatment control (data not shown).

The potential for accumulation of fermentation intermediates in the landfill ecosystem has been documented.[25] In our study, elevated benzoate concentrations inhibited phenol degradation by the MR-P enrichment culture. By keeping an elevated acetate concentration, the rate of phenol degradation in a methanogenic sewage sludge was reduced with an accumulation of benzoate.[26] While the carboxylation of phenol to benzoate is energetically favorable, acetogenic degradation of benzoate requires syntrophic removal of acetate and hydrogen based on the thermodynamics illustrated in Equations 1 and 2.[23]

$$\text{phenol} + HCO_3^- + H^+ + H_2 \text{ ® benzoic acid} + 2H_2O \; G^o = 45.6 \text{ kJ/mol} \quad (1)$$

$$\text{benzoic acid} + 7H_2O \text{ ® } 3 \text{ acetate}^- + 4H^+ + HCO_3^- + 3H_2 \; G^o = 50.5 \text{ kJ/mol} \quad (2)$$

The results presented here demonstrate that phenol degradation can be controlled by efficient benzoate degradation which requires active methanogenic removal of acetate and hydrogen.

## CONCLUSION

This study showed that decomposing refuse from landfills harbors microbial populations capable of toluene, phenol, and p-cresol degradation. Results suggested that phenol and p-cresol degrading microorganisms are widespread in the landfill ecosystem. It was also demonstrated that both neutral pH and efficient removal of acetate are beneficial for phenol degradation by an anaerobic consortium enriched from decomposing refuse. In the early stage of refuse decomposition in landfills, carboxylic acids rapidly accumulate, resulting in a pH decrease.[4] Before the development of an active methanogenic population in landfills, a long lag is generally required to overcome this acidic pH. One technique for controlling methane production in landfills is leachate recycle.[27] Thus, the development of effective landfill management strategies to maximize trace organic contaminant degradation in landfills may be possible.

## REFERENCES

1. Carra, J.S. and R. Cossu, Municipal Solid Waste and Sanitary Landfilling in the United States of America, *International Perspectives on Municipal Solid Wastes and Sanitary Landfilling*, Academic Press, Inc., New York, 1990, pp. 1–14.
2. Lesage, S., R.E. Jackson, M.W. Priddle, and P.G. Riemann, "Occurrence and fate of organic solvent residues in anoxic groundwater at the Gloucester landfill, Canada," *Environ. Sci. Technol.*, 24, pp. 559–566, 1990.
3. Sawhney, B.L. and R.P. Kozloski, "Organic pollutants in leachates from landfill sites," *J. Environ. Qual.*, 13, pp. 349–352, 1984.
4. Barlaz, M.A., D.M. Schaefer, and R.K. Ham, "Bacterial population development and chemical characteristics of refuse decomposition in a simulated sanitary landfill," *Appl. Environ. Microbiol.*, 55, pp. 55–65, 1989.
5. Grbic-Galic, D. and T.M. Vogel, "Transformation of toluene and benzene by mixed methanogenic cultures," *Appl. Environ. Microbiol.*, 53, pp. 254–260, 1987.
6. Kuhn, E.P., J. Zeyer, P. Eicher, and R.P. Schwarzenbach, "Anaerobic degradation of alkylated benzenes in denitrifying laboratory aquifer columns," *Appl. Environ. Microbiol.*, 54, pp. 490–496, 1988.
7. Beller, H.R., D. Grbic-Galic, and M. Reinhard, "Microbial degradation of toluene under sulfate-reducing conditions and the influence of iron on the process," *Appl. Environ. Microbiol.*, 58, pp. 786–793, 1992.
8. Cossu, R. and R. Serra, Codisposal of Difficult Wastes and MSW in Sanitary Landfill, The Fifth International Conference on Solid Waste Management and Secondary Materials, Philadelphia, PA, 1989.
9. Knox, K. and J. Gronow, "A reactor based assessment of co-disposal," *Waste Manage. Res.* 8, pp. 255–276, 1990.

10. Reinhart, D.R. and F.G. Pohland, "The assimilation of organic hazardous wastes by municipal solid waste landfills," *J. Indus. Microbiol.* 8, pp. 193–200, 1991.
11. Watson-Craik, I.A. and E. Senior, "Treatment of phenolic wastewaters by co-disposal with refuse," *Water Res.*, 23, pp. 1293–1303, 1989.
12. Sulisti Watson-Craik, I.A. and E. Senior, "Studies on the co-disposal of o-cresol with municipal refuse," *J. Chem. Tech. Biotechol.* 65, pp. 72–80, 1996.
13. Dolfing, J. and J.M. Tiedje, "Acetate inhibition of methanogenic, syntrophic benzoate degradation," *Appl. Environ. Microbiol.*, 54, pp. 1871–1873, 1988.
14. Healy, J.B. and L.Y. Young, "Catechol and phenol degradation by a methanogenic population of bacteria," *Appl. Environ. Microbiol.*, 35, pp. 216–218, 1978.
15. Eleazer, W.E., W.S. Odle, Y.-S. Wang, and M.A. Barlaz, "Biodegradability of municipal solid waste components in laboratory-scale landfills," *Environ. Sci. Technol.*, 31, pp. 911–917, 1997.
16. Wang, Y.-S., C.S. Byrd, and M.A. Barlaz, "Anaerobic biodegradability of cellulose and hemicellulose in excavated refuse samples using a biochemical methane potential assay," *J. Indus. Microbiol.*, 13, pp. 147–153, 1994.
17. Willson, B.H., G.B. Smith, and J.F. Rees, "Biotransformations of selected alkylbenzenes and halogenated aliphatic hydrocarbons in methanogenic aquifer material: A microcosm study," *Environ. Sci. Technol.*, 20, pp. 997–1002, 1986.
18. Aelion, C.M., C.M. Swindoll, and F.K. Pfaender, "Adaptation to and biodegradation of xenobiotic compounds by microbial communities from a pristine aquifer," *Appl. Environ. Microbiol.*, 53, pp. 2212, 1987.
19. Characterization of Municipal Solid Waste in the United States: 1994 Update, EPA/530-R-94-042, United States Environmental Protection Agency, Washington, D.C., 1994.
20. Chian, E.S.K and F.B. DeWalle, "Characterization of soluble organic matter in leachate," *Environ. Sci. Technol.*, 11, pp. 158–163, 1977.
21. Hess, T.F., S.K. Schmidt, J. Silverstein, and B. Howe, "Supplemental substrate enhancement of 2,4-dinitrophenol mineralization by a bacterial consortium," *Appl. Environ. Microbiol.*, 56, pp. 1551–1558, 1990.
22. Rozich, A.F. and R.J. Colvin, "Effects of glucose on phenol biodegradation by heterogeneous populations," *Biotechnol. Bioeng.*, 28, pp. 965–971, 1986.
23. Knoll, G. and J. Winter, "Degradation of phenol via carboxylation to benzoate by a defined, obligate syntrophic consortium of anaerobic bacteria," *Appl. Microbiol. Biotechnol.*, 30, pp. 318–324, 1989.
24. Beaudet, R., J.G. Bisaillon, M. Ishaque, and M. Sylvestre, "Isolation of an anaerobic bacterial consortium degrading phenolic compounds-assay in swine waste," *Agric. Wastes*, 17, pp. 131–140, 1986.
25. Barlaz, M.A., R.L. Ham, and D.M. Schaefer, "Methane production from municipal refuse: A review of enhancement techniques and microbial dynamics," *Crit. Rev. Environ. Control*, 19, pp. 557–584, 1990.
26. Knoll, G. and J. Winter, "Anaerobic degradation of phenol in sewage sludge," *Appl. Microbiol. Biotechnol.*, 25, pp. 384–391, 1987.
27. Townsend, T.G., W.L. Miller, and J.F.K. Earle, "Leachate-recycle infiltration ponds," *ASCE J. Environ. Eng. Div.*, 121, pp. 465, 1995.

# 17 EVALUATION OF TOXIC ORGANICS REMOVAL BY SIMULTANEOUS ADSORPTION AND BIODEGRADATION

**Frieda Orshansky** and **Nava Narkis**
Environmental and Water Resources Engineering
Technion—Israel Institute of Technology, Haifa 32000, Israel

## ABSTRACT

Simultaneous adsorption and biodegradation processes were investigated. Three toxic organic compounds, nitrobenzene, aniline, and phenol were selected to be studied individually and in mixtures as a model of polluted toxic industrial wastewaters. Their behavior was assessed separately in adsorption, biotreatment, as well as in combined systems, by physicochemical and respirometric methods. Powdered activated carbon (PAC) was used as an adsorbent in these experiments. In the adsorption systems, in single, bi- and trisolute solutions, the limiting adsorption capacities, $Q^o$ and b, the energy of adsorption related constants, were determined from the linearized forms of Langmuir adsorption isotherms. $Q^o$ values for aniline and phenol were found to be similar and approximately 2.5 times higher for nitrobenzene. The energy related coefficient b, was similar for nitrobenzene and phenol, and almost 5 times higher for aniline. The presence of mineral nutrients, needed for biological processes, significantly decreased the energy related constants for phenol and aniline, but did not affect this parameter for nitrobenzene. Kinetic characteristics of those three compounds' biodegradation by acclimated microbial cultures were determined in biological systems. The degree of bioresistance of the organics was estimated. PAC addition to the biological systems affected differently the biooxidation of the three compounds. The presence of PAC considerably enhanced the microbial respiration in the bioreactor with nitrobenzene, distinctly intensified that of phenol, but reduced it in the reactor with aniline. Different organic removal mechanisms were suggested in simultaneous adsorption and biodegradation of each of the three compounds, due to their different adsorption characteristics. It was found that the strong inhibiting effect of the three compounds in highly concentrated mixture on microbial respiration was completely wiped out by adding PAC to the biological reactor. As a result, the initial COD 3350 mg $O_2$/L was reduced to 27.6 mg $O_2$/L. The optimal PAC dose was determined by using respirometric data.

The results obtained can be used for the evaluation of simultaneous adsorption and biodegradation as a method for effective removal of toxic organics from industrial wastewaters, under appropriate conditions. The method can be applied for the industrial wastewater treatment alone as well as for pretreatment, before discharging to the municipal biological treatment plant. The process is expected to decrease the harmful effect on conventional activated sludge processes and to prevent the disposal of hazardous materials into the environment.

## INTRODUCTION

The problem of the environmental protection from industrial toxic, teratogenic, and carcinogenic pollutants is becoming more and more acute. The traditional approach to industrial wastewaters purification is treating them together with large amounts of domestic sewage in a biological treatment plant, with the idea of their manifold dilution. However, most of the industrial contaminants are nonbiodegradable at such conditions and have a deteriorating effect on municipal wastewater treatment plant operation and effluents' quality. Thus, the above method provides "cosmetic" purification and results in incomplete removal of the contaminants. To obtain a higher and stable treated water quality for a wide range of uses, advanced treatment processes are required. Activated carbon adsorption proved to be an effective method for residual toxic or-

*52nd Purdue Industrial Waste Conference Proceedings*, 1997, Ann Arbor Press, Chelsea, Michigan 48118. Printed in U.S.A.

ganics removal.[1–4] However, this method has significant disadvantages due to its high costs and its preferable application to low concentrated water treatment.

Significant contributions to industrial wastewater purification technology and an improvement of purified water quality and costs reduction were achieved when activated carbon was used in biological treatment plants.[5–9] Although the Powdered Activated Carbon Treatment (PACT) has been widely used, its mechanisms concerning the relations between biodegradation and carbon adsorption, in the simultaneous processes, remain obscure. The aim of the present research is to develop proper approaches for the evaluation of adsorption and biodegradation for removal of toxic organics in high concentrations by simultaneous processes.

## EXPERIMENTAL

A modified PACT, simultaneous adsorption and biodegradation process was studied in systems containing relatively high concentrations of organic compounds and Powdered Activated Carbon (PAC) and relatively low concentrations of biomass. Such a combination of constituents was expected to facilitate biokinetic investigations and to give a more precise quantitative evaluation of the separate roles of adsorption and biodegradation in the combined process.

Low biodegradability, toxicity, and significant harmful effects on the environment were considered as high priority characteristics in the selection of the organic substances to be studied in the research.[10–13] In addition, the effect of the organic molecule chemical structure on the adsorbability, biodegradability, and the kinetic aspects of these processes was also taken into consideration. The following toxic aromatic compounds, which are included in the EPA list of the toxic priority pollutants, were selected: nitrobenzene—slowly biodegradable, toxic, and teratogenic organic substance; aniline—toxic and teratogenic organic compound, which besides being bioresistant was described as an inhibitor to other organics biodegradation in wastewater treatment plant;[13] and phenol—a widely-spread toxicant, common in many industrial wastewaters. All three compounds have different charge and polarity of the molecules which are supposed to affect their behavior in adsorption, biological, and combined processes. These compounds were studied individually and in mixtures as a model of concentrated toxic industrial wastewaters. Physicochemical and respirometric methods were used to evaluate adsorption, biological, and their simultaneous treatment.

## MATERIALS

The following chemicals were used in the research: nitrobenzene, analytical grade, manufactured by Fluka Chemika; extra pure aniline, manufactured by Merck; analytical grade phenol, manufactured by Riedel-de-Haen; powdered activated carbon, product of BDH Chemicals, Ltd.; 4-aminoantipyrine, product of Janssen; and 2-napthol-3.6-disulfonic acid disodium salt, manufactured by Fluka Chemika. All the additional mineral analytical grade reagents used were manufactured by Merck and Riedel-de-Haen.

## DESCRIPTION OF EXPERIMENT

Adsorption isotherms were determined for individual compounds nitrobenzene, aniline, and phenol, and for their mixtures. The adsorption was studied in distilled water and in aqueous solutions containing mineral nutrients, needed for the biological processes. The composition of the mineral ingredients is given in Table I.

Batch adsorption experiments were carried out as described by Orshansky and Narkis.[9,15] Linearized Langmuir isotherms were obtained from which the limiting adsorption capacity, $Q^o$, and b, the energy related constant of adsorption and the correlation coefficient, were calculated.

### Biodegradation

For the biodegradation studies acclimated mixed microbial cultures were obtained from raw sewage by 2 to 4 months adaptation period separately to nitrobenzene, aniline, and phenol in batch or continuous flow reactors. Organic compound concentrations in feeding media were gradually increased, as their stable bioremoval had been achieved. The concentrations ranged from 100 to 600 mg/L of nitrobenzene and from 100 to 1000 mg/L of aniline and phenol.

**Table I. Composition of Mineral Ingredients in Aqueous Solutions of Organic Compounds**

| | | Concentration | | |
|---|---|---|---|---|
| Ingredient | Units | Nitrobenzene | Aniline | Phenol |
| $(NH_4)_2 SO_4$ | mg/L | 135 | 125 | 500 |
| $MgSO_4 \cdot 7H_2O$ | mg/L | 200 | 200 | 200 |
| Phosphate buffer pH 7.2 (for BOD *Standard Method*[14]) | mL/L | 5 | 30 | 30 |
| $CaCl_2$ | mg/L | 27.5 | 27.5 | 27.5 |
| $FeCl_3 \cdot 6H_2O$ | mg/L | 0.5 | 0.5 | 0.5 |

All organic substrate biooxidation parameters were determined separately in shaken batch reactors, containing mineral nutrients solutions. Initial concentrations of acclimated biomass were between 40 and 90 mg/L and organics from 100 to 500 mg/L. Control of biodegradation was carried out by using specific analytical methods, biomass concentration monitoring, and respirometric measurements. A Sapromat electrolytic respirometer, B6 VOITH was used for the biooxidation control. Biokinetic constants calculations were performed according to Gaudy's approach.[16,17]

### Simultaneous Adsorption and Biodegradation

The combined processes were investigated in shaken batch reactors and in respirometer reactors. Solutions in reactors contained mineral ingredients (Table I) and nitrobenzene, aniline, or phenol individually and in mixtures. The concentration was 500 mg/L of each organic substance in single, bi-, and in trisolute solutions. PAC dosage was based on the concept that every new dose of carbon should adsorb approximately 100 mg of organic substance, according to the experimentally determined $Q^o$ values. The PAC concentrations in nitrobenzene solutions ranged from 0.31 g/L to 1.5 g/L and from 0.74 g/L to 3.68 g/L for phenol and aniline solutions. For mixtures of the organics PAC concentrations were calculated according to their $Q^o$ determined in their mixtures.

Acclimated microbial cultures were added to each reactor to support the initial biomass concentrations of about 90 mg/L for nitrobenzene and 40 to 50 mg/L for phenol and aniline. In the respirometric investigations of simultaneous adsorption and biodegradation treatment, one batch reactor did not contain PAC and served as a basic biological treatment control.

### Analytical Control

The nitrobenzene concentration in the single solute systems was controlled by direct UV spectrophotometric measurements at l = 208 nm and in the trisolute solutions at l = 310 nm. Phenol concentration in the single, bi-, and trisolute systems was determined by 4-amynoantipyrine colorimetric method.[14] Aniline concentration in all systems was determined by diazocoupling with 2-naphthol-3.6-disulfonic acid disodium salt.[18] The spectrophotometer used was a Milton Roy Spectronic 601.

All the data were representative of repeated experiments. The calculated standard deviations of the results, including oxygen uptake respirometric measurements, were less than ± 2.5%, excluding nitrobenzene adsorption isotherm, in trisolute solution, which was higher, up to ±7%.

## RESULTS AND DISCUSSION

The example of the linearized form of the Langmuir isotherms, for nitrobenzene adsorption in distilled water and mineral nutrients saline solution, is given in Figures 1 and 2. The adsorption isotherms for aniline and phenol in single solute saline solutions were described earlier.[19,20] Figure 3 shows the linear form isotherms of the aniline and phenol adsorption from bisolute saline solution, and Figure 4 shows the isotherms of each of the three compounds in their mixture in saline solution. In bisolute and trisolute solutions, the isotherm of each compound was found to obey the Langmuir equation. Hence, the isotherms were used to calculate $Q^o$ and b.

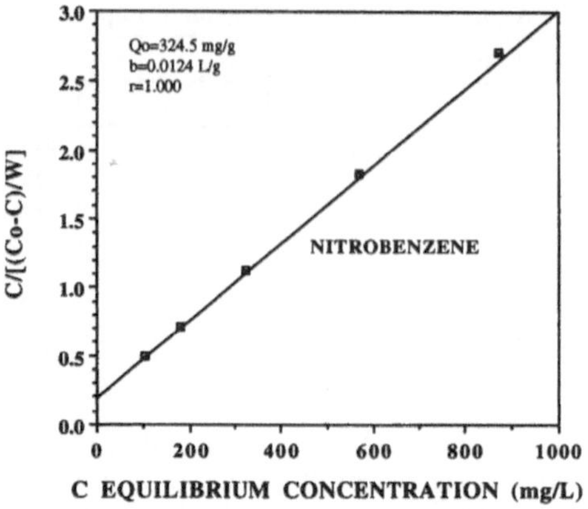


**Figure 1.** Linearized Langmuir isotherm of nitrobenzene adsorption in distilled water on powdered activated carbon.

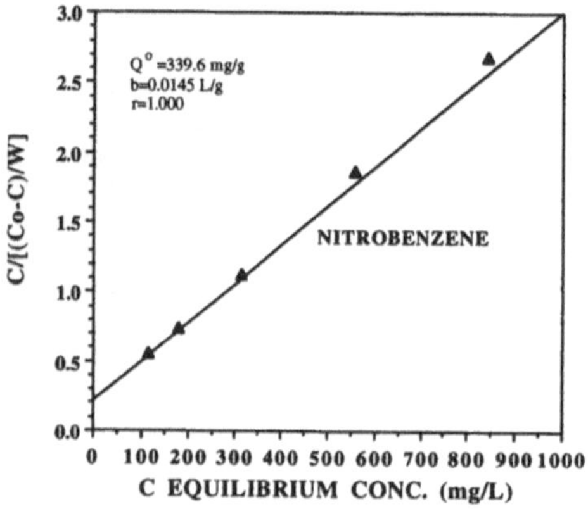


**Figure 2.** Linearized Langmuir isotherm of nitrobenzene adsorption in mineral nutrients saline solution on powdered activated carbon.

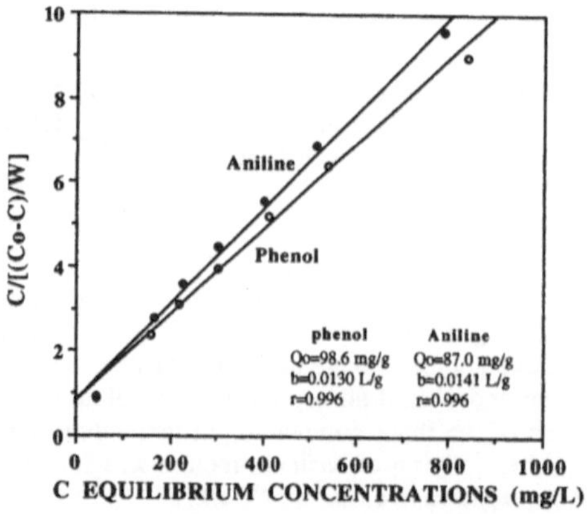


**Figure 3.** Linearized Langmuir isotherm of phenol and aniline in bisolute saline solution on powdered activated carbon.

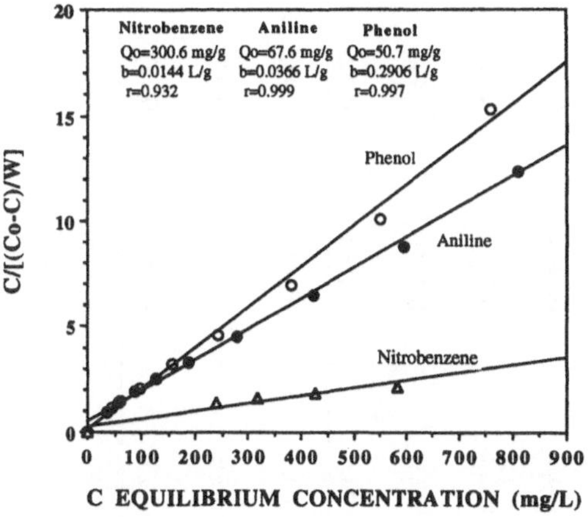


**Figure 4. Linearized Langmuir isotherm of nitrobenzene, phenol and aniline in trisolute saline solution on powdered activated carbon.**

The isotherms equations, the calculated values of $Q^o$, limiting adsorption capacities, and b, the energy related Langmuir constants are summarized in Table II. The results show that the adsorbability of phenol and aniline were similar, while for nitrobenzene it was about 2.5 times higher. Similar values of the energy related constants are obtained for phenol and nitrobenzene, while it was 5 times higher for aniline. The presence of the mineral nutrients was found to reduce considerably the energy of adsorption of phenol and aniline and had no effect on the b value of nitrobenzene. Although the energy of adsorption of aniline was reduced in saline solutions, it still remained about twice as high as that of phenol and nitrobenzene. The PAC adsorption capacity in the tricomponents system was found to be in the ratio of nitrobenzene, aniline, and phenol as 6.0 : 1.2 : 1.0, and the energy related coefficient as compared to single solute saline solutions remained unchanged for nitrobenzene, while it became 1.3 times higher for aniline and increased 24 times for phenol.

## Biotreatment

The processes of the substrates' biooxidation individually and in mixtures were studied in two parallel series of batch reactors. In the first series, the reactors were placed in the thermostatic shaker, and the microbial growth and residual liquid phase organics concentrations were determined. In the second series, the oxygen uptake, during substrates' biooxidation, was determined in respirometric batch reactors. "$O_2$ surrogate" technique, along with biomass growth control, was used for the biokinetics determination.[21]

Figure 5 shows the oxygen uptake during biodegradation of various initial nitrobenzene concentrations in the batch reactors. Distinct lag phases are observed at the nitrobenzene biooxidation in each concentration investigated. This phenomenon was not observed in aniline and phe-

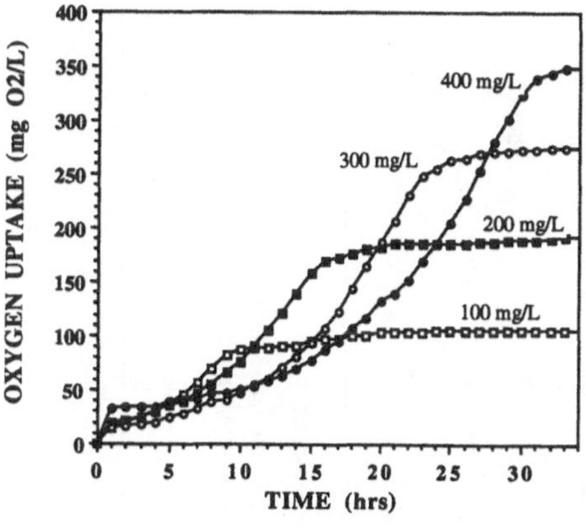


**Figure 5. Oxygen uptake during biodegradation of various initial nitrobenzene concentrations in batch biological reactors.**

**Table II. Adsorption Characteristics of Nitrobenzene, Aniline, and Phenol in Single, Bi, and Trisolute Systems in PAC**

| Medium | Sorbate | Langmuir Linearized Isotherm Equation $C/q_e = \frac{1}{bQ^o} + \frac{C}{Q^o}$ | Fig No. | $Q^o$ $\frac{\text{mg organic}}{\text{p PAC}}$ | b L/G | r Correlation coefficient |
|---|---|---|---|---|---|---|
| Single solute distilled water solution | Nitrobenzene | $C/q_e = 0.1860 + 2.8959 \times 10^{-3}C$ | Fig. 1 | 324.5 | 0.0124 | 1.000 |
| | Aniline | $C/q_e = 7.4740 \times 10^{-2} + 7.9019 \times 10^{-3}C$ | Orshansky & Narkis, 1997, Fig. 2 | 126.6 | 0.1057 | 0.996 |
| | Phenol | $C/q_e = 0.3603 + 7.3690 \times 10^{-3}C$ | Orshansky & Narkis, 1997, Fig. 1 | 135.7 | 0.0204 | 0.999 |
| Single solute mineral nutrients saline solution | Nitrobenzene | $C/q_e = 0.2093 + 2.9445 \times 10^{-3}C$ | Fig. 2 | 339.6 | 0.0145 | 1.000 |
| | Aniline | $C/q_e = 0.3187 + 7.9319 \times 10^{-3}C$ | Narkis & Orshansky, 1995, Fig. 14 | 126.1 | 0.0249 | 0.997 |
| | Phenol | $C/q_e = 0.5003 + 6.0117 \times 10^{-3}C$ | Narkis & Orshansky, 1993, Fig. 7 | 166.3 | 0.0121 | 0.995 |
| Bisolute aniline-phenol saline solution | Aniline | $C/q_e = 0.8169 + 1.1497 \times 10^{-2}C$ | Fig. 3 | 87.0 | 0.0141 | 0.996 |
| | Phenol | $C/q_e = 0.7815 + 1.0147 \times 10^{-2}C$ | Fig. 3 | 98.6 | 0.0130 | 0.996 |
| Trisolute nitrobenzene, aniline phenol saline solution | Nitrobenzene | $C/q_e = 0.2305 + 3.6967 \times 10^{-3}C$ | Fig. 4 | 300.6 | 0.0144 | 0.932 |
| | Aniline | $C/q_e = 0.4340 + 1.4585 \times 10^{-2}C$ | Fig. 4 | 67.6 | 0.0336 | 0.999 |
| | Phenol | $C/q_e = 6.7878 \times 10^{-2} + 1.9351 \times 10^{-2}C$ | Fig. 4 | 50.7 | 0.2906 | 0.997 |

**Table III. Specific Microbial Growth Rates as a Function of Nitrobenzene, Aniline and Phenol Initial Concentrations in Batch Reactors**

| Initial substrate concentration, mg/L | μ, Specific Growth Rate, $h^{-1}$ | | |
|---|---|---|---|
| | Nitrobenzene | Aniline | Phenol |
| 100 | 0.032 | 0.064 | 0.172 |
| 200 | 0.029 | 0.050 | 0.112 |
| 300 | 0.027 | 0.034 | — |
| 400 | 0.024 | 0.025 | 0.063 |
| 500 | 0.018 | 0.020 | 0.043 |

nol biodegradation studies at the same conditions.[9] A special investigation of batch microbial nitrobenzene biodegradation contributed to the appropriate explanation of this fact. It was found that two main microbial cultures were responsible for the nitrobenzene biodegradation. One of them, poorly flocculating, executed the initial transformation of the molecule without oxygen consumption, preparing it for further oxidative metabolism. This stage is probably displayed as lag period of the oxygen uptake in the respirometric results. The other culture, well flocculating, performed the oxidative part of the substrate metabolism, which is displayed as the logarithmic phase of the oxygen uptake. Hence, nitrobenzene removal by biodegradation can be considered as a two-stage process, with two different approaches to kinetics calculation.

Table III summarizes the values of the specific microbial growth rates, μ, as a function of nitrobenzene, aniline, and phenol initial concentrations in the batch reactors. The time Δt from the beginning of the process up to the asymptotic oxygen uptake reached was used for μ calculations. For example, m value for 100 mg/L nitrobenzene biooxidation stage alone was found to be 0.145 $h^{-1}$, as compared to 0.032 $h^{-1}$ for the entire process.

Oxygen uptake asymptotics, as a function of initial nitrobenzene concentrations, are presented in Figure 6. The equation was used for BOD determination and interpretation of the respirometric results. The calculated BOD is about 0.83 mg $O_2$ per mg nitrobenzene biodegraded, as compared to 1.34 mg $O_2$ per mg phenol and 1.59 mg $O_2$ per mg aniline, obtained previously.[9]

Bisolute aniline/phenol biotreatment processes were controlled by monitoring the organics concentrations in liquid, as well as by microbial oxygen uptake measurements. The microbial culture used was previously acclimated to the mixture of these substrates. For example, the biodegradation of a highly concentrated mixture of 500 mg/L of each compound is presented in Figure 7 which shows the residual organics concentrations and the oxygen uptake as a function of process time. After 48 hours the total oxygen uptake value was found to be equal to the sum of the two substrates' BOD. The complete biooxidation of 400 mg/L aniline alone in a single solute solution, required 48 hours.[9] It is suggested that the presence of the two compounds has a synergistic effect on microbial biodegradation of each substrate. This may be accounted for by the prehistory of the microorganisms cultivated in batch reactors containing high concentrations of each compound.

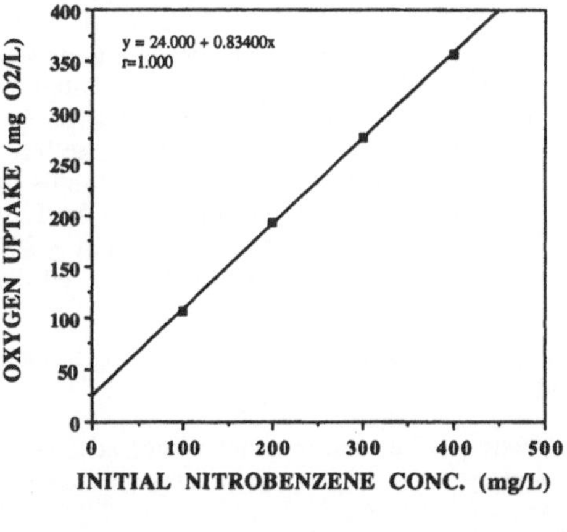


**Figure 6. Oxygen uptake asymptotic values as a function of the initial nitrobenzene concentrations in batch reactors.**

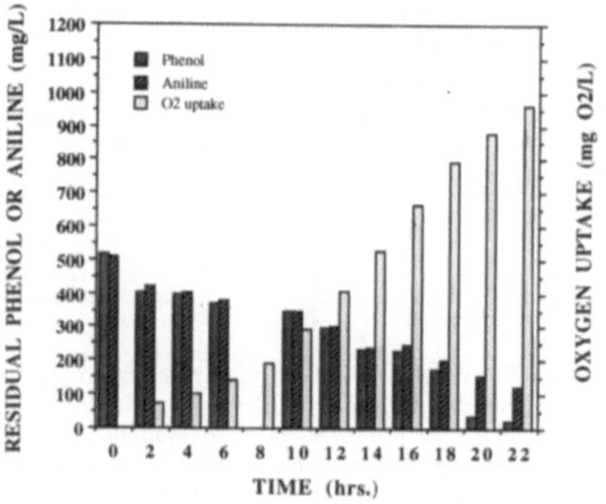


**Figure 7. Residual phenol and aniline in liquid and oxygen uptake during biological treatment of 500 mg/L phenol and 500 mg/L of aniline in bisolute solutions.**

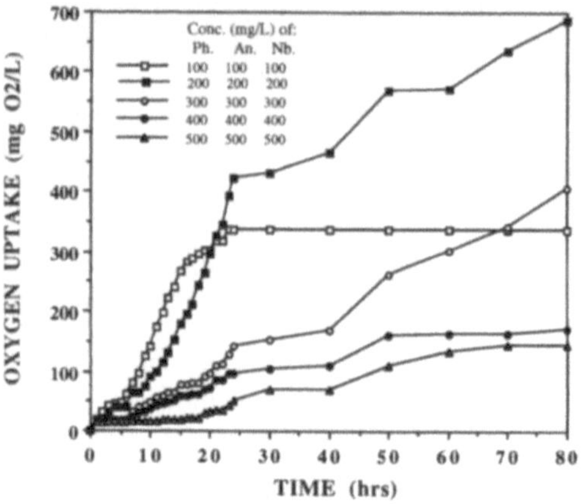


**Figure 8. Oxygen uptake during biodegradation of mixtures of nitrobenzene, aniline, and phenol at various initial concentrations.**

For tricompound systems the microbial culture was obtained by mixing equal volumes of the three cultures, previously acclimated to each compound separately. The oxygen uptake during biooxidation of various initial concentrations of the organic substances, in mixtures containing nitrobenzene, aniline, and phenol, as a function of time is given in Figure 8. From this figure it can be seen that about 90% of the organics from the mixture containing initial concentration of 100 mg/L of each component are biooxidized after 23 hours. In the mixture of 200 mg/L of each organic compound, about 90% are bioremoved after 80 hours only. The saw-like curves of the oxygen uptake indicate a step-by-step biodegradation of these substrates in the mixture. In the mixture of 300 mg/L of each organic compound, the inhibiting effect of high concentrations was likely to be overcome after 40 hours, but only 36% of the total organics was biooxidized after 80 hours. The higher the organics concentration, the more distinct was the inhibiting effect on microbial respiration. Residual concentrations of aniline and phenol, in the mixtures of the three compounds, as a function of biodegradation time are shown in Figure 9. Similar rates of biodegradation of aniline and phenol were obtained in bisolute system containing 500 mg/L of each substrate (Fig. 7), and trisolute solution of aniline, phenol, and nitrobenzene containing 200 mg/L of each component (Fig. 9). This indicates that 200 mg/L nitrobenzene inhibited the process of aniline and phenol biodegradation even by acclimated microorganisms.

## Simultaneous Adsorption and Biodegradation

Constant concentrations of 500 mg/L of each substrate were used in simultaneous adsorption and biodegradation studies, in single, bi-, and trisolute systems. All the experiments were carried

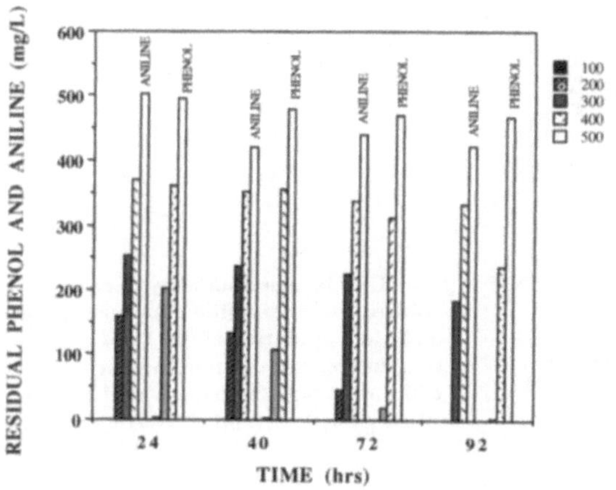


**Figure 9. Residual concentrations of aniline and phenol after biodegradation of the trisolute solution as a function of time.**

out in parallel batch reactors on a shaker and in a Sapromat respirometer, using acclimated microbial cultures.

Figure 10 shows the oxygen uptake by microorganisms during PACT simultaneous adsorption and biodegradation of 500 mg/L nitrobenzene in a single solute solution, in reactors containing various PAC concentrations. As can be seen, PAC addition distinctly accelerated the biooxidation process. In the absence of PAC, total biooxidation was achieved after approximately 48 hours. A minimum dose of carbon 0.31 g/L shortened this period to 38 hours. The highest acceleration was observed with PAC concentrations from 0.62 to 1.54 g/L. The PAC dose of 0.62 g/L can be considered as an optimum from both kinetic and economical points of view.

Opposite results were obtained in previous studies of aniline in simultaneous adsorption and biodegradation process.[9] Figure 11 shows that in the linear part of the curves similar slopes or rates of aniline biooxidation were observed, irrespective of the presence or absence of various PAC concentrations. A lag phase of 12 hours was observed in the presence of each PAC dose, while in the absence of PAC, no lag period was observed. Differences in PAC doses did not substantially affect microbial respiration. These facts indicate that PAC presence had no promotive effect on aniline biooxidation in simultaneous adsorption and biological treatment. It was analytically confirmed that the nonbiodegraded aniline remained strongly attached to the PAC surface. Phenol behavior was found to be closer to that of nitrobenzene with less manifested effect of PAC addition.

The different effects of PAC addition on the organics biooxidation can be explained as a result of different adsorption characteristics, like $Q^o$ and the energy of adsorption.

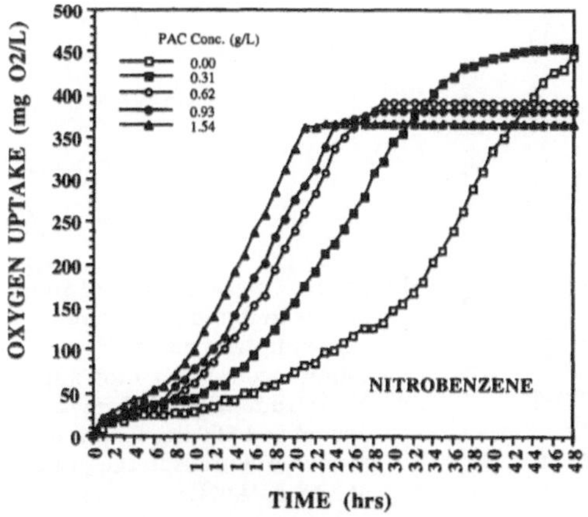


**Figure 10. Oxygen uptake by microbial culture during simultaneous adsorption and biodegradation of 500 mg/L nitrobenzene in batch reactors containing various PAC concentrations.**

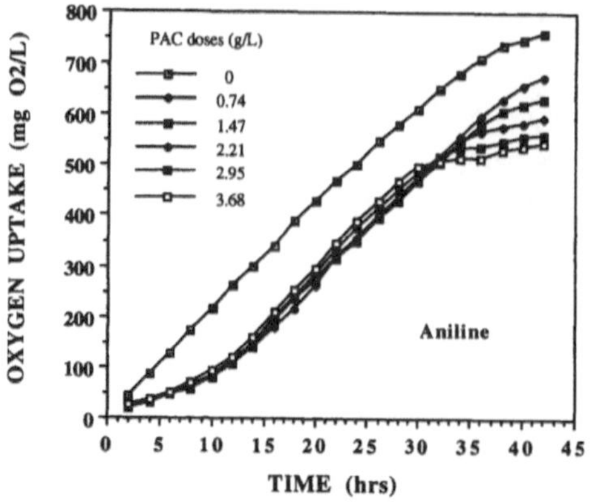


**Figure 11. Oxygen uptake by microbial culture during simultaneous adsorption and biodegradation of 500 mg/L aniline in batch reactors containing various PAC concentrations (Orshansky and Narkis, *Water Res.*, 31 (3), pp. 399–410, 1997.**

Figure 12 shows the residual aniline and phenol in the liquid phase and the oxygen uptake by microbial culture during simultaneous adsorption and biodegradation in bisolute solution.

The rates of aniline and phenol bioremoval from the liquid phase were similar in both processes, in biotreatment alone (Fig. 7) and in simultaneous adsorption and biodegradation (Fig. 12). This was reflected in the same slopes of the liquid phase organics concentration (Figs. 7 and 12). On the other hand, the oxygen uptake in PAC presence is significantly accelerated. For example, 960 mg $O_2$/L of oxygen uptake was recorded after 22 hours of biotreatment alone (Fig. 7) and after 16 hours in the simultaneous process (Fig. 12). This can be explained as an enhancement of the biooxidation of both organics adsorbed on PAC surface due to decreased energy of adsorption of aniline in bisolute system.

Figure 13 shows the microbial oxygen uptake, during simultaneous adsorption and biodegradation, in the mixture of nitrobenzene, aniline, and phenol of 500 mg/L of each component. A very strong effect of PAC addition on biooxidation of such a concentrated mixture was observed. Practically no oxygen uptake was recorded in the reactor in the absence of PAC. Even 1.5 g/L PAC brought about biooxidation of 67% of the total organic compound after 48 hours. At higher PAC concentrations, 75% of biooxidation of the total organics was achieved after 40 hours. No lag phase in oxygen uptake was observed for immediate and very fast phenol biooxidation in the liquid phase, as was analytically confirmed, while for nitrobenzene, it seems that biooxidation took place on PAC surface.

Table IV summarizes the results of the chemical analysis of the residual concentrations of the individual organic compounds in the liquid phase. The calculated oxygen uptake rates were

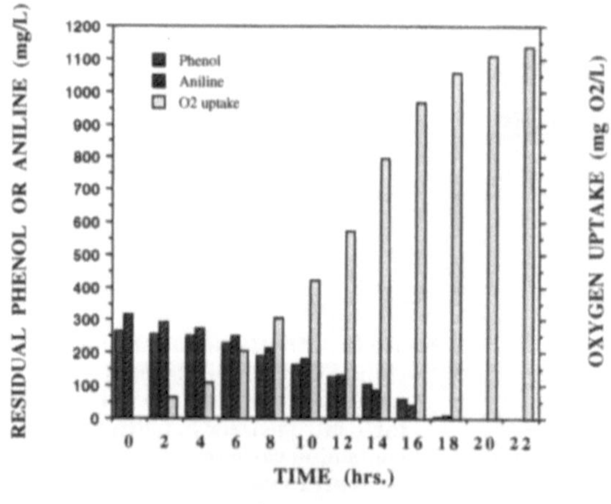


**Figure 12. Residual phenol and aniline in liquid phase and oxygen uptake during simultaneous adsorption and biodegradation of 500 mg/L phenol and 500 mg/L aniline in bisolute solutions in the presence of 3.68 g PAC/L.**

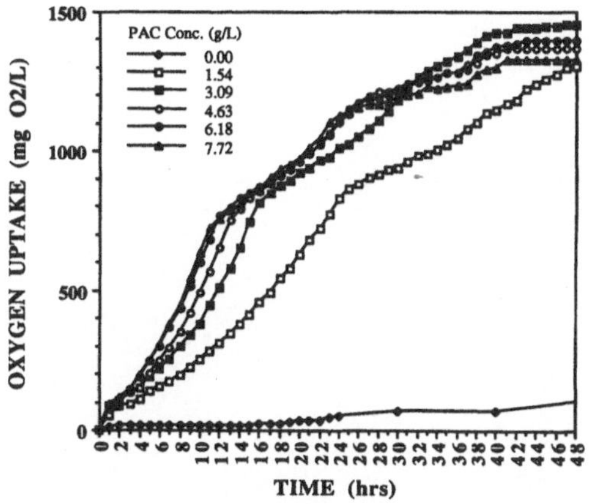


**Figure 13. Oxygen uptake by acclimated microbial culture during simultaneous adsorption and biodegradation of nitrobenzene, aniline, and phenol in mixture of 500 mg/L of each.**

**Table IV. Residual Organics Concentrations in the Liquid Phase After Simultaneous Adsorption and Biological Treatment of the Three Compunds Mixture in the Presence of Various PAC Concentrations**

| Time of Treatment, hr | PAC Concentration g/L | Residual Concentrations of Organics in the Liquid Phase in mg/L | | |
|---|---|---|---|---|
| | | Nitrobenzene | Aniline | Phenol |
| | 0.00 | 500,6 | 491.9 | 514.9 |
| | 1.54 | 355.0 | 425.9 | 471.1 |
| 0.4 | 3.09 | 99.0 | 375.5 | 399.1 |
| | 4.64 | 35.0 | 311.2 | 318.4 |
| | 6.18 | >0.1 | 262.0 | 244.6 |
| | 7.72 | >0.1 | 211.6 | 167.4 |
| | 0.00 | — | 486.9 | 476.4 |
| | 1.54 | 171.3 | 435.5 | 341.2 |
| 12.0 | 3.09 | — | 356.2 | 170.9 |
| | 4.64 | — | 225.9 | 52.9 |
| | 6.18 | 12.0 | 163.9 | 5.2 |
| | 7.72 | 4.7 | 100.6 | 1.5 |
| | 0.00 | — | 422.6 | 448.3 |
| | 1.54 | — | 387.3 | 208.1 |
| 18.0 | 3.09 | 42.0 | 299.5 | 2.5 |
| | 4.64 | 21.8 | 201.2 | 1.7 |
| | 6.18 | — | 130.9 | 2.4 |
| | 7.72 | 1.5 | 74.9 | 0.6 |
| | 0.00 | — | 462.3 | 457.1 |
| | 1.54 | 310.0 | 398.0 | 37.5 |
| 24.0 | 3.09 | 20.0 | 288.7 | 3.5 |
| | 4.64 | >0.1 | 169.5 | 2.2 |
| | 6.18 | >0.1 | 90.7 | 1.3 |
| | 7.72 | >0.1 | 32.9 | 0.6 |
| | 0.00 | 316.9 | 428.0 | 444.8 |
| | 1.54 | 102.0 | 79.9 | 1.2 |
| 48.0 | 3.09 | 7.4 | >0.01 | >0.001 |
| | 4.64 | >0.1 | >0.01 | >0.001 |
| | 6.18 | >0.1 | >0.01 | >0.001 |
| | 7.72 | >0.1 | >0.01 | >0.001 |

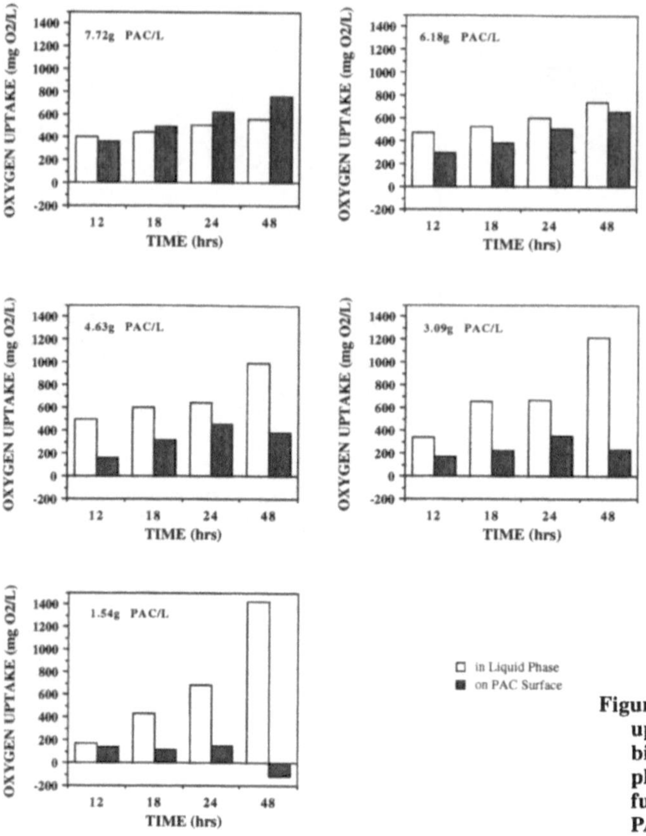


**Figure 14. Calculated oxygen uptake consumed for organics biooxidation in the liquid phase and on PAC surface as a function of process time and PAC concentration.**

found to be higher than the rates of the organics' bioremoval from the liquid phase. It seems that intensive biooxidation occurred directly on the PAC surface simultaneously with the liquid phase substrates' biodegradation. Figure 14 shows the calculated oxygen uptake consumed for the organics biooxidation in the liquid phase and on the PAC surface as a function of process time and PAC concentration.

From the data obtained in the trisolute solution regarding $Q^o$ and b, adsorption, oxygen uptake, and organics concentrations balance, it can be assumed that surface biooxidation of nitrobenzene occurred much quicker than in the liquid phase. Hence, a simultaneous adsorption and biodegradation process can be recommended for complete removal of the toxic organics from highly concentrated trisolute solutions. About 75% were removed just by biodegradation. The initial COD of 3350.8 mg $O_2$/L was reduced to residual 27.6 mg $O_2$/L. When 3.09 g PAC per liter treated water was added, no residual organics concentrations were detected after 48 hours of treatment.

## CONCLUSIONS

Simultaneous absorption and biodegradation process can be used for effective removal of the toxic organics, such as nitrobenzene, aniline and phenol, from highly concentrated single, bi- and trisolute solution.

It was found that these compounds had mutual effects on adsorption, biological and combined processes parameters. PAC adsorption characteristics can be used for prediction of its role in the simultaneous adsorption and biological treatment. Addition of PAC reduced significantly the inhibitory effect on biodegradation of these toxic organics, particularly in highly concentrated mixtures. Respirometric method can be recommended for optimal PAC dose determination.

## ACKNOWLEDGMENTS

The research was supported by The Ministry of Science and Arts, The Program for Immigrant Absorption and the Center for Absorption in Science, The Ministry of Immigrant Absorption, and The Ministry of the Environment, State of Israel. The grants from these ministries are gratefully acknowledged.

## REFERENCES

1. Asakawa, T. and Ogino, K. "Removal of trace organic compounds dissolved in water. VII Specific adsorption of phenol on activated carbon." *Yukagaku*, 33 (6), pp. 384–387, 1984.
2. McKay, G., Bino, M.J., and Altameni, A.R. "The adsorption of various pollutants from aqueous solutions onto activated carbon." *Water Res.*, 19 (4), pp. 491–495, 1985.
3. Smith, E.H. "Evaluation of multicomponent adsorption equilibria for organic mixtures onto activated carbon." *Water Res.*, 25 (2), pp. 125–134, 1991.
4. Vidic, R.D. and Suidan, M.T. "Adsorption of phenolic compounds by GAC." *Hazard. Ind. Wastes*, 23, pp. 374–385, 1991.
5. Hutton, D.G. and Robertaccio, F.L. Wastewater Treatment USA du Pont de Nemours, E.J. and Co., U.S. 3904518 750909, U.S. 702846 700114, 1975.
6. Galil, N. and Rebhun, M. "Powdered activated carbon combined with biological treatment to remove organic matter containing cresols and xylenols." *J. Environ. Sci. Health*, Part A, A (1), pp. 199–218, 1992.
7. Brenner, A., Belkin, S., et al. "Evaluation of activated carbon adsorption capacity by a toxicity bioassay." *Water Res.*, 27 (10), pp. 1577–1583, 1993.
8. Narkis, N. and Orshansky, F. "Upgrading of Biodegradation Process for the Removal of Toxic Organics from Polluted Waters by Addition of Activated Carbon." Annual Report, Ministry of Science and Arts, Israel, 1995.
9. Orshansky, F. and Narkis, N. "Characteristics of organics removal by PACT simultaneous adsorption and biodegradation." *Water Res.*, 31 (3), pp. 399–410, 1997.
10. U.S. Environmental Protection Agency. Effect of Hazardous Materials Spill on the Biological Treatment Processes. EPA-600/2-77-239, 1977.
11. Callahan, M.A. et al. Water Related Environmental Fate of 129 Priority Pollutants. EPA-44014079-029a, b, NTIS, 1977.
12. Keth, L.H. EPA's Priority Pollutants: Where They Come From—Where They Are Going. Water—1980, AICHE Symp. Ser. 77, 249–258, 1981.
13. *Industrial Water Pollution Control*, McGraw-Hill, New York, 1989, p. 177.
14. *Standard Methods for the Examination of Water and Wastewater* 18th ed. Am. Public Health Assoc. Am. Wat. Wks. Assoc. and Wat. Pollut. Control. Fed., 1992.
15. Narkis, N. and Orshansky, F. Optimal Combinations of Adsorption and Intensive Biological Treatment of Highly Concentrated Industrial Wastewaters and Hazardous Wastes. Annual Report, Ministry of the Environment, Israel (November), 1996.
16. Rozich, A.F., Gaudy, A.F. and Colvin, R.J. Application of Respirometry for Operating Activated Sludge Processes Treating Conventional and Hazardous Wastes, 60th Annual Conference Water Pollution Control Federation, WPCF, PA, 1987.
17. Gaudy, A.F., Jr., Rozich, A.F., Garniewski, S., Morran, N.R. and Ekambaram, A. Methodology for Utilizing Respirometric Data to Assess Biodegradation Kinetics. Proc. 42nd Annual Industrial Waste Conference, Purdue University, 1988, p. 573.
18. Korenman, J.M. *Photometric Analysis*, Chimija (Russ), 1975, p. 417.
19. Narkis, N. and Orshansky, F. Upgrading of Biodegradation Processes for the Removal of Toxic Organics from Polluted Waters by Addition of Activated Carbon. Annual Report, The Ministry of Science and Technology, Israel, March, 1993.
20. Narkis, N. and Orshansky, F. Upgrading of Biodegradation Processes for the Removal of Toxic Organics from Polluted Waters by Addition of Activated Carbon. Annual Report, The Ministry of Science and Technology, Israel, December, 1993.
21. Gaudy, A.F., Ekambaram, A. and Rozich, A.F. (1988). A Respirometric Method for Biokinetic Characterization of Toxic Wastes. 43rd Proc. Ind. Waste. Conf., 35–44.

# Section Five
# BIOLOGICAL: HIGH STRENGTH WASTES

## 18 PILOT TREATABILITY TESTING OF FOOD PROCESSING WASTEWATERS USING THE ANAFLUX™ ANAEROBIC FLUIDIZED-BED REACTOR


**Ravinder Menon**, Principal Engineer
Industrial Systems Group, Infilco Degremont, Inc.
Richmond, Virginia 23294

**Claude Delporte**, Research Engineer
Degremont SA, 78290 Croissy-sur-Seine, France

**Douglas S. Johnstone**, Regional Manager
Infilco Degremont, Inc., Scotch Plains, New Jersey 07076



### ABSTRACT

This chapter reports on the pilot testing conducted to evaluate the feasibility of using the ANAFLUX anaerobic fluidized-bed reactor technology toward treating high-strength wastewaters generated in a food processing plant. The reactor's ability to accommodate variations in influent hydraulic and organic loading as well as the time required to achieve and sustain optimum volumetric loading rates were examined.


### INTRODUCTION

A food processing plant in northern New Jersey (Plant F) produces canned fruits in syrup, jams, jellies, canned vegetables, and condiments. Wastewater (WW) is generated primarily from the various washing and rinsing operations performed on foodstuff and equipment.

Plant F has an existing wastewater treatment system consisting of screening, anaerobic pretreatment, followed by an aerobic activated sludge process. The anaerobic step consists of a packed-bed filter, installed in 1989. Changes in the product-mix and manufacturing practices have resulted in changes in the wastewater quantity and quality from the plant. The wastewater plant is consequently not able to handle the pollution load from the plant adequately at all times. Only a portion of the plant WW is sent to the anaerobic phase, the rest being bypassed to the aerobic step. Plant F is interested in investigating new technology in this area, which could treat the wastewaters produced at a greater efficiency.

The ANAFLUX anaerobic fluidized-bed reactor is a relatively novel concept in anaerobic treatment. The reactor uses mineral support media for bacterial attachment. The media is kept in a fluidized state by an upflow velocity of the forward feed plus recycle. Plant F was interested in a pilot test program to study the feasibility of using this technology for treating the wastewaters generated.

The objectives of the pilot study were threefold:

- To verify the feasibility of treatment using the ANAFLUX reactor
- To evaluate the reactor's ability to accommodate variations in influent characteristics
- To develop full-scale sizing parameters, in order to generate a proposal for a new ANAFLUX reactor system.


*52nd Purdue Industrial Waste Conference Proceedings*, 1997, Ann Arbor Press, Chelsea, Michigan 48118. Printed in U.S.A.

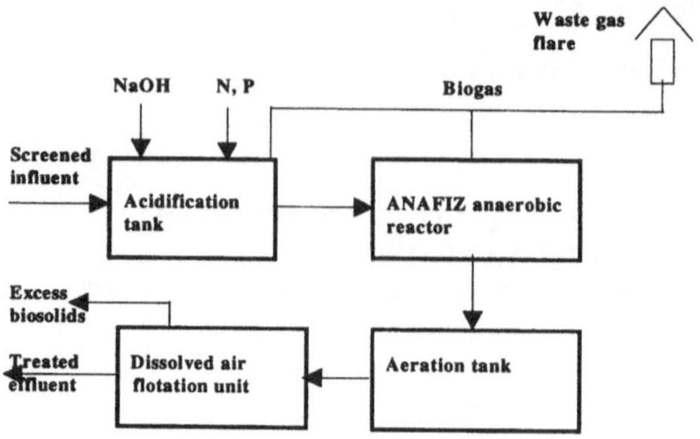


**Figure 1. Schematic of existing WWT system at Plant F.**

## Existing Wastewater Treatment Plant

The WW from the plant is screened and pumped into the acidification tank, where alkalinity and nutrients are added. Some flow equalization is also obtained in this step. The two primary phases of anaerobic treatment, hydrolysis and acidogenesis are accomplished in the acidification tank. The acidified WW then is pumped to the ANAFIZ™ packed-bed anaerobic reactor. This reactor is where methanogenesis takes place and the volatile fatty acids produced in the acidogenesis phase are ultimately converted into methane gas. Figure 1 illustrates the existing WWT plant at Plant F.

### *Influent Characteristics*

Average influent characteristics after screening are as follows:

| | |
|---|---|
| Flow Rate | 250 $m^3$/day |
| COD | 15000 mg/L (3750 kg/day) |
| $BOD_5$ | 10000 mg/L (2500 kg/day) |
| TSS | 1000 mg/L (250 kg/day) |
| Temperature | 38°C |

The influent is deficient in both nitrogen and phosphorus. Nutrient supply (N and P) is needed to maintain biological activity under optimum conditions.

### *Effluent Characteristics*

The treated effluent (DAF subnatant) averages the following results:

| | |
|---|---|
| COD | 400 mg/L |
| $BOD_5$ | 220 mg/L |
| TSS | 150 mg/L |
| pH | 6.5 to 8.6 |

Treated effluent is discharged into the city sewer system for further treatment by the POTW (publicly owned treatment works).

## DESCRIPTION OF THE ANAFLUX PROCESS:

The ANAFLUX process, which involves attached growth on a mobile medium, is particularly well suited to moderately strong wastewaters. The key advantages of the process are:

- no danger of clogging of the reactor, due to the fluidization of the granular medium
- rapid startup

- compact installation
- no danger of biomass entrainment in treated effluent
- accommodates wide variation in flow, within the velocity range acceptable for fluidization.

## Principle

Based on the fluidized bed principle, this unit consists of a reaction zone in which the raw influent passes through a mass of granular material, which it expands by its upward flow.

The medium, Biolite(, was selected for its specific characteristics:

- porous structure with a high specific surface area
- esistance to attrition
- manufactured under strictly controlled conditions.

The bacteria grow on the granular medium and form a biofilm ranging in thickness from 5 to 70 μm. This type of reactor allows concentrated growth of large populations of active bacteria, and therefore is designed for the highest volumetric loadings (30 to 80 kg of COD/$m^3$ of active biomass/day).

The attached biomass can reach concentrations of 40 to 50 kg of dry solids (DS) per $m^3$ (40,000–50,000 mg/L) of expanded medium; the limited thickness of the biofilm indicates that virtually all biomass is involved in the biological treatment process which, accordingly, allows a significant reduction in the mass loading (kg COD/kg DS/day) and results in high COD removal efficiencies of up to 90%. Figure 2 illustrates the principle of operation of the ANAFLUX reactor. Acidified influent and recycle are pumped by the fluidization pump to the bottom of the reactor. The upflow velocity inside the reactor is adequate to maintain the Biolite and the attached biomass in a fluidized state. The three-phase separator at the top separates gas, liquid, and solids. Biogas, consisting mainly of methane and carbon dioxide is withdrawn from the top of the reactor, whereas the treated effluent is withdrawn from an internal launder. Part of the effluent withdrawn goes back to the suction of the fluidization pump, along with the forward feed, in order to maintain the required velocity for fluidization. Media collects at the bottom of the cone and may be withdrawn from time to time and run through a media attrition pump, in order to maintain proper specific gravity and appropriate biofilm thickness.

The gradual colonization of the granular medium is controlled to prevent excessive losses and to allow wasting of excess sludge.

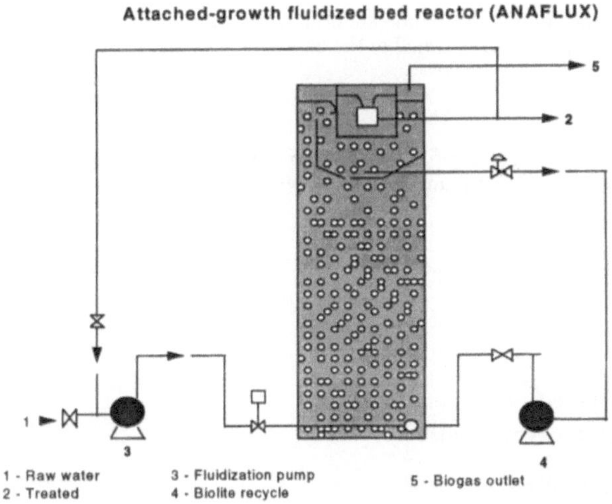


**Figure 2. Schematic of the ANAFLUX fluidized-bed reactor.**

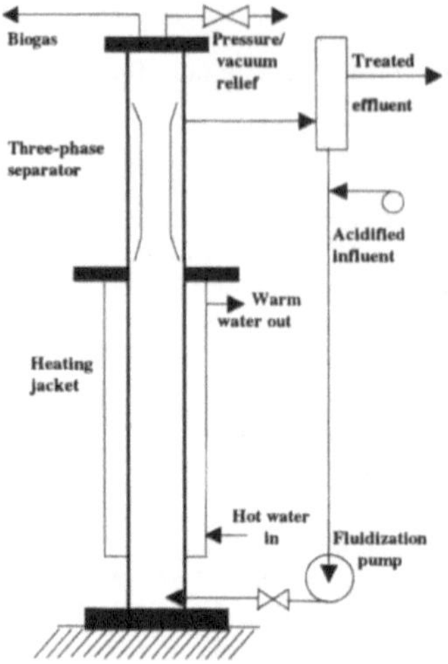


**Figure 3. Schematic of ANAFLUX pilot reactor used in the testing.**

## PILOT TESTING

### Description of the ANAFLUX Pilot Unit

Figure 3 presents a schematic of the ANAFLUX pilot reactor used in the testing. The reactor consisted of a 100 mm diameter acrylic column, approximately 2.5 meters in overall height. Biolite media was contained only in the bottom 1.8 meters of the cylinder, which was provided with a heating jacket to maintain proper temperature control at about 35–37° C. The acidified influent along with recycle was pumped to the bottom of the unit by the fluidization pump, which was a peristaltic pump (0.8 L/hour).

The WW used in the testing was acidified plant WW, which was already dosed with adequate quantities of nutrients and alkalinity. The acidified influent was stored in a plastic container, fitted with a stirrer. The forward feed pumps (one operating + one standby) pumped the acidified influent into the recirculation line, where it mixed with the effluent recycle and was fed into the fluidization pump. The fluidization pump produced a velocity inside the reactor, which was adequate to maintain an expansion of the Biolite media in the range of 30–50%. As the effluent flowed upward within the reactor, organic constituents were degraded with the resultant production of methane gas. Toward the top of the reactor, the three-phase separator separated treated effluent, biogas, and any entrained media granules with attached biomass. Treated effluent was withdrawn and partially recycled to the fluidization pump suction. Biogas produced was metered prior to discharge and any entrained Biolite was returned to the anaerobic treatment zone.

### Process Control/Analysis

Control analyses were performed on average samples in order to monitor the system and optimize process loadings and detention times. Pilot plant operations were monitored on the basis of the following daily measurements:

- acidified influent flow rate
- recycle flow rate
- temperature
- expansion of the Biolite.

The following parameters were analyzed on a regular basis:

$COD_{tot}$: Total chemical oxygen demand
pH
$M_{alk}$: Methyl orange alkalinity

| | |
|---|---|
| VFA: | Volatile fatty acids (by titration) |
| SS (TSS): | Total suspended solids |

The biogas produced was supposed to be metered, but due to problems with the instrumentation, no metering was done. All analyses were conducted according to the latest edition of the *Standard Methods*.

### Testing Schedule

| | |
|---|---|
| May 7 to May 8 1996 | Installation and startup of pilot unit |
| May 9 to May 15, 1996 | Acclimatization of the biomass |
| May 15 to May 28, 1996 | Increase in the organic loading: 13.8 → 17.6 kg of COD/$m^3$ of expanded Biolite/day |
| May 29 to May 30, 1996 | Sharp increase in loading → 39 kg of COD/ $m^3$ of expanded Biolite/day |
| May 31 to June 10, 1996 | Stabilization of loading at 30 kg of COD/ $m^3$ of expanded Biolite/day |
| June 12 to June 19, 1996 | Stabilization of loading at 35 kg of COD/ $m^3$ of expanded Biolite/day |
| June 20 to June 28, 1996 | Stabilization of loading at 40 kg COD/ $m^3$ of expanded Biolite/day |
| → June 24, 1996 | 0.4 liter of glue added to the influent |

## RESULTS

### Period from May 15–May 28, 1996

- Increase in organic loading: 13.8 → 17.6 kg COD/ $m^3$ expanded Biolite/day
- Average organic loading: 16.5 kg COD/ $m^3$ expanded Biolite/day
- Test period: 14 days

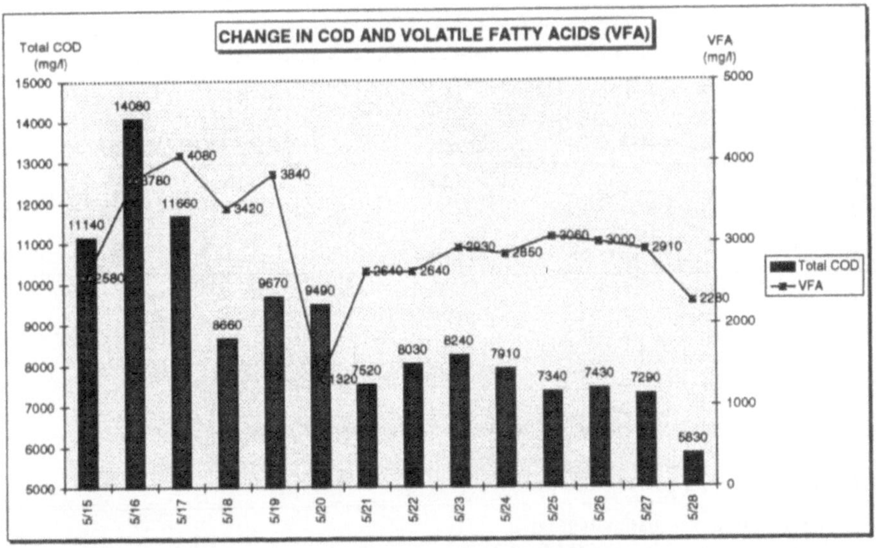


**Figure 4. ANAFLUX reactor influent characteristics (May 15–28, 1996).**

**Table I. ANAFLUX Reactor Influent Characteristics (May 15–28, 1996)**

| mg/L | pH | $M_{alk}$ mg/L ($CaCO_3$) | VFA mg/L ($CH_3COOH$) | $COD_{tot}$ mg/L |
|---|---|---|---|---|
| Minimum | 6.5 | 3200 | 1320 | 5830 |
| Maximum | 7.9 | 5500 | 4080 | 14080 |
| Mean | 7.35 | 4378 | 2952 | 8879 |

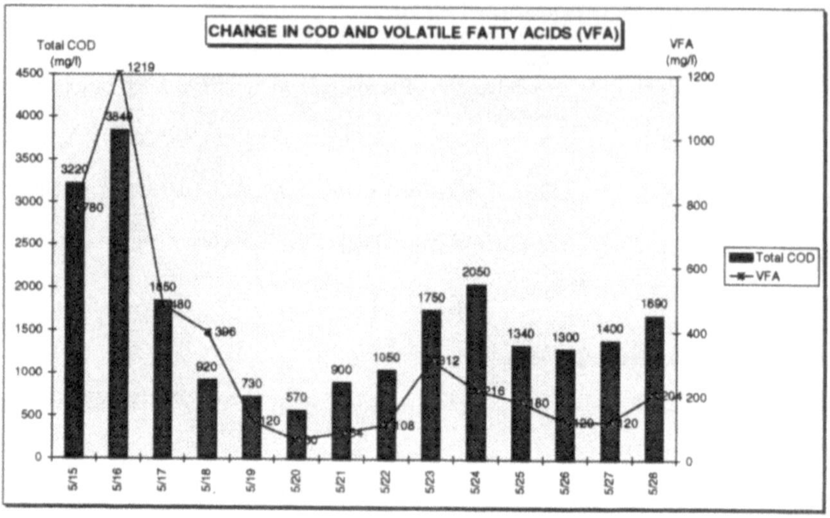


**Figure 5. ANAFLUX reactor effluent characteristics (May 15–28, 1996).**

**Table II. ANAFLUX Reactor Effluent Characteristics (May 15–28, 1996)**

| mg/L | pH | $M_{alk}$ mg/L ($CaCO_3$) | VFA mg/L ($CH_3COOH$) | $COD_{tot}$ mg/L |
|---|---|---|---|---|
| Minimum | 7 | 4200 | 60 | 570 |
| Maximum | 7.9 | 5560 | 1219 | 3840 |
| Mean | 7.55 | 5050 | 314 | 1615 |

The average COD removal efficiency during this period was 81.8%.

## Period from May 31 to Oct. 6, 1996

- Stabilization of organic loading:
  → 30 kg COD/ $m^3$ expanded Biolite/day
- Test period: 10 days.

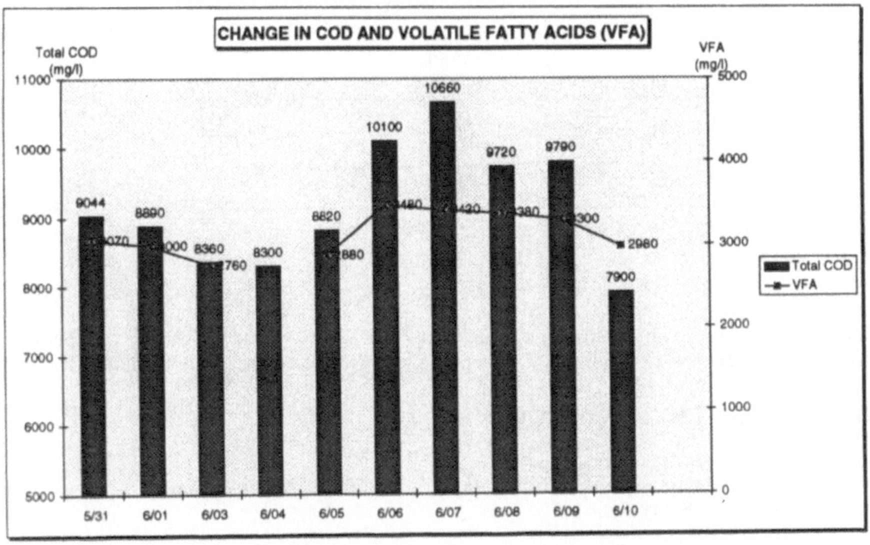


**Figure 6.** ANAFLUX reactor influent characteristics (May 31–June 10, 1996).

**Table III.** ANAFLUX Reactor Influent Characteristics (May 31–June 10, 1996)

| mg/L | pH | $M_{alk}$ mg/L ($CaCO_3$) | VFA mg/L ($CH_3COOH$) | $COD_{tot}$ mg/L |
|---|---|---|---|---|
| Minimum | 7.15 | 5100 | 2760 | 7900 |
| Maximum | 8 | 5400 | 3480 | 10660 |
| Mean | 7.6 | 5260 | 3141 | 9158 |

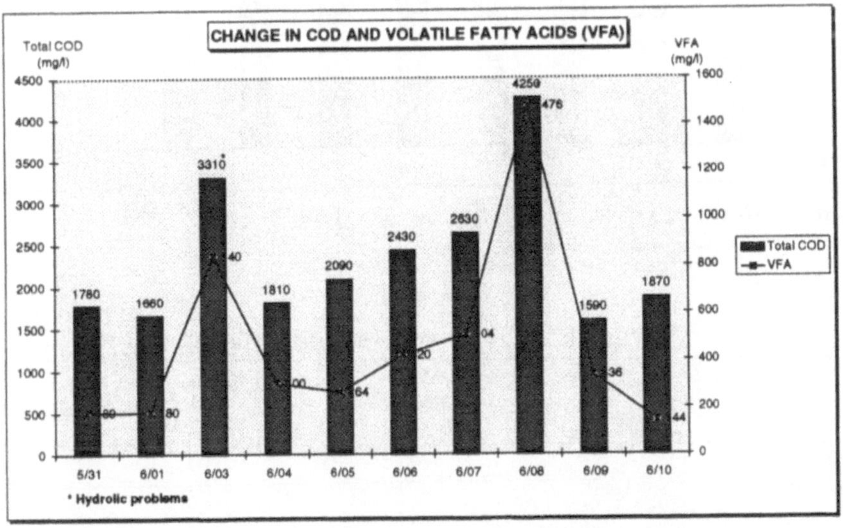


**Figure 7.** ANAFLUX reactor effluent characteristics (May 31–June 10, 1996).

**Table IV. ANAFLUX Reactor Effluent Characteristics (May 31-June 10, 1996)**

| mg/L | pH | $M_{alk}$ mg/L ($CaCO_3$) | VFA mg/L ($CH_3COOH$) | $COD_{tot}$ mg/L |
|---|---|---|---|---|
| Minimum | 7.6 | 5500 | 180 | 1590 |
| Maximum | 7.9 | 6610 | 1476[a] | 4250[a] |
| Mean | 7.7 | 5708 | 464 | 2342 |

[a]These values correspond to the malfunction.

The average COD removal efficiency during this period was 74.4 %. During this period, hydraulic problems related to fluidization disrupted the operation of the pilot installation.

## Period from June 12 – June 19, 1996

- Stabilization of the organic loading
  → 35 kg COD/$m^3$ expanded Biolite/day
- Test period → 8 days

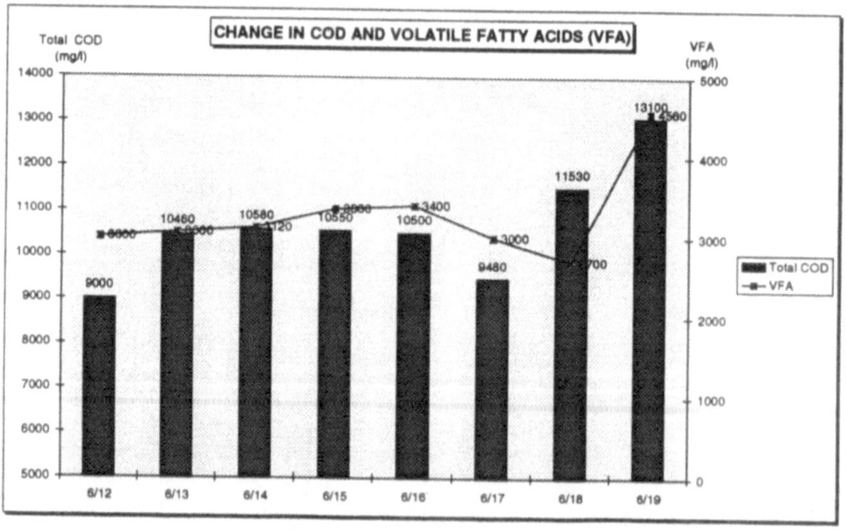


**Figure 8. ANAFLUX reactor influent characteristics (June 12–June 19, 1996).**

**Table V. ANAFLUX Reactor Influent Characteristics (June 12–June 19, 1996)**

| mg/L | pH | $M_{alk}$ mg/L ($CaCO_3$) | VFA mg/L ($CH_3COOH$) | $COD_{tot}$ mg/L |
|---|---|---|---|---|
| Minimum | 7 | 4700 | 2700 | 9000 |
| Maximum | 8.1 | 7000 | 4560 | 13100 |
| Mean | 7.45 | 5710 | 3275 | 10650 |

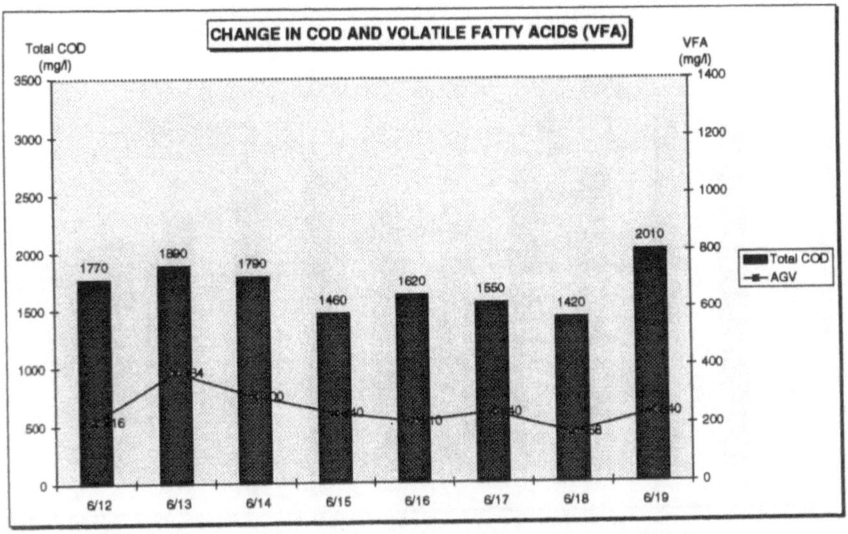


**Figure 9. ANAFLUX reactor effluent characteristics (June 12–June 19, 1996).**

**Table VI. ANAFLUX Reactor Effluent Characteristics (June 12–June 19, 1996)**

| mg/L | pH | $M_{alk}$ mg/L ($CaCO_3$) | VFA mg/L ($CH_3COOH$) | $COD_{tot}$ mg/L |
|---|---|---|---|---|
| Minimum | 7.3 | 5250 | 168 | 1420 |
| Maximum | 7.8 | 7400 | 384* | 2010 |
| Mean | 7.5 | 6054 | 250 | 1689 |

The average COD removal efficiency during this period was 84.1 %.

## Period from June 20–June 28, 1996

- Stabilization of the organic loading
  → 40 kg COD/m$^3$ expanded Biolite/day
- Introduction of glue into raw influent on June 24, 1996
- Test period → 9 days.

**Table VII. ANAFLUX Reactor Influent Characteristics (June 20–June 28, 1996)**

| mg/L | pH | $M_{alk}$ mg/L ($CaCO_3$) | VFA mg/L ($CH_3COOH$) | $COD_{tot}$ mg/L |
|---|---|---|---|---|
| Minimum | 7 | 6400 | 4200 | 12,220 |
| Maximum | 8.2 | 7800 | 6360 | 15,090 |
| Mean | 7.35 | 7070 | 4850 | 13,398 |

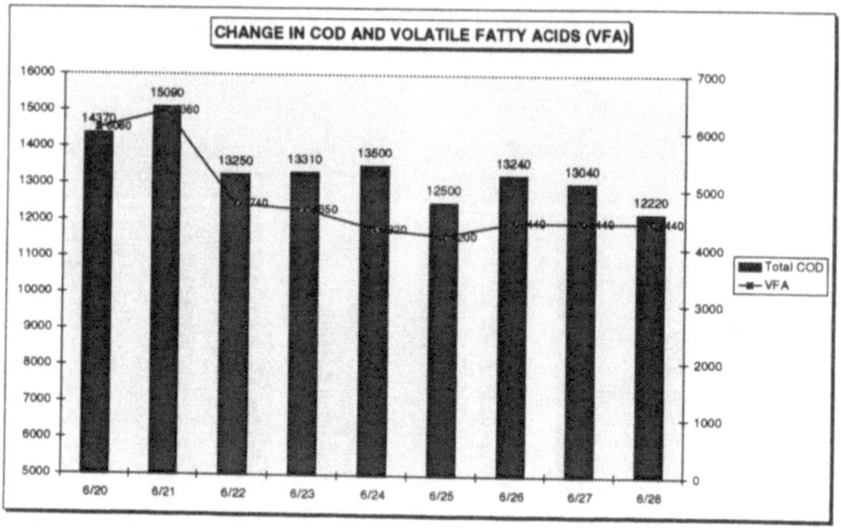


**Figure 10. ANAFLUX reactor influent characteristics (June 20–June 28, 1996).**

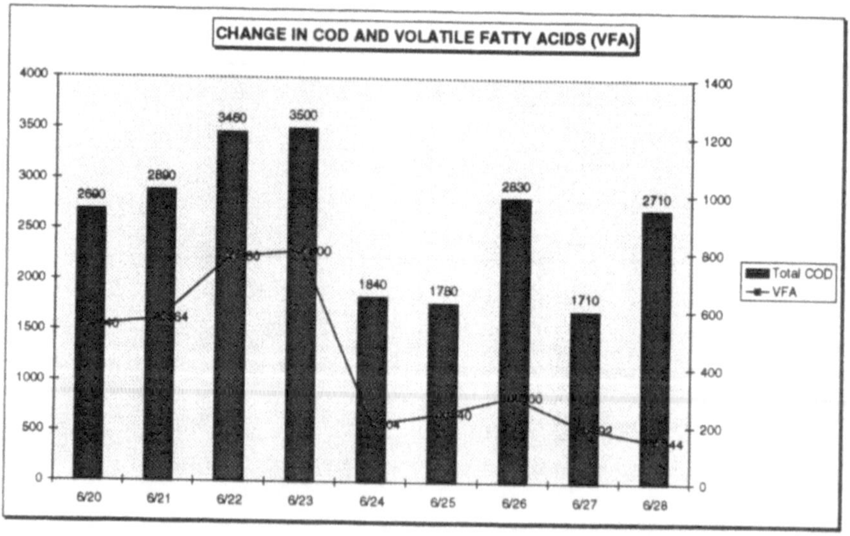


**Figure 11. ANAFLUX reactor effluent characteristics (June 20–June 28, 1996).**

**Table VIII. ANAFLUX Reactor Effluent Characteristics (June 20–June 28, 1996)**

| mg/L | pH | $M_{alk}$ mg/L ($CaCO_3$) | VFA mg/L ($CH_3COOH$) | $COD_{tot}$ mg/L |
|---|---|---|---|---|
| Minimum | 7.35 | 7400 | 144 | 1710 |
| Maximum | 7.9 | 8750 | 800 | 3500 |
| Mean | 7.7 | 7978 | 418 | 2601 |

**Table IX. Results of Glue Addition on ANAFLUX Pilot Reactor**

| | | ANAFLUX OUTLET | |
|---|---|---|---|
| | | VFA mg/L $CH_3COOH$ | $COD_{tot}$ |
| Prior to addition of glue | June 24 '96 | 204 | 1840 |
| After addition of glue | June 25 '96 | 240 | 1780 |
| | June 26 '96 | 300 | 2830 |
| | June 27 '96 | 192 | 1710 |

**Table X. Results of Sudden Spikes in Organic Loading on the ANAFLUX Pilot Reactor**

| Organics Loading kg COD/m$^3$ Expanded Biolite/day | VDFA mg/L $CH_3COOH$ | $COD_{tot}$ mg/L |
|---|---|---|
| 17.6 | 204 | 1690 |
| 35.4 | 180 | 1550 |
| 39 | 24 | 1880 |

The average COD removal efficiency during this period was 80.5%.

On June 24, 1996, 400 mL of glue were added to the raw influent to check the impact on anaerobic treatment. Table VII presents the results of this addition on the performance of the ANAFLUX pilot reactor.

The one-time addition of glue representing 0.5% of the volume had no significant impact on COD removal.

## Influence of a Sudden Rise in Organics Loading

On May 29 and May 30, 1996, the organic loading rose from 17.6 kg to 35.4 kg COD/m$^3$ expanded Biolite/day and from 35.4 to 39 kg COD/ m$^3$/day.

There is no significant difference in results for these two days.

The increase in biogas production caused an accumulation of condensate water in the flow meter, leading to an increase in reactor pressure with the ensuing problems. This is what caused the hydraulic problems observed during this period.

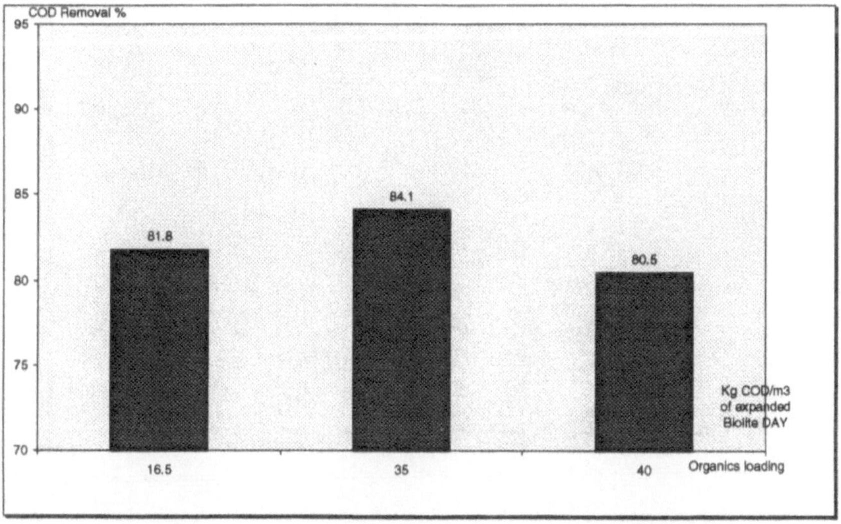


**Figure 12. Changes in COD removal efficiencies as a function of organics loading.**

### Changes in COD Removal Efficiencies as a Function of Organics Loading

No significant difference in the COD removal efficiency was observed over the loading range studied.

## CONCLUSIONS

The characteristics of the food processing wastewaters tested are compatible with the ANAFLUX anaerobic fluidized-bed reactor.

Tests of anaerobic treatment showed that COD removal rates of greater than 80% can be obtained for volumetric loading rate of 40 kg COD/m$^3$ of expanded Biolite/day. Variations in the volumetric loading rate (from 17–35 kg COD/m$^3$/day) due to the fluctuations in the COD content of the influent had no significant impact on COD removal rates.

An accidental discharge of glue in a proportion of 0.5% of the total flow does not disrupt the operations of the ANAFLUX reactor.

## REFERENCE

*Standard Methods for the Examination of Water and Wastewater*, 19th ed., A.P.H.A., A.W.W.A., and W.E.F., 1995.

# 19 INVESTIGATION OF SOLID WASTE DRYING BY AEROBIC BIOLOGICAL TREATMENT

**Joachim Brummack**, Dr.-Ing.
**Bernhard Gemende**, Dr.-Ing., speaker
**Peter Loetzsch**, Professor Dr.-Ing. habil.
**Silke Paar**, Dipl.-Ing.
Dresden University of Technology,
Process and Environmental Engineering Institute

## ABSTRACT

The application of biological prior to thermal treatment of municipal solid waste is very promising to facilitate separation of inert materials and homogenization of municipal solid waste. In addition, a remarkable energetic effect can be achieved. By using the biological treatment as a drying step, the input water content for a subsequent incineration could be considerably reduced and optimized. Corresponding pilot-scale tests in fixed bed and rotary reactors are presented. The experimental results are checked by energy balances and compared to the theoretical drying potential. Based on these experiences, optimal conditions for the biological process and parameters for its control can be derived.

## INTRODUCTION

In the last years, many efforts have been made to introduce biological treatment processes for municipal solid waste (MSW) prior to disposal. With these processes, the amount as well as the environmental problems related to MSW disposal could be remarkably reduced. Meanwhile, mechanical-biological waste treatment (MBWT) has become an alternative to mere disposal. Even with simple technological means, the process is very effective. The large-scale application proved a considerable limitation of degradation processes in the disposal body and a significant increase of the waste compaction density (up to the two-fold or even somewhat more) due to improved separation of materials. Whereas also the emissions during the aerobic degradation in open windrows cause no extreme enhancement with concentrations in the order, of or less than, the disposal background.[1] A critical and often controversial issue discussed is the natural (and much more technical) limit for the reduction of the organic content. By long-term biological treatment, the minimal achievable content of organic matter, expressed in terms of ignition loss, is about 20% of mass. This organic content may still implicate some disposal problems. Only by thermal treatment processes (e.g., incineration) the organic content could be further reduced to values of about 5%, which are, for example, prescribed by German disposal law. However, one of the major disadvantages of that biological treatment is that the energy (or organic content) contained in the waste is not used at all.

The combined application of biological and thermal treatment is also very promising with respect to an energetic utilization of MSW. By using the biological treatment as a drying step, the input water content for a subsequent incineration could be considerably reduced and optimized. Thus the energy efficiency compared to direct waste incineration is improved. To investigate the drying effect by aerobic biological treatment, several laboratory and pilot-scale tests were carried out. Main objectives of these tests were to determine the amount of the reaction heat generated and the air needed for a controlled biological drying process. Furthermore, a first assessment should be made upon the possible reduction of the waste moisture content and the implicated enhancement of the calorific value. For the experiments, both a static fixed bed and a rotary reactor

*52nd Purdue Industrial Waste Conference Proceedings*, 1997, Ann Arbor Press, Chelsea, Michigan 48118. Printed in U.S.A.

were used. The experimental results were checked by energy balances and compared to the theoretical drying potential. Based on these experiences, optimal conditions for the biological process and parameters for its control can be derived.

## APPLICATION OF MECHANICAL-BIOLOGICAL TREATMENT PRIOR TO INCINERATION

In former contributions, it was demonstrated that MBWT is very successful in reducing waste disposal risks and volumes.[1] However, even by sophisticated biological treatment, the organic content can only be reduced to about 20% in terms of the ignition loss or to about 7 to 10% of total organic carbon (TOC). This organic content still represents a long-term risk for waste disposal. Therefore, legislators have limited the organic content of waste to be disposed. Only by thermal treatment can a drastic reduction of the organic content be achieved. Nevertheless, this does not mean that biological processes will be obsolete in waste treatment.

### Separation of a High Calorific Fraction After Mechanical-Biological Treatment

A properly controlled mechanical-biological treatment facilitates separation of inert materials and the recovery of recyclables. As discussed with technological schemes in a previous publication,[13] such a separation can lead to an enriched high calorific waste fraction; e.g., for a subsequent incineration. The preceding biological treatment (usually with temperatures up to 70°C) detaches agglomerates or even compound materials and causes a homogenization of the waste mixture. A following mechanical separation and sorting-out of foreign materials generates an incineration input material with more uniform caloric characteristics which does not tend to block the transport mechanisms (e.g., grates, conveyors). Besides that path (I.), Figure 1 indicates another possibility for a combined application of biological and thermal waste treatment: the utilization of the heat generated by the biological reaction for waste drying.

### Drying by Aerobic Biological Treatment

By a short-term aerobic biological treatment, the waste mixture can be dried. Especially the initial intensive phase of this treatment with its high temperatures and larger heat generation can be effectively used to dry out the substrate. The process heat of the exothermal biodegradation of

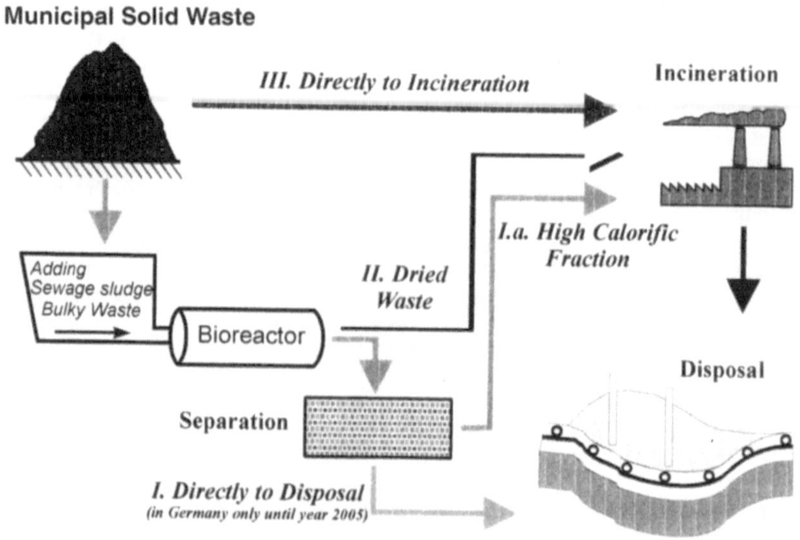


**Figure 1. Treatment paths for municipal solid waste.**

**Table I. Literature Citations for Water Content (in mass %) in Aerobic Biodegradation**

| Substrate | Minimum | Optimum | Maximum | Reference |
|---|---|---|---|---|
| Residual waste & sewage sludge | < 30 | 50 | 60 | 2 |
| Municipal solid waste & kitchen waste | 20 | 50 | 70 | 3 |
| Organic waste | | 40–60 | 70 | 3 |
| Organic waste | 20 | | | 4 |
| Organic waste | 25–35 | | | 5 |
| Organic waste | 25–30 | | | 6 |

organic substance delivers the energy to vaporize the water in the waste mixture, thereby decreasing moisture content and increasing the calorific value of the waste. However, if the moisture content is too low to support the transport and living processes of the microorganisms, the biological treatment will stop or at least the degradation rate will be remarkably reduced. Thus, a sort of auto-inhibition will take place, connected with a sanitization of the mixture. That is why this effect is sometimes also called thermo-stabilization of the substrate. In Table I some data are cited about this minimum moisture content, together with the moisture range given in some references. Obviously the water content is very much dependent on the substrate characteristics, like porosity, particle size distribution, material composition, and maximum water holding capacity. But it can be derived that the minimal moisture content is well below 30 mass % and dependent on the different substrate characteristics in the range of about 20 to 25 mass %. A further drying effect can only be achieved by other driving forces and energy sources; e.g., convection drying by fresh unsaturated air. That was obviously the case for the figures of less than 15 mass % given in Reference 7 for thermally stabilized waste mixtures.

In the initial intensive phase, mainly the easily degradable fraction of the organic substance is consumed in the bioreaction, because of its higher degradation rates. This portion and its energetic equivalent will be lost for a subsequent energetic (or thermal) utilization. Thus, the calorific heat of the dry substance (the combustion heat) will decrease. However, especially during longer storage periods (after renewed wetting by atmospheric influences), this easily degradable fraction is always subject to further decay, also with minor rates. It will usually be lost anyway for the energetic exploitation.

The aim of the biological drying is to have a net energetic and technological effect. It is important to effectively control the bioreaction/stabilization. The stabilization effect and the increase of the calorific value by the reduced moisture content should well balance the decrease in calorific value by the degradation of organic substance. Therefore, all process parameters (e.g., air supply, temperature regime, usuable reaction heat, mixture turning) have to be controlled and optimized. In the presented investigations, some of the main important process characteristics and parameters are studied. The discussed pilot tests and calculations are the basis for a technological utilization of the process. This will be the main focus of this study, though the process is likewise accompanied by the previously mentioned improvements of transport and separation characteristics (although not so emphasized, as after longer biotreatment).

## TESTS WITH A STATIC FIXED BED REACTOR

The main objective of the primary tests was to determine the heat generated by the biodegradation reaction. In addition, some further questions of the heat transfer to the air, as well as the heat loss, had to be experimentally tested. For these tests a static fixed bed reactor was used. It is a cylindrical reactor with a diameter of 573 mm and a height of 800 mm. The reactor has a perforated support plate (at a height of 336 mm) to hold the waste mixture and to separate it from the leachate and condensate, thus offering a maximum waste load of about 114 L. The air input is provided from the bottom of the reactor by a controlled fan. Besides the waste parameters (mass, water, and organic content), the temperature distribution and air velocities across and at different heights of the reactor were also measured. The influence of air supply and organic content on the reaction heat and the heat loss were studied.[8] Together with some issues of measuring and analytical devices, this should form the basis for the further pilot tests and the design of the process control.

To determine the maximal heat generated by the bioreaction, different waste mixtures were tested. Main substrate was source-separated organic waste from the household collection of the

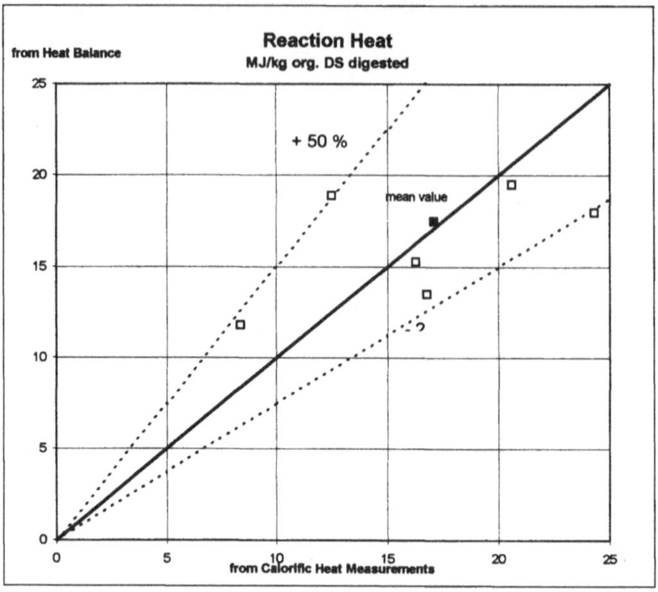


**Figure 2. Comparison of reaction heats determined from heat balance and from calorific heat measurements.**

city of Dresden. As mentioned, the main source for metabolism in the intensive initial phase are the easily degradable substances. Thus, that source should produce the maximum heat values under realistic conditions. The constituents were in different compositions:

- organic household waste (kitchen waste, vegetables, coffee filters, peels);
- yard waste (grass trimmings, leafs, branches, soil);
- paper;
- about 5 % inert and foreign materials (plastics, bags, metals).

To get a porous matrix and to alter the organic substrate content according to the test scheme, wood chops were added as structuring material.

The reaction heat was determined by two methods. At first, derived from the measured caloric (enthalpy, specific heat, temperatures) and mass parameters (including water content) of air and waste at input as well as at output (or throughout the process). For this, the heat loss had to be considered over the process time of 3 or 10 days. Secondly, samples were taken before and after the tests to experimentally measure the heat of combustion in a calorimeter. Some experimental results are given in Figure 2. The values for the reaction heat upon both methods are compared for the same degradation time and waste mixture. The values are not completely equal. However, they are in the same order (even for changing input substrate conditions) and within a range of –25 % to +50 %, with the more accurate values from the calorimeter measurements. But these values also include a measuring error of about 20 % mainly, resulting from the errors caused by the sampling-out of the inhomogeneous waste mixture.

The obtained mean value of about 16.5 to 17 MJ/kg degraded organic substance well coincides for both methods and is in good correspondence with data given in the literature for comparable substrates (see Table II). In general, the larger values are given for organic rich substrates, with the exception of kitchen waste.[10] Obviously, a number of parameters will influence the heat generated. First of all, whether or not optimal conditions for the bioreaction could be established; i.e., which reaction rate was prevailing during the measurements. This was also exactly the main outcome of the static fixed bed reactor tests. Those tests with inhomogeneous temperature profiles across the reactor (probably caused by local limitations in air supply directly to the organic substrate and the lack of mixing) ended up in lower values.

Similar values for the reaction heat (based on the calorific heat of the wet mixture) are also gained by estimation from the average biochemical reactions. As already noted, for this calcula-

**Table II. Literature Citations for Reaction Heat (in MJ/kg digested dry substance) for Aerobic Biodegradation**

| Substrate | Equipment, Duration [in days] | Reaction Heat $Q_R$ in MJ/(kg DS digested) | Ref. |
|---|---|---|---|
| Kitchen & yard waste | composting box, 14 d | 20 | 9 |
| Mixed organic waste | reactor, 10 d | 12.9 | 10 |
| Yard waste (leafs, branches, meadow trimmings) | reactor, 10 d | 17.1–24.1 | 10 |
| Kitchen waste | reactor, 10 d | 12–18.5 | 10 |
| Fine fraction of mixed municipal solid waste | reactor, 10 d | 20.5 | 10 |
| Municipal solid waste (residues after source separation) | mechanical-biological treatment | 14 | 11 |

**Table III. Estimation of Reaction Heat after BOIE,[12] Assuming Representative Substances for the First Intensive Biodegradation Phase[8]**

| Characteristic Representative | Chemical Formula | $Q_H$ in MJ/kg | Assumed Mass Fractions of Representatives | | | | |
|---|---|---|---|---|---|---|---|
| Hydrocarbon (mono- & polysaccharide) | $(C_6H_{10}O_6)_X$ | 13.5 | $x_{HC}$ | 0.6 | 0.7 | 0.8 | 0.4 |
| Fat (palmitic acid) | $CH_3(CH_2)_{14}COOH$ | 36.5 | $x_F$ | 0.2 | 0.2 | 0.15 | 0.1 |
| Protein (amino acid) | $CH_2(NH_2)COOH$ | 14.0 | $x_P$ | 0.1 | 0.1 | 0.05 | 0.1 |
| (Fresh primary) sewage sludge | $C_{22}H_{39}O_{10}N$ | 23.5 | $x_{SS}$ | 0.1 | — | — | 0.4 |
| Calculation<br>Equation from BOIE[12] with mass fractions of elements: C, H, N, O | | | $\Sigma x$ | 1.0 | 1.0 | 1.0 | 1.0 |
| $Q_H = 34.8C + 93.7H + 6.3N + 10.5S - 10.8O$ | | | $Q_H(x)$ | 19.1 | 17.5 | 17.0 | 19.8 |

tion in the intensive first phase (of about 3 days) of biodegradation, only the easily degradable substances have to be considered; whereas taking characteristic representatives for these substances and assuming a composition of them at the start of the process.[8] Then the reaction heat could be estimated by an equation of BOIE[12] from the portion of the individual elements in the substance. The main assumptions and calculations are summarized in Table III. Another simpler estimation could be made by assuming the composition of these easily degradable components in the mixture and calculating the reaction heat with the individual average calorific values given in the literature; i.e., hydrocarbons (14–17 MJ/kg), proteins (14 MJ/kg), and fats (36–40 MJ/kg). The obtained values correspond with the experimental results.

As Figure 3 indicates, during the tests a decrease of the (gross) calorific value for the dry substance was observed, mainly in the initial three days. It is due to a remarkable degradation of the organics accompanied by a reduction of the organic content by about 17 to over 30 %. This was expected with respect to the specified test objective of a maximum heat generation by the bioreaction, though this is not the objective for the application of the process itself as a drying step with a positive calorific effect. This effect was investigated in a rotary reactor.

## Tests with a Rotary Reactor

The rotary reactor has much better features to intensify the bioreaction by mixing the substrate, while exploiting the generated heat for drying. That is why it was chosen for the first pilot tests. In Figure 4 the rotary reactor is schematically sketched. The reactor has a diameter of 1200 mm and a length of 2920 mm. The thickness of the metal wall is 10 mm without insulations. The rotational speed can be controlled in the range from 0.1 to 7.2 revolutions per minute. Air input and output are provided through flanges at the end walls. Measuring devices for temperatures, air

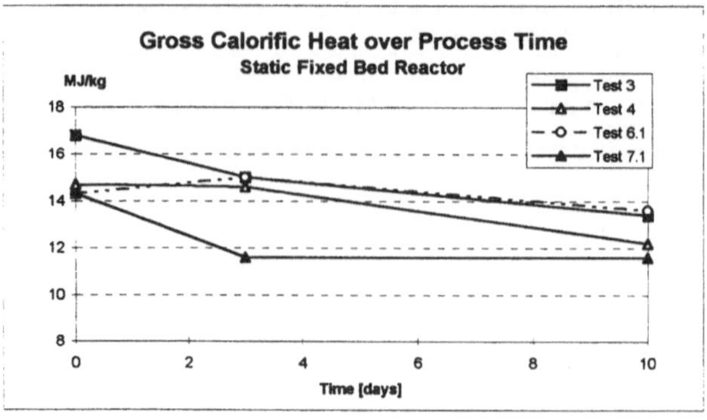


**Figure 3. Gross calorific heat over process time for fixed bed reactor tests.**

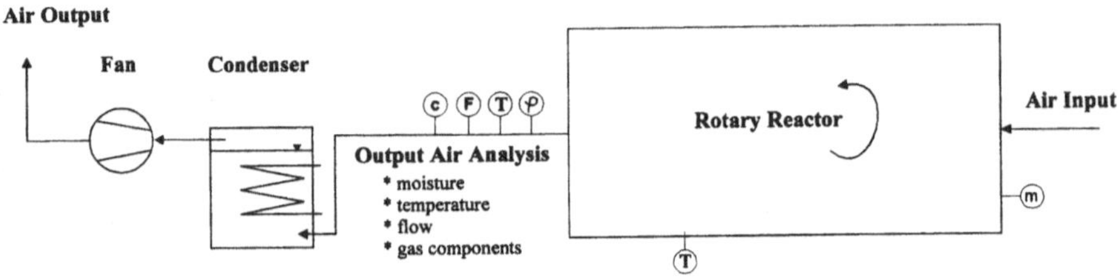


**Figure 4. Schematic sketch of rotary test reactor.**

flow, moisture contents, gas components, and mass permit a proper process monitoring and control.

In the pilot tests the following issues were investigated:

- finding appropriate parameters for an optimal drying process and its control;
- proving the effectiveness of stabilization under real conditions;
- determining water uptake by circulating air;
- determining air demand;
- deriving conclusions for the large-scale technological and equipment design as well as suitable parameters for a stable process.

To study a wide range of waste input, different mixture of municipal solid waste, source-separated organic waste and sewage sludge were tested. Table IV indicates the percentage range of the individual constituents at input. In Table V an example is given for the change of the main input parameters compared to the output after 4 days. The given experimental calorimeter values

**Table IV. Percentage Range of Input Waste Constituents**

| Main Constituents | Range [mass %] |
|---|---|
| Organic waste | 20–40 |
| Plastics | 10–20 |
| Textiles | 5–25 |
| Paper & cardboard | 15–25 |
| Metals & inert materials | 5–10 |
| Sewage sludge | 0–40 |

**Table V. Example for Main Parameters of an Aerobic Biodegradation Test in the Rotary Reactor**

| Parameter | Input | Output |
|---|---|---|
| Mass (kg) | 500.2 | 386.3 |
| Reactor filling ratio (volume %) | 60.0 | |
| Moisture (mass %) | 48.3 | 36.2 |
| Ignition loss (mass %) | 65.3 | 68.7 |
| (Gross) calorific heat of dry substance in the continuous phase (MJ/kg) | 13.6 | 13.5 |
| (Net) calorific heat of wet substance in the continuous phase (MJ/kg) | 5.9 | 7.7 |

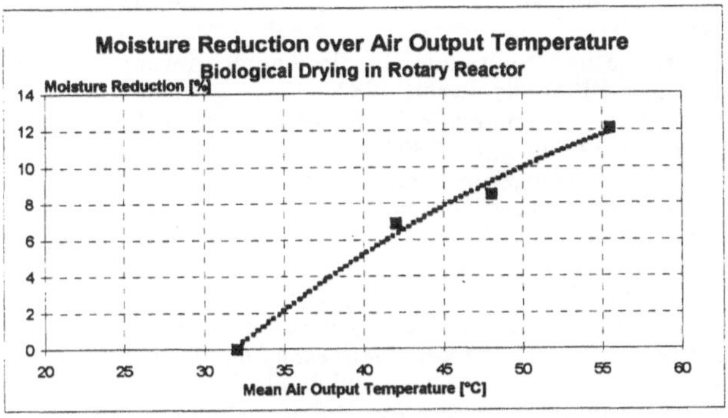


**Figure 5. Moisture reduction over air output temperature for rotary reactor tests.**

for the gross and net calorific heats are for the continuous phase. There the bulky materials as well as the larger, calorific-rich components; e.g., plastics, bags, have been separated. These substances are not determining the biodegradation. In addition they cause sampling problems, because it is rather difficult to take a representative sample from those larger components with respect to the mass content of the different fractions.

As Figure 5 shows, the moisture reduction is strongly depending on the air output temperature with its capacity to hold the moisture. Again, that temperature is mainly determined by the air flow, the generated reaction heat and the heat loss to the environment (and to heat up the substrate). These relationships can be modeled by heat and mass balances. A comparison of these calculations and the experimental results give a good correspondence.[8]

## SUMMARY AND CONCLUSIONS

The presented first tests and accompanying calculations demonstrate that the utilization of aerobic biodegradation as a drying step can substantially reduce the input water content and can potentially increase the calorific value of a waste mixture prior to incineration. Thus, a combined application of biological and thermal treatment is also very promising with respect to an energetic utilization of MSW. After optimizing the biological drying, the energy efficiency compared to direct waste incineration can be improved. The value of the reaction heat generated by the biotreatment is in the range of 15 to 20 MJ/kg with mean values of 16.5 to 17 MJ/kg, depending on the process and input waste parameters. This reaction heat can be effectively used to reduce the water content. The moisture reduction in the rotary reactor is strongly depending on the air output temperature. For an optimal moisture reduction, an air output temperature of about 65 °C or more could be derived from the calculations, supported by the experiments. From these experiments, mainly those with the rotary reactor, and the calculated balances, a set of optimal process parameters can be fixed. All experimental findings are in good correspondence with calculations resulting from balances or models. Besides the already mentioned values, the air demand for the bioreaction and the uptake of the reaction heat and the vaporized water was determined. The air demand for the biological drying is about 3 to 5 times greater than for the biotreatment itself.

Furthermore, the optimal filling ratio for the reactor can be derived, depending on the input waste characteristics. Appropriate parameters for monitoring and controlling the process progress and effect were tested and determined. The experimental results support the practical and theoretical drying potential of aerobic bioreactions. Some preliminary tests were carried out to determine the influence of the heat loss to the environment and through the output air and waste streams on the process effect. Corresponding improvements on equipment insulation and heat recovery possibilities were developed and are actually tested. The achieved results are very promising for an effective large-scale application.

## REFERENCES

1. Wotte, J., J. Brummack, S. Paar, and B. Gemende, Aeration Optimization and Emission Control of Biological-Mechanical Waste Treatment, In: *Proceedings of the 51th Industrial Waste Conference, May 6–8, 1996*, Purdue University, West Lafayette, IN. R.F. Wukasch, Ed. Ann Arbor Press, Chelsea, MI, 1996, p. 371.
2. Thomé-Kozmiensky, K.J., Systeme und Aggregate für die Kompostierung, In: *Kompostierung und Brennstoffgewinnung*, EF-Verlag, Berlin, 1983.
3. Bidlingmaier, W., Das Wesen der Kompostierung von Siedlungsabfällen, In: *Kumpf, Maas, Strub., "Handbuch Müll- und Abfallbeseitigung*, KZ 5305, Berlin, 1983.
4. Schuchardt, F., *Verlauf von Kompostierungsprozessen in Abhängigkeit von technisch-physikalischen und chemischen Randbedingungen*, Essen, 1988.
5. Fricke, K., Rottesteuerung und Qualtiätssicherung, In: *Grundlagen der Kompostierung*, EF-Verlag, Berlin, 1990.
6. Wiemer, K., *Mechanisch-Biologische Restabfallbehandlung nach dem Trockenstabilatverfahren*, Abfall-Wirtschaft-Neues aus Forschung und Praxis, Witzenhausen, 1996.
7. Wiemer, K. and M. Kern, *Mechanisch-Biologische Restabfallbehandlung nach dem Trockenstabilatverfahren*, Abfall-Wirtschaft-Neues aus Forschung und Praxis, Witzenhausen, 1995.
8. Hieke, J.R., Untersuchungen zur Konditionierung von Restabfällen. Diploma Thesis, Dresden University of Technology, Process and Env. Eng. Institute, 1997.
9. Wiegel, U., Optimierung der Vorrotte von getrennt gesammelten Küchen- und Gartenabfällen durch prozeßgesteuerte Rotteführung in geschlossenen Reaktoren, Research Report, ITU, Berlin, 1989.
10. Krogmann, U., *Kompostierung: Grundlagen zur Einsammlung und Behandlung von Bioabfällen unterschiedlicher Zusammensetzung*, Economica Verlag, Berlin, 1994.
11. Reinhardt, T. and J. Jager, Schadstoffbelastung der Abluft bei mechanisch-biologischer Restabfallbehandlung und anschließender Deponierung, In: Wiemer, K. and M. Kern, *Biologische Abfallbehandlung 3*, Abfall-Wirtschaft-Neues aus Forschung und Praxis, Witzenhausen, 1996.
12. Boie, W., *Vom Brennstoff zum Rauchgas*, Teubner Verlagsgesellschaft, Leipzig, 1957.
13. Wotte, J., J. Brummack, S. Paar, and B. Gemende, Biological-Mechanical Waste Treatment Tests at a Landfill in East Germany, in *Proceedings of the 50th Industrial Waste Conference, May 8–10, 1995*, Purdue University, West Lafayette, IN. R.F. Wukasch, Ed. Ann Arbor Press, Chelsea, MI, 1996, p. 319.

# 20 CASE STUDY FOR FRUIT FLY FACILITY WASTEWATER MANAGEMENT: AGRICULTURAL PROCESSING FOR LAND LIMITED CONDITIONS

**J. Tao**, Graduate Research Assistant

**P. Y. Yang**, Professor
Dept. of Biosystems Engineering
University of Hawaii
Honolulu, Hawaii 96822

## ABSTRACT

A fruit fly facility on the island of Oahu, Hawaii was investigated for its wastewater characteristics and treatment/disposal alternatives. Based on wastewater characteristics and bench-scale study, an on-site intermittently-aerated batch operated treatment was installed to treat the fruit fly wastewater. With F/M ratios maintained at the level of 0.2 g COD/g MLSS/day or below, a more than 90% removal of influent COD can be achieved. The system also achieved average total nitrogen, total phosphorus, $BOD_5$, TSS removal efficiencies of 86%, 60%, 98%, and 92%, respectively. The overall system performance and cost analysis showed that this treatment system had very stable and excellent operational performance while physically it was very simple, economical, and land saving for an on-site small wastewater treatment system. This treatment system integrated with the existing irrigation system successfully provides the most effective wastewater treatment/reuse management solution among all alternatives currently available to the fruit fly facility.

## INTRODUCTION

Agricultural processing wastewater can be high or low volume of wastewater flow with high organic content compared to domestic sewage. Management of the agricultural processing wastewater with land-limited conditions is rather difficult, especially when it cannot be discharged into the existing domestic sewer system and only limited land is available for final disposal. The Fruit Fly Rearing Facility, owned by the California Department of Food and Agriculture (CDFA), located on a limited farm land on the island of Oahu, Hawaii, produces sterilized fruit flies for biological control of fruit fly infestation. It also creates high organic content of wastes and wastewater.

The facility experienced great difficulties in managing the wastewater disposal ever since the wastewater was no longer allowed to be discharged into the sewer system. Drip irrigation of the untreated wastewater for a 7-acre farm area was initially approved by Hawaii State Department of Health. After the operation period of over a year, a final approval of this practice was still unattainable because the performance of the installed irrigation systems could not meet the state regulatory requirements. The high concentrated wastewater caused the development of anaerobic conditions in the soil, rock, and pipelines as well as the storage tanks. This generated tremendous obnoxious odor and attracted flies in the field and around the irrigation system, not to mention other adverse environmental impacts and maintenance problems. It is required that an appropriate management of the wastewater generated from the fruit fly facility needs to be developed for the protection of the environmental quality.

This paper explains how a biological wastewater treatment system was integrated into the existing irrigation system for an effective fruit fly wastewater management over the period of September, 1995–August, 1996.

*52nd Purdue Industrial Waste Conference Proceedings*, 1997, Ann Arbor Press, Chelsea, Michigan 48118. Printed in U.S.A.

## WASTEWATER CHARACTERISTICS

The CDFA fruit fly facility operated two shifts daily. During the morning shift, the wastewater was originated from general utensil, facility, and floor wash. At nights, the used fruit fly diet was scraped off the trays and collected for animal feed. The trays were then washed for use at the next morning shift. The two shifts contributed almost equally to the total amount of the wastewater. The fruit fly media used mostly contained wheat mill, grain, sugar, and torula yeast, indicating that the media should contain a major portion of organic matter with a minor portion of inorganic material.

The wastewater originated primarily from washing and cleaning, and discharged to a concrete collecting tank as shown in Figure 1. From September 28, 1995 to June 19, 1996, grab samples were collected once every weeks from the collection tank and analyzed for overall characteristics. Significant variations in wastewater quality were experienced and it was decided to prepare the cumulative frequency of occurrence graphs. Table I presents 50 and 90 percentile probability values of wastewater characteristics based on 44 samples. The wide fluctuation in the wastewater characteristics is apparent.

There were wide fluctuations in the wastewater characteristics. This was mainly due to the solid content of the wastewater. With good inplant cleaning procedures and solid screening, the solid content of wastewater in the collecting tank was mainly soluble. The soluble COD to total COD ratio was as high as 83.3% (50 percentile level). The suspended solid content of 450 mg/L (50 percentile level) reasonably represented the average suspended solid content of the wastewater, as did the total COD of 2400 mg/L and other parameters. The biodegradability of the wastewater was about 90%, based on the laboratory experiment of soluble COD removal test as shown in Figure 2. Therefore, biological processes would be appropriate for treatment of this wastewater.

The wastewater also had a wide variation in daily flowrates. The wastewater flows were largely determined by the intensity of fruit fly production. Figure 3 shows the monthly 50 and 90 percentile values of the daily wastewater flows from January to August, 1996.

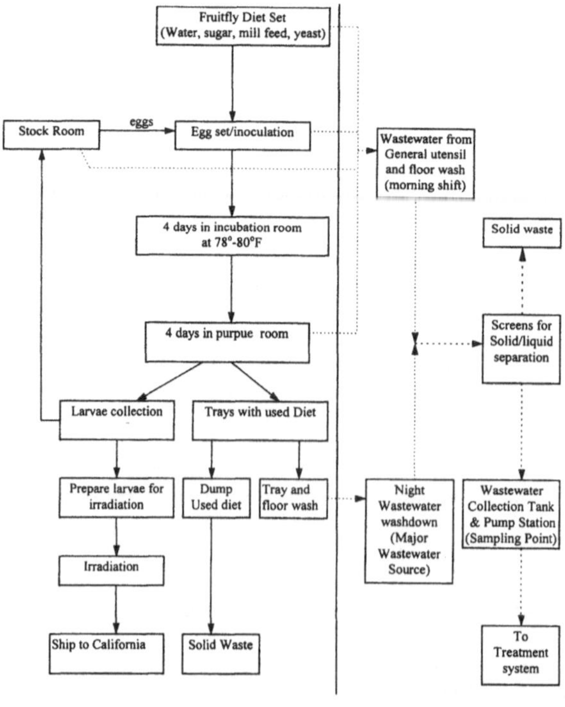


**Figure 1. The fruit fly facility process flow, sources of wastewater and sampling point (collecting tank).**

**Table I. Fruit Fly Wastewater Characteristics (sample taken from 9/28/95 to 6/19/96)**

| Characteristics | Range | 50 Percentile Level | 90 Percentile Level |
|---|---|---|---|
| pH | 3.2–5.87 | 4.1 | 4.75 |
| Total solids mg/L | 425–5100 | 1750 | 3350 |
| Total suspended solids mg/L | 90–2570 | 450 | 2100 |
| Total COD, mg/L | 734–6100 | 2400 | 5200 |
| Soluble COD, mg/L | 661–4444 | 2000 | 2800 |
| Total $BOD_5$, mg/L | 535–2700 | 1400 | 2120 |
| TKN, mg/L | 6–190 | 86 | 142 |
| $NH_4^+$-N, mg/L | 1–35 | 9.1 | 19.7 |
| $NO_3^-$-N, mg/L | 0.5–15 | 3.2 | 9 |
| Total phosphorus as mg P-$PO_4$/L | 4.9–113 | 32 | 50 |

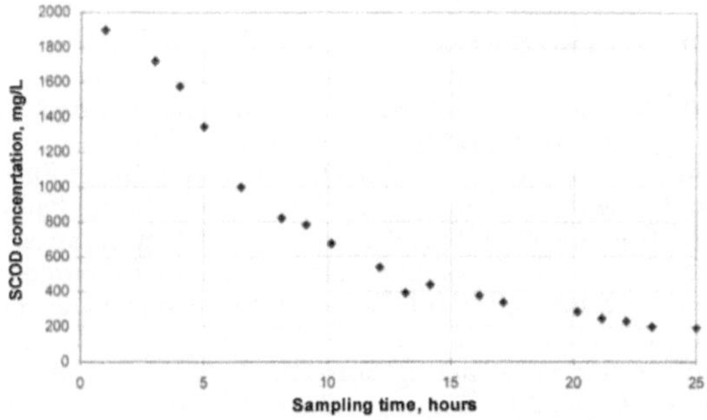


**Figure 2. Fruit fly wastewater biodegradability curve.**

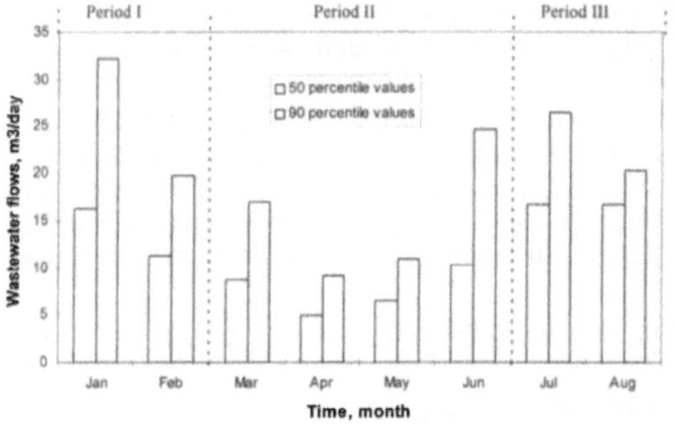


**Figure 3. Monthly 50 and 90 percentile values of wastewater flow.**

The pattern of wastewater flow shown in Figure 3 reflects the intensity of fruit fly production for the nine months period. According to the fruit fly facility, January, June, July, and August were the full fruit fly production period. After January, the fruit fly production and wastewater flows decreased until June, when the full production was resumed and would continue.

The high flow in January was significantly contributed by wet weather. Based on the wastewater flow study above, it was estimated that during the normal production period, the average daily wastewater flow was about 17 $m^3$ (4500 gallons). The maximum daily wastewater flow was over 24.6 $m^3$ (6500 gallons).

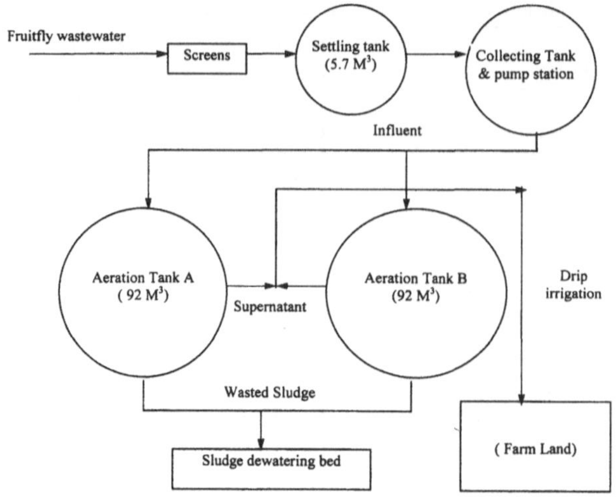


**Figure 4. Flow diagram of the fruit fly wastewater treatment/reuse system.**

## DESCRIPTION OF ON-SITE WASTEWATER TREATMENT SYSTEM

Based on wastewater characteristics and bench-scale treatment study, an intermittent-aeration batch-operated activated sludge treatment system was designed and installed at the CDFA fruit fly rearing facility. The treatment process is very similar to a traditional Sequencing Batch Reactor[1] except that the aeration is intermittent during the reaction period. Two 92 $m^3$ (7.3 m diameter × 2.4 m height) tanks receive wastewater from the 6′ (1.8 m) diameter × 5′ (1.5 m) reinforced concrete collecting tank/pump station. A 5 HP air blower supplied air with an air flow rate of 4.7 $m^3$ per minute (165 cfm). The air discharged through 250 holes (4 mm in diameter) evenly drilled in a 1.8-m diameter concentric circle of pipe near each tank floor. The blower was operated intermittently by a timer for different aeration ratios during different study periods. At the end of each treatment cycle, the aeration was stopped for settling, followed by discharging the supernatant for drip irrigation. The discharge pipe outlet in the aeration tank was adjusted so that the discharged supernatant accounted for only about half of the liquid volume contained in the aeration tanks. After the discharge, the equal amount of raw wastewater was then pumped to the tanks for another treatment cycle. The treatment cycle was normally two days to maintain a HRT of four days for the aeration tank. The treated effluent was distributed by a pump for drip irrigation. A plastic filter with a disk element (equivalent 155-mesh screen) was positioned after the pump. This filter could further remove solids to avoid plugging of irrigation pipe lines.

The treatment units with the two reactors were completely set up and integrated into the existing irrigation system in late December, 1995 at California fruit fly rearing facility. Since then, wastewater and microbial culture were fed into the two reactors. The whole treatment/disposal system consists of solid/liquid separation units, a collection tank/pump station, two aeration tanks/reactors, an air blower, a dissolved oxygen monitoring unit and irrigation pipelines as shown in Figure 4.

After system startup, intermittent aeration with three different aeration ratios were used in three periods during the nine-month study.

Period I: January 20, 1996–February 29, 1996. Aeration Ratio: 3:2 (6 hours air on, 4 hours off).

Period II: March 1, 1996–July 7, 1996. Aeration Ratio: 1:1 (2 hours air on, 2 hours off).

Period III: July 12, 1996–August 31, 1996. Aeration Ratio: 2:1 (4 hours air on, 2 hours off).

## SYSTEM PERFORMANCE

The operation of the on-site treatment system including two parallel reactors (Tanks A and B) started in the last week of 1995. After system startup, the aeration time was adjusted and three

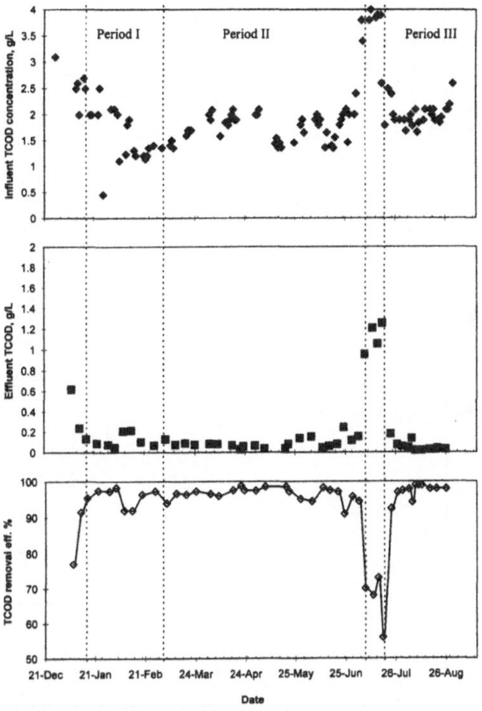


**Figure 5. Influent and effluent COD concentration and COD removal efficiency for Tank A.**

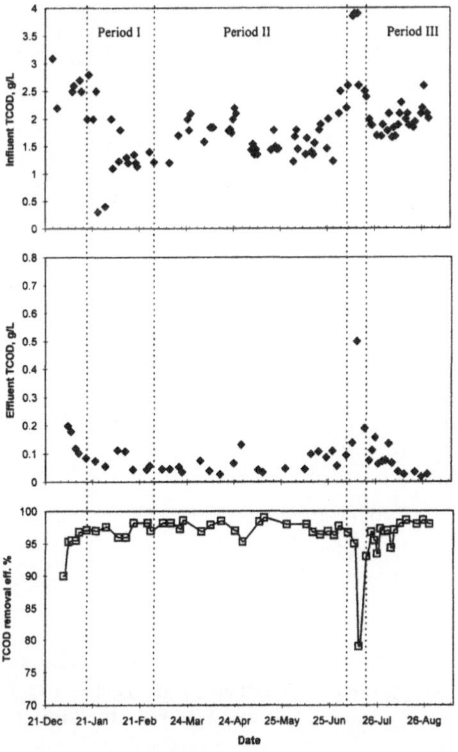


**Figure 6. Influent and effluent COD concentration and COD removal efficiency for Tank B.**

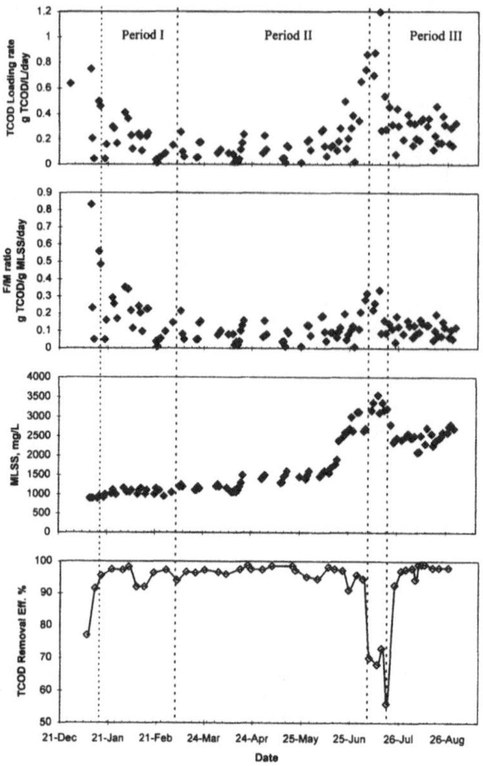


**Figure 7. TCOD removal based on COD loading rate and F/M ratio for Tank A.**

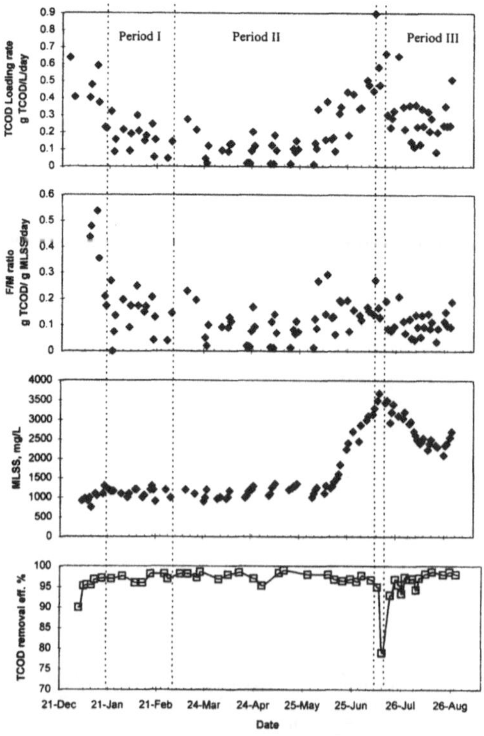


**Figure 8. TCOD removal based on COD loading rate and F/M ratio for Tank B.**

aeration ratios were studied. Figures 5 and 6 show influent and effluent characteristics and TCOD removal efficiency of Tanks A and B during the nine-month study period. Figures 7 and 8 show the TCOD removal efficiency and effluent quality based on TCOD loading rates and F/M ratios for both tanks.

## Startup Period

The startup period for both tanks lasted 2–3 weeks from the last week of 1995 to mid-January, 1996. In the initial startup, the wastewater was pumped to the two reactors. Since the daily wastewater flow during the startup period was 15 to 30 $m^3$ and the reactor volume was 90 $m^3$ each, it took about a week to fill the two reactors to 2/3 full. The reactors were continuously aerated since the wastewater was filled. At the beginning of the startup, enormous foaming was noticed. This is considered a common occurrence when intensive aeration is used.[2,3] The reactors were maintained and operated less than 2/3 full. This prevented the foam and adhered microbial flocs from creeping over the sides of the reactors and onto the ground. The foam gradually decreased and almost disappeared at the end of the startup period of about 2 weeks. As showed in Figure 5, the effluent TCOD was reduced to a level of 0.1 g/L after startup periods.

## Period I

Period I was from January 20 to February 29, 1996, when the aeration time was controlled by a timer with an aeration ratio of 3:2 (6 hours on, 4 hours off). The decanting outlet was initially built in the reactors so that about half of the full liquid volume (42 $m^3$) could be discharged. In the full-scale treatment system, the reactors could not be fully filled daily because the daily wastewater volume was less than 40 $m^3$. The wastewater was intermittently filled in a treatment cycle. Depending on the wastewater generation, it took two to three days to fill the reactor up during Period I. After being filled up, the reactors were continuously aerated for another day before the settling for discharge. Therefore, the treatment cycle for Period I was about 3 to 4 days, with a HRT of 6–8 days.

As shown in Figures 5 and 6, the effluent TCOD was below 0.15 g/L, with a TCOD removal efficiency higher than 90%. The TCOD loading rate was below 0.4 g TCOD/L/day. However, due to the low MLSS concentration level of 1.0g/L, the F/M ratio was up to 0.4 g TCOD/g MLSS/day.

## Period II

Period II was from March 1 to July 7, 1996, when the aeration ratio was adjusted to 1:1 (2 hours on, 2 hours off). The daily aeration time was 12 hours; 25% less than Period I. The fruit fly production was slowed down in this period until early June 1996. Before early June 1996 in Period II, the wastewater generated was less than half of normal fruit fly production period. The low wastewater volume resulted in even longer treatment cycles (4–7 days) and HRTs (8–14 days) as compared to Period I. This long SBR cycle and HRTs had an advantage of biosolids minimization. Since the operation of this system, the sludge has not been wasted and disposed. The high HRTs (HRTs > 4 days ) not only favored the treatment efficiency, but also avoided the troublesome sludge disposal problem. In Figures 7 and 8, the loading rates were all below 0.4 g TCOD/L/day with averages of 0.1 TCOD/L/day. The overall F/M ratios were below 0.2 g TCOD/g MLSS/day. The effluent TCOD was averaged 0.080 g/L with a TCOD removal efficiency of over 95%. The system had an excellent performance under a very low organic loading during the three-month low fruit fly production period.

Since June, 1996, the full fruit fly production was resumed with high volume of wastewater. The SBR cycles were therefore reduced to 2–3 days with HRT of 4–6 days. As shown in Figures 5 and 6, influent TCOD was greatly increased, as was the volume. It was later found that during that period, night shift workers did not scrap off and separate the fruit fly diet from trays in order to save labor and time. This practice not only increased both the suspended and soluble solid content of the wastewater, but also required more washing water. The TCOD values of the wastewater reached over 4 g/L, almost doubling the average TCOD values. The high volume and high concentrated wastewater were not equally pumped to the two reactorS. Instead, Tank A received the wastewater most of the time, resulting in more serious overloading than Tank B. As shown in Figure 7, the loading rate increased from 0.4 g/L/day to 1.2 g/L/day. The F/M ratios reached the level of 0.3gTCOD/gMLSS/day, or even higher. A significant increase in the MLSS concentration was also apparent in Figure 7. The increase of the MLSS value may be partly due to higher suspended solid content of the influent wastewater. Therefore, the actual ratio of food to microorganism would be even higher than the F/M values shown in Figure 7. The higher organic loading and F/M ratio explained the increased effluent TCOD value and eventually caused the system to overload. The two reactors were then restarted with continuous aeration and were back in normal within a week.

### Period III

Period III was from the system restartup in July 8 to August 30, 1996. After the restartup, the aeration ratio was adjusted to 2:1 (4 hours on, 2 hours off). The daily aeration time was increased to 16 hours; 25% more than Period II.

In this period, an inplant wastewater minimization program was enforced by the fruit fly rearing facility. More night shift workers were assigned to scrap off the fruit fly diet before cleaning the trays. As showed in Figure 3, the maximum wastewater (90 percentile value) was cut down by 20% in August. More importantly, it decreased the wastewater organic concentration. It also suggested avoiding daily wastewater being pumped into one single tank. This helped to reduce the risk of overloading. Since then, the system has been running very successfully again.

As shown in Figure 7, there was a decrease in MLSS concentration which was maintained at 2000 mg/L level. The loading rates were reduced to below 0.45 g TCOD/L/day. The F/M ratios were controlled under 0.2 g TCOD/g MLSS/day. The system was run with a 3-day treatment cycle with HRT of 6 days. Under these conditions, the treatment system produced effluent with average TCOD value of 0.1 g/L. The TCOD removal efficiency reached over 95%. In comparing the overall process performance of the three operation periods, it is apparent that the operation of Period III is appropriate to maintain the treatment efficiency.

## OVERALL EFFLUENT QUALITY

The two reactors had a similar pattern of treatment performance and effluent qualities. Before July 1996, the treatment system was very successful with average effluent $BOD_5$ and TCOD of 29 mg/L and 83 mg/L, respectively, and TCOD removal efficiency of over 95%. The record showed a couple of very high daily wastewater flows of more than 38 $m^3$ (10,000 gallons) in January. This was due to heavy rain. The high volume and low concentration of the wastewater during the heavy rain period did not impose any serious problem to the proper functioning of the treatment system. However, a temporary system failure for both reactors occurred in July, as shown in Figures 5–8. The system failure was found to be caused by system overloading. Despite this, the treatment system has been operating very successfully since the startup in late 1995. Table II shows the average effluent quality from mid-January to August 31, 1996.

**Table II. Effluent Quality of Tank A and B[a]**

| Parameters | Influent (50 Percentile Values) | Effluent of Tank A | Effluent of Tank B |
|---|---|---|---|
| TCOD, mg/L | 2400 | 82.9 ± 49.7 | 83.1 ± 42.3 |
| $BOD_5$, mg/L | 1400 | 28.9 ± 21.2 | 25.6 ± 16.1 |
| TKN, mg/L | 86 | 8.3 ± 9.2 | 8.9 ± 7.7 |
| $NH_4^+$-N, mg/L | 9.1 | < 3 | < 3 |
| $NO_3^-$-N, mg/L | 3.2 | 1.9 ± 2.8 | 2.9 ± 4.1 |
| T-P, mg/L | 32 | 13 ± 5.3 | 12.4 ± 8.2 |
| pH | 4.1 | 7.2 ± 0.3 | 7.1 ± 0.4 |
| TS, mg/L | 1750 | 350 ± 34 | 305 ± 7.1 |
| TDS, mg/L | 1300 | 310 ± 43.8 | 274 ± 26.7 |

[a]Totally, 40 samples for each tank (excluding 4 samples from overloading period) were used for the calculation of mean value and standard deviation.

The system achieved high treatment efficiency for both reactors. The 50 percentile value of the influent TCOD, $BOD_5$ TSS,TKN, $NH_4^+$-N, $NO_3$—N and Total phosphorous were 2400, 1400, 450, 86, 9.1, 3.2 and 32 mg/L, respectively. After the treatment, the average values of TCOD, BOD5, TKN, and total phosphorous for Tank A were reduced to 82.9, 28.9, 38, 8.3, 3, 1.9 and 13 mg/L, respectively. Average TCOD removal efficiency of over 95% was achieved. The system also achieved average total nitrogen, total phosphorus, $BOD_5$, and TSS removal efficiencies of about 86%, 60%, 98% and 92%, respectively. The biosolids have excellent settleability, with sludge SVI values of 105 or less. The cause of overloading in mid-July of 1996 was identified and the problem was solved. The information is very useful in avoiding future overloading.

The pH was very consistent in the effluent. Before treatment, the pH of the wastewater ranged from 3 to 5. The low pH of the wastewater did not show any adverse effect from the treatment process. After the treatment, the pH increased to a neutral, while the TCOD of the effluent decreased to an average of 83 mg/L. The correlation of the pH value and TCOD of the effluent could be very useful for on-site monitoring. Since there was no on-site chemical analysis lab available for this small wastewater treatment system, the pH value of the effluent can be used as an indicative parameter for the effluent quality.

## EFFLUENT DISCHARGE AND REUSE

Currently, the treated wastewater effluent is reused for drip-irrigation of the nonfood plantation in the farm. For this current practice, the treated water is classified as R-3 reclaimed water according to the Guidelines for the Treatment and Use of Reclaimed Water in the State of Hawaii.[4] R-3 water is defined as undisinfected secondary reclaimed water. Under the requirement of the Department of Heath of the State of Hawaii, lysimeters were installed in the field. $BOD_5$ and $NO_3$—N concentration for the effluent and soil solution of the irrigation were analyzed and monitored.

The $BOD_5$ (<30 mg/L) and $NO_3$—N (<5 mg/L) concentrations of the effluent were satisfactory for drip or surface irrigation. The $BOD_5$ values of soil solution (average $BOD_5$ value of 13 mg/L) were also very satisfactory. The nitrate-N concentration, however, is somewhat higher than the effluent and the state requirement. The nitrate-N concentration of soil solution ranged from 10 to 50 mg/L, with an average of 23 mg/L. The untreated wastewater was used for irrigation for more than nine months until the wastewater project started. The long-term effect of the previous poor wastewater disposal practice and use of fertilizers may contribute to the high level of nitrate in the soil. The significant achievements of the project, compared with the previous treatment/disposal practice, were that the obnoxious odor has been completely eliminated and the previous frequent drip line clogging problem solved. The treatment process and effluent disposal practice satisfactorily meet the requirement of the Hawaii Department of Health.

## COMPARISON OF WASTEWATER MANAGEMENT ALTERNATIVES

The three fruit fly wastewater management alternatives were compared and presented in Table III. Among these three alternatives, Alternative 1 was a backup plan for the facility to have the excess wastewater hurled away when the previous irrigation system could not handle high flows during high production periods. Alternative 1 alone is least desirable for managing all the wastewater. Although this alternative requires minimum investment cost, it would cost \$115,200–\$201,600 annually, which makes this alternative very expensive in both the short and long run. Alternative 2 is a practical and feasible option. However, the installation and operation of such an irrigation system proved not simple and effective, as expected. Odor, potential underground water contamination, irrigation line clogging, and maintenance problems would be overwhelming. The established irrigation system already costs over \$100,000 and still cannot guarantee handling the wastewater disposal problem and meeting the regulations by the Hawaii State Department of Health. Expanding the storage capacity and improving the performance of this irrigation system are required, and can be costly. Alternative 3 achieves the best results among the wastewater disposal and reuse alternatives.

The wastewater treatment system of Alternative 3 was easily integrated into the existing irrigation system. The treatment system achieved over 90% BOD removal efficiency. The effluent met the Hawaii standard of R-3 reclaimed water, and its use for drip or surface irrigation met the guidelines for the treatment and use of reclaimed water in Hawaii. The batch-operated intermittent aeration treatment system provides stable and excellent operation performance for the treat-

**Table III. Comparisons of the Wastewater Management Alternatives**

| Management Alternatives | Cost | Performance |
|---|---|---|
| Alternativbe 1:<br>Disposal by commercial tankers | $0.08/gal<br>or<br>**$201,600/yr** | – Unreliable<br>– Odor problem<br>– Secondary disposal problem |
| Alternative 2:<br>Drip irrigation with untreated wastewater | | – Serious odor problem<br>– Ground water contamination<br>– Operation & maintenance problems |
| **Installed irrigation system** | $100,911 | – Large storage capacity required<br>– Unstable and unreliable |
| **Required Expanding and improvement** | $35,000 | – Final approval questionable |
| Total: | ***$135,911*** | |
| Alternative 3:<br>Treatment for R-3 reclaimed water reuse | | – Excellent organic and nutrient removal<br>– Odor completely eliminated |
| **Installed drip irrigation system** | $70,911<br>(Excluding $30,000 for the unused three storage tanks) | – Easily operated & maintained<br>– Cost-effective |
| **Established treatment system** | $45,775 | – R-3 reclaimed water reuse approved |
| **Further Improvement** | $28,995 | |
| Total: | ***$145,181*** | |

ment of the fruit fly wastewater, while physically it is very simple, economical, and land saving. The operation of the on-site fruit fly wastewater treatment system also requires little maintenance and labor to produce an acceptable effluent for the wastewater reuse. The capital cost for the treatment system is less than $50,000. The overall capital cost for the new integrated wastewater treatment and reuse system is $118,466, compatible with the old irrigation system. Even with further expansion and improvement including a third reactor, a backup blower, and a sludge dewatering bed, the total capital cost will be below $150,000, while Alternative 2 with expansion and improvement could cost up to $140,000. Clearly, Alternative 3 is very cost-effective. This cost-effective alternative is a good example, and can be easily adopted by other small food and agro-industry practices in Hawaii for the wastewater treatment/reuse in the land-limited condition.

## CONCLUSION

An on-site intermittent aerated batch-operated treatment system was constructed to treat the wastewater from the CDFA fruit fly facility. The performance of this on-site full-scale treatment system was excellent. At the F/M ratio of less than 0.2 g COD/g MLSS/day, a more than 90% COD removal efficiency could be achieved. The system also achieved the average removal efficiencies of total nitrogen, total phosphorus, $BOD_5$, and TSS of about 86%, 60%, 98%, and 92%, respectively.

Despite the variations in wastewater flow and concentration, the intermittent-aeration batch-operated treatment system proved to have stable and excellent operation performance for treatment of the fruit fly wastewater, while physically it was very simple, economical, and land saving. The construction and operation of the system is simple and very cost-effective. The integration of the present treatment process with the irrigation system provides the most effect approach in managing the fruit fly wastewater disposal in a land-limited condition among all other available alternatives. This cost-effective alternative is a good example and could be readily adopted by other small food and agro-industry practices for the wastewater treatment/reuse in the land-limited condition.

## ACKNOWLEDGMENT

Funding for the research was supported, in part, by the California Department of Food and Agriculture (CDFA), and the College of Tropical Agriculture and Human Resources of the University of Hawaii. Both administrative and field assistance provided by CAFA, Mr. Nori Tanaka, and Harris Chang, are appreciated.

## REFERENCES

1. Ketchum, L.H., Jr. Design and Physical Features of SBR Reactors. Proceedings of the First IAWQ Specialized Conference on Sequencing Batch Reactor Technology. Munich Germany, March, 1996. pp. 7–14.
2. Loehr, R.C. *Pollution Control for Agriculture*, 2nd ed. Academic Press, Orlando, FL, 1984.
3. Merkel, R. *Managing Livestock Wastes*. AVI Publishing Company, Inc., Westport, CT, 1981.
4. Guidelines for the Treatment and Use of Reclaimed Water. Hawaii State Department of Health, Honolulu, HI, 1993.

# 21 USE OF A BENCH-SCALE SEQUENCING BATCH REACTOR TO SIMULATE A FULL-SCALE PLUG-FLOW REACTOR FOR TREATMENT OF A HIGH-STRENGTH INDUSTRIAL WASTE STREAM ISOLATED FROM A MUNICIPAL WASTEWATER

**A. K. Umble** and **L.H. Ketchum, Jr.**
Department of Civil Engineering & Geological Sciences
University of Notre Dame, Notre Dame, Indiana 46556

## ABSTRACT

A bench-scale sequencing batch reactor (SBR) was used first to test the treatability of an isolated, high-strength industrial waste stream that enters the Elkhart Wastewater Treatment Plant. Because the SBR is the ideal plug-flow reactor (PFR), it was then also used to evaluate the feasibility of isolating one of the plant's PFRs and operating it as a side-stream treatment train for this industrial stream. The SBR results showed that the waste stream was readily treatable within the 12-hour reaction cycle applied. Additionally, the SBR's simulation of the full-scale PFR demonstrated that the industrial waste stream could be treated to acceptable levels within the hydraulic configuration of one of the existing PFR tanks isolated from the remainder of the system. This study demonstrates the utility of the bench-scale SBR in effectively studying complex, full-scale treatment problems, at negligible cost in equipment and operator time. Results such as these provide plant managers and facility planners with specific treatment data to make sound judgments on capital facility improvements and process enhancements to address various regulatory compliance issues. The study method also provides an excellent educational tool for plant operators to broaden their knowledge base for biological waste treatment.

## INTRODUCTION

The City of Elkhart, Indiana, is an industrial town of 40,000 residents and home to more than 50 industries permitted to discharge their wastewaters to the municipality's publicly owned treatment plant. The facility treats an average daily flow of 66,500 $m^3$, of which nearly 12% is made up from the discharge of one industry alone. The organic and ammonia loadings within this single industrial waste stream are significant and often impact the process control strategies imposed by plant operations personnel. Currently, this waste stream is blended with the raw municipal wastewater at the plant's headworks facility.

The plant utilizes a conventional activated sludge process for removal organics and nitrification of ammonia. The biological process includes five parallel plug-flow reactors (PFR) and eight final clarifiers. Each reactor contains a volume of nearly 3,900 $m^3$. Return activated sludge (RAS) is introduced at the head end of each reactor at rates varying between 65% and 100% of the total primary effluent flow. Normal operation of the plant targets a mixed liquor suspended solids (MLSS) concentration of 3,500 mg/L with wastage rates set to meet a sludge retention time (SRT) of 7 days. A schematic of the plant's primary and secondary treatment layout is given in Figure 1.

Because the industrial waste stream in question is conveyed to the plant site independent of the remaining municipal flow, thought has been given to developing a side-stream treatment train solely for this flow. In so doing, the plant could be buffered from shock organic and ammonia loadings common to this discharge, while still achieving acceptable treatment of the isolated stream. Conceptually, one alternative is to utilize one of the five PFRs as a side-stream treatment

*52nd Purdue Industrial Waste Conference Proceedings*, 1997, Ann Arbor Press, Chelsea, Michigan 48118. Printed in U.S.A.

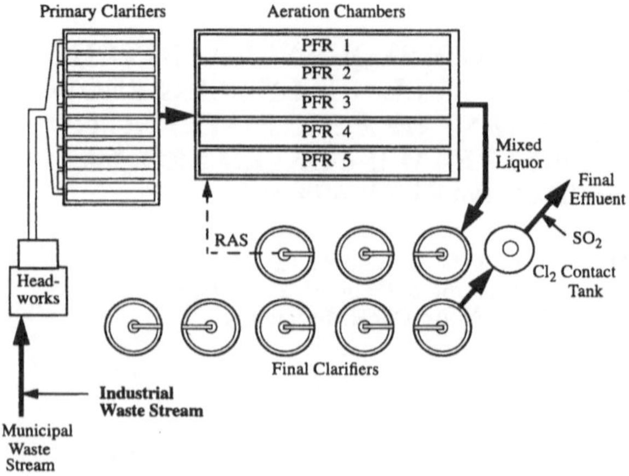


**Figure 1. Plant layout - existing.**

train solely for the treatment of this industrial stream. Considering the average daily volumetric inflow rate of the industrial stream and the working volume of one of the PFR tanks, an average hydraulic retention time (HRT) of between 8 to 12 hours could be attained, depending on the rate of RAS imposed. Though operated hydraulically independent from the other plant inflow, this isolated reactor would receive RAS in its current manner. By adjusting the gate valves controlling the RAS rates entering a PFR, the operator has flexibility in controlling the MLSS concentrations. Therefore, the MLSS of the isolated tank, though operated with a different HRT than the other tanks, could be maintained in equivalency with the other. The question for investigation is whether such physical modifications to the plant could achieve the treatment objectives for the industrial waste stream while operated in parallel to the treatment of the remaining municipal wastewaters.

An effective tool for investigating this question is a bench-scale Sequencing Batch Reactor (SBR). The bench-scale SBR is a simply operated, inexpensive tool for evaluating full-scale system responses to particular influent characteristics, process modifications, or combinations of both. The operation of a bench-scale SBR requires minimal attention of the Plant Operator. The primary tasks are attending to decanting and refilling the reactor at the end of each treatment cycle, and adhering to specific sampling schedules.

The batch treatment mode of the SBR operates in five basic sequences. These are: FILL, REACT, SETTLE, DRAW, and IDLE. By manipulating the times associated with each of these sequences, all aspects of an activated sludge treatment process can be achieved within the confines of a single vessel. The details on these sequences have been fully described by Irvine and Ketchum.[1]

The SBR is, by definition, the ideal plug-flow reactor. In a review of the mathematical development for the SBR, Irvine and Ketchum[1] describe how the SBR cycle translates into time that the plug-flow reactor achieves in space. Therefore, by setting a reaction cycle in the SBR equal in duration to the hydraulic retention time of the PFR, and establishing equivalency in biomass concentration and sludge retention time, the SBR can provide valuable insight in the PFR's operational performance and efficiencies.

The primary questions evaluated in this project were two-fold. First, is the industrial waste stream treatable within an 8–12 hour reaction period utilizing the microbial consortiums commonly present in the full-scale plant? The selection of MLSS and SRT parameters for this question do not necessarily need to correspond to the full-scale plant. Secondly, would the 8–12 hour HRT of the existing PFR coupled with normal operating parameters of MLSS and SRT be adequate to consistently achieve acceptable effluent quality in terms of organics and ammonia removals from this waste stream?

The purpose in this two-phased study is as follows. If the initial treatability study shows that a 12-hour reaction cycle is inadequate for acceptable treatment, then it would be safe to assume that an isolated PFR also would not achieve acceptable performance within its 12-hour HRT. If the results of the treatability study show the waste stream is treatable within the 12-hour time frame, then the second study is warranted. The subsequent study is designed to simulate not only the reaction period (i.e., HRT), but the normal operating parameters of organism concentration and age within the PFR as well. If it is found that these parameters in the SBR do not permit acceptable treatment performance, then other alternatives for the isolated side-stream treatment train (such as a full-scale SBR system) would be necessary. Finally, if the results of the simulation study are positive, the data generated by the study would be sufficient to begin concept designs, preliminary designs, and cost estimates for the necessary modifications for retrofitting the plant's hydraulic infrastructure to accommodate an isolated PFR side-stream treatment train.

## METHODS

The bench-scale SBR used was a 15 L glass vessel. The maximum working volume for both studies was 10 L. Aeration was supplied by two aquarium-size air pumps fitted with air diffuser stones. The batch treatment sequences were controlled using a simple home appliance timer. Figure 2 shows the sequencing of the 12-hour reaction cycle. Eleven hours of this cycle were dedicated to aerated REACT. A dump FILL was used to begin each cycle. Supernatant was decanted with a siphon hose in dump fashion following a SETTLE period of between 45 minutes to 1 hour. No IDLE sequence was utilized. The average waste stream characteristics during the two study periods are given in Table I. The ammonia and COD levels are much higher than typical for municipal waste streams.

Waste-activated sludge taken from the Return Activated Sludge (RAS) line of the full-scale plug-flow system was used to start the SBR for this initial treatability study. Enough RAS was added to attain an initial mixed liquor concentration of 5000 mg/L. The FILL volume was set to 3 L per cycle. Minimal wasting was conducted in the first several cycles to allow the MLSS con-

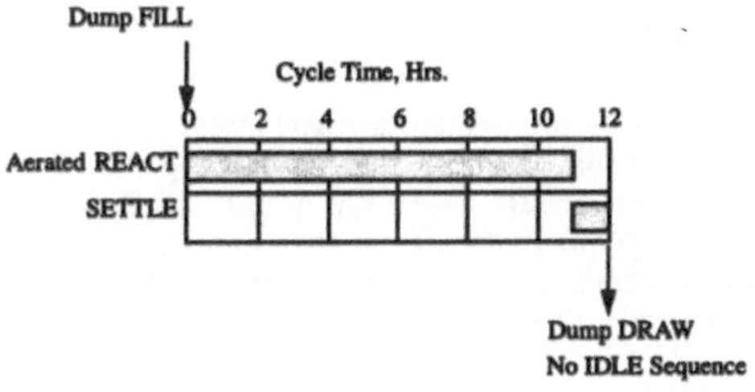


**Figure 2. SBR cycle.**

**Table I. Average Industrial Waste Stream Influent Characteristics[a]**

| Parameter | Treatability Period Conc., mg/L | PFR Simulation Period Conc., mg/L |
|---|---|---|
| COD[b] | 1410<br>(2273, 993) | 1390<br>(1662,1144) |
| TSS[c] | 188<br>(340, 85) | 455<br>(1160, 252) |
| $NH_3$-N[d] | 155<br>(222, 95) | 120<br>(228, 71) |
| Total P[e] | 11.2<br>(22.8, 7.5) | 11.6<br>(12.8, 8.8) |

[a]Average values are reported with maximum and minimum values in parentheses.
[b]Chemical Oxygen Demand.
[c]Total Suspended Solids.
[d]Ammonia nitrogen.
[e]Total Phosphorus.

centration to increase to approximately 9200 mg/L. At that time, wasting of mixed liquor commenced at between 200–250 mL daily. This wastage rate was sufficient to maintain a relatively constant MLSS concentration near 9000 mg/L and a SRT of 40 days. At these operational conditions, the average loading rate to the reactor was 0.14 mg COD-mg of MLVSS$^{-1}$-day$^{-1}$. A month of operation was provided as an acclimation period for the biomass. Then a track study was conducted on the 11-hour REACT sequence to monitor the reductions and transformations in the organic and major inorganic constituents.

A second SBR study was conducted to simulate the potential for the full-scale PFR to treat this waste stream while operating under the normal plant operational targets for MLSS and SRT of 3300 mg/L and 7 days, respectively. The reactor was restarted using 5 L of RAS seed from the full-scale activated sludge PFR. To achieve the desired operational targets in the SBR, higher loadings of substrate were needed than used in the initial study to stimulate higher biological growth rates. This also required higher wastage rates (i.e., 1200–1400 mL of mixed liquor-day$^{-1}$). Therefore, a FILL volume of 5 L was selected for this study. This resulted in an average loading rate of 0.32 mg COD-mg MLVSS$^{-1}$-day$^{-1}$. Following 12 cycles of operation, a period within which the biomass would still be representative of the full-scale system, a track study was conducted to evaluate the efficiency of the 11-hour REACT sequence. It should be noted that no chemical additives were introduced to precipitate phosphorus in either study, though this is the current practice in the full-scale plant. All phosphorus removal observed in the SBR studies was attributed to assimilatory mechanisms for metabolism and intercellular storage.

To monitor the SBR's performance, effluent samples from both studies were taken 3 times per week and analyzed for COD, $NH_3$-N, Total P, TSS, and VSS (volatile suspended solids) according to *Standard Methods*.[2] The mixed liquor was also sampled three times per week and analyzed for total and volatile suspended solids content. Finally, a microscopic review of the biological activity was conducted for each study according to standard procedures practiced in the full-scale plant operations. These were compared to observations of samples taken from the full-scale PFR during the same treatment period.

## RESULTS

The 12-hour reaction cycle proved adequate for treating the industrial waste stream in both the initial treatability study and in the PFR simulation. Table II shows the reactor's performance under both operational strategies. For each of the parameters listed, the removal efficiencies were acceptable. The improved efficiency for TSS removal in the PFR simulation study is actually a result of significantly higher influent solids loadings during the study period. The removal efficiency of near 90% for total P was achieved solely through biological processes since no chemical additions were introduced.

Settling performance of the biological floc during both studies was excellent as assessed from the Sludge Volume Index (SVI) method. The average SVI for the treatability study was 51 mL-L$^{-1}$g MLSS$^{-1}$, and 40 mL-L$^{-1}$g MLSS$^{-1}$ for the PFR simulation study. In bench-scale SBR studies, Borgatti[3] found that SVIs in this range were evidence of good settling characteristics. In this range, a relative balance between floc-forming and filament-forming organisms is present, resulting in optimum effluent quality in terms of suspended solids. The sludge settling performance resulting from the 11 hours of continuous aerated REACT also reflects what Palis and Irvine[4] found in that better settling occurred in SBR systems that received continuous aeration.

**Table II. SBR Average Performance for Treatability and PFR Simulation Studies**

| Parameter | Treatability Study Conc., mg/L | Treatability Study Removal Efficiency, % | PFR Simulation Conc., mg/L | PFR Simulation Removal Efficiency, % |
|---|---|---|---|---|
| COD | 110 | 92.2 | 163 | 88.3 |
| TSS | 12 | 93.6 | 19 | 95.8 |
| $NH_3$-N | 0.50 | 99.7 | 0.18 | 99.9 |
| Total P | 1.2 | 89.3 | 1.6 | 86.2 |

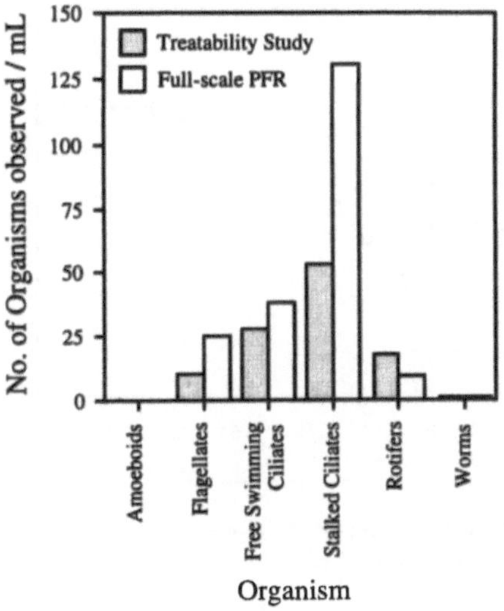


**Figure 3. Microbiological evaluation of SBR for treatability study and of full-scale PFR at same sample time.**

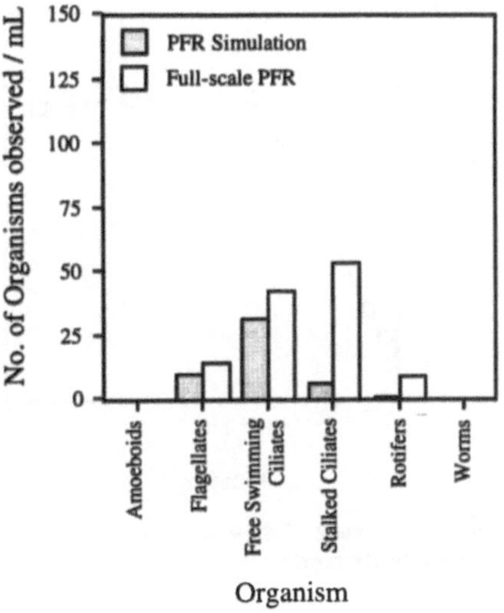


**Figure 4. Microbiological evaluation of SBR for treatability study and of full-scale PFR at same sample time.**

It is normal procedure in the operation of an activated sludge process to assess the biological activity based on the presence and relative dominance of various protozoan and invertebrate animals. The microscopic evaluations of the biomass for both studies and their comparisons to the full-scale PFR for the same sampling times are shown in Figures 3 and 4. During the initial study, the protozoan community in both the SBR and the full-scale PFR biomass was dominated by stalked and free-swimming ciliates. Relative dominance by these organisms is an indication of a healthy, good settling biomass. This is interesting, despite the substantially longer SRT applied in the SBR.

In the PFR simulation test, the protozoan community in the SBR biomass was dominated by free-swimming ciliates, while the full-scale PFR biomass continued to be dominated by stalked ciliates. The shift in the SBR protozoan community from the initial study was likely related to shifts at the bacterial level. Though the SRT was the same as the PFR in the latter study, these shifts were likely a result of differences in loading condition and substrate types, which greatly influence the presence or absence of those bacterial populations that become food sources for the various competing protozoan populations. As the bacterial food sources change, so too will the protozoan populations.

Figures 5 through 14 compare the SBR's biological response to the substrate in both studies over the REACT period of the cycle. In each of these graphs, the first sample point occurs just prior to the dump FILL and indicates time zero. This is followed immediately by the start of the REACT period. The SETTLE sequence begins at time = 11 hours.

## DISCUSSION

The results of the treatability study clearly show that the components of the waste stream are readily biodegradable. In fact, oxidative reactions for organics and ammonia were essentially complete in less than half of the allocated REACT time (Figures 7 and 9). The majority of the COD removed appears to have occurred within the first 15 minutes of the REACT period. Additional removals continued to take place but at a much slower rate for an additional 3 hours. This is likely the period during which breakdown of the more complex organics occurred.

Rapid uptake of organic carbon is evidence of several conditions. First, as Goronzy and Rigel[6] suggested, the rate of uptake is mediated by rapid absorption and storage kinetics within the cells and biological floc. This stored carbon is then metabolized at a later time during the endogenous

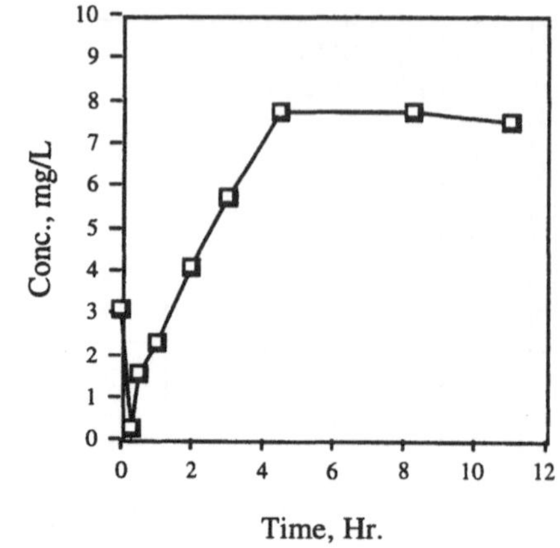


**Figure 5. Track of dissolved oxygen: treatability study.**

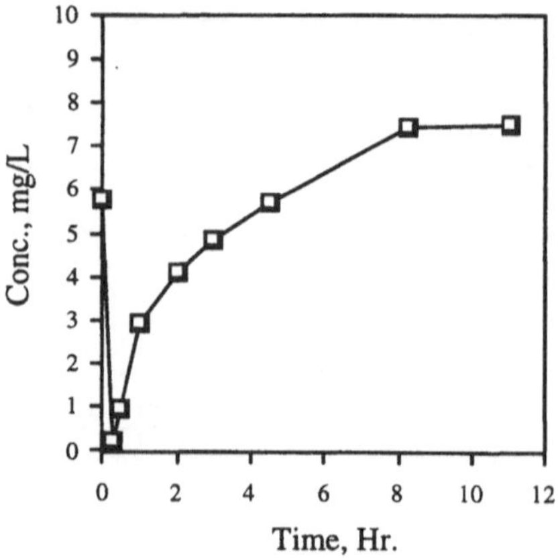


**Figure 6. Track of dissolved oxygen: PFR simulation.**

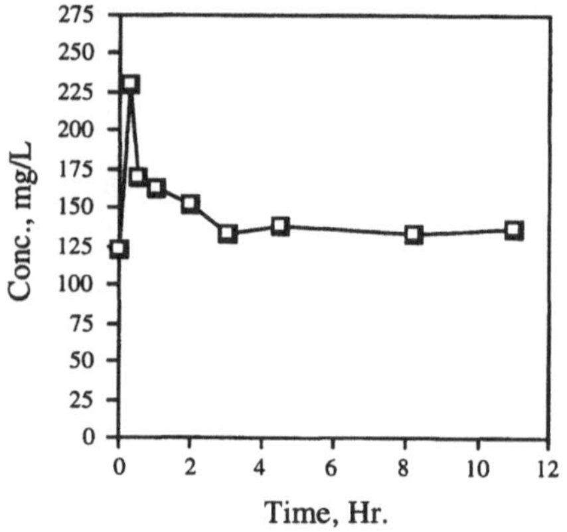


**Figure 7. Track of COD: treatability study.**

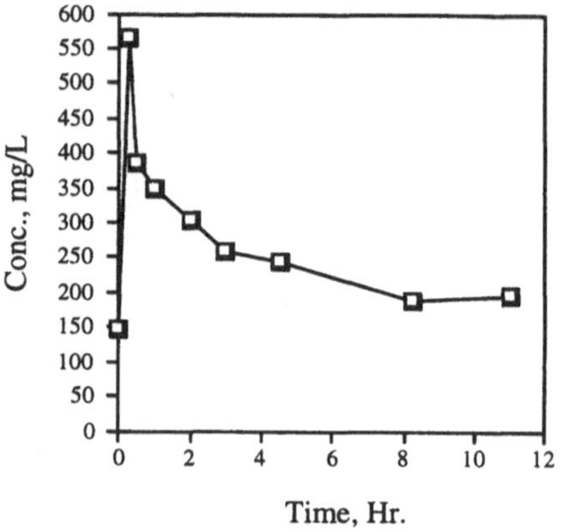


**Figure 8. Track of COD: PFR simulation.**

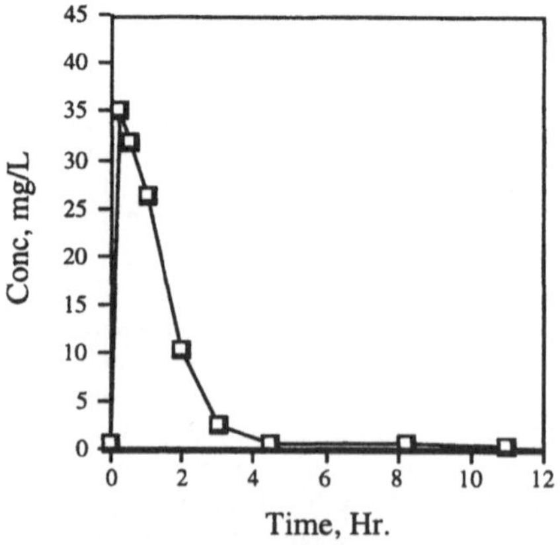


**Figure 9. Track of $NH_3$-N: treatability study.**

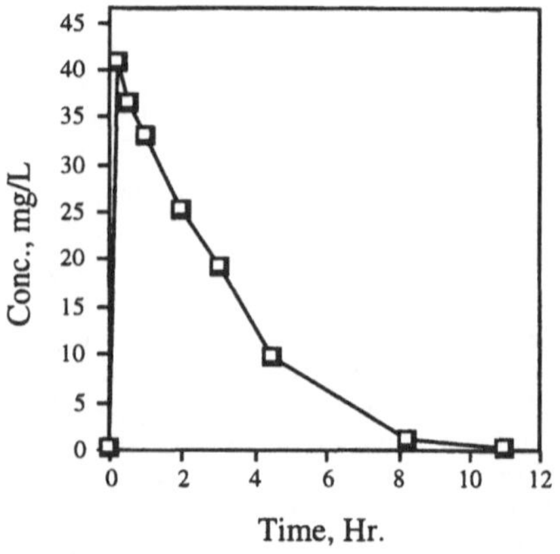


**Figure 10. Track of $NH_3$-N: PFR simulation.**

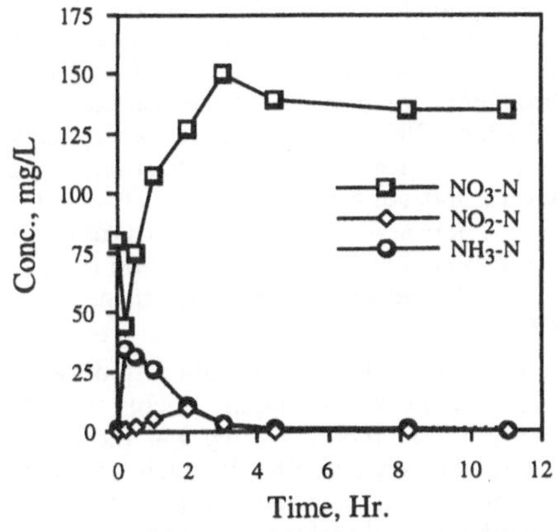


**Figure 11. Track of inorganic-N: treatability study.**

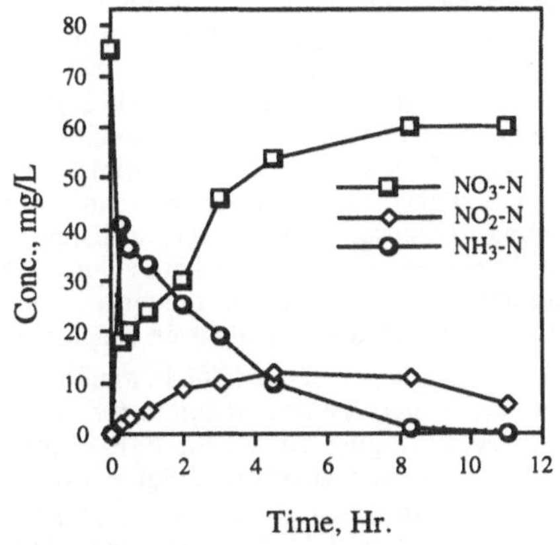


**Figure 12. Track of inorganic-N: PFR simulation.**

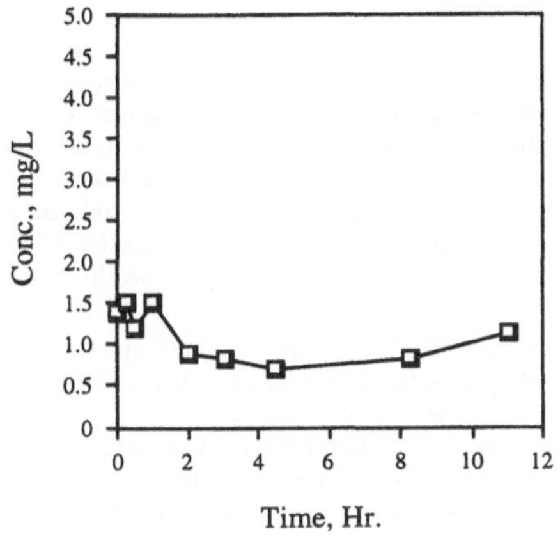


**Figure 13. Track of total P: treatability study.**

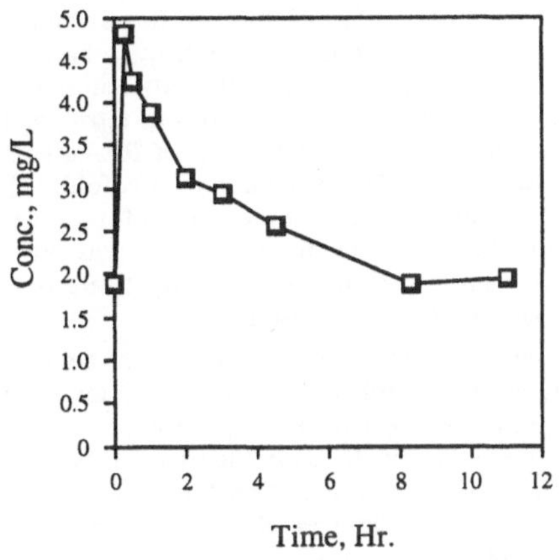


**Figure 14. Track of total P: PFR simulation.**

respiration phase which likely occurred between 4.5 and 11 hours of the REACT period that coincided with high DO levels. Secondly, rapid uptake is an indication that the influent COD may be dominated by one or more highly labile constituents. This has been confirmed from preliminary analyses of this industrial waste stream's COD using gas chromatography techniques, which showed that the primary constituents are neutralized organic acids. Thirdly, rapid uptake is also reflective of the high concentration of viable organisms present in the biomass at the time of substrate introduction.

By comparison, in the PFR simulation study, removal of the major fraction of the influent COD also occurred within the initial 15-minute reaction period (Figure 8). However, the rate of oxidation of the more complex organics was less than in the initial study, and a lower portion of these complex organics was ultimately removed. This result indicates that the lower concentration of organisms in the mixed liquor may have limited a more complete COD oxidation. Therefore, in a full-scale PFR side-stream application operated with similar MLSS concentration levels and under a 12-hour HRT, a higher concentration of the less degradable COD would probably be observed in the mixed liquor exiting that PFR.

It is widely known that nitrification of ammonia is accomplished when sufficiently long periods of dissolved oxygen (DO) levels greater than 1 mg/L are present and a relatively low organism growth rate is permitted.[7,8] For both the initial treatability study and the subsequent PFR simulation study, the high DO levels sustained throughout the majority of the reaction periods (Figures 5 and 6) satisfy the first condition, while the long SRTs (40 and 7 days) satisfy the second. Therefore, we would expect these conditions to allow for adequate nitrification provided that inhibitory compounds or environments are absent. Indeed, this is what was observed. According to Figures 9 and 10, the oxidation of ammonia appears to commence concurrently with the absorption of organics in both studies. The implication here is that the oxidation of organic material seems not to inhibit this nitrification activity. This verifies observations made by Alleman and Irvine[9] in their studies on bench-scale SBRs.

The performance of the SBR in ammonia oxidation during the PFR simulation study was acceptable, though the time required for the ammonia removal was essentially double that of the initial study (Figure 10). In the PFR simulation study, the ammonia loading was increased by nearly 25% over that of the initial study. Additionally, the concentration of organisms was decreased by two-thirds. But since nitrifying organisms are known to account for only 2–4% of the bacterial community,[9] the slower rate of ammonia oxidation is therefore attributable to a limitation in the sheer number of organisms rather than in substrate or oxygen. This is evidenced by the dissolved oxygen levels throughout the REACT sequence, as indicated by Figure 6. The continual increase in DO levels indicate that oxygen did not limiting the system.

Nitrification is known to occur at its highest rate when dissolved oxygen levels are greater than 2 mg/L.[10] Additionally, as Bishop, et al.[10] suggested, the rate of nitrification follows two distinctly different rates, the first occurring at 2 mg/L of DO followed by a more rapid rate when levels reach 4–5 mg/L. Though this characteristic was not observed in the initial treatability study, it was present in the PFR simulation study. As depicted in Figure 11, the transformation of ammonia into the oxidized forms of nitrite and nitrate in the treatability study occurred at a maximum rate during the first hour of REACT. But during this time, dissolved oxygen levels were below 2 mg/L. The explanation is twofold: (1) the nitrifying organisms experienced little competition from heterotrophic organisms for oxygen after 15 minutes into the reaction period and any available oxygen in the build fluid was utilized immediately for nitrification activity, and (2) there were more than enough nitrifying organisms in the biomass to metabolize the limited amount of ammonia present.

Figure 12 depicts the nitrification of ammonia for the PFR simulation. Here, the most rapid nitrification rate did occur once the DO levels reached 4–5 mg/L, and follows well the characteristic trait described by Bishop, et al..[10] In the simulation study, the degradation of COD is not completed as rapidly as in the treatability study (see Figure 8). Therefore, the nitrifiers likely experienced some competition for oxygen as the heterotrophs completed the organic oxidations. Also, the sheer number of nitrifying organisms needed to oxidize the higher ammonia loadings was much less because of the lower SRT. Despite the lower rate of nitrification in the PFR simulation, it is clear that complete nitrification of the applied ammonia did occur within the reaction period provided. It should be noted, however, that the reaction period does not seem to be adequate for the complete transformation of nitrite to nitrate. Only within the final three hours of the REACT sequence is a decline in nitrite observed. But its complete transformation to nitrate remains incomplete, leaving residual nitrites in the effluent. If excessive concentrations of nitrite accumulate, this could be of concern to the operator if chlorine is used as the primary disinfection process for the final plant effluent. Nitrite is readily oxidized by chlorine and therefore creates an increase in the chlorine dosage that must be applied to complete disinfection.

The comparison of the total phosphorus reductions between the two studies is depicted in Figures 13 and 14. Clearly, the rate of phosphorus uptake is more rapid in the treatability study than in the PFR simulation. Final effluent concentration levels were reached within 4 hours of aerated REACT. Because no operational modifications were applied to develop alternating anaerobic/aerobic/anoxic conditions, a technique recognized by Shin and Jun[11] as a means to enhance phosphorus removal, uptake occurred during oxic conditions. This uptake was likely a result of intercellular storage under aerobic metabolism. It is interesting to note that in the initial treatability study, some phosphorus was released in the latter period of the REACT sequence. The time of the start of this release corresponds with the time of peak dissolved oxygen concentration in the reactor, after which the dissolved oxygen declined slightly. This is mostly an indicator of cellular lysis resulting from endogenous respiration activity. This phosphorus release was not observed in the PFR simulation study.

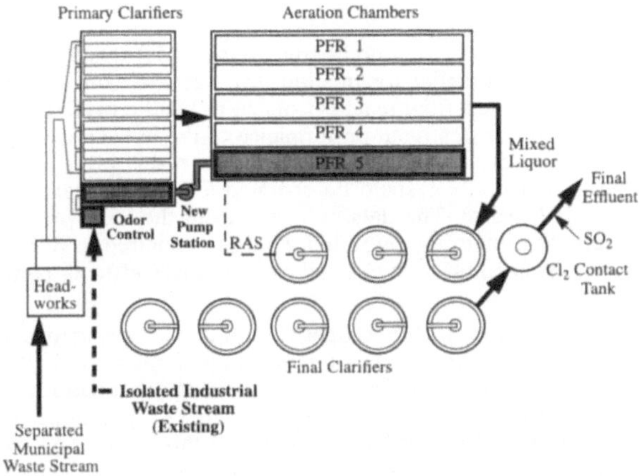


**Figure 15. Plant layout for side stream treatment concept.**

There are implications for the full-scale PFR plant stemming from the results of this project. The studies show that the industrial waste stream in question is highly degradable and readily treatable by the consortiums of organisms present in the facility's activated sludge system using the current operational targets. But the degradability efficiency is related to a need for ample reaction time. The 11 hours of aerobic reaction environment evaluated in the PFR simulation study is an adequate reaction duration.

By isolating one of the PFR tanks in the five-tank full-scale system for this waste stream, a hydraulic retention time of 10–12 hours can be established. The specific HRT chosen would depend on the operator's choice of the RAS rate. A concept of how the plant might be retrofitted to accommodate this isolated side-stream treatment is illustrated in Figure 15. Following the separation of the flow streams at the plant's headworks, the isolated industrial stream would bypass the headworks facility. Because this industrial stream does not contain grits, sands, or other debris, preliminary treatment through the headworks is unnecessary. Historically, the isolated industrial wastewater stream in question has created malodorous conditions at the plant site. Consequently, the flow could first be routed through a new small mixing chamber where pre-aeration or chemical additives for odor control could be applied, as necessary. From there the flow would be discharged through a short, newly constructed channel into one of the existing primary clarifiers. This clarifier would be sealed off from its existing municipal influent channel by closing and sealing existing gates. The effluent collection channel at the end of the isolated clarifier would then be closed off from the other municipal flows by a concrete wall. The primary effluent would then be lifted to PFR Tank No. 5 by a new, duplex submersible pump, lift station.

The results of the PFR simulation study indicated that the isolated waste stream could be successfully treated utilizing the microbial consortiums present in the full-scale system operating under the normal MLSS concentration and plant SRT.H Maintaining a consistent MLSS concentration in the isolated PFR requires the plant operator to make periodic adjustments to the rate of the RAS entering the isolated PFR tank. This is done simply by adjusting the existing RAS air-operated gate valves located at the head end of the reactors. Mixed liquor from the isolated tank would then be combined with that of the other tanks in the effluent channel at the end of the plug-flow length and sent to the final clarifiers for settling.

Because of the PFR simulation study, separating the organisms from the municipal waste treatment train by isolating a separate final clarifier is not necessary. This is critical to the economic feasibility of this plant retrofit alternative because substantial piping modifications and rerouting of flows would be required in isolating a final clarifier. An additional advantage to this blending of biomass is that the high ammonia levels introduced into the isolated PFR provide opportunity for the sheer number of nitrifiers to be increased within the system. This allows for enhanced ammonia removal in the municipal train as well as the organisms are returned and redistributed throughout the five reactors.

## CONCLUSIONS

The use of a bench-scale SBR to test the treatability of a high-strength industrial waste stream and to simulate a full-scale plug-flow reactors potential for treating that waste stream proved successful. This study showed that a full-scale plug-flow reactor isolated to handle only the industrial flow could achieve an acceptable level of treatment performance of the industrial waste stream under the current influent characteristics and the tank hydraulics. In this project, the bench-scale SBR demonstrated that the full-scale PFR could be operated as a side-stream isolated treatment reactor for this high strength waste. The data collected from the SBR provide substantial preliminary design parameters. Minimal pipe rerouting would be required to retrofit the existing PFR, but the installation of an isolated influent channel and a primary effluent pump station would be required.

The bench-scale SBR is a simple-to-operate, powerful investigative tool that allows an operator or wastewater plant manager to study and evaluate complex treatment problems at negligible equipment cost and without the high costs of outside environmental engineering consultations. In addition, the time required of the plant operator to carry out the FILL and DRAW operations and then necessary sampling is minimal (averaged only 30 minutes per day in this study) and can easily be fit into the operator's normal routines in the full-scale operation. Finally, the educational experience gained by the operator from participating in this type of simulation study is invaluable in improving the operator's working knowledge of the activated sludge treatment system.

## ACKNOWLEDGMENTS

The authors wish to thank the City of Elkhart, Indiana, for its cooperative participation in this project by providing facilities, personnel, utilities, and analytical services. Specifically, the efforts of the Primary and Secondary Operators at the Wastewater Treatment Plant facility are recognized and appreciated. Also, acknowledgment is given to Matt Tatarko, Ph.D., of the University of Notre Dame for his consultation on the gas chromatography aspects of this project. This work was funded in part through a grant from the University of Notre Dame and through a grant funded by the City of Elkhart.

## REFERENCES

1. Irvine, R.L. and Ketchum, L.H. "Sequencing batch reactors for biological wastewater treatment." *CRC Crit. Rev. Environ. Control*, 18, pp. 255–294, 1989.
2. *Standard Methods for the Examination of Water and Wastewater*, 19th ed., Franson, M.A.H., Eaton, A.D., Clesceri, L.S., and Greenberg, A.E., Eds. APHA, AWWA, WEF, 1995.
3. Borgatti, D. Controlling Filamentous Growth in SBRs by Alternating Process Controls. Ph.D. Thesis, University of Notre Dame, Notre Dame, IN, 1981.
4. Palis, J.C. and Irvine, R.L. "Nitrogen removal in a low-loaded, single tank SBR." *J. Wat. Poll. Control Fed.*, 57, pp. 82–86, 1985.
5. Kerri, K.D., Dendy, W.B., Brady, J., and Crooks, W., Eds. *Operation of Wastewater Treatment Plants*, 4th ed., Volume II: A Field Study Training Program. California State University, Sacramento Foundation, 1996, p. 96.
6. Goronzy, M.C. and Rigel, D. "Biological phosphorus removal in a fed-batch reactor without anoxic mixing sequences." *Res. J. Wat. Poll. Control Fed.*, 63, pp. 248–258, 1991.
7. Jones, W.L., Wilderer, P.A., and Schroeder, E.D. "Operation of a three-stage SBR system for nitrogen removal from wastewater." *Res. J. Wat. Poll. Control Fed.*, 62, pp. 268–274, 1990.
8. Manning, J.F. and Irvine, R.L. "The biological removal of phosphorus in a SBR." *J. Wat. Poll. Control Fed.*, 57, pp. 87–94, 1985.
9. Alleman, J.E. and Irvine, R.L. "Nitrification in the sequencing batch biological reactor." *J. Wat. Poll. Control Fed.*, 52, pp. 2747–2753, 1980.
10. Bishop, D.F., Heidman, J.A., and Stamberg, J.B. "Single stage nitrification/denitrification." *J. Wat. Poll. Control Fed.*, 48, pp. 520–532, 1976.
11. Shin, H.S. and Jun, H.B. "Development of excess phosphorus removal characteristics in a sequencing batch reactor." *Wat. Sci. Tech.*, 25, pp. 433–440, 1992.

# Section Six
# BIOLOGICAL: ANIMAL AND DAIRY WASTES

## 22 RECOVERY OF PROTEIN, FAT AND OIL FROM SLAUGHTERHOUSE PROCESS WATER WITH THE DISSOLVED AIR CONCENTRATION PROCESS

**Louis D. le Roux**, Senior Sales Technologist
**Jelte Lanting**, Technology Manager
Biothane Corporation
2500 Broadway\Dwr # 5
Camden, New Jersey 08104
**P. Borup Sorensen**, President
The Aminodan Group
P.O. Box 43, Vestmole 1
DK-9990, Skagen, Denmark

### ABSTRACT

This chapter discusses the Biothane-Aminodan (B-A) process which was developed for protein and fat recovery from slaughterhouse process water where the recovered product is used as animal feed. The basic principles and techniques involved in protein and fat coagulation and flocculation are discussed in relation to their removal with the dissolved air concentration system. Operating data of a full-scale protein, fat, and oil recovery installation at a poultry slaughterhouse is also presented.

The B-A process uses FDA edible food approved additives for soluble protein and emulsified fat coagulation and flocculation, as compared to metal salt coagulants and acrylamide based flocculants that are used in many dissolved air flotation systems for slaughterhouse wastewater treatment. The product recovered in the B-A system does not contain any toxic metal salts or cancer-causing acrylamide and can safely be rendered into animal feed.

The B-A system uses dissolved air concentration (DAC©) water for protein and fat removal from process water. During DAC© water addition, many small, homogeneously sized air bubbles (≈30 μm in diameter) are formed in the process water which attach to the flocs. The air-floc particles have enough buoyancy to float to the surface of the concentrator. Only part of the recycled water stream is used as DAC© water, which results in substantial savings in system operating (energy) cost.

Typically, the overall BOD removal efficiency of the B-A system varies between 70 and 85%. BOD removal efficiency is highest when process water soluble BOD consists mostly of proteins and fats. More than 95% total FOG and suspended solids are removed from process water.

### INTRODUCTION

Food processing companies generate an estimated 18 million tons of organic waste each year.[1] These wastes consist mostly of unused raw agricultural materials, fermentation and microbial residues, biosolids (biomass) and "float."[2] Float, also known as skimmings, results from the physical/chemical removal of proteins and fats from food processing plant wastewater (process

*52nd Purdue Industrial Waste Conference Proceedings*, 1997, Ann Arbor Press, Chelsea, Michigan 48118. Printed in U.S.A.

water) with dissolved air flotation (DAF) systems. Disposal of these solids is a significant environmental problem. Traditional disposal methods, such as land application, landfills, and incineration, are increasing in cost and becoming more restricted in use.[3] For example, landfilling of biosolids and float are no longer allowed in North Carolina and New Jersey, and banning of residuals disposal in landfills is likely to occur in more states as the number of landfills and their effective capacities continue to decrease.[4] As a result, many wastewater treatment agencies will be forced to modify their residuals management practices to comply with new regulatory requirements and to mitigate public concerns.

Float constitutes a significant portion of the total solid wastes generated by food processing companies. During calendar year 1994, meat processing companies slaughtered a total of 35 million cattle, 96 million hogs, and 7.3 billion chickens.[5] As a result, they produced an estimated total of 122,850 million gallons of wastewater containing 1,720 million pounds of biochemical oxygen demand (BOD). During this same time period, approximately 1351 million gallons of float was produced as a result of treatment of these wastewaters with DAF systems. Disposal of float is not only expensive, but also results in a loss of valuable products and revenue for slaughtering and meat processing companies.

Slaughterhouse process water BOD consists mostly of soluble and insoluble proteins, as well as free and emulsified fats and oils. Most of these materials can be recovered with conventional DAF systems and have the potential to be reused as dog and cat food, animal food, ingredients for detergents, and lubricants. However, many DAF systems use metal salts (mostly iron and aluminum based) for protein and fat coagulation prior to flocculation with synthetic, organic polymers. The backbone monomer most widely used in polymer synthesis is acrylamide, which has been shown to cause cancer in laboratory animals.[6] Furthermore, many synthetic polymers in use for animal feeding purposes have been labeled "generally recognized as safe" (GRAS), even though none are approved by the Food and Drug Administration (FDA) as animal food additives.1 Therefore, recycling of float containing GRAS polymers and metal salts as animal feed is questionable and still needs FDA approval.

There exists a need for a process that uses FDA-approved additives for protein and fat recovery from meat processing plant process water. This chapter discusses the Biothane-Aminodan (B-A) process which was developed for protein and fat recovery from process water with the use of edible food approved (EFA©c) additives. The basic principles and techniques involved in protein and fat coagulation and flocculation are discussed in relation to their removal with the dissolved air concentration (DAC©) system. Operating data of a full-scale protein, fat, and oil recovery installation at a poultry slaughterhouse will also be presented.

## BIOTHANE-AMINODAN PROCESS

### Process Description

The B-A process was developed by The Aminodan Group in Denmark specifically to recover proteins and fats from slaughterhouse and meat processing plant wastewater with the use of EFA© additives. No ferric or alum based coagulants or GRAS polymers are used in the B-A system; therefore, the recovered product does not contain any cancer-causing compounds and can safely be rendered into animal feed. The B-A process typically consists of three stages or sections thereof (Figure 1):

Stage One— Primary Process Water Purification,
Stage Two—Secondary Process Water Purification, and
Stage Three—Product Recovery and Dewatering.

#### *Stage One: Primary Process Water Purification*

Raw process water coming from a slaughterhouse or meat processing plant is first screened before it is transferred to the sedicyclone (Figure 1). Several screens are available for coarse solids removal, ranging from static, concave-shaped screens to internally or externally fed rotary drum screens. Screen opening size varies between 0.01 and 0.05″, with 0.02″ opening size the most commonly used. Screened solids, which consist mostly of bone and meat pieces, feathers (in the case of poultry slaughtering), and hair (in the case of cattle slaughtering), are further processed into meat and bone meal at a rendering plant.

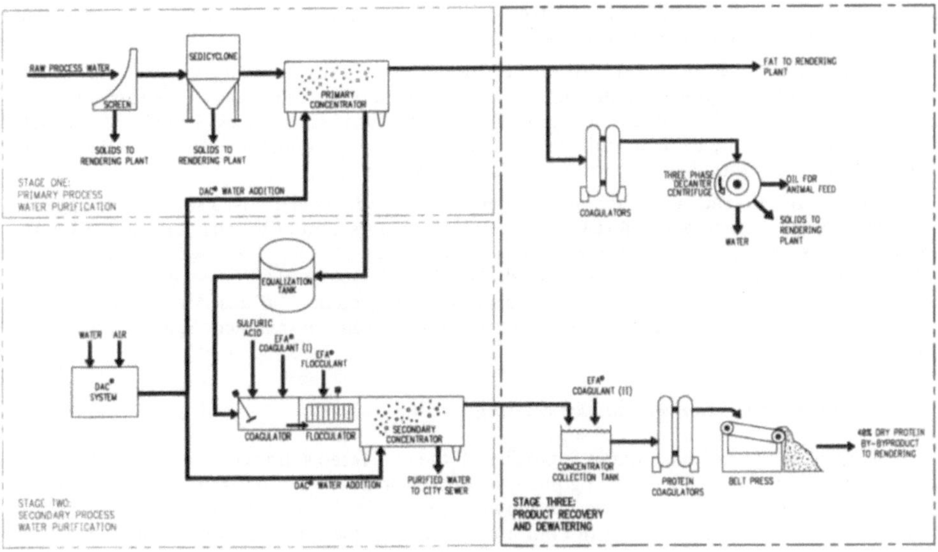

**Figure 1. Flow diagram of the Biothane-Aminodan process.**

Small, settleable suspended solids are removed from process water in the sedicyclone before it is forwarded to the primary concentrator (Figure 1). The sedicyclone operates similar to hydrocyclones; it uses centrifugal force to enhance suspended solids separation from process water. The recovered solids are sent to a renderer for further processing into meat and bone meal. At the inlet of the primary concentrator, DAC© water is added to the process water, during which many small air bubbles are formed in the process water stream. The air bubbles attach to the suspended solids, which consist mostly of small proteinaceous and free floatable fat particles, and cause enough buoyancy to float them to the top of the primary concentrator (Figure 1).

The removal efficiency of suspended solids and free fat from process water in the primary concentrator depends on several factors, such as: (1) fraction of fat in suspended solids, (2) air bubble size, (3) air-to-solids ratio, (4) concentrator solids loading, and (5) concentrator hydraulic loading. Of these factors, air bubble size and fraction of fat in suspended solids directly affect suspended solids removal when no coagulants and flocculants are used. Highest suspended solids and fat removal efficiencies are achieved when most of the suspended solids are free fat and when the air bubble size ranges between 20 and 50 μm. The Biothane-Aminodan DAC© system is designed to produce air bubbles of homogeneous size with an average diameter of 30 um. More than 95% of free floatable fat is removed in the primary concentrator.

*Stage Two: Secondary Process Water Purification*

After prepurification in the primary concentrator, the process water is forwarded to the secondary process water purification stage (Stage Two, Figure 1). In stage two, residual suspended solids, soluble proteins and emulsified fats, oils, and grease (FOG) are removed. Process water first enters the coagulator during which soluble proteins and fats are coagulated through the addition of sulfuric acid and EFA© coagulant I. During coagulation, the electrostatic surface potential of the protein and oil "particle" is reduced such that van der Waals attraction forces dominate and cause the particles to coagulate (see section: "Theory of Coagulation and Flocculation").[7] After coagulation, the process water is forwarded to the flocculator where an EFA© flocculant is added. During flocculation, the coagulated fat and protein "particles" attach to the flocculant to form large flocs (see section: "Theory of Coagulation and Flocculation").[7] Both the coagulator and flocculator are designed with sufficient reaction time between the additives and protein and fat particles for optimum coagulation and flocculation.

Process water flows by gravity from the flocculator to the secondary concentrator. DACc water is added to the process water between the flocculator and secondary concentrator during which small air bubbles (≈30 μm in diameter) form and attach to the flocs. The air-floc particles have enough buoyancy to float to the top of the secondary concentrator from where they are removed and collected in the concentrate collection tank (Figure 1).

The removal efficiency of soluble proteins and emulsified fats and oils, also expressed as soluble biochemical oxygen demand (BOD), depends on factors similar to those mentioned for suspended solids and fat removal in the primary concentrator. Typically, the overall BOD removal efficiency for both stages in series varies between 70 and 85%. BOD removal efficiency is highest when process water-soluble BOD consists mostly of proteins and fats. More than 95% total FOG and suspended solids are removed from the process water in treatment Stages One and Two (Figure 1).

*Stage Three: Product Recovery and Dewatering*

*Processing of Solids Recovered in Stage One*: If the recovered solids consist mostly of fat, then they are send directly to the rendering plant for further processing into yellow grease (in the case of poultry slaughtering or processing) or brown grease (in the case of cattle slaughtering or meat processing). If the solids are high in both protein and fat, then they are first separated into oil, protein, and water in a three-phase decanter centrifuge before being processed into animal feed. In order to enhance protein and fat separation in the centrifuge, the solids are coagulated in protein coagulators (Figure 1). After centrifugation, the oil is rendered into yellow or brown grease and the proteinaceous solids are rendered into meat and bone meal. Many slaughterhouses and meat processing plants own rendering plants for solids processing; however, companies which do not own rendering plants sell their solids to independent rendering companies.

*Processing of Solids Recovered in Stage Two:* The proportion of protein and fat in the recovered product (called concentrate) from the secondary concentrator depends on the ratio of soluble protein to emulsified fat in the process water. Ideally, the wastewater treatment system should be designed so that the process water undergoes minimum turbulence in the primary process water treatment stage (Stage One) to reduce mechanical dispersion. Since most of the fat and oil will then be in the free floatable form, most of it will be removed in Stage One. The remaining soluble proteins and emulsified FOG are removed in Stage Two.

The protein content of secondary concentrate from chicken slaughterhouses varies typically between 55 and 65% (dry basis). Before the concentrate is rendered into meat and bone meal, it first is dewatered to 35% TS. Concentrate dewatering results in lower rendering cost because less moisture has to be evaporated to dry the concentrate to 90% TS. During concentrate dewatering, the solids are first conditioned in the concentrate collection tank through the addition of EFA© coagulant II (Figure 1). EFA© coagulant II enhances the coagulation of proteinaceous solids in the protein coagulators, necessary to dewater the concentrate with a belt filter. After solids dewatering, the solids are send to the rendering plant for further processing into meat and bone meal.

## Theory of Coagulation and Flocculation in Relation to the B-A Process

*Coagulation*

Soluble protein and emulsified fat have a net electrostatic surface charge in their natural state (Figure 2).7 As a result of this surface charge, which is typically negative, an electrical double layer exists at the solids-water interface.[8] The negatively charged protein and fat "particles" repel each other, resulting in a stable suspension (Figure 2). Particles do not separate quickly in a stable suspension due to the "energy barrier" that has to be overcome in order for coagulation to take place (Figure 3[a]).[7]

Destabilization of protein and fat particles can be achieved through several mechanisms, such as compression of the electrical double layer, surface charge neutralization, or enmeshment in a precipitate.[9] In the B-A process, protein and fat coagulation are accomplished via two mechanisms: (a) particle surface charge neutralization through sulfuric acid addition, and (b) compression of the electrical double layer surrounding each particle through the addition of an EFA© coagulant. If it is assumed that the surface charge of proteins and fats results from the presence of

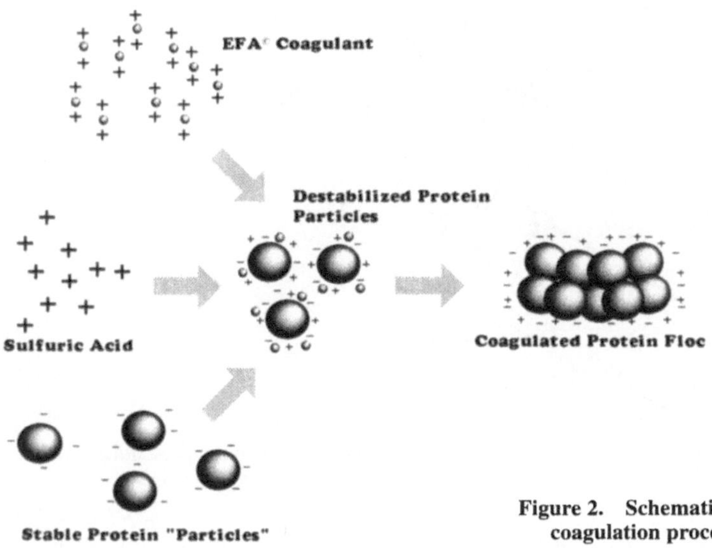


**Figure 2. Schematic presentation of the coagulation process.**

carboxyl (-COOH) and amino (-NH2) groups, then the fat or protein molecule (NH2-R-COOH) can react as follows:

$$\text{NH3+-R-COOH} \qquad \text{NH}_3^{+}\text{-R-COO}^{-} + \text{H}^{+} \tag{1}$$

$$\text{NH}_3^{+}\text{-R-COO}^{-} \qquad \text{NH}_2\text{-R-COO}^{-} + \text{H}^{+} \tag{2}$$

In these reactions, the surface charge on the molecule depends on the concentration of hydrogen ions (H+) in solution. As pH increases, Equations 1 and 2 shift to the right and the particle (protein or fat) surface charge becomes increasingly negative. Since process water pH is normally between 6 and 8, addition of sulfuric acid reduces the pH to the iso-electric point of the molecules, at which point it is in the zwitterion state (net zero charge). At the zwitterion state, the "energy barrier" is reduced and protein/fat particles with sufficient kinetic energy to over-

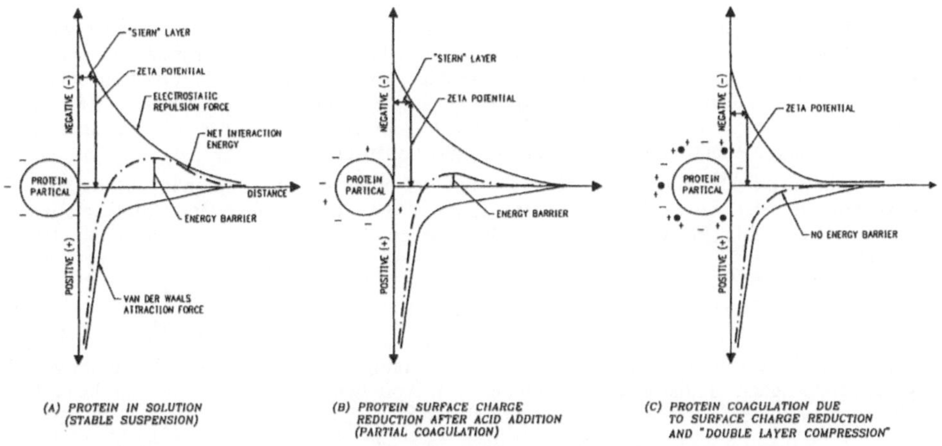


**Figure 3. Interaction energy of protein/fat particles at different stages of coagulation.**

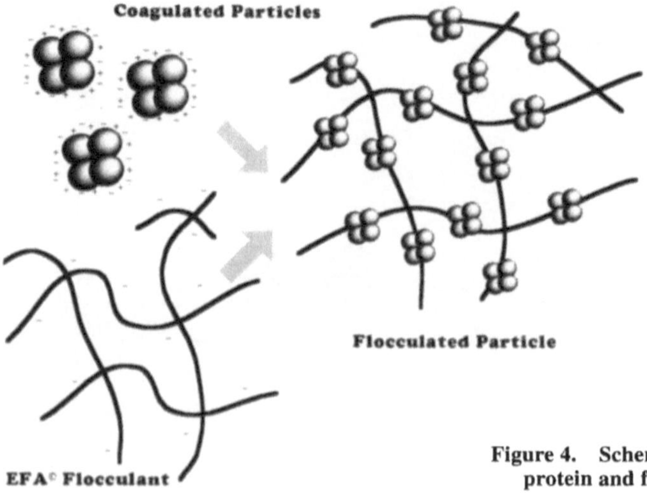


**Figure 4. Schematic presentation of protein and fat flocculation.**

come this lower "energy barrier" will coagulate (Figure 3[b]). In order to further reduce the zeta-potential (measure of the electrostatic repulsion force, see Figure 3) of the protein/fat particles, EFA© coagulant I is added. EFA© coagulant I results in the compression of the electrical double layer of the particles (lowers the zeta-potential) and results in the elimination of the "energy barrier" (Figure 3[c)].9 At this point, coagulation is complete and the protein particles appear in the process water as small "pin-point flocs" (Figure 2).

*Flocculation*

After coagulation, the pin-point flocs are flocculated into larger flocs in the B-A flocculator through the addition of an EFA© flocculant (Figure 4). The flocculant contains negatively charged sites onto which the coagulated particles attach to form larger flocs that can be separated from the wastewater with air. During flocculation, the process water is gently agitated to provide sufficient contact between the pin-point flocs and flocculant for optimum flocculation to occur. After flocculation, DAC© water is added to the process water during which small air bubbles are formed which attach or get entrapped in the floc-matrix and float to the surface of the secondary concentrator (Figure 1).

## The DAC© System

The B-A DAC© system is designed so that it supersaturates a portion of the treated water, recycled from the secondary concentrator, with air under high pressure (>80 psi). Air is mixed with recycled process water in the ejector, from where it is transferred to the dissolved air concentration tank to provide sufficient retention time for complete saturation to take place. After saturation, dissolved air concentration (DAC©) water is made available for solids concentration through pressure reduction over a specially designed diffuser.

Air bubble size distribution and quantity of bubbles released from solution are critical to solids concentration performance; hence, so is the design of the diffuser. The overall solids removal efficiency of a bubble (Ec) was shown to be the product of collision (E1) and attachment efficiencies ($E_2$):10

$$E_c = E_1 \quad E_2 \tag{3}$$

Considering a spherical bubble of diameter $d_b$ rising vertically at constant velocity and a spherical particle of diameter $d_p$, Reay and Ratcliff[10] predicted for $d_b < 100$ μm that the collision efficiency between air bubbles and solid particles to be:

$$E_1 \quad (d_p/d_b)^2 \tag{4}$$

Equation 4 shows that for a given air consumption, the recovery of fine particles through flotation is enhanced by the presence of small air bubbles. This is due to their large number density and longer residence time in solution which increase the probability of collision between a solid particle and air bubble. It is also clear from Equation 4 that if an increase in particle size can be achieved for a given bubble size, then flotation should be enhanced. This leads to the practice of soluble protein and fat coagulation and flocculation.

The B-A diffuser is designed so that it produces air bubbles of homogeneous size (~30 μm), which results in high solids collision efficiency. The DACc water is introduced through more than one diffuser, to enhance solids collision efficiency for improved process water purification.

## Advantages of B-A System

The B-A system offers the following advantages to slaughterhouse and meat processing plants:

- recovers salable products from process water, which results in savings in wastewater treatment cost and payback of capital equipment
- uses Edible Food Approved (EFA©) additives for protein and fat coagulation and flocculation. The recovered product does not contain any cancer-causing compounds and can safely be rendered into animal feed
- reduces rendering cost due to high TS in the dewatered (35% TS) concentrate
- increases life expectancy of rendering equipment due to the elimination of corrosive metal salts in the concentrate
- decreases or completely eliminates municipal surcharges due to the large reduction in process water BOD, TSS, and FOG
- decreases the size of, or eliminates the need for primary wastewater treatment equipment due to reduced effluent BOD, TSS, and FOG

# FULL-SCALE SYSTEM PERFORMANCE: POULTRY SLAUGHTERHOUSE

## Background

The first B-A system in the poultry industry to use EFA© additives was installed in 1982 at DANPO Poultry, Denmark. Since then, several B-A systems have been installed worldwide, of which Devon Crest Poultry is one. Devon Crest Poultry slaughters 48,000 broilers (between 46 and 48 days old) per day and produces mostly whole chicken and chicken portions.

The main source of process water arises from chicken slaughtering, cooking, and processing with some contribution from crate washing and general cleaning of the plant and premises. The plant uses on average 10.2 L (2.7 gal) per bird (gpb), which is substantially lower than the 4 and 7 gpb used by similar U.S. slaughtering companies. Process water TSS varies between 250 and 1,000 mg/L and COD varies between 900 and 2,800 mg/L throughout the day (Figure 5). Highest TSS and COD occur between noon and 4 p.m. daily, which corresponds to initial rinsing of the processing equipment before cleaning. Process water flow increases gradually throughout the day from 8 $m^3$/hr to 35 $m^3$/hr.

## System Layout and Performance

### *System Layout*

The process water from the plant is screened with a 0.02″ opening size, static screen before it is collected in the equalization tank. The equalization tank serves as a buffer to equalize the flow as well as pollution load before the process water is purified in the B-A system. Due to the low fat content of the process water, only a secondary process water purification system was installed (Stage Two, see Figure 1).

Process water pH is first reduced to the iso-electric point of the proteins (normally between 4 and 5) through the addition of sulfuric acid, a process which is called acidulation. After soluble protein and emulsified fat coagulation, an EFA© flocculant is added for flocculation. EFA© flocculant I is not used at this plant since the process water purification performance obtained with sulfuric acid and EFA© flocculant I alone, was sufficient. After flocculation, the flocs are re-

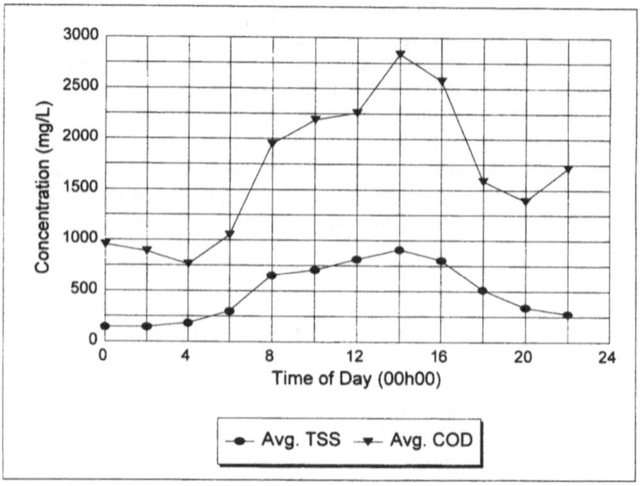


**Figure 5. Process water TSS and COD concentration at different times of the day (April, 1993: Devon Crest Poultry, England).**

moved from the secondary concentrator through DAC© water addition. The recovered protein concentrate is processed in a protein coagulator and then dewatered with a belt press before it is sold to a local renderer for inclusion in dog and cat food.

*System Performance*

The B-A system reduces process water COD and TSS (on average) by 71% and 80%, respectively. Results of process water analysis performed during May 1996 showed that process water COD was reduced from 4,320 to 1,240 mg/L and TSS was reduced from 1,050 to 223 mg/L.

As a result of the installation of the B-A system, Devon Crest Poultry was not only able to save on municipal surcharges, but also completely eliminated sludge disposal cost, since the recovered product is sold to a renderer.

## ACKNOWLEDGMENTS

We wish to thank Mr. Des Morris, chief engineer at Devon Crest Poultry Ltd., for his assistance in providing us with operating data on the B-A process water treatment system.

## REFERENCES

1. le Roux, L.D., J.B. Litchfield, R.L. Belyea, and M.E. Tumbleson, Effect of Acid/Base Treatment on the Dewaterability and Nutritional Quality of Biosolids as Animal Feed, *Proceedings of the 51st Industrial Waste Conference,* 51, May 6–8, 1996, pp. 573–583, Ann Arbor Press, Inc., Chelsea, MI.
2. Belyea, R.L., T.E. Clevenger, D.L. Van Dyne, S.E. Eckhoff, M.A. Wallig, and M.E. Tumbleson, Using Biosolids from Agricultural Processing as Food for Animals, *First Biomass Conference Americas: Energy, Environment, Agriculture and Industry* (Volume II), NREL, Golden, CO, 1993, pp. 1416–1426.
3. Garvey, D., C. Guarino, and R. Davis, "Sludge disposal trends around the globe," *WATER/Eng. Manage.*, December 1993, pp. 17–20.
4. Scott, C. and D. Kuchenrither, "Wastewater residuals plan management," *Environ. Protection,* 4(7), pp. 12–16, 1993.
5. Watt Poultry Group: 1997 Marketing Guide, Watt Publishing, 122 S. Wesley Ave., Mt. Morris, IL 61054.

6. Molak, V., A Review of the Literature, DHHS (NIOSH) Publication No. 91-115, U.S. Department of Health and Human Services, Public Health Service, National Institute for Occupational Safety and Health.
7. le Roux, L.D., Dewatering Biosolids from a Milk Processing Plant: Agricultural Fibers as Flocculant Aids and Ultrafiltration Membrane Dewatering, Ph.D. Thesis, University of Illinois at Urbana-Champaign, Urbana, IL, 1996.
8. Zita, A. and M. Hermansson, "Effect of ionic strength on bacterial adhesion and stability of flocs in a wastewater activated sludge system," *Appl. Environ. Microbiol.*, 60(9), pp. 3041–3048, 1994.
9. Amirtharajah, A. and C.R. O'Melia, Chapter 6: Coagulation Processes: Destabilization, Mixing and Flocculation, In: *Water Quality and Treatment: A Handbook of Community Water Supplies*, 4th ed., F.W. Pontius, Ed., McGraw-Hill, Inc., New York, 1990, pp. 269–361.
10. Gochin, R.J., Chapter 19: Flotation, In: *Solids-Liquid Separation*, 3rd ed., Svarovsky, Ladislav, Ed., Butterworth & Co., University Press, Cambridge, UK, 1990.

# 23 ASBR TREATMENT OF BEEF SLAUGHTERHOUSE WASTEWATER

**Damon Morris**, Graduate Research Assistant

**Shihwu Sung**, Assistant Professor of Environmental Engineering

**Richard R. Dague**, Professor of Environmental Engineering
Department of Civil Engineering
Iowa State University
Ames, Iowa 50011

## ABSTRACT

This research focused on the treatment of wastewater from the Excel Corporation beef slaughterhouse located at Schuyler, Nebraska. Research was conducted using the Anaerobic Sequencing Batch Reactor (ASBR) process developed at Iowa State University by Dr. Richard R. Dague and coworkers (U.S. Patent No. 5,189,079).

This research was intended to show the feasibility for application of the wastewater to the ASBR system. Research was conducted using two ASBR units over a 13-month period. The first unit was seeded with granular sludge obtained from the Heileman brewery in Lacrosse, Wisconsin and the second unit was seeded with flocculent sludge from the municipal wastewater treatment plant in Ames, Iowa. The units were operated at 30°C to simulate the wastewater temperature at the Excel facility.

This research has shown that the ASBR system is effective in the treatment of wastewater from the Schuyler Excel Corporation slaughterhouse. The system remained effective in terms of removals and operational parameters over a range of organic and solids loadings. Periods of sharp increase in organic loading (shock loading) due to the sampling and feeding procedure were also evaluated and the results shows the system performed well. System performance depended significantly on hydraulic retention time (HRT), with lower HRTs resulting in lower removal efficiencies. Consistent results can be attained at a 30-hour HRT. At this loading, the granular achieved average removals of over 70% based on $BOD_5$ and 50% based on COD.

Granular and flocculent seed were both effective in treatment of the wastewater. Granular seed was more easily held in the reactor, resulting in better startup and higher removals early in the study. MLVSS concentrations in the granular reactor ranged from approximately 10 to 16 g/L under steady state conditions. After establishment, flocculent seed performed as well as granular seed at lower HRTs with more consistent removals even though it operated at much lower biomass levels. MLVSS concentrations in the flocculent reactor ranged from approximately 2 to 4 g/L under steady state conditions.

## INTRODUCTION

### Background

This research dealt with the treatment of beef slaughterhouse wastewater using the Anaerobic Sequencing Batch Reactor (ASBR) process developed at Iowa State University by Dr. Richard R. Dague and coworkers (U.S. Patent No. 5,189,079).[6,7,9,15] The research is based on laboratory studies of the Anaerobic Activated Sludge process developed by Dague, McCarty, and Pfeffer in 1966.[4] The ASBR is a batch-fed system which treats wastewater by anaerobic decomposition of organic matter. This system functions as a high rate process in the treatment of waste through the

*52nd Purdue Industrial Waste Conference Proceedings*, 1997, Ann Arbor Press, Chelsea, Michigan 48118. Printed in U.S.A.

metabolism of organic matter. An anaerobic bacterial consortia including methanogens convert complex carbohydrates, proteins, and fats to carbon dioxide and methane. The system is sealed from the outside environment to maintain anaerobic conditions. This allows the system to operate without odor emissions and allows for the collection of biogas, a valuable fuel, consisting of carbon dioxide and methane.

Research and development of the ASBR has been conducted at ISU using a variety of substrates and under various conditions. Wastes which have been successfully treated in lab-scale operations include swine waste, pork slaughterhouse wastewater, pet food processing waste, and biochemical and enzyme production waste.[2,5,10,13] The ASBR system has been run under various conditions including differing seed sources and concentrations, a range of temperatures, and as a part of multiple stage systems.[1,7]

Previous application of slaughterhouse wastewater to the ASBR system is limited to a single study by Calaghan.[2] Beef slaughterhouse wastewater seems to be an acceptable substrate for use in the ASBR. Slaughterhouse waste is high in organic matter, which is ideal for biological treatment. The presence of protein allows the system to be run without addition of supplemental alkalinity because it is naturally produced through protein ammonification to form ammonium bicarbonate during the treatment process. The ASBR system can also handle the high solid concentrations typically present in slaughterhouse wastewater.

Presently the wastewater at the Excel slaughterhouse is initially treated in uncovered anaerobic lagoons prior to activated sludge treatment. Following anaerobic pretreatment, the wastewater undergoes aerobic treatment and clarification before discharge. The ASBR system would be designed to replace the lagoon pretreatment step allowing the aerobic secondary treatment and clarification steps to remain as they presently exist.

The ASBR would allow pretreatment to operate at higher rates, require less area, reduce odors, and allow for recovery of methane. The ASBR can operate at low HRTs because of the effective biosolid and liquid separation enabling a longer SRT. Because the system operates at high rates, wastewater is processed more quickly and a much lower reactor volume is required for the system.

### Objectives and Scope of Study

The purpose of this research was to determine the treatability of the Excel Corporation beef slaughterhouse wastewater when applied to the ASBR system. Research was conducted using two ASBR units over a 13-month period. The first unit was seeded with granular sludge obtained from the Heileman brewery in Lacrosse, Wisconsin. The second unit was seeded with flocculent sludge obtained from a conventional primary anaerobic digester at the municipal wastewater treatment plant in Ames, Iowa. The units were operated at 30°C to simulate wastewater temperature at the Excel treatment facility.

The study was intended to assess application of the ASBR system to the wastewater and any differences in performance between the system seeded with granular biomass and the system seeded with flocculent biomass. Performance was tested with respect to removals based on total chemical oxygen demand (COD), soluble COD, and the five-day biological oxygen demand ($BOD_5$) tests. Biogas was sampled to determine its quantity and composition. Other parameters were monitored routinely to determine typical operating conditions. These included solids, alkalinity, pH, volatile fatty acids (VFA), ammonia, and production of hydrogen sulfide in the biogas. These data were collected in order to characterize the operation of the overall system as well as to detect upsets in day-to-day performance. After collection and analysis of data from the ASBR study, optimum design parameters for the full-scale system were recommended.

## EXPERIMENTAL APPROACH

Two identical reactors were used for the duration of the study. The reactors were constructed from Plexiglas and had a total volume of 11.5 L. The inside diameter of the body was 5.0 inches. The body of the reactors was cylindrical and made with 1/4 inch thick walls and had a height of 35 inches. The reactors were operated at a volume of 9 L. This volume allowed adequate headspace for biogas leaving the liquid phase, foam produced by the reactor, and to avoid accidental overflows from reaching the gas collection system.

**Table I. HRT and Corresponding Time Periods for the Granular Reactor**

| Days of Study | HRT, Granular Reactor |
|---|---|
| 5–53 | 36 |
| 53–105 | 30 |
| 105–143 | 24 |
| 143–199 | 18 |
| 199–340 | 24 |
| 340–02 | 30 |

**Table II. HRT and Corresponding Time Periods, Flocculent Reactor**

| Day of Study | HRT, Flocculent Reactor |
|---|---|
| 5–133 | 36 |
| 133–264 | 30 |
| 264–340 | 24 |
| 340–419 | 18 |

Feed was obtained from the splitter box at the Excel beef slaughterhouse in Schuyler, Nebraska prior to primary treatment. Due to the presence of large particulate matter such as rocks, hair, and straw the feed was strained through a number 20 sieve before being fed to the reactors.

The reactor systems were operated with eight cycles per day throughout the study. This resulted in a complete decant, feed, and react cycle every three hours. This number of cycles was chosen because it seemed to provide an adequate react period while maintaining reasonable feed and decant volumes.

Over the course of the research, the reactors were operated over a range of hydraulic retention times. A summary of the HRTs and their corresponding time periods for each reactor system are given in Tables I and II. The decant cycle ranged from approximately three to five minutes over the course of the study. The feed cycle ranged from approximately 10 to 15 minutes over the study. The extra feeding time may have allowed biomass to adjust to increased feed concentration in the reactor.

Periodic mixing inside the reactors was provided by the gas recirculation mixing system. Mixing was provided every ten minutes during the react portion of the cycle for a period of 30 seconds to evenly distribute substrate. Additional mixing was provided during the last mixing period before decanting to remove trapped biogas from the settled biomass. The purpose of this extra mixing was to aid in settling due to the tendency for biogas to be released when the partial pressure was lessened from the declining liquid level during the decant cycle.

The biogas collection system served to collect biogas leaving the reactor, measure the quantity of biogas produced, and allow sampling to analyze the quality of the biogas. Components of the biogas collection system include the gas bag, an observation bottle, a sulfide scrubber, a sampling port, and a gas meter. The layout of the experimental setup for the reactor systems is given in Figure 1.

Methane, carbon dioxide, and nitrogen levels in the biogas were determined with the use of a Gow-Mac Instrument Company series 69-350 gas chromatograph with a thermal conductivity detector. Hydrogen sulfide levels in the biogas were determined using a Drager gas detector pump with Drager hydrogen sulfide 100/a colorimetric indicator tubes, Model No. 67-19-001. Alkalinity, VFA, solids levels, COD, and BOD were determined using methods found in Standard Methods.[14] Alkalinity and VFA were determined using the titration method and distillation method, respectively. The COD test used for this research was the closed reflux, titrimetric method, and BOD was measured using the five-day biochemical oxygen demand ($BOD_5$) test.

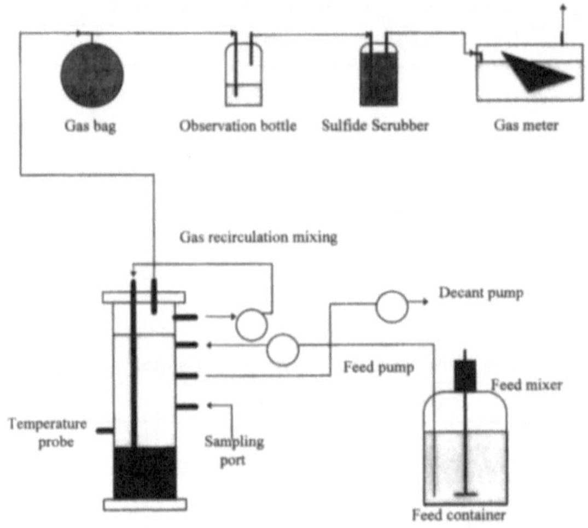


**Figure 1. ASBR experimental set-up.**

## RESULTS AND DISCUSSION

### Lab Scale Operations

This research has shown that the ASBR system is effective in the treatment of wastewater from the Schuyler Excel Corporation slaughterhouse. The system remained stable over a range of organic loadings. Periods of sharp increase in organic loading (shock loading) due to the sampling and feeding procedure were also handled well by the reactor.

Stability of the lab scale systems was indicated by operational parameters such as alkalinity, VFA, pH, and ammonia. The alkalinity remained at sufficient levels throughout the study to buffer increases in VFA and drops in pH without the use of supplemental alkalinity. The range for alkalinity in the reactors was from 900 to 1580 mg/L as $CaCO_3$. The alkalinity for both reactor systems throughout the study is shown in Figures 2 and 3. Levels of VFA never reached inhibitory levels for either system during the study. The maximum levels in the systems were approximately 180 mg/L for the granular reactor and 750 mg/L for the flocculent reactor. High VFA levels were only reached during unsteady transitional periods as shown in Figures 4 and 5.

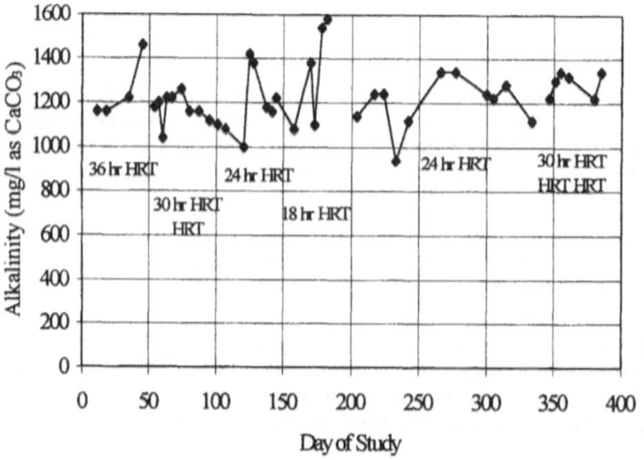


**Figure 2. Alkalinity in the granular reactor.**

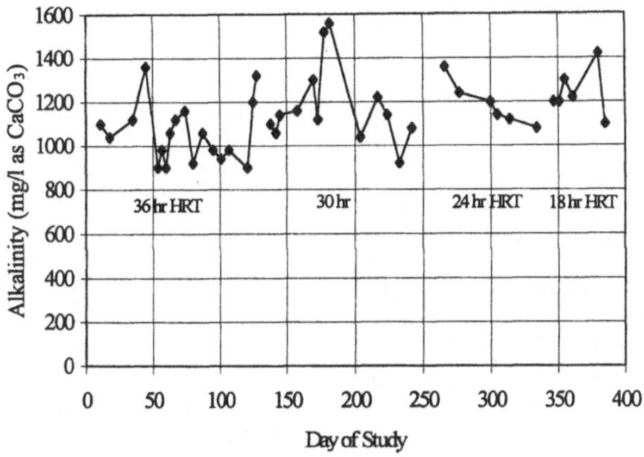


**Figure 3. Alkalinity in the flocculent reactor.**

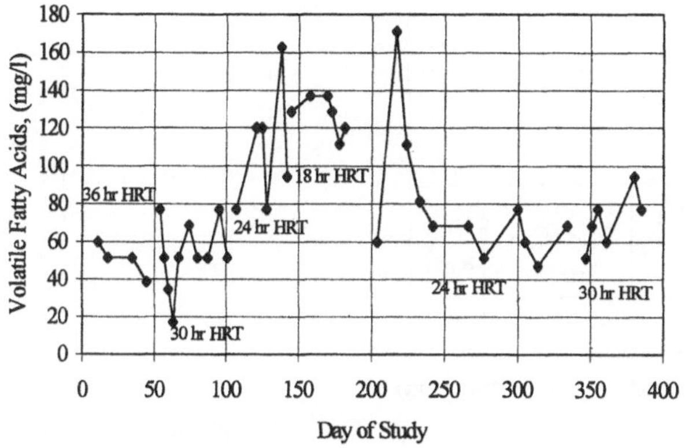


**Figure 4. Volatile fatty acids in the granular reactor.**

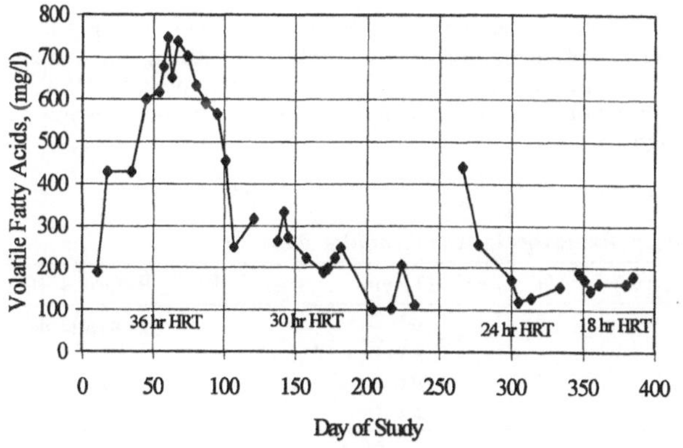


**Figure 5. Volatile fatty acids in the flocculent reactor.**

Levels were very low over most of the study, indicating excellent conversion of acids to biogas. The pH level in the reactor system remained in the range suitable for growth of methanogens, although occasional drops below the optimum level of 6.8 TO 7.2 were present.24 The pH for each reactor system is presented in Figures 6 and 7. The production of ammonia from protein degradation was monitored over the study. It was found that ammonia levels were less than inhibitory levels for both systems throughout the study.

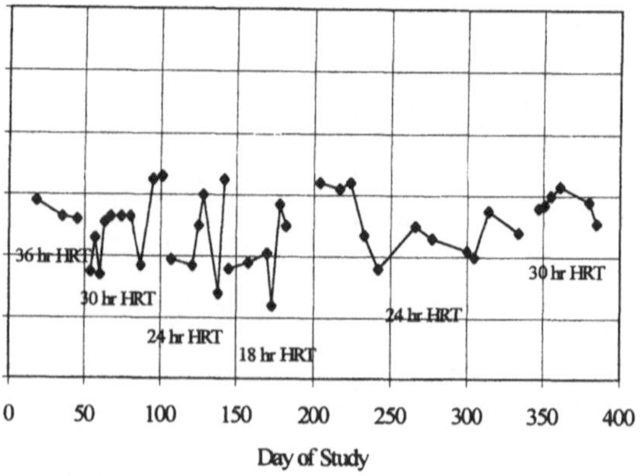


**Figure 6. pH in the granular reactor.**

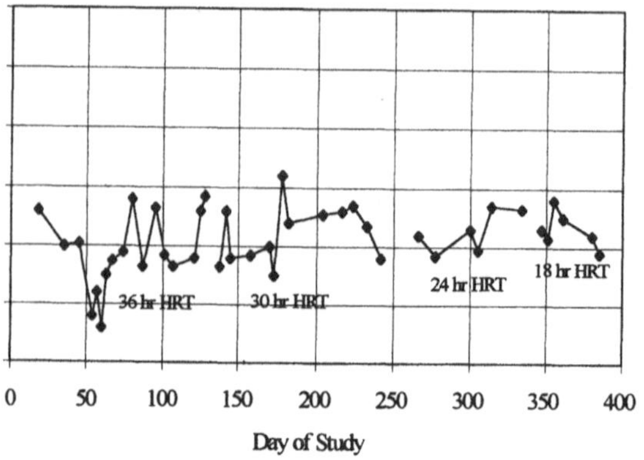


**Figure 7. pH in the flocculent reactor.**

Throughout the study, higher HRTs showed better removals. As the HRTs were brought to 18 hours and occasionally at 24 hours, periods of instability were present in the system showing significant variation in removals from one testing period to another. Average removals by HRT are given in Tables III and IV. Removals based on COD, SCOD, and BOD5 for individual testing periods are presented in Figures 8 through 13.

**Table III. Average Removals for the Granular Reactor**

| HRT (hours) | Total COD Removal, (%) | Soluble COD Removal, (%) | $BOD_5$ Removal, (%) |
|---|---|---|---|
| 36 | 63 | 94 | no data available |
| 30 | 53 | 95 | 77 |
| 24 | 35 | 91 | 50 |
| 18 | 27 | 89 | 37 |

**Table IV. Average Removals for the Flocculent Reactor**

| HRT (hours) | Total COD Removal, (%) | Soluble COD Removal, (%) | $BOD_5$ Removal, (%) |
|---|---|---|---|
| 36 | 57 | 85 | 76 |
| 30 | 44 | 91 | 54 |
| 24 | 36 | 90 | 47 |
| 18 | 34 | 85 | 46 |

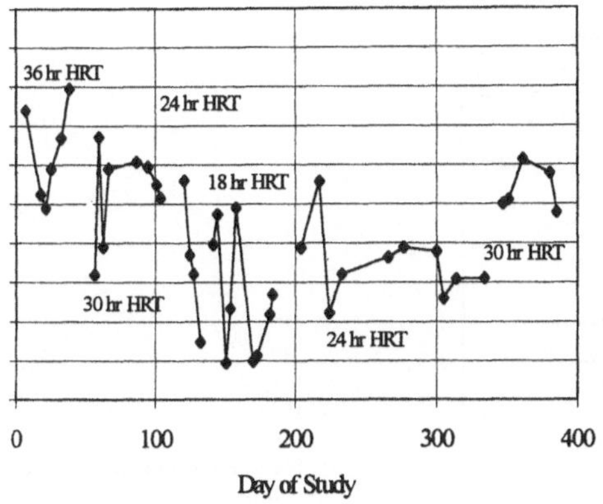


**Figure 8. COD removals in the granular reactor.**

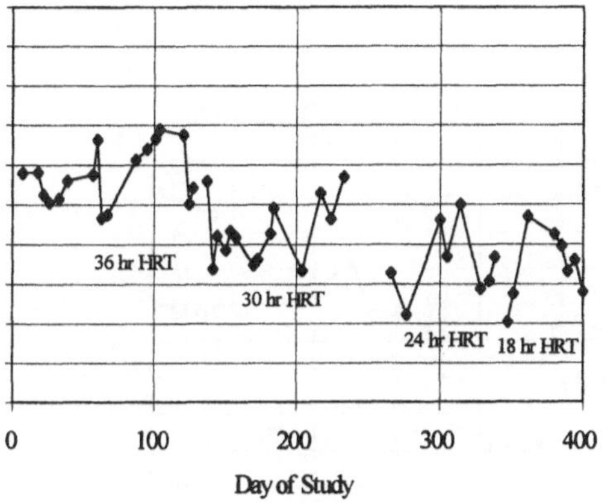


**Figure 9. COD removals in the flocculent reactor.**

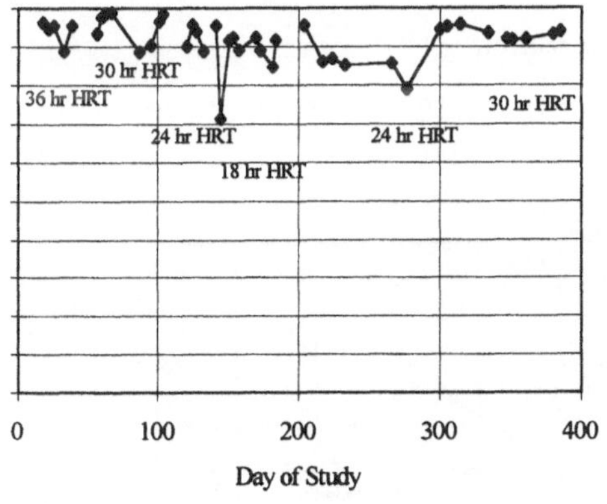


**Figure 10. SCOD removals in the granular reactor.**

The high fraction of solids present in the wastewater and their resistance to degradation likely resulted in rate-limiting hydrolysis and liquefaction reactions. This is shown by the low levels of VFA and high wastewater solids concentrations in the effluent. Similar results were found in studies of slaughterhouse wastewater treatment by Sayed et al.[11,12] Early in the study, the granular reactor performed better than the flocculent reactor, but as the flocculent reactor matured it

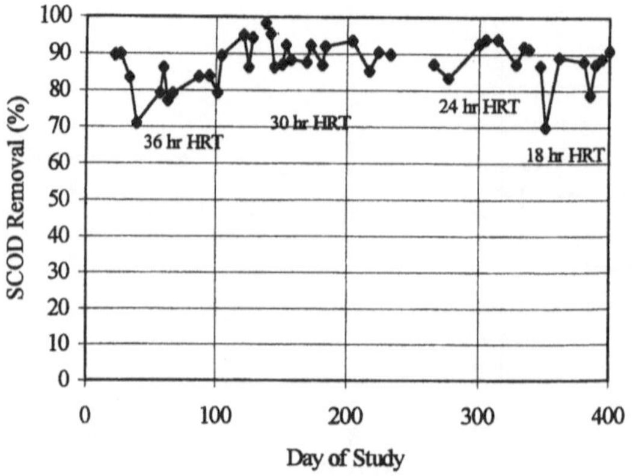


**Figure 11. SCOD removals in the flocculent reactor.**

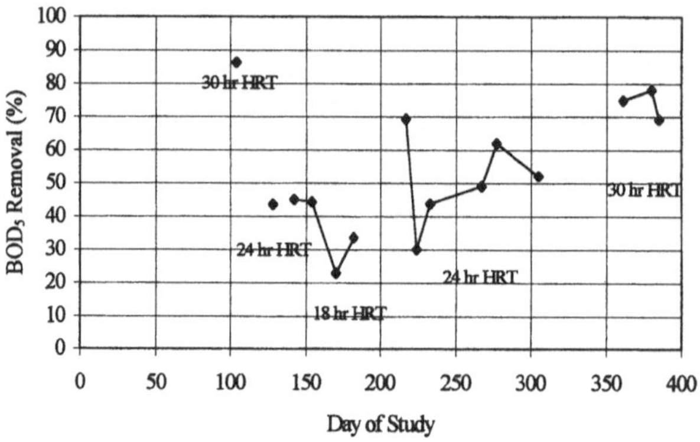


**Figure 12. $BOD_5$ removals in the granular reactor.**

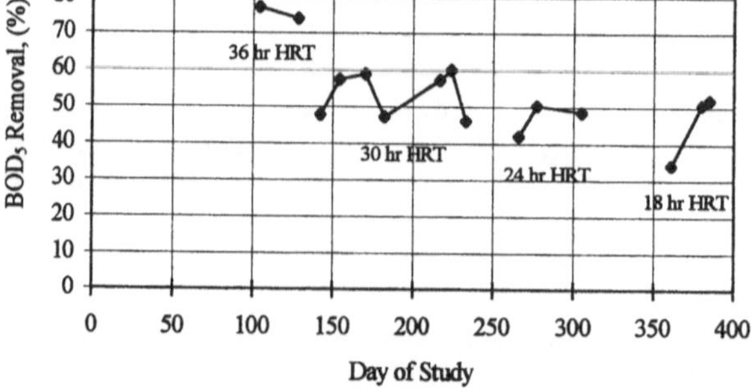


**Figure 13. $BOD_5$ removals in the flocculent reactor.**

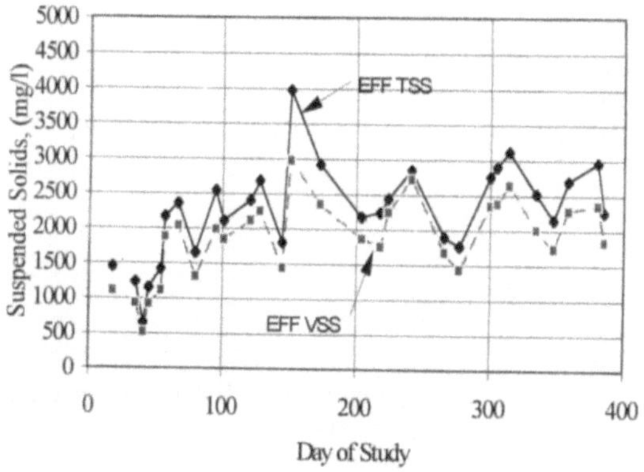


**Figure 14. Effluent suspended solids in the granular reactor.**

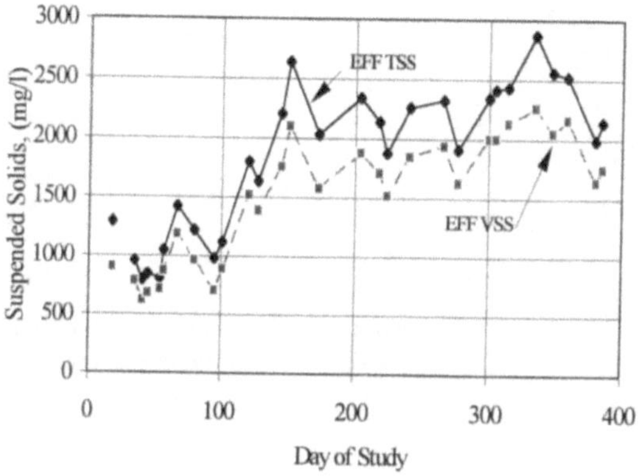


**Figure 15. Effluent suspended solids in the flocculent reactor.**

achieved higher removals. Solids destruction appeared to be higher in the flocculent reactor, probably due to the large surface area of the biomass. Influent and effluent solids concentrations for the reactor systems are shown in Figures 14 through 16.

Levels of mixed liquor suspended solids over the study for the reactor systems are given in Figures 17 and 18. The granular seed was retained better in the reactor due to larger particle size and higher particle density as shown in Figure 17. This allowed for a better reactor startup and earlier reductions in HRT. The flocculent reactor could not retain solids as well as the granular reactor and operated at much lower levels, as shown in Figure 18. Although the startup phase took a longer time and achieved lower removals than the granular reactor, the higher activity of the flocculent biomass and ability to degrade solids allowed better performance at lower HRTs with lower levels of biomass.

Levels of methane, carbon dioxide, and nitrogen remained fairly constant over the duration of the study, as shown in Figures 19 and 20. Methane levels of approximately 77% should be consistently recovered in the biogas. Average hydrogen sulfide concentrations in the biogas were 1040 ppm in the granular reactor and 1310 ppm in the flocculent reactor. The highest measured level over the course of the study was 1800 ppm in the flocculent reactor. All gases which contain over 1000 ppm hydrogen sulfide need to be treated before use.[3] The biogas produced during the treatment of this waste stream would require scrubbing to remove hydrogen sulfide before it could be used in the generation of power.

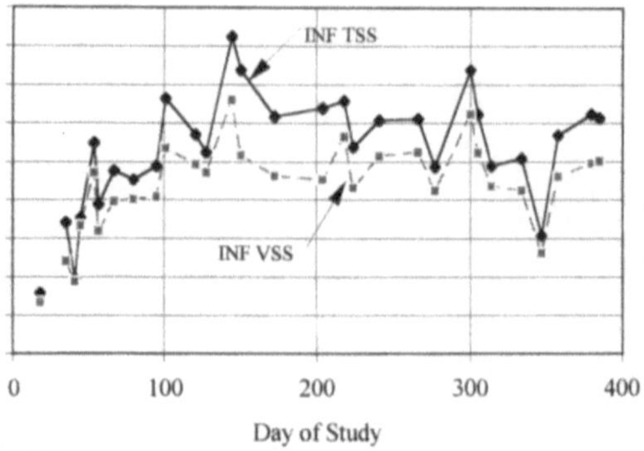


**Figure 16. Influent suspended solids.**

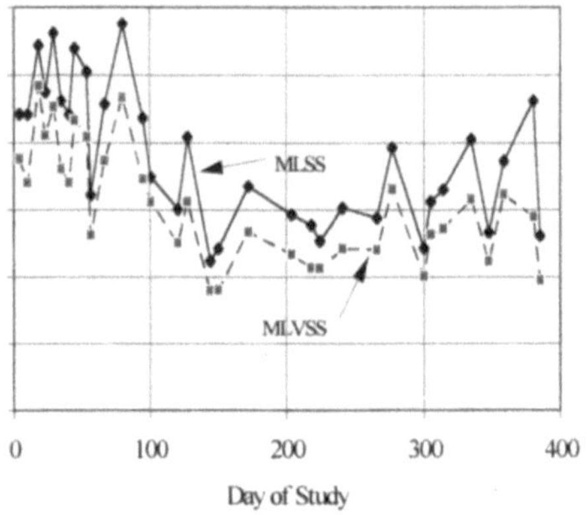


**Figure 17. Mixed liquor suspended solids in the granular reactor.**

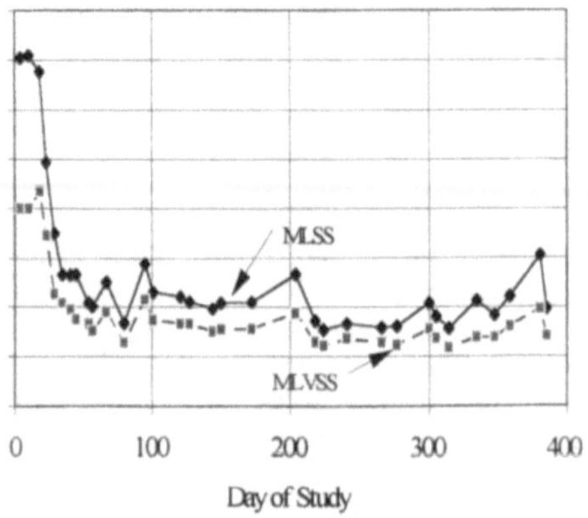


**Figure 18. Mixed liquor suspended solids in the flocculent reactor.**

## Engineering Significance

Recommendation for design of a full-scale system based on this research are to operate at a 30-hour HRT. The system seems to function best at a 30 HRT to achieve high loading rates and still ensure consistent removals. At this loading, the granular reactor achieved removals of over 70% based on BOD5 and 50% based on total COD. As the reactors mature, lower HRTs may be attainable to allow for expansion and increased wastewater flows.

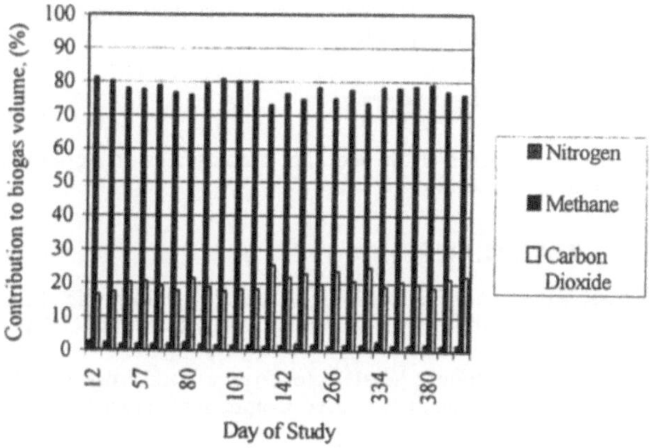


**Figure 19. Biogas composition in the granular reactor.**

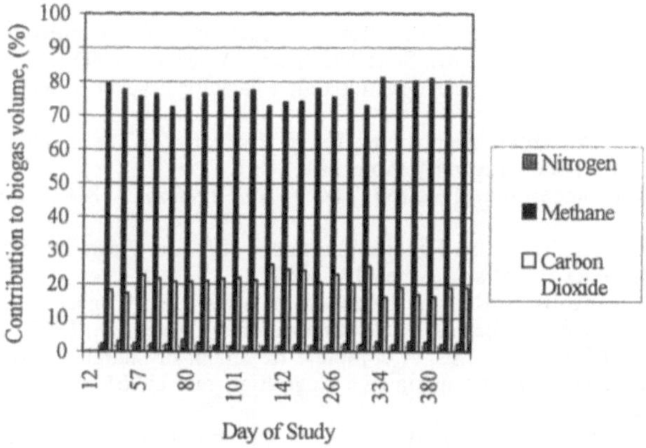


**Figure 20. Biogas composition in the flocculent reactor.**

## CONCLUSIONS

The following conclusions were drawn based on the results of this research:

1. The use of the ASBR was effective in the treatment of beef slaughterhouse wastewater obtained from the Excel corporation in Schuyler, Nebraska. Removal efficiencies are highly dependent on HRT, with longer residence times resulting in higher removals. Total COD removals averaged from 27 to 63% by HRT and BOD5 average removals ranged from 37 to 77% by HRT.
2. Hydrolysis and liquefaction reactions involving the degradation of suspended solid organics appeared to be the rate limiting step in treatment.
3. The ASBR system can be expected to produce biogas containing methane in the range of 73 to 81%. This biogas also contains high levels of hydrogen sulfide (up to 1800 ppm) which will need to be reduced prior to use as a fuel source.
4. The ASBR system is capable of treating the wastewater without the addition of alkalinity. Natural production of alkalinity through the degradation of protein is sufficient to maintain acceptable levels.
5. Granular and flocculent seed were both effective in treatment of the wastewater. Granular seed was more easily held in the reactor, resulting in better startup and higher removals early in the study. The flocculent seed showed formation of small granules as early as 100 days into the study. After establishment, flocculent seed

performed as well as granular seed at lower HRTs with more consistent removals even though it operated at much lower biomass levels.

## ACKNOWLEDGMENTS

Support for this research was provided by a grant from the U.S. Department of Agriculture through the Iowa Biotechnology Byproducts Consortium.

## REFERENCES

1. Banik, G.C. and Dague, R.R. ASBR Treatment of Dilute Wastewater at Psychrophilic Temperatures. Water Environment Federation, 69th Annual Conference and Exposition, Dallas, TX, 1996.
2. Calaghan, R.P. The Pretreatment of Slaughterhouse Wastewater from a Pork Processing Facility with an Anaerobic Sequencing Batch Reactor. Creative Component, Iowa State University, Ames, IA, 1996.
3. Caterpillar Gas Engine Power Systems General Catalog, Ziegler Power Systems, Ames, IA, 1996.
4. Dague, R.R., McKinney, R.E., and Pfeffer, J.T. "Anaerobic activated sludge." *J. Water Pollut. Control Fed.*, 38(2), pp. 220–226, 1966.
5. Fernandez-Garcia, S.D. Anaerobic Sequencing Batch Reactor Treatment of a Wastewater from Bioenzyme Production. Master's Thesis, Iowa State University, Ames, IA, 1995.
6. Habben, C.E. Initial Studies of the Anaerobic Sequencing Batch Reactor. Master's Thesis, Iowa State University, Ames, IA, 1991.
7. Kaiser, S.K. Initial Studies of the Anaerobic Sequencing Batch Reactor at a Thermophilic Temperature. Master's Thesis, Iowa State University, Ames, IA, 1991.
8. Novaes, R.F.V. "Microbiology of anaerobic digestion." *Water Sci. Technol.*, 18(12), pp. 1–14, 1986.
9. Pidaparti, S. and Dague, R.R. Sequencing Batch Reactor Treatment of Swine Waste at 35°C and 25°C. Master's Thesis, Iowa State University, Ames, IA, 1991.
10. Rowzee, M.V. Anaerobic Sequencing Batch Reactor Treatment of a High Fat, High Protein Waste from Pet Food Processing. Master's Thesis, Iowa State University, Ames, IA, 1996.
11. Sayed, S., van der Spoel, H., and Truijen, G.J.P. "A complete treatment of slaughterhouse wastewater combined with sludge stabilization using two stage high rate UASB process." *Water Sci. Technol.*, 27(9), pp. 83–90, 1993.
12. Sayed, S., van der Zanden, J., Wijffels, R., and Lettinga, G. "Anaerobic degradation of the various fractions of slaughterhouse wastewater." *Biological Wastes*, 23, pp. 117–142, 1988.
13. Schmitt, C.G. Anaerobic Sequencing Batch Reactor Treatment of Swine Wastes at 20°C, 25°C, and 35°C. Master's Thesis, Iowa State University, Ames, IA, 1992.
14. *Standard Methods for the Examination of Water and Wastewater.* 16th ed. American Public Health Association, Washington D.C., 1995.
15. Sung, S. and Dague, R.R. Fundamental Principles of the Anaerobic Sequencing Batch Reactor. Presented at the 47th Industrial Waste Conference, Purdue University, West Lafayette, IN, 1992.

# 24 DAIRY WASTEWATER TREATMENT WITH A TEMPERATURE-PHASED ANAEROBIC SEQUENCING BATCH REACTOR SYSTEM

**Prince Dugba**, Research Assistant

**Ruihong Zhang**, Assistant Professor
Biological and Agricultural Engineering Department
University of California
Davis, California 95616

**Richard R. Dague**, Professor
Civil and Construction Engineering Department
Iowa State University
Ames, Iowa 50010

## ABSTRACT

Three two-stage anaerobic sequencing batch reactor (ASBR) systems were used to treat dairy wastewater in the laboratory to evaluate a two-stage thermophilic-mesophilic system against a two-stage mesophilic-mesophilic system. The total working volume of 15 liters for each system was divided between the first- and second-stage reactors in the volume ratios of 1:4, 1:4, and 1:2 for the three systems, respectively. All three systems were automated to obtain precision feeding, decanting, and data collection using a computer-based control and data acquisition system. Organic loading rates of 2, 3, and 4 g VS/L/day were tested at the system hydraulic retention times of 3 and 6 days. The volatile (VS) removal varied from 26 to 44% with the thermophilic-mesophilic systems destroying 26–50% more VS than the mesophilic-mesophilic system. No fecal coliforms were detected in the effluents from the thermophilic-mesophilic systems compared to only two log reduction in fecal coliforms with the mesophilic-mesophilic system.

## INTRODUCTION

Farm animals in the U.S.A. generate an estimated 2 billion tons of solid waste annually,[1] some nine times that generated from American homes. Traditionally, this waste is disposed of through land application. In California, for example, about 90% of animal manure is applied to the land with little or no pretreatment. However, increasing environmental awareness and dwindling agricultural land force farmers to look for efficient waste treatment alternatives. One promising alternative is the use of an anaerobic digester system to stabilize the waste for odor control and to recover energy and nutrients.

Most conventional digesters are continuously-stirred tank reactors (CSTR) with hydraulic retention time (HRT) equal to the solid retention time (SRT). In a CSTR, active biomass is easily washed out with the effluent when the HRT is too short. The HRT has to be long enough (15 to 20 days) to maintain a viable bacteria population necessary to complete the anaerobic digestion process. High-rate reactors such as the upflow anaerobic sludge blanket (UASB), the anaerobic biofilter, and the fluidized bed reactor developed to improve biomass retention have shown better treatment of dilute wastes compared to the CSTR. The HRTs can be on the order of several hours to a few days compared to 15 to 20 days for the CSTR. Most of these high-rate reactors, however, require inert media placed in the reactors to support the bacterial growth. Uniform feed distribution requires special manifold designs that are difficult to maintain with high-solid wastewater. Clogging of the sludge blanket or the media matrix could be a problem when treating a

---

*52nd Purdue Industrial Waste Conference Proceedings*, 1997, Ann Arbor Press, Chelsea, Michigan 48118. Printed in U.S.A.

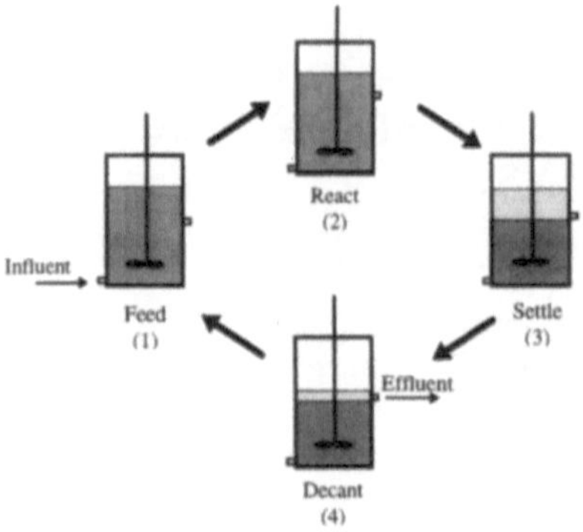


**Figure 1. Four phases of the anaerobic sequencing batch reactor (ASBR).**

high-solid waste such as animal waste. New high-rate reactors capable of handling high-solid influent and simple to operate would be the most promising alternatives. One such system, known as the Anaerobic Sequencing Batch Reactor (ASBR) (U.S. Patent No. 5,185,079) was developed in 1992 by Sung and Dague[2] at Iowa State University. The ASBR operates in four distinct phases during each batch cycle. Figure 1 shows a schematic of a typical ASBR cycle.

At the beginning of the cycle, the reactor is fed at a given load and mixed to start the react phase. It is then mixed intermittently during the react phase until the settle phase. By this time, biogas production is minimum, allowing for efficient settling and retention of microbes to the bottom of the reactor. During the forth and final phase, the decant phase, the supernatant is decanted with minimum washout of bacteria mass. Feeding follows to start another batch cycle. This kind of digester operation provides an efficient way to maintain an optimum food to microbial ratio (F/M) with no requirement for an elaborate feed distribution manifold.

Although there is no previous study on the efficiency of the use of the ASBR to treat dairy wastewater, reports from the studies using this technology on other high-solid wastewaters are promising. Schmit and Dague[3] achieved 40–60% VS removal from swine waste at the organic loading rates of 0.9 to 5.0 gVS/L/d at temperatures between 20 to 35°C using the ASBR. For the VS loading rates of 0.8 to 6.5 gVS/L/day, no significant difference was observed between the performance of an ASBR operated on a 3-day HRT scheme and that at a 6-day HRT scheme.[4] Other projects underway, including a scale-up study of the ASBR, have shown promising results in the application of this technology to animal waste treatment.

Harris and Dague[5] observed that thermophilic biofilters were able to handle the organic loads of synthetic substrate (nonfat dry milk) at the rate of 2.5 times that of the mesophilic biofilters. They also observed improved dewaterability of the effluents from thermophilic systems. This is particularly useful for high-solid wastewater such as animal manure. However, thermophilic digester effluent was found to have higher concentrations of volatile fatty acids (VFAs).[5–7] Feeding this effluent into a second-stage mesophilic reactor will polish the final effluent by further converting the VFAs to biogas. A temperature-phased two-stage anaerobic treatment system consisting of two reactors in series with the first reactor operated at a thermophilic temperature of 55°C, followed by the second reactor operated at a mesophilic temperature of 35°C, combines the advantages of both types of reactors while eliminating their disadvantages.

One major advantage of thermophilic digestion is the destruction of pathogenic microbes, thereby making the biosolids from the digestion systems suitable for land application with less risk of disease transmission through groundwater contamination. Supernatant liquid from such systems can also be used for flushing animal buildings with less risk of disease transmission. A recent work by Han and Dague[8] reported a near total (99.9998%) destruction of fecal coliforms with thermophilic treatment of domestic sewage. The highest coliform count which they recorded for the effluent from a temperature-phased CSTR system was 310 MPN/g TS. This is well below the CFR 40, Part 503 coliform requirement (not to exceed 1000 MPN/g TS) for Class A sludge used for land application. They also recorded 18% higher VS destruction and 16% more methane production for the temperature-phased CSTR system compared to a CSTR.

This chapter presents laboratory test results from the use of a two-stage temperature-phased anaerobic sequencing batch reactor system to treat screened dairy manure.

## OBJECTIVES

The specific objectives of this study were to:

- evaluate the performance of a thermophilic-mesophilic temperature-phased anaerobic sequencing batch reactor (ASBR) system in treating dairy wastewater.
- determine the optimum design parameters and operational schemes for an ASBR system for treating dairy wastewater.

## MATERIALS AND METHODS

### Experimental Design

Three parallel systems—I, II and III—each consisting of two ASBR in series were tested for the effects of temperature, VS loading rate, and hydraulic retention time (HRT). Each system has a total volume of 15 L divided between the first- and second-stage reactors, R1 and R2, in the ratios 1:4, 1:4, and 1:2 for systems I, II, and III, respectively. The first-stage reactor of system I and all the second-stage reactors of system I, II, and III were maintained at 35°C. The first-stage reactors of systems II and III were maintained at 55°C. Table I shows the operating parameters of individual reactors and overall systems. Figure 2 shows a three dimensional view of the system.

**Table I. Operational Parameters**

| | System I | | | System II | | | System III | | |
|---|---|---|---|---|---|---|---|---|---|
| Operational Parameters | R1 | R2 | System | R1 | R2 | System | R1 | R2 | System |
| Reactor Volume (L) | 3 | 12 | 15 | 3 | 12 | 15 | 5 | 10 | 15 |
| Temperature (°C) | 35 | 35 | | 55 | 35 | | 55 | 35 | |
| HRT (days) | 0.6 | 2.4 | 3 | 0.6 | 2.4 | 3 | 1 | 2 | 3 |
| | 1.2 | 4.8 | 6 | 1.2 | 4.8 | 6 | 2 | 4 | 6 |
| VS Loading Rate | 10.0 | 2.5 | 2 | 10.0 | 2.5 | 2 | 6.0 | 3.0 | 2 |
| (gVS/L/d) | 15.0 | 3.8 | 3 | 15.0 | 3.8 | 3 | 9.0 | 4.5 | 3 |
| | 20.0 | 5.0 | 4 | 20.0 | 5.0 | 4 | 12.0 | 6.0 | 4 |

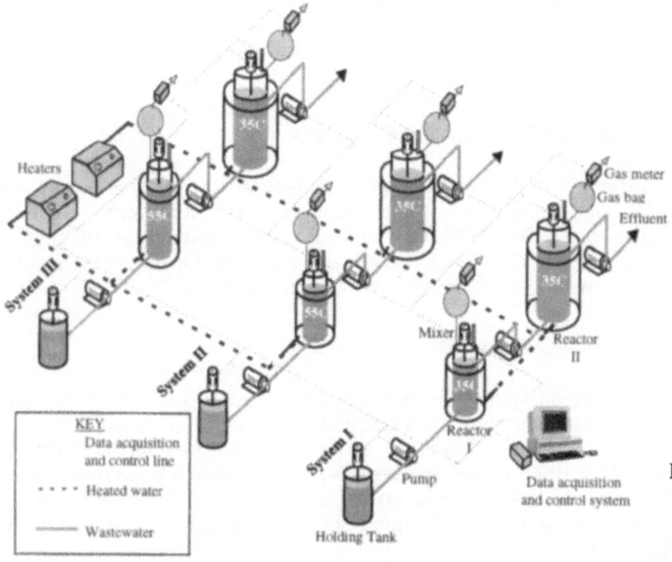


**Figure 2. Temperature-phased ASBR systems for treating dairy wastewater.**

Three system VS loading rates of 2, 3, and 4 gVS/L/day were tested with each system at the system HRTs of 3 and 6 days. Note from Table I that the individual reactor VS loading rates were higher than the system loading rates because each reactor volume was only a fraction of the system volume. For the same reason, the reactor HRTs were shorter than the system HRTs.

All the reactors and holding tanks are made of Plexiglas. Each reactor has an external water jacket which was heated with a thermostatically controlled flowthrough heating recirculator to maintain each reactor at a desired temperature. A wet tip gas meter was used to measure the volume of biogas produced by each of the six reactors.

Using the ASBR operating cycle shown in Figure 1, the substrate from the holding tank was fed to the first-stage reactor while its effluent was fed to the second-stage reactor. The operation of the entire system was automated using a computer-based data acquisition and control system. A custom-made level sensor was used to ensure precision feeding and decanting. The computer control and sensor system for the bioreactor systems is described in detail by Dugba and Zhang.[9]

### Reactor Startup and Operation

All the reactors were seeded with mesophilic anaerobic digester sludge from the Municipal Wastewater Treatment Plant at Davis, CA. They were all started at 1% initial mixed liquor volatile suspended solids concentration (MLVSS). The temperature of the thermophilic reactors was raised from 35°C to 55°C over a one-hour period. All reactors were allowed to acclimatize for 24 hours before being fed at the rate of 0.5 gVS/L/day for one week with dairy manure screened through a sieve with 2 x 2 mm openings. The organic loading rates of all six reactors were then raised to 1.0 gVS/L/day. Feeding was maintained at this loading rate until all the reactors started producing gas. It took about a week for the mesophilic reactors to start producing gas, and about five weeks for the thermophilic reactors to start producing gas. System loading rates of 2, 3, and 4 gVS/L/day were then tested from June 1996 through March 1997, first at 6-day HRT and then at 3-day HRT.

### Reactor Performance Evaluation

Samples from the reactorsÕ influent and effluent, and their mixed liquor were analyzed for total solids (TS), volatile solids (VS), volatile suspended solids (VSS), volatile organic acids (VOA), ammonia nitrogen ($NH_3$-N), total Kjeldahl nitrogen (TKN), total phosphorous (P), total potassium (K), pH and alkalinity, using *Standard Methods.*[10] Biogas was collected and measured with the Rebel wet tip gas meters. Gas chromatographic (GC) analysis was done on the gas samples from individual reactors to determine their methane ($CH_4$) and carbon dioxide ($CO_2$) contents. Except for the wet tip gas measurement that was continuous, all measurements were repeated three times for each organic loading rate and HRT after reactors had reached a quasi-steady-state which was characterized by a less than 5% variation in biogas production during a one-week period.

## RESULTS AND DISCUSSION

There was no significant difference in the composition of the biogas from the six reactors. The biogas contained 62–66% $CH_4$ and 28–31% $CO_2$. Solid retention times (SRT) were also similar for all the reactors at both the 3- and 6-day HRTs. For the same organic loading rate, the raw influent VS concentration for the 3-day HRT is twice that at the 6-day HRT. Operating the reactors in the ASBR mode achieved 13–18 day SRTs for both the 3- and 6-day HRTs.

### Volatile Solids Removal and Biogas Production

Table II and Figure 3 show the VS removals at the system loading rates of 2, 3, and 4 gVS/L/day and 6 days HRT. The corresponding methane productions are also shown in Table II. For the organic loading rates of 2, 3, and 4 gVS/L/day and 6 days HRT, system I (1:4 volume, two-stage mesophilic-mesophilic ASBR system) achieved 29, 31, and 26% VS removal; system II (1:4 volume, thermophilic-mesophilic ASBR system) achieved 37, 42, and 39% VS removal; and system III (1:2 volume, two-staged thermophilic-mesophilic ASBR system) achieved 33, 35, and 31% VS removal, respectively.

**Table II. Volatile Solids Removal and Methane Production at 6-Days HRT**

| | VS Load, g/L/day | | | | | | | | |
|---|---|---|---|---|---|---|---|---|---|
| | 2 Reactor System | | | 3 Reactor System | | | 4 Reactor System | | |
| | I[a] | II[b] | III[c] | I | II | III | I | II | III |
| Reactors | VS Removal, % | | | | | | | | |
| R1 | 9.4 | 13.2 | 20.2 | 12.6 | 18.4 | 23.9 | 14.6 | 17.4 | 18.1 |
| R2 | 19.7 | 24.0 | 12.7 | 18.4 | 23.3 | 10.7 | 11.5 | 21.6 | 12.8 |
| System | 29.1 | 37.1 | 32.9 | 31.0 | 41.7 | 34.6 | 26.1 | 38.9 | 30.9 |
| | Methane Production, L/L/day | | | | | | | | |
| R1 | 0.44 | 0.61 | 0.56 | 0.87 | 1.28 | 1.00 | 1.36 | 1.61 | 1.01 |
| R2 | 0.23 | 0.28 | 0.18 | 0.32 | 0.41 | 0.22 | 0.27 | 0.50 | 0.36 |
| System avg. | 0.27 | 0.34 | 0.25 | 0.43 | 0.58 | 0.38 | 0.48 | 0.72 | 0.49 |

[a]1:4 volume ratio, two-staged mesophilic-mesophilic ASBR system.
[b]1:4 volume ratio, two-staged thermophilic-mesophilic ASBR system.
[c]1:2 volume ratio, two-staged thermophilic-mesophilic ASBR system.

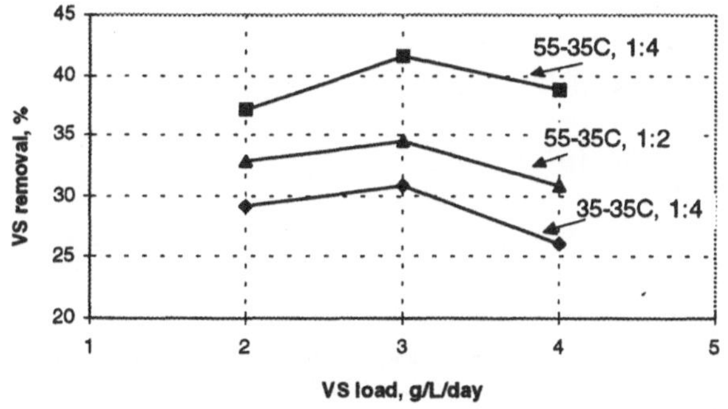


**Figure 3. Volatile solids removal at 6 days HRT.**

Table III and Figure 4 show the VS removal and methane production at 3-days HRT. For the organic loading rates of 2, 3, and 4 gVS/L/day and 3-days HRT, system I (1:4 volume, two-stage mesophilic-mesophilic ASBR system) achieved 29, 35, and 30% VS removal; system II (1:4 volume, thermophilic-mesophilic ASBR system) achieved 44, 44, and 44% VS removal; and system III (1:2 volume, two-staged thermophilic-mesophilic ASBR system) achieved 29, 40, and 37% VS removal, respectively.

At both 6- and 3-days HRTs, a significantly higher VS removal was achieved with the thermophilic-mesophilic ASBR systems (II and III) than the mesophilic-mesophilic system (I). Compared to system I, system II removed 27–50% more VS, while system III removed up to 23% more. About the same amount of VS was removed by both systems I and III for the VS loading rate of 2 gVS/L/d at 3 days HRT. Both systems I and II had the same 1:4 first-stage to second-stage reactor volume ratio, and therefore the VS removal and methane production of the individual reactors in Tables II and III can be compared directly. Note that the thermophilic first-stage reactor in system II removed 10–46% more VS than the mesophilic first-stage reactor in system I. The second-stage reactor in system II also outperformed its counterpart in system I. This could be attributed to the fact that a significant amount of the substrate fed into the second-stage reactor of the thermophilic-mesophilic system was already partially hydrolyzed and converted to volatile fatty acids (VFAs). The VFA concentration in the thermophilic effluents (1056–1900

**Table III. Volatile Solids Removal and Methane Production at 3-Days HRT**

| | VS Load, g/L/day | | | | | | | | |
|---|---|---|---|---|---|---|---|---|---|
| | 2 Reactor System | | | 3 Reactor System | | | 4 Reactor System | | |
| | I[a] | II[b] | III[c] | I | II | III | I | II | III |
| Digester | VS Removal, % | | | | | | | | |
| R1 | 10.5 | 12.9 | 12.9 | 12.9 | 14.3 | 20.9 | 12.8 | 16.2 | 18.8 |
| R2 | 18.7 | 31.0 | 16.4 | 21.9 | 29.6 | 19.0 | 17.4 | 28.0 | 18.5 |
| System | 29.3 | 43.9 | 29.3 | 34.8 | 43.8 | 39.9 | 30.2 | 44.2 | 37.3 |
| | Methane Production, L/L/day | | | | | | | | |
| R1 | 0.49 | 0.60 | 0.36 | 0.90 | 0.99 | 0.87 | 1.19 | 1.50 | 1.04 |
| R2 | 0.22 | 0.36 | 0.23 | 0.38 | 0.51 | 0.40 | 0.40 | 0.65 | 0.52 |
| System avg. | 0.27 | 0.41 | 0.25 | 0.48 | 0.61 | 0.49 | 0.56 | 0.82 | 0.62 |

[a]1:4 volume ratio, two-staged mesophilic-mesophilic ASBR system.
[b]1:4 volume ratio, two-staged thermophilic-mesophilic ASBR system.
[c]1:2 volume ratio, two-staged thermophilic-mesophilic ASBR system.

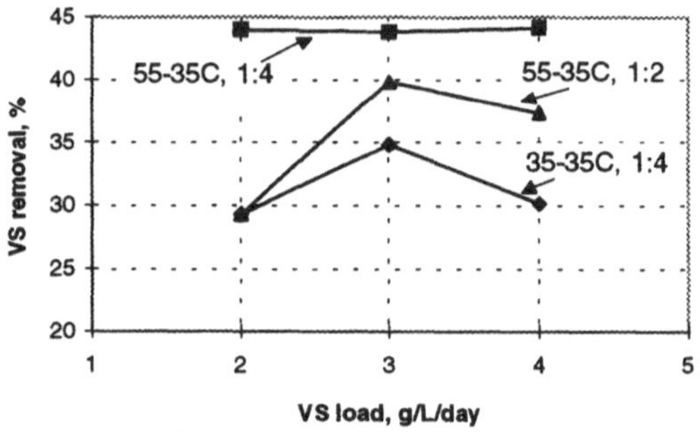


**Figure 4. Volatile solids removal at 3-days HRT.**

mg/L) was about three times higher than the VFA concentration in the mesophilic effluents (330–630 mg/L). The individual reactors in system III cannot be compared directly to their counterparts in system I since the reactor volumes are different. After adjusting the reactor performance to unit volume basis, however, it can be seen that the combined performance of the two reactors in system III was better than that in system I, except for the 2 gVS/L/d at 3 days HRT treatment.

Although system III (1:2 volume ratio) had a larger thermophilic volume compared to system II, the former performed consistently not as good as the latter. This may be due to difference in individual reactor HRTs as the first-stage reactor in system III (1:2 volume ratio) had 5/3 times longer HRTs than the first-stage reactor in system II (1:4 volume ratio). This also translated into higher VS loading rates for the first-stage reactor in systems II compared to system III. Table I lists the VS loading rates for the individual reactors.

Overall, there was no significant difference in the VS removals between 3- and 6-days HRTs for the organic loading rates of 2, 3, and 4 gVS/L/day. This result is significant because it translates into 50% reduction in reactor volume at 3-days HRT compared to the reactor volume at 6 - days HRT.

Note that the methane production per unit reactor volume of the first-stage reactors was significantly higher than the second-stage reactors in each of the three systems. The trend is illustrated

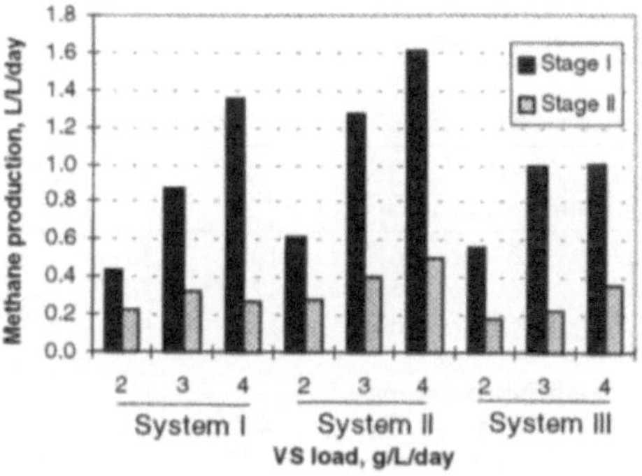


**Figure 5. Methane production per unit reactor volume.**

in Figure 5. This indicates a less-than-optimum organic loading to the second-stage reactors. Increase in organic loading rates will be further investigated in subsequent studies.

## Fecal Bacteria Destruction

No fecal coliform CFU was detected in the effluents from the thermophilic-mesophilic systems, indicating their near-complete destruction, compared to only about 1-2 log reductions in CFU with the mesophilic-mesophilic system. This is significant, especially in the dairy wastewater treatment system where the effluent water may be used to flush the confinement buildings. The coliform destruction also reduces the risk of animal disease transmission due to animal waste disposal.

## Nutrient Transformation

Table V shows the nutrient composition of the wastewater samples. Each of nitrogen, phosphorous, and potassium concentration was proportional to the organic loading rates. Ammonia nitrogen concentration in the effluent streams was generally higher than that in the influent streams due to mineralization. Phosphorous concentration in the effluents was significantly lower than in the influents due to cellular fixation and solids settling. The pH of all samples was fairly constant at 6.7 to 7.3.

**Table IV. Fecal Coliform Colony Forming Unit (CFU) Counts of the Wastewater Samples**

| | VS Load, g/L/day | | | | | | | | |
|---|---|---|---|---|---|---|---|---|---|
| | 2 Reactor System | | | 3 Reactor System | | | 4 Reactor System | | |
| | I[a] | II[b] | III[c] | I | II | III | I | II | III |
| Sample | Fecal Coliform, MPN/L | | | | | | | | |
| Influent | 2.8e+7 | 2.8e+7 | 2.8e+7 | 5.3e+8 | 5.3e+8 | 5.3e+8 | 2.7e+7 | 2.7e+7 | 2.7e+7 |
| R1, effluent | 2.3e+7 | <20[d] | <20[d] | 3.6e+7 | <20[d] | <20[d] | 6.2e+5 | <20[d] | <20[d] |
| R2, effluent | 1.2e+7 | <20[d] | <20[d] | 7.1e+6 | <20[d] | <20[d] | 8.1e+4 | <20[d] | <20[d] |

[a]1:4 volume ratio, two-staged mesophilic-mesophilic ASBR system.
[b]1:4 volume ratio, two-staged thermophilic-mesophilic ASBR system.
[c]1:2 volume ratio, two-staged thermophilic-mesophilic ASBR system.
[d]no fecal coliform colony forming unit was detected in these samples.

**Table V. Nutrient Composition of the Wastewater Stream**

| | VS Load, g/L/day | | | | | | | | |
|---|---|---|---|---|---|---|---|---|---|
| | 2 Reactor System | | | 3 Reactor System | | | 4 Reactor System | | |
| | I[a] | II[b] | III[c] | I | II | III | I | II | III |
| Sample | Total Kjeldahl Nitrogen (TKN), mg/L | | | | | | | | |
| Influent | 496 | 496 | 496 | 673 | 673 | 673 | 1357 | 1357 | 1357 |
| R1, effluent | 632 | 604 | 682 | 800 | 608 | 667 | 1270 | 1388 | 1429 |
| R2, effluent | 524 | 570 | 501 | 773 | 697 | 606 | 1436 | 1123 | 1397 |
| | Ammonia Nitrogen, mg/L | | | | | | | | |
| Influent | 295 | 295 | 295 | 314 | 314 | 314 | 479 | 479 | 479 |
| R1, effluent | 322 | 368 | 352 | 461 | 370 | 461 | 595 | 678 | 678 |
| R2, effluent | 402 | 384 | 352 | 482 | 461 | 576 | 772 | 807 | 808 |
| | Total Potassium (K), mg/L | | | | | | | | |
| Influent | 403 | 403 | 403 | 621 | 621 | 621 | 1083 | 1083 | 1083 |
| R1, effluent | 426 | 441 | 468 | 300 | 301 | 348 | — | 327 | 534 |
| R2, effluent | 458 | 429 | 440 | 546 | 478 | — | 396 | 656 | 264 |
| | Total Phosphorous (P), mg/L | | | | | | | | |
| Influent | 181 | 181 | 181 | 236 | 236 | 236 | 411 | 411 | 411 |
| R1, effluent | 150 | 171 | 158 | 219 | 171 | 116 | 437 | 209 | 170 |
| R2, effluent | 82 | 52 | 77 | 97 | 95 | 335 | 276 | 315 | 263 |

[a] 1:4 volume ratio, two-staged mesophilic-mesophilic ASBR system.
[b] 1:4 volume ratio, two-staged thermophilic-mesophilic ASBR system.
[c] 1:2 volume ratio, two-staged thermophilic-mesophilic ASBR system.

## CONCLUSIONS

All three anaerobic systems that have been evaluated in this study: I (two-stage 1:4 volume ratio mesophilic-mesophilic ASBR), II (two-stage 1:4 volume ratio thermophilic-mesophilic ASBR) and III (two-stage 1:2 volume ratio thermophilic-mesophilic ASBR), demonstrated stable treatment performance with dairy wastewater.

The following conclusions can be drawn from this study:

1. A relatively higher solids retention time (SRT) of 13-18 days was achieved at both 3- and 6-days HRTs. There was no significant difference in the SRTs between the 3- and 6-days HRT treatment.
2. The VS removal for the three systems tested varied from 26 to 44%. There was no significant difference in VS removals between the 3- and 6-days HRTs. This will translate into a significantly smaller reactor capable of handling high organic loads at short HRTs.
3. The thermophilic-mesophilic systems (II and III) produced more biogas than the mesophilic-mesophilic system (I) for the same organic loading rates (2, 3, and 4 gVS/L/day). Compared to system I, system II removed 27–50% more VS, while system III removed up to 23% more.
4. The thermophilic reactors reached a steady state performance faster than the mesophilic reactors when VS loading rate was increased from a lower to a higher level. This indicates a better buffering capability at the thermophilic temperature when reactors are exposed to shock loads. Subsequent studies at higher system organic loads will be needed to determine systems' stability performance.
5. No fecal coliform CFU was detected in the effluents from the thermophilic-mesophilic systems, compared to only about 1-2 log reductions in CFU with the mesophilic-mesophilic system.

## REFERENCES


1. USDA-Soil Conservation Service (USDA-SCS). Agricultural Waste Management Field Handbook. National Engineering Handbook. Issued in April 1992. Chapters 1–11.
2. Sung, S. and R.R. Dague. Fundamental Principles of the Anaerobic Sequencing Batch Reactor Process. In: *Proceedings of the 47th Purdue Industrial Waste Conference*. Lewis Publishers, Inc., Chelsea, MI, 1992.
3. Schmit, C.C. and R.R. Dague. Anaerobic Sequencing Batch Reactor Treatment of Swine Wastes at 20°C, 25°C, and 35°C. In: *48th Purdue Industrial Waste Treatment Conference Proceedings*. Lewis Publishers, Chelsea, MI, 1993.
4. Zhang, R.H. and R.R Dague. Treatment of Swine Wastes by the Anaerobic Sequencing Batch Reactor System. Proceedings of the Symposium on Agricultural and Food Processing Waste, 1995, p. 301.
5. Harris, W.L. and R.R. Dague. "Comparative performance of anaerobic filters at mesophilic and thermopilic temperatures." *Water Environ. Res.*, 65(6), pp. 764–771, 1993.
6. Buhr, H.O. and J.F. Andrews. "Review paper: The thermophilic anaerobic digestion process." *Water Res.*, 11, pp. 129–143.
7. Kaiser, S.K. and R.R. Dague. "The temperature-phased anaerobic biofilter process." *Water Sci. Technol.*, 29(9), pp. 213–223, 1994.
8. Han, Y. and R.R. Dague. Laboratory Studies on the Temperature-Phased Anaerobic Digestion of Domestic Wastewater Sludge. Presented at the 68th Annual Conference of the Water Environment Federation, Miami Beach, FL, October 21–25, 1995.
9. Dugba, P.N. and R.H. Zhang. Automation of a Prototype Biological Treatment System for Agricultural and Food Processing Waste. Paper No. 96-3041: ASAE International Meeting, July 14–18, 1996, Phoenix, AZ. ASAE, 2950 Niles Rd., St., Joseph, MI 49085-9659.
10. *Standard Methods for the Examination of Water and Wastewater*, 16th ed. American Public Health Association, American Water Works Association, Water Pollution Control Federation. 16th Ed., Washington, D.C. 20005, 1985.

# Section Seven
# BIOLOGICAL: PHENOLIC WASTE TREATMENT

## 25 FATE OF AMMONIA, THIOCYANATE, PHENOLICS, AND CYCLIC AROMATICS WITHIN ACTIVATED SLUDGE COKING PLANT WASTE TREATMENT

**Clifford R. Lange**, Assistant Professor
**David E. Chambers**, Graduate Student
Department of Civil Engineering, Auburn University
Auburn, Alabama 36849-5337

### INTRODUCTION

Coal distillation processes, such as coal gasification and coal coking produce complex wastewaters. The coking process involves the destructive distillation of coal in an anoxic environment. The volatile compounds released during the coking process are either condensed or scrubbed out in the gas main or primary coolers. Tar is decanted from the condensate and the remaining aqueous waste, known as the flushing liquor or weak ammonia liquor (WAL), is the major source of coking plant wastewater.

Coke plant wastewaters include a complex mixture of organic and inorganic compounds, including: ammonia, cyanide, thiocyanate, monohydric phenols (70% phenol, 25% m-cresol, and 5% o-cresol), heterocyclic nitrogenous compounds (HNCs), and polyaromatic hydrocarbons (PAHs). The concentrations of these wastewater constituents are a function of: coal type; coking temperature; and process technology. The ammonia and cyanide concentrations of the WAL are typically diminished by air stripping processes to make the wastewater more amenable to biological treatment. Dilution water may be added to the waste to reduce the concentrations of other inhibitory compounds prior to biological treatment.

The goal of biological treatment of coking process wastewater is to mineralize organic contaminants and to oxidize the ammonia and thiocyanate to nitrate. One of the most commonly employed biological treatment processes for coking plant wastwaters is activated sludge. These processes are typically operated with long mean cell residence times ($\Theta_C$s) and can achieve high degrees of ammonia nitrification, thiocyanate conversion, and organic removals. However, the presence of inhibitory compounds in the wastewater may cause process upsets.

Nitrification is the biological oxidation of ammonia ($NH_4^+$) to nitrite ($NO_2^-$), and ultimately to nitrate ($NO_3^-$). The specialized population of bacteria which mediate this oxidation are known as nitrifiers. Although there are seven known genera of nitrifiers, the species *Nitrosomonas* and *Nitrobacter* perform most of the nitrification that occurs in wastewater treatment. *Nitrosomonas europea* has been identified as the only organism capable of oxidizing ammonia to nitrite in coking plant biological processes, while several species of *Nitrobacter* are known to oxidize nitrite to nitrate. Nitrifiers have low growth rates and cell yields and are easily out-competed by heterotrophs under high growth rate conditions. Consequently, nitrification occurs only in low growth rate processes and is slow to recover from process upsets.

Nitrifiers are inhibited by a number of compounds present in coking plant wastewater. Excessive levels of free ammonia were reported by Wong-Chong and Loehr[1] to be inhibitory to *Nitrosomonas* (10 mg/L) and *Nitrobacter* (0.1 mg/L). Similarly, free nitrous acid ($HNO_2$) concentrations in excess of 0.22 mg/L are toxic to nitrifying populations. Thiocyanate concentrations of

*52nd Purdue Industrial Waste Conference Proceedings*, 1997, Ann Arbor Press, Chelsea, Michigan 48118. Printed in U.S.A.

100 mg/L caused 12% inhibition of nitrification.[2] Phenol can inhibit nitrification at concentrations of 4 mg/L or higher (Stafford[3] and Blok[4] demonstrated that ammonia oxidation is inhibited by other phenolics, including: o-cresol (4.4 mg/L), m-cresol (4.0 mg/L), and aniline (0.5 mg/L). Becarri[2] showed that concentrations of quinoline above 5.0 mg/L can cause inhibition of nitrification. Known heavy metal inhibitors of nitrification include: Cd (10 ppm); Cr (1 ppm); Cu (20 ppm); Ni (1 ppm), and Zn (10 ppm).

Biological conversion of other organic and inorganic compounds present in coking plant wastewater has been well documented. High degrees of thiocyanate, cyanide, and phenols removal in an activated sludge process was described by Shvaraman et al.[5] Thiocyanate is biologically hydrolized to form ammonia in the activated sludge process.[6] Phenolic compounds are easily biodegraded, while the heterocyclic nitrogenous compounds and polyaromatic hydrocarbons are poorly degraded.[7] Biological oxidation efficiencies of 99% or greater have been reported for activated sludge processes treating phenols.[8–10] Conversely, the overall PAH levels in activated sludge remain largely unaffected by biodegradation, and may be increased by partitioning to the sludge in long residence time processes.[9] The biodegradation of heterocyclic compounds is relatively slow compared to phenol biodegradation. Cooper et al.[11] found the biodegradation of pyridine and quinoline in coking wastewater to be both slow and incomplete.

To assess the feasibility of operating a two-stage, activated sludge process for treatment of coal coking wastewater, pilot processes were constructed and operated at Gulf States Steel, Inc., located in Gadsten, Alabama. The overall wastewater flow at the Gulf States Steel facility varies from 0.3 to 0.5 million gallons per day. This flow is an amalgamation of three sources: (1) ammonia stripper effluent; (2) blast furnace blowdown; and (3) dilution water. The flow from the ammonia stripper is approximately 100 gpm and is the primary source of carbon as well as heat in the system. Quantitative measure of blowdown water from the blast furnace cooling system is difficult since the system is pumped sporadically. However, this flow is significantly less than the ammonia stripper flow. The blowdown water has a low organic strength, but contains up to 240 mg/L ammonia. Because of the inhibitory nature of the wastewater, approximately 0.4 MGD of river water is used for dilution and as a temperature control aid.

The existing biological treatment process at Gulf State Steel consists of two 28,000 $ft^3$ aeration basins, which are connected in parallel. Recycling of biological solids is not employed in the existing process. The wastewater is cooled before entering the first basin. Because the wastewater is essentially devoid of phosphorous, sufficient potassium phosphate was added to the wastewater to yield a total phosphate concentration of 15 mg/L prior to biological treatment. The process performs well as a method of organic carbon removal, but does not significantly reduce the ammonia concentration.

This chapter presents the results of a study conducted to assess the feasibility of utilizing a two-stage, low $\Theta_C$, nitrifying activated sludge process for treatment of coking plant wastewater. Using pilot processes, the effects of varying $\Theta_C$ from 1.5 to 15 days was investigated. The specific objectives of the study were to:

- Determine the fate of ammonia, thiocyanate, and total nitrogen in activated sludge processes treating coking waste
- Assess the biodegradation of phenolics, HNCs, and PAHs in the process
- Assess the degree of accumulation of phenolics, HNCs, and PAHs in the biosolids.

## MATERIALS AND METHODS

Two-stage activated sludge pilot processes were constructed at Gulf State Steel, Inc. for use in this study. A schematic of the system is shown in Figure 1. Each system consisted of two aeration basins which were fabricated from 55-gallon drums and were operated in series. The pilot processes had an operating volume of 6.64 $ft^3$ in the first basin and 5.8 $ft^3$ in the second. The aeration basins were mixed and aerated using sintered glass diffusers. Clarifiers were fabricated from 30-gallon steel cans with a sheet metal cone fitted through a hole in the can bottom. Metering pumps were utilized for sludge recycle and wasting. Influent flow to the process was achieved by gravity flow.

The pilot-scale aeration basins were seeded with acclimated biomass obtained from the plant's current treatment system. Influent was taken from a mixing basin at the head of the operating treatment process. The influent flow rate was 540 mL/min, yielding a hydraulic residence time

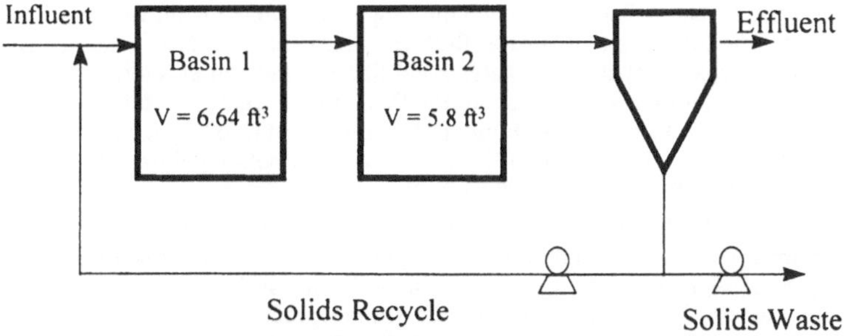


**Figure 1. Schematic of the pilot-scale two-stage activated sludge process.**

($\Theta_H$) of 0.5 days. The mean cell residence time ($\Theta_C$) was chosen as the process variable and was used to evaluate overall performance of the pilot systems. The $\Theta_C$s were controlled by varying the sludge wasting rate, and ranged from 1.5 to 15 days in this investigation. The systems were operated until steady-state BODs, DOCs, and VSS values were achieved.

Performance of the pilot systems was assessed daily based on measurements of pH, alkalinity, dissolved oxygen (DO), dissolved organic carbon (DOC), mixed liquor suspended solids (MLSS), biochemcial oxygen demand (BOD), ammonia, nitrite, nitrate, and thiocyanate. Each of these parameters was measured in accordance with *Standard Methods for the Examination of Water and Wastewater*, 19th edition.[12] Effluent and wasted sludge were sampled for phenolic, heterocyclic nitrogenous compound (HNC), and polyaromatic hydrocarbons (PAH) quantification. Phenolics, HNCs, and PAHs were identified and quantified using a Hewlett Packard GCD GC/MS system and methods outlined in *Standard Methods*.[12]

## RESULTS OF PILOT STUDY

The average values for chemical composition of the coking plant wastewater influent and pilot process effluents are presented in Table I. The influent was variable on a day-to-day basis, had a pH of 7.1, an alkalinity of 172 mg/L as $CaCO_3$, a VSS of 25 mg/L, a BOD of 127.8 mg/L , and a

**Table I. Average Parameter Values for Influent and Effluent from Pilot Activated Sludge Units With $\Theta_C$s of 1.5, 7.5, 10, and 15 Days**

| | Influent | 1.5 Day | 7.5 Day | 10 Day | 15 Day |
|---|---|---|---|---|---|
| BOD, mg/L | 127.8 | 39.7 | 24.3 | 15.6 | 7.9 |
| DOC, mg/L | 47.9 | 22.1 | 14.6 | 14.4 | 14.8 |
| MLVSS, mg/L | 25 | 144 | 567 | 720 | 980 |
| pH | 7.1 | 7.1 | 7.2 | 7.2 | 7.5 |
| Alkalinity, mg/L as $CaCO_3$ | 172 | 141.6 | 91.2 | 71.2 | 60.3 |
| $NH_3$, mg/L | 49.2 | 44.2 | 17.1 | 8.4 | 3.0 |
| SCN, mg/L | 74.7 | 33.2 | 7.3 | 7.7 | 6.2 |
| $NO_3^-$, mg/L | 0 | 0 | 37.2 | 69.1 | 86.9 |
| $NO_2^-$, mg/l | 0 | 0 | 20.2 | 25.8 | 32.6 |
| Total N, mg-N/L | 58.4 | 44.4 | 29.6 | 31.9 | 33.4 |
| Al, ppm | 2.10 | 1.94 | 1.71 | 2.02 | 1.80 |
| Cd, ppm | – | – | – | – | – |
| Cr, ppm | 0.95 | 0.97 | 0.73 | 0.76 | 0.79 |
| Cu, ppm | – | – | – | – | – |
| Fe, ppm | 3.62 | 3.32 | 2.98 | 2.03 | 1.85 |
| Mn, ppm | 0.54 | 0.46 | 0.41 | 0.46 | 0.38 |
| Ni, ppm | 0.06 | 0.06 | 0.05 | 0.05 | 0.06 |
| Pb, ppm | – | – | – | – | – |
| Zn, ppm | 0.66 | 0.61 | 0.54 | 0.49 | 0.47 |

DOC of 47.9 mg/L. The influent contained 49.2 mg/L of ammonia and 74.2 mg/L of thiocyanate, but had no measurable nitrate or nitrite. Both the ammonia and thiocyanate concentrations are relatively low for coking plant wastewaters, and likely reflect an efficient stripping process and the large amount of dilution water added to the flow stream. Heavy metals present in the influent included: aluminum (2.10 ppm), chromium (0.95 ppm), iron (3.62 ppm), manganese (0.54 ppm), nickel (0.06 ppm), and zinc (0.66 ppm). These values are below published values for inhibition of nitrification and heterotrophic metabolism.

The reduction in BOD and DOC are indicators of organic removal by aerobic heterotrophic microorganisms. As shown in Table I, the average effluent BOD decreased from 39.7 mg/L to 7.9 mg/L as $\Theta_C$ increased from 1.5 to 15 days. This indicates an increased efficiency of biodegradable organic carbon removal as the activated sludge process $\Theta_C$ was increased. The average DOC was 22.1 mg/L for the process with a 1.5 day $\Theta_C$ and remained consistent at higher $\Theta_C$ s. As expected, the average reactor mixed liquor volatile suspended solids (MLVSS) increased as $\Theta_C$ increased (see Table I). The minimum MLVSS of 144 mg/L was observed at a $\Theta_C$ of 1.5 days. The low biomass concentration was accompanied by settling problems. The maximum MLVSS of 980 mg/L was measured for the process with a $\Theta_C$ of 15 days, which exhibited excellent settling properties. No significant reduction in the effluent concentrations of heavy metals, except iron, were observed in any process, indicating that only iron is being significantly accumulated in the biosolids.

As shown in Table I, the removal of thiocyanate and ammonia increased as the process $\Theta_C$ increased. The average thiocyanate concentration was reduced from 74.7 mg/L to 33.2 mg/L in the 1.5 day $\Theta_C$ process. Greater than 90% reductions of thiocyanate occurred at the higher $\Theta_C$s. The average ammonia concentration for the process with a 1.5 day $\Theta_C$ was 44.2 mg/L, which is only slightly lower than the influent concentration. No detectable nitrite or nitrate was present at this residence time, indicating that nitrification did not occur. As the process $\Theta_C$ increased, a decrease in ammonia concentration occurred, with a concomitant increase in nitrite and nitrate concentrations. These results demonstrate that a significant degree of nitrification occurred at the 7.5 day $\Theta_C$, with nitrification efficiency increasing to over 94% at a $\Theta_C$ of 15 days.

Reduction of the total soluble nitrogen was observed in all of the pilot processes (see Table I). In the activated sludge process with a $\Theta_C$ of 1.5 days, the total nitrogen was reduced from 58.4 mg/L to 44.4 mg/L. This reduction in total nitrogen was attributed to incorporation into biomass, and represents a BOD:N ratio of slightly less than 10:1. The highest reduction of total nitrogen occurred at the 7.5 day $\Theta_C$. The resulting BOD:N ratio was 10:2.2, which is higher than typically reported for biomass synthesis. The processes with higher $\Theta_C$s had slightly lower reductions in total nitrogen, with BOD:N ratios of approximately 10:2. These higher BOD:N ratios may reflect a change in the microbial population due to the presence of the nitrifier population. Alternatively, the increase in the nitrogen removal may be due to a small amount of denitrification occurring in the clarifier. Gas bubbles, observed in the clarifiers of the processes with $\Theta_C$s of 7.5 days or greater, are consistent with the hypothesis of denitrification.

The concentrations of specific phenolic compounds in the process influent and effluents are shown in Table II. The total phenols concentration of the influent averaged 21.9 mg/L and was composed of: phenol (4.43 mg/L), ortho-cresol (3.21 mg/L), meta-cresol (4.26 mg/L), 2,4-dimethyl phenol (0.92 mg/L), 3,5-dimethyl phenol (5.38 mg/L), and aniline (5.38 mg/L). All of the phenolic compounds were detectable, but were removed by greater than 99% in the effluent from activated sludge process with a 1.5 day $\Theta_C$. In the processes with higher $\Theta_C$s, only phenol was present at detectable levels.

The HNCs present in the influent included: indole (4.32 mg/L), quinoline (5.25 mg/L), isoquinoline (4.08 mg/L), acridine (1.09 mg/L), and carbazole (2.41 mg/L). The average effluent concentrations of all HNCs decreased as $\Theta_C$ increased. For the activated sludge process with a $\Theta_C$ of 1.5 days, only isoquinoline (0.41 mg/L) and acridine (0.18 mg/L) were less than 90% removed from the wastewater. At a $\Theta_C$ of 7.5 days, all HNCs were reduced to concentrations below 0.1 mg/L except isoquinoline (0.27 mg/L).

The average total PAH concentration of the influent wastewater was 0.55 mg/L. Individual PAHs in the influent included: naphthalene (0.05 mg/L), pyrene (0.12 mg/L), anthracene (0.15 mg/L), and fluoranthene (0.08 mg/L). The total PAH concentration of the process with a 1.5 day $\Theta_C$ was 0.18 mg/L, of which anthracene comprised the largest fraction. As the $\Theta_C$ increased, the concentration of total PAHs and the individual PAH compounds decreased, with removals exceeding 90%.

**Table II. Concentrations (mg/L) of Specific Organic Compounds in the Influent and Effluent from Pilot Activated Sludge Units With $\Theta_C$s of 1.5, 7.5, 10, and 15 Days**

| | Influent | 1.5 Day | 7.5 Day | 10 Day | 15 Day |
|---|---|---|---|---|---|
| Phenol | 4.43 | 0.04 | 0.03 | 0.01 | – |
| o–Cresol | 3.21 | 0.03 | – | – | – |
| m–Cresol | 4.26 | 0.02 | – | – | – |
| 2,4 DMP | 0.92 | 0.001 | – | – | – |
| 3,5 DMP | 5.38 | 0.004 | – | – | – |
| Aniline | 3.31 | 0.02 | – | – | – |
| Indole | 4.32 | 0.06 | 0.02 | 0.01 | 0.003 |
| Quinoline | 5.25 | 0.10 | 0.06 | 0.03 | 0.01 |
| Isoquinoline | 4.08 | 0.41 | 0.27 | 0.04 | 0.01 |
| Acridine | 1.09 | 0.18 | 0.07 | 0.02 | 0.01 |
| Carbazole | 2.41 | 0.16 | 0.05 | 0.03 | 0.01 |
| Naphthalene | 0.05 | 0.003 | 0.002 | 0.001 | 0.001 |
| Pyrene | 0.12 | 0.008 | 0.004 | 0.003 | 0.002 |
| Fluoranthene | 0.08 | 0.006 | 0.003 | 0.002 | 0.001 |
| Anthracene | 0.15 | 0.08 | 0.03 | 0.01 | 0.01 |
| Total PAHs | 0.55 | 0.18 | 0.05 | 0.03 | 0.02 |

**Table III. Concentrations (mg/g VSS) of Specific Organic Compounds in Biomass from Pilot Activated Sludge Units With $\Theta_C$s of 1.5, 7.5, 10, and 15 Days**

| | 1.5 Day | 7.5 Day | 10 Day | 15 Day |
|---|---|---|---|---|
| Phenol | 0.41 (0.44%) | 0.08 (0.07%) | — | — |
| o-Cresol | 7.03 (4.8%) | 6.46 (7.6%) | 2.88 (3.2%) | 2.12 (2.2%) |
| m-Cresol | 4.19 (4.7%) | 5.27 (4.7%) | 7.44 (6.3%) | 8.78 (6.7%) |
| 2,4 DMP | 0.88 (4.6%) | 0.75 (3.1%) | 0.72 (2.8%) | 0.65 (2.2%) |
| 3,5 DMP | 5.36 (4.8%) | 4.03 (2.8%) | 3.61 (2.4%) | 3.99 (2.3%) |
| Aniline | 3.27 (4.8%) | 3.25 (3.7%) | 3.38 (3.7%) | 4.39 (4.1%) |
| Indole | 1.81 (2.0%) | 2.29 (2.0%) | 3.98 (3.3%) | 3.86 (2.8%) |
| Quinoline | 5.17 (4.9%) | 7.28 (5.4%) | 8.07 (5.6%) | 8.86 (5.6%) |
| Isoquinoline | 4.05 (5.3%) | 4.59 (4.5%) | 5.37 (4.8%) | 7.18 (5.7%) |
| Acridine | 3.54 (18.5%) | 3.73 (13.7%) | 4.87 (16.2%) | 7.71 (23.1%) |
| Carbazole | 1.90 (4.1%) | 2.42 (3.9%) | 3.35 (5.1%) | 3.71 (5.0%) |
| Naphthalene | 1.4 (100%) | 1.4 (100%) | 1.6 (100%) | 1.5 (98%) |
| Pyrene | 3.4 (100%) | 4.5 (100%) | 5.2 (100%) | 5.4 (100%) |
| Anthracene | 9.2 (100%) | 10.4 (100%) | 11.9 (100%) | 13.5 (100%) |
| Fluoranthene | 3.5 (100%) | 3.7 (100%) | 4.1 (100%) | 4.4 (100%) |
| PAHs | 18.4 (100%) | 19.0 (100%) | 22.1 (100%) | 25.4 (100% |

The concentrations of phenols in biomass from the activated sludge processes are presented in Table III. Phenol was present in small amounts in the processes with $\Theta_C$s of 1.5 days (0.41 mg/g VSS) and 7.5 days (0.08 mg/g VSS), but was nondetectable at higher $\Theta_C$s. These results, coupled with the low effluent phenol concentrations, indicate that phenol is almost completely biodegraded in the process, regardless of $\Theta_C$. At a $\Theta_C$ of 1.5 days significant amounts of o-cresol (7.03 mg/g), m-cresol (4.19 mg/g), 2,4-dimethyl phenol, 3,5-dimethyl phenol (5.36 mg/g), and aniline (3.27 mg/g) were partitioned into the biomass. The biomass concentrations of each phenolic compound, except m-cresol, decreased as $\Theta_C$ increased, indicating an increased degree of biodegradation. However, significant amounts of these compounds were present at all $\Theta_C$s. The biomass concentration of m-cresol increased from 4.19 mg/g VSS to 8.78 mg/g VSS as $\Theta_C$ increased. Based on these findings, it appears that sorption plays a significant role in activated sludge removal of phenolics from coking plant wastewater.

The concentrations of HNCs in activated sludge biomass are presented in Table III. For activated sludge with a $\Theta_C$ of 1.5 days, large concentrations of indole (1.81 mg/g VSS), quinoline (5.17 mg/g VSS), isoquinoline (4.05 mg/g VSS), acridine (3.54 mg/g VSS), and carbazole (1.90

mg/g VSS) were present in the biomass. As the process $\Theta_C$ increased, the concentrations of HNCs in the biomass increased. At the 15 day $\Theta_C$, the biomass concentrations of the HNCs were: indole (3.86 mg/g VSS), quinoline (8.86 mg/g VSS), isoquinoline (7.18 mg/g VSS), acridine (7.71 mg/g VSS), and carbazole (3.71 mg/g VSS). Based on these results, adsorption of HNCs onto the biomass appears to play an important role in removal by the pilot activated sludge processes.

The biomass total PAH levels of the pilot processes increased from 18.1 mg/g VSS at a $\Theta_C$ of 1.5 days to 25.4 mg/g VSS at a $\Theta_C$ of 15 days (see Table III). The concentration of all individual PAHs increased with increasing $\Theta_C$. These levels were significantly higher than the influent values, and indicate that a majority of the influent PAHs are accumulated on the biomass. These trends are consistent with the findings of other investigators.

The percentage of phenols, HNCs, and PAHs that are removed by sorption was calculated as:

$$\frac{[\text{MLVSS}] \times [\text{Biomass}] \times \Theta_H \times 100\%}{[\text{influent}] \times \Theta_C} \qquad (1)$$

The percentages of removal due to sorption are indicated in parentheses in Table III. Less than 1% of the phenol removed by the activated sludge was attributed to sorption and less than 8% of the other phenolic compounds was removed by sorption. This indicates that the majority of the total phenols removal by activated sludge was attributable to biodegradation. Similarly, only a small percentage of most HNCs was removed by sorption, indicating that biodegradation was the predominant mechanism of HNC removal. Acridine, however, had between 13.7 and 23.1% of its removal attributable to sorption. The percent removal by sorption to biomass was calculated for PAHs and was found to exceed 100% of the influent mass. Therefore, it was concluded that PAHs were removed by sorption and were not significantly biodegraded.

## CONCLUSIONS

Based on the results of the pilot study, it was concluded that a two-stage activated sludge process can be successfully employed to treat coking plant wastewater. Removals of BOD, phenols, HNCs, and PAHs from coking wastewater were acceptable for processes operated with a $\Theta_C$ of 1.5 days. However, ammonia and thiocyanate removals were small, and no significant nitrification occurred. At higher $\Theta_C$s, ammonia and thiocyanate removals exceeded 90% and nitrification occurred. The highest process $\Theta_C$ (15 days) yielded the highest degree of overall treatment.

The ultimate fate of phenol in the activated sludge processes appears to have been biodegradation. The other phenolic compounds were mostly biodegraded, and with less than 10% of the influent phenolic mass being adsorbed onto the biomass. The fate of HNCs in activated sludge treatment of coking wastewater was also predominantly attributable to biodegradation. Only a small amount of bioadsorption of HNCs was observed. PAHs had much higher biomass concentrations, and appear to have been removed exclusively by adsorption. The high degree of bioconversion observed for phenolics and HNCs makes the two-stage activated sludge process an attractive alternative for treating coking plant wastes.

## REFERENCES


1. Wong-Chong, G. and R. Loehr, *Water Res.*, 12, p. 605, 1978.
2. Beccari, M., et al. *Environ. Tech. Lett.*, 1, p. 245, 1980.
3. Stafford, D. *J. Appl. Bact.*, 37, p. 75, 1981.
4. Blok, J. *$H_2O$*, 14, (11), p. 242, 1981.
5. Shivaraman, N. et al., *Environ. Pollut.*, 39, p.141, 1985.
6. Wong-Chong, G. and J. Fittipaldo, *Iron Steel Eng.*, October, 1994.
7. Yi, Q. et al. *Water Res.*, 28, (3), p. 701, 1994.
8. Kostenbader, S. *JWPCF*, 41, (2), p. 199, 1969.
9. Klein, J. and D. Lee, *Biotechnol. Bioeng. Symp.* 8, p.379, 1978.
10. Tsai, K. et al. *Proc. 38th Purdue Ind. Waste Conf.*, West Lafayette, IN, 1983, p. 529.
11. Cooper, R. and J. Catchpole, *Water Res.*, 7, p. 1137, 1973.
12. *Standard Methods for the Examination of Water and Wastewater,* 16th ed., APHA, 1995.

# 26 SORPTION CHARACTERISTICS OF STABILIZED BIOLOGICAL SLUDGE

**Noah I. Galil**, Senior Lecturer
Department of Civil Engineering
Technion - Israel Institute of Technology
Technion City, Haifa 32000, Israel

**Shaul Mevorah**, Environmental Engineer
"Balasha - Jalon" Consulting Engineers, Ltd.
Haifa, Israel.

## ABSTRACT

Cadmium was sorbed by sludge solids and had the tendency to achieve saturation at equilibrium concentrations above 15 mg/L. Sorption of phenol was relatively low along equilibrium concentrations of up to 10 mg/L, and then sorption substantially increased. Comparing the sorption of cadmium on sludges which have been stabilized at similar levels, a small advantage could be observed for sludge treated by anaerobic digestion. No significant differences between anaerobic and aerobic sludges were found for phenol sorption. The stabilization level appears to be an important factor in the sorption of cadmium and phenol on sludges originating from both aerobic and abaerobic digestion processes. Both contaminants have increased their sorption characteristics as the concentration of dissolved organic matter in the aqueous solution was higher. Phenol, especially, exerted low sorption in low organic solutions and increased significantly its sorption characteristics in the presence of dissolved organics. Soluble organics matter, originating from the sludge, was found as an important factor controlling the sorption of cadmium and phenol by sludge, which released dissolved organic matter proportionally to the stabilization level.

## INTRODUCTION

Studies on sorption characteristics of sludges, originating from biological treatment processes of wastewater, are based on different methods in the preparation of the biomass, before its contact with the specific compounds which are supposed to be sorbed from the solution. Selvakumar and Hsieh,[1] Bell and Tzesos[2] and others based their experimental work on sludge pretreatment and its transformation into dry particles or granular structures. Kennedy et al.[3] and others performed their experiments using sludge in wet form, as it could be produced by the treatment processes.

Studies on sorption of heavy metals onto sludge and different types of biomass have been performed by Cheng et al.,[4] Koers and Mavinic,[5] Nelson et al.,[6] Lo and Lin,[7] Gourdon at al.,[8,9] and Venkobahar,[10] indicated that biomass could behave as sorbing media. Other studies reported sorption of organic matter onto biomass: phenol, chlorophenol, chlorobenzene, ethylbenzene,[1,11] methylchloride, chloroform, di-chloroethylene, di-nitro- o-cresol, tetrachloetylene, lindane,[12] pentachlorophenol, 2-chlorobiphenol,[13] trichloroethane,[14] chlorophenol, di-chlorophenol, trichlorophenol, pentachlorophenol.[3]

Studies reported that the Freundlich isotherm was found to be the most appropriate mathematical relationship to describe the sorption of either heavy metals or organic matter onto sludge from suspended growth biological reactors, such as activated sludge. The process itself was characterized by a rapid phase, usually the first minutes of the exposure, followed by a slow phase, which could last for 24 hours or more. Most of the materials, about 80%, were sorbed during the first phase.

*52nd Purdue Industrial Waste Conference Proceedings*, 1997, Ann Arbor Press, Chelsea, Michigan 48118. Printed in U.S.A.

Like in other cases, sorption onto biomass sludge is reported to be influenced by the general characteristics of the solution (pH and temperature), the properties of the compounds which are supposed to be sorbed (hydrophobicity, octanol/water partitioning coefficient), and the sorbing media, which in these cases is represented by different types of sludge.

The process which is being utilized for sludge stabilization can be either aerobic or anaerobic, and this would influence the chemical composition of the biosolids.[15] Another important factor, which could affect sorption characteristics of sludge, would be the level of stabilization, since this parameter controls the content of organic biodegradable material in the biosolids.

The objectives of this study were to observe sorption characteristics of sludge in wet form, at different levels of stabilization, obtained by separate aerobic and anaerobic processes. The target compounds, utilized separately for each type of sludge, included cadmium and phenol.

## MATERIALS AND METHODS

The source of sludge used in the experiments was an activated sludge plant in Haifa, Israel. The sludge was supplied from the pumping station which recycles settled sludge from the sedimentation tanks to the aeration reactors. The sludge residence time in the process is usually 5 to 7 days. Most of the experimental work was based on a 500-liter batch of sludge containing an initial concentration of 8,400 mg/L suspended solids. The initial content of volatile suspended solids was 81.6%. After a gravitational thickening of eight hours, the suspended solids content increased to 18,300 mg/L, the supernatant was carefully removed, and the remaining thickened sludge was transferred into two different tanks. Both tanks were equipped with mechanical paddle mixers. One tank was closed and operated as anaerobic digester, while the other tank, used for aerobic digestion, was open and included an air diffusion supply system located at the bottom. Sludge temperatures were kept in the range of 21–24°C in the aerated tank and 25–29°C in the other one.

Sludge was taken for the sorption experiments after different periods of stabilization and was previously filtered by a Buchner funnel. Filtered sludge, still contained about 70 to 80% of water, therefore careful measurements and calculations have been made in order to ensure that the same amount of dry solids has been added to each flask used in the experiments.

The aqueous solutions were prepared in tap water shortly before the experiments and contained different controlled concentrations of either cadmium or phenol; pH was kept at 7.2; tap water was characterized by 780 mg/L total dissolved solids and electrical conductivity of 820 µmho/cm.

The experiments were performed in Erlenmeyer flasks of 250 mL, placed on magnetic stirrers. The contact time was three hours. During the experiments, flasks were hermetically closed. At the end of each experiment, samples were filtered and taken to analyses. Cadmium was analyzed using a Perkin Elmer atomic absorption apparatus. Phenol, COD, and suspended solids analyses were performed according to the *Standard Methods for Examination of Water and Wastewater* (1992).

## RESULTS AND DISCUSSION

### Sorption Experiments

Sorption experiments were performed contacting separately cadmium and phenol with solids originating from filtered sludges. These sludges have been preliminarily stabilized at different levels by either aerobic or anaerobic processes.

The experimental results obeyed Freundlich-type isotherms; however, different sorption characteristics were obtained for each one of the target compounds used. The mathematical expression for this process is provided by the Freundlich isotherm:

$$q = K\,C^n \qquad (1)$$

where
- $q$ = the amount of target compound connected by sorbant media, (mg/gr)
- $C$ = equilibrium concentration of the target compound in the solution, (mg/L)
- $K, n$ = coefficients of the process.

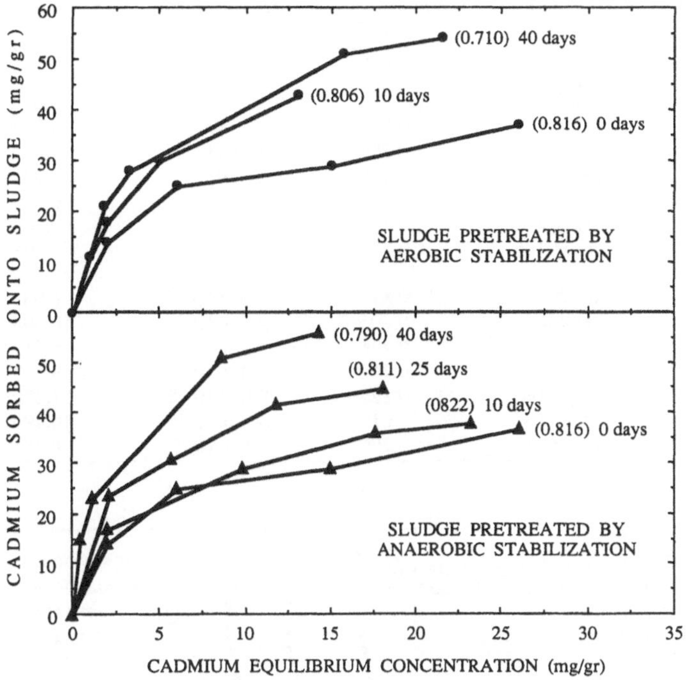


**Figure 1. Sorption of cadmium on sludges pretreated to different levels of aerobic or anaerobic stabilization.**

Cadmium was sorbed by all sludges and showed tendencies to achieve saturation values (Fig. 1). This type of sorption may occur when the sorbent can bind to certain sites of the sorbant media and, as these sites become more and more "taken," sorption would become more limited. In the case of cadmium, the exponent n always had values lower than 1 for all the sludges in the study, as summarized in Table I.

Sorption of phenol onto sludges was low along a range of equilibrium concentrations up to 10–15 mg/L and then steeply increased (Fig. 2). This type of sorption, where the exponent n values are higher than 1, could indicate that sludge solids might sorb phenol after developing special binding sites. These sites might be, apparently, few at low equilibrium concentrations and could develop when phenol concentrations in the solution exceeded 10 mg/L.

This type of sorption may create severe problems in cases of unexpected increased phenol concentrations which may be accidentally spilled in the influent of a biological reactor, based on suspended biomass. As shown in Figure 2, for the initial sludge characterized by volatile/total ratio of 0.816 (this sludge could be considered as similar to biomass found in a typical bioreactor), an increased phenol concentration from normal values of 0.1 to 0.2 mg/L, to more than 10 mg/L, would substantially increase the amounts of phenol accumulated on the biomass. This sorption characteristic might explain disturbances to the bioprocess caused by increased loads of phenol reported in the literature.[16–18]

## The Type of Sludge Stabilization Process

The initial sludge, obtained from the activated sludge plant, has been separately stabilized before performing sorption experiments, by either aerobic or anaerobic processes. The results were expressed in terms of the ratio between volatile/total suspended solids found in the sludge (V/T), as described in Figure 3. This ratio could reflect the different chemical properties, caused by the different total organic contents, achieved at different stabilization levels. The aerobic sta-

**TABLE 1. Sorption Coefficients for the Freundlich Isotherm Obtained in the Experimental Work**

| Target Pollutant | Sludge Stabilization | V/T ratio | K | n | $R^2$ |
|---|---|---|---|---|---|
| Cadmium | aerobic | 0.816 | 11.589 | 0.359 | 0.956 |
| Cadmium | aerobic | 0.806 | 11.803 | 0.530 | 0.982 |
| Cadmium | aerobic | 0.710 | 17.298 | 0.382 | 0.994 |
| Cadmium | anaerobic | 0.822 | 13.521 | 0.335 | 0.998 |
| Cadmium | anaerobic | 0.811 | 18.578 | 0.310 | 0.987 |
| Cadmium | anaerobic | 0.790 | 20.557 | 0.398 | 0.992 |
| Phenol | aerobic | 0.816 | $1.9 \times 10^{-10}$ | 8.037 | 0.865 |
| Phenol | aerobic | 0.806 | $4.5 \times 10^{-13}$ | 10.398 | 0.996 |
| Phenol | aerobic | 0.776 | $1.2 \times 10^{-6}$ | 6.219 | 0.998 |
| Phenol | aerobic | 0.710 | $1.4 \times 10^{-4}$ | 3.873 | 0.924 |
| Phenol | anaerobic | 0.822 | $6.0 \times 10^{-14}$ | 12.070 | 0.958 |
| Phenol | anaerobic | 08.11 | $4.2 \times 10^{-12}$ | 10.498 | 0.910 |
| Phenol | anaerobic | 0.790 | $4.0 \times 10^{-9}$ | 8.385 | 0.811 |

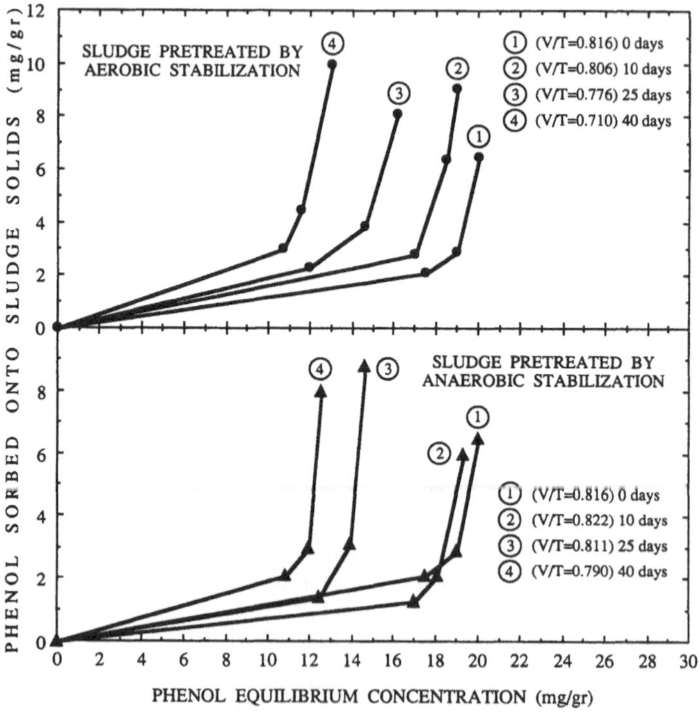


**Figure 2. Sorption of phenol on sludges pretreated to different levels of aerobic or anaerobic stabilization.**

bilization was somehow more efficient, probably due to the fact that the temperature levels were relatively low for the anaerobic stabilization, since this was performed at room temperature.

Comparing the sorption of cadmium on sludges, which have been stabilized at similar ratios of V/T , one can see a small advantage for the anaerobic digestion (Fig. 4A). As for phenol sorption, no significant differences could be observed, as caused by the type of sludge stabilization (Fig. 4B).

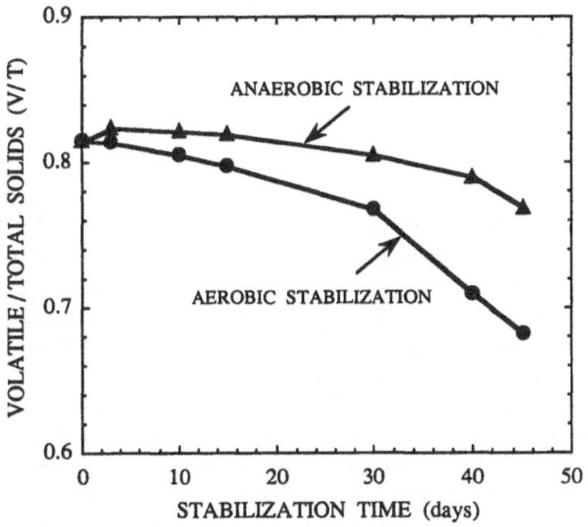


**Figure 3. The stabilization levels achieved by aerobic or by anaerobic stabilization in terms of volatile to total solids ratio versus the stabilization time.**

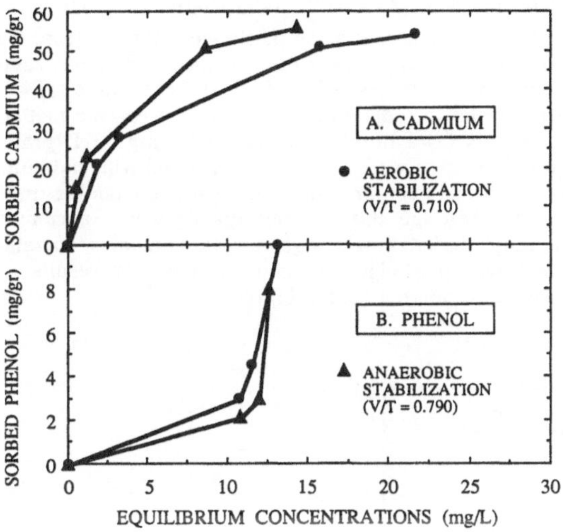


**Figure 4. Comparison of the sorption of cadmium and phenol on sludges pretreated aerobically or anaerobically at similar stabilization levels.**

## Sludge Stabilization Levels

Samples of sludge have been taken from the digestion tanks at three different stabilization levels, after 10, 25, and 40 days. The sludge of each batch was contacted with cadmium and with phenol, in separate experiments.

The stabilization levels appeared to be an important factor in the sorption of cadmium and phenol on sludges originating from both aerobic and anaerobic digestion processes. As shown in Figures 1 and 2, the sorption of both contaminants, cadmium and phenol, improved as the sludge stabilization level was higher (the V/T ration was lower).

The coefficients K and n , summarized in Table I, refer to experiments which have been performed on the same batch of sludge (500 liters) after different stabilization processes at different stabilization levels. Since sludge characteristics may fluctuate, as influenced by many factors, these values should be considered for comparative information, rather than as absolute constants. For example, comparing the experimental coefficients, one can see that the lower the V/T ratio, the higher were the values of K. In the case of phenol, a change of about 4% in the V/T ratio (from 0.822 to 0.790) was followed; in the case of anaerobic stabilization, by an increase of five orders of magnitude in the values of K.

The values of the exponent n in the case of cadmium did not seem to be affected by the V/T ratio; however, in the sorption of phenol these values were lower as the V/T ratio decreased, except one case of aerobic sludge at V/T = 0.806 .

## Effects of Soluble Organics

The fact that the stabilization level of sludge affected the sorption process could be explained by chemical and biochemical changes which may occur on the surfaces of the sludge solids. This study focused on the amount of dissolved organic matter present in the solution during the sorption process. The source for these organics could be the soluble organic matter released to the aqueous solution by the sludge itself and this could be quantified by performing contact experiments, similar to sorption experiments, including only sludge and water, without the addition of cadmium or phenol. The results are presented in Figure 5 and indicate that while the raw initial sludge released about 25 mg COD/gram of sludge solids, both stabilization processes enabled the sludge to release a progressing amount of organic matter, up to 190 and 150 mg/gram, released by sludges which have been stabilized during 40 days by either aerobic and anaerobic processes; accordingly, indicating that the more stabilized the sludge, the higher was the amount of soluble organic matter released in the aqueous solution. This would lead to the conclusion that at least some of the different sorption characteristics obtained in the experiments summarized in Figures 1 and 2 could be caused by different levels of soluble organic matter released by the sludges.

An additional experiment was performed, based on a new batch of sludge, supplied from the same source and stabilized by aerobic and anaerobic processes. After a period of 36 days, samples were taken from both digestors and exposed to a process of washing and centrifuge separation. This process was repeated five times in order to reduce the "reservoir" of organic matter; therefore, both sludges could release after this ÒwashingÓ only about 12 mg COD/gram. Washed solids from aerobic and anaerobic stabilization were separately contacted with cadmium and with phenol, using aqueous solutions which had been previously prepared, containing three different levels of organic matter: (a) tap water, here the only soluble organics were contributed by the washed sludge; (b) tap water "artificially" enriched by sludge extracted organics, containing 355 mg/L; (c) tap water enriched to 663 mg/L of sludge extracted organics. The results for sorption experiments with cadmium and phenol are summarized in Figures 6 and 7.

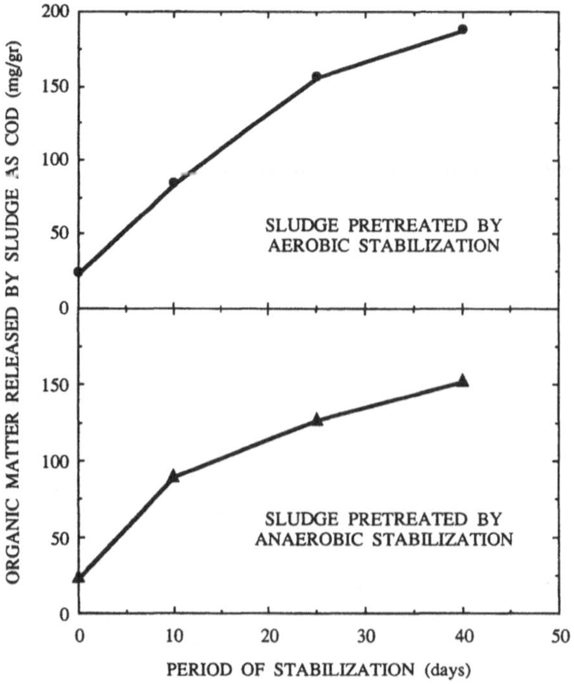


**Figure 5. Organic matter released by sludge at different periods of stabilization.**

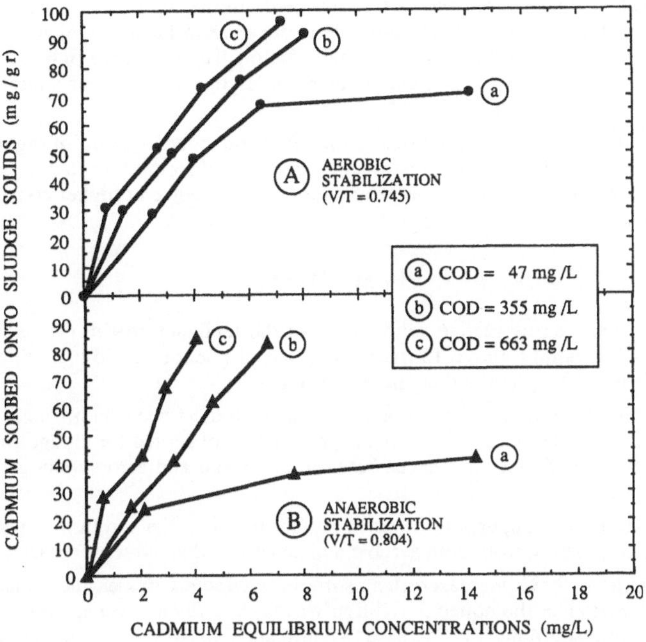


**Figure 6 . Sorption of cadmium on sludge at different levels of dissolved sludge organic matter.**

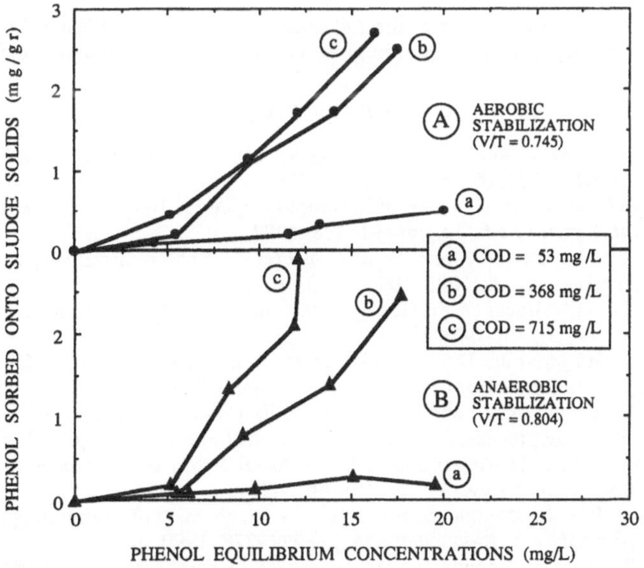


**Figure 7. Sorption of phenol on sludge at different levels of dissolved sludge organic matter.**

Since the sludge solid phase used as sorbant media in the reported experiments was identical in all the three solutions used, it was possible to separate influences which may have been exerted by sludge solids at different levels of stabilization. Therefore, the only changing parameter which could control the sorption of contaminants in sludge was the concentration of dissolved organic matter in the solution.

Both contaminants have increased their sorption, as the concentration of organic matter in the aqueous solution was higher. Phenol exerted relatively low sorption in low organic solution and increased significantly its sorption characteristics in the presence of higher concentrations of dissolved organics.

## SUMMARY

Cadmium was sorbed by sludge solids and had the tendency to achieve saturation values at equilibrium concentrations above 15 mg/L. Sorption of phenol was low along equilibrium concentrations of up to 10 mg/L and then increased steeply.

Comparing the sorption of cadmium on sludges, which have been stabilized at similar ratios of volatile to total solids (V/T), a small advantage could be observed for sludge treated by anaerobic digestion. No significant differences between anaerobic and aerobic sludges were found for the phenol sorption.

The stabilization level appears to be an important factor in the sorption of cadmium and phenol on sludges originating from both aerobic and anaerobic digestion processes.

Both contaminants have increased their sorption characteristics as the concentration of dissolved organic matter in the aqueous solution was higher. Phenol, especially, exerted low sorption in low organic solutions and increased significantly its sorption characteristics in the presence of dissolved organics. Soluble organic matter, originating from the sludge, was found as an important factor controlling the sorption of cadmium and phenol by sludge solids, which released dissolved organics proportionally to its stabilization level.

A future step in the study of sludge sorption characteristics would have to focus on the sludge solids after different types and different levels of stabilization, separating the influences of dissolved organics in the solution.

## ACKNOWLEDGMENT

The study has been carried out at the Laboratory for Industrial Wastewater Treatment and Water Renovation, Department of Civil Engineering, Technion - Israel Institute of Technology.

## REFERENCES

1. Selvakumar, A and Hsieh, H.N. Adsorption of organic compounds by microbial biomass, *J. Env. Sci. Health,* A23, (6), pp. 543–557, 1988.
2. Bell, J.P. and Tsezos, M. Removal of hazardous organic pollutants by biomass adsorption. *J. Water Pollut. Control Fed.*, 59, pp. 191–198, 1987.
3. Kennedy, K.J., Lu, J., and Mohn, W.W. Biosorption of chlorophenols by anaerobic granular sludge. *Water Res.*, 26, pp. 1085–1092, 1992.
4. Cheng, M.H. and Minear, R.A. Heavy metals uptake by activated sludge. *J. Water Pollut. Control Fed.*, 47, pp. 362–376, 1975.
5. Koers, D.A. and Mavinic, D.S. Aerobic digestion of waste activated sludge at low temperatures. *J. Water Pollut. Control Fed.*, 48, pp. 460–468, 1977.
6. Nelson, P.O., Chung, A.K., and Hudson, M.C. Factors affecting the fate of heavy metals in the activated sludge process. *J. Water Pollut. Control Fed.*, 53 III, pp. 1323–1333, 1981.
7. Lo, S.L. and Lin, C.Y. Adsorption of heavy metals from wastewater with waste activated sludge. *J. Chinese Inst. Eng.*, 12, pp. 451–461, 1989.
8. Gourdon, R., Rus, E., Bhende, S. and Sofer S.S. Mechanism of cadmium uptake by activated sludge. *Applied Microb. Biotechnol.*, 34, pp. 274–278, 1990.
9. Gourdon, R., Rus, E., Bhende, S., and Sofer, S.S. A comparative study of cadmium uptake by free and immobilized cells from activated sludge. *J. Env. Sci. Health*, A25, pp. 1019–1036, 1990.

10. Venkobachar, C. Metal removal by waste biomass to upgrade wastewater treatment plants. *Water Sci. Technol.*, 22, pp. 319–320, 1990.
11. Selvakumar, A. and Hsieh, H.N. Adsorption of organic compounds by microbial biomass. *Int. J. Environ. Studies*, 30, pp. 313–319, 1987.
12. To come later
13. Bell, J.P. and Tsezos, M. Removal of hazardous organic pollutants by adsorption on microbial biomass. *Water Sci. Technol.*, 19, pp. 409–416, 1987.
14. Tsezos, M. and Seto, W. The adsorption of chloroethanes by microbial biomass. *Water Res.*, 20, pp. 851–858, 1986.
15. Frasinetti, S., Citterio, S., and Nappi, P. Chemical changes in sludge stabilization. *Biocycle*, 31, pp. 50–52, 1990.
16. Galil, N., Rebhun, M., and Brayer, Y. Disturbances and inhibition in biological treatment of wastewater from an integrated oil refinery. *Water Sci. Technol.*, 22, pp. 21–29, 1988.
17. Galil, N. and Rebhun, M. Powdered activated carbon combined with biological treatment in removal of organic matter containing cresols and xylenols. *J. Environ. Sci. and Health,* 24, pp. 199–218, 1992.
18. Rebhun, M. and Galil, N. Inhibiton by hazardous compounds in an integrated oil refinery. *J. Water Pollut. Control Fed.*, 60, pp. 1953–1959, 1988.

# 27 BIODEGRADATION OF PHENOLIC COMPOUNDS WITH A SEQUENCING BATCH BIOFILTER

**Germán Buitrón**, Research Leader
**José Ortiz**, Engineering Student
Environmental Bioprocesses Department
Institute of Engineering
National University of Mexico
C. U., Apartado Postal 70-472
04510 Mexico, D.F., Mexico

## ABSTRACT

Sequencing batch reactor systems have been used to manipulate both the organisms' distribution established in the reactor and the physiological state of the organisms developed, in particular they have been used to remove specific organic compounds present in industrial effluents. It has been observed that a way to improve the bioreactor activity is maintaining a high biomass concentration; i.e., using a biofilm system. This chapter presents the study of a sequencing batch biofilter packed with a low cost material for toxic wastewater treatment. The packing material was a porous volcanic rock (puzolane) with a mean diameter of 2.5 cm and 75% of porosity. The water in the reactor was maintained at the laboratory temperature (15°C). A synthetic effluent mixture of chlorophenols (60–400 mg/L) was used as the toxic wastewater model (phenol, 4-chlorophenol, 2,4-dichlorophenol and 2,4,6-trichlorophenol, 25% each). The mixture of contaminants was eliminated with an efficiency superior to 93% as total organic carbon and 99% as total phenols. A maximal applied load of 3.6 kg COD/$m^3$-d was achieved. The kinetic of phenols degradation under starvation period was examined.

## INTRODUCTION

Toxic compounds present in industrial wastewater are problematic to treat, due to the fact that their concentration fluctuates with time (for example, during the washing of the reaction tanks in the production unit). As a result, the microorganisms in a biological treatment are frequently subject to shock loading (and changes in concentration). In a continuous system, the microorganisms are unable to acclimate to such fluctuations. Sequencing batch reactor systems have been used to manipulate both the organism distribution established in the reactor and the physiological state of the organisms developed; in particular, they have been used to remove specific organic compounds present in industrial effluents. It has been demonstrated that the periodicity imposed by the operation of the system can overcome the impact of variations in the rate of electron donor input to the system.[1]

A way to improve the bioreactor activity is maintaining a high biomass concentration; i.e., using a biofilm system.[2,3] In the biofilm processes the inner layer of microorganisms is protected, tolerating the toxic shock loading.[4] One example of these processes is the fluidized bed. However, the disadvantage of such systems is their high energy consumption in order to maintain in suspension the support, as well as the high costs associated with the optimized bed, generally protected by a patent. Kaballo et al.[5] compared periodic and continuous packed bed reactors treating 4-chlorophenol. Organisms were more evenly distributed through the reactor operated periodically. As a result, the sequencing batch biofilter was better able to treat large shock loads (5-fold increase in the specific elimination rate per biomass unit) than the continuous flow process.

*52nd Purdue Industrial Waste Conference Proceedings*, 1997, Ann Arbor Press, Chelsea, Michigan 48118. Printed in U.S.A.

The objective of this work was to the study a sequencing batch biofilter, SBB, packed with a low-cost material (porous volcanic rock) for industrial wastewater treatment. A synthetic effluent mixture of chlorophenols was used as the industrial wastewater model.

## EXPERIMENTAL PROCEDURE

### SBB System

The system consisted of a 7-liter, aerated batch biofilter formed by a column of 60 cm high and 15 cm internal diameter. The packing material (5.5 liters) was a porous volcanic rock (puzolane) with a mean diameter of 2.5 cm and 75% of porosity (packed material). The water in the reactor was maintained at the laboratory temperature (15°C). Each cycle of the SBB consisted of four periods: fill (10 min), reaction (variable), settle (30 min) and draw (10 min), controlled by a timer. The volume exchange in each cycle was 75% of the total volume. Fill and draw were done by peristaltic pumps (Figure 1).

Activated sludge from a municipal wastewater treatment plant was used as inoculum (1.7 gVSS/l). A mixture of phenol (25%), 4-chlorophenol (25%), 2,4-dichlorophenol (25%), and 2,4,6-trichlorophenol (25%) was utilized as representative of the toxic compounds in the range of concentration of 60 to 400 mg total phenols/L, and it was the sole source of carbon and energy. Mineral salts (N, P, and oligoelements) were added to the reaction medium.[6]

### Operational Conditions

The main advantage of the batch processes is that they are time-oriented, so when microorganisms are put in contact with the toxic compounds, enough time can be provided to accomplish the acclimation.[7] In contrast, when a continuous system is used, a dilution rate must be fixed. There is a possibility that such a dilution rate is higher than the specific growth rate of microorganisms; thus, the specific degraders will not be preserved in the reactor.

In order to acclimate the microorganisms, the reaction phase of the first cycles of the batch biofilter was maintained until 80% of the initial organic carbon was degraded (or 90% of the total phenols was degraded). After the acclimation, the duration of the cycles was maintained in 24 hr. When the removal efficiencies were constant and higher than 90%, an operational condition was changed. First, the initial concentration was maintained constant and the duration of the cycle was reduced. For the concentration of 133 mg of total phenols/L, the duration of the cycle

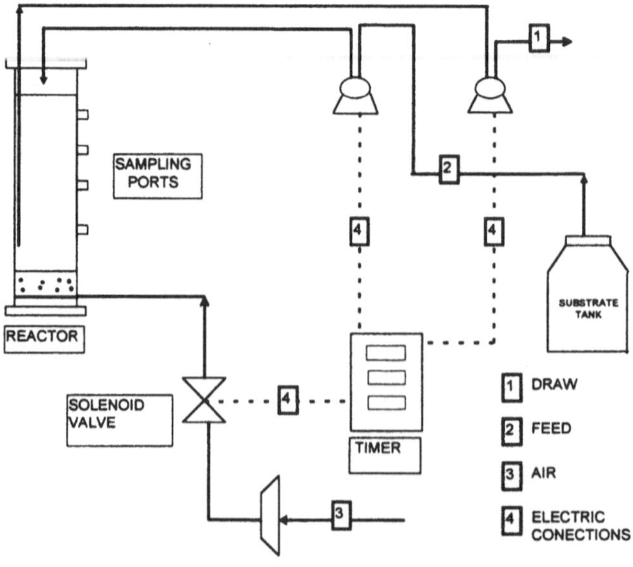


**Figure 1. Experimental setup.**

was stepwise reduced from 24 to 12, 8, 6, and 4 hours. The same procedure was applied for the initial concentrations of 266 and 400 mg of total phenols/L. In this manner, the reactor was operated with a constant efficiency, regardless the degradation time or the initial toxic concentration.

### Analytical Procedure

Residual phenols concentration was determined by using a HPLC (Waters 510) coupled to an integrator (Spectra Physics SP 400). A radial column (Resolve C-18, Radial Pack) was utilized for separation. Detection was performed at 280 nm with an isocratic mixture of water-ammonium acetate 0.3M/acetonitrile/methanol (56:34:10, v/v/v). Total organic carbon, TOC, was determined in samples with a Shimadzu TOC-5050 carbon analyzer. Suspended solids, chemical oxygen demand (COD), and total phenols (4-aminoantipyrine method) were determined according to the *Standard Methods*.[8]

## RESULTS AND DISCUSSION

### Acclimation and Startup of the Reactor

Fresh activated sludge was acclimated into the packed reactor to a concentration around 60 mg/L of total phenols. In the first cycle, the initial total organic carbon concentration, TOC, was 59 mg/L, composed of 26.7 mg TOC/L of chlorophenols and 32.3 mg TOC/L of residual organic carbon from wastewater. Next cycles were feed only with the organic carbon done by the mixture of phenols. The monitoring of the first SBB cycle shows that it took 16 days to degrade more than 90% of initial substrate, in an organic carbon basis (Figure 2). The second cycle took 5 days and the third cycle took 2 days. From the fourth cycle the reactor was operated every 24 hours.

### Colonization of the Packing Material

After the acclimation phase, the duration of the SBB was maintained for 24 hours and initial chlorophenols concentration was raised to 133 mg/L. The exchange volume was 3/4; thus, the concentration in the tank, at the beginning of the degradation, was around 100 mg/L because of the dilution with the remaining water (1/4 of the total volume). The reactor was operated under this condition for 40 days. In Figure 3 we can observe that suspended solids concentration in the effluent decreased from 80 mg/L to around 20 mg/L after a period of 20 days. During this period the biomass started to colonize the porous material. Also, some microorganisms that were unable to adapt to the phenols were lost with the decanted water, as suspended material.

During the 165 days of operation of the reactor there was not a significant change in the effluent suspended solids concentration, nor in the clogging of the biofilter. The microorganisms acclimated seem to have a low yield and for this reason the backwash of the filter was not necessary during this period (more than 400 cycles and 5 months of operation). As the initial concentration was raised from 267 to 400 mg of total phenols/L (day 110), a slight increase of suspended solids in the effluent was observed, probably because of the toxicity of the influent, but the quality of the effluent was restored after 10 cycles.

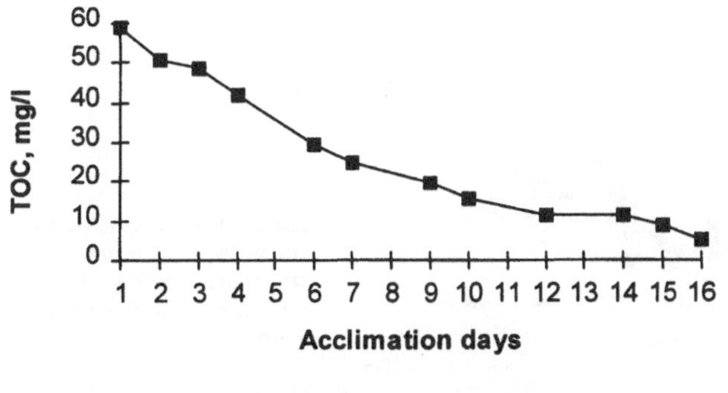


**Figure 2. Biomass acclimation to the mixture of chlorophenols in the SBR bioreactor. First cycle**

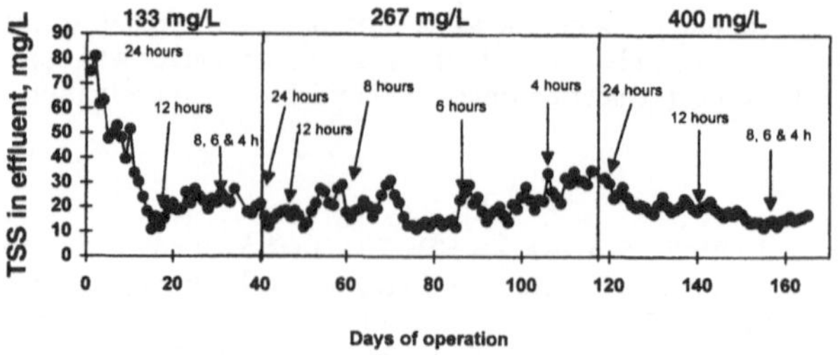


**Figure 3. Evolution of TSS in the effluent as a function of operation time.**

The biomass attached on the packed material was determined by a gravimetric manner. A porous rock sample was dried at 103°C, weighted and passed at 550°C to burn the organic matter. The difference in weight was considered the volatile solids. An average of 2.8 mg volatile solids per $cm^2$ of rock surface (evaluated assuming a spherical shape) was found in the reactor after a month of operation.

## Reactor Performance

The evolution of the phenols' removal efficiency is shown in Figure 4. For all the conditions studied, the biofilter always eliminated more than 90% of the initial phenols. Removal efficiencies varied from 93 to 100% (99.1 ± 1.3% in average). In days 19, 84, and 117, the removal efficiency was lower than 97% because in these days there was a change in the operational conditions (see operation conditions in Figure 3); i.e. a diminution of the degradation time (days 19 and 84) or increase in the effluent concentration (day 117). Nevertheless, the removal efficiency was recovered after one operation cycle.

HPLC analysis indicated that. after the treatment, 0.1 mg 2,4-dichlorophenol/L and 0.8 mg 2,4,6-trichlorophenol/L were present in the effluent (average for the best conditions obtained for the initial concentration of 133 phenols/L). All the phenol and 4-chlorophenol were eliminated by the microorganisms.

The performance of the SBB as a function of the total organic carbon in the effluent is shown in Figure 5. In general, an average of 92% of the initial organic carbon was eliminated by the sequencing batch biofilter. Residual concentrations averaged 7.4 mg/L, varying between 2 and 22 mg/l. As in the case of phenols removal, the TOC peaks in the effluent were observed when a change in an operational condition was effectuated. The greatest variations were observed when the reactor was operated with 400 mg of total phenols/L.

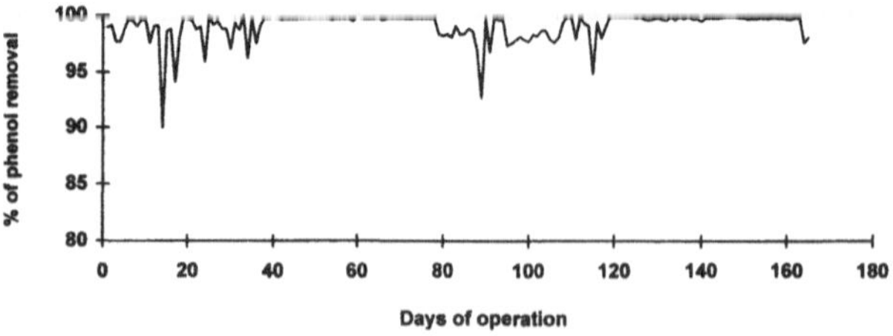


**Figure 4. Total phenols removal efficiencies as a function of the operation days.**

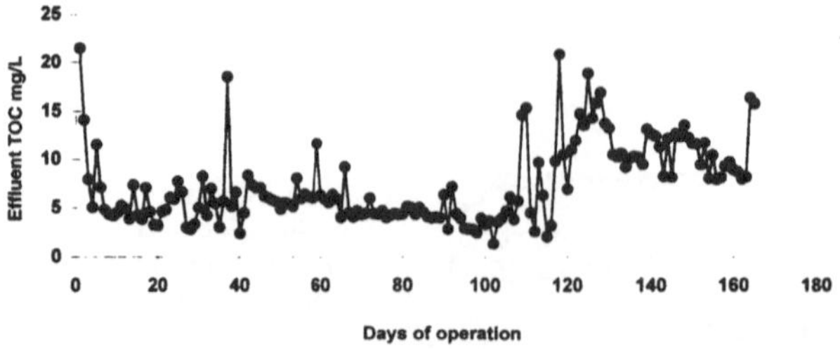


**Figure 5. Organic carbon in the effluent as a function of the operation days. For operational changes see Figure 2.**

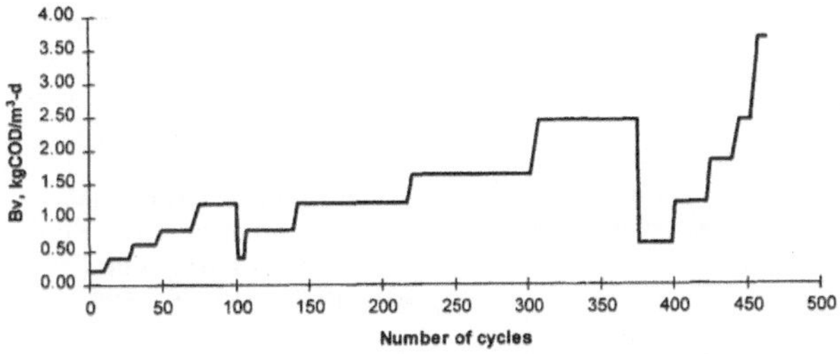


**Figure 6. Applied volumetric load as a function of the number of cycles.**

## Volumetric Load

The applied volumetric loads during the reactor operation varied from 0.2 to 3.65 kg $COD/m^3{}_{liquid}$-day. As the duration of the cycle was reduced, this parameter increased, reaching the highest value for the influent with 400 mg of total phenols/L and 4 hours of reaction time. Figure 6 depicts the variation of the volumetric load as a function of the operation cycles.

## Influence of Starvation Periods

Buitrón et al.[9] discussed the negative effect of starvation periods on 4-chlorophenol degradation for an activated sludge SBR system. In that study, the aeration was prolonged 20 to 23 hours after the toxic degradation had been completed. It was found that the activity of the acclimated population decreased drastically as a result of starvation. Degradation time for 100 mg of 4-chlorophenol/L increased six times (from 0.7 to 4.5 hr). The loss in the microbial capacity was attributed to a decline in both the enzymatic activity and the viability of the suspended cells.

To study the influence of starvation periods on the biofilm microbial activity, the reaction (aeration) phase of the biofilter was prolonged without adding exogenous substrate. During the cycles of 4 hours, a 24-hour cycle was introduced. This procedure was applied for both 100 and 200 mg total phenols/L in the reactor. Thus, after the degradation the microorganisms were maintained in endogenous respiration. The kinetic of the phenols biodegradation was followed before and after the starvation period (Figures 7 to 10).

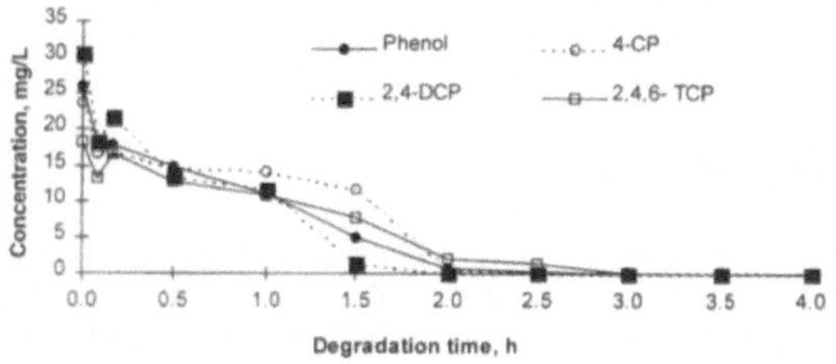


**Figure 7. Kinetics of phenols degradation before starvation. 100 mg total phenols/L; cycle No. 99.**

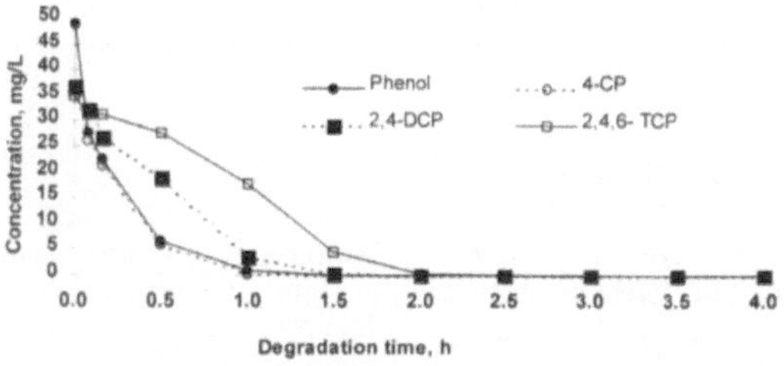


**Figure 8. Kinetics of phenols degradation before starvation. 200 mg total phenols/L; cycle No. 374.**

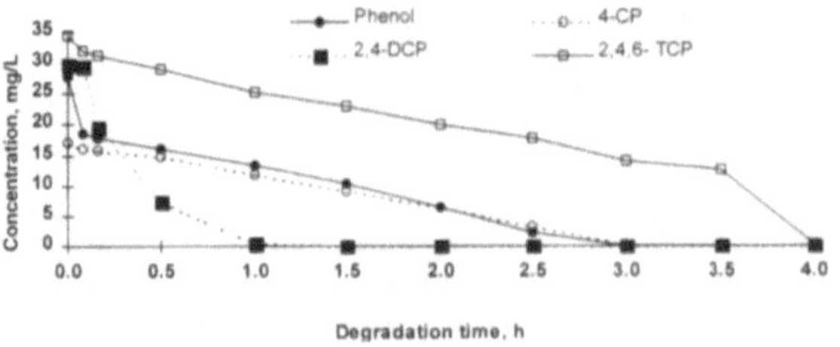


**Figure 9. Kinetics of phenols degradation after starvation (24 hours). 100 mg total phenols/L; cycle No. 100.**

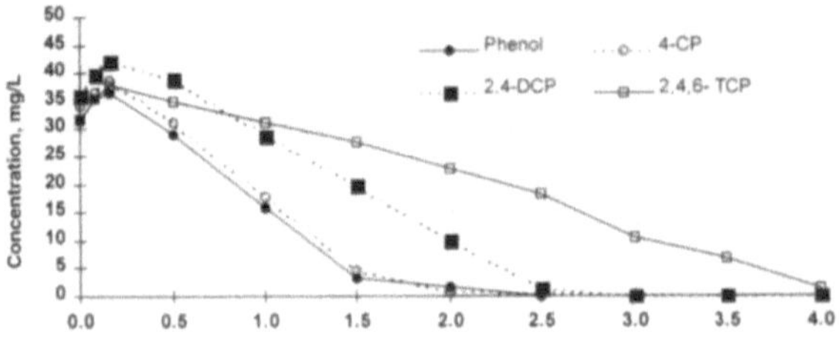


**Figure 10. Kinetics of phenols degradation after starvation (24 hours). 200 mg total phenols/L; cycle No. 375.**

Before starvation time, degradation was achieved in 2 to 3 hours, regardless of the toxic initial concentration. For 100 mg phenols/L, the degradation of the four compounds was simultaneous but for 200 mg/L it was sequential: phenol and 4-chlorophenol were degraded first, followed by 2,4 dichlorophenol and 2,4,6-trichlorophenol (Figures 7 and 8).

After the starvation period (Figures 9 and 10), a change in the degradation kinetics was observed. The degradation of trichlorophenol was the more affected; it took nearly four hours to complete the oxidation of this compound. But, in spite of the changes in the kinetic behavior, the overall phenols' degradation was not affected, and only a slight diminution of TOC removal was observed after the starvation period. For the following cycles, the reactor performance was recovered. The amount of phenols degraded was superior to 87% as TOC and 99% as total phenols.

The toxics elimination was not affected because the biofilm buffered the negative effect of starvation. Also, it is possible that in the porous material some quantity of phenols remained during starvation, so the microorganisms were always in contact with the toxic, producing the induced enzymes necessary to degrade the phenols and maintaining the viability of the biofilm.

## CONCLUSIONS

The degradation of a mixture of chlorophenols (phenol, 4-chlorophenol, 2,4-dichlorophenol and 2,4,6-trichlorophenol) in a SBB packed with a porous volcanic rock was investigated. The mixture of contaminants was eliminated with an efficiency superior to 93% as total organic carbon and 99% as the total phenols, regardless of the initial toxics concentration (varied from 60 to 400 mg/L). A maximal volumetric eliminated load of 3.6 kg COD/$m^3$-d was achieved. The process showed an excellent behavior against the operational problems due to the combination of the properties present in the SBR operation and the attached biomass systems. In particular, it was observed that the influence of starvation periods had no adverse effect on the reactor performance.

## ACKNOWLEDGMENT

This research was supported by the Consejo Nacional de Ciencia y Tecnolog'a, through a grant CONACYT 1117 PA. The technical assistance provided by Mrs. Leticia Garc'a for HPLC analysis is acknowledged.

## REFERENCES


1. Irvine R.L. and L.H. Jr. Ketchum, "Sequencing batch reactors for biological wastewater treatment," *CRC Crit. Rev. Environ. Control*, 18(4), pp. 225–294, 1989.
2. Grady C.P.L., Jr., "Biodegradation of toxic organics: status and potential," *J. Environ. Eng.*, 116, pp. 805–828, 1990.
3. Puhakka J.A. and K. Järvinen, "Aerobic fluidized-bed treatment of polychlorinated phenolic wood preservative constituents," *Water Res.*, 26, pp. 765–770, 1992.
4. Gantzer, C. J. "Inhibitory substrate utilization by steady-state biofilms," *J. Environ. Eng.*, 115, pp. 302–319, 1989.
5. Kaballo H.P., Y. Zhao, and P.A. Wilderer, "Elimination of p-chlorophenol in biofilm reactors-A comparative study of continuous flow and sequenced batch operation," *Water Sci. Technol.*, 31(1), pp. 51–60, 1995.
6. AFNOR "Evaluation en milieu aqueux de la biodégradabilité aérobie "ultime" des produits organiques solubles," *Normalisation française, NFT 90-312*, 1985.
7. Buitrón G. and B. Capdeville, "Enhancement of the biodegradation activity by the acclimation of the inoculum," *Environ. Technol.*, 16, pp. 1175–1184, 1995.
8. *Standard Methods for Examination of Water and Wastewater*,18th ed., APHA, AWWA, WPCF, New York, 1992.
9. Buitrón G., B. Capdeville, and P. Horny, "Improvement and control of the microbial activity of a mixed population for degradation of xenobiotic compounds," *Water Sci. Technol.*, 29(7), pp. 317–326, 1994.

## REFERENCES

# 28 REMOVAL OF PHENOL FROM INDUSTRIAL WASTEWATERS USING *ARTHROMYCES RAMOSUS* PEROXIDASE IN A CONTINUOUS FLOW SYSTEM

**M.S. Ibrahim**, Assistant Lecturer

**H.I. Ali,** Professor
Faculty of Engineering
Ain Shams University
Cairo, Egypt

**K.E. Taylor**, Professor
Faculty of Science
University of Windsor
Windsor, Ontario, Canada N9B 3P4

**N. Biswas**, Professor

**J. K. Bewtra**, Professor
Faculty of Engineering
University of Windsor
Windsor, Ontario Canada, N9B 3P4

## ABSTRACT

Various aromatic compounds, in particular phenols and aromatic amines, are present in wastewaters of numerous industries such as coal conversion, petroleum refining, organic chemicals, and dyes. Most of these compounds are toxic and some have been determined to be human carcinogens; therefore, the removal of such compounds from industrial aqueous effluent is of great practical significance.

An enzymatic method for the removal of phenols from industrial aqueous effluent has been developed in the past several years. In this method, peroxidase enzymes catalyze the oxidation of phenol with hydrogen peroxide, generating phenoxy radicals. These radicals diffuse from the active site of the enzyme into solution and react nonenzymatically to eventually form higher oligomers and polymers which can be removed from wastewater by sedimentation or filtration.

In this study, *Arthromyces ramosus* peroxidase (ARP) was applied to remove 1.0 mM phenol in synthetic wastewater in a continuous flow system. The system consisted of a mixing tank where hydrogen peroxide was added to the mixture of phenol, enzyme, and polyethylene glycol (PEG) to initiate the reaction. A plug flow reactor where the reaction takes place was followed by a flocculation tank where the alum was added, and finally the polymers formed were settled in a sedimentation tank and removed from the system.

The removal efficiency of the system was dependent on the enzyme dose, molar ratio between hydrogen peroxide and phenol, pH, and PEG concentration. These parameters were optimized first using batch experiments, and then applied to the continuous experiments.

## INTRODUCTION

Phenolic compounds are priority pollutants which occur in the effluent of many industries such as coal conversion, wood preservation, pulp and paper mills, metal casting, petroleum refining, resins, plastics, iron and steel, textile and organic chemical manufacturing.[1] The majority of phe-

*52nd Purdue Industrial Waste Conference Proceedings*, 1997, Ann Arbor Press, Chelsea, Michigan 48118. Printed in U.S.A.

nolic compounds are toxic;[2] some have been classified as hazardous substances, and some are known or suspected carcinogens.[3] Typical concentrations of phenols in industrial wastewaters range between 100 and 1000 mg/L.[4]

Several existing biological, chemical, or physical treatment processes can be used to remove phenols from contaminated aqueous mixtures.[5] Biological treatment systems can be most effective but may not remove specific chemicals to low levels consistently.[6] Many chemicals in industrial wastes can be inhibitory to mixed culture biological processes, and phenol is a well-known example.[7]

Chemical and physical processes are not appropriate for treating moderate to high strength wastewaters, since they are not highly selective of the contaminants removed during treatment. Thus, large amounts of oxidant or adsorbent would be required to remove targeted compounds from high strength wastewaters, even when the targeted compounds are present in relatively low concentrations. These methods can be designed to achieve low final concentrations of pollutants as polishing steps.

The concept of using enzymes to remove specifically targeted pollutants was developed in the 1970s.[8] For this purpose, enzymes isolated from their parent organisms are preferred over intact organisms containing the enzymes because the isolated enzymes are highly specific for their substrates, their potency can be better standardized, they are easier to handle and store,[9] and their concentration is easier to control since it does not depend on microbial growth.[10]

Enzymes also offer a number of advantages over conventional methods.[11] These advantages are: the removal of a broad range of compounds, including biorefractory compounds and those which are toxic to microorganisms; selective removal of a single class of contaminant; quick response to shock loadings; effectiveness at low hydraulic retention times; online addition with a minimal delay (no biomass growth and acclimation); simpler process control due to the use of isolated enzymes; less sludge production (since no biomass is invoiced); effectiveness over a broad range of pH, temperature, and salinity; less requirement of oxidant than conventional chemical oxidation methods; equally effective at high and low substrate concentrations; and operation under milder, less corrosive conditions than chemical oxidation methods.

Horseradish peroxidase (HRP) has demonstrated an ability to catalyze the polymerization and precipitation of aromatic compounds from aqueous solution.[1] Recently, peroxidase from *Arthromyces ramosus* (Fungi Imerfecti) has been shown to have specific activity and substrate profiles similar to horseradish peroxidase.[12]

## Mechanism of *Arthromyces ramosus* Peroxidase

The one-electron oxidation of aromatic substrate is given by the Chance-George mechanism[13] and can be depicted by the following mechanism:

$$E + H_2O_2\ E_i + H_2O \tag{1}$$

$$E_i + AH_2\ E_{ii} + {}^{\bullet}AH \tag{2}$$

$$E_{ii} + AH_2\ E + {}^{\bullet}AH + H_2O \tag{3}$$

Native enzyme (E) reacts with peroxide to form "Compound I" which is oxidized by two electrons above the native enzyme and carries the peroxidic oxygen. Compound I accepts an aromatic compound in its active site and carries out its oxidation. The oxidized aromatic, now a free radical, is released from the catalytic site while the enzyme is reduced to "Compound II." The catalytic cycle is completed when Compound II oxidizes a second aromatic molecule, resulting in the release of a second free radical into the solution and returning the enzyme to its native state. The free radicals couple to form dimers and some of the free radical coupling products are substrates for the enzyme. Yu et al.[14] detected ten soluble products; five stable dimers were identified as *o,o′*-biphenol, *p,p′*-biphenol, *o,p′*-biphenol, *p*-phenoxyphenol, and *o*-phenoxyphenol. They found that *p*-phenoxyphenol and *p,p′*-biphenol reacted with the enzyme faster than the parent phenol, and this provided an explanation for the 1:1 observed molar ratio between hydrogen peroxide and phenol.

While Reactions 1, 2, and 3 give the catalytic cycle of peroxidase, peroxide, and phenol, a number of side reactions take place. Some of these reactions are responsible for the inactivation of the enzyme resulting in a finite catalyst lifetime. Experiments showed that apparent enzyme inactivation during phenol polymerization was mainly caused by end-product polymer, which adsorbed the enzyme molecules and hindered the access of substrate to the enzyme's active site.[15] In the presence of proteins or hydrophilic synthetic polymers, the enzyme adsorption was suppressed and the amount of enzyme required to achieve complete removal was reduced. Polyethylene glycol, gelatin,[4,15] talc,[16] and a cationic polymer coagulant[17] have been shown to be effective in reducing the inactivation of enzymes. The best additive of these is polyethylene glycol due to its low environmental impact and as a nontoxic organic compound, it has been declared fit for human consumption.[18]

In the presence of excess hydrogen peroxide, Compound II can be oxidized to Compound III,[19] according to:

$$E_{ii} + H_2O_2 \ E_{iii} + H_2O \tag{4}$$

Compound III is catalytically inactive and its formation does not a terminal inactivation since Compound III decomposes to native peroxidase according to:

$$E_{iii} \rightarrow E + O_2^- \tag{5}$$

The return to the native enzyme state is sufficiently slow that, once in the Compound III form, the enzyme is severely hampered in carrying out the catalytic oxidation of aromatic substrates. Therefore, any accumulation of enzyme in the Compound III state represents a loss in catalytic efficiency.

In the absence of reductant substrate, and with excess of $H_2O_2$, peroxidase shows the kinetic behavior of suicide inactivation, $H_2O_2$ being the suicide substrate. The process involves both a reversible inactivation pathway leading to Compound III as above, and a pathway leading to an irreversible inactivated intermediate, Compound $P_{670}$.[19,20]

The nature of the soluble and insoluble products formed in the polymerization process must be characterized before the enzymatic treatment process can be implemented on an industrial scale. Aitken et al.[21] demonstrated that the product characteristics (molecular mass and toxicity) varied, depending on the parent phenol being treated, the source of the enzyme, and the reaction pH. The authors concluded that it may be possible to achieve detoxification of phenols through appropriate reaction engineering. In addition, they indicated that the nature of the products which would be produced in treating a specific wastewater cannot be predicted. Rather, well-designed feasibility tests must be performed on each specific wastewater to assess the effects of enzymatic oxidation on phenol removal and reaction product characteristics, including toxicity. So, separation of the reaction products from the solution is of great importance. The disposal of these solid products has yet to be determined and depends primarily upon their toxicity and mobility in the environment; however, incineration and disposal in landfills are options to be considered.

This chapter demonstrates the use of *Arthromyces ramosus* peroxidase in the removal of phenol in a continuous flow system and the removal of the reaction products from the wastewater.

## MATERIALS AND METHODS

*Arthromyces ramosus* peroxidase (ARP) with an RZ $\approx$ 0.5 was a gift from the Biotech Environmental Inc. of a development preparation of Novo Nordisk. Enzyme activity was determined by a phenol/AAP assay that contained the following in 0.05 M phosphate buffer pH 7.4 in a final volume of 1.0 mL: 10 mM phenol, 2.4 mM 4-aminoantipyrine (AAP), 0.2 mM hydrogen peroxide, and 20–50 µL of appropriately diluted enzyme sample. The rate of reaction was monitored with a Hewlett Packard Spectrophotometer (Model 8452A) at 510 nm and calculated using an extinction coefficient of 6000 $M^{-1}cm^{-1}$ based on $H_2O_2$. One unit of activity is defined as the number of micromoles of $H_2O_2$ converted per minute at pH 7.4 and 25 $^0$C. Enzyme concentration are expressed as U/mL in this study. The concentration of the enzyme was determined by measuring the absorbance of the haem at 404 nm and a peroxidase extinction coefficient of 102,000 $M^{-1}cm^{-1}$. For this stock of enzyme, it was found that 1 nM gave an activity of 0.0194 U/mL.

Phenol was purchased from Aldrich Chemical Co., Milwaukee, WI, and had a purity of 99% or greater. Phenol concentration was determined using a colorimetric assay in which the phenolic compounds within a sample reacted with 2.08 mM AAP in the presence of 8.34 mM potassium ferricyanide reagent. The extent of the color-generating reaction was proportional to phenol concentration. After the full development of the color, the absorbance at 510 nm was converted to total phenol concentration in the cuvette using a calibration curve.

Hydrogen peroxide (30%) was purchased from BDH Inc., Toronto, Ontario. Peroxide solutions for the analytical assays were prepared daily. The concentration of $H_2O_2$ was determined using an end-point assay, based on AAP/phenol system. The concentration of ARP in the microcuvette was 0.1 mg/L, while those of other reactants (except peroxide) were as cited above for the ARP assay. The extent of color development was proportional to hydrogen peroxide concentration. The absorbance at 510 nm after ten minutes reaction time was converted to peroxide concentration using a calibration curve. Polyethylene glycol (average molecular mass = 3350) was purchased from Sigma Chemical Co., St. Louis, MO.

The batch reactors consisted of vials containing 30 mL of a mixture of phenol, ARP, and PEG in a dechlorinated tap water. The pH was adjusted to 7.0 by using either hydrochloric acid or sodium hydroxide. The reaction was initiated by adding hydrogen peroxide to the mixture. The reaction solution was agitated with a magnetic stirrer and Teflon coated bar. The retention time in all the experiments was 3 hours to allow the reaction to go to completion. Reaction was stopped by the addition of concentrated catalase enzyme to consume hydrogen peroxide. The mixture was treated by adding alum to a final concentration of 100 mg/L, and was allowed to coagulate for 20 minutes and then a sample was taken to centrifuge at 3000 rpm for 30 min. The supernatant was analyzed for residual phenol.

The continuous system consisted of:

- An inlet tank where phenol, enzyme, and PEG were mixed together in the tank and the activity of the enzyme was determined before and after the experiment.
- A mixing tank where the hydrogen peroxide was injected with a syringe pump (Harvard 22, purchased from Harvard Apparatus Canada, Saint Laurent, Quebec; Becton Dickson inert and nonadsorbing plastic syringes were purchased from Fisher Scientific of Montreal). The mixing time was 5 min.
- A plug flow reactor where the enzymatic reaction took place for 2 hours.
- A flocculation tank where the alum and sodium hydroxide were added just ahead of the tank to allow for alum mixing and coagulation for 30 min.
- A sedimentation tank with a retention time of 3 hours, where the settled solids were removed from the bottom of the tank.

## RESULTS AND DISCUSSION

A series of batch reactions were conducted to determine the best operating conditions. The optimum values were as follows:[22] a phenol concentration of 1 mM and a removal efficiency of more than 95% required the minimum enzyme dose = 0.6 U/mL, optimum PEG dose = 30 mg/L, optimum pH = 7.0, optimum $H_2O_2$/phenol ~ 1.0, and the minimum reaction time = 1.0 hour. In these batch reactor studies, it was demonstrated that no removal of phenol occurred in the presence of either hydrogen peroxide or enzyme alone. Alum treatment alone did not result in any removal. Also, no phenol removal was observed in the presence of PEG alone or in the presence of PEG and $H_2O_2$ and PEG and enzyme together. Therefore, removal of phenol was attributed to the combined action of ARP and $H_2O_2$. It was also concluded that the PEG enhanced the efficiency of the catalytic reaction, without causing any additional removal of phenol by side-reactions.

The removal of phenol in the continuous flow system was evaluated by measuring the percentage of the removed phenol from the solution and the clearance of the final effluent after sedimentation tank.

The continuous flow system (CFS) was first operated as shown in Figure 1. The reaction conditions were: phenol = 1 mM, ARP dose = 1.75 U/mL, PEG dose = 100 mg/L, and $H_2O_2$ dose = 1.9 mM. After adding hydrogen peroxide to the mixture of phenol, ARP, and PEG, a brown color was developed. This color had a peak at 400 nm and it was attributed to the oxidation of *p,p'*-biphenol to *p*-diphenoquinone.[23] No precipitate was observed in the sedimentation tank and the color was not removed by filtration.

The removal ratio of phenol after treatment with alum and centrifugation was 91%. It was concluded from this that alum was necessary to remove the color and by-products from the solution but the removal efficiency was dependent on where alum was added.

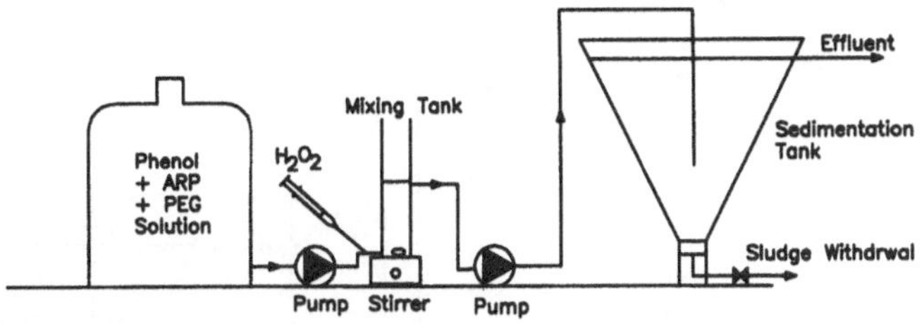


**Figure 1. The continuous flow system setup #1.**

A separate experiment was conducted to determine if the enzyme was immobilized on the alum floc while being still active. This was done by preparing enzyme stock solutions with a measured activity. It was observed that after adding alum and completing coagulation, the enzyme was still active with almost the same activity. However, after filtering the coagulated enzyme stock solutions, using a 0.2 _m syringe filter, the measured enzyme activity dropped to 15% of its initial activity. This clearly demonstrated that the enzyme was immobilized on the alum floc and was removed on the filter with the floc.

Since there was not a remarkable decrease in the enzyme activity when it was immobilized on the alum floc, the experimental setup was modified to add alum to the reaction in the mixing tank along with the hydrogen peroxide (Fig. 2). The operating conditions for the best run were: phenol = 1 mM, ARP = 2.25 U/mL, PEG = 100 mg/L, $H_2O_2$ = 1.5 mM, and alum dose = 50 mg/L. When alum was added in the mixing tank, good floc was formed and a clear supernatant in the sedimentation tank was obtained, but it did not give good results in removing phenol. The low removal efficiency was attributed to two phenomena: (i) as the tap water had a lower buffering capacity than any chemical buffer, alum decreased the pH of the reaction mixture and shifted the pH to a lower value than the optimum pH; (ii) as the floc entered the sedimentation tank and began to settle, the enzyme was also removed from the reaction mixture with the settled floc. The removal efficiency of phenol in this run was 91%.

Subsequently, sodium hydroxide was added in the mixing tank to adjust the pH value. The same removal efficiency was obtained as before by using a lower enzyme dose of 1.5 U/mL but this enzyme dose was still high as compared to the batch reactor dose of 0.6 U/mL. This high enzyme dose was required because 85% of the enzyme activity was removed from the solution in the sedimentation tank.

The best results were obtained by dividing the treatment scheme into two parts: removal of phenol by enzymatic reaction, followed by the removal of soluble reaction products by conventional coagulation and sedimentation process. This is shown in Figure 3.

Table I shows the operating condition of the CFS and the corresponding phenol removal efficiency. The removal efficiency of the phenol was dependent on the enzyme dose, influent pH of the reaction, the molar ratio of hydrogen peroxide and phenol, PEG dose and alum concentration, and the use of sodium hydroxide to adjust the pH in the coagulation process. The optimum values that had been determined from the batch experiments gave a phenol removal efficiency of 94%. The effluent suspended solids was 6.4 mg/L.

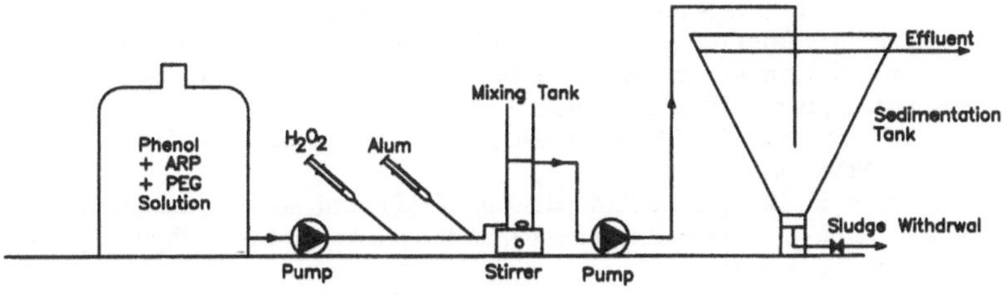


**Figure 2. The continuous flow system setup #2.**

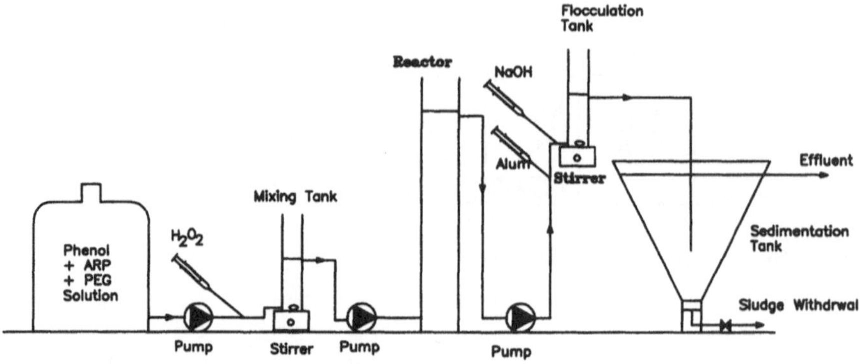


**Figure 3. The proposed continuous flow system setup.**

**Table I. The Operating Condition of the Continuous Flow System and the Phenol Removal Efficiency**

| ARP Dose U/mL | PEG dose mg/L | $H_2O_2$ dose mM | Alum and (NaOH) Doses mg/L | Influent pH | Removal Efficiency %[a] |
|---|---|---|---|---|---|
| 0.43 | 400 | 1.2 | 50 (0) | 7.55 | 63 |
| 0.55 | 400 | 1.1 | 50 (0) | 7.4 | 70 |
| 0.60 | 100 | 1.2 | 50 (0) | 7.7 | 90 |
| 0.60 | 75 | 1.1 | 50 (0) | 7.7 | 90 |
| 0.63 | 100 | 1.1 | 25 (2.5) | 7.7 | 90 |
| 0.60 | 30 | 1.1 | 25 (2.5) | 7.0 | 94 |

[a]Based on the sedimentation tank effluent analysis without any further treatment.

## CONCLUSIONS

The results from these experiments showed that:

- The enzyme-catalyzed polymerization and precipitation is an effective method for the removal of phenol from wastewater.
- Based on batch reactor studies,[22] polyethylene glycol (PEG) was effective in improving the removal efficiency by 28–70% for 1 and 10 mM phenol solution, respectively.
- The optimum operating condition obtained from batch reactor studies can be applied to a continuous flow system and obtain the same removal ratio providing a good sedimentation facility that give a clear supernatant..

## REFERENCES

1. Klibanov, A.M., Alberti, B.N., Morris, E.D., and Fleskin, L.M. "Enzymatic removal of toxic phenols and anilines from wastewater." *J. Appl. Biochem.*, 2, pp. 414–421, 1980.
2. Klibanov, A.M. "Enzymatic removal of hazardous pollutants from industrial aqueous effluent." *Enzyme Eng.*, 6, pp. 319–323, 1982.
3. Verschueren K. *Handbook of Environmental Data on Organic Chemicals.* Van Nostrand Reinhold Co., New York, 1977.
4. Wu, J., Taylor, K.E., Bewtra, J.K. and Biswas, N. "Optimization of the reaction conditions for enzymatic removal of phenol from water in the presence of polyethylene glycol." *Water Res.*, 27(12), pp. 1701–1706, 1993.
5. Patterson, J.W. *Industrial Wastewater Treatment Technology*, 2nd ed. Butterworths, Boston, MA, 1985.

6. Grady C.P.L. "Biodegradation of toxic organics: Status and potential." *J. Environ. Eng.*, 116, pp. 805–828. 1990.
7. Allsop P.J., Moo-Young M. and Sullivan G.R. "The dynamics and control of substrate inhibition in activated sludge." *Crit. Rev. Environ. Control*, 20, pp. 115–167, 1990.
8. Aitken, M.D. "Waste treatment applications of enzymes: Opportunities and obstacles." *J. Chem. Eng.* 52, pp. B49–B58, 1993.
9. Vieth, W.R. and Venkatasubramanian, K. "Enzyme Engineering: Part I. The utility of supported enzyme systems." *Chemtech.* Nov., pp. 677–684, 1973.
10. Bailey, J.E. and Ollis, D.F. *Biochemical Engineering Fundamentals*. 2nd ed. McGraw-Hill Book Company, New York, 1986.
11. Nicell, J.A., Al-Kassim, L., Bewtra, J.K., and Taylor, K.E. "Wastewater treatment by enzyme catalysed polymerization and precipitation."*Biodeterioration Abstracts*, 7(1), pp. 1–8, 1993.
12. Kjalke, M., Anderson, M.B., Schneider, P., Christensen, B., Schulein, M., and Welinder, K.G. "Comparison of structure and activities of peroxidases from *Coprinus cinereus*, *Coprinus macrorhizus* and *Arthromyces ramosus*." *Biochim. Biophys. Acta*, 1120, pp. 248–256, 1992.
13. Job, D. and Dunford, H.B. "Substituent effect on the oxidation of phenols and aromatic amines by horseradish peroxidase compound." *Eur. J. Biochem.*, 66, pp. 607–614, 1976.
14. Yu, J., Taylor, K.E., Zou, H., Biswas, N., and Bewtra, J.K., "Phenol conversion and dimeric intermediates in horseradish peroxidase-catalyzed phenol removal from water." *Environ. Sci. Technol.*, 28, pp. 2154–2160, 1994.
15. Nakamoto, S. and Machida, N. "Phenol removal from aqueous solutions by peroxidase-catalyzed reaction using additives." *Water Res.*, 26, pp. 49–54, 1992.
16. Arseguel, D., and Baboulene, M. "Removal of phenol from coupling of talc and peroxidase. Application for depollution of waste water containing phenolic compounds." *J. Chem. Technol. Biotechnol.*, 61, pp. 331–335, 1994.
17. Tatsumi, K., Ichikawa, H., and Wada, S. "Dephenolization from aqueous solution by treatment with peroxidase and a coagulant." *Water Sci. Tech.*, 30, pp. 79–86, 1994.
18. Harris, J.M. Introduction to Biotechnical and Biomedical Applications of Poly(ethylene Glycol). In: *Poly(ethylene glycol) Chemistry: Biotechnical and Biomedical Applications*. Plenum Press, New York, 1992, pp. 1–14.
19. Arnao, M.B., Acosta, M., del Rio, J.A., Varon, R., and Garcia-Canovas, F. "A kinetic study on the suicide inactivation of peroxidase by hydrogen peroxide." *Biochim. Biophys. Acta*, 1041, pp. 43–47. 1990.
20. Baynton, K.J., Bewtra, J.K., Biswas, N., and Taylor, K.E. "Inactivation of horseradish peroxidase by phenol and hydrogen peroxide: a kinetic investigation." *Biochim. Biophys. Acta*, 1206, pp. 272–278. 1994.
21. Aitken, M.D., Massey, I.J., Chen, T., and Heck, P.E. "Characterization of reaction products from the enzyme catalyzed oxidation of phenolic pollutants." *Water Res.*, 28(9), pp. 1879–1889, 1994.
22. Ibrahim, M.S., Ali, H.I., Taylor, K.E., Biswas, N. and Bewtra J.K. Optimization of the Removal of Phenol from Wastewater Catalyzed by *Arthromyces ramosus* Peroxidase. paper submitted to the '97 CSCE-ASCE Environmental Engineering Conference. Edmonton, Alberta, Canada. July 23–25, 1997.
23. Sawahata, T. and Neal, R.A. "Horseradish peroxidase-mediated oxidation of phenol." *Biochem. Biophys. Res. Commun.*, 109, pp. 988–994, 1982.

# Section Eight
# BIOLOGICAL: ANAEROBIC SYSTEMS

## 29 FEASIBILITY AND BENEFITS OF METHANOGENESIS UNDER OXYGEN-LIMITED CONDITIONS

**D.H. Zitomer**, Assistant Professor

**J.D. Shrout**, Research Assistant
Civil and Environmental Engineering
Marquette University
Milwaukee, Wisconsin 53201

### ABSTRACT

Methanogenic and aerobic (or microaerophilic) biological processes are often considered mutually exclusive and separated as biological wastewater treatment options. However, under oxygen-limited conditions, both aerobic respiration and methanogenesis may be practically accomplished by a single mixed culture. This chapter describes sustained batch culture, oxygen-limited methanogenic serum bottle and bench-scale systems. Serum bottle cultures exhibited methanogenic activity similar to or greater than that of a strictly anaerobic culture maintained in parallel. The COD removal efficiencies of anaerobic, oxygen-limited, and aerobic bench-scale reactors receiving 30,000 mg/L of sucrose were comparable (94%), but a system receiving 1 g $O_2/L_R$-day achieved a lower final effluent COD than the strictly anaerobic reactor. After a shock-load of sucrose, the pH recovered in low-aeration batch reactors in 28 to 34 days, whereas anaerobic pH did not recover after 52 days of observation. In the future, methanogenesis under limited-aeration may be employed as an energy-efficient treatment option to achieve low final COD concentrations, minimal biosolids generation, and mineralization of a broad range of specific organic chemicals.

### INTRODUCTION

Oxygen, which is classically considered to be the nemesis of methanogenic processes, may potentially enhance methanogenesis under some conditions. Although even low levels of dissolved oxygen are considered to be extremely toxic to methanogens, methanogens have been found to survive short periods in the presence of dissolved oxygen and to coexist with aerobic or microaerophilic organisms. For example, methane production by the methanogen *Methanothrix soehngenii* did not decline after oxygen was administered for 48 hours and then removed.[1] Other methanogens, including *Methanosarcina barkeri*,[2] *Methanobacterium bryantii*,[3] *Methanothrix soehngenii*,[1] *Methanobacterium thermoautotrophicum*, and *Methanobrevibacter arboriphilus*[4] all exhibit limited tolerance to low oxygen levels.

Mixed cultures of methanogenic, acetogenic, and obligate aerobic microorganisms have been described, and methane production has been sustained over a wide range of oxygen supply rates.[5] Obligate aerobes in these systems have been found to oxidize acetate and propionate produced by acetogenic anaerobes during lactose fermentation,[6] and to transform acetamide in synergistic association with methanogens.[7] Stoichiometric equations for propionate- and ethanol-fed methanogenic, oxygen-limited batch cultures have been developed using electron equivalent balances.[8] The oxygen utilized by various cultures ranged from 10 to 30% of the added COD, whereas the remainder of removed COD was converted to methane.

---

*52nd Purdue Industrial Waste Conference Proceedings*, 1997, Ann Arbor Press, Chelsea, Michigan 48118. Printed in U.S.A.

Survival of methanogens in oxygen-limited mixed cultures is often attributed to formation of reduced microniches in otherwise oxic environments.[4] For instance, facultative microorganisms on the surface of granular sludge particles from an upflow anaerobic sludge blanket reactor (UASBR) consumed oxygen before it diffused into the inner-particle region.[9] Similarly, porous support media such as calcium alginate beads have been employed to co-culture strict aerobes (outer-bead region) and anaerobes (inner-bead region).[10] This oxygen shielding effect may explain the viability of a full-scale methanogenic bioreactor (Hormel Packing Plant, Austin, MN) that has been operating since 1960 which achieves effluent BOD concentrations consistently below 20 mg/L even though biomass is aerated prior to thickening and recycle.[11] Shielding may also explain the presence of methanogens in aerobic activated sludge[12] and the cultivation of anaerobic granular sludge in a reactor seeded with aerobic activated sludge.[13]

It is significant that both reductive and oxidative biotransformations may occur concomitantly under oxygen-limited conditions. The ability of anaerobic systems to reduce highly chlorinated compounds that are relatively recalcitrant to aerobic biotransformation is documented.[14] On the other hand, the ability of aerobic processes to oxidize aromatic organics resistant to anaerobic transformation is also apparent. At the same time, aerobic cometabolic oxidation of chlorinated organics may occur in the presence of methane and some other primary substrates.[15]

Similarly, the mineralization of DDT, polycyclic aromatic hydrocarbons,[4] highly chlorinated solvents, and other organics often proceeds through a sequence of anaerobic, aerobic, or anoxic biotransformations.[14] For example, the mineralization rate of a chlorinated insecticide, Methoxychlor, was significantly increased when a single mixed culture was incubated under sequential anaerobic/aerobic conditions.[16] Mineralization was not apparent under strictly anaerobic nor aerobic conditions alone. When considering in-situ technologies for soil/groundwater remediation, it may not be feasible to divide the system into two locales containing different cultures. Consider that PCB congeners (Aroclor 1242) in sediment are biotransformed when biologically active soil samples are incubated under anaerobic conditions, and then under aerobic conditions.[17] The same initial culture performs both reductive and oxidative steps. It has also been reported that transition from anaerobic to aerobic conditions is required for particular cultures to reductively dechlorinate trichloroethylene and tetrachloroethylene.[18]

Other possible benefits of methanogenic, oxygen-limited systems relate to gross pollutant removal in wastewater treatment. Combined systems are potentially much more energy-efficient than conventional aerobic systems, requiring less energy for blower operation and producing significantly less biosolids to be handled, transported, and disposed. On the other hand, combined systems may achieve lower final biochemical oxygen demand (BOD) than conventional anaerobic processes and recover from organic shock-loads more quickly.

In the study reported herein, anaerobic and oxygen-limited methanogenic cultures were maintained under stable conditions in both serum bottle and bench-scale batch reactors. The methanogenic activity, COD removal rates, biomass yields, and ability to recover from shock COD loads are described.

## METHODS

### Serum Bottle Studies

Seed organisms for the mixed cultures were obtained from anaerobic digesters of the South Shore Wastewater Treatment Plant, Milwaukee, Wisconsin. Fifty millileters of anaerobic sludge were added to 160-mL serum bottles. The bottles were then flushed with methane gas and sealed with rubber septa.

All cultures contained 50 mL of active volume, and were maintained in a batch mode having a two-day wasting/feeding cycle. A solids retention time (SRT) of 10 days was maintained by wasting 10 mL of culture every two days. Subsequently, 10 mL of medium containing approximately 50 mg of COD in the form of sucrose, ammonia-nitrogen, phosphorus, and other trace nutrients were added. An oxidation-reduction indicator dye, resazurin, was also added. Nitrate-nitrogen, nitrite-nitrogen, and sulfate concentrations were very low, so that COD oxidation associated with reduction of these electron acceptors was negligible.

A factorial approach was used in which two oxygenation conditions and three oxygen doses were maintained. Cultures maintained under the first oxygenation condition received oxygen at the time of medium addition (aerobic/anaerobic cultures), whereas cultures maintained under the second condition received oxygen 1 day into the 2-day feeding cycle (anaerobic/aerobic cultures). Pure oxygen was injected into serum bottles using syringes fitted with needles. Oxygen volumes injected to various cultures included 4, 12, and 48 mL every two days. These addition rates correspond to 10, 30, and 125% of the added COD, respectively. A strictly anaerobic control was also maintained. Headspace gas was wasted daily using a 50-mL glass syringe with a wetted barrel. Microorganisms were acclimated for at least three SRTs (30 days) before data were collected. All culture conditions were maintained at 35 C on a shaker table in the dark.

Methanogenic activity assays were performed by withholding oxygen the day of the assay (to maintain anaerobic conditions) and adding 400 mg/L of acetate (as calcium acetate) to each mixed culture. Culture wasting and medium addition was not performed for 7 days, and gas production was measured daily using a 50-mL glass syringe with a wetted barrel. Under the studied conditions, the concentration of extraneous electron acceptors, such as nitrate and sulfate, were negligible. Therefore, under anaerobic conditions biogas was generated, theoretically containing 30% carbon dioxide and 70% methane.

Methanogenic activity was estimated as the initial rate of gas production determined from the initial slope of a plot of cumulative gas production versus time. Relative activity (R) of a culture was calculated as the estimated initial rate of gas produced by the culture of interest, divided by the initial rate of gas production exhibited by the strictly anaerobic control.

## Bench-Scale Batch Reactor Studies

Two bench-scale batch reactors were seeded with a mixture containing anaerobic digester sludge and aerobic mixed liquor from the South Shore Wastewater Treatment Plant, Milwaukee, Wisconsin. Reactors were operated under oxygen-limited conditions. Air was introduced by means of a timer-actuated solenoid valve which was open for 15 seconds every half-hour. When open, the solenoid valves provided an air flow of 900 mL/minute (at 35°C) to one reactor, and 95 mL/minute to the other, corresponding to 1 and 0.1 grams of oxygen per liter of reactor per day (g $O_2/L_R$-day). Oxygen transfer rates were significantly lower than the addition rates since some oxygen escaped in off-gas. In addition to oxygen-limited reactors, conventional anaerobic (methanogenic) and aerobic (dissolved oxygen greater than 2.0 mg/L) systems were operated in parallel for comparison.

All four systems were 3.8-liter vessels containing 3.0 liters of culture (Figure 1) maintained at approximately 35 C in a temperature-controlled room. All SRTs were maintained at 20 days by wasting 150 mL of reactor contents every day and replacing it with a basal medium and sucrose (5 to 80 g/L) as the primary substrate. The basal medium contained the following in Milwaukee, Wisconsin tap water: yeast extract (2 g/L); $Fe(NH_4)_2 \cdot 6H_20$ (2 mg/L); $NiCl_2 \cdot 6H_2O$ (4 mg/L); $CoCl_2 \cdot 6H_2O$ (4 mg/L); $(NH_4)_2 HPO_4$ (60 mg/L); $Na_2 S \cdot 9H_20$ (60 mg/L), and $Na_2HCO_3$ (4000 mg/L). The daily organic loading rate (OLR) was the same for all reactors, but varied over time from 0.25 to 4 grams of COD per liter of reactor per day ( g $COD/L_R$-day) (Figure 2).

Soluble chemical oxygen demand (COD) was measured by first filtering the sample through a 0.45 µm glass fiber filter, and then measuring COD according to *Standard Methods*[19] Mixed liquor volatile suspended solids (MLVSS) were determined according to *Standard Methods*.[19] The pH of reactor content aliquots was measured using a probe and pH meter.

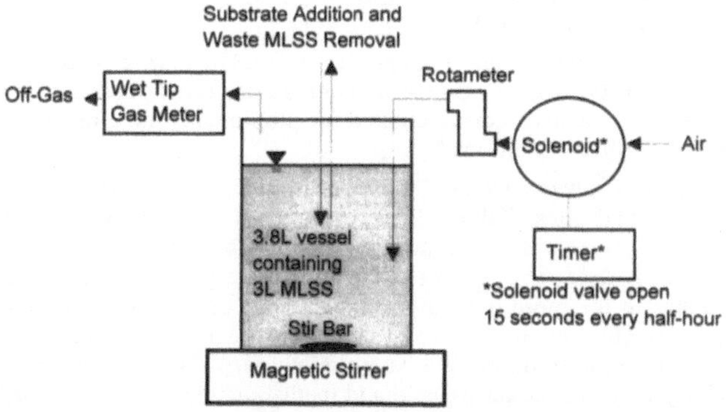


**Figure 1. Typical bench-scale batch reactor. Oxygen-limited reactors were constructed as shown. The strictly anaerobic reactor did not include the air line, solenoid valve, timer, and rotameter. The strictly aerobic reactor did not include the timer, solenoid valve, and wet tip gas meter.**

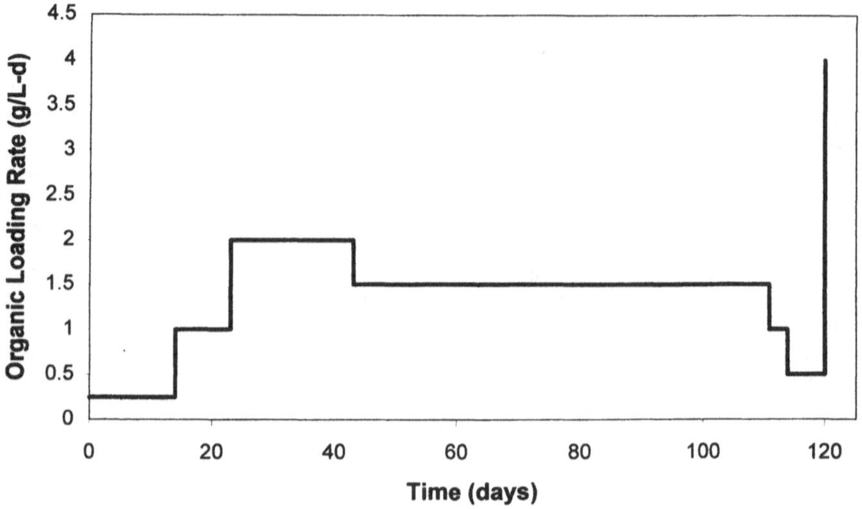


**Figure 2. Bench-scale batch reactor organic loading rate (OLR) versus time.**

The four reactors were shock-loaded by increasing the organic loading rate to 4 g COD/$L_R$-day to investigate process stability and recovery (Figure 2). During the subsequent 60 days, wasting and substrate addition was not performed and the pH of reactor contents was measured.

## RESULTS AND DISCUSSION

### Serum Bottle Studies

Oxygen-limited and anaerobic serum bottle cultures were maintained for over five months under sustainable conditions. The dissolved oxygen concentration in systems receiving oxygen doses of 0, 10, and 30% of the added COD was typically zero, as indicated by a lack of pink color exhibited by the resazurin dye in the basal medium. During brief periods (approximately 30 minutes) immediately after oxygen addition, however, dissolved oxygen was present. In addition, the systems receiving an oxygen dose of 125% of the COD typically contained dissolved oxygen at all times, yet approximately 0.1% methane was detected in the headspace of these systems. More importantly, the results of methanogenic assay tests indicate that a significant population of acetotrophic methanogens was present in all cultures (Figures 3 and 4), and that the specific methanogenic activity (R) was highest in systems which received the most oxygen (Table I). Therefore, oxygen addition did not prevent the growth of methanogens, but increased their initial activity under the specific conditions studied. The relatively low percent of methane in the headspace in comparison to the relatively high methanogenic activity may be attributable to aerobic oxidation of methane (methanotrophic reactions).

Maximum biomass yields of oxygen-limited cultures ranged from 0.13 to 0.08 g VSS/g sucrose as COD (Table I), and are more typical of strictly methanogenic, rather than aerobic processes. Lower-than-expected yields have also been calculated for propionate- and ethanol-fed methanogenic cultures under oxygen-limited conditions.[8]

Higher COD removal efficiencies and lower residual COD concentrations were exhibited by batch systems under oxygen-limited conditions (Table I). Sucrose may have been oxidized directly by aerobic or microaerophilic processes. It is also possible that aerobic oxidation of intermediates such as propionic acid, acetic acid, and hydrogen occurred. The possibility of lower concentrations of these intermediates in the more oxic environments may also explain the relatively high activity of methanogens in cultures that received the most oxygen. Elevated concentrations of these intermediates may potentially inhibit methane production.[11] Other reactor configurations, such as UASBR and separate, staged acetogenic and methanogenic processes would

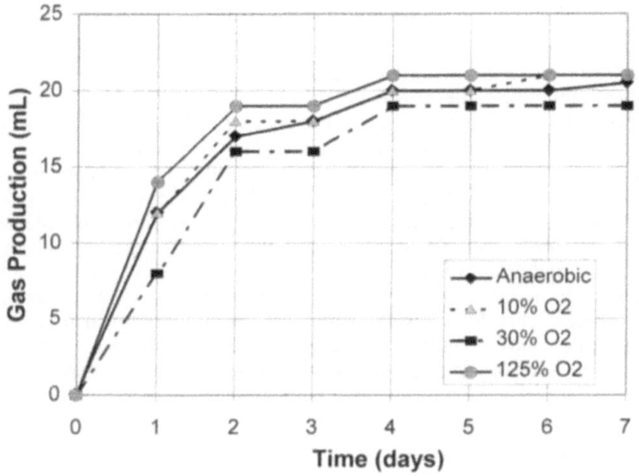


**Figure 3. Results of methanogenic activity tests for anaerobic-aerobic serum bottle cultures.**

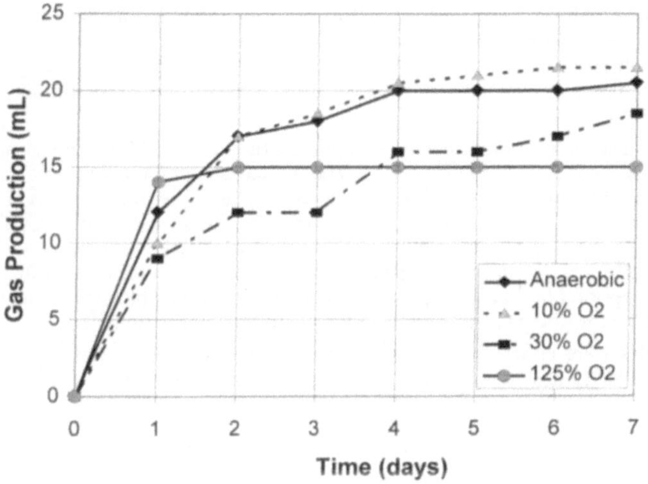


**Figure 4. Results of methanogenic activity tests for aerobic-anaerobic serum bottle cultures.**

**Table 1. Sucrose Biotransformation Under Methanogenic, Oxygen-Limited Conditions**

| Culture | Condition | Oxygen Added (% COD) | pH | MLVSS (mg/L) | Effluent COD (mg/L) | COD Removal (%) | $Y_{obs}$ (gVSS/gCOD) | R[a] |
|---|---|---|---|---|---|---|---|---|
| 1 | anaerobic | - | 6.8 | 330 | 1970 | 61 | 0.11 | 1.0 |
| 2 | aerobic/anaerobic | 10 | 6.7 | 310 | 1300 | 74 | 0.08 | 0.80 |
| 3 | aerobic/anaerobic | 30 | 6.9 | 270 | 1290 | 74 | 0.07 | 0.80 |
| 4 | aerobic/anaerobic | 125 | 7.1 | 520 | 540 | 89 | 0.12 | 1.2 |
| 5 | anaerobic/aerobic | 10 | 6.8 | 290 | 1330 | 74 | 0.08 | 1.0 |
| 6 | anaerobic/aerobic | 30 | 6.8 | 430 | 1300 | 74 | 0.12 | 0.70 |
| 7 | anaerobic/aerobic | 125 | 7.1 | 535 | 590 | 90 | 0.13 | 1.2 |

[a]R= Relative Rate of Methanogenic Activity.

likely be much more efficient. Methanogenesis with limited oxygen under these conditions should be investigated.

Although survival of methanogens under low-aeration conditions is usually attributed to formation of anaerobic microniches, the cultures described herein existed as dispersed, suspended growth. Large flocs or pellets were not apparent, and a predominance of free-swimming bacteria was observed microscopically. Therefore, exceptional spatial variation in redox conditions is unlikely. Others have reported ephemeral methane production by pure cultures of methanogens in the presence of dissolved oxygen.[20] Although a decline in production was evident, it was a "relatively slow process" under some conditions. Similarly, some researchers have considered methanogenesis under aerobic (or microaerophilic) conditions when investigating the aerobic marine environment, whereas more traditional concepts, such as the presence of anaerobic microenvironments, were described as unlikely, based upon conservative estimates.[21] The possibility of sustained methanogenesis under very low dissolved oxygen conditions is even more intriguing when considering that superoxide dismutase (an oxygen "detoxifying" enzyme) has been detected in methanogens,[3] and other oxygen-adapting phenomenon have been reported for methanogens.[22]

## Bench-Scale Batch Reactor Studies

The four bench-scale reactors were maintained for over seven months under stable conditions. Initially, all cultures were black. After a month of operation, the strictly anaerobic culture remained black, whereas the limited oxygen and aerobic cultures were brown and light brown, indicating less reduction of sulfate and iron to ferrous sulfide. The anaerobic reactor headspace gas contained approximately 47% methane, and methane was detected in the off-gas from both oxygen-limited reactors.

Soluble COD removal efficiencies of approximately 94% were exhibited by all systems at an OLR of 1.5 g COD/$L_R$-day (influent COD = 30,000 mg/L). The limited-aeration conditions studied did not result in reduced COD removal as compared to the anaerobic or aerobic systems. In addition, the oxygen-limited reactor receiving 1 g $O_2/L_R$-day exhibited the lowest residual COD concentration of 1,400 mg/L (Figure 7), whereas the aerobic reactor exhibited the highest COD concentration of 2,400 mg/L (Figure 8), probably in part due to release of soluble microbial products (SMP). Others have reported rapid COD uptake in the first 2 to 3 hours of aerobic batch culture reactions, followed by a slow increase in soluble COD due to SMP formation.[23]

Initial COD removal kinetics were approximately zero-order for all systems (Figures 5, 6, 7, and 8). Rates were estimated using linear regression based upon data from the first 1.5 hours of

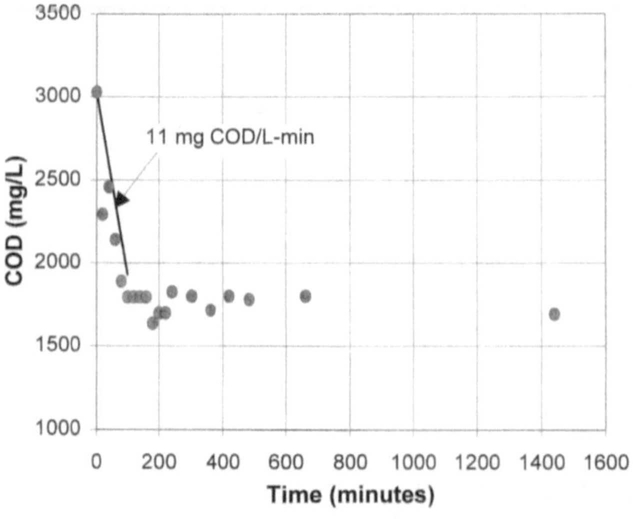


**Figure 5. COD vs. time in anaerobic bench-scale reactor.**

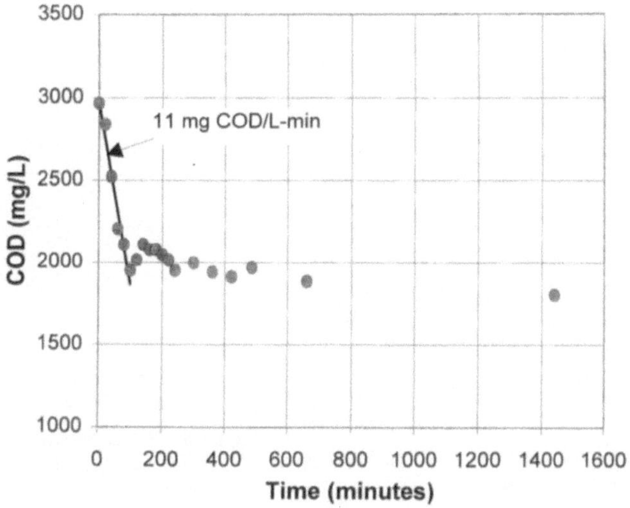


**Figure 6. COD vs. time in $0.1g\ O_2/LR$-day oxygen-limited bench-scale reactor.**

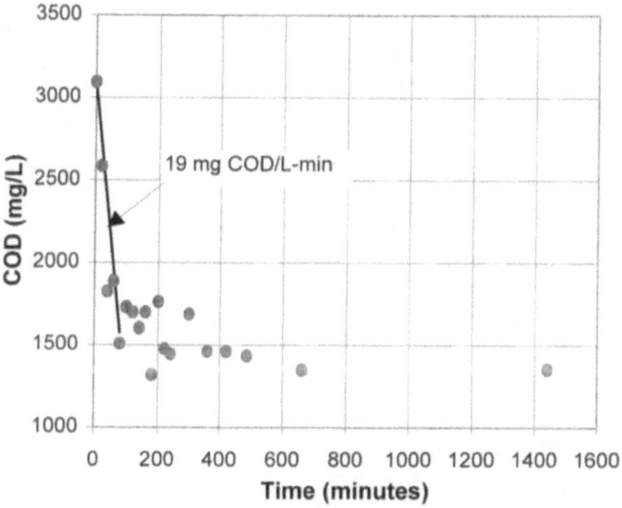


**Figure 7. COD vs. time in $1.0g\ O_2/LR$-day oxygen-limited bench-scale reactor.**

reaction. When compared to strictly anaerobic and aerobic systems, limited oxygen addition did not cause a decrease in the initial rate of COD removal. The 1 g $O_2/L_R$-day-oxygen-limited reactor exhibited the highest initial rate of 19 mg COD/L-minute, whereas the strictly anaerobic culture removal rate was 11 mg COD/L-minute. The aerobic and 0.1 g $O_2/L_R$-day-oxygen-limited cultures exhibited initial rates of 9 and 11 mg COD/L-minute, respectively. Increased rates of methane production under limited aeration conditions is not without precedent. Others have found that the addition of 4 mg $O_2/L_R$-day to essentially anaerobic cultures doubled methane production when algae was the primary substrate.[24]

The pH in reactors after the shock load of 4 g COD/$L_R$-day was administered is presented in Figure 9. The pH in all reactors decreased from approximately 7 to 5 after the shock load, and the pH of the strictly anaerobic reactor did not recover to the initial pH after 52 days. However, the 1 g $O_2/L_R$-day- and 0.1 g $O_2/L_R$-day-oxygen-limited reactors returned to the initial pH condi-

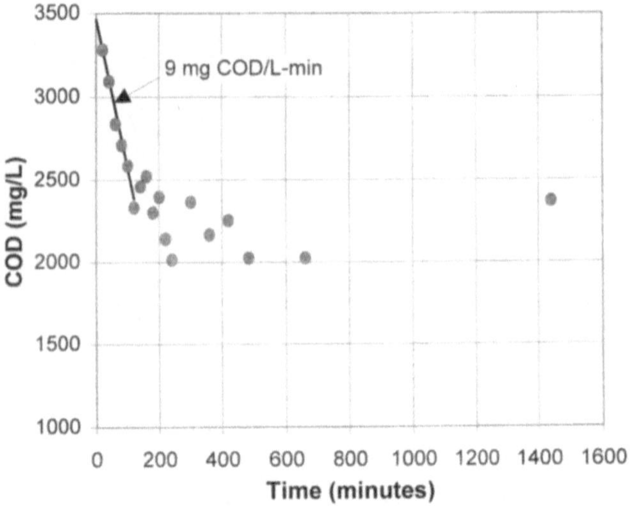


**Figure 8. COD vs. time in aerobic bench-scale reactor.**

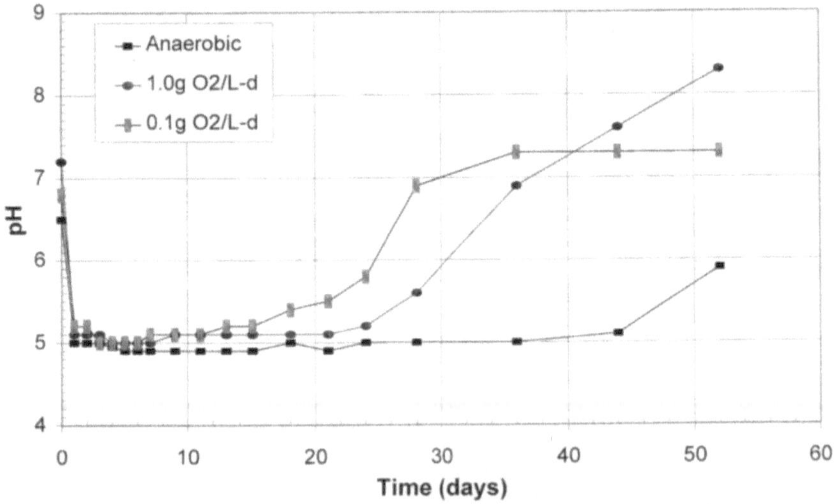


**Figure 9. The pH vs. time in bench-scale reactors after organic shock-load increase to 4 g COD/LR-day.**

tion after approximately 34 and 28 days, respectively. It is possible that aerobic (or microaerophilic) metabolism of volatile fatty acids and stripping of carbon dioxide and hydrogen ($H_2$) increased the rate of pH recovery in the systems under low aeration.

## CONCLUSIONS

Methanogenesis is sustainable in oxygen-limited cultures. Even though dissolved oxygen is present in bulk liquid at all times, methane may still be produced and methanogenic activity can be significant. Under the studied conditions, relative methanogenic activity was higher in some cultures that received oxygen than it was in parallel, strictly anaerobic cultures.

Low aeration rates do not necessarily cause a decrease in COD removal in essentially anaerobic processes. Overall COD removal efficiencies for oxygen-limited, strictly anaerobic, and strictly aerobic cultures were comparable under the complete-mix, suspended growth conditions investigated. In one case, effluent COD values were lower for a methanogenic system under low aeration. It may be possible to achieve the low COD values associated with aerobic biological processes while producing the minimal waste biomass associated with anaerobic systems. Other reactor configurations, such as UASBR and separate, staged acetogenic and methanogenic processes would likely be more efficient. Methanogenesis with limited oxygen under these conditions should be investigated.

The pH in the complete-mix methanogenic reactors recovered more quickly under low-aeration conditions. Similarly, low aeration rates may prevent pH decreases in other highly loaded systems since volatile acids are potentially oxidized and carbon dioxide and hydrogen are stripped out.

In the future, methanogenesis under limited-aeration may be employed as an energy-efficient treatment option to achieve low final COD concentrations, minimal biosolids generation, and mineralization of a broad range of organics. Anaerobic and aerobic are only two concepts in an infinite spectrum of possibilities.

## ACKNOWLEDGMENT

Financial support was provided by the Marquette University Water Quality Center and Department of Civil and Environmental Engineering. Laboratory guidance and some analyses were performed by Mr. Don Gamble, Water Quality Center Laboratory Supervisor, and Ms. Bindu George, Graduate Student. The authors thank Dr. R.E. Speece for his helpful suggestions during the preparation of this paper.

## REFERENCES

1. Huser, A.A., Wuhrmann, K., and Zehnder, A.J.B., "*Methanothrix soehngenii* gen. nov. sp. nov., a new acetotrophic non-hydrogen-oxidizing methane bacterium," *Arch. Microbiol.*, 132, pp. 1–9, 1982.
2. Kiener, A., and Leisinger, T., "Oxygen sensitivity of methanogenic bacteria," *Syst. Appl. Microbiol.*, 4, pp. 305–312, 1983.
3. Kirby, T.B., Lancaster, J.R., and Fridovich, I., "Isolation and characterization of the iron-containing superoxide dismutase of *Methanobacterium bryantii*," *Arch. Biochem. Biophys.*, 210, pp. 140–148, 1981.
4. Field, J., Stams, A., Kato, M., and Schraa, G., "Enhanced biodegradation of aromatic pollutants in cocultures of anaerobic and aerobic bacterial consortia," *Antonie van Leeuwenhoek*, 67, pp. 47–77, 1995.
5. Gerritse, J., Schut, F., and Gottschal, J.C., "Mixed chemostat culture of obligately aerobic and fermentative or methanogenic bacteria grown under oxygen-limiting conditions," *FEMS Microbiol. Lett.*, 66, pp. 87–94, 1990.
6. Gerritse, J., Schut, F., and Gottschal, J.C., "Modeling mixed chemostat cultures of and aerobic bacterium, *Comamonas testosteroni*, and an anaerobic bacterium, *Veillonella alcalescens*: Comparison with experimental data," *Appl. Environ. Microbiol.*, 58(5), pp. 1466–1476, 1992.
7. Guyot, J.P., Ferrer, H., and Florina, R., "Methane production from acetamide in an upflow anaerobic sludge-blanket reactor based on a synergistic association between an aerobic rod and methanogens," *Appl. Microbiol. Biotechnol.*, 43, pp. 1107–1111, 1995.
8. Zitomer, D.H., "Stoichiometry of Combined Aerobic and Methanogenic COD Transformation Accomplished by a Single Mixed Culture," manuscript under review by *Water Res.*, 1996.
9. Kato, M.T., Field, J.A., and Lettinga, G., "High tolerance of methanogens in granular sludge to oxygen," *Biotechnol. Bioeng.*, 42, pp. 1360–1366, 1993.
10. Kurosawa, H. and Tanaka, H., "Advances in immobilised cell culture: Development of co-immobilised mixed culture system of aerobic and anaerobic micro-organisms," *Process Biochem. Int.*, December, pp. 189–196, 1990.
11. Speece, R.E., *Anaerobic Biotechnology for Industrial Wastewaters*, Archae Press, Nashville, TN, 1996.

12. Lens, P.N., DePoorter, M.-P., Cronenberg, C.C., and Verstraete, W.H., "Sulfate reducing and methane producing bacteria in aerobic wastewater treatment systems," *Water Res.*, 29(3), pp. 871–880, 1995.
13. Wu, W., Hu, J., Gu, X., Zhao, Y., Zhang, H., and Gu, G., "Cultivation of anaerobic granular sludge in UASB reactors with aerobic activated sludge seed," *Water Res.*, 21, pp. 789–799, 1987.
14. Zitomer, D.H., and Speece, R.E., "Sequential environments for biotransformation of aqueous contaminants," *ES&T*, 27(2), pp. 227–244, 1993.
15. Oldenhuis, R., Vink, R.L.J.M., Janssen, D.B., and Witholt, B., "Degradation of chlorinated aliphatic hydrocarbons by *Methylosinus trichosporium* OB3b expressing soluble methane monooxygenase," *Appl. Environ. Microbiol.*, 55(11), pp. 2819–2826, 1989.
16. Fogel, S., Lancione, R.L., and Sewall, A.E., "Enhanced biodegradation of methoxychlor in soil under sequential environmental conditions," *Appl. Environ. Microbiol.*, 44(1), pp. 113–120, 1982.
17. Anid, P.V. and Vogel, T.M., "Sequential Anaerobic-Aerobic Biodegradation of PCBs in the River Model," In: Proceedings, *On-Site Bioreclamation-Processes for Xenobiotic and Hydrocarbon Treatment*, R.E. Hinchee and R.F. Olfenbuttel, Eds., Butterworth-Heinemann, Boston, MA, pp. 428–436, 1991.
18. Kastner, M., "Reductive dechlorination of tri- and tetrachloroethylene depends on from aerobic to anaerobic conditions," *Appl. Environ. Microbiol.*, 57(7), pp. 2039–2046, 1991.
19. *Standard Methods for the Examination of Water and Wastewater*, 19th. ed., APHA (American Public Health Assoc.), AWWA (American Water Works Assoc.), and WEF (Water Environment Federation), 1995.
20. Roberton, A.M. and Wolfe, R.S., "Adenosine triphosphate pools in *Methanobacterium*," *J. Bacteriol.*, 102(1), pp. 43–51, 1970.
21. Rudd, J.W.M. and Taylor, C.D., "Methane cycling in aquatic environments," *Adv. Aquatic Microbiol.*, 2, pp. 77–149, 1980.
22. Zinder, S.H., Physiological Ecology of Methanogens, In: *Methanogenesis*, J.G. Ferry, Ed., Chapman and Hall, New York, 1993, p. 136.
23. Pribyl, M., Tucek, F., Wilderer, P.A., and Wanner, J., "Amount and nature of soluble refractory organics produced by activated sludge micro-organisms in sequencing batch and continuous flow reactors," *Water Sci. Technol.*, 35(1), pp. 27–34, 1997.
24. Pirt, S.J. and Lee, Y. K., "Enhancement of methanogenesis by traces of oxygen in bacterial digestion of biomass," *FEMS Microbiol. Lett.*, 18, pp. 61–63, 1983.

# 30 TREATABILITY OF TWO DIFFERENT INDUSTRIAL WASTEWATER STREAMS USING THE ANAEROBIC MIGRATING BLANKET REACTOR

**James J. Flamming**, Graduate Research Assistant
**Largus T. Angenent**, Graduate Research Assistant
**Shihwu Sung**, Adjunct Assistant Professor
**Richard R. Dague**, Professor (Deceased)
Iowa State University, Ames Iowa 50011

## ABSTRACT

Treatability studies were conducted on two different industrial waste streams using a new compartmentalized, high-rate, anaerobic reactor system known as the anaerobic migrating blanket reactor (AMBR). During the first experimental run, studies were conducted on a 4:1 mixture of cardboard recycling plant wastewater and industrial grade bioenzyme plant wastewater. Later, a second experimental run was made using the cardboard recycling plant wastewater by itself. Soluble chemical oxygen demand (SCOD) removals of 78% were achieved at a COD loading rate of 8 g/L/day while treating the combined waste stream, and 72% at a COD loading rate of 10 g/L/day while treating solely the cardboard recycling wastewater. Additional studies suggest that compartmentalized reactors such as the AMBR can provide additional sites for the metabolism of higher energy organics that accumulate during shock-loading conditions.

## INTRODUCTION

Most municipalities utilize aerobic biological processes in the treatment of large, relatively dilute combinations of domestic and industrial wastewater streams. Anaerobic treatment used is usually in the form of digesters for final stabilization of resulting sludge streams. However, the treatment of industrial waste streams by the same methods can become costly. In general, industrial wastewaters are higher in concentration than domestic, or domestic/industrial blends that occur at municipal wastewater treatment plants. This increased organic concentration results in increased aeration requirements, increased reactor size, and increased production of waste sludges, which themselves require further treatment and stabilization. In addition, industrial waste streams can vary greatly with respect to waste strength and quantity of flow. These fluctuations can also result in large variations in the oxygen requirements and sludge production of the aerobic systems. Fluctuations can occur rapidly over a short period of time or slowly over an entire season, and depend on the productivity and use of raw materials by a particular industry.

In some cases, biological alternatives must be discarded altogether due to the inhibitory or even toxic nature of the waste. In such cases, physical/chemical treatment methods may be used, but many times these can be quite costly to operate. Thus, many industries simply decide not to investigate industrial pretreatment methods at all, but rather, opt to pay high sewage use fees. As a result, the municipality becomes responsible for the ultimate stabilization of the waste. By discharging their waste to the sewers, the strength of these streams becomes diluted, and the effects of high strength/high variability/inhibitory constituents are minimized.

While many industries feel forced to live by this method, cost-effective wastewater treatment alternatives do exist in the form of anaerobic biological systems. These systems can undergo high organic loading rates typical of industrial waste streams without large increases in costs. In fact, the higher the strength of the waste stream, the more potential profit is available due to

*52nd Purdue Industrial Waste Conference Proceedings*, 1997, Ann Arbor Press, Chelsea, Michigan 48118. Printed in U.S.A.

methane generation. Anaerobic systems are one of the few pretreatment alternatives with the potential to pay for themselves over time. In recent years, many of the presumed disadvantages of these systems, such as process instability, and slow startup, have been overcome by increased amounts of research, operational know-how, and implementation of new techniques. High-rate systems, such as the widely used upflow anaerobic sludge blanket (UASB),[1] and more recently, the anaerobic sequencing batch reactor (ASBR),[2] along with others, have opened new doors to the application of anaerobic technologies. High-rate systems in general can take higher loadings and higher degrees of variability due to the high level of biomass retention. While understanding that these systems usually do not provide complete treatment, experts feel there are very few drawbacks of anaerobic systems relative to other treatment methods,[3,4] and that these methods must be considered as innovative technologies.[5]

Anaerobic biodegradation consists of three steps, each governed by a different trophic group. The first step is the hydrolytic fermentation step, where complex organics are fermented by acidogenic bacteria into simpler organic acids. These acids are then used by syntrophic acetogenic bacteria to form $H_2$, $CO_2$, acetate, and formate. Methane is then formed from $H_2$ and $CO_2$ via reductive methane formation, from acetate via a decarboxylation, or from formate degradation. Approximately 70% of the methane produced in anaerobic digesters originates from acetoclastic methanogenesis. In cases where sulfate levels are high, sulfate reducing bacteria (SRB) can coexist with or outcompete the methanogens in the terminal reaction.[6]

The way these three biodegradation steps have been utilized in digesters has varied. The majority of anaerobic treatment systems have either consisted of a single tank, multiple trophic group design (such as the UASB and ASBR) or a dual tank, phase-separated design, where the hydrolytic step is separated from the acetogenic/methanogenic steps. Recently, a new generation of plug-flow, compartmentalized anaerobic reactors have been developed that can make use of what Fox and Pohland[7] described as partial phase separation. Lettinga[3] referred to these new systems as staged anaerobic reactor systems. Some of the reactor designs that could be described as staged reactors include: the Anaerobic Baffled Reactor,[8] the Multiplate Anaerobic Reactor,[9] the Upflow Staged Sludge Bed Reactor,[10] the Plate-Column Reactor,[11] the Rotating-Stationary Fixed-Film Reactor,[12] the Internal Circulation Reactor,[13] the Biogas Tower Reactor,[14] and some designs of the Anaerobic Rotating Biological Contactor.[15] Some authors have developed modifications to the ABR design.[16,17] Other variations of compartmentalized or baffled reactors are also found in the literature,[18–20] as well as reversing flow compartmentalized systems.[21,22]

While phase separation is sometimes promoted in plug flow reactors,[7] compartmentalized reactors can be operated so that in each of the compartments all phases of the anaerobic degradation process remain present. In this way, each compartment can develop a different consortia of bacteria to handle the different substrate and environmental conditions that exist within that compartment. Systems such as these can provide higher treatment efficiencies because intermediates that are difficult to degrade, such as propionate, will be degraded in compartments that develop optimum environments for the required bacteria. These systems can also have a higher process stability.[3]

Angenent and Dague[23] and coworkers at Iowa State University have been developing the anaerobic migrating blanket reactor (AMBR), a high-rate, compartmentalized reactor configuration that receives wastewater on a continuous flow basis. The liquid enters the initial compartment and is mixed with the biomass through the intermittent use of mechanical mixers. Next, the liquid flows between a set of baffles into the adjacent compartment where it once again comes in contact with biomass. This continues until the liquid reaches the final compartment and eventually flows out through the effluent baffle system (see Figure 1). As the liquid flows through the reactor, biomass migrates with it. Flocculent biomass migrates through the reactor faster than granular biomass and thus a selection process for better settleable granular biomass occurs. As the liquid continues to be fed to the initial compartment, the volatile fatty acid (VFA) concentration increases, which causes the pH in the initial compartment to decline. In addition, if the granular biomass is allowed to migrate through the system, the concentration of biomass in the initial compartment also decreases. Meanwhile, the concentration of VFAs in the final compartment remains low. This results in low amounts of biogas production and allows the final compartment to act as a clarifier. Eventually, a point is reached when the pH decreases to 6.5 or the biomass approaches a limiting concentration in the initial compartment. The flow is then step fed to the next compartment or is reversed entirely. In this manner, the wastewater always comes in contact with a high population of biomass in an environment that does not allow total phase separation. Reversing the flow direction also allows the AMBR to operate without an effluent recycle stream, which helps to promote plug-flow conditions in the reactor. Furthermore, the reversing flow scheme allows the compartments of an AMBR to work under feast/famine conditions.[23]

To this point, initial studies and granulation studies have been conducted using the AMBR.[23,24] The initial studies have shown that the AMBR could achieve 99% soluble COD removal at a load of 23 gCOD/L/d and a hydraulic retention time (HRT) of 12 hours when fed nonacidified sucrose as a synthetic waste stream. Another AMBR was able to maintain high levels of biomass (16 g MLVSS/L) at high COD loading rates up to 30 g/L/d. The AMBR has been found to successfully produce, maintain, and grow granular biomass. Unlike many of the compartmentalized systems that depend on an upflow velocity for selection and granulation, the AMBR operates on a flow pattern that is horizontal overall. The results shown in these studies support the findings of Wirtz and Dague[25] and Vanderhaegen et al.[26] that an upflow velocity is not needed for granulation to occur. The study that is described below is the first application of the AMBR to a nonsynthetic waste stream.

## GENENCOR INTERNATIONAL AND CEDAR RIVER PAPER COMPANY

The City of Cedar Rapids, Iowa, approached the Biotechnology Application Group at Iowa State University concerning investigations into the anaerobic pretreatment of two industrial waste streams. One of the companies, Genencor International, is an industrial grade bioenzyme producer that produces mostly cellulases, proteases, pectinase and lipase enzymes for the food processing industry. The plant discharges approximately 0.5 million gallons per day (MGD) of wastewater, which varies dramatically in its characteristics, and has been known to have nitrogen levels higher than those desired for anaerobic treatment. This waste stream was originally the subject of a treatability study conducted by Fernandez-Garcia[27] which utilized an ASBR operating at a 36-hr HRT and COD loading rates up to 3 g/L/d. This study met with limited success however, primarily due to the large variation in waste characteristics and the possibility of inhibitory constituents.

Another factory nearby, Cedar River Paper Company (CRPC), was built shortly after Genencor, and produces corrugated medium and linerboard from recycled corrugated containerboard and mixed waste paper. The factory discharges over 2.0 MGD of wastewater, which is relatively consistent in its characteristics, but tends to have lower than desired nitrogen levels. The City felt that, when mixed, the larger amount of CRPC water would level out the variability of the Genencor waste and dilute any inhibitory constituents, while Genencor's waste could provide adequate nitrogen levels for the treatment of both waste streams.

## SCOPE AND OBJECTIVES

This study is an extension of the work done by Fernandez-Garcia.[27] In this case, though, treatability studies were run on two different waste streams, CRPC's wastewater, and a 4:1 mixture of CRPC's and Genencor's wastewaters. The studies were conducted at 35° Celsius, using a 20 L AMBR.

Initially, two trial runs were conducted using the 4:1 mixture. These were done in order to optimize the reactor design. After this initial trial run period, the study was split into two experimental runs which correspond to the two different waste streams. The objectives of run 1 included: (1) Determine whether the 4:1 combined waste stream can be successfully pretreated; (2) Determine what the optimal operating conditions are for this system; and (3) Investigate any toxic or inhibitory characteristics of this waste stream. The objectives for run 2 were: (1) To determine the treatability of the CRPC wastewater; (2) Provide further insight into the fundamental performance characteristics of the AMBR; and (3) Determine the effects of hydraulic shockloads on the AMBR.

## MATERIALS AND METHODS

Figure 1 shows a schematic of the reactor. A four compartment AMBR was constructed out of 0.25 inch Lexan for this study. The reactor had a working volume of 20 L, and was operated in a constant temperature room at 35 ± 1°C. All the electrical equipment was controlled by programmable timers (ChronTrol Timer Controlers, ChronTrol Corporation, Model XT). Each compartment was connected to a separate peristaltic feed pump (Masterflex Peristaltic Pumps, Cole Palmer Instrument Company, Model 7520-35), and was mixed intermittently by a separate timer-

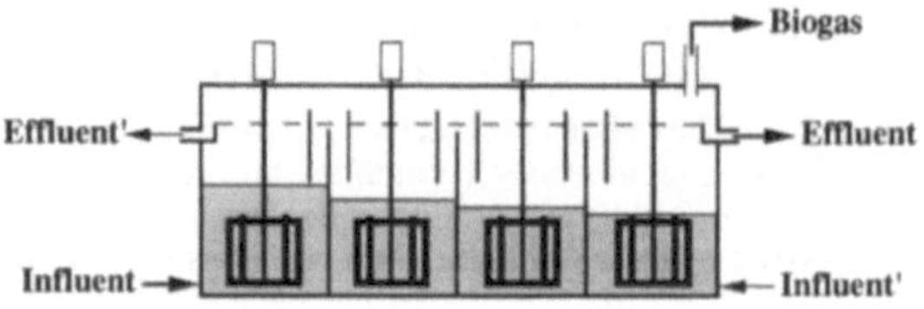


**Figure 1. AMBR schematic.**

controlled mixer (Eastern Mixers, EMI, Model 5VB-C). Paddles were used to ensure gentle mixing.

The effluent from the reactor flowed over the effluent weir into an effluent trough. Effluent then flowed through timer-controlled, ball valve actuators (True Blue Electric Actuators, Model EBV 65, BLDL, Inc., Totowa, NJ), to a clarifier. For all of run 1 and most of run 2 the effluent then flowed into a backwash holding tank, which held a reserve of effluent that could be used to backwash the effluent troughs. The troughs were backwashed four times a day for 15 seconds per backwash using peristaltic pumps (Masterflex Pump Controller, Cole Palmer Instrument Company, Model 7553-50). Due to the limited size of the lab scale reactor, backwashing was necessary to keep the effluent troughs clean and prevent plugging of the 3/8 inch inner diameter effluent ports. The total quantity of liquid used to backwash the troughs each day was less than 1% of the total daily flow, even at the highest HRT. Therefore, the amount of water reintroduced to the reactor was considered insignificant. Toward the end of run 2 the liquid backwashing was replaced with biogas recirculation. This allowed for more cleaning cycles per day, which cut down on plugging. Finally, the effluent exited the backwash holding tank through an overflow port and was directed to the sewer drain.

The resultant biogas flowed out of the reactor into a gas bag, which provided a gas reserve for when the pressure dropped in the reactor after the flow direction was changed. Then the biogas went through a observation bottle/water trap, passed through a gas sampling septum port, and into a hydrogen sulfide scrubbing bottle that was filled with steel wool. Finally, the biogas flowed through a Wet Test Meter (GCA/Precision Scientific) which recorded the total gas production, and then was vented through the laboratory hood.

Effluent samples were drawn at the midpoint between reversing the flow (6 hours into a 12-hour cycle) so as to be representative of overall reactor performance. Total and soluble COD, total VFA, alkalinity, total and volatile suspended solids (TSS and VSS), sulfate, and five-day biochemical oxygen demand ($BOD_5$) tests were conducted according to *Standard Methods*.[28] Ammonia and total Kjehldal nitrogen tests were conducted according to EPA Methods. Biogas composition was determined using gas chromatography (Gow Mac Model 350 with thermal conductivity detector; column pack stainless steel Poropak Q 80/100 mesh, length = 6 ft; diameter = 1/8 in., temperature = ambient; injector port temperature = 100°C; detector temperature = 150°C; outlet temperature = 70°C; helium carrier gas, flowrate = 60 mL/min; sample volume = 0.9 mL). $H_2S$ concentration in the gas stream was determined using Drager Tubes (Dragerwerk Aktiengesellschaft, Model CH 28101). Total sulfide levels in the aqueous phase were determined using the method described in Truper and Schlegel[29] and a spectrophotometer at 660 nm (Spectronic Spectrophotometer, Model 501). Specific methanogenic activity (SMA) tests were conducted as described by Rinzema et al.[30] The scanning electron microscope (SEM) was used to describe and characterize the granular biomass (as described in Banik et al.[31]).

As stated previously, the data presented below were collected over two trial runs and two experimental runs. Since a number of changes were being made to the reactor design during the two trial runs, a smaller number of operating parameters were monitored. After the necessary modifications were made in the reactor design, a larger number of operating parameters was monitored.

Waste samples were collected from the factories and transported back to the Biotechnology Applications Laboratory at Iowa State University. During the two trial runs and run 1, the waste samples were mixed at a 4:1 ratio of CRPC and Genencor wastes, respectively. For experimental run 2, only CRPC waste was fed to the reactor. The samples were stored over the long term (maximum 14 days) and while being fed into the reactor in refrigeration units that kept the sample at 4°C ± 1°. Sodium bicarbonate alkalinity was added as needed to the waste samples to keep the pH in the initial compartment above 6.5. In addition, nutrients had to be added to the samples during run 2. A nutrient stock solution was made and added to the samples at varied levels in order to determine the required uptake (primarily for ammonia). The nutrient stock solution was made of 290 mL/L of $NH_4OH$ and 22.6 g/L of $K_2HPO_4$.

For trial run 1, the AMBR was seeded with mesophilic granules which were collected from an UASB reactor in Lacrosse, Wisconsin, that is treating brewery wastewater. During both trial runs the reactor plugged occasionally due to problems in the lab-scale design, and granules washed out of the reactor into the clarifier. When this happened, the granules were collected and stored in a refrigerator. Eventually, these granules were used to reseed the AMBR for trial run 2 and experimental run 1. For experimental run 2, the reactor was reseeded with granules collected from the UASB reactor in Lacrosse. However, these granules required grinding with magnetic stir bars and continuously mixing in the AMBR at high rpms. This breakup of granules was needed to prevent floating problems due to the sudden change in waste streams.

## TREATABILITY RESULTS

Table I shows some of the influent characteristics of the individual waste streams and of the 4:1 combined waste stream. The waste stream from Genencor was very difficult to characterize, as can be seen by the large range of values shown for most of the parameters. Genencor utilized several batch processes in the manufacture of its products. This resulted in a large degree of variability in effluent quality. On the other hand, the CRPC waste stream was relatively consistent in its effluent characteristics. By mixing these two waste streams at a 4:1 ratio, the high degree of variability inherent to the Genencor waste stream evened out, and the low levels of nitrogen inherent to the CRPC waste stream were raised.

The results of both the trial runs and runs 1 and 2 are shown in Figures 2 to 7. While the number of analyses conducted during the trial runs was not as extensive as during experimental runs 1 and 2, the results which are presented in the figures below are significant to the treatability study. The time scale which is shown at the bottom of Figures 2 to 5 is an absolute time period for the entire study (the four different runs were made in sequential order).

Figure 2 shows the variation in COD removal and effluent total VFA concentration during each of the runs. It should be noted that the $BOD_5$/COD ratio for the effluent in all four runs was approximately 0.30. While changes in the reactor design were being made during trial runs 1 and 2 to avoid plugging problems, significant results were obtained. During trial run 1, the system performed well for the first 40 days, during which the system achieved total COD (TCOD) removals of approximately 60% and soluble COD (SCOD) removals in excess of 80%. Furthermore, Figure 2 shows that the calculated TCOD removal, which was determined from the methane production as described in Angenent and Dague,[23] ranged from approximately 30 to 50%. The remaining COD removal was assumed to be the result of sulfate reduction. Table I shows that the influent waste streams varied in sulfate concentration between 500 mg/L and 1300 mg/L. Moreover, $H_2S$ concentrations in the biogas stream also indicated a significant degree of COD removal due to SRBs. $H_2S$ consistently accounted for 0.3 to 1.5% of the total biogas

**Table I. Wastewater Characteristics of Individual Waste Streams and 4:1 Mixture**

| Waste Characteristics | Units | Genencor | CRPC | 4:1 mixture |
|---|---|---|---|---|
| Plant Flow | MGD | 0.3–0.7 | 2.0–3.5 | 2.0–4.0[a] |
| COD | g/L | 1.0–11.0 | 2.8–4.2 | 2.5–5.0 |
| Average COD | g/L | 5.6 | 3.4 | 3.7 |
| Standard deviation, $\sigma$ | g/L | 2.8 | 0.6 | 0.7 |
| TSS | mg/L | 200–7,000 | 200–500 | 200–1,000 |
| VSS | mg/L | 150–5,500 | 150–350 | 100–1,000 |
| Sulfate | mg/L as $SO_4^{-2}$ | 700–1,300 | 500–850 | 500–750 |
| TKN | mg/L | 500–850 | 5–10 | 200–300 |
| NH3 | mg/L as N | 100–1,200 | > 1 | 25–200 |
| Alkalinity | mg/L as $CaCO_3$ | 200–1,700 | 200–300 | 150–400 |
| Runs used in | N. A. | Fernandez-Garcia[27] | Run 2 | Trial runs 1&2, and Run 1 |

[a]Estimated value.

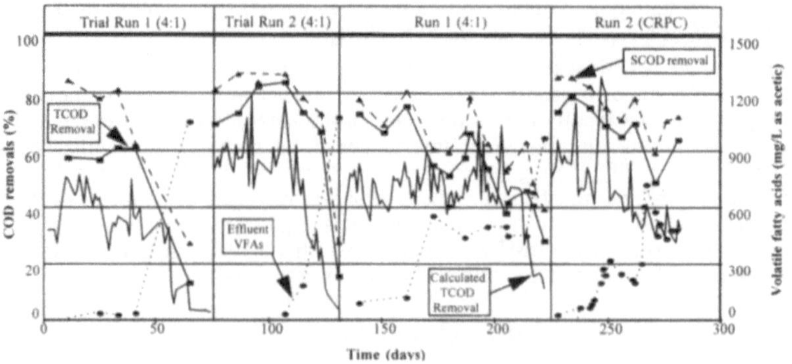


**Figure 2. Variation of COD removal and effluent total VFA concentration during each of the runs.**

stream in all four of the runs, with the majority of the readings being greater than or equal to 1.0%. During run 2, the sulfate removal rate was measured at 97% at an eight-hour HRT.

After the first 40 days of operation, however, the system failed rapidly. The effluent VFA concentration rose from less than 100 mg/L to over 1000 mg/L in a period of approximately 15 days. It was suspected that some form of inhibitory compound was present in the waste stream and tests were conducted in accordance with recommendations made by Fernandez-Garcia.[27] Nevertheless, nothing could be concluded from the results as to the identity of the inhibitor.

During trial run 2, overall system removals improved to a point where the TCOD and SCOD removals both exceeded 80%. Calculated TCOD removal hovered at approximately 60%, and the system performed relatively well, until a sample containing very high levels of total suspended solids (approximately 7 g/L TSS) was obtained from Genencor. It was initially thought that the dilution resulting from the mixture with CRPC wastewater would prevent any major problems from occurring within the reactor. Mixing the waste streams did result in a lower TSS level (around 1000 mg/L), but this was not enough. The refractory solids were retained in the reactor, which soon became completely filled with these solids. In spite of this, total and soluble COD removals both remained above 70%, but gas production decreased and mixing of the biomass was impossible. Even after replacing the waste stream, the solids did not leave the reactor. Attempts were made to drain out the solids manually, but this resulted in the loss of active biomass. The VFA levels began to rise and once again the reactor failed.

Once the changes in the reactor design were finalized, a more intensive effort was made to monitor system performance and achieve optimum loading conditions. Run 1 did not achieve as high TCOD and SCOD removals as trial run 2 did, but TCOD and SCOD removals did exceed 75% and 80%, respectively. SMA results also showed that average activity of the biomass increased from an initial level of 0.10 gCOD/gVSS/d to a high of 0.41 gCOD/gVSS/d at day 150. At around day 170, the reactor started to exhibit results similar to those observed during trial run 1. The TCOD removal decreased to 50% which correlated with the removal attributable to methanogenesis only. The effluent VFA levels also rose from less than 100 mg/L to over 500 mg/L. System COD removals did recover in about 15 days to 65% TCOD removal and 79% SCOD removal, but shortly after that they dropped again and leveled off at around 40% TCOD removal before eventually dropping below 30% TCOD removal. Furthermore, VFA levels rose in excess of 1000 mg/L, and calculated TCOD removal decreased to 10%. SMA analysis showed the average specific methanogenic activity to be less than 0.10 gCOD/gVSS/d. Once again, tests were run to determine what caused the inhibition, but these proved inconclusive. Ammonia levels in the last Genencor sample rose over an order of magnitude from what was typically seen (from about 100 mg/L as N to over 1000 mg/L as N), but according to Bhattacharya and Parkin,[32] this was not inhibitory.

The remainder of the studies focused on the treatment of the CRPC waste stream alone, due to unknown inhibition of the Genencor waste and the limited success of previous treatability studies using this waste.[27] The last part of Figure 2 shows the results of run 2, in which the AMBR

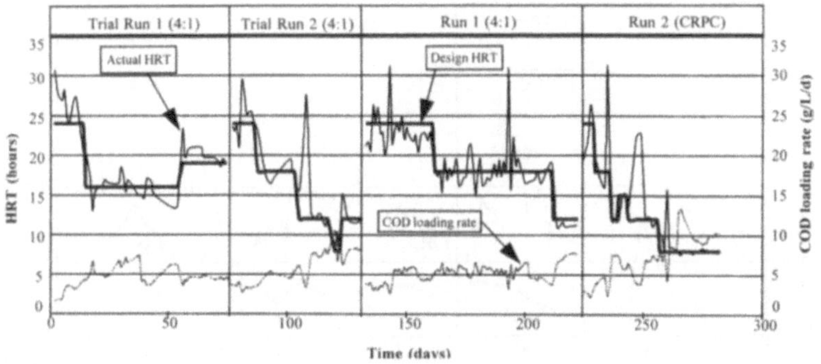


**Figure 3. Variation in design and actual HRT and COD loading rate during each of the runs.**

maintained SCOD removals in excess of 70% for the majority of the study. The initial SMA of the seed granules was 0.59 gCOD/gVSS/d. SCOD and TCOD removals temporarily declined around day 270, but as will be seen later, this was due to a spike in influent COD. The calculated TCOD removals paralleled the measured TCOD removals for the most part, and the increased effluent VFAs were a result of increased COD loading rates. At the end of run 2, the average SMA of the biomass was 0.45 gCOD/gVSS/d.

Figure 3 shows the variation in the design and actual HRTs and the COD loading rate with time for all four runs. For the most part, the lowest HRT achieved while treating the 4:1 waste stream mixture was 12 hours. An 8-hour HRT was attained for only a short period of time during trial run 2, but that was done only as an attempt to wash out some of the solids that had accumulated in the reactor. Had it not been for these accumulated solids, the reactor probably would have done well at an 8-hour HRT (based on COD removal data, see Figure 2), barring any inhibition. The COD loading rate was also higher during trial run 2, where loadings of approximately 9 g COD/L/d were maintained for some time. Hence, the AMBR maintained high removals (78% SCOD removal) at COD loadings as high as 8 g/L/d.

The AMBR showed significant removals while treating the CRPC waste stream alone. An 8-hour HRT was maintained for some time, which resulted in COD loading rates in excess of 10 g/L/d. At these loadings, the system still maintained SCOD removals greater than 70%, and at its highest COD load (approximately 14 g/L/d) the SCOD removals remained near 60%.

Mixed liquor concentrations and effluent volatile suspended solids (VSS) levels versus time are shown in Figure 4. The mixed liquor suspended solids (MLSS) and mixed liquor volatile sus-

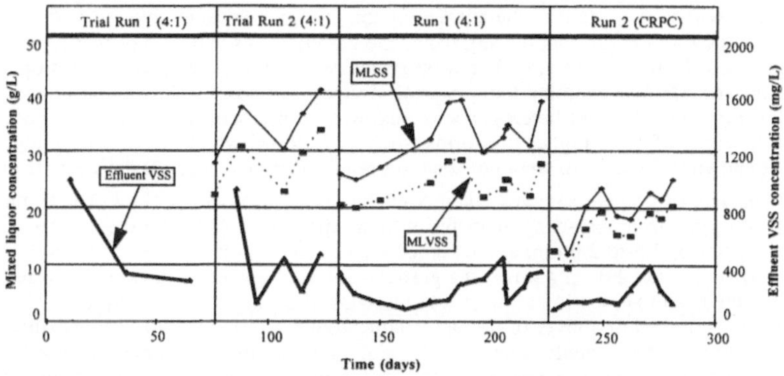


**Figure 4. Mixed liquor concentrations and effluent VSS levels during the individual runs.**

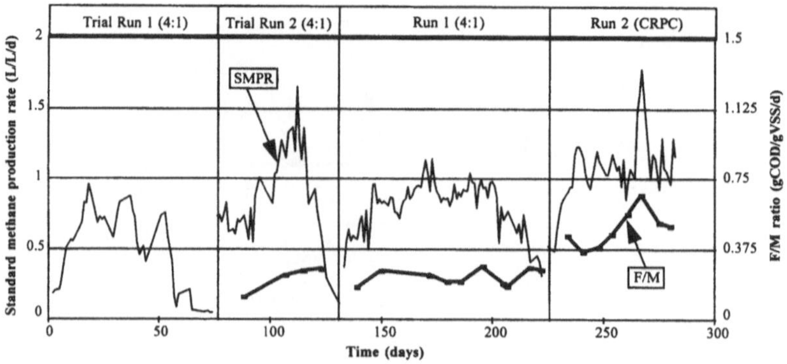


**Figure 5. Variation in standard methane production rate (SMPR) and F/M ratio during the individual runs.**

pended solids (MLVSS) concentrations were not monitored during trial run 1. Figure 4 shows an overall increase in MLSS and MLVSS throughout the other three experimental runs. This shows that the lab-scale AMBR system was capable of supporting and growing viable granular biomass, which supports similar findings of Angenent and Dague.[23] In this study, MLSS concentrations as high as 40 g/L and MLVSS levels as high as 34 g/L were obtained. This is because the effluent VSS levels remained low, usually less the 400 mg/L. However, higher values of effluent VSS were seen during startup.

Figure 4 does not show the mixed liquor levels for the end of trial run 2 when the reactor filled with refractory solids. However, it was observed at the time that the reactor was completely filled with solids. Run 2 was operated at a lower level of biomass (MLVSS < 20 g/L) to provide for better mixing, and to see how the AMBR would perform at lower mixed liquor levels. As was described previously, the AMBR performed well at these levels.

The variation of the standard methane production rate (SMPR) and the food-to-microorganism ratio (F/M) over time are shown in Figure 5. As stated above, the variation in F/M ratio during trial run 1 was not recorded because no mixed liquor suspended solids measurements were taken. The maximum F/M ratio in the reactor which treated the 4:1 mixture was 0.25 $d^{-1}$, while the F/M ratio of the reactor treating CRPC waste during run 2 was significantly higher (never below 0.35 $d^{-1}$). This reflects the lower MLVSS concentration and higher COD loading rates that were used during run 2, and is an indication of better performance. The higher SMPRs during trial run 2 and run 2 indicate the higher levels of organic degradation that were obtained by the methanogens. The SMPR during trial run 2 and run 2 was usually above 1.0 L/L/d (liters of methane at STP per liter of reactor volume per day).

SEM analysis of granules taken at the ends of runs 1 and 2 are shown in Figures 6 (a–f). Granules used during run 1 generally ranged in diameter from 1.0 to 1.5 mm throughout the duration of the run. Shortly after startup, the surface of the granules became covered with a grayish-tan coating. As shown in Figure 6b, the surface of the granules was made up primarily of rod and filamentous bacteria along with granule matrix structures. Granules were also sliced open and analyzed using the SEM to examine their interior structure. Figure 6c shows that granules taken at the end of run 1 consisted of heterogeneous mixtures of microcultures of rod and cocci bacteria, and the skeletons of other higher life forms, such as diatoms, which are a form of algae. The diatoms, as shown in Figure 6d, were not seen in granule samples taken 30 days earlier in run 1.

SEM analysis of granule samples taken over the duration of run 2, show that granule size increased with time, from a starting diameter of approximately 1.0 mm for the crushed granules, to a final diameter of 1.5 to 2.0 mm at the end of the run (see Figure 6e). As opposed to the heterogeneous structure found during run 1, the granules in run 2 tended to develop a layered structure. Figure 6f shows a cross section view of a granule taken during run 2 on day 253. The surface of the granule is covered in a layer of rod and filamentous bacteria extending approximately 50 um into the interior of the granule. The rest of the granule depth was composed of a homogenous mixture of rod and cocci bacteria and matrix material. The amount of rod and cocci bacteria in the interior of the granule appeared to increase over time.

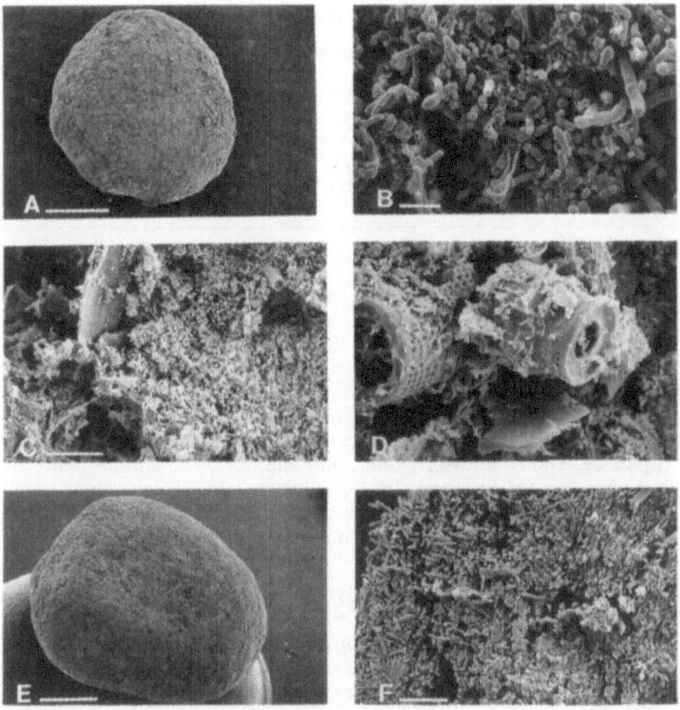


**Figure 6. SEM photographs of granule structures: (a) view showing size of granule taken at end of run 1 (bar equals 0.5 μm); (b) view of surface of granule taken at end of run 1 (bar equals 2 μm); (c) view of interior of granule taken at end of run 1 (bar equals 20 μm); (d) closeup view of diatom skeletons found in interior of granule taken at end of run 1 (bar equals 10 μm); (e) view showing size of granule taken at end of run 1 (bar equals 0.5 μm); (f) cross section view of layered structure in granule taken during run 2 (day 253, bar equals 20 μm).**

## PROFILES AND SHOCK-LOADING STUDIES

Analyses were carried out in addition to the treatability studies to provide additional information on how the AMBR process works. As wastewater passes through a continuous feed, anaerobic, compartmentalized reactor, a fraction of the total organic material is removed in each compartment. In this way, plug-flow kinetic profiles prevail over the length of the reactor,[23,25,33] Table II shows results similar to those shown by Angenent and Dague,[23] along with data on pH and dissolved hydrogen sulfide. Percent SCOD remaining are given for 18- and 8-hour HRTs and two different pH profiles taken at 8-hour HRTs are also given. The hydrogen sulfide concentration in the liquid was only measured at an 8-hour HRT.

In addition, individual VFA concentrations were determined within each of the four compartments over a 12-hour cycling time to show how these acids profiled within the reactor. This was done at an 8-hour HRT to determine a baseline from which other loadings could be compared. At time zero, the flow direction of the reactor had just been reversed and feed was introduced into compartment 1. This continued for 10 hours, after which the liquid was step-fed to compartment 2 for the last two hours of the cycle. A sample was also collected 24 hours later for comparison purposes.

Baseline VFA results are shown in Figure 7. Acetic acid was by far the most prevalent acid present in any of the compartments, followed by propionic acid and finally butyric acid. Valeric acid was also monitored, but its concentration never exceeded 15 mg/L, and thus it was not considered significant. Soon after the flow direction changed at time zero, acetate concentration rose

**Table II. Profiles of Percent SCOD Remaining, pH, and Hydrogen Sulfide During Run 2**

| Parameter | HRT | Influent | Compartment 1 | Compartment 2 | Compartment 3 | Compartment 4 |
|---|---|---|---|---|---|---|
| % SCOD remaining | 18 | 100 | 59.2 | 30.8 | 6.3 | 4.7 |
| % SCOD remaining | 8 | 100 | 59.8 | 45.5 | 37.2 | 29.7 |
| pH | 8 | 7.45 | 6.71 | 6.80 | 6.96 | 6.98 |
| pH | 8[a] | 7.30 | 6.60 | 6.70 | 6.84 | 6.86 |
| H2S in liquid (mg/L) | 8[a] | N.A. | 122 | 106 | 103 | 100 |

[a]Taken at the same time.

quickly in compartment 1. After two hours, the rate of production of acetate slowed, but still increased until a maximum was reached at 10 hours. It then decreased, as the flow was step-fed to compartment 2, and compartment 1 was given a chance to metabolize some of the high acid concentrations. Propionate, and to a lesser extent, butyrate, followed similar patterns. As the liquid moved through the reactor, the concentrations of all three volatile acids decreased in a plug-flow manner, similar to that shown in the previous works mentioned above. Samples taken at 24 hours showed that acid concentrations returned to similar levels as before, which indicate the system was at steady state conditions.

Figure 8 shows the results of a similar study conducted under shock-loaded conditions. The flowrate was increased by 50% for a period of 6 hours, going from an 8-hour HRT to a 5.3-hour HRT. This was done to simulate the variation of flow that can occur at the CRPC plant. With the exception of higher acid concentrations at time zero, the acid profiles in compartment 1 were almost identical to those seen during the baseline measurements. It was in compartments 2, 3, and 4 that a noticeable difference was noticed. The acetate concentrations in each of these compartments increased from the levels seen in the baseline analysis, but plug-flow conditions continued to prevail across the reactor. The 24-hour sample showed an increase in acid concentrations compared to the 0 hour sample, which would imply that the reactor had not yet completely recovered from the shock load after 18 hours of normal operation.

## DISCUSSION

The results did not show conclusively whether the AMBR process could successfully treat a 4:1 mixture of CRPC and Genencor wastewaters. Successful treatment was observed for some

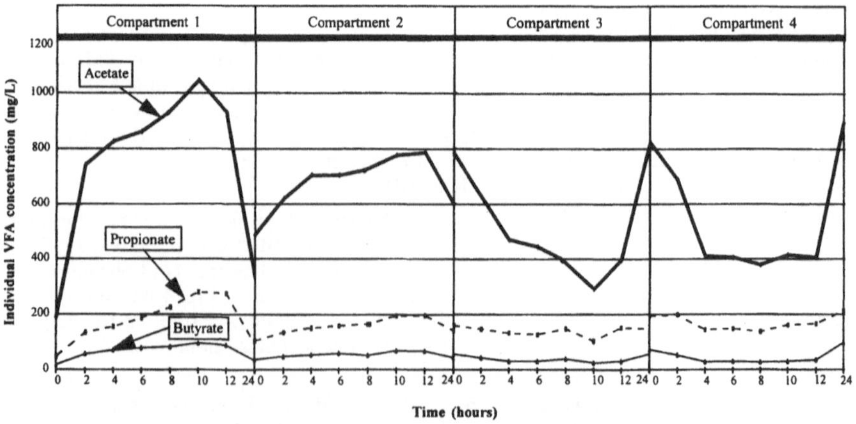


**Figure 7. Individual VFA concentrations within each compartmet over a 12-hour flow cycle at an 8-hour HRT.**

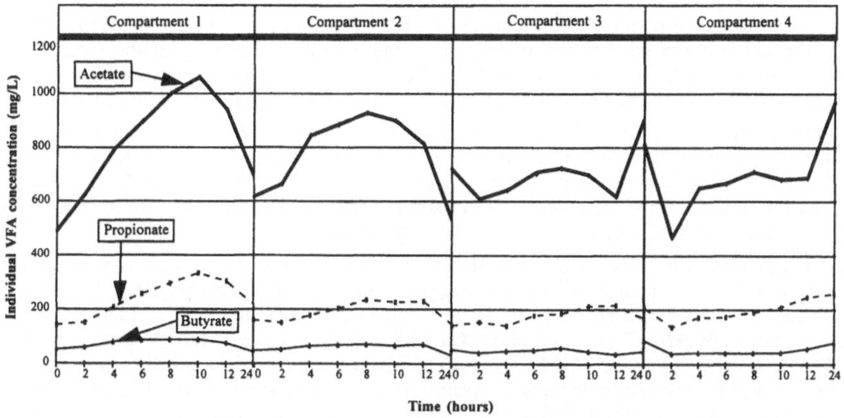


**Figure 8. Individual VFA levels in each compartment when exposed to a hydraulic shockload (duration of shockload: hour 0 to hour 6; intensity: immediate 50% increase in flow at hour 0, then immediate return to normal flow conditions at hour 6; resultant HRT: 8-hour HRT to 5.3-hour HRT, and then returning to an 8-hour HRT).**

time, but with the periodic inhibitory nature of the Genencor waste, long-term success of the reactor was not achieved. Mixing the waste streams at a 4:1 ratio limited the amount of variability seen in the waste, but did not dilute it enough to prevent potential inhibition. In addition, the variability of Genencor's waste made identifying the inhibitory constituents difficult. Successful treatment of the 4:1 mixture followed by rapid failure showed that the inhibitor was either periodically introduced to the waste stream, or was constantly present and only periodically occurred at inhibitory concentrations. Parkin and Speece[34] found that periodic exposure to inhibitory substances were handled by plug-flow anaerobic systems that separated solids retention time (SRT) and HRT and thus retained high levels of biomass. Consequently, this would apply to the AMBR system. Furthermore, Parkin et al.[35] showed that systems acclimated to inhibitory constituents through a variety of different mechanisms. However, in order to acclimate the biomass to the inhibitor, repeated exposures may be necessary. Thus, reactors that are being acclimated to inhibitory constituents must be carefully monitored and operated, to ensure that overexposure and failure does not occur.

When comparing the measured and the calculated COD removals, it was concluded that both SRB and methane producing bacteria (MPB) were responsible for degrading COD. The 97% sulfate removal observed during run 2 showed that SRB probably use organic carbon faster than MPB, up to a point where the sulfate concentration became limiting. A significant fraction of the COD remained and was utilized by the MPB.[36] CRPC wastewater was successfully treated at $H_2S$ concentrations as high as 122 mg/L.

The difference between the measured and calculated COD removals might give some insight into the nature of the inhibition observed. In run 1, the measured COD removal eventually dropped to levels that could be accounted for by the methane production. This might show that the SRB are inhibited first, which could help to limit the search for possible inhibitors. A review article prepared by Hao et al.[37] listed some of the possible inhibitors of SRB. Sulfide is known to be toxic to SRB at relatively low concentrations. However, run 2 had similar $H_2S$ levels in the gas stream compared to run 1 without experiencing similar inhibition. Therefore, sulfide was probably not the main inhibitor. However, a spike in sulfide levels in the reactor might have occurred, and Hilton and Oleszkiewicz[38] found that inhibitory levels of sulfide were lower for SRB compared to MPB. In addition, Maillacheruvu and Parkin[39] found that inhibition coefficients for acetate-utilizing SRB were much lower than those for aceticlastic MPB. Postgate[40] listed several SRB inhibitors including selenite ion, molybdate ion, microbicides and bacteriostatic substances, along with others.

Further investigation is needed into the identity of the inhibitor(s) which were present in the Genencor waste stream before application of pilot or full-scale treatment systems. As stated above, the ammonia concentration of the waste increased significantly in the final inhibitory sample of run 1. While the ammonia itself did not cause the inhibition, perhaps it could serve as a clue as to the origin of the toxicity and perhaps as a warning signal for potential inhibition.

Moreover, efforts are required to prevent spikes in suspended solids concentration from occurring in the Genecor waste. Solids levels as high as in trial run 2 cannot be allowed. In fact, for the lab scale AMBR, it was observed that solids levels in excess of 600 mg/L TSS caused rather rapid accumulation within the reactor. These solids settled out readily, and thus, clarification or some other form of solids separation would be very effective.

The treatment of CRPC wastewater at HRTs of 8 hours in the AMBR was successful in terms of stability and COD removal. Additional alkalinity and nutrients were required, but the alkalinity was added at an as-needed basis. At its most unstable conditions (while changing to a lower HRT), the lab scale AMBR required $NaHCO_3$ additions of 2.5 g/L of feed to keep the pH in the initial compartment above 6.5. As the reactor stabilized to new loading conditions, the amount of bicarbonate addition required decreased. Ammonia was added at a variety of levels to determine the amount of uptake required. Regardless of the amount of ammonia added to the sample, the uptake consistently amounted to between 20 and 30 mg/L as N.

However, some additional aspects need to be noted about the treatment of CRPC wastewater. As with the 4:1 mixture, the $H_2S$ levels in the biogas stream averaged in excess of 1.0%, and a similar grey-tan coating developed on the granule surface. This grey-tan coating might ultimately be a result of the fine solid particles within the CRPC waste. While having a relatively low level of TSS (< 250 mg/L), the solids present in the waste stream tended to form clumps of solids and granules, particularly in quiescent areas of the reactor (between baffles). Most of the solids were probably paper fibers, but part of these solids were soluble in hexane, which could imply that some of the solids might be glue originating from between the layers of the corrugated cardboard. The method used for processing SEM samples could have dissolved these glue particles and thus, they would not appear in the SEM photographs (Figures 6 a–f).

The granules used to treat the 4:1 mixture tended to develop a mixture of different bacteria throughout the granule structure, whereas the granules treating the CRPC waste stream alone, developed a layered structure. This is similar to the results shown by Guiot et al.,[41] who stated that granules with layered structures tended to form when treating primarily carbohydrate waste streams. The diatom skeletons which were observed at the end of run 1 probably did not grow inside the granules. Instead, they may have grown on the side of the clear, Lexan reactor, where light was available, and microniches of dissolved oxygen could be developed and used. They also could have entered the reactor in the waste stream. Aside from providing nuclei upon which granules could develop and grow, they probably played an insignificant role within the granular biomass.

Some interesting conclusions can be drawn from the profiles and shock-loading studies. Van Lier[42] theorized that hydrogen sulfide could be stripped out of waste streams in the initial compartment of staged reactors due to pH profiles that occur between compartments. This would allow higher sulfate wastewaters to be treated without concern for sulfide toxicity. This was theorized, but never verified. The results in this paper show only a small profile in $H_2S$ concentration in the liquid, but does not disprove the theory. If, at the time of the experiment, the difference in pH values between the compartments would have been larger, as it often was, the profile in $H_2S$ levels would have been more significant. At pHs around 6.5, the percentage of unionized $H_2S$, usually the most toxic form, is 75%, compared to less than 50% at pH values above 7.0.[38] In addition, if the headspace for each compartment would have been separate, higher levels of hydrogen in the initial compartment could have been removed from the reactor. As stated by Van Lier,[42] this would help to maintain pH profiles across the reactor, due to improved thermodynamic conditions for acidification in the initial compartment and acetogenesis in later compartments. Based on these results the theory seems possible, but requires further study to verify.

The individual VFA data showed that plug-flow conditions existed within the AMBR, similar to other compartmentalized systems, and that partial phase separation can occur in the AMBR. The extent of the staging is probably not as extensive as in unidirectional, compartmentalized systems, since the direction of flow is reversed and biomass is allowed to migrate through the reactor. This can be disadvantageous because specific populations of bacteria within the individual compartments might not as readily form as in other systems. On the other hand, this could be a significant advantage. Extensive staging has led to different populations of biomass in the individual compartments that have had different growth and decay rates. This difference could make maintaining the correct balance of staged bacterial populations difficult. Problems such as these have been encountered in unidirectional reactors, and withdrawal of biomass from initial compartments or periodic recycle of biomass from final to initial compartments might be required to maintain stable operation.[9,42] As stated previously, reversing the flow direction in the AMBR produces feast-famine conditions.[23] This allows for less extensive partial phase separation, while the reactor configuration still maintains the kinetic advantages of plug-flow.

Finally, the individual VFA data showed that under shock-loading the AMBR maintained plug-flow conditions, and that regardless of the loading conditions, the initial compartment carried out the bulk of the conversion of organic material. McCartey and Mosey[43] described the carbon and electron flow through an anaerobic system in an analogous "rain barrel model." When a reactor is exposed to an increase in flow over a short period of time, the equilibrium in the system shifts, so that some of the carbon and electron flow is channeled through intermediate acids, alcohols, and lactate, rather than the usual acetate, $CO_2$, and $H_2$. This was observed by comparing the baseline and shock-load data, in which the individual VFA concentrations in the last 3 compartments increased upon shock-loading. During shock-loading conditions, intermediate acids (and presumably other higher energy organics) formed in the initial compartment, and were later converted to acetate, $CO_2$, and $H_2$ in the following compartments. Thus, during shock-loading, the multiple compartments used in staged reactors such as the AMBR could provide additional sites for the metabolism of these higher energy organics. This is a major difference in comparison to single pass reactors which, with the exception of those that utilize recycle streams, have only one chance to metabolize the waste products. Ultimately, this is one of the major advantages of staged or compartmentalized systems.

## CONCLUSIONS

The following conclusions can be drawn from these studies:

1. The 20-liter AMBR achieved SCOD removals of 78% at a COD loading rate of 8 g/L/day and an HRT of 12 hours while treating a 4:1 mixture of cardboard recycling plant wastewater and industrial grade bioenzyme plant wastewater. However, inhibitory constituents in the Genencor waste stream were responsible for sudden failures in the reactor operation.
2. The 20-liter AMBR achieved SCOD removals of 72% at a COD loading rate of 10 g/L/day and an HRT of 8 hours while treating the cardboard recycling wastewater alone.
3. The lab-scale AMBR was able to maintain and grow granular biomass while treating these waste streams, obtaining high mixed liquor concentrations, even in excess of 35 g/L MLVSS. During run 2, biomass concentrations were limited to 25 g/L MLVSS for operating reasons.
4. Compartmentalized reactors such as the AMBR can provide additional sites for the metabolism of higher energy organics that accumulate during shock-loading conditions.

## ACKNOWLEDGMENTS

This research could not have been possible were it not for the vision of Dr. Richard R. Dague, who unfortunately passed away October 16, 1996. This paper is dedicated to his memory. This research was supported by a grant from the U.S. Department of Agriculture, contract number 91-34188-5943 through the Iowa Biotechnology Byproducts Consortium.

## REFERENCES

1. Lettinga, G., A.F.M. van Velsen, S.W. Hobma, W. de Zeeuw, and A. Klapwijk, "Use of the upflow sludge blanket (USB) reactor concept for biological wastewater treatment, especially for anaerobic treatment," *Biotechnol. Bioeng.*, 22, pp. 699–734, 1980.
2. Sung, S., and R.R. Dague, "Fundamental principles of the anaerobic sequencing batch reactor process," *Proceedings, 47th Purdue Industrial Waste Conference,* Lewis Publishers, Chelsea, MI, 1992, pp. 393–407.

3. Lettinga, G., "Anaerobic digestion and wastewater treatment systems," *Antonie van Leeuwenhoek*, 67, pp. 3–28, 1995.
4. Speece, R.E., Advances in Anaerobic Biotechnology for Industrial Wastewater Treatment, In: *Anaerobic Treatment of Industrial Wastewaters*, Torpy, M.F., Ed., Noyes Data Corporation, Park Ridge, NJ, 1988, pp. 1–6.
5. Switzenbaum, M.S., "Obstacles in the implementation of anaerobic treatment technology," *Bioresource Technol.*, 53, pp. 255–262, 1995.
6. Bhattacharya, S.K., V. Uberoi, and M.M. Dronamraju, "Interaction between acetate fed sulfate reducers and methanogens," *Water Res.*, 30 (10), pp. 2239–2246, 1996.
7. Fox, P., and F.G. Pohland, "Anaerobic treatment application and fundamentals: substrate specificity during phase separation," *Water Environ. Res.*, 66(5), pp. 716–724, 1994.
8. Bachmann A., V.L. Beard, and P.L. McCarty, "Comparison of fixed-film reactors with a modified sludge blanket reactor," *Proceedings of the First International Conference on Fixed-Film Biological Processes*, 2, pp. 1192–1211, 1982.
9. Guiot, S.R., B. Safi, J.C. Frigon, P. Mercier, C. Mulligan, R. Tremblay, and R. Samson, "Performance of a full-scale novel multiplate anaerobic reactor treating cheese whey effluent," *Biotechnol. Bioeng.*, 45, pp. 398–405, 1995.
10. Van Lier, J.B., F. Boersma, M.M.W.H. Debets, and G. Lettinga, "'High rate' thermophilic anaerobic wastewater treatment in compartmentalized upflow reactors," *Proceedings, Seventh International Symposium on Anaerobic Digestion*, Cape Town, South Africa, 1994.
11. Tanaka, K., E. Mikami, and T. Suzuki, "Anaerobic treatment of starch-containing wastewater using a plate-column reactor," *J. Ferment. Technol.*, 64(3), pp. 259–263, 1986.
12. Noyola, A., B. Capdeville, and H. Roques, "Anaerobic treatment of domestic sewage with a rotating-stationary fixed-film reactor," *Water Res.*, 22(12), pp. 1585–1592, 1988.
13. Hack, P.J.F.M., S.H.J. Vellinga, and L.H.A. Habets, Growth of Granular Sludge in the BIOPAQ IC-Reactor, In: *Granular Anaerobic Sludge; Microbiology and Technology, Proceedings of the GASMAT-Workshop*, G. Lettinga, A.J.G. Zehnder, J.T.C. Grotenhuis, and L.W. Holshoff Pol, Eds., Pudoc, Wageningen, The Netherlands, 1988, pp. 211–215.
14. Reinhold, G., S. Merrath, F. Lennemann, and H. Markl, "Modelling the hydrodynamics and the liquid-mixing behavior of a biogas tower reactor," *Chem. Eng. Sci.*, 51(17), pp. 4065–4073, 1996.
15. Tait, S.J. and A.A. Friedman, "Anaerobic rotating biological contactor for carbonaceous wastewaters," *J. Water Pollut. Control Fed.*, 52, pp. 2257–2269, 1981.
16. Boopathy, R. and A. Tilche, "Anaerobic digestion of high strength molasses wastewater using hybrid anaerobic baffled reactor," *Water Res.*, 25 (7), pp. 785–790, 1991.
17. Yu, H. and G.K. Anderson, "Performance of a combined anaerobic reactor for municipal wastewater treatment at ambient temperature," *Resour. Conserv. Recycling*, 17, pp. 259–271, 1996.
18. Stadlbauer, E.A., R. Achenbach, D. Doll, B. Jehle, B. Kufner, L. Oey, and J. Quurck, "Design and performance of pulsed anaerobic digesters," *Water Sci. Technol.*, 25(7), pp. 351–360, 1992.
19. Beeftink, H.H. and P. Staugaard, "Structure and dynamics of anaerobic bacterial aggregates in a gas-lift reactor," *Appl. Environ. Microbiol.*, Nov., pp. 1139–1146, 1986.
20. Yang, P.Y. and C.Y. Chou, "Horizontal-baffled anaerobic reactor for treating diluted swine wastewater," *Proceedings, 39th Purdue Industrial Waste Conference*, Butterworth Publishers, Stoneham, MA, 1984, pp. 585–597.
21. Basu, S.K., "Innovative wastewater treatment using reversing anaerobic upflow system (RAUS)," *Proceedings, 50th Purdue Industrial Waste Conference*, Lewis Publishers, Chelsea, MI, 1995, pp. 379–383.
22. Beyen, G.P., P. Delaplace, S. Ghekiere, M. Mertens, and L. Vriens, "The anaerobic-aerobic unitank system for advanced wastewater treatment: pilot- and full-scale experience on brewery wastewater," *Mededelingen van de Faculteit Landbouwwetenschappen: Rijksuniversiteit Gent*, 53 (4b), pp. 1927–1938, 1988.
23. Angenent, L.T. and R.R. Dague, "Initial studies on the anaerobic migrating blanket reactor," *Proceedings, 51st Purdue Industrial Waste Conference*, Ann Arbor Press, Inc., Chelsea, MI, 1996, pp. 271–288
24. Angenent, L.T., S. Sung, and R.R. Dague, "Granulation and start-up of the anaerobic migrating blanket reactor process," In: *8th International Conference on Anaerobic Digestion (AD-97) Poster Papers*, Sendai, Japan, May 25–29, 1997.

25. Wirtz, R.A. and R.R. Dague, "Enhancement of granulation and start-up in the anaerobic sequencing batch reactor," *Proceedings, 67th Annual Conference of the WEF,* Chicago, IL, 1994, pp. 241–252.
26. Vanderhaegen, B., E. Ysebaert, K. Favere, M. Van Wambeke, T. Peeters, V. Panic, V. Vandenlangenbergh, and W. Verstraete, "Acidogenesis in relation to in-reactor granule yield," *Water Sci. Technol.*, 25(7), pp. 21–30, 1992.
27. Fernandez-Garcia, S.D., Anaerobic Sequencing Batch Reactor Treatment of a Wastewater from Bioenzyme Production, Master's thesis, Iowa State University, Ames, IA, 1995.
28. *Standard Methods for the Examination of Water and Wastewater,* 19th ed., American Public Health Association, Washington, D.C., 1995.
29. Truper, H.G. and H.G. Schlegel, "Sulphur metabolism in Thiorhodaceae I. Quantitative measurements on growing cells of *Chromatium okenii,*" *Antonie van Leeuwenhoek,* 30, pp. 225–238, 1964.
30. Rinzema, A., J. van Lier, and G. Lettinga, "Sodium inhibition of acetoclastic methanogens in granular sludge from a UASB reactor," *Enzyme Microb. Technol.*, 10, pp. 24–32, 1988.
31. Banik, G.C., T.G. Ellis, and R.R. Dague, "Structure and methanogenic activity of granules from an ASBR treating dilute wastewater at low temperatures," In: *8th International Conference on Anaerobic Digestion (AD-97),* Sendai, Japan, May 25–29, 1997.
32. Bhattacharya, S.K. and G.F. Parkin, "The effect of ammonia on methane fermentation processes," J. Water Pollut. Control Fed., 61, pp. 55–59, 1989.
33. Bachmann, A., V.L. Beard, and P.L. McCarty, "Performance characteristics of the anaerobic baffled reactor," *Water Res.*, 19 (1), pp. 99–106, 1985.
34. Parkin, G.F. and R.E. Speece, "Anaerobic Biological Waste Treatment," *Chem. Eng. Prog.*, December, pp. 55–58, 1984.
35. Parkin, G.F., R.E. Speece, C.H.J. Yang, and W.M. Kocher, "Response of methane fermentation systems to industrial toxicants," *J. Water Pollut. Control Fed.,* 55(1), pp. 44–53, 1983.
36. Maillacheruvu, K.Y., G.F. Parkin, C.Y. Peng, W. Kuo, Z.I. Oonge, and V. Lebduschka, "Sulfide toxicity in anaerobic systems fed sulfate and various organics," *Water Environ. Res.*, 65(2), 1993, pp. 100–109.
37. Hao, O.J., J.M. Chen, L. Huang, and R.L. Buglass, "Sulfate-reducing bacteria," *Crit. Rev. Environ. Sci. Technol.*, 26(1), pp. 155–187, 1996.
38. Hilton, B.L. and J.A. Oleszkiewicz, "Toxicity of sulfides to the anaerobic treatment process," *Proceedings, 42nd Purdue Industrial Waste Conference,* Lewis Publishers, Chelsea, MI, 1987, pp. 739–747.
39. Maillacheruvu, K.Y. and G.F. Parkin, "Kinetics of growth, substrate utilization and sulfide toxicity for propionate, acetate, and hydrogen utilizers in anaerobic systems," *Water Environ. Res.*, 68(7), pp. 1099–1106, 1996.
40. Postgate, J.R., *The Sulphate Reducing Bacteria,* 2nd ed., Cambridge University Press, Cambridge, England, 1984.
41. Guiot, S.R., A. Pauss, and J.W. Costerton, "A structured model of the anaerobic granule consortium," *Water Sci. Technol.*, 25(7), pp. 1–10, 1992.
42. Van Lier, J.B., "Limitations of thermophilic anaerobic wastewater treatment and the consequences for process design," *Antonie van Leeuwenhoek,* 69, pp. 1–14, 1996.
43. McCarty, P.L. and Mosey, F.E., "Modelling of anaerobic digestion processes (A discussion of concepts)," *Water Sci. Technol.*, 24(8), pp. 17–33, 1991.

# 31 RECIRCULATION IMPACT BY INFLUENT METHANOGENIC SLUDGES DURING TWO-PHASE ANAEROBIC WASTE DIGESTION

**J. Hermana**, Research Associate

**G.K. Anderson**, Professor of Environmental Engineering
Department of Civil Engineering
University of Newcastle-upon-Tyne
NE1 7RU, United Kingdom

## ABSTRACT

An attempt to improve process efficiency was carried out by applying recirculation of methanogenic sludge to the influent of an acidogenic reactor during two-phase anaerobic digestion of a soluble organic wastewater. It was intended to investigate its impact as well as to determine its performance at different hydraulic retention times (HRTs). The two-phase system used consisted of an anaerobic packed bed reactor (APBR) as the acidogenic reactor and an upflow anaerobic reactor (UAR) as the methanogenic reactor.

## INTRODUCTION

Two-phase anaerobic digestion has been proposed as an alternative for treating industrial wastewaters since the 1970s.[1,2] It is considered to be more advantageous than a single-phase process because it is more stable (particularly during variable process conditions); has a shorter recovery time after failures caused by organic shock loadings; is more resistant to inhibitory conditions; has a higher treatment efficiency, and less total reactor volume is required.[3–6]

The disadvantage of phase separation, however, is that the acidogenic organisms may cause the settleability of granular sludge to deteriorate and also adversely affect the formation and growth of granular sludges when they are not removed from the effluent of the acidogenic reactor.[7] Currently, a staging process in anaerobic digestion is proposed instead of phase separation for treating slightly or nonacidified wastewaters, because it provides a higher treatment efficiency and a greater process stability.[7–9] This study was based on the assumption that a high treatment efficiency can be achieved by maintaining the immobilization of a balanced microbial community inside the reactor, so that the concentration of intermediate products will be sufficiently low.

Nevertheless, taking into account the potential of an acidogenic reactor to act as a buffer to either toxic effects or discontinuous wastewater flow in a separated reactor, the significance of the two-phase anaerobic process still has to be considered.[5]

Although there have been many publications dealing with two-phase anaerobic digestion, information on the possibility of enhancing its application is scarce. Understanding the reaction kinetics as well as metabolism has not been confirmed. An appreciation of the limiting factors, hydrogen flow mechanisms, types and concentration of intermediate products, and interspecies transfer between the microorganisms becomes of importance for the overall improvement of a two-phase process.[10]

The separation of the acidogenic and methanogenic processes results in the accumulation of hydrogen due to the suppression of growth of methanogens in the acidogenic reactor.[11] This may lead to an alteration of the fermentation pattern tending toward increased utilization of those metabolic routes that do not employ $H^+$ reduction as a means of disposing of electrons but in-

*52nd Purdue Industrial Waste Conference Proceedings*, 1997, Ann Arbor Press, Chelsea, Michigan 48118. Printed in U.S.A.

stead employ organic electron acceptors.[12] As a result, the concentration of propionate and butyrate, the most common intermediate products, will increase. It is well known that the product distribution will determine the level of hydrogen partial pressure buildup inside the reactor. The formation of propionate will reduce and butyrate will increase hydrogen concentration in the acidogenic reactors, respectively.[11]

Considering that regulating the intermediate products in the acidogenic reactor is able to enhance the methanogenesis process,[5] the concentration of propionate has to be kept at a minimum due to the methanogenesis of propionate being the slowest during the second phase.[13] A possible way of achieving this could be by recirculating some of the methanogenic sludge back to the acidogenic reactor. The lack of direct interaction between both groups of bacteria during phase separation, which may in turn alter the formation of propionate, may be reduced due to the presence of the methanogenic bacteria. Moreover, this recycled methanogenic sludge could also utilize hydrogen which would eventually reduce the partial pressure of hydrogen in the first reactor. Nevertheless, to satisfy two-phase anaerobic conditions, the level of recirculation should be adjusted in such a way that the predominance of the acid-formers in the first reactor and the methane-formers in the second reactor is maintained during the whole process.

The aims of this research were therefore to improve and control process efficiency by applying recirculation of the methanogenic sludge to the influent of the acidogenic reactor during two-phase anaerobic digestion of a soluble organic wastewater at different hydraulic retention times (HRTs). The two-phase system used consisted of an anaerobic packed bed reactor (APBR) as the acidogenic reactor and an upflow anaerobic reactor (UAR) as the methanogenic reactor.

## MATERIAL AND METHODS

### Laboratory-Scale Two-Phase Anaerobic Digestion System

The reactor unit was originally developed by Armfield Ltd. but modified as deemed necessary for the purpose of the research program, and is illustrated in Figure 1. It consisted of two anaerobic reactors, each consisting of a Plexiglas column, approximately 32.0 cm high and with an inside diameter of 13.8 cm. They were mounted on a stainless steel base equipped with a monitoring

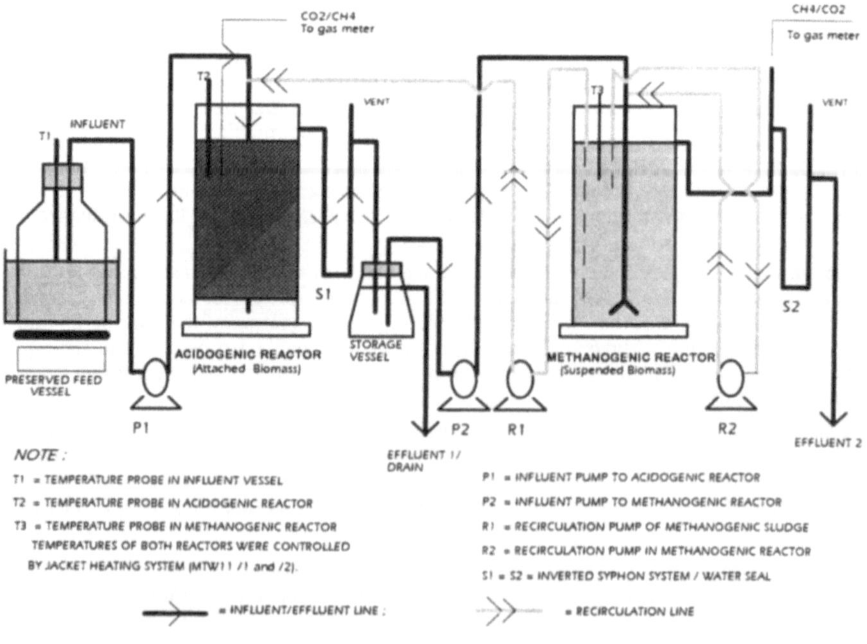


**Figure 1. Schematic diagram of laboratory setup.**

and control board. Both reactors were operated in an upflow mode with each reactor temperature being maintained at (34 ± 1)°C by MTW11/1 and /2 jacket heating controllers. The feeding and recirculation pump systems were all of the Watson Marlow peristaltic type and were adjusted to achieve the required operational conditions. The effluent was discharged via an inverted siphon system in order to prevent biogas escaping from the reactor. The biogas produced was collected in a water-displacement aspirator with gas sampling ports fitted to both anaerobic reactors.

## Anaerobic Packed Bed Reactor (APBR)

The first-phase reactor was an anaerobic packed bed reactor (APBR). It was randomly packed with 130 nonporous Etapak pal rings confined within metal grids up to 76% of its height. The pal rings had a diameter of 2.54 cm, a height of 2.75 cm, with a specific surface area of 100 $m^2/m^3$ and approximately 95–97 % void space. The effective volume of the acidogenic reactor was 4.55 liters. The liquid level was maintained at 3.2 cm above the packing unit to provide sufficient moisture for the attachment of biomass, while the gas space above the liquid level was about 4 cm.

## Upflow Anaerobic Reactor (UAR)

The methanogenic reactor was an upflow suspended growth system (UAR). Although the UAR had the same capacity as the APBR, excluding the pal-rings media; i.e., 4.84 liter, its effective volume was only 4.75 liter. Throughout the course of experiments, the liquid volume in the APBR was controlled in order to maintain a higher hydraulic retention time (HRT) in the methanogenic reactor than that of the acidogenic reactor. A storage vessel of 250 mL was installed between both reactors for this purpose. The flow rate ratio between the influent of the UAR and the influent of the APBR was within the range of 40 to 70%, which corresponds to a HRT ratio of 1.5–2.7 : 1.

## Operational Parameters

For the purpose of the research program, three independent variables were determined; namely, hydraulic retention time (HRT), COD concentration of influent, and recirculation ratio of the methanogenic sludge, as shown in Table I. For the latter case, methanogenic sludge was recirculated from the methanogenic reactor to the influent of the acidogenic reactor with ratios of 0%, 10%, and 30%. These values were calculated based on the ratio between recirculation flow rate and the total influent flow rate to the acidogenic reactor. Since the HRT of the acidogenic reactor for the equal recirculation ratio system had to be constant, the calculation was as follows:

$$Q_R = R\,(Q_{in} + Q_R) \tag{1}$$

where: $Q_R$ = recirculation flow rate of methanogenic sludge
$Q_{in}$ = influent flow rate
$R$ = recirculation ratio

## Wastewater Characteristics

The composition of the synthetic wastewater used, with glucose as the main energy and carbon source, is presented in Table II. The tap water which was used to make the feed was left at room

**Table I. Operational Variables Applied During the Experiment**

| Variables | Range |
|---|---|
| Organic loading (kgCOD/$m^3$.day | 3.2–68.6 (to acidogenic reactor only) |
| | 1.6–33.5 (to both anaerobic reactors) |
| HRT (h) | 2.8–12.3 (in acidogenic reactor) |
| | 6.0–19.6 (in methanogenic reactor) |
| Feed COD concentration (mg/L) | 1600–8000 |
| Methane-sludge recirculation ratio (%) | 0, 10 and 30 |

**Table II. Composition of Synthetic Wastewater Used**

| Parameter | Concentration (mg/L) |
|---|---|
| Glucose, $C_6H_{12}O_6$ | 8000 |
| Ammonium hydrogen carbonate, $NH_4HCO_3$ | 2500 |
| Potassium dihydrogen phosphate, $KH_2PO_4$ | 400 |
| Sodium hydrogen carbonate, $NaHCO_3$ | 400 |
| Trace metal solution: | |
| $MgSO_4 \cdot 7H_2O$ | 5 |
| $FeCl_3$ | 5 |
| $CaCl_2$ | 5 |
| KCl | 5 |
| $CoCl_2$ | 1 |
| $NiCl_2$ | 1 |

temperature for 24 hours prior to use, so the effect of its residual chlorine on anaerobic bacteria was minimized. The feed vessel was kept in a refrigerator at a constant temperature, below 4°C while its contents were continuously mixed by a magnetic stirrer.

### Seed Sludge

Seed anaerobic sludge was collected from a local municipal WWTP. It was sieved through a mesh with a pore size of 0.1 mm and was allowed to stand for 24 hours before being introduced into both reactors. Initial seed concentrations for the acidogenic and methanogenic reactors were 6000 and 8000 mgVSS/L, respectively. The sludge in both digesters was warmed to 34 ± 1°C and left for 24 hours before the start of the acclimatization.

### Analytical Methods

The analytical methods were carried out according to *Standard Methods*[14] and consisted of COD, SS, VSS, alkalinity, and pH of the influent and effluent samples. Volatile fatty acids (VFAs) were analyZed by Unicam 610 Gas Chromatograph and biogas production was measured by a water-displacement system. Gas composition was determined using a Becker Model 403 Gas Chromatograph.

## RESULTS AND DISCUSSION

The impact of recirculation of the methanogenic sludge to the acidogenic reactor was evaluated by investigating operational conditions in two series, as summarized in Table III. The first experiment examined the performance of both anaerobic reactors at recirculation ratios of 0%, 10%, and 30%, with different COD influent concentrations. Throughout this series of experiments with CODs of 1600 mg/L and 5000 mg/L, the mixing in the methanogenic reactor was achieved by recirculating its content through the influent line at the bottom part of the reactor, using a flowrate of 93 L/d corresponding to an upflow velocity ($V_{up}$) of 6.24 m/d. The second series of experiments examined the effect of HRT at a 30% recirculation ratio. The influent COD concentration was fixed at 8000 mg/L and the HRT was decreased in steps from 5.2 to 2.8 hours for the acidogenic reactor and from 9.3 to 6.8 hours for the methanogenic reactor. As illustrated in Table III, the organic loading rates (OLRs) were subsequently increased after changing the HRT from approximately 37 $kgCOD/m^3.d$ to a maximum of 69 $kgCOD/m^3.d$ for the acidogenic reactor.

### Impact of Methanogenic Sludge Recirculation

The COD removal efficiencies under steady state conditions are summarized in Table IV. The sample means and their corresponding standard deviations are shown in order to indicate the significant differences between each run. The COD removal efficiency was affected by both the recirculation ratio and the influent COD concentration. It was always greater in the acidogenic reactor fed with recirculated methanogenic sludge than in the reactor fed without recirculation.

**Table III. Operational Conditions During the Experiments**

| Series | Run | Recirculation (%) | COD influent (mg/L) | HRT (hours) Acid-R[a] | HRT (hours) Methane-R[b] | OLR ($kgCOD/m^3.d$) to Acid-R[a] |
|---|---|---|---|---|---|---|
| I | I.1 | 0 | 1600 | 11.9 | 19.6 | 3.2 |
| | I.2 | 10 | 1600 | 11.9 | 19.6 | 3.2 |
| | I.3 | 30 | 1600 | 11.9 | 19.6 | 3.2 |
| | I.4 | 0 | 5000 | 8.9 | 16.3 | 13.5 |
| | I.5 | 10 | 5000 | 8.9 | 16.3 | 13.5 |
| | I.6 | 30 | 5000 | 8.9 | 16.3 | 13.5 |
| | I.7 | 0 | 8000 | 12.3 | 19.0 | 15.6 |
| | I.8 | 10 | 8000 | 12.3 | 19.0 | 15.6 |
| | I.9 | 30 | 8000 | 12.3 | 19.0 | 15.6 |
| | I.10 | 0 | 8000 | 2.8 | 6.8 | 68.6 |
| | I.11 | 10 | 8000 | 2.8 | 6.8 | 68.6 |
| | I.12 | 30 | 8000 | 2.8 | 6.8 | 68.6 |
| II | II.1 | 30 | 8000 | 5.2 | 9.3 | 37.0 |
| | II.2 | 30 | 8000 | 3.8 | 8.3 | 50.5 |
| | II.3 | 30 | 8000 | 3.5 | 7.5 | 54.9 |
| | II.4 | 30 | 8000 | 3.4 | 9.1 | 56.5 |
| | II.5 | 30 | 8000 | 2.8 | 6.8 | 68.6 |

[a] Acid-R = acidogenic reactor.
[b] Methane-R = methanogenic reactor.

**Table IV. Removal Efficiency at Steady State During First Series of Experiments**

| Recirculation (%) | CODinfluent (mg/L) | COD effluent (mg/L) Acid-R[a] | COD effluent (mg/L) Methane-R[b] | COD Removal, (%) Acid-R[a] | COD Removal, (%) Methane-R[b] | COD Removal, (%) Total |
|---|---|---|---|---|---|---|
| 0 | 1600 | 1096 ± 102 | 556 ± 53 | 32 | 48 | 65 |
| 10 | 1600 | 764 ± 35 | 171 ± 35 | 52 | 78 | 89 |
| 30 | 1600 | 665 ± 75 | 65 ± 4 | 58 | 90 | 96 |
| 0 | 5000 | 3440 ± 165 | 2210 ± 156 | 31 | 36 | 56 |
| 10 | 5000 | 2183 ± 160 | 449 ± 50 | 56 | 79 | 91 |
| 30 | 5000 | 2468 ± 79 | 599 ± 66 | 51 | 76 | 88 |
| 0 | 8000 | 2687 ± 355 | 670 ± 70 | 66 | 75 | 92 |
| 10 | 8000 | 1510 ± 180 | 458 ± 50 | 81 | 70 | 94 |
| 30 | 8000 | 1365 ± 78 | 220 ± 29 | 83 | 84 | 97 |
| 0 | 8000 | 5400 ± 113 | 2620 ± 14 | 32 | 51 | 67 |
| 10 | 8000 | 4856 ± 250 | 2503 ± 234 | 39 | 48 | 69 |
| 30 | 8000 | 4117 ± 29 | 2010 ± 120 | 49 | 51 | 75 |

[a]Acid-R = acidogenic reactor.
[b]Methane-R = methanogenic reactor.

This condition was different in the methanogenic reactor, especially when fed with the high strength influent (8000 mg/L); nevertheless, the total COD removal was consistently higher. The increased COD removal with the 30% recirculation ratio was approximately 1.5 times greater in the high strength feed acidogenic reactor, and 1.8 times greater in the medium and low strength feed reactors, than in the reactor without recirculation.

In all systems, the overall COD removal generally decreased with increasing feed concentration of the influent, but it was not the case for those with a similar organic loading rate (OLR). At feed concentrations of 5000 and 8000 mgCOD/L, with the corresponding OLR of 13.5 and

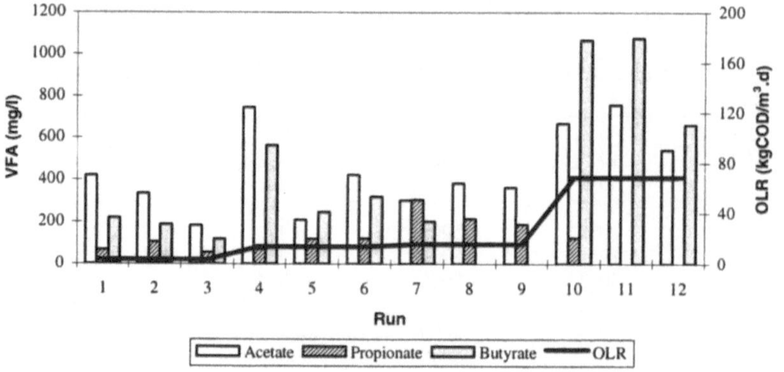


**Figure 2. VFA distribution in acidogenic reactor during first series of experiments.**

15.6 kgCOD/m$^3$.d, respectively, it appears that the higher COD concentration was more favorable to the system. Furthermore, increasing the OLR with a constant COD concentration of 8000 mg/L resulted in a decrease in COD removal.

With respect to COD removal, the optimum conditions for two-phase anaerobic digestion was at a feed concentration of 8000 mg/L COD at an OLR of 15.6 kgCOD/m$^3$.d. It applied to all runs with the COD removal being at least 92% in the reactor fed without recirculation. The maximum COD removal was achieved by the reactor with 30% recirculation of methanogenic sludge.

*Impact on Acidogenesis*

The impact of recirculation of the methanogenic sludge to the acidogenic reactor can be seen by the volatile fatty acids (VFA) distribution for different recirculation ratio values and COD influent concentrations (Figure 2). The VFA distribution in both acidogenic and methanogenic reactors is also shown in Table V. Since only the three main acids became of significant interest, other acids are not considered during the discussion but they are included in the total VFAs in Table V. They consisted of isobutyric acid, isovaleric acid, and valeric acid in very small concentration, but their presence during the process was also rare and inconsistent.

From Figure 2, acetic acid was the most common VFA, followed by propionic and butyric. In the lower strength feed reactor (COD 1600 mg/L) and the medium strength feed reactor (COD 5000 mg/L), the product distribution was dominated by acetic acid and butyric acid, while the propionic acid concentration was lower. In the higher strength feed reactor (COD 8000 mg/L), the VFA distribution varied, depending on the OLR. At an OLR of 15.6 kgCOD/m$^3$.d, acetic acid and propionic acid were dominant with no residual butyric acid (runs I.7–I.9). At the higher OLR (68.6 kgCOD/m$^3$.d), no propionic acid was found in the reactors fed with methanogenic sludge recirculation, while the dominant VFA products were acetic and butyric acid. Contrary to the VFA distribution in the low and medium strength feed reactors, the butyric acid concentration was higher than that of acetic (runs I.10–I.12). Comparing this with previous studies, this is in agreement with the fact that butyric acid becomes dominant at high dilution rates and at mesophilic temperatures for high sludge loadings, while propionic and acetic acids are the major product at lower dilution rates.[15–16] Using dilution rates of (0.51–0.71) h$^{-1}$, at pH 5.8, these authors found that the product distribution was relatively constant.

Throughout this series of experiments, the total VFA concentration was lower in the reactor with methanogenic sludge recirculation than that in the reactor without recirculation. With respect to this, not all three main VFAs decreased with increasing recirculation ratio. As shown in Table V, except in the case of the medium strength feed reactor, only the 30% recirculation ratio was able to reduce most acids during the acidogenesis process. It was clear, therefore, that recirculation of the methanogenic sludge was only able to reduce the overall VFA concentration, not particular individual acids at the same time.

**Table V. Main VFA Distribution During First Series of Experiments**

| Run | Acetate (mg/L) | | Propionate (mg/L) | | Butyrate (mg/l) | | VFA total, mg/L | |
|---|---|---|---|---|---|---|---|---|
| (No.) | Acid-R[a] | Methane-R[b] | Acid-R[a] | Methane-R[b] | Acid-R[a] | Methane-R[b] | Acid-R[a] | Methane-R[b] |
| I.1 | 421 (28) | 271 (53) | 66 (3) | 51 (21) | 220 (38) | 54 (2) | 707 | 376 |
| I.2 | 334 (2) | 54 (2) | 104 (5) | 20 (5) | 188 (17) | 17 (4) | 628 | 91 |
| I.3 | 189 (29) | 0 | 54 (15) | 0 | 117 (37) | 0 | 353 | 0 |
| I.4 | 745 (127) | 413 (18) | 68 (4) | 78 (11) | 563 (117) | 248 (18) | 1431 | 734 |
| I.5 | 210 (52) | 150 (61) | 117 (32) | 0 | 247 (38) | 0 | 682 | 150 |
| I.6 | 424 (34) | 96 (18) | 120 (18) | 50 (13) | 320 (23) | 0 | 696 | 146 |
| I.7 | 303 (28) | 0 | 305 (17) | 0 | 202 (70) | 0 | 975 | 0 |
| I.8 | 384 (42) | 0 | 213 (19) | 0 | 0 | 0 | 579 | 0 |
| I.9 | 363 (16) | 0 | 188 (16) | 0 | 0 | 0 | 551 | 0 |
| I.10 | 668 (18) | 1532 (18) | 122 (4) | 158 (11) | 1068 (46) | 618 (251) | 2143 | 2485 |
| I.11 | 758 (96) | 555 (72) | 0 | 106 (61) | 1078 (79) | 418 (73) | 2151 | 1079 |
| I.12 | 542 (18) | 602 (88) | 0 | 0 | 662 (4) | 202 (4) | 1372 | 804 |

[a]Acid-R = acidogenic reactor.
[b]Methane-R = methanogenic reactor.
( ) standard deviation of sample mean.

*Impact on Methanogenesis*

The effects of the VFA distribution obtained in the acidogenic reactor on the performance of methanogenesis are shown in Figure 3. At the optimum operating conditions of the experiment (runs I.7 - I.9), no VFA was found in the effluent of methanogenic reactor. It is interesting to note also that this result was achieved from the acidogenic reactor effluent which mainly contained acetic acid and propionic acid, rather than acetic acid and butyric acid. It is widely known that propionic acid is the most undesirable acid for methanogenesis due to its low degradation rate; consequently, its concentration should be ideally kept at a minimum.[13]

The product distribution after acidogenesis was considered to be the main factor affecting final effluent quality, and based on the free energy calculation, it was stated that either lactate or ethanol would be preferable as the acidogenesis product because of the available free energy which would then be able to provide a maximum potential energy for the methanogens. Another author reported that the formation of propionate from lactate could be avoided by controlling the dilution rate and pH value between 0.1–0.15 $h^{-1}$ and 4.5–6.0, respectively. Acetate was formed from lactate which was faster than the conversion of butyrate.[13] It has also been reported that lactate was only available under imbalanced conditions, and it will eventually decrease or disappear after equilibrium of the digestion system has been reached.[15]

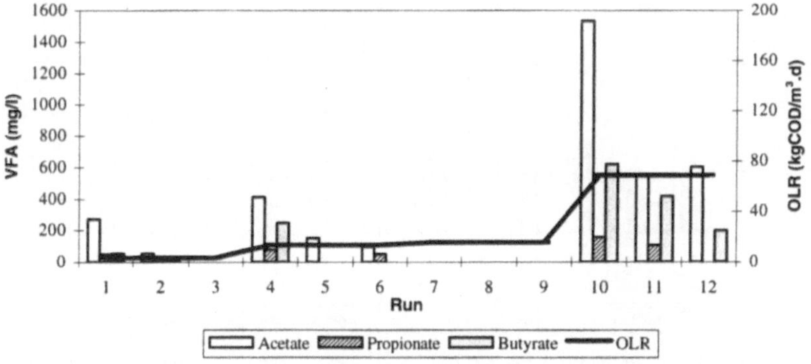


**Figure 3. VFA distribution in methanogenic reactor during first series of experiments.**

In these experiments, methanogenesis of the effluent from the acidogenic reactor, which consisted primarily of acetic and butyric acids had different results. When acetic acid was more dominant than butyric acid, methanogenesis produced a greater removal efficiency than that in the reverse conditions. It was likely that it was not the concentration of the acid products which affected the efficiency of methanogenesis but rather the type of product distribution and its composition. Acetate and butyrate have been found to be consumed at a higher rate than propionate,[15] but it was still found during these experiments that COD removal in the reactor containing acetate-propionate was better than that from the reactor containing acetate-butyrate.

During runs I.7–I.9, no volatile fatty acids were found. Probably during acidogenesis, some glucose was converted into forms other than acids. It has been reported that having glucose as the main carbon source ethanol was found to be a major product from an acidogenic reactor.[3] Similar work had also indicated that end products from glucose during acidogenesis within a similar range of temperature and pH values, were lactate, ethanol, butyric acid, propionic acid, and acetic acid.[15–16]

*Biogas, pH and Alkalinity*

The biogas proportions in both acidogenic and methanogenic reactors are shown in Figure 4, while pH and alkalinity are shown in Figure 5. The overall results are tabulated in Tables VI and VII.

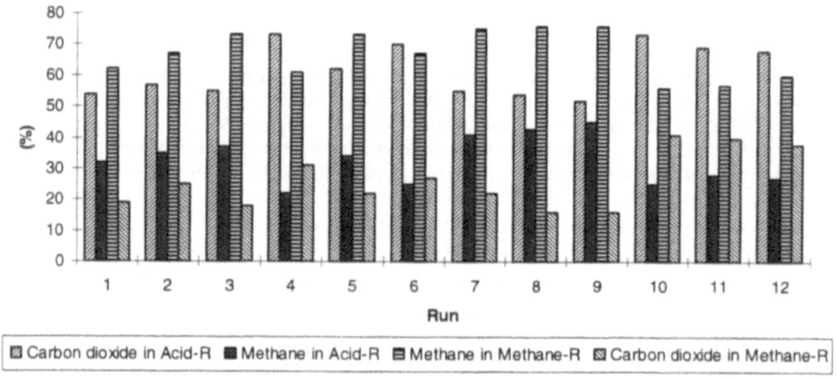


**Figure 4. Biogas composition in both reactors during first series of experiments. (Acid-R = acidogenic reactor, Methane-R = methanogenic reactor).**

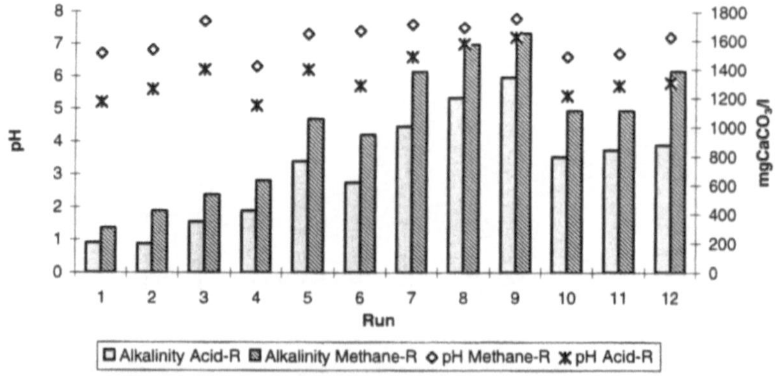


**Figure 5. pH and alkalinity in both reactors during first series of experiments. (Acid-R = acidogenic reactor, Methane-R = methanogenic reactor).**

**Table VI. Reactor Performance at Steady State During First Series of Experiments (COD concentration - 1600 mg/L and 5000 mg/L)**

| | 1600 mg/l | | | 5000 mg/L | | |
|---|---|---|---|---|---|---|
| Effluent | Run I.1 0%[a] | Run I.2 10%[a] | Run I.3 30%[a] | Run I.4 0%[a] | Run I.5 10%[a] | Run I.6 30%[a] |
| **Acidogenic Reactor :** | | | | | | |
| pH | 5.2 (0.1) | 5.6 (0.2) | 6.2 (0.1) | 5.1 (0.1) | 6.2 (0.1) | 5.7 (0.1) |
| Alkalinity (as $CaCO_3$ ) (mg/L) | 203 (27) | 198 (11) | 349 (22) | 423 (45) | 763 (45) | 615 (90) |
| Gas production (L/d) | 0.7 (0.1) | 0.9 (0.1) | 1.0 (0.1) | 8.6 (0.9) | 13.6 (3.3) | 8.9 (0.7) |
| % $CO_2$ | 54 | 57 | 55 | 73 | 62 | 70 |
| %$CH_4$ | 32 | 35 | 37 | 22 | 34 | 25 |
| **Methanogenic Reactor :** | | | | | | |
| pH | 6.7 (0.3) | 6.8(0.2) | 7.7 (0.1) | 6.3 (0.2) | 7.3 (0.2) | 7.4 (0.1) |
| Alkalinity (as CaCO3) (mg/L) | 306 (27) | 423(2) | 535 (20) | 632 (20) | 1053 (15) | 945 (13) |
| Gas production (L/d) | 0.8 (0.1) | 1.8(0.2) | 1.6 (0.2) | 3.6 (0.8) | 6.2 (0.7) | 5.4 (0.1) |
| % $CH_4$ | 62 | 67 | 73 | 61 | 73 | 67 |
| % $CO_2$ | 19 | 25 | 18 | 31 | 22 | 27 |
| $m^3$ $CH_4$/kg COD utilized | 0.17 | 0.35 | 0.34 | 0.25 | 0.37 | 0.28 |

[a]Methanogenic sludge recirculation.

**Table VII. Reactor Performance at Steady State During First Series of Experiment (COD concentration - 8000 mg/L)**

| | 8000 mg/L | | | | | |
|---|---|---|---|---|---|---|
| Effluent 0%[a] | Run I.7 10%[a] | Run I.8 30%[a] | Run I.9 0%[a] | Run I.10 10%[a] | Run I.11 30%[a] | Run I.12 |
| **Acidogenic Reactor :** | | | | | | |
| pH | 6.6 (0.2) | 7.0 (0.1) | 7.2 (0.1) | 5.4 (0.0) | 5.7 (0.2) | 5.8 (0.0) |
| Alkalinity (as CaCO3) (mg/L) | 1003 (35) | 1200 (29) | 1340 (39) | 795 (92) | 844 (59) | 880 (28) |
| Gas production (L/d) | 22.3 (3.5) | 25.1 (2.8) | 25.8 (1.3) | 7.5 (1.8) | 51.8 (2.0) | 46.6 (0.5) |
| % $CO_2$ | 55 | 54 | 52 | 73 | 69 | 68 |
| %$CH_4$ | 41 | 43 | 45 | 25 | 28 | 27 |
| **Methanogenic Reactor :** | | | | | | |
| pH | 7.6 (0.1) | 7.5 (0.1) | 7.8 (0.0) | 6.6 (0.0) | 6.7 (0.1) | 7.2 (0.0) |
| Alkalinity (as CaCO3) (mg/L) | 1380 (61) | 1570 (32) | 1648 (45) | 1110 (71) | 1112 (41) | 1385 (304) |
| Gas production (L/d) | 5.9 (1.0) | 3.7 (0.4) | 3.0 (0.2) | 26.8 (0.4) | 28.8 (2.5) | 22.8(2.4) |
| % $CH_4$ | 75 | 76 | 76 | 56 | 57 | 60 |
| % $CO_2$ | 22 | 16 | 16 | 41 | 40 | 38 |
| $m^3$ $CH_4$/kg COD utilized | 0.37 | 0.44 | 0.33 | 0.32 | 0.42 | 0.39 |

[a]Methanogenic sludge recirculation.

Carbon dioxide was dominant in the acidogenic reactor, between 52 and 73%, while the methane content was between 22 and 45%. Although there was no significant difference in the overall biogas composition in each run, increasing recirculation tended to decrease the $CO_2$ content in the acidogenic reactor, but increased the $CH_4$ content in the methanogenic reactor. A similar result was obtained by Zoetemeyer et al.,[16] who found that increasing the pH reduced gas production and percentage $CO_2$.

Biogas composition during the optimum operating procedures (runs I.7–I.9) was approximately 41–45% methane and 52–55% carbon dioxide, and resulted in more methane than in the other systems. This composition of biogas also resulted in more methane and less carbon dioxide in the methanogenic reactor when compared to others; i.e., 75–7% methane and only 16–22% of carbon dioxide. The biogas composition in the methanogenic reactor for the other operating procedures varied between 56–73% for methane and between 19–41% for carbon dioxide.

Methane production was a maximum in the reactor with 10% recirculation ratio in all operating procedures, and the average methane production was greater in the high strength feed reactor than in the medium and low strength feed reactors. It varied from 0.32–0.44 $m^3CH_4$/kgCOD utilized.

Zoetemeyer[15] reported that gas production was about 80% $CO_2$ during acidogenesis. In this present study, that figure was quite close to the carbon dioxide content in the medium and high strength feed reactors without methanogenic sludge recirculation (Tables VI and VII). As indicated in Table IV, this condition gave less COD removal than that for the other operating procedures.

An indication of the stability of anaerobic systems has been reported to be the alkalinity:volatile fatty acids ratio.[3,17] The higher the ratio, the more stable the anaerobic system would be. The ratio in a separated system, for instance, was 60% higher than in the single-phase system.[3] Applying the alkalinity:volatile fatty acids ratio to this study, it was found that the ratio in the system with recirculation was approximately three times greater than the ratio in the system without recirculation (Figure 5). The increase in buffering capacity of the acidogenic reactor, due to the impact of effluent recirculation instead of sludge recirculation, from a methanogenic reactor had been reported before.[18] In general, the recirculation of methanogenic sludge had a higher ratio value than that in the reactor without recirculation. Hence the recirculation of methanogenic sludge improved the stability of the two-phase anaerobic digestion process.

The operating pH at each operating procedure varied between 5.1 and 7.2 in the acidogenic reactor and 6.3 and 7.8 in the methanogenic reactor. Similar to the alkalinity, the operating pH in each system increased with increasing recirculation ratio. The pH values during acidogenesis reflected the VFA distribution. At the lower pH values (5.1–5.8), the dominant acids were acetic and butyric, while at the higher pH values (6.6–7.2) the dominant acids were acetic and propionic. This finding seemed to be in accordance with previous investigations that butyric acid was the major acid product of glucose degradation when the reactor was operated at pH 5.7. However, at pH greater than 6.0, lactic acid was dominant, leading to the formation of acetic acid, formic acid, and ethanol.[16] It has also been found that low pH (<6) and short HRT (<6 hours) brought about the production of more n-butyric acid rather than propionic acid.[5] The effect of pH was significant in determining end-product distribution, but not total VFA production. VFA production was more dependent upon temperature, while its product distribution varied depending upon pH and temperature.[18,19]

Recirculation of the methanogenic sludge was able to improve buffering capacity in the acidogenic reactor as well as improve its stability, increase gas production and its composition, and improve the overall COD removal efficiency.

## Effect of HRT on the Performance of Anaerobic Reactors

In the second series of experiments, the COD loading rate for the acidogenic reactor was set at 37.0 $kg/m^3$.d and was increased over time to a maximum of 68.6 $kg/m^3$.d with a feed concentration of 8000 mg/L COD. Figure 6 shows the performance of both reactors at different OLRs due to the decreased HRT. During the initial period with an organic loading rate of 37.0 $kg/m^3$.d; i.e., an HRT of 5.2 hours, the COD removal increased in the acidogenic reactor. When the HRT was changed instantaneously to 3.8 hours (i.e., an OLR of 50.5 $kg/m^3$.d), the COD removal increased to its peak value. It should be noted that during the first day of this period (day 11) the acidogenic reactor feed pump delivered only one-third of the proposed capacity due to a technical problem. This meant that the HRT of the acidogenic reactor was about 11 hours instead of 3.8

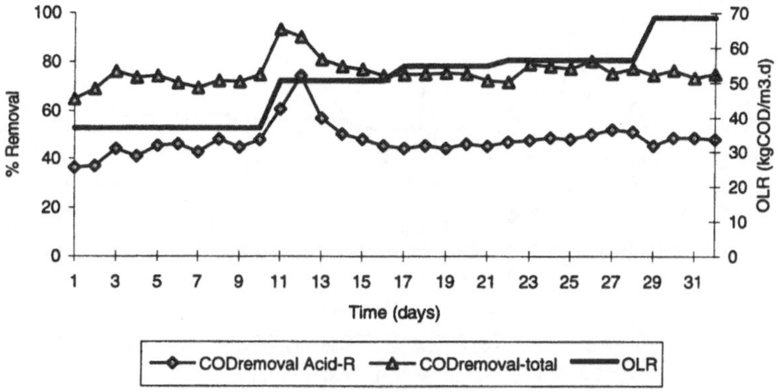


**Figure 6. Effect of HRT on removal efficiency during second series of experiments. (Acid-R = acidogenic reactor).**

hours, which is not shown in Figure 6. There was no significant difference between the COD removal at either the low (2.8 hr) or the high (5.2 hr) HRT. It was also found that the total COD removal at the low HRT was slightly greater than that at the higher HRT.

In Figure 7 it can be seen that the VFA distribution was dominated by acetic acid and butyric acid throughout the course of experiments. Again, acetic acid was the major acid product in the higher HRT reactor; i.e., the lower OLR (37.0 kg/m$^3$.d), while butyric acid was dominant in the lower HRT reactor; i.e., the higher OLR. During a pump failure, both acetic and butyric acids were reduced to a minimum but increased again with increased OLR. Propionic acid was noted shortly after the OLR changed from (37.0 to 50.5) kg/m$^3$.d, but was not found with increased loading rate. This was contradictory to a report which has shown that propionic acid increased with increased loadings, while the butyric acid concentration was low and relatively unaffected by the organic load using a feed concentration of 12000 mg/L COD in the series of CSTR and AFBR, as acidogenic and methanogenic reactors, respectively.3

Throughout this study, the system was able to retain its performance level after power failure and during uncontrolled shock-loadings. With respect to this, product distribution in the acidogenic reactor was more constant than that in the methanogenic reactor. During the absence of feed during power failure, acetic acid was converted completely within a very short period and butyric acid was converted at a similar rate (Figures 7 and 8).

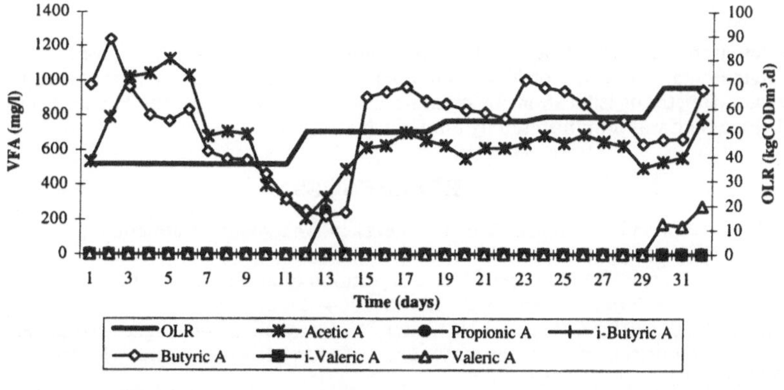


**Figure 7. Effect of HRT on VFA distribution in the acidogenic reactor during second series of experiments.**

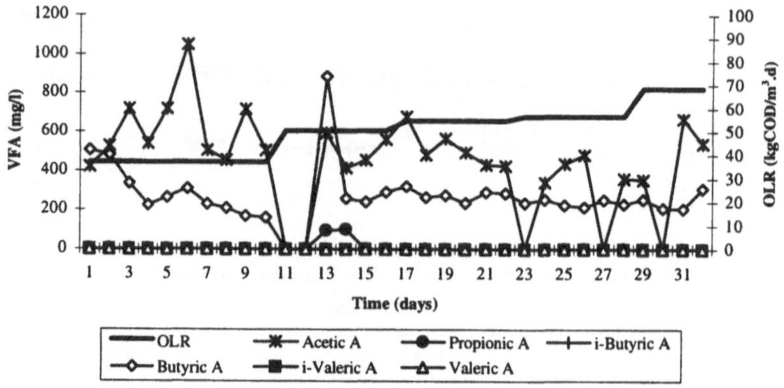


**Figure 8. Effect of HRT on VFA distribution in the methanogenic reactor during second series of experiments.**

## CONCLUSION

The optimum operating procedure for two-phase anaerobic digestion during this study was at a feed concentration of 8000 mg/L COD and at the organic loading rate of 15.6 kgCOD/m$^3$.d. COD removal was affected by both recirculation and influent concentration, with the maximum being achieved with the reactors fed with 30% recirculation of methanogenic sludge.

The impact of recirculation of methanogenic sludge to the influent of the two-phase anaerobic system under study was:

- increased COD removal efficiency,
- reduced total VFA concentration, but individual acids were not reduced at the same time,
- increased stability of the system, indicated by the high alkalinity:volatile acids ratio,
- increased pH value in the acidogenic reactor and in the methanogenic reactor,
- increased biogas production,
- reduced concentration of $CO_2$ in the acidogenic reactor and increased concentration of $CH_4$ in the methanogenic reactor,
- maintenance of performance during low HRT; i.e., 2.8 hours for the acidogenic reactor and 6.8 hours for the methanogenic reactor, and
- ability to recover from either power failure or during uncontrolled shock loadings.

## ACKNOWLEDGMENT

This research was carried out in the Environmental Engineering laboratory, Department of Civil Engineering, University of Newcastle upon Tyne, UK. The first author wishes to thank the Indonesian Ministry of Education for granting a scholarship through the Higher Education Project (HEP), which was administered by the British Council.

## REFERENCES

1. Pohland, F.G., and S. Ghosh, "Developments in Anaerobic Treatment Processes," *Biotech. Bioeng. Symp.*, 2, pp. 85–106, 1971.
2. Massey, M.L., and F.G. Pohland, "Phase separation of anaerobic stabilization by kinetic controls," *J. Water Pollut. Control Fed.*, 50, pp. 2204–2222, 1978.
3. Bull, M.A., R.M. Sterritt, and J.N. Lester, "An evaluation of single- and separated-phase anaerobic industrial wastewater treatment in fluidised bed reactors," *Biotechnol. Bioeng.*, XXVI, pp. 1054–1065, 1984.
4. Dinopoulou, G. and J.N. Lester, "Optimization of a two-phase anaerobic digestion system treating a complex wastewater," *Environ. Technol. Lett.*, 10, pp. 799–814, 1989.

5. Hanaki, K., T. Matsuo, M. Nagase, and Y. Tabata, "Evaluation of effectiveness of two-phase anaerobic digestion process degrading complex substrate," *Water Sci. Technol.*, 19, pp. 311–322, 1987.
6. Jeyaseelan, S. and T. Matsuo, "Effects of phase-separation in anaerobic-digestion on different substrates," *Water Sci. Technol.*, 31, 9, pp. 153–162, 1995.
7. Lettinga, G., "Anaerobic-digestion and waste-water treatment systems," *Antonie van Leeuwenhoek International J. Gen. Molecular Microbiol.*, 67, 1, pp. 3–28, 1995.
8. Vanlier, J.B., F. Boersma, M.M.W.H Debets, and G. Lettinga, "High-rate thermophilic anaerobic waste-water treatment in compartmentalized upflow reactors," *Water Sci. Technol.*, 30, 12, pp. 251–261, 1994.
9. Vanlier, J.B., J.L.S. Martin, and G. Lettinga, "Effect of temperature on the anaerobic thermophilic conversion of volatile-acids by dispersed and granular sludge," *Water Res.*, 30, 1, pp. 199–207, 1996.
10. Verstraete W., L. de Baere, and A. Rozzi, "Phase separation in anaerobic digestion: Motives and method," *Trib. CABEDEAU*, 453–454 (3), pp. 367–375, 1981.
11. Mossey, F.E., "New developments in the anaerobic treatment of industrial wastes," *Effluent Water Treat. J.*, pp. 85–93, 1983.
12. Cohen A., A.M. Breure, J.G. van Andel, and A. van Deursen, "Influence of phase separation on the anaerobic digestion of glucose I. maximum COD-turnover rate during continuous operation," *Water Res.*, 14, pp. 1439–1448, 1980.
13. Kisaalita, W.S., K.V.Lo, and K.L. Pinder, "Influence of dilution rate on the acidogenic phase products distribution during two-phase lactose anaerobiosis," *Biotechnol. Bioeng.*, 34, pp. 1235–1250, 1989.
14. Greenberg, A.E., R.R. Trussell, and L.S. Clisceri, Eds., *Standard Methods for Examination of Water and Wastewater*, 16th ed., Public Health Association, Washington, D.C, 1985.
15. Zoetemeyer, R.J., P. Arnoldy, A. Cohen, and C. Boelhouwer, "Influence of temperature on the anaerobic acidogenic of glucose in a mixed culture forming part of a two-stage digestion process," *Water Res.*, 16, pp. 313–321, 1982.
16. Zoetemeyer, R.J., J.C. van den Heuvel, and A. Cohen, "pH influence on acidogenic dissimilation of glucose in an anaerobic digestor," *Water Res.*, 16, pp. 303–311, 1982.
17. Fongastitkul, P., D.S. Mavinic, and K.V. Lo, "A two-phased anaerobic digestion process: Concept, process failure and maximum system loading rate," *Water Environ. Res.*, 66, 3, 1994.
18. Romli, M., P.F. Greenfield, and P.L. Lee, "Effect of recycle on a two-phase high-rate anaerobic wastewater treatment system," *Water Res.*, 28, 2, pp. 475–482, 1994.
19. Kasapgil, B., O. Ince, and G.K. Anderson, "Determination of operating conditions in an anaerobic acid-phase reactor treating dairy wastewater," *Proceedings of the 50th Purdue Industrial Waste Conference, Purdue University*, 1995, pp. 669–681.
20. Ruzicka, Marek, "An extension of the Mosey model," *Water Res.*, 30, 10, pp. 2444–2446, 1996.

# 32 EMPIRICAL AND KINETIC MODELING OF SOLID SUBSTRATE ANAEROBIC DIGESTION (DASS) FOR JOINT MUNICIPAL-INDUSTRIAL WASTE STABILIZATION

**H.M. Poggi-Varaldo**
Ecatepec Institute of Technology-TESE and P3 Consulting Engineers
P.O. Box 75-202, Mexico D.F., 07300, Mexico

**F. Esparza-García**
Centre for Advanced Studies and Research (CINVESTAV)
Dept. Biotechnology & Bioengineering
Mexico City, Mexico

**G. Fernández-Villagómez**
Centre for Disaster and Hazards Prevention (CENAPRED)
Mexico City, Mexico

**S. Caffarel- Méndez**
Ecatepec Institute of Technology-TESE and P3 Consulting Engineers
P.O. Box 75-202, Mexico D.F., 07300, Mexico

**E. Arce-Medina**
ESIQIE del IPN, Graduate School of Chemical Engineering
Mexico City, Mexico

## ABSTRACT

Dry anaerobic digestion of solid wastes (DASS for the Spanish abbreviation of solid substrate anaerobic digestion) is receiving much attention worldwide as an alternative for industrial and municipal waste stabilization and reclaiming. Although some efforts were made in DASS process modeling, they were concentrated on the thermophilic DASS and only addressed empirical or semiempirical models. This chapter, then, aimed at modeling the mesophilic steady-state DASS process with both empirical and kinetic models.

Bench-scale, semicontinuous, mesophilic reactors were operated at six mass retention times (MRT, 15, 18, 21, 25, 30, and 40 days) by duplicate. Process performance was evaluated in terms of volatile solids efficiency removal, biogas productivity, methane content in biogas, volatile organic acids contents in mixed solids, etc. Methanogenic biomass was quantitated by coenzyme F 420 which is a factor practically specific of methanogenic bacteria. The feedstock was a mixture of lignocellulosic wastes, food wastes, and biosolids intended to simulate the codigestion of industrial and municipal wastes. Empirical models based on Levenspiel kinetics were applied to the efficiency and the unit removal rate of volatile solids. Another kinetic model based on the conventional anaerobic digestion concept (as a series process hydrolysis/acidogenesis, volatile organic acids uptake rate, and methanogenesis) was adapted and fitted to data.

The efficiency increased in the range 60% to 83% (biodegradable volatile solids base) with increasing MRT (18 to 40 days MRT). However, the efficiency was very poor (27%) at 15 days MRT. Unit removal rate and biogas productivity increased steadily from 40 to 18 days MRT, with a sudden drop of both responses at 15 days MRT. The overall performance indicated a partial process inhibition at 15 days MRT.

*52nd Purdue Industrial Waste Conference Proceedings*, 1997, Ann Arbor Press, Chelsea, Michigan 48118. Printed in U.S.A.

Empirical models of efficiency vs. MRT and unit removal rate vs. loading rate fitted adequately the data and predicted process cessation for MRT between 13.3 to 14.5 days. The kinetic model based on the conventional anaerobic digestion concept, modified for inhibition of the hydrolysis rate and volatile acids net uptake rate, also fitted the experimental results. A model with fractional order in volatile solids (approx. 0.4) and a Levenspiel-like inhibition factor gave the best fitting of the hydrolysis rate. For the volatile organic acids uptake rate, a Haldane model was the best representation. The kinetic coefficient values were lower than those reported for similar kinetic models of low total solids anaerobic digestion (suggesting the influence of mass transfer retardation effects), while the stoichiometric coefficients (i.e., the yield synthesis) gave comparable values.

## INTRODUCTION

Mexico faces increasing pressures regarding treatment and disposal of municipal solid wastes (MSW), industrial solid wastes, and wastewater solids. Nearly 52,000 tonne/day of MSW, 370,000 tonne/day of solid industrial wastes, and 1,900,000 $m_3$/day of wastewater solids are generated in Mexico.[1] The main disposal option is sanitary landfills, but only one-third of the generated MSW is disposed of there. Incineration is only applied to infectious wastes in sanitary landfills. Although there is a positive trend and measures are being taken in order to solve the MSW problem, the number of sites properly designed as sanitary landfills is scarce (nearly 40 so far).[2] Thus, there is a need for finding environmentally-friendly alternatives for the treatment, reclaiming, and disposal of the solid wastes generated in Mexico.

Solid substrate anaerobic digestion (DASS) has become an attractive technology for stabilizing agricultural, food, yard, and municipal solid wastes.[3–8] In spite of the intense research activity on DASS, some areas need more research:[5] fast and reliable process startup (type of inocula and overall startup strategies); process modeling; nonhazardous industrial solid waste stabilization; process response to toxic substances; anaerobic compost post-treatment requirements and alternative uses. So far, DASS modeling has embraced the application of semiempirical models based on biogas production and kinetic approaches based only on substrate consumption, which did not take into account the active methanogenic biomass.[9–11] The present work reports the advances made on modeling the mesophilic DASS process in steady state.

## METHODOLOGY

### Model Development

#### *Empirical Models*

A first approach to modeling continuous mesophilic DASS was to apply empirical models based on either the efficiency or the removal rate as response variables. As it is exhibited in Table I, the biodegradable efficiency $_{-b}$ (efficiency based on biodegradable volatile solids) is a function of retention time following a modified Levenspiel relationship. The second model relates the organic matter removal rate (RR) and the biodegradable organic matter loading rate ($LR_b$). It is first order in $LR_b$ and includes a Levenspiel-like inhibition factor in terms of the $LR_b$ and a critical loading rate ($LR_{b,c}$). The rationale behind this approach was that at low loading rates, the process is limited by the hydrolysis and the removal is proportional to the loading, but at higher loading rates the process is limited by methanogenesis and its inhibition. Regarding the latter, the $LR_{b,c}$ is a critical value of the loading rate beyond which the DASS process is completely inhibited and the removal rate drops to zero.

#### *Kinetic Model*

*1. Assumptions.* The kinetic model was based on the conventional concept of the anaerobic digestion as a biochemical series process hydrolysis/acidogenesis/methanogenesis.[12,13] Other assumptions were:

- digesters are in steady state
- complete mix reactors, and complete mix
- the mass retention time is equal to the cellular retention time

**Table I. Empirical Models for Mesophilic Continuous DASS**

| Response | Model | Equation |
|---|---|---|
| Efficiency (based on biodegradable VS) | Levenspiel (modified) | $\eta_b = [1 - (\theta_c/\theta)]^a$ |
| Organic matter removal rate | First order modified with Levenspiel | $RR = q*LR_b*[1 - (LR_b/LR_{b,c})]^a$ |

Notes:
- a exponent
- $LR_b$ biodegradable organic matter loading rate
- $LR_{b,c}$ critical biodegradable organic matter loading rate
- q empirical kinetic coefficient
- $\theta$ mass retention time
- $\theta_c$ critical mass retention time

- the mass transfer effects are eventually lumped in the kinetic coefficients and reaction orders;[14–17]
- the principle of the limiting rate (or bottleneck rate) in a series of reactions applies. If $r_h$, $r_a$ y $r_s$ are the hydrolysis, the acidogenesis and methanogenesis rate respectively, then

$$r_a \gg r_h \cong r_s \tag{1}$$

- the methanogenic bacteria concentration in the feed is negligible when compared to that inside the digester
- the hydrolysis rate can be modeled either as a first order in the complex substrate (B, biodegradable volatile solids), or fractional order to take into account diffusion effects,[16] or the latter modified by an inhibition factor depending on the retention time to additionally take into account both differential degradabilities of different fractions of the volatile solids[18] and the inhibitory effects of acids and pH (Poggi-Varaldo, unpublished data), see Table II.
- the methanogenesis kinetic can be represented either by Monod model with endogenous decay, or by models displaying inhibition by substrate (Table III), where the limiting substrate is the pool of volatile organic acids (VOA) expressed as acetic acid.[12,19,20] One reason to support this approach is that 70% of the methane in waste

**Table II. Kinetic Models for the Hydrolysis Rate ($r_h$)**

| Response | Model | Equation |
|---|---|---|
| $r_h$ | Levenspiel | $r_h = k_h*B*[1 - (B/B_c)]^a$ |
| $r_h$ | Haldane | $r_h = k_h/[1 + (K_h/B) + (B/K_i)]$ |
| $r_h$ | Levenspiel-Poggi | $r_h = k'_h*(B)^n*[1 - (\theta_c/\theta)]^a$ |

Notes:
- a exponent in Levenspiel and Levenspiel-like model
- B concentration of biodegradable complex organic matter, expressed as biodegradable volatile solids
- $B_c$ critical concentration of biodegradable complex organic matter
- $K_h$ affinity coefficient in the Haldane kinetic
- $k_h$ specific hydrolysis coefficient for a first order kinetic
- $k'_h$ specific hydrolysis coefficient for a fractional order kinetics
- $K_i$ inhibition coefficient in Haldane model
- n exponent or order of reaction of substrata B
- $\theta$ mass retention time
- $\theta_c$ critical mass retention time

**Table III. Kinetic Models for the Specific Uptake Rate of Volatile Organic Acids ($u_p$)**

| Response | Model | Equation |
|---|---|---|
| $u_p$ | Monod | $u_p = k_m*S*[K_s + S]$ |
| $u_p$ | Levenspiel | $u_p = k_m*[S/(K_s + S)]*[1 - (S/S_c)]^a$ |
| $u_p$ | Haldane | $u_p = k_m/[1 + (K_s/S) + (S/K_i)]$ |
| $u_p$ | Haldane gral. | $u_p = k_m/[1 + (K_s/S) + (S/K_i)^p]$ |

Notes:
- a exponent
- $k_m$ specific uptake rate coefficient
- $K_i$ inhibition coefficient
- $K_s$ affinity coefficient
- p exponent
- $S_c$ critical VOA concentration beyond which the uptake rate is zero

digesters is thought to be effected via the aceticlastic pathway[12,21] of acetic acid produced during the fermentation of complex substrates and the cleavage of C3 to C5 organic acids by syntrophic bacteria.[22]

Under these assumptions, the model is based on three main mass balances:

- mass balance of the biodegradable complex organic matter (B)
- mass balance of the VOA (S), and
- mass balance of methanogenic bacteria ($X_m$).

The general form of the mass balances is:

$$\text{Accumulation} = \text{In} - \text{Out} + \text{Generation} - \text{Consumption} \tag{2}$$

*2a. Mass balance for the complex biodegradable organic matter.* The consumption term in this balance is the hydrolysis rate $r^h$ of the substrate biodegradable volatile solids B. The hydrolytic/fermentative biomass associated to this process is not modeled because of its heterogeneity and the difficulty to find a representative parameter in DASS.[23] The balance takes the following form:

$$0 = \frac{d(MB)}{dt} = \dot{m}_o B_o - \dot{m}B - r_h M \tag{3}$$

where
- $m_o$ = feed mass flowrate (mass/time)
- m = effluent mass flowrate (mass/time)
- $B_o$ = concentration of biodegradable VS in the feed (mass bVS/mass feed)
- B = concentration of biodegradable VS in the effluent mixed solids (mass bVS/mass effluent)
- M = mass of mixed solids inside the reactor (mass reactor).

*2b. Mass balance of volatile organic acids.* If the total VOA organic concentration is represented by S and expressed on acetic acid (COD-HAc) basis, the balance takes the following form:

$$0 = \frac{d(MS)}{dt} = \dot{m}_o S_o - \dot{m}S + r_h - fM - r_s M \tag{4}$$

where
- f = factor mgCOD/mgbVS
- $S_o$ = VOA concentration in the feed (mass COD-HAc/mass feed)
- S = VOA concentration in the effluent mixed solids (mass COD-HAc/mass effluent)

*3. Mass balance of methanogenic bacteria.* The overall mass balance for methanogenic bacteria is given by Equation 5:

$$0 = \frac{d(MX_m)}{dt} = \dot{m}_o X_{m,o} - \dot{m}X_m + r_c M \tag{5}$$

where $X_{m,o}$ = methanogenic bacteria concentration in the feed (mass methanogenic bacteria/mass feed)
$X_m$ = methanogenic bacteria concentration in the effluent mixed solids (mass methanogenic bacteria/mass mixed solids effluent)
$r_c$ = methanogenic bacteria growth rate (mass of methanogenic bacteria/time).

The methanogenic bacteria growth rate ($r_c$) can be modeled with a term expressing the synthesis of biomass from VOA consumption and a component related to the endogenic decay of the biomass:[19,24]

$$r_c = Yr_s - k_d X_m \tag{6}$$

where $Y$ = yield or synthesis coefficient of the methanogenic bacteria (mass methanogenic bacteria/mass VOA)
$k_d$ = endogenous decay coefficient (1/time).

Equation 6 is further transformed into Eq. 7 dividing by the methanogenic bacteria concentration $X_m$

$$w_p = Yu_p - kd \tag{7}$$

where $w_p$ = specific methanogenic bacteria growth rate (1/time), and
$u_p$ = specific VOA uptake rate (1/time).

## Experimental Design

Digesters 3-L capacity were operated at MRT 15, 18, 21, 25, 30, and 40 days, by duplicate, at 39°C, and fed twice a week in a draw-and-fill mode. All reactors received the same feedstock (a mixture of newspaper, food waste, and biosolids) at feed rates determined by the operating MRT. Main process responses were monitored and analyzed as described elsewhere.[25] Volatile solid reduction efficiency was calculated based on mass balances.

## Procedures and Analysis

Biogas volume and composition were measured as outlined elsewhere.[6] Preparation of samples for measuring pH, alkalinity, and volatile organic acids (VOA) was based on leaching refuse and compost samples according to Kasali and Senior,[26] and further modified by Poggi-Varaldo and Rinderknecht.[27] Solids were analyzed as previously described.[7] Cellulose and lignin were determined by the Goering and Van Soest's method.[28] Total Kjeldahl nitrogen and ammonia-N were determined on 3-g homogeneized wet samples, while the substrate chemical oxygen demand was analyzed in 1-g wet sample according to *Standard Methods*.[29] The VOA were determined by gas chromatography.[6] The methanogenic biomass was quantitated by coenzyme F-420 analysis, the latter was determined by HPLC.[30] A factor of 781 μg F-420/g cell was used for calculating methanogenic bacteria concentration from coenzyme F-420 determinations.[12,21,31,32]

# RESULTS AND DISCUSSION

The feedstock was a lignocellulose-rich type of waste (9.91% lignin and 33.35% cellulose). The feedstock ultimate anaerobic degradability (UAD) as determined by two independent methods[25] gave similar results: 62.4% VS (3.0% std. dev.) from extrapolating the steady stage reactor efficiency data and 61.9% VS (3.7% std. dev.) in the batch assay, respectively. These values sug-

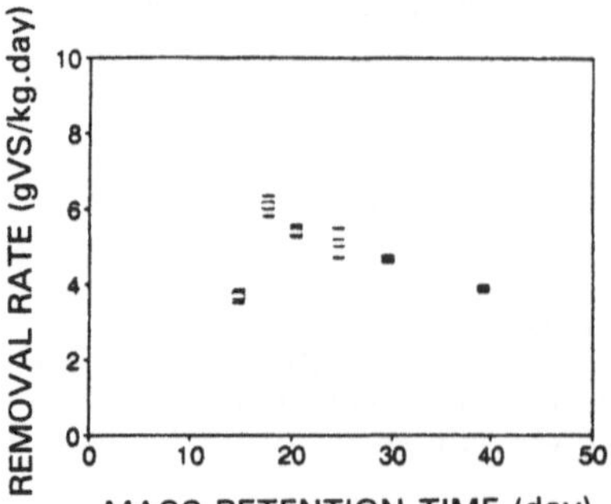


**Figure 1. Organic matter removal rate of DASS reactors at different mass retention times. The bars indicate the standard deviation of duplicate reactors.**

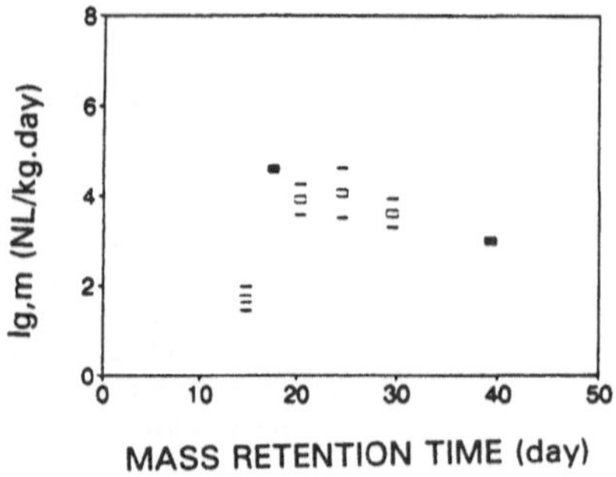


**Figure 2. Biogas productivity of DASS reactors at different mass retention times. The bars indicate the standard deviation of duplicate reactors.**

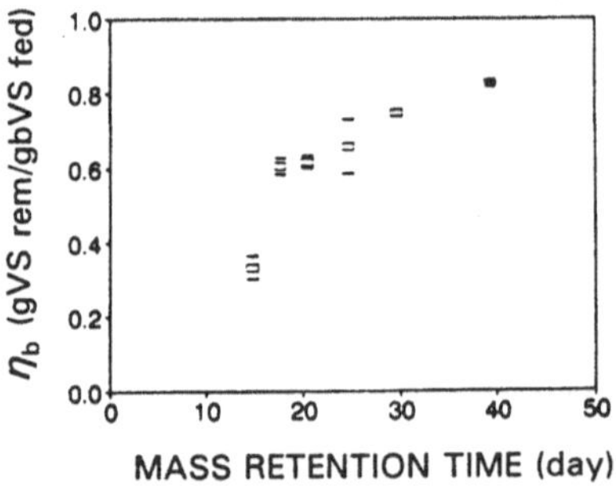


**Figure 3. Volatile solid removal efficiency (based on biodegradable VS) of DASS reactors at different mass retention times. The bars indicate the standard deviation of duplicate reactors.**

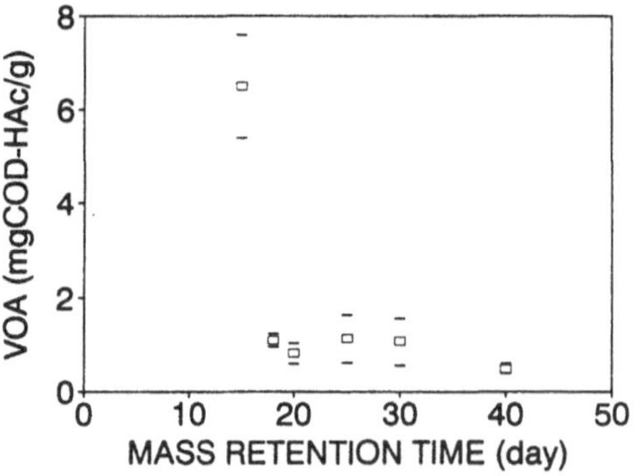


**Figure 4. Volatile organic acid concentration in DASS reactors at different mass retention times.**

gested a low-to-moderate degradability, as compared to values up to 90% VS of segregated wastes reported in the European practice.[3] On the other hand, our results were close to that found for a lignocellulose-rich substrate in thermophilic pilot trials.[33]

## Digester Performance

The overall performance of the reactors is discussed in terms of the organic matter removal rate, volatile solid removal efficiency, biogas productivity, and VOA contents inside the reactors, as shown in Figures 1 to 4. The $\eta_b$ increased in the range 60% to 83% with increasing MRT (18 to 40 days MRT). However, the $\eta_b$ was very poor (27%) at 15 days MRT (see Fig. 3). The overall performance indicated a partial process inhibition at 15 days MRT, as it is shown by the lower biogas productivity (Fig. 2) and higher VOA contents (Fig. 4) in these reactors. Also, reactors operated at 15 day MRT had the lowest content of coenzyme F-420 (data not shown).

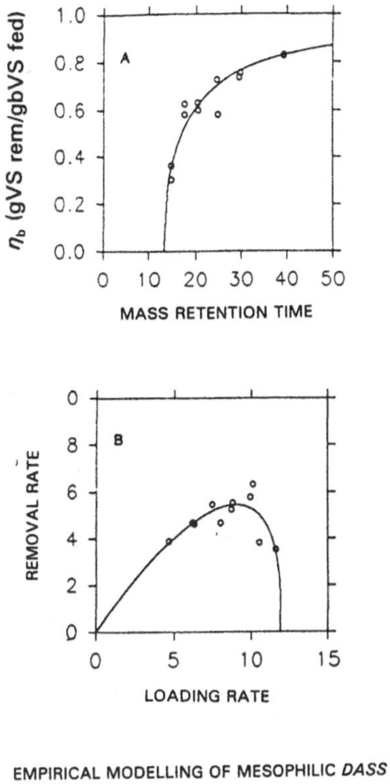


**Figure 5. Fitting of empirical models to continuous mesophilic DASS performance. (A). Biodegradable volatile solids removal efficiency versus mass retention time. (B). Organic matter removal rate versus loading rate (based on biodegradable volatile solids).**

## Empirical Model Fitting

As was mentioned above, models based on Levenspiel analogy were applied. A first model related the efficiency vs. the MRT; the second related the removal rate vs. loading rate. Both models fitted the results reasonably well (Fig. 5 and Table IV). Process complete inhibition was predicted for a 13.3 days critical MRT and a critical loading rate $LR_{b,c}$ of 11.9 g bVS/kg.day (which translates into a MRT of 14.5 days), respectively.

**Table IV. Parameter Values of the Empirical Models Fitted to Continuous Mesophilic DASS**

| Response | Model | $\Theta_c$ [1] | $LR_{b,c}$ [2] | a [3] | Norm |
|---|---|---|---|---|---|
| $\eta_b$ | Levenspiel (modified) | 13.3 | — | 0.451 | 0.180 |
| Removal rate | First order plus Levenspiel | — | 11.94 | 0.327 | 1.831 |

Notes:
[1]in days
[2]in g bVS/kg.day
[3]dimensionless

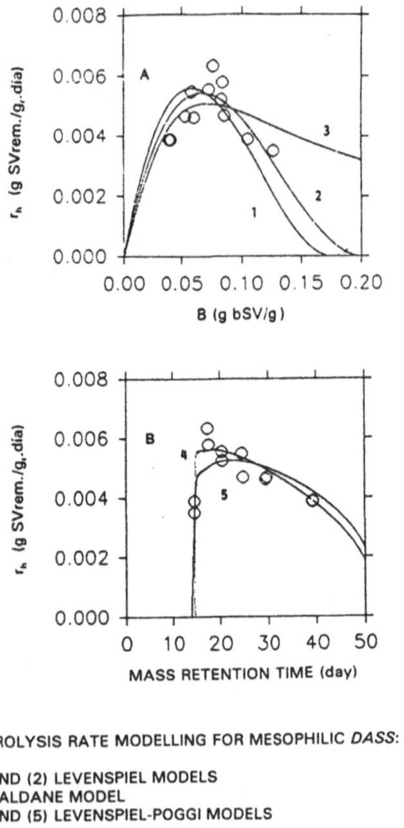


**Figure 6. Model fitting for the hydrolysis rate. (A) Levenspiel (1,2) and Haldane (3) models. (B). Levenspiel-Poggi (4,5) models.**

## Kinetic Model Fitting

Haldane and Levenspiel-like models were fitted to the hydrolysis rate $r_h$ (see Fig. 6). The best fit was obtained with the Levenspiel-like ones (Table V) with fractional order in volatile solids B (approx. 0.5), which strongly suggested the importance of mass transfer effects.[13] An examina-

**Table V. Parameter Values of the Different Kinetic Models Fitted to the Hydrolysis Rate of DASS**

| Model | $k_h$[1] | $B_c$[2] | $K_s$[2] | $K_i$[2] | $k'_h$[3] | $\Theta_c$[4] | n[5] | a[5] | Norm |
|---|---|---|---|---|---|---|---|---|---|
| Levenspiel (1)[6] | 0.216 | 0.175 | — | — | — | — | 1 | 2 | 0.0029 |
| Levenspiel (2) | 0.184 | 0.20 | — | — | — | — | 1 | 2 | 0.0033 |
| Haldane (3) | 2.00 | — | 13.93 | 0.00036 | — | — | — | — | 0.0022 |
| Levenspiel-Poggi (4 | — | — | — | — | 0.019 | 14.75 | 0.464 | 0.063 | 0.0019 |
| Levenspiel-Poggi (5) | — | — | — | — | 0.016 | 14.00 | 0.384 | 0.168 | 0.0027 |

Notes:
1. in 1/day
2. in g bVS/kg wm
3. in (g bVS/kg wm)$^{1-n}$/día
4. in day
5. dimensionless
6. the number after the model name in the Model column refers to the curves in Figure 6.

**Table VI. Values of the Hydrolysis Rate Kinetic Coefficient from Several Works in Anaerobic Digestion**

| | | Hydrolysis Coefficient | | |
|---|---|---|---|---|
| Process[1] | Subtrate | $k_h$(1/d) | $k'_h$ ($gbSV^{1-n}/g^{1-n}.d$) | Reference |
| Batch DAN líq.[2] | cellulose | 0.17 in sludge | — | 19 f. |
| Batch DAN f. líq. | cellulose | 0.12 | — | 37 |
| Batch DAN f. líq. | pig manure and straw | 0.02 | — | 38 |
| Cont. DAN f. líq. | cellulose in cattle manure | 0.05 | — | 18 |
| Cont. DAN f. líq. | cellulose | 0.28–0.52 | — | 34 |
| Cont. DAN f. líq. | primary sludge | 0.20 | — | 39 |
| Cont. DAN f. líq. | primary sludge | 0.37–0.74 | — | 35 |
| Batch DAN f. semilíq.[3] | newspaper and cattle manure | 0.026 | — | 36 |
| Batch DASS | newspaper and cattle manure | 0.013 | — | " |
| Cont. DASS | cattle manure and straw | 0.023[4] | — | " |
| Cont. DASS | MSW and sludge | 0.18–0.22 | 0.016–0.019 | this work |

Notes:

1. all processes are mesophilic
2. DAN = digestión anaerobia, f.l'q. = liquid or slurry mode
3. f. semil'q = semi-liquid, at 10% TS
4. estimated from biogas production assuming density 800 kg/m$^3$, 20% reactor VS, substrate estimate degradability 60%, and biogas yield 0.72 NL/gVS rem.

tion of Table VI allows for the comparison of the values of hydrolysis coefficients obtained in this work with those reported by other researchers. It is apparent that the values from continuous DASS are two to three times lower than those found by Speece and McCarty[34] and O'Rourke[35] for cellulose and primary sludges in slurry-mode anaerobic digestion. On the other hand, the coefficient values for continuous DASS favorably compared with those reported by Wujcik[36] for batch DASS of newspaper and cattle manure mixtures.

Figure 7 and Table VII show that the volatile organic acid uptake rate $u_p$ best fit was given by a Haldane model (inhibition by substrate S; i.e., the total volatile organic acids concentration). The inhibition coefficient value for the Haldane kinetics in this work (16300 mg COD/kg) was of the same order of those reported in the literature[20] (Table VII). However, an examination of Table VIII shows that the value of the maximum specific VOA uptake coefficient (0.92 mg COD/mg cell.day) was much lower than those found in the literature for liquid or slurry-mode anaerobic digestion. This again suggests that mass transfer retardation effects lumped into the biokinetic coefficient could be responsible for the low values.

The methanogenic bacteria growth rate $w_p$ was fitted by a linear model in $u_p$ (Fig. 8 and Table VIII). The yield and the endogenous decay coefficient were Y = 0.049 mg cell/mg COD-HAc

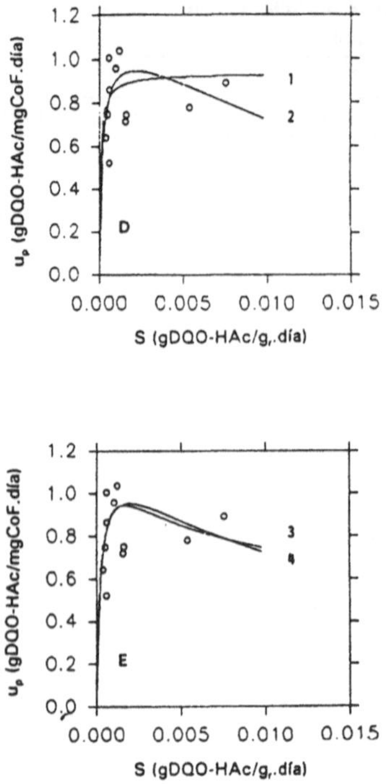


VOLATILE ORGANIC ACIDS UPTAKE RATE MODELLING FOR *DASS*:

(1) MONOD; (2) LEVENSPIEL
(3) HALDANE; (4) GENERALIZED HALDANE

**Figure 7. Model fitting for the specific VOA uptake rate. (A) Monod (1) and Levenspiel (2) models. (B) Haldane (3) and generalized Haldane (4) models.**

and $k_d$ = 0.001 1/day, respectively (Fig. 9, Table IX, and Table X). The yield value fell in the middle of the range reported for yields of both pure culture methanogenic bacteria and anaerobic consortia (Table IX); however, the endogenic decay coefficient $k_d$ was 10 to 20 times lower than the corresponding values found for liquid phase anaerobic digestion (Table X).

**Table VII. Parameter Values of the VOA Specific Uptake Rate Fitted to Continuous Mesophilic DASS**

| Model | $k_m{}^1$ | $K_s{}^2$ | $K_i{}^2$ | $S_c{}^2$ | $p^3$ | $a^3$ | Norm |
|---|---|---|---|---|---|---|---|
| Monod | 0.933 | 0.000775 | — | — | — | — | 0.711 |
| Haldane | 1.18 | 0.000232 | 0.01630 | — | — | — | 0.686 |
| Haldane | 2.00 | 0.000521 | 0.00289 | — | 0.40 | — | 0.675 |
| Levenspiel | 1.124 | 0.000210 | — | 0.0525 | — | 2.0 | 0.683 |

Notes:
1. in g COD/mg CoF.day
2. in g COD-HAc/g wm
3. dimensionless

**Table VIII. Values of the VOA Uptake Rate Kinetic Coefficients from Several Works in Anaerobic Digestion**

| Model | $k_m$[1] | $K_s$[2] | $K_i$[2] | $S_c$[2] | p | Reference |
|---|---|---|---|---|---|---|
| Monod | 8.7 | 165 | — | — | — | 19 |
| Monod | 2.6-5.1 | NR[3] | — | — | — | 40 |
| Monod | 8.5 | 185 | — | — | — | 41 |
| Monod | 2.6-11.6 | 11-421 | — | — | — | 13 |
| Haldane | NR | NR | 7980[4] | — | — | 20 |
| Haldane | NR | NR | 59860[5] | — | — | 42 |
| Haldane generalized[6] | 0.19 | 1667 | 2300 | — | 4 | 43 |
| Monod[7] | 0.73 | 775 | — | — | — | this work |
| Levenspiel[7] | 0.88 | 210 | — | 52500 | — | ″ |
| Haldane[7] | 0.92 | 232 | 16300— | — | ″ | ″ |
| Haldane[7] generalized | 1.56 | 6521 | 2890 | — | 0.40 | ″ |

Notes:
1. in mg COD/mg VS.day OR mg COD/mg cell.day (DASS)
2. in mg COD/L or mg COD/kg (DASS)
3. NR = not reported
4. estimated from reported value 40 mg HAc (unionized)/L, assuming pH 7 and acetic acid $pK_a$ of 4.7
5. estimated from reported value 300 mg HAc (unionized)/L and similar to Note 4.
6. the substrate was phenol, 2.38 mg COD/mg phenol
7. $k_m$ values estimated from values in Table VII and an average content of 781 μg CoF-420/g cell

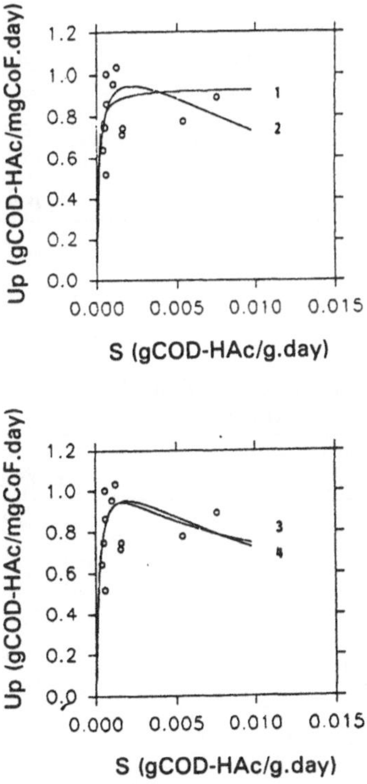


VOLATILE ORGANIC ACIDS UPTAKE RATE MODELLING FOR *DASS*:

(1) MONOD; (2) LEVENSPIEL
(3) HALDANE; (4) GENERALIZED HALDANE

**Figure 8. Model fitting for the specific VOA uptake rate.**

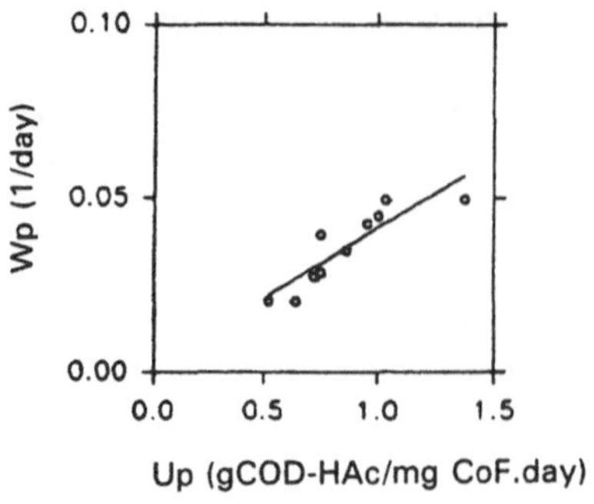


**Figure 9. Model fitting for the specific methanogenic growth rate.**

**Table IX. Values of the Methanogenic Bacteria Yield from Several Works in Anaerobic Digestion**

| | Synthesis Yield Coefficient | | |
|---|---|---|---|
| Culture or Process[1] | g/mol Methane | g cell/g COD | Reference |
| *Methanosarcina barkeri* | 1.1–1.9 | 0.046–0.079 | 44 |
| *Methanosarcina barkeri* str. TM1 | 1.6–2.0 | 0.067–0.083 | " |
| Mixed, cont. DAN f. líq.[2] | | 0.041–0.054 | 19 |
| Mixed, cont. DAN f. líq. | | 0.040 | 41 |
| Mixed, cont. DAN f. líq. | | 0.010–0.054 | 13 |
| Mixed, cont. DASS | | 0.049 | this work |

Notes:
1. all cultures and processes are mesophilic
2. DAN = anaerobic digestion, f.líq. = liquid or slurry mode

**Table X. Endogenous Decay Coefficient Values of Methanogenic Biomass in Several Anaerobic Digestion Processes**

| Culture or Process[1] | Endogenous Decay Coefficient (1/day) | Reference |
|---|---|---|
| Mixed, cont. DAN f. líq.[2] | 0.015 | 19 |
| Mixed, cont. DAN f. líq. | 0.036 | 41 |
| Mixed, cont. DAN f. líq. | 0.004–0.037 | 13 |
| Pure and continuous, *M. barkeri* | 0.004 | 45 |
| Mixed, cont. DASS | 0.0004–0.008 | this work |

Notes:
1. all cultures and processes mesophilic
2. DAN = anaerobic digestion, f.líq. = liquid or slurry mode

## CONCLUSIONS

The findings from our efforts on modeling the mesophilic continuous DASS of solid wastes and biosolids can be summarized as follows:

- empirical models fitted the steady state efficiency and removal rate of mesophilic DASS reasonably well; they predicted process cessation at MRT between 13 and 14.75 days
- the kinetic model of the mesophilic DASS based on the series hydrolysis/acidogenesis/methanogenesis concept, modified for inhibition, also gave an adequate representation of the experimental results
- for the hydrolysis rate, a model with fractional order in volatile solids (approx. 0.4) and a Levenspiel-like inhibition factor in terms of the mass retention time gave the best fitting
- the Haldane model was the best representation of the volatile organic acids uptake rate
- the methanogenic bacteria-specific growth rate was proportional to the volatile organic uptake rate, except for a departure of the reactors operated a 15 day MRT
- the lower values found for the hydrolysis rate coefficient and the maximum specific VOA uptake kinetic coefficient in this work, as compared with literature data from liquid anaerobic digestion, seemed to indicate the influence of mass transfer retardation effects.

## ACKNOWLEDGMENTS

Dr. R. Thauer, Dr. G.D. Vogels, and Dr. L.G.M. Gorris supplied coenzyme F-420 standards and helpful hints for its HPLC analysis. Their contributions are greatly appreciated.

## REFERENCES

1. Poggi-Varaldo, H.M., Valdés, L., Esparza-García, F., and Fernández-Villagómez, G. "Solid substrate anaerobic co-digestion of paper mill sludge, biosolids and municipal solid wastes." *Water Sci. Technol.* 35(2–3), pp. 197–204, 1996.
2. CINVESTAV. Project Proposal. Solid Substrate Anaerobic Digestion of Municipal and Industrial Wastes. México, D.F., 1994. (In Spanish).
3. Cecchi, F., Traverso P.G., Mata-Alvarez J., Clancy J., and Zaror, C. "State of the art of R&D in the anaerobic digestion process of municipal solid waste in Europe." *Biomass*, 16, pp. 257–284, 1988.
4. de Jong, H.B.A, Koopmans, W.F., van der Knijff, A., and Jumelet, H.J. Conversion Techniques for Vegetable, Garden and Fruit Waste. Haskoning Report, Nijmegen, The Netherlands, 1993.
5. Poggi-Varaldo, H.M., Yu, A., and Rinderknecht, N. "Dry anaerobic digestion: A review and feasibility study," *R'95 Recovery, Recycling, Re-integration*, Feb. 1–4, Geneve, Switzerland, 1995.
6. Poggi-Varaldo, H.M. and Valdéz-Ledezma, L. "Optimization of dry anaerobic digestion start-up," *17th Canadian Waste Management Conference* , Sept. 11–14, Quebec City, Canada, 1995.
7. Poggi-Varaldo, H.M. and Oleszkiewicz, J. "Anaerobic co-composting of municipal solid waste and waste sludge at high total solids levels," *Environ. Technol.*, 13(3), pp. 409–421, 1992.
8. Rivard, C.J., Himmel, M.E., Vinzant, T.B., Adney, W.S., Wyman, C.E., and Grohmann, K. "Anaerobic digestion of processed municipal solid waste using a novel high solids reactor: maximum solids levels and mixing requirements," *Biotechnol. Lett.*, 12(3), pp. 185–190, 1990.
9. Cecchi, F., Marcomini, A., Pavan, P., Fazzini, G., and Mata-Alvarez, J. "Mesophilic digestion of the organic fraction of refuse: performance and kinetic study," *Waste Manage. Res.* 8, pp. 33–44, 1990.
10. Pfeffer, J.T. "Temperature effects on anaerobic fermentation of domestic refuse," *Biotechnol. Bioeng.* 26, pp. 771–787, 1974.

11. Wujcik, W.J. and Jewell W.J. "Dry anaerobic fermentation," *Biotechnol. Bioeng. Symp.*, 10, pp. 43–65, 1980.
12. McCarty, P.L. "Anaerobic waste treatment fundamentals. III." *Public Works*, 95, pp. 91–94, 1964.
13. Pavlostathis, S.G. and Giraldo-Gómez, E. "Kinetics of anaerobic treatment." *Water Sci. Technol.*, 24(8), pp. 35–61, 1991.
14. Contois, D.E. "Kinetics of bacterial growth: relationship between population density and specific growth of continuous cultures." *J. Gen. Microbiol.*, 21, pp. 40–50, 1959.
15. Harremöes, P. Biofilm Kinetics, In: Mitchell R., Ed., *Water Pollution Microbiology*, Vol. 2, Chapter 4, John Wiley, New York, 1978, pp. 71–111.
16. Henze, M. and Harremöes, P. "Anaerobic treatment of wastewater in fixed-film reactors-a literature review." *Water Sci. Technol.*, 15, pp. 1–101, 1983.
17. Suidan, M.T., Ritmann, B., and Traeguer, V.K. "Criteria of establishing biofilm kinetics." *Water Res.*, 21, pp. 481–498, 1987.
18. Singh, R., Jain, M.I., and Tauro, P. "Rate of anaerobic digestion of cattle waste." *Agric. Wastes*, 4, pp. 267–272, 1982.
19. Lawrence, A.W. and McCarty, P.L. "Kinetics of methane fermentation in anaerobic treatment." *J. Water Pollut. Control Fed.*, 41, pp. R1–R17, 1969.
20. Andrews, J.F. and Graef, S.P. Dynamic Modelling and Simulation of the Anaerobic Digestion Process. In: Pohland F.G., Ed., *Anaerobic Biological Treatment Processes*, Adv. Chem. Series, 105, American Chemical Society, Washington, D.C., 1971, pp. 126–162.
21. Tilche, A. and Shan, Y.X. Light and Scanning Electron Microscope Observations on the Granular Biomass of Experimental SBAF and HABR Reactors. *International Workshop Granular Anaerobic Sludge-Microbiology and Technology.* GASMAT, Lunteren, Netherlands, 25–27 October, 1987.
22. McCarty, P.L. and Mosey, F.E. "Modelling of anaerobic digestion processes—a discussion of concepts." *Water Sci. Technol.*, 24(8), pp. 17–35, 1991.
23. Mata-Alvarez, J. and Cecchi, F. Kinetic Models Applied to the Anaerobic Biodegradation of Complex Organic Matter. In: Kamely, D., Chakrabarty, A., and Omenn, G.S. "Biotechnology and biodegradation." *Adv. Appl. Biotechnol. Ser.*, 4, pp. 318–344, 1989.
24. Metcalf & Eddy. *Wastewater Engineering: Treatment, Disposal and Reuse,* 3rd. edition, McGraw-Hill Co., New York, 1991.
25. Poggi-Varaldo, H.M., Rodríguez-Vázquez, R., Fernández-Villagómez, G., and Esparza-García, F. "Inhibition of mesophilic solid-substrate anaerobic digestion by ammonia nitrogen." *App. Microbiol. Biotechnol.* 47(3), pp. 284–291, 1997.
26. Kasali, G.B., and Senior, E. "Effects of temperature and moisture on the anaerobic digestion of refuse," *J. Chem. Technol. Biotechnol.*, 44(1), pp. 31–41, 1989.
27. Poggi-Varaldo, H.M. and Rinderknecht, N. Mini-Handbook of Solid Waste Analysis, CINVESTAV del IPN, México D.F., México, 1990. (In English, available upon request).
28. Goering, H.K. and Van Soest, P.J. Forage Fiber Analysis, U.S. Dept. of Agriculture Handbook No. 379. 1970.
29. *Standard Methods for the Examination of Water and Wastewaters,* 16th edition, APHA-WPCF-AWWA, Washington, D.C., 1985.
30. Gorris, L.G.M., van der Drift, C., and Vogels, G.D. "Separation and quantification of cofactors from methanogenic bacteria by high-performance liquid chromatography: Optimum and routine analysis," *J. Microbiol. Methods,* 8, pp. 175–190, 1988.
31. Gorris, L.G.M. and van der Drift, C. Methanogenic Cofactors in Pure Cultures of Methanogens in Relation to Substrate Utilization. In: H.C. Dubourguier, G. Albagnac, J. Montreuil, C. Romond, P. Sautiere, and J. Guillaume, Eds.. *Biology of Anaerobic Bacteria.* Elsevier Sci. Publ. B.V., Amsterdam, 1986, pp. 144–150.
32. Peck, M.W. "Changes in concentrations of coenzyme F 420 analogs during batch growth of *Methanosarcina barkeri* and *Methanosarcina mazei.*" *Appl. Environ. Microbiol.*, 55(4), pp. 940–945, 1989.
33. Kayhanian, M., and Tchobanoglous, G. "Innovative two-stage process for the recovery of energy and compost from the organic fraction of solid wastes," *Water Sci. Technol.*, 27(2), pp. 133–140, 1993.
34. Speece, R.E. and McCarty, P. "Nutrient requirements and biological solids accumulation in anaerobic digestion." In: W.W. Eckenfelder, Ed.. *Advanced Water Pollution Research,* Vol. 2, Pergamon Press, New York, 1964.
35. O'Rourke, J.T. Kinetics of Anaerobic Treatment at Reduced Temperatures. Ph.D. thesis, Stanford University, Stanford, CA, 1968.

36. Wujcik, W.J. Dry Anaerobic Fermentation to Methane of Organic Residues. Ph.D. thesis, Cornell University, Ithaca, NY, 1980.
37. Greco, R.L., Coto, J.M., Dentel, S.K., and Gossett, J.M. Aluminum-Organic Interactions Influencing Anaerobic Digestion of Coagulated Substrates. Technical Report, Cornell University, Environ. Eng. Dept., Ithaca, NY, 1983.
38. Llabres-Luengo, P. and Mata-Alvarez, J. "Kinetic study of anaerobic digestion of straw-pig manure mixtures." *Biomass*, 14, pp. 129–142, 1987.
39. Eastman, J.A. and Ferguson, J.F. "Solubilization of particulate organic carbon during the acid phase of anaerobic digestion". *J. Water Pollut. Control Fed.*, 53, pp. 352–366, 1981.
40. van den Berg, L. "Effect of temperature on growth and activity of a methanogenic culture utilizing acetate." *Can. J. Microbiol.*, 23, pp. 898–902, 1977.
41. Kugelman, I.J. and Chin, K.K. Toxicity, Synergism, and Antagonism in Anaerobic Waste Treatment Processes. In: F.G. Pohland, Ed.. *Anaerobic Biological Treatment Processes*. Adv. Chem. Series, American Chemical Society, Washington D.C., 1971, pp. 55–90.
42. Hill, D.T. and Barth C.L. "A dynamic model for simulation of animal waste digestion." *J. Water Pollut. Control Fed.*, 49, pp. 2129–2143, 1977.
43. Neufeld, R.D., Mack, J.D., and Strakey, J.P. "Anaerobic phenol biokinetics." *J. Water Pollut. Control Fed.*, 52, pp. 2367–2377, 1980.
44. Taylor, G.T. The Methanogenic Bacteria. In: Bull, M.J., Ed. *Progr. Ind. Microbiol.*, Elsevier, UK, 16, pp. 233–329, 1982.
45. Wandrey and Aivisidis. "Continuous anaerobic digestion with *Methanosarcina barkeri.*" *Ann. New York Acad. Sci.*, 413, pp. 489–500, 1983.

## LIST OF ABBREVIATIONS

| | |
|---|---|
| a | exponent in Levenspiel or Levenspiel-like kinetic |
| B | concentration of complex organic matter expressed on biodegradable volatile solids basis |
| $B_c$ | critical concentration of complex organic matter expressed on biodegradable volatile solids basis |
| $B_o$ | concentration of complex organic matter in the feed, expressed on biodegradable volatile solids basis |
| bVS | biodegradable volatile solids |
| COD | chemical oxygen demand |
| CoF420 | coenzyme F-420 |
| DAN | anaerobic digestion |
| DASS | solid-substrate anaerobic digestion |
| f.líq. | at low total solid contents, liquid or slurry mode |
| HAc | acetic acid |
| HBu | butyric acid |
| HPLC | high pressure liquid chromatography |
| HPr | propionic acid |
| HVal | valeric acid |
| $I_{g,m}$ | biogas productivity (NL biogas/kg mixed solids.day) |
| $k_d$ | endogenous decay coefficient of methanogenic bacteria |
| $K_h$ | affinity coefficient in the hydrolysis rate |
| $k_h$ | hydrolysis specific coefficient in first order kinetics on substrate B |
| $k«_h$ | hydrolysis specific coefficient in fractional order kinetics on substrate B |
| $K_i$ | inhibition coefficient in Haldane kinetics |
| $k_m$ | maximum specific coefficient of VOA uptake in the methanogenesis rate |
| $K_s$ | affinity constant in the methanogenesis rate |
| LR | loading rate |
| $LR_b$ | loading rate based on bVS |
| $LR_{b,c}$ | critical loading rate based on bVS |
| m | mass flow rate of effluent mixed solids |
| $m_o$ | feed mass flow rate |
| M | mass content inside reactor |
| MRT | mass retention time |
| MSW | municipal solid waste |
| n | order of bVS in the hydrolysis kinetics |
| NKT | Kjeldahl total nitrogen |

| | |
|---|---|
| NL | volume of biogas, in liters (corrected at 0°C and 102 bar) |
| p | exponent in the generalized Haldane kinetics |
| P | phosphorus |
| q | empirical kinetic coefficient |
| $r_c$ | growth rate of methanogenic bacteria |
| $r_h$ | hydrolysis rate |
| $r_s$ | VOA consumption rate |
| RR | organic matter removal rate |
| S | concentration of VOA (expressed on COD-HAc basis) in the effluent mixed solids |
| $S_c$ | critical concentration of VOA (expressed on COD-HAc basis) in the Levenspiel kinetics for the VOA uptake rate |
| $S_o$ | concentration of VOA (expressed on COD-HAc basis) in the feed |
| TS | total solids or dry matter |
| $u_p$ | specific uptake rate of VOA |
| VOA | volatile organic acids |
| $w_p$ | specific growth rate of methanogenic bacteria |
| $X_m$ | concentration of methanogenic biomass |
| Y | yield synthesis coefficient of methanogenic bacteria |

**Greek Characters:**

| | |
|---|---|
| $\eta_b$ | organic matter removal efficiency (based on biodegradable VS) |
| $\theta$ | mass retention time |
| $\theta_c$ | critical mass retention time |

# Section Nine
# PHYSICAL-CHEMICAL: GENERAL

## 33 IMMOBILIZATION OF Pb(II) AND Cd(II) IN BLASTING WASTES BY PORTLAND CEMENT COMPOUNDS

**Brian A. Dempsey, Charles S. Tricou, Wenjun Liu, and Matthew Hemming**
Department of Civil & Environmental Engineering
The Pennsylvania State University
212 Sackett Building
University Park, Pennsylvania 16802-1408

### ABSTRACT

This report addresses the effectiveness of Portland Cement-Soluble Silicates (PCSS) blasting additives for reduction of TCLP metals and for reduction of long-term liabilities upon disposal. The work was performed on blasting wastes from the Marine Corps Logistics Base (MCLB) in Albany, Georgia. TCLP Cd exceeded TCLP limits in half of Black Beauty samples without PCSS or with only 10% (w/w) PCSS. Cd exceedances were eliminated with 15 or 20% PCSS. Pb exceeded TCLP limits on 15% of regular blasting wastes, but 10% PCSS eliminated all Pb exceedances. Steel blasting wastes had 83% exceedances for TCLP Cd, and plastic or glass wastes frequently exceeded TCLP Pb and Cd.

Leaching experiments were performed on each blasting waste with and without laboratory addition of PCSS. If pH was continuously adjusted with rapid mixing so that the leaching environment never rose above pH 5, then every sample (including those with PCSS) exceeded TCLP standards for Cd and there were several exceedances for Pb.

These results are consistent with thermodynamic principles modified by chemical kinetic considerations. High temporary pH permits instantaneous precipitation of $PbSiO_3(s)$, $CdSiO_3(s)$, or calcium silicate hydrate (CSH). These materials then dissolve very slowly. PCSS is very alkaline, keeping the TCLP test pH well above 5 and preventing immediate acid titration for all reasonable disposal scenarios.

PCSS was not physically separated from the blasting materials during dry sieving, and inconsequential separation of PCSS materials from blasting wastes are expected during accumulation, storage, transfer, and disposal operations. This was confirmed by scanning electron microscopy (SEM) and energy dispersive spectroscopy (EDS).

Recommendations included: (1) segregation of wastes likely to result in TCLP exceedances; (2) reducing contamination through selection of blasting materials and through masking operations; (3) use of PCSS on wastes that may result in TCLP exceedances; (4) mono-fill disposal to maintain alkaline; and (5) studies of the long-term leachability of TCLP metals under disposal conditions.

### INTRODUCTION

The objectives were to determine the short- and long-term stability of blasting wastes with and without addition of PCSS. Short-term refers to RCRA regulations. Long-term refers to Superfund regulations, liabilities, and environmental responsibilities.

The work included an initial site visit, evaluation of historic data, critical evaluation of PCSS materials, theoretical investigation of the effects of PCSS on the reactivity and mobility of trace

*52nd Purdue Industrial Waste Conference Proceedings*, 1997, Ann Arbor Press, Chelsea, Michigan 48118. Printed in U.S.A.

metals in blasting wastes, and experimental evaluation of the effects of PCSS on trace metals in blasting wastes. The experimental work included measurements of total and leachable Pb, Cd, and Cr for three blasting wastes (Black Beauty, garnet, and plastic) with and without addition of 15% (w/w) PCSS and also evaluation of the possible physical separation of PCSS from the blasting waste during storage and transportation.

## MATERIALS AND METHODS

Blasting wastes from plastic, Black Beauty, and garnet blasting operations were analyzed for total and extractable metal concentrations. These samples plus a PCSS material (Blastox©) were analyzed by Scanning Electron Microscopy (SEM), Energy Dispersive Spectroscopy (EDS), Atomic Absorption Spectroscopy (AAS), and colorimetric techniques according to conventional techniques. Leachable metals were determined by techniques that are described below. Total extractable metals were determined using concentrated $HNO_3$ and HCl at boiling temperatures.

Dry sieving was performed using standard U.S. mesh screens, stacked on a conventional shaking device. The fraction of PCSS in sieved samples was determined by discontinuous acidimetric titration, which took advantage of the observation that more than 99% of total alkalinity was due to PCSS in mixtures that contained 85% blasting waste and 15% PCSS.

## REVIEW OF HISTORIC DATA

Historic data are summarized in Table I, which describes the number of MCLB samples that exceeded TCLP standards for each category of blasting material with and without preblast addition of PCSS. Cd was the most frequent problem. Addition of 10% PCSS reduced Pb to below TCLP levels, and 15% PCSS reduced Cd to below TCLP levels. PCSS did not reduce TCLP Cr.

Data in Table II indicates that long-term release of Cd would not be a concern for the Black Beauty wastes, since 96% of the total extractable Cd in Black Beauty wastes without PCSS was leached during a single TCLP test, which left only 4% of total extractable Cd for all subsequent leaching. Addition of 10% PCSS reduced TCLP Cd to 66% of total extractable Cd, due to the precipitation of sparingly soluble precipitates of $CdSiO_3(s)$ or due to encapsulation of Cd within Calcium Silicate Hydrate (CSH). Addition of 15% or 20% PCSS reduced the TCLP Cd to less than 3% of the total extractable Cd.

**Table I. Samples Characterized as Hazardous Due to Excessive TCLP Extractable Metals**

| Blasting waste | Hazardous TCLP Pb | Hazardous TCLP Cd | Hazardous TCLP Cr |
|---|---|---|---|
| TCLP Standard | 5 ppm | 1 ppm | 5 ppm |
| Black Beauty | 8 out of 56 samples | 25 out of 56 samples | 2 out of 56 samples |
| BB w/10% PCSS | 0 out of 10 samples | 8 out of 10 samples | 0 out of 10 samples |
| BB w/15% PCSS | 0 out of 10 samples | 0 out of 10 samples | 0 out of 10 samples |
| BB w/20% PCSS | 0 out of 10 samples | 0 out of 10 samples | 1 out of 10 samples |
| Steel | 0 out of 6 samples | 5 out of 6 samples | 0 out of 6 samples |
| Plastic | 2 out of 3 samples | 1 out of 3 samples | 1 out of 3 samples |
| Glass | 2 out of 3 samples | 3 out of 3 samples | 0 out of 3 samples |
| Total | 12 out of 98 samples | 42 out of 98 samples | 4 out of 98 samples |

**Table II. TCLP Extractable Metals as Percentage of Total Extractable Metal (Historic Samples from MCLB)**

| Blasting Waste | TCLP Pb as % of Total Extractable Pb | TCLP Cd as % Total Extractable Cd | TCLP Cr as % of Total Extractable CR |
|---|---|---|---|
| Black Beauty | 2.1 | 96 | 7.4 |
| BB w/10% PCSS | 1.8 | 64 | 6.5 |
| BB w/15% PCSS | <1.0 | <3 | 10.0 |
| BB w/20% PCSS | <1.0 | <3 | 17.3 |

**Table III. Examples of PCSS Solidification Processes (from Conner, 1990)**

| | |
|---|---|
| Chemfix® Process | Chemfix Technologies, Inc., Kenner, Louisiana |
| Fujimasu Process, SoliRoc | Fujimasu Synthetic Chemical Laboratory Company, Ltd. |
| ETA Process | Environmental Technologies Associates, Lima, Ohio |
| SolidTek EnvironGuard/Protek/Profix™ | SolidTek, Inc., Atlanta, Georgia |
| Blastox® | The TDJ Group, Inc., Cary, Illinois |
| N® and Portland Cement | PQ Corporation, Valley Forge, Pennsylvania |

Some additional problems were identified through analysis of historic data.

- Steel blast resulted in high total concentrations of Pb, Cd, and Cr. This could be due to contamination in the steel itself, and causes concern about long-term liability due to potentially increased long-term leaching of trace metals.
- Use of plastic or glass media resulted in an increased percentage of samples that failed the TCLP test, compared to Black Beauty blasting wastes.
- More PCSS resulted in increased leaching of Cr. This could be due to release of the anionic chromates at the higher pH values that are caused by addition of PCSS.

## Use of PCSS to Stabilize Waste Material

Some PCSS products are listed in Table III. Other formulations have been patented or described in the literature (Conner, 1990).

PCSS products contain a mixture of silicate materials. Hock et al.(1996) reported that Blastox© contained 65.5% CaO, 22.1% SiO2, 4.6% Al2O3, 3.55% MgO, 2.1% Fe2O3, and less than 0.5% of several other constituents. This formulation is similar to Type I cement.

PCSS immobilize metals by precipitation, adsorption, or encapsulation. Precipitation means that a new solid phase is formed having a fixed atomic ratio between the removed metals and silicate. Adsorption means that metals become attached to the surface of silica or a silicate material. Encapsulation is often considered the primary mechanism for solidification and stabilization processes and occurs when contaminants are trapped inside CSH precipitates. CSH forms after calcium silicate (e.g., portland cement) and water are mixed and has the generic chemical formula 3CaO:2SiO2:3H2O.

When soluble silicates are combined with water, silica gels or activated silicas are formed. Adsorption of trace metals onto silica has been studied by many investigators. Significant adsorption occurs at one to two pH units below the pH required for precipitation of the metal hydroxides, and removal of metals usually increases from less than 10% to greater than 90% within one to two pH units. Pb adsorption occurs at lower pH than Cd adsorption. As the pH is increased, metals may be first adsorbed and then precipitated as a new phase.

Figure 1 demonstrates the theoretical solubility of PbSiO3(s) and CdSiO3(s) if the total dissolved Si is either 0.01 or 0.001 M. The flat lines on the left side of the curves mean all of the added Pb or Cd was dissolved at these acidic pH values. Precipitation of PbSiO3(s) and CdSiO3(s) is sufficient to achieve compliance with TCLP standards when pH is greater than about 7.5 for Pb, or greater than about 8.5 for Cd. As the pH is increased above these levels, the precipitated removals increase.

When water is added to blasting waste that contains PCSS, the pH rises well above the pH values that permit precipitation of Pb or Cd silicates or CSH. Although the TCLP is nominally performed at pH 5, the pH of the test rapidly rises above 10 due to the dissolution of PCSS, so that precipitation of Pb and Cd silicates is likely during the TCLP test. Subsequent dissolution of these solids is slow so that these precipitates could persist even if the pH changes to make the precipitates undersaturated.

Disposal in a dedicated monofill should be considered for these blasting wastes, especially if some waste has been treated with PCSS. Under these conditions the pH will remain high due to the silicate materials and due to lack of acid generation from microbial decomposition of co-filled organic materials.

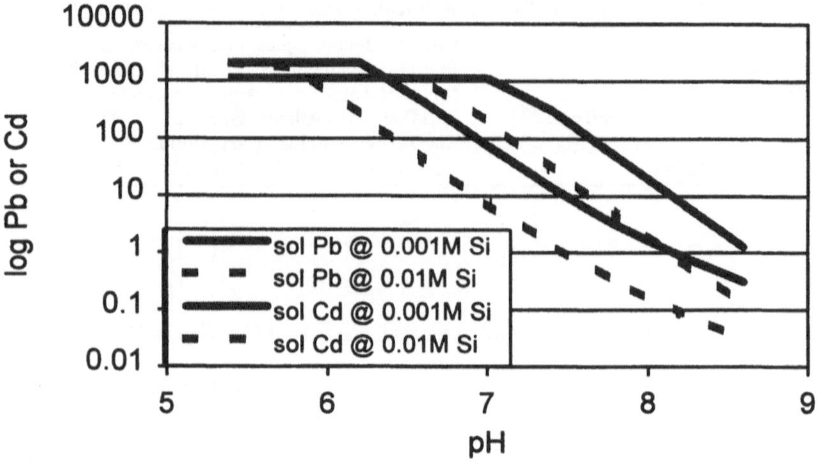


**Figure 1. Solubility of PbSiO3(s) and CdSiO3(s) when total monomeric silica is 0.001 M or 0.01 M.**

## Leaching of Trace Metals from MCLB Blasting Wastes

The TCLP test requires selection of an appropriate leaching solution, which will either be a mixture of acetic acid and acetate if the pH after 24 hours (with this solution) remains slightly acidic, or will be acetic acid only if the sample tends to increase the pH substantially. However, the TCLP test does not require subsequent adjustment of the pH, even if the value rises well above the target pH 5 value. Since PCSS amendments are highly alkaline, these materials may result in pH much higher than 5 during the TCLP test.

We conducted two sets of experiments to evaluate the effects of leachate pH on the extraction of Pb, Cd, and Cr from the MCLB blasting wastes and from the MCLB blasting wastes plus 15% Blastox.

- The first set of experiments involved intermittent adjustments of pH (at 0.04, 0.4, 3, and 5 days) to target values (pH 3, 5, 7, and 9). These tests were similar to the TCLP tests, which would allow long periods of high pH conditions which could be appropriate for the formation of metal silicates.
- The second set of experiments involved constant high-intensity mixing and continuous adjustment to maintain pH 5 over 12 hours. These tests were designed to rapidly impose a constant pH 5, so that there was no exposure to a higher pH that might result in precipitation of metal silicates.

The total extractable concentrations of the materials supplied by MCLB are reported in Table IV.

**Table IV. Total Extractable Concentrations of Pb, Cd, and Cr in Materials Supplied by MCLB**

| Total Concentrations | Pb (ppm) | Cd (ppm) | Cr (ppm) |
|---|---|---|---|
| Blastox© | 257 | 10 | 188 |
| Black Beauty | 360 | 47 | 520 |
| Plastic | 335 | 10 | 80 |
| Garnet | 405 | 78 | 117 |

*Extractions with Intermittent Adjustment of the pH*

Results of extractions with intermittent adjustment of pH to the designated values are shown in Tables V to VII for Black Beauty, Plastic and Garnet blasting wastes, respectively. BDL means below the detection limit. Pb, Cd, or Cr concentrations greater than the TCLP standard are shaded. Although the pH was adjusted to the reported value at the beginning of the experiment and at each of the reported times, the pH rose in between these times due to the slow dissolution of alkaline materials such as PCSS. For samples with 15% PCSS, the pH rose above 10. For brevity, results from 1 and 120 hours are not reported.

These conditions resulted in very effective removal of Pb, even in the absence of PCSS. Three samples exceeded TCLP Cd standards and four samples were between 75 and 100% of TCLP Cd standards, but addition of PCSS lowered Cd to undetectable levels. All of the Pb and Cd exceedances were for ultimate pH of 3 or 5, while all of the Cr exceedances were for pH 7 or 9.

**Table V. Intermittent pH Adjustment—Pb, Cd, and Cr (in mg/L) for 100 g of Black Beauty in 2L of Solution**

| T (hr) | pH | Pb | Pb(B) | Cd | Cd(B) | Cr | Cr(B) |
|---|---|---|---|---|---|---|---|
| 9 | 3 | 0.01 | 0.04 | 0.47 | BDL | 1.49 | 1.54 |
| 72 | 3 | 0.14 | 0.04 | 1.00 | BDL | 0.17 | BDL |
| 9 | 5 | 0.4 | | 1.11 | 0.03 | 1.07 | 1.12 |
| 72 | 5 | 0.04 | 0.04 | 0.95 | BDL | BDL | 1.47 |
| 9 | 7 | 0.12 | 0.05 | 0.11 | BDL | 3.67 | 2.06 |
| 72 | 7 | 0.03 | 0.03 | 0.12 | BDL | 4.03 | 2.1 |
| 9 | 9 | 0.03 | 0.02 | 0.05 | BDL | 3.85 | 0.47 |
| 72 | 9 | 0.02 | 0.02 | 0.01 | BDL | 3.58 | 2.71 |

(B) Means that the Black Beauty Contained 15% (w/w) of PCSS.

**Table VI. Intermittent pH Adjustment—Pb, Cd, and Cr (in mg/L) for 100 g of Plastic Waste in 2L of Solution**

| T (hr) | pH | Pb | Pb(B) | Cd | Cd(B) | Cr | Cr(B) |
|---|---|---|---|---|---|---|---|
| 9 | 3 | 0.03 | 0.03 | 0.45 | BDL | 1.93 | 1.02 |
| 72 | 3 | 0.02 | 0.06 | 0.95 | BDL | 2.84 | 1.3 |
| 9 | 5 | — | 0.02 | 0.28 | BDL | 2.1 | 1.29 |
| 72 | 5 | — | 0.03 | 0.86 | BDL | 1.26 | 2.21 |
| 9 | 7 | 0.07 | 0.05 | 0.12 | BDL | 2.73 | 1.63 |
| 72 | 7 | 0.02 | 0.07 | 0.18 | BDL | 2.94 | 1.79 |
| 9 | 9 | 0.01 | BDL | 0.1 | BDL | 3.25 | 1.9 |
| 72 | 9 | 0.03 | 0.10 | BDL | BDL | 2.8 | 1.9 |

(B) means that the Plastic contained 15% (w/w) of PCSS.

**Table VII. Intermittent pH Adjustment—Pb, Cd, and Cr (in mg/L) for 100 g of Garnet Waste in 2L of Solution**

| T (hrs) | pH | Pb | Pb(B) | Cd | Cd(B) | Cr | Cr(B) |
|---|---|---|---|---|---|---|---|
| 9 | 3 | 0.02 | 0.05 | 0.11 | 0.01 | BDL | BDL |
| 72 | 3 | BDL | 0.03 | 0.86 | BDL | BDL | 0.1 |
| 9 | 5 | BDL | 0.05 | 0.29 | BDL | 0.17 | 0.16 |
| 72 | 5 | 0.02 | 0.06 | 0.21 | BDL | 1.49 | 0.55 |
| 9 | 7 | BDL | 0.06 | 0.2 | BDL | 0.5 | 0.74 |
| 72 | 7 | BDL | 0.10 | 0.1 | BDL | 0.4 | 0.58 |
| 9 | 9 | 0.06 | 0.04 | BDL | BDL | 0.97 | 0.98 |
| 72 | 9 | 0.03 | 0.07 | 0.05 | BDL | 0.45 | 0.63 |

(B) means that the Garnet contained 15% (w/w) of PCSS.

**Table VIII. Continuous pH Adjustment—Pb, Cd, and Cr (in mg/L) for 100 g of Black Beauty Waste in 2L of Solution**

| T (hr) | Pb | Pb(B) | Cd | Cd(B) | Cr | Cr(B) |
|---|---|---|---|---|---|---|
| 1 | 1.25 | 0.75 | 1.10 | 1.00 | 3.43 | 2.55 |
| 3.5 | 1.41 | 0.31 | 1.17 | 1.13 | 3.52 | 1.40 |
| 5.5 | 1.67 | 0.26 | 1.27 | 1.04 | 3.61 | 1.02 |
| 12 | 1.73 | 0.20 | 1.35 | 1.04 | 4.26 | 0.71 |

(B) means that the Black Beauty contained 15% (w/w) of PCSS.

**Table IX. Continuous pH Adjustment—Pb, Cd, and Cr (in mg/L) for 100 g of Plastic Waste in 2L of Solution**

| T (hr) | Pb | Pb(B) | Cd | Cd(B) | Cr | Cr(B) |
|---|---|---|---|---|---|---|
| 1 | 3.55 | 2.41 | 1.41 | 1.19 | 1.96 | 1.69 |
| 3.5 | 4..95 | 2.60 | 1.48 | 1.19 | 2.13 | 1.81 |
| 5.5 | 5.16 | 2.97 | 1.53 | 1.28 | 2.09 | 1.76 |
| 12 | 5.31 | 2.15 | 1.56 | 1.24 | 2.21 | 1.43 |

(B) means that the Plastic contained 15% (w/w) of PCSS.

**Table X. Continuous pH Adjustment—Pb, Cd, and Cr (in mg/L) for 100 g of Garnet Waste in 2L of Solution**

| T (hr) | Pb | Pb(B) | Cd | Cd(B) | Cr | Cr(B) |
|---|---|---|---|---|---|---|
| 1 | 0.40 | 0.31 | 1.95 | 1.48 | 0.73 | 0.38 |
| 3.5 | 0.71 | 0.23 | 2.09 | 1.60 | 0.90 | 0.26 |
| 5.5 | 0.78 | 0.22 | 2.12 | 1.55 | 0.94 | 0.19 |
| 12 | 1.10 | 0.26 | 2.31 | 1.57 | 1.27 | 0.29 |

(B) means that the Garnet contained 15% (w/w) of PCSS.

*Extractions with Continuous Acid Additions to Maintain pH 5*

In these experiments, acid was continuously added as needed to keep the pH from rising to more than 5.2. The demand for acid was greatest at the beginning of the experiments. In addition, these experiments were intensely mixed to permit rapid reaction between the acid and the solid materials.

Results are reported in Tables VIII through X, and are significantly different than for intermittent addition of acid. Every sample, with or without PCSS, produced soluble Cd in excess of the TCLP Cd standards. Pb concentrations were also significantly higher in these experiments compared to the experiments with intermittent additions of acid. For Pb and Cd, the soluble concentrations continued to increase with time in the absence of PCSS, but concentrations were lower and stable over time in the presence of Blastox. Thus, PCSS provided benefits even at a constant pH 5, but was not sufficient to prevent leachable Cd and Pb in excess of TCLP standards.

Maintenance of the low pH decreased solubility of Cr compared to tests where the pH was allowed to rise in between acid additions.

## INVESTIGATIONS OF PHYSICAL SEPARATION OF PCSS FROM BLASTING WASTES

Separation between blasting wastes and PCSS during collection, storage, transportation, or disposal operations could result in inadequate treatment over the short time-frame of RCRA testing and also over the longer time-frame of Superfund liability.

**Table XI. Enrichment Factors for PCSS After Dry Sieving of Mixtures**

| Mesh Size | Black Beauty Plus 15% Blastox© | Plastic Media Wastes Plus 15% Blastox© | Garnet Wastes Plus 15% Blastox© (w/w) |
|---|---|---|---|
| 60 (0.098–0.42 m) | 2.64 | 1.10 | 1.18 |
| 100 (0.059–0.098 m) | 0.78 | 0.85 | 0.60 |
| Fines (<0.059 m) | 0.94 | 0.98 | 1.09 |

This issue is related to the regulatory requirement that water should not be added directly to solid wastes without a treatment permit, even though addition of water is an important part of the PCSS stabilization process. Although leachates will contact the waste materials after deposition in a landfill, all of the preliminary processes (blasting, collection and storage of the fines, transfer to a dumpster, transport to the landfill site, and on-site dumping) should occur without the addition of water. The TDJ Corporation (1994) reported that the EPA has sanctioned the use of Blastox©, even though the immobilization of metals might only occur after the addition of water, due to the fact that blasting wastes are not listed, and therefore the potentially hazardous characteristic of blasting wastes can only be determined after the addition of water for the TCLP test.

However, if separation were to occur at any time before contact with leachate, then the protections offered by the PCSS amendments might be lost, and this could result in long-term liabilities. We studied a "worst case" scenario for separation of PCSS from the blasting waste materials. A mixture of blasting waste (Black Beauty, garnet, or plastic waste) and Blastox© (added as neat product @ 15% w/w) was ground by mortar and pestle, and therefore the mixture was not as intimately mixed as would occur if PCSS were added prior to blasting operations. We then used vibrating mesh screens (20 through 100 mesh) in an attempt to separate the dry, recently combined materials. The accumulated mass on each sieve was weighed. One gram of each sample was analyzed by discontinuous titration using 0.1N HCl; i.e., an excess of acid was added to the solid sample, one day was allowed for dissolution reactions, and then each sample was back-titrated to pH 5 using 0.1N NaOH. The mass of PCSS in each sample was calculated by the net amount of acid that was required. The method was calibrated using samples of Blastox© (which completely reacted with 0.1N HCl within two hours) and samples of the blasting wastes (which did not consume any acid after one day).

The PCSS "enrichment factors" are shown in Table XI, where a value of 1.00 means that the reported fraction retained exactly the fraction of Blastox© as was in the original mixture. An "enrichment factor" approaching zero would indicate that Blastox© was completely removed from the sieve fraction. Scanning electron microscopy (SEM) photographs demonstrated intimate mixing of Blastox© with the Black Beauty and the garnet blasting materials, but very little attachment of Blastox© with the plastic media.

Both the sieving/acid titration experiments and the SEM photographs demonstrated that Blastox, once incorporated with blasting media, did not separate even under a worst-case scenario.

## CONCLUSIONS

- The most frequent problems in MCLB blasting wastes were for Cd (42 out of 98 samples exceeded TCLP levels). There have been several TCLP exceedances for Pb (12 out of 98 samples) and very few TCLP violations for Cr (4 out of 98 samples).
- Based on historic records, PCSS at 10% (w/w) reduced TCLP Pb to below TCLP levels, and PCSS at 15% (w/w) reduced Cd to below TCLP levels. PCSS did not reduce TCLP Cr.
- Mass balances showed that Cd will not present a long-term problem after disposal of these materials, since 96% of the total extractable Cd in straight Black Beauty blasting wastes was leached during a single TCLP test, leaving only 4% of extractable Cd for all subsequent TCLP extractions.

  Steel blast contained high total Pb, Cd, and Cr, even when TCLP concentrations were below limits.
- Plastic and glass blasting wastes exceeded TCLP limits test more often than other blasting wastes.

- TCLP Cr increased after the addition of PCSS, perhaps due to decreased sorption of anionic Cr(VI) at the higher pH values.
- Even with final pH of 5, leachable Pb, Cd, and Cr were quite low when poor mixing and slow addition of acid resulted in intermittently high pH values that were conducive to the formation of silicate precipitaters. PCSS reduced the concentrations of Pb and Cd to very low values for these conditions.
- On the other hand, intense mixing and instantaneous addition of acid to maintain the pH at 5 resulted in high Pb, Cd, and Cr concentrations, even when PCSS had been added to the blasting wastes. Every sample tested, with or without PCSS, exceeded TCLP Cd standards for these conditions.
- Both theory and experiments indicate the need to temporarily maintain pH at alkaline values to obtain the benefit of the PCSS reagents. A temporary pH greater than about 7.5 is needed for good removal of Pb and greater than about 8.5 for good removal of Cd. After lowering the pH to 5, these precipitates were inert and did not dissolve during the time frame of the TCLP test.
- PCSS was not physically separated from the blasting waste, even after imposition of a worst-case scenario for separation according to size.

## RECOMMENDATIONS

- Inventory the amount of Cd-plated material that is blasted during each operation. PCSS should be considered for use at 15% (w/w) for blasting jobs where Cd could be a TCLP characteristic contaminant. These blasting wastes should be separated from other blasting wastes.
- Consider disposal of all blasting wastes in a mono-fill that does not receive co-fill containing high organic content or mineral acids. If the leachate remains alkaline, then the leachable concentrations of metals will be low, especially if some of the blasting waste includes PCSS.
- Cd is the most important of the TCLP metals for these blasting wastes. However, since untreated blasting materials release more than 50% of total Cd during one leaching cycle, therefore the reservoir of Cd is insufficient to produce a long-term problem. Mass balances and the multiple extraction leaching (MEL) procedure can predict long-term effects.
- Plastic, glass, and steel blast wastes should be monitored carefully and the source of metals in these blasting wastes should be investigated.

## BIBLIOGRAPHY

Conner, J.R. *Chemical Fixation and Solidification of Hazardous Wastes.*
Van Nostrand Reinhold, New York, 1990.

Hock, V.F., et al. Demonstration of Lead-Based Paint Removal and Chemical Stabilization Using Blastox (Draft Research Report FEAP TR 96/20). Alexandria, VA: U.S. Army Construction Engineering Research Laboratory, 1996.

Mchedlov-Petrossyan, O.P. *Thermodynamics of Silicates*. Springer-Verlag, New York, 1985.

The PQ Corporation. Treatability Study Manual for Solidification/Stabilization. Valley Forge, PA.

TDJ Group, Inc. Performance Evaluation of Blastox In-Situ on Various Structures. S.G. Pinney & Associates, Inc., P.O. Box 9220, Port St. Lucie, FL 34985-9220, 1994.

# 34 THE EFFECTS OF A SURFACTANT ON POST-EXTRACTION TREATMENT OF PCE CONTAMINATED WATER

**Stefanie R. Lange**, Research Associate
**Clifford R. Lange**, Assistant Professor
Department of Civil Engineering
Auburn University
Auburn, Alabama 36830

## INTRODUCTION

Remediation of contaminated groundwater has long been a concern of environmental engineers and scientists. One category of environmentally important contaminants is chlorinated hydrocarbons. Chlorinated hydrocarbons in the environment are generally present in the form of Dense Non-Aqueous Phase Liquids (DNAPLs) because of their high molecular weight and relatively low solubility in water. High densities combined with low viscosities make remediation of chlorinated solvents difficult.[1] One specific chlorinated hydrocarbon of interest is tetrachloroethylene, also known as perchloroethylene (PCE), which is a solvent used in dry cleaning and degreasing. PCE has been a commonly encountered contaminant at remediation sites due to industrial discharges, municipal waste treatment plant discharges, and spills.

Pump and treat extraction has been the most widely used form of aquifer remediation for PCE contaminated groundwater sites. However, large pumping volumes and long pumping times are required to effectively remediate these sites. Recently, investigations have been conducted to assess the feasibility of surfactant enhanced pump and treat techniques. These investigators have demonstrated that Surfactant Enhanced Aquifer Remediation (SEAR) can greatly reduce the volume of water required to flush PCE from a contaminated groundwater site.[1]

Surfactants are able to increase the apparent solubility and mobility of PCE.[2] This allows the PCE to go into solution and decreases liquid surface tension causing the PCE to be more readily transported out of the groundwater through pumping. However, little research has been done concerning the treatment of these wastes containing surfactants after they have been pumped out of the groundwater.

The presence of surfactants in a wastewater solution complicates the mechanisms controlling the adsorption of PCE onto granular activated carbon. It is hypothesized that competition exists between the carbon and the surfactant micelles for PCE adsorption. PCE bound in surfactant micelles can pass through a carbon column without being adsorbed. The surfactants also block the pore structure of the activated carbon, resulting in reduced absorbent surface area. It has been observed that the rate of PCE adsorption by sand is controlled by intraparticle-pore diffusion,[3] indicating that the potential for fouling of pores by surfactants can have a negative affect. Additionally, the surfactant can coat the surface of the carbon, which will alter the surface properties of the absorbent.[4]

Surfactant in wastewater solutions also present problems for air stripping due to changes in solubility and the generation of foaming. Further, surfactants alter the rate of mass transfer into the gas bubble, reducing the mass transfer coefficient,[5] and thereby slowing the air stripping process.

*52nd Purdue Industrial Waste Conference Proceedings*, 1997, Ann Arbor Press, Chelsea, Michigan 48118. Printed in U.S.A.

## MATERIALS AND METHODS

To assess the effects of the surfactant, Tween-80, on Granular Activated Carbon (GAC) adsorption of PCE, batch isotherm tests and packed column tests were performed. Test solutions containing PCE only and PCE with Tween-80 at concentrations above the critical micelle concentration or CMC (1% and 4% by volume) were used. Calgon React-A activated carbon was chosen for this study based on the manufacturers' recommendations.

Isotherms were constructed using 40 mL amber EPA vials which were charged with GAC masses ranging from 0.001 to 3.33 grams. The vials were subsequently filled to capacity with test solution. Agitation of the samples was conducted using a rotary shaker for 24 hours. The aqueous phase PCE concentrations were quantified using gas chromatography. The data obtained were analyzed using the standard Freundlich isotherm format.

Column studies were conducted using 10-cm long, 1.5-cm diameter glass columns. These columns were operated in the upflow mode. A peristaltic pump was used to deliver a flow rate of 2.5 $cm^3$/min, yielding a loading rate of approximately 8 bed volumes per hour. Samples of column effluent were collected and analyzed for PCE using gas chromatography. The surfactant, Tween-80, was quantified using capillary electrophoresis.

To assess the effects of Tween-80 on air stripping of PCE, tests were performed using 500 mL gas washing bottles. Test solutions containing PCE only and PCE with Tween-80 at concentrations above CMC (1% and 4% by volume) were examined in the air stripping study. Solution volumes of 250 mL were added to the 500 mL washing bottles and compressed air at a gas flow rate of 20 mL/minute was bubbled through the mixtures. PCE stripped from each test solution was trapped by passing the off-gas through liquid hexane. Samples of the hexane solution were subsequently analyzed for PCE concentration over time using gas chromatography.

## GRANULAR ACTIVATED CARBON ADSORPTION

### Isotherm Experiments

Batch isotherm experiments were run in duplicate for the 200 mg/L PCE solutions, the 200 mg/L PCE in a 1% Tween-80 solutions, and the 200 mg/L PCE in a 4% Tween-80 solutions. Freundlich isotherms were constructed to analyze the batch data. The Freundlich isotherm representation was selected due to the recommendation by the carbon manufacturer (Calgon) and since it is the commonly used model for hydrocarbon adsorption onto granular activated carbon.[6] The manufacturer recommended the pulverization of the carbon before using it in batch experiments. The pulverization alters the behavior of the carbon by changing the surface area and altering mass transfer limitations. Thus, the carbon would act as a powdered activated carbon and would show adsorption capacity more favorable than those obtained using the granular form.

The Freundlich isotherm that models adsorption at equilibrium is in the form of,[7]

$$X/M = K_f \, C^{1/n} \qquad (1)$$

where
- $X$ = mass of adsorbate adsorbed on adsorbent, mg
- $M$ = mass of adsorbent, g
- $K_f$ = equilibrium constant indicative of adsorptive capacity
- $C$ = solution concentration at equilibrium after adsorption, mg/L
- $n$ = constant indicative of adsorption intensity.

The Freundlich isotherms developed for the 200 mg/L PCE solutions are shown in Figure 1. The concave nature of the PCE-only isotherm suggests that nonideal adsorption of PCE by the GAC occurred.[8] This means that X/Ms are lower than desired at low concentrations and the carbon cannot achieve low effluent concentrations in batch processes. The nonideal nature of adsorption does not preclude the carbon's use in columns, where the carbon is in equilibrium with the influent concentration and X/M would be much higher.

The Freundlich isotherm for the saturated PCE in a 1% Tween-80 solution appeared to show two phases of adsorption. The change in isotherm shape indicates that the mechanisms of adsorption have been altered. The first phase, occurring at equilibrium concentrations ranging from 10 to 100 mg/L, is characterized by low X/Ms, which indicates that adsorption is unfavorable. The second phase of the isotherm, occurring at concentrations above 100 mg/L, shows that adsorp-

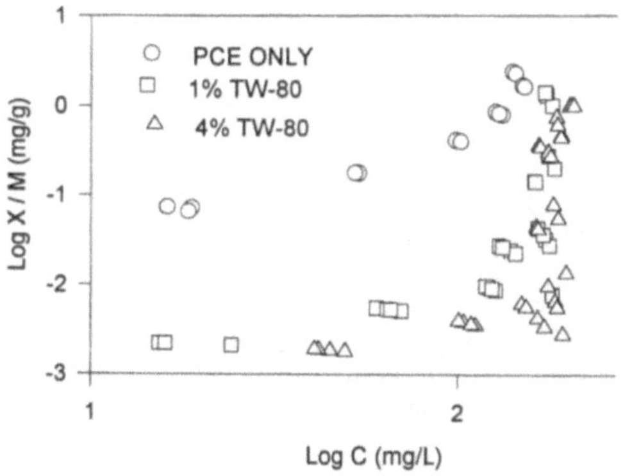


**Figure 1. Freundlich isotherms for PCE with and without Tween-80.**

tion is favorable as demonstrated by the X/Ms.[8] In general, X/Ms in the presence of 1% Tween-80 were significantly lower than those observed for PCE only.

The Freundlich isotherm for the saturated PCE in a 4% Tween-80 solution generated a trend similar to the 1% Tween-80 solution. Again two phases appeared, one phase at a low X/M and another at a high X/M. However, X/M is lower for the 4% case than for the 1% case. Therefore, it appears that adsorption of PCE is significantly reduced at the higher surfactant concentration.

## Column Studies

Column experiments were performed using 200 mg/L PCE solutions. A representative breakthrough curve for PCE is presented in Figure 2. A sharp breakthrough curve was observed, and is common behavior for a single organic contaminant in a simple matrix.[9] Breakthrough occurred after 1500 pore volumes, and exhaustion occurred after 2500 pore volumes.

Based on the PCE-only column runs, it is apparent that Calgon React-A carbon performs well at treating PCE. The length of the adsorption zone can be estimated by,

$$L = [(V_{ex} - V_b) / V_{ex}] * h \tag{2}$$

where $L$ = length of the adsorption zone
$V_{ex}$ = volume treated at exhaustion
$V_b$ = volume treated at breakthrough
$h$ = height of column (10 cm).

By substituting the geometrical values of the columns, the length of the adsorption zone is found to be approximately 4 cm. The mass of PCE adsorbed by the column was determined by integrat-

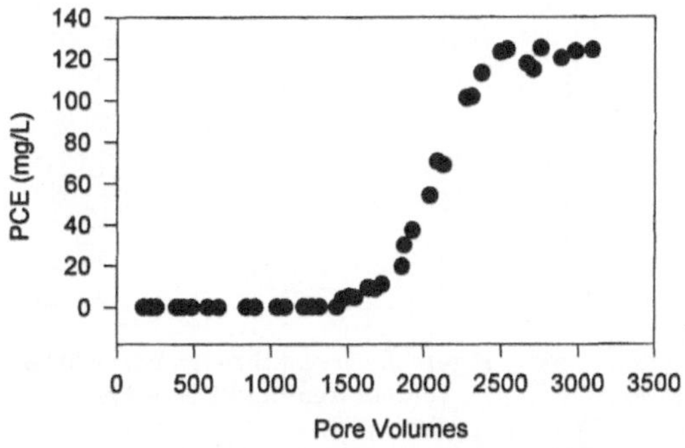


**Figure 2. Breakthrough curve for a GAC column treating PCE.**

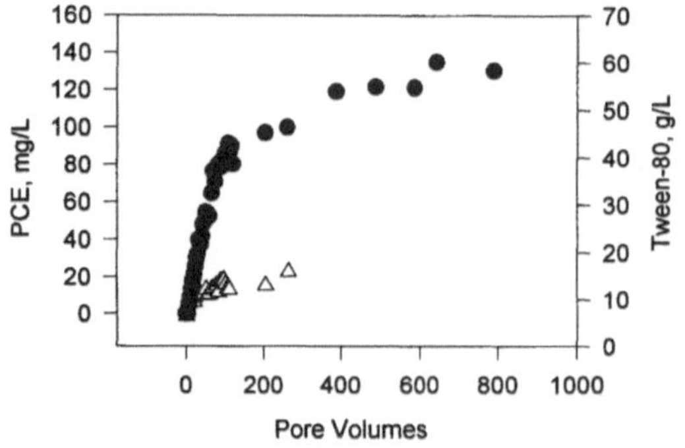


**Figure 3. Breakthrough curve for a column treating PCE in 1% Tween-80.**

ing the area above the breakthrough curve. For the PCE-only solution, the mass adsorbed is 4,230 mg.

A representative breakthrough curve for 200 mg/L PCE in a 1% Tween-80 solution is presented in Figure 3. PCE concentrations in the column effluent are indicated by circles, while Tween-80 concentrations are shown as triangles. The sharp curve observed for the PCE-only samples was not observed for PCE in a 1% Tween-80 solution. Breakthrough of PCE occurred at 8 pore volumes, while exhaustion occurred after 200 pore volumes. Breakthrough of Tween-80 occurred soon after one pore volume, indicating that Tween-80 behaves like a conservative tracer.

The presence of 1% Tween-80 in the solutions appears to greatly diminish the efficacy of Calgon React-A carbon for removing PCE. The length of the adsorption zone is approximately 10 cm long and is not distinct. This indicates that Tween-80 is interfering with adsorption equilibrium. The mass of PCE adsorbed by the carbon, as determined by integration of the area above the curve, was 403 mg. This is an order of magnitude lower than the adsorption observed than when no Tween-80 was present.

When comparing the breakthrough curves for PCE and Tween-80 for the 1% solution, two observations were made. First, the Tween-80 breaks through more rapidly than PCE. Second, the concentration of Tween-80 increases to a maximum more quickly than does PCE. Thus, the ratio of PCE to Tween-80 in the effluent increases during the run. It is apparent that PCE is initially being drawn out of the surfactant micelles since the Tween-80 breaks through before PCE is detected. It is further hypothesized that as more of the 1% Tween-80 solution is applied to the column, less PCE is lost by the micelles, causing the increased ratio of PCE to Tween-80.

The results of a representative column run using 200 mg/L PCE in a 4% solution are shown in Figure 4. Breakthrough of PCE occurred after about 1.6 pore volumes, and exhaustion occurred at 40 pore volumes. Breakthrough of Tween-80 occurred after one pore volume. The concentra-

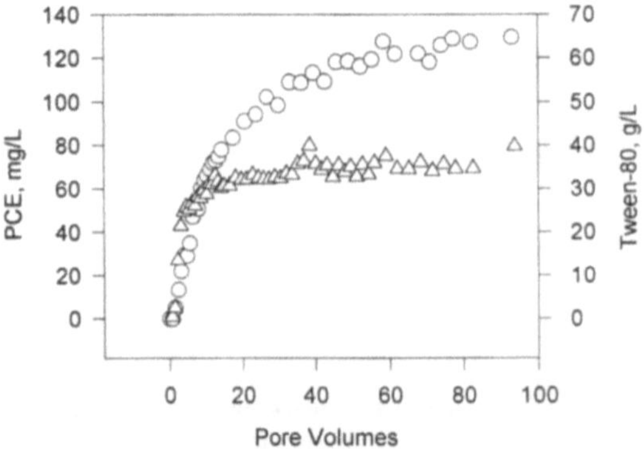


**Figure 4. Breakthrough curve for a column treating PCE in 4% Tween-80.**

tion of Tween-80 increased more rapidly than did PCE, reaching 95% of influent values after 25 pore volumes.

The presence of 4% Tween-80 interfered with GAC adsorption of PCE to a greater extent than was observed for 1% Tween-80. The mass of PCE adsorbed was calculated to be 38 mg, which is two orders of magnitude less than for PCE only. Once again, the rapid breakthrough and increase of Tween-80 concentrations indicates that complex partitioning between GAC and surfactant micelles is occurring in the column.

Based on the results of the experimental investigation, the following conclusions were made:

1. The presence of Tween-80 in a PCE solution alters the adsorption by GAC. This was demonstrated by the two-phase nature of the Freundlich isotherms obtained after surfactant addition.
2. Tween-80 concentrations above the CMC can greatly reduce the efficiency of GAC columns in treating PCE.
3. Partitioning of PCE between the aqueous phase, GAC, and Tween-80 micelles occurs.

It can be concluded that the use of GAC adsorbers to directly remove PCE from Surfactant Enhanced Aquifer Remediation (SEAR) wastewater is not feasible. The presence of Tween-80 would result in short column runs and the inefficient use of carbon. Ongoing studies are underway in the Department of Civil Engineering at Auburn University to examine the effects of separation and recovery of Tween-80 prior to adsorption.

## AIR STRIPPING

Preliminary studies were conducted to assess the feasibility of using air stripping as a means of post-extraction treatment of SEAR wastewater. As a baseline for comparison, tests were conducted using a saturated (200 mg/L) PCE solution without surfactant. The results of a typical test are presented in Figure 5. The cumulative mass of PCE stripped from the wastewater increased rapidly over the first 30 minutes and reached a plateau after 60 minutes. The total mass of PCE recovered in the hexane was approximately half of the initial mass of PCE. Extraction of the wastewater, however, confirmed that no measurable PCE remained in the water after 60 minutes.

The results of an air stripping experiment using a solution containing PCE and 1% Tween-80 are presented in Figure 6. Previous attempts at stripping the waste resulted in an excessive amount of foaming. To reduce foaming, 0.04 gram of pulverized lime was added to the solutions containing surfactant. This amount of lime was determined to be the minimum concentration capable of reducing foam production. The addition of surfactant and lime did not appear to significantly alter the total mass of PCE stripped from the wastewater. However, the time required to achieve this stripping of PCE was increased from 30 minutes to 100 minutes.

The results of an air stripping experiment using a solution containing PCE, 4% Tween-80, and 0.04 gram of lime, are shown in Figure 7. The lime addition was successful in minimizing foaming. As observed for the 1% Tween-80 solution, the entire mass of PCE was stripped from the

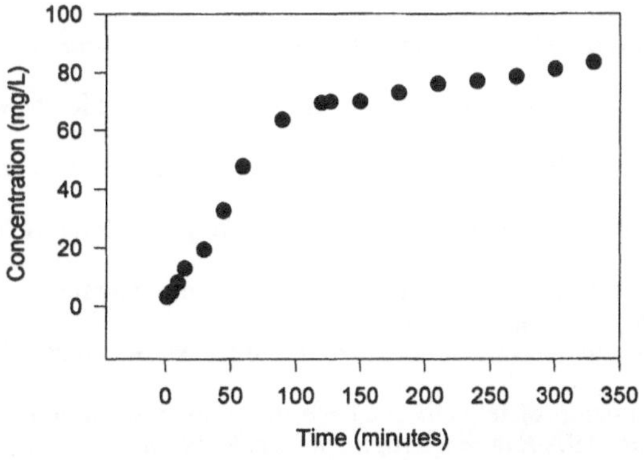


**Figure 5. Results of air stripping experiment using a PCE-only test solution.**

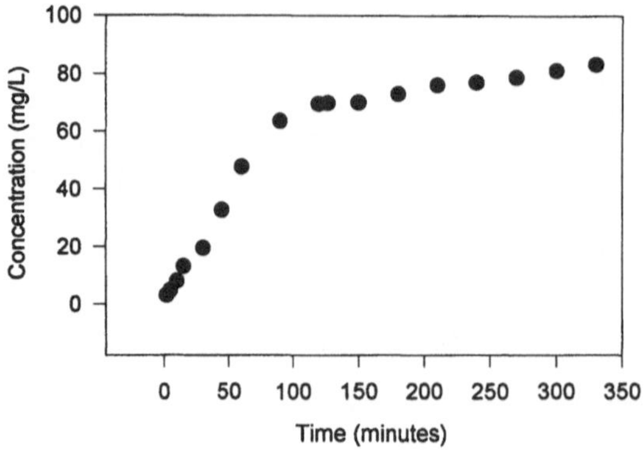


**Figure 6. Results of air stripping experiment using a PCE in 1% Tween-80.**

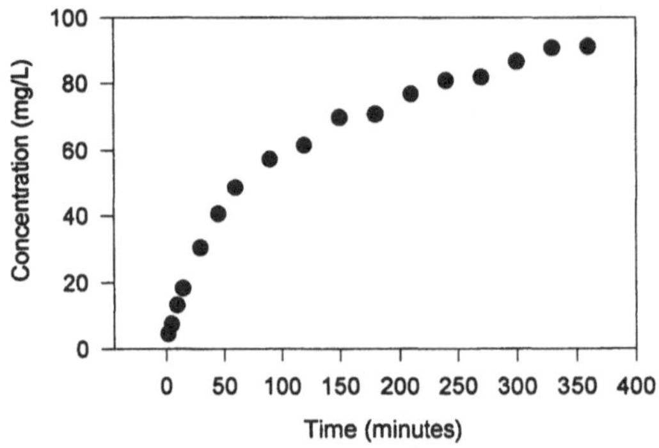


**Figure 7. Results of air stripping experiment using a PCE in 4% Tween-80.**

wastewater. The presence of 4% Tween-80 increased the time required to strip the PCE to 400 minutes.

Based on the results of preliminary experiments, air stripping appears to be a feasible method for removing PCE from post-extraction groundwater and industrial wastewater. The detection times required to achieve PCE stripping increased as surfactant concentration increases. Foaming, due to the presence of surfactant, poses a problem for the use of air stripping. However, the addition of cations, such as calcium, can reduce foaming to acceptable levels.

## BIBLIOGRAPHY

Allred, B. and G.O. Brown. "Surfactant-induced reductions in soil hydraulic conductivity," *GWMR,* Summer 1994, p. 174, 1994.

Biswas, N., et al. "Model for predicting PCE desorption from contaminated soils," *Water Environ. Res.,* 64(2), p. 170, 1992.

Brown, C.L., et al. "Simulation of surfactant-enhanced aquifer remediation," *Water Resour. Res.,* 30 (11), p. 2959, 1994.

Eckenfelder, W.W., *Industrial Water Pollution Control,* Second edition, McGraw-Hill, Inc., New York, 1989.

Jeng, J.J., et al. "Surface effects and mass transfer in bubble column," *Ind. Eng. Chem. Process Des. Dev.* 25(4), p. 974, 1986.

Weber, W.J., *Physiochemical Processes for Water Quality Control,* John Wiley and Sons, New York, 1972.

Weber, W.J., et al. "Sorption phenomena in subsurface systems: Concepts, models, and effects on contaminant fate and transport," *Water Res.,* 25(5), p. 499, 1991.

West, C.C. and J.H. Harwell, "Surfactants and subsurface remediation," *Environ. Sci. Technol.,* 26(12), p. 2324, 1992.

Young, D.F. and W.P. Ball, "A *priori* simulation of tetrachloroethene transport through aquifer material using an intraparticle diffusion model," *Environ. Prog.,* 13(1), p. 9, 1994.

# 35 IMPROVED SCRUBBING OF $NO_x$ EMISSIONS FROM AN ALUMINUM ANODIZING BRITE-DIP SYSTEM USING ALKALINE PERMANGANATE

**Hugh G. Allerton**
TRW Vehicle Safety Systems Inc.
Washington, Michigan

**Rodney Todd**
TRW Vehicle Safety Systems Ltd.
Midland, Ontario

**Jerome H. Jacobs and Richard Prober**
Metcalf & Eddy, Inc.
Cleveland, Ohio

## ABSTRACT

This chapter describes the scrubbing system for control of NOx emissions from aluminum anodizing. Changing from an alkaline sodium sulfide scrubbing liquor to alkaline potassium permanganate, with continuous bleed and replenishment of scrubbing liquor, has eliminated incidents of visible brown plumes or acid mists.

## INTRODUCTION

The former Decor Products Division of TRW located in Midland, Ontario (now KLECO) produces extruded aluminum products such as railings for automatic passive track seat belts and picture frames. During the anodizing process, the parts undergo Brite-Dip treatment by sequential immersion in tanks containing concentrated sulfuric acid, phosphoric acid, and nitric acid. The Brite-Dip treatment generates fumes containing gaseous oxides of nitrogen (NOx), principally $NO_2$, as well as nitric acid vapor and entrained droplets of the acid solutions. Exhaust air containing these acidic fumes discharged directly to the atmosphere until about 1980, when a packed-column wet scrubber was installed.

Figure 1 is a schematic flow diagram of the existing air scrubbing system for the Brite-Dip process. An exhaust air hood collects and conveys the fumes to the scrubber, an 8-ft diameter Fiberglas column containing 8 feet of 2-inch Tripack packing. The air flow (14,000 CFM) is countercurrent to scrubbing liquor (240 US gpm), which is injected through nozzles above the packing. Scrubbed air discharges into the atmosphere through a 3 foot, 8-inch diameter stack containing a chevron-type mist eliminator. Scrubbing liquor recirculates from a 1700 US gallon sump. Spent scrubbing liquor is neutralized in the plant wastewater pretreatment system, which discharges into the Town of Midland sanitary sewer system. The municipal waste water treatment plant (WWTP) is located within 1 km of the Decor facility.

## ORIGINAL SCRUBBING CHEMISTRY

Before 1993, the scrubbing solution was an alkaline liquor containing sodium sulfide. It was made up in 500 US gallon batches from 50% caustic soda and technical grade sodium sulfide. As indicated schematically in Figure 2, the principal liquid phase reactions were as follows:

*52nd Purdue Industrial Waste Conference Proceedings*, 1997, Ann Arbor Press, Chelsea, Michigan 48118. Printed in U.S.A.

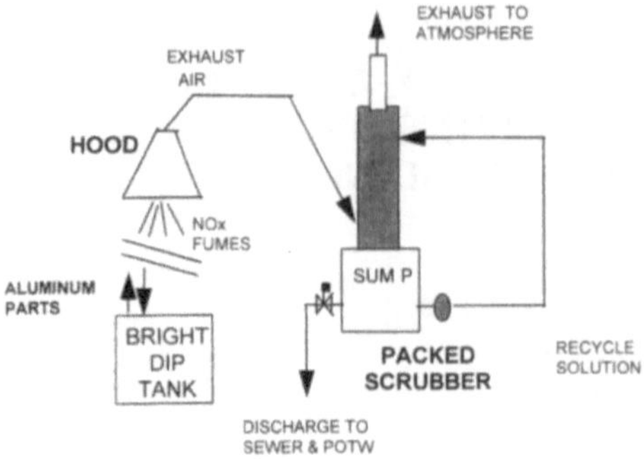


**Figure 1. Brite-Dip and air scrubbing schematic.**

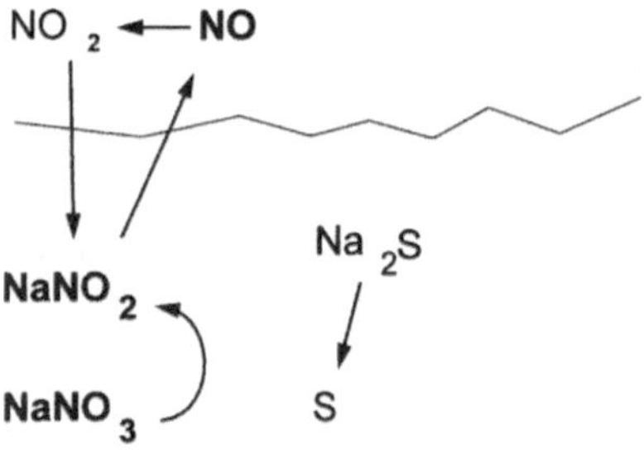


**Figure 2. Old chemistry.**

- absorption of $NO_2$: $2\ NaOH + 2\ NO_2 \rightarrow NaNO_2 + NaNO_3 + H_2O$ (1)
- nitrite reduction: $2\ NaNO_2 + 2\ H_2O + Na_2S \rightarrow 2\ NO + 4\ NaOH + S$ (2)
- nitrate reduction: $NaNO_3 + Na_2S + H_2O \rightarrow NaNO_2 + 2\ NaOH + S$ (3)

These reactions remove brownish $NO_2$ from the gas phase but return much of it as colorless NO. The scrubber usually alleviated visible plume emissions, but was not effective in totally eliminating NOx emissions. At times, when weather conditions caused an inversion to take place, a "detached" brownish plume was observed; that is, the exhaust air was colorless at the stack, but it became visibly brownish some distance away from the stack. That was the result of a gas phase reaction that converted some of the NO to $NO_2$.

- oxidation of NO: $NO + 1/2\ O_2 \rightarrow NO_2$ (4)

The existing system also was not totally effective in capturing the nitric acid vapor and acidic droplets, despite contacting alkaline scrubbing liquor in the scrubbing column. There were periodic complaints of "acid biting" due to low pH (<1 S.U.) of the stack gas in the localized area downwind.

Spent scrubbing solution containing residual alkalinity and unreacted sodium sulfide was discharged periodically into the plant wastewater treatment system. This produced a shock loading of alkalinity and Chemical Oxygen Demand (due to the residual sulfide), which then discharged into the sanitary sewer system. This loading on the Midland WWTP had the potential to upset the activated sludge system.

## REHABILITATION

TRW Vehicle Safety Systems initiated upgrading of the Brite-Dip Scrubbing System in 1992. The objectives were as follows:

- Emissions of NOx shall be in compliance with Ontario Ministry of Environment (MOE) limit of 500 ug/$m^3$ at the property boundary
- Eliminate incidents of a visible plume, and
- Resolve acid biting.

Stack sampling results in July 1992 showed a maximum NOx concentration of approximately 60 ppm by volume (about two-thirds $NO_2$ and one-third NO). The property boundary concentrations were determined to be about 310 μg/$m^3$, well within the MOE limit.

However, the visible plume and acid biting episodes still had to be addressed. Metcalf & Eddy, Inc. (M&E) conducted an evaluation of the existing system, including inspection of the scrubber column internals, a review of the current system loading and process chemistry, and a review of available performance data. It was determined that corrective maintenance was needed to restore the operation of the scrubber to the manufacturer's specifications. At this time the scrubbing process chemistry was also questioned. A phased rehabilitation program was recommended to alleviate the problems.

The initial phase included repairs to restore the original system capacity. The scrubbing liquor inlet nozzles were replaced, to ensure uniform distribution of the scrubbing liquor flow over the column cross section. The fouled packing in the column was replaced, to restore the original absorption efficiency. However, these necessary repairs proved to be inadequate to achieve the objectives in brief trials using the original scrubbing chemistry. Occasional incidents of acid biting and visible plumes still occurred.

The second phase upgraded the scrubbing liquor system, including new chemical feed pumps and process control instrumentation, as necessary for continuous bleed of spent scrubbing liquor. Also, the scrubbing chemistry was changed to use higher alkalinity (pH > 12.5 S.U.) and an oxidizer—potassium permanganate.

A third phase of upgrading was proposed, to eliminate acid biting incidents by replacing the existing chevron-type mist eliminators with a larger unit. This has not been necessary.

## NEW SCRUBBING CHEMISTRY

Use of sodium sulfide—caustic soda solutions as the scrubbing liquor was not entirely successful at elimination of the visible plume, as discussed above. The sulfide (reducing agent) enhanced release of absorbed $NO_2$ back into the gas phase as NO; i.e., Reactions 2 and 3. This chemistry proved difficult to manage in practice. Other alkaline scrubbing solutions (without the reducing agent) were tried, with no significant effect on the stack gas composition.

An alternative scrubbing liquor composition was recommended, using an oxidizer—potassium permanganate—to stabilize the absorbed NOx and more concentrated caustic soda to ensure neutralization of nitric acid vapor and/or entrained acidity. The principal chemical reactions with the new scrubber liquor composition are as follows:

- absorption of $NO_2$: $2\,NaOH + 2\,NO_2 \rightarrow NaNO_2 + NaNO_3 + H_2O$ (1)
- oxidation of $NaNO_2$: $3\,NaNO_2 + 2\,KMnO_4 + H_2O \rightarrow 3\,NaNO_3 + 2\,MnO_2 + 2\,KOH$ (5)

As shown schematically in Figure 3, these reactions retain the absorbed $NO_2$ in the scrubber liquor, unlike the original chemistry that allows for release back into the gas phase as NO. Therefore, the overall NOx removal is more effective. However, precipitation of manganese dioxide in the scrubbing liquor could foul the packing. Therefore, continuous bleed of spent scrubber liquor is required.

## PERFORMANCE

All three objectives have been realized, using this new chemistry, with appropriate instrumentation and continuous bleed of spent scrubbing liquor. Since the change of chemicals in August 1993 and implementation of continuous bleed in November 1993, there have been no visible plume incidents and no complaints of "acid biting" by plant personnel. Plant personnel have learned that an accumulation of solids in the liquor must be controlled through preventive maintenance to preclude the discharge of scrubber solution droplets through the stack.

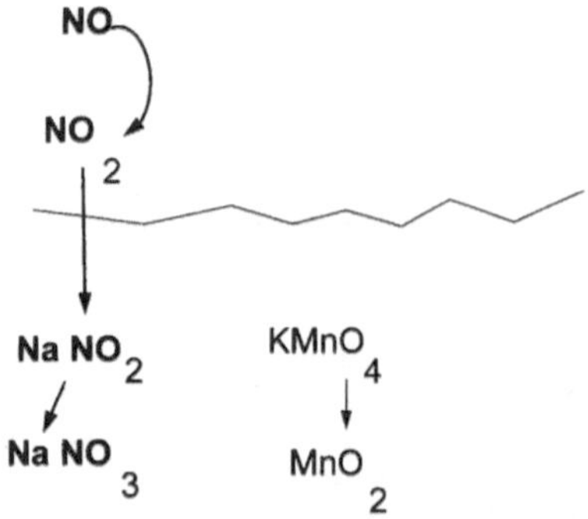


**Figure 3. New chemistry.**

The system performance was monitored closely in the first five months after startup. Scrubber liquor samples were analyzed daily for nitrite, nitrate, and potassium permanganate concentrations using colorimetric field test kits. Scrubber liquor sample concentrations during this period were recorded ranging as high as the following:

| | |
|---|---|
| $NO_3$-N | 33,000 mg/L |
| $NO_2$-N | 825 mg/L |
| $KMnO_4$ | 0.7 % |
| pH | 13 S.U. |

Figure 4 summarizes the scrubber system performance from the startup after Phase 2 improvements. Total removal of NOx during this period amounted to more than 4,500 kg as $NO_2$. There were three distinct periods of operation, as summarized in Table I. The system loading was light (31 kg/day average, based on actual NOx absorbed) in the first six weeks. It increased considerably (to 72 kg/day) and then returned to light levels (32 kg/day) for the last four weeks. Potassium permanganate consumption decreased steadily over the five month trial. It averaged 13.9

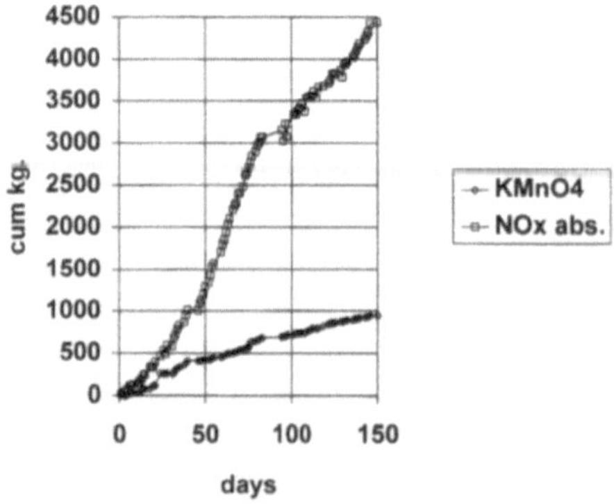


**Figure 4. Scrubber performance.**

**Table I. Chemical Requirements (kg/day)**

| Period | NOx Absorption | $KMnO_4$ Addition |
|---|---|---|
| Period I | 31 | 13.9 |
| Period 2 | 72 | 9.4 |
| Period 3 | 32 | 6.3 |

kg/day in the initial light loadings, then decreased to 9.4 kg/day during the heavy loading and decreased further to 6.3 kg/day for the final light loadings.

Theoretically, the weight ratio of potassium permanganate ($KMnO_4$) usage to NOx absorption to should be 1.1. The observed ratio was 0.45 in the initial period, and it decreased to about 0.16 after continuous bleed was implemented. In all three operation periods, $KMnO_4$ usage was much less than the stoichiometric quantity. This may be due to a more complicated chemistry than represented above. For example, potassium permanganate may be acting as a catalyst rather than as a direct oxidizing agent consumed by Reaction 5. Alternatively, it could be reacting to form a stable soluble manganese species such as the hexavalent $MnO_3^{=}$ which has been observed at pH of about 13 S.U.[1]

Precipitation of manganese dioxide ($MnO_2$) has been controlled by the continuous bleeding of spent liquor. In addition, the scrubbing liquor sump was cleaned at two to three month intervals, at which times the liquor was replaced with fresh solution. Small quantities of dark, granular material believed to be $MnO_2$ were found in the sump before each cleaning.

The change in chemistry has been beneficial in terms of impacts on the plant wastewater treatment system. Discharges of spent scrubber solution, even with solids present, has not overloaded the system. Residual potassium permanganate, which is not removed there, can beneficially assist the Town of Midland wastewater treatment facility in reducing oxygen demands from other sources.

The system has been in operation for more than three and one-half years, with no visible plume or acid mist incidents. It has achieved cost savings and eliminated upsets of the plant wastewater pretreatment system.

## REFERENCES

1. Stumm, W. and Morgan, J. J., *Aquatic Chemistry*, Wiley—Interscience, New York, 1970, p. 533.

# 36 REMOVAL OF 2-CHLOROPHENOL FROM WASTEWATER WITH THE ULTRASONIC/FENTON PROCESS

**Jih-Gaw Lin**
Institute of Environmental Engineering
National Chiao Tung University
75 Po-Ai Street
Hsinchu, Taiwan, ROC

**Allen C. Chao**
Department of Civil Engineering
North Carolina State University
Raleigh, North Carolina

**Ying-Shih Ma**
Institute of Environmental Engineering
National Chiao Tung University
75 Po-Ai Street
Hsinchu, Taiwan, ROC

## ABSTRACT

Using the ultrasonic/Fenton oxidation to remove 2-chlorophenol from wastewater is presented this 354paper. The combining ultrasonic and Fenton treatment enhances the decomposition efficiency of 2-chlorophennol and reduced the reaction time. In previous studies, the optimal conditions for carrying out the proposed ultrasonic/Fenton process are: pH maintained at 3, ionic strength of the solution adjusted at 0.1 N, ultrasonic amplitude maintained at 120 mm, and the reaction temperature controlled at 25°C. In this study, the influent of Fenton reagent concentration and ratios of $Fe^{2+}$ to $H_2O_2$ on the removal efficiency of 2-chlororphenol is investigated. Laboratory results indicate that the percentage of 2-chlorophenol decomposition increases for higher concentrations of Fenton reagent applied. If the concentration of $H_2O_2$ exceeds 50 mg/L, more than 100% of 2-chlorophenol is decomposed with more than 80% mineralized. The kinetics of the 2-chlorophenol decomposition rate can be fitted with a pseudo first-order kinetic equation, and the reaction rate constants are dependent upon the concentration of Fenton reagent applied. With increasing Fenton reagent concentrations, the observed reaction rate constants also increase. The oxidation/reduction potential (ORP) of the reaction is also monitored. Variations of the monitored ORP values are related to the oxidation of 2-chlorophenol.

## INTRODUCTION

Chlorinated organic substances that are thought to be carcinogenic have been detected in raw wastewater and also in the chlorinated drinking water as well as wastewater effluent. They are formed in natural environments as metabolites of agricultural pesticides and in some industrial plants. For example, chlorophenols, chlorinated enol lactones and cyclopenten-1, 3diones were formed during chlorination of softwood kraft pulp. In water or wastewater treatment plants, the chlorination process is often used for disinfection. Natural water often contains humic materials (Smeds et al., 1995) which are precursors to the formation of carcinogens. The applied chlorine may react with these organic substances to form the various chlorinated by-products (Karlsson et al., 1995; Smeds et al., 1995). These chlorinated by-products have been reported to be carcino-

*52nd Purdue Industrial Waste Conference Proceedings*, 1997, Ann Arbor Press, Chelsea, Michigan 48118. Printed in U.S.A.

genic (Minero et al., 1995; Karlsson et al., 1995; Muftikian et al., 1995) in addition to causing taste and odor problems. Smeds et al. (1995) also reported that the mutagenic activity of a chlorinated compound seemed to increase with an increasing number of chlorine atoms in the molecules.

Using ultrasound to accelerate conventional chemical reactions and redox processes began in 1927. In recent years, research efforts have focused on applying the ultrasonic energy to removing toxic and hazardous organic compounds from contaminated water or wastewater. The ultrasonic process has been used for remediating contaminated soil and groundwater; and the organic compounds treated with ultrasonic process are converted into highly biodegraded products (Orzechowska et al., 1995). Several parameters affecting the treatment efficiency of the ultrasonic process include the ultrasound frequency or power supply, saturated gas, pH, temperature, ionic strength, catalyst, and the initial concentration of contaminant (Mason et al., 1992; Lin et al., 1996a, b; Petrier et al., 1992; Okouchi et al., 1992; Kotronarou et al., 1991). The most influential parameter to the degradation or mineralization of organic substances is the pH value of the solution (Lin et al., 1996b; Ma and Lin, 1996). Chemical oxidation of contaminated surface water and groundwater represents an important method for the destruction of biodegradable or nonbiodegradable pollutants (Minero et al., 1995). Advanced oxidation processes (AOPs) using the various chemical oxidants have been reported to be effective to various extents for removing chlorinated organic substances from water and wastewater. Degradation of organic substances is accomplished by the formation of hydroxyl radicals which are super active oxidants (Ollis et al., 1991; Werderitsch et al., 1996). The organic pollutants reported to be effectively removed include pesticides (Werderitsch et al., 1996; Arnold et al., 1995), phenol (Scheck and Fermmel, 1995; Ding et al., 1995; Canonica et al., 1995), chlorophenol (Minero et al., 1993; Lin et al., 1995; Lin et al., 1996a, b; Minero et al., 1995), acetic acid (Savage and Smith, 1995), selenite (Gruebel et al., 1995), and chlorinated hydrocarbons (Orzechowska et al., 1995; Martin et al., 1995). Among the many AOP methods, the Fenton process is often used because of its simplicity. The decomposition of chlorinated organic substances using the Fenton process is accomplished through a mechanism of dechlorination in the presence of Fe3+ and $H_2O_2$. In aqueous solution, hydrogen peroxide and ferrous or ferric ions produce OH radicals which can readily oxidize most organic substances nonselectively. The oxidants used in Fenton process are commonly available, inexpensive, and no special equipment and reaction vessels are required (Arnold et al., 1995). This process is also reported to enhance the biodegradability of organic substances, thus it is often used as a pretreatment of wastewater prior to biological treatments. Muftikian et al. (1995) indicated that several chlorinated hydrocarbons could be rapidly hydro-dechlorinated in aqueous solution to ethane in the pelletized iron system, however, no intermediate reaction was observed to take place. The Fenton process has also been used in degrading the waste organic substances in industrial wastewater (Elizardo, 1991; Huang et al., 1995). Major parameters affecting the Fenton process were reported to include pH, concentrations of ferrous or ferric ions, species of iron, and demands of $H_2O_2$ (Kochany et al., 1992; Sun and Pignatello, 1993; Takemura et al., 1994; Muftikian et al., 1995; Huang et al., 1995). For enhancing the treatment efficiency, many reports showed that the pH value of the solution was usually controlled between 2 and 4, especially at less than 3 (Kochany et al., 1992; Huang et al., 1995).

The decomposing rate of the organic matter in advanced oxidation process is generally fitted with either a first-order model (Kotronarou et al., 1991; Okouchi et al., 1992; Savage et al., 1995; Martin et al., 1995; Gruebel et al., 1995), or a pseudo first-order kinetic model (Mascolo et al., 1995; Lin et al., 1995; Lin et al., 1996a, b; Ma and Lin, 1996). The Arrhenius model offers a good correlation for experiment data obtained at different reaction temperature levels, implying that the acetic acid concentration plays the most important role in acetic acid oxidation under critical condition (Savage et al., 1995). Martin et al. (1995) indicated that the kinetics equations for the degradation rate of 4-chlorophenol is a function of [4-cp], $[ClO_3^-]$, and $[O_2]$. Gruebel et al. (1995) proposed that the change of selenite to selenate by UV oxidation in the presence of oxygen, titania, and light also follow the first-order law. The rate of oxidation is not a simple function of pH, but is also related to the variation of light intensity, ionic strength, oxygen partial pressure, and selenite concentration. Thus, a pseudo first-order model is often used as in the case of the degradation of herbicide prometryne with sodium hypochlorite (Mascolo et al., 1995).

Using only a single AOP method is less efficient than the combination of more than one AOP methods or catalysts for removal of organic matter.[2,4,13,14] Glaze[19] proposed that ozone can be combined with other processes or agents to be more effective for water and wastewater treatment than using the ozonation process alone. Lin et al. (1996b) reported that the degradation of 2-

chlorophenol using an ultrasonic process at an amplitude of 120 mm in the presence of 200 mg/L $H_2O_2$ was more effective than the ultrasonic oxidation process alone.

## OBJECTIVE

The objective of this study is to investigate the use of a combined ultrasonic and Fenton process for removal of 2-chlorophenol from synthetic wastewater samples. The efficiency of 2-chlorophenol removal using different concentrations of Fenton reagent is investigated. Specific objectives of this study also include (I) to investigate the transformation of 2-cp and to determine rate constants and (II) to identify and determine the pathway of 2-cp decomposition with the ultrasound/Fenton catalytic process.

## METHODOLOGY

### Materials

Table I summarized the physical and chemical of the commercial 2-chlorophenol (Merck 802253) used in preparing the synthetic 2-chlorophenol samples. This chemical has a pKa value of 8.49 that affects the type of 2-chlorophenol in solution. Other chemicals used in this study include hydrogen peroxide (Merck 8597), sodium perchlorate monohydrate (Merck 6564), perchloric acid, (FEARK Art.-Nr. 11935), sodium hydroxide (YAKURI, Test No. 31511601), iron(II) sulfate, (FEARK Art.-Nr. 00686), 6-chloro-m-cresol (Chem. Service CAS No. 59-50-7), 6-chloro-m-phenol (Chem. Service CAS No. 615-74-7), 2, 6-dichloro-*p*-benzoquinone (Chem. Service CAS No. 697-91-6). These chemicals can be directly used without further purification. The stock 2-chlorophenol solution (10,000 mg/L) was prepared with deionized water from a millipore purification system, and reserved in a brown Pyrex bottle. It is futher diluted to a final concentration of 100 mg/L with deionized water for the oxidation study. $NaClO_4$ was used to control the ionic strength at 0.1 N. The pH was controlled at 3 with HClO4 and NaOH. All solutions were prepared with deionized water from a millipore purification system.

### Apparatus and Experimental Conditions

Figure 1 shows the schematic drawing of the experimental apparatus. All oxidation reactions were carried in a 1-liter reactor made of Pyrex glass. The reactor content is continuously circulated, at a rate of 500 mL/min, into a sonication cell to receive the ultrasonic treatment. The ultrasonic wave is generated using a Microson XL2020 sonicator (Heat system, Ltd., USA) operated at 20 kHz. The double amplitude at the tip (Part. No. 419) of the standard horn (Part. No. 200, 1/2" d) was adjusted to 120 mm (corresponding to a power input of 165 W). Pure oxygen was continuously bubbled through the reactor content to maintain a dissolved oxygen level at 30–32 mg/L. The reaction temperature of the flask was kept constant at 25°C using a thermostatically controlled outer water jacket. The oxidation studies were carried out using a combination

**Table I. Physical and Chemical Characteristics of 2-cp**

| Molecular Formula | $C_6H_5ClO$ |
|---|---|
| Structure | 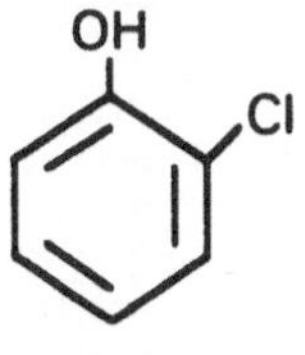 |
| Molecular weight | 128.56 |
| Melting point | 281.85°K |
| Boiling point | 448.15–449.15°K |
| Density | 1.26 Kg/L |
| pKa | 8.49 (25°C) |

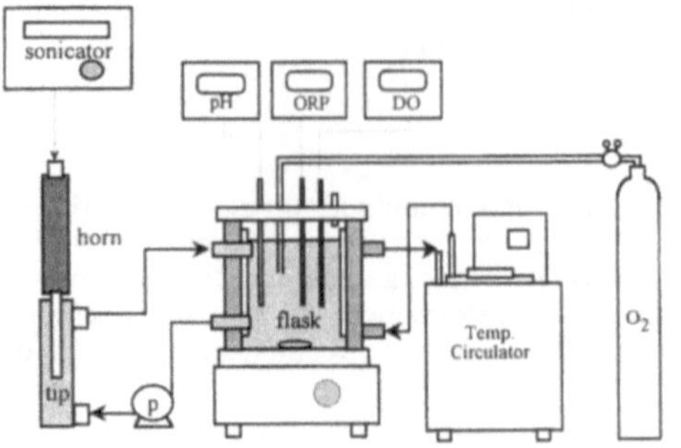


**Figure 1. Schematic diagram of experimental apparatus.**

of 0, 10, 50, 100, 200, and 500 mg/L of $H_2O_2$ and 0.5, 1, 2, 5, and 10 mg/L of $Fe^{2+}$. Based on the results obtained from previous oxidation studies (Lin et al., 1996a, b), the optimal conditions for conducting the ultrasonic/Fenton oxidation process were used to carry out all oxidation reactions in this study. These conditions are pH 3 and ionic strength 0.1 N with pure oxygen being continuously bubbled through the reactor content. All oxidation reactions were allowed to proceed for 100 minutes. Samples of the reactor content were removed at regular time intervals for analyses. Variations of the pH and ORP values were continuously monitored using a pH meter and an ORP meter. Concentrations of 2-chlorophenel and other intermediate products were analyzed using a gas chromatography/mass spectrometry (HP 1800A GCD with DB-5 column, 30 m × 1/2" i.d.).

## RESULTS AND DISCUSSIONS

### Effects of the Fenton Reagent Concentrations on ORP Values

The initial ORP value for samples is 395 mV. It varies with concentrations of $Fe^{2+}$ and $H_2O_2$ applied as well as the reaction time. Figures 2(a) and 2(b) show variations of the ORP ratio which is defined as the ratio of the ORP value measured at time t ([ORP]t) to the original ORP valued ([ORP]0) or ([ORP]t/[ORP]0) versus the various concentrations of $H_2O_2$ for $Fe^{2+}$ concentrations of 0.5 and 10 mg/L, respectively.

According to the results plotted in Figures 2(a) and 2(b), the variation of [ORP]t/[ORP]0 at six different $H_2O_2$ concentrations can be divided into two groups. The first group including the $H_2O_2$ concentrations of 200 and 500 mg/L showed a greater oxidation potential during the overall reaction. The sample with 0 mg/L $H_2O_2$ is not considered to have Fenton reaction and should be excluded in this group. The [ORP]t/[ORP]0 values for samples with 200 and 500 mg/L in this group were always greater than 1. The second group including the $H_2O_2$ concentrations of 10, 50, and 100 mg/L have lower oxidation potential values. The major difference between these two groups was caused by the amounts of $H_2O_2$ applied to the samples. Since the oxidation power increased with an increasing amount of oxidant, the [ORP]t/[ORP]0 value of a high Fenton reagent concentration is always greater than that of samples with low Fenton reagent concentrations ap-

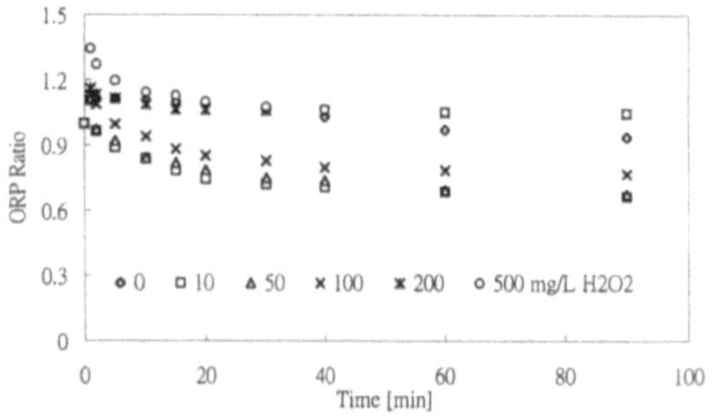


**Figure 2(a). The ORP values profiles at 0.5 mg/L $Fe^{2+}$ and various $H_2O_2$ dosages with the ultrasound/Fenton process.**

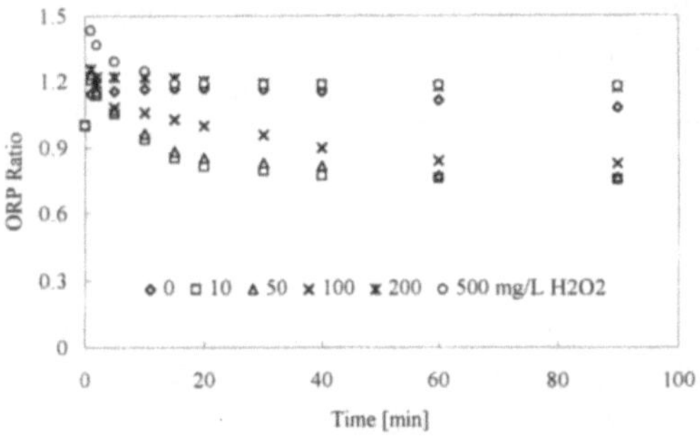


**Figure 2(b). The ORP values profiles at 10 mg/L $Fe^{2+}$ and various $H_2O_2$ dosages with the ultrasound/Fenton process.**

plication. In the presence of both $H_2O_2$ and $Fe^{2+}$ in the aqueous solution, $Fe^{2+}$ could be transferred to $Fe^{3+}$, and produced OH radicals and hydroxide ion shown in Eq. 1.

$$Fe^{2+} + H_2O_2 \rightarrow Fe^{3+} + \cdot OH + OH^- \tag{1}$$

$Fe^{3+}$ also can react with $H_2O_2$ and produce an intermediate as the Fe-$O_2H^{2+}$. Fe-$O_2H^{2+}$ could be transferred to $Fe^{2+}$ and $O_2H$ radical with the sonication by ultrasound or UV light shown in Eq. 2.

$$Fe^{3+} + H_2O_2 \xrightarrow{-H^+} Fe - O_2H^{2+} \xrightarrow{sonication} Fe^{2+} + \cdot O_2H \tag{2}$$

However, the Fe3+ also reacts with $O_2H$ radical and produces $Fe^{2+}$. Hence, the $Fe^{2+}$ produced by other chain reactions reacted with $H_2O_2$ again, and increased the amount of the OH radical. In Figures 2(a) and 2(b), the [ORP]t/[ORP]0 ratios increased with an increasing initial concentration of $H_2O_2$. The greater [ORP]t/[ORP]0 ratio indicates the greater amount of the reactive OH radicals. Therefore, the pollutant could be decomposed easily. The [ORP]t/[ORP]0 ratios dropped after the reaction duration for the first 1–2 min, then keeping constant. In Figures 2(a) and 2(b), we investigated that the initial increment of [ORP]t/[ORP]0 ratio with $Fe^{2+}$ addition only was lower than that with both $Fe^{2+}$ and $H_2O_2$ additions. The production of the OH radicals by the $Fe^{2+}$ reacting with $H_2O$ was lower than that by $Fe^{2+}$ reacting with $H_2O_2$. Thereby, the enhancement of [ORP]t/[ORP]0 ratio by $Fe^{2+}$ addition only was insignificant. Furthermore, the changing ratios of [ORP]t/[ORP]0 were between 0.9 to 1.1. Figures 2(a) and 2(b) also investigate the effects of $Fe^{2+}$ initial concentration on the variety of ORP values The results indicate that the ratio of [ORP]t/[ORP]0 was enhanced with an increasing initial concentration of $Fe^{2+}$. The ORP ratio for different Fenton reagent concentrations measured at 1, 30, 60, and 90 min reaction times is summarized in Table II. The increasing ratio of ORP values within the initial 1 minute indicates that the initial variation of ORP values is caused by the addition of oxidant. The subsequent increasing ORP ratios as observed at reaction times of 30, 60, and 90 minutes represent the oxidation power induced by the residual $Fe^{2+}$ and $H_2O_2$. In Table II, the changing ratios of ORP values are all positive within the initial reaction 1 minute. This observation suggests a positive effect of oxidants addition on the increasing oxidation power. Furthermore, a negative trend of ORP changing ratios occurs at 30, 60, and 90 minutes for lower $H_2O_2$ concentrations (10, 50, and 100 mg/L $H_2O_2$) in the second group. This is in contrast to the aforementioned positive change of ORP values with higher $H_2O_2$ concentration of 200 and 500 mg/L. Based on only the ORP ratios, the ultrasonic/Fenton oxidation should be carried out with $H_2O_2$ dosages greater than 100 mg/L. In reality, the changing ratios of ORP values are also dependent on the concentrations of $Fe^{2+}$ applied as discussed in the following sections.

## Effects of Fenton Reagent Concentrations on 2-cp Decomposition

In this study, simulated wastewater with an initial concentration of 100 mg/L 2-chlorophenole was used in the oxidation study. Variations of the remaining 2-chlorophenol with the ultrasound/Fenton process at $Fe^{2+}$ 0.5 and 10 mg/L versus reaction time are shown in Figures 3(a)

**Table II. The Effects of Different Fenton Reagent Concentrations on ORP Changing Ratios $([ORP]_t-[ORP]_0)/[ORP]_0$**

| Concentration of $H_2O_2$ | Concentration of $Fe^{2+}$ | $([ORP]_t-[ORP]_0)/[OEP]_0$ | | | |
|---|---|---|---|---|---|
| | | 1 min | 30 min | 60 min | 90 min |
| No $H_2O_2$ addition | $FE^{2+}$=0.5 mg/L | 0.113 | 0.066 | –0.028 | –0.061 |
| | $FE^{2+}$=1.0 mg/L | 0.127 | 0.091 | 0 | –0.038 |
| | $FE^{2+}$=2.0 mg/L | 0.134 | 0.099 | 0.035 | 0.008 |
| | $FE^{2+}$=5.0 mg/L | 0.139 | 0.114 | 0.073 | 0.066 |
| | $FE^{2+}$=10.0 mg/L | 0.142 | 0.162 | 0.114 | 0.081 |
| 10 mg/L $H_2O_2$ addition | $FE^{2+}$=0.5 mg/L | 0.104 | –0.276 | –0.291 | –0.334 |
| | $FE^{2+}$=1.0 mg/L | 0.109 | –0.261 | –0.289 | –0.301 |
| | $FE^{2+}$=2.0 mg/L | 0.170 | –0.241 | –0.276 | –0.286 |
| | $FE^{2+}$=5.0 mg/L | 0.197 | –0.230 | –0.256 | –0.268 |
| | $FE^{2+}$=10.0 mg/L | 0.205 | –0.208 | –0.241 | –0.246 |
| 50 mg/L $H_2O_2$ addition | $FE^{2+}$=0.5 mg/L | 0.113 | –0.248 | –0.304 | –0.322 |
| | $FE^{2+}$=1.0 mg/L | 0.124 | –0.233 | –0.278 | –0.299 |
| | $FE^{2+}$=2.0 mg/L | 0.177 | –0.211 | –0.248 | –0.281 |
| | $FE^{2+}$=5.0 mg/L | 0.208 | –0.180 | –0.243 | –0.261 |
| | $FE^{2+}$=10.0 mg/L | 0.230 | –0.170 | –0.230 | –0.238 |
| 100 mg/L $H_2O_2$ addition | $FE^{2+}$=0.5 mg/L | 0.129 | –0.169 | –0.213 | –0.230 |
| | $FE^{2+}$=1.0 mg/L | 0.149 | –0.139 | –0.179 | –0.203 |
| | $FE^{2+}$=2.0 mg/L | 0.215 | –0.129 | –0.172 | –0.190 |
| | $FE^{2+}$=5.0 mg/L | 0.215 | –0.110 | –0.152 | –0.182 |
| | $FE^{2+}$=10.0 mg/L | 0.256 | –0.040 | –0.159 | –0.172 |
| 200 mg/L $H_2O_2$ addition | $FE^{2+}$=0.5 mg/L | 0.159 | 0.061 | 0.053 | 0.048 |
| | $FE^{2+}$=1.0 mg/L | 0.159 | 0.081 | 0.058 | 0.056 |
| | $FE^{2+}$=2.0 mg/L | 0.192 | 0.111 | 0.084 | 0.076 |
| | $FE^{2+}$=5.0 mg/L | 0.225 | 0.124 | 0.106 | 0.101 |
| | $FE^{2+}$=10.0 mg/L | 0.256 | 0.190 | 0.177 | 0.172 |
| 500 mg/L $H_2O_2$ addition | $FE^{2+}$=0.5 mg/L | 0.344 | 0.078 | 0.056 | 0.048 |
| | $FE^{2+}$=1.0 mg/L | 0.354 | 0.096 | 0.071 | 0.068 |
| | $FE^{2+}$=2.0 mg/L | 0.382 | 0.122 | 0.096 | 0.089 |
| | $FE^{2+}$=5.0 mg/L | 0.403 | 0.134 | 0.116 | 0.114 |
| | $FE^{2+}$=10.0 mg/L | 0.435 | 0.185 | 0.182 | 0.180 |

and 3(b). The residual amounts of 2-cp are dependent on the quantity of $H_2O_2$ and $Fe^{2+}$ as well. It is obvious that more $H_2O_2$ can induce a greater amount of OH radical production as described in Eq. 3, and hence a stronger oxidation power.

$$H_2O_2 + ultrasound \rightarrow 2 \cdot OH \quad (3)$$

The organic substance could be decomposed either to some intermediates or to final products, and the process is shown in Eq. 4. The amount of OH radicals produced from $H_2O_2$ cleavage influences the potential of 2-cp decomposition. With more $H_2O_2$ added into the solution, more OH radicals are produced and more 2-cp can be oxidized into the final products.

$$Organic\ Compound + \cdot OH \rightarrow Intermediates\ or\ Products \quad (4)$$

Hence, the experimental results indicate that the effects of $H_2O_2$ concentrations on 2-cp decomposition are significant. In Figure 3(a), the residual amounts of 2-cp are 57.1, 31.1, 16.1, 3.8, 0.7, and 0.2 mg/L after 90 minutes oxidation duration with the ultrasound/Fenton process at $Fe^{2+}$ 0.5 mg/L and 0, 10, 50, 100, 200, and 500 mg/L $H_2O_2$, respectively. When the $H_2O_2$ concentration exceeds 200 mg/L, 2-cp decomposition is not significantly enhanced by additional increase

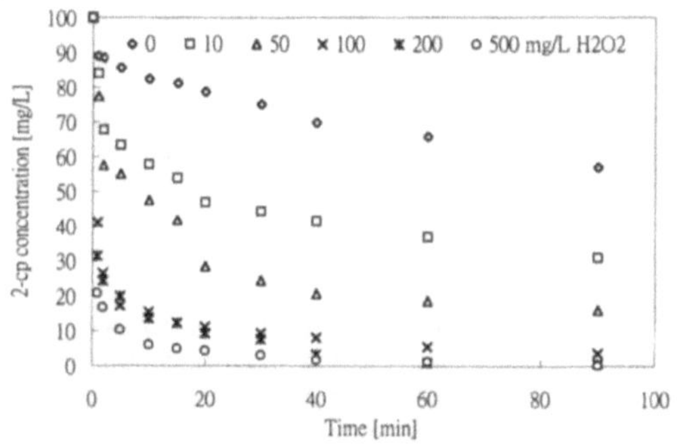


**Figure 3(a). The 2-cp decomposition profiles at 0 mg/L $Fe^{2+}$ and various $H_2O_2$ dosages with the ultrasound/ Fenton process.**

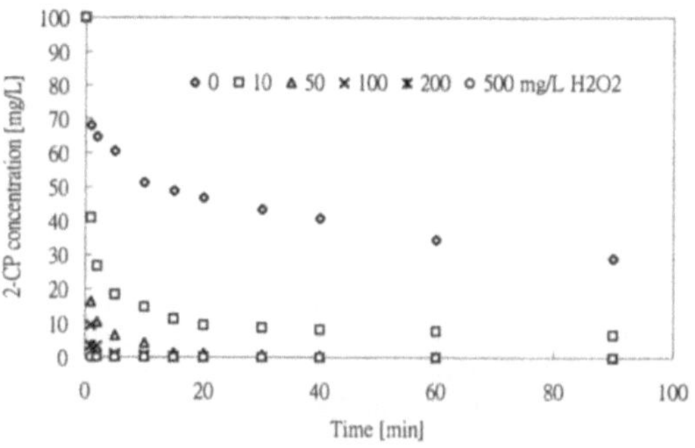


**Figure 3(b). The 2-cp decomposition profiles at 10 mg/L $Fe^{2+}$ and various $H_2O_2$ dosages with the ultrasound/ Fenton process.**

of $H_2O_2$ dosage. In Figure 3(b), the residual amounts of 2-cp are 29, 6.7, 0, 0, 0, and 0 mg/L after 90 minutes oxidation duration with the ultrasound/Fenton process at $Fe^{2+}$ 10 mg/L and 0, 10, 50, 100, 200, and 500 mg/L $H_2O_2$. Hence, addition of $H_2O_2$ and $Fe^{2+}$ concentrations also affects the efficiency of 2-cp decomposition. This is evidenced by the observation that the residual amounts of 2-cp decreases with an increasing $Fe^{2+}$ concentration [Figures 3(a) and 3(b)].

If the $H_2O_2$ concentrations are kept above 50 mg/L and $Fe^{2+}$ concentrations above 10 mg/L, the residual amounts of 2-cp after the Fenton treatment will be lower than the detection limit of 0.22 mg/L. In literature, Sedlak and Andren (1991) developed a Fenton process to oxidize chlorobenzene at an initial pH of 3. Chlorobenzene disappeared within the first 2 hours of the reaction by addition of $H_2O_2$ at a rate of 5 mM/h. Their results are close to the results obtained in this study.

## Effects of the Fenton Reagent Concentration on $H_2O_2$ Utilization

In this study, the concentrations of $H_2O_2$ were analyzed by a real-time sampling and analyzing during the 2-cp decomposition with the ultrasound/Fenton process. The amount of $H_2O_2$ consumed during the reaction period is shown in Figures 4(a) and 4(b). The concentrations of $Fe^{2+}$ in Figures 4(a) and 4(b) are 0.5 and 10 mg/L, respectively. Utilization of $H_2O_2$ is dependent on the amounts of $Fe^{2+}$ applied. More $Fe^{2+}$ applied can react with $H_2O_2$ to produce greater amounts of OH radicals in the solution according to Eq. (1). At a $H_2O_2$ dosage of 100 mg/L, the utilization ratios at $Fe^{2+}$ 0.5 and 10 mg/L were 56.4 and 72.8% after 90 min, respectively [Figures 4(a) and 4(b)]. When the concentrations of $H_2O_2$ applied are greater than 100 mg/L, only 25–67.5% of $H_2O_2$ could be utilized in pollutant destruction. This observation suggests that more $H_2O_2$ added into the aqueous solution will yield greater 2-cp decomposition ratios [Figures 3(a) and 3(b)]. However, the trend of $H_2O_2$ utilizing does not parallel the decomposition of 2-chlorophenol. The utilization ratios of $H_2O_2$ decrease with increasing initial concentration of $H_2O_2$. For treating wastewater containing 100 mg/L 2-cp, if the concentrations of $H_2O_2$ are below 50 mg/L, 72–92% of $H_2O_2$ will be consumed and 84–100% of 2-cp will be decomposed. This phenomenon suggests that a significantly removing ratio of 2-cp decomposition can be obtained at a low con-

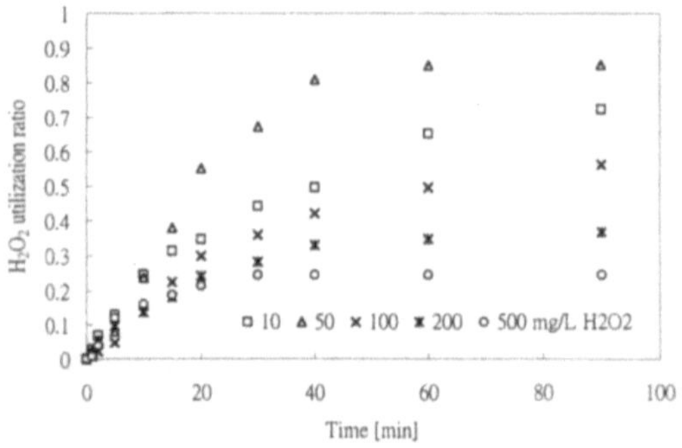


**Figure 4(a). The $H_2O_2$ utilization profiles at 0.5 mg/L $Fe^{2+}$ and various $H_2O_2$ dosages with the ultrasound/Fenton process**

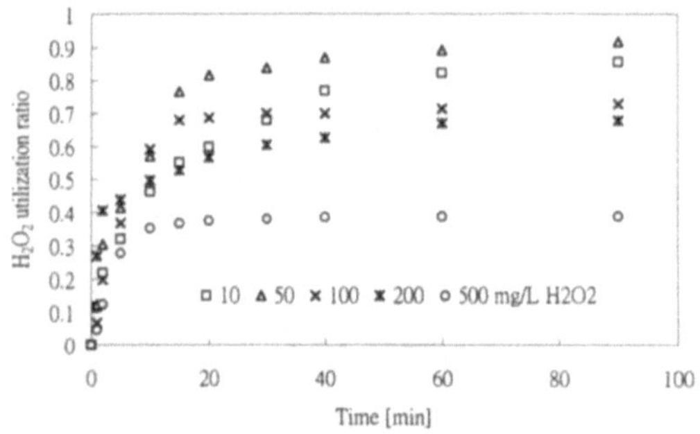


**Figure 4(b). The $H_2O_2$ utilization profiles at 10 mg/L $Fe^{2+}$ and various $H_2O_2$ dosages with the ultrasound/Fenton process.**

centration of $H_2O_2$. Arnold et al. (1995) showed that lowering the $FeSO_4/H_2O_2$ ratio should improve the reaction efficiency because there would be less $Fe^{2+}$ competing with atrazine for OH radicals. In this study, the decomposition of 2-cp per unit $H_2O_2$ utilized was 2.27 mg 2-cp/mg $H_2O_2$. In addition, as the initial concentration of $H_2O_2$ increased to 500 mg/L, the ratio would be decreased to 0.66 mg 2-cp/mg $H_2O_2$.

## Effects of the Fenton Reagent Concentration on 2-cp Mineralization

The major goal of this study is to mineralize the hazardous 2-cp into nonhazardous final products, e.g., $CO_2$. The production of inorganic carbon such as $CO_2$ as well as the reduction of TOC (total organic carbon) can be used to indicate mineralization of the organic substance (Ding et al., 1995). The removal of TOC is used to indicate the amount of 2-cp mineralized during Fenton treatment. Figure 5 plots the variation of 2-cp mineralization at various concentrations of $H_2O_2$ and $Fe^{2+}$ applied. The mineralization ratios are noted to increase with increasing concentrations of $H_2O_2$ and $Fe^{2+}$ applied. With the addition of 10 mg/L $Fe^{2+}$ and 500 mg/L $H_2O_2$, more than 86% of 2-cp is mineralized. Furthermore, at these dosage levels, all 2-cp is observed to be completely decomposed at the end of the first 1 minute reaction time. But the $H_2O_2$ utilization rate is only 25% as shown in Figure 4(b). In this study, the raw sample contains 100 mg/L 2-cp which is much higher than the concentration of 2-cp found in most water and wastewater samples. The 86% mineralization may be more than needed for treating natural water and wastewater samples. For the oxidation process to be cost-effective, it is not necessary to use the highest dosages of 10 mg/L $Fe^{2+}$ and 500 mg/L $H_2O_2$ or running the reaction at the longest reaction time of 90 minutes. But the results demonstrate that the ultrasonic/Fenton process will yield an efficiency sufficient to meet most treatment goals in the real world.

## Reaction Kinetics

A number of reports found in literatures proposed that the decomposition of organics with AOPs follows the pseudo first-order kinetics.[4,7,22,32] The reduction of 2-cp is fitted with a pseudo

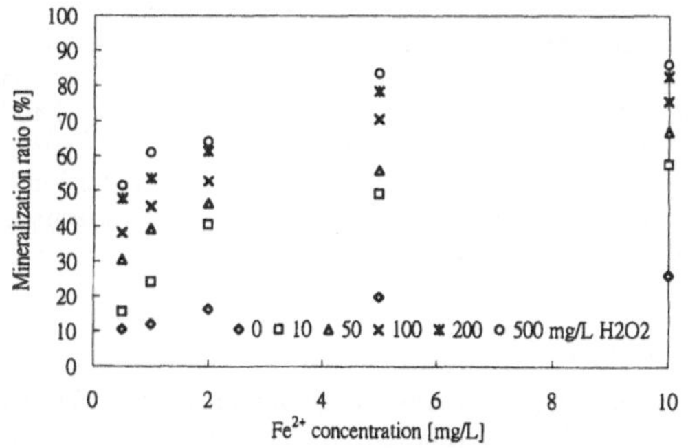


**Figure 5. The 2-cp mineralization profiles at various concentrations of $H_2O_2$ and $Fe^{2+}$ with the ultrasound/Fenton process.**

first-order kinetics model. Values of the rate constant, k, in Eq. (4), are calculated using Eq. 5 which relates the k value to concentrations of 2-cp and OH radicals.

$$\frac{-d[2-cp]}{dt} = k_{obs}^{2-cp}[2-cp][\cdot OH] \tag{5}$$

$$\frac{-d[H_2O_2]}{dt} = k_{obs}^{2-cp}[H_2O_2] \tag{6}$$

The experimental data can be fitted well by the pseudo first-order kinetics, with R2 values being greater than 0.9. Figures 6(a) and 6(b) show the variations of the pseudo first-order rate constants of 2-cp decomposition and $H_2O_2$ utilization. In Figure 6(a), the 2-cp decomposition

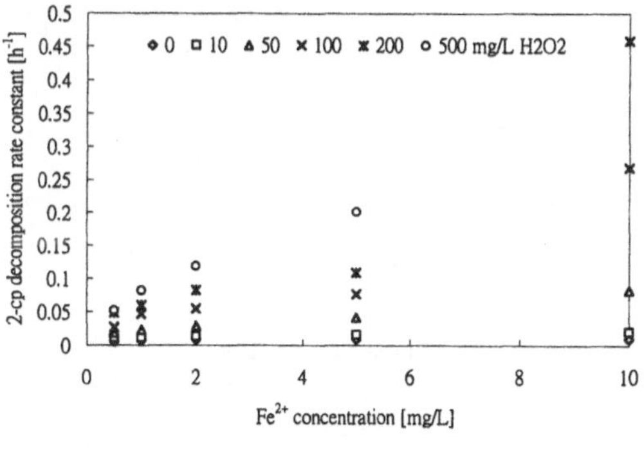


**Figure 6(a). The 2-cp decomposition rate constant profile at various concentrations of $H_2O_2$ and $Fe^{2+}$ with the ultrasound/Fenton process.**

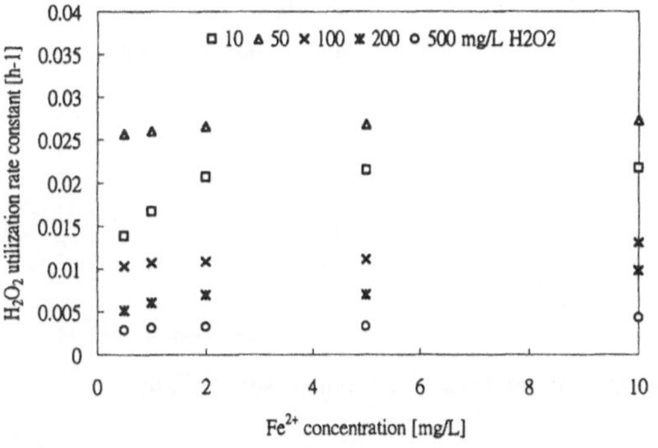


**Figure 6(b). The $H_2O_2$ utilization rate constant profile at various concentrations of $H_2O_2$ and $Fe^{2+}$ with the ultrasound/Fenton process.**

rate constants are observed to depend on the initial concentrations of $H_2O_2$ and $Fe^{2+}$. However, the profiles of $H_2O_2$ utilization rate constants are seen to depend on only the initial concentrations of $Fe^{2+}$, but independent of $H_2O_2$. The experimental data shown in Figure 6(a) suggests that the enhancement of rate constant for increasing $Fe^{2+}$ dosages at lower $H_2O_2$ concentration, i.e., 0 or 10 mg/L, is insignificant but is significant at higher $H_2O_2$ concentrations. Thus a greater amount of $Fe^{2+}$ was more effective on 2-cp decomposition if used with a higher concentration of $H_2O_2$. In contrast to the results presented in Figure 6(a), Figure 6(b) suggests variations of $H_2O_2$utilization rate constant with various $Fe^{2+}$ concentrations are insignificant.

## Intermediates and Reaction Pathway

In this study, the liquid-liquid extraction procedure following the Standard Method 6420 B was to analyze the 2-cp decomposition and intermediates formation. Seven different kinds of intermediates included 4-chloro-1,3-benzendiol, 2-chloro-5-methyl-phenol, 2-chloro-6-methyl-phenol, 4-chloro-3-methyl-phenol, 2-chloro-p-hydroquinone, 2-chloro-p-benzoquinone, and 2,6-dichloro-2,5-cyclohexadiene-1,4-dione. 2-chloro-p-benzoquinone was detected as the major product. Two possible destruction mechanisms are proposed to account for the production of these seven species. These mechanisms are shown in Figures 7(a) and 7(b). Huang et al. (1995) proposed that the reaction between OH radicals and phenol were taking place in the ortho- and para- positions. However, based on the results obtained in this study, the OH radicals react with phenol in the para- and meso-positions during 2-cp decomposition. If the OH radical reacts with 2-cp at the meso-position, 2-cp could be connected with an OH radical and produce 4-chloro-1,3-benzendiol. After a meso- OH radical addition, a methylation takes place and leads to a substitution at the OH functional group site. Thus, 4-chloro-1,3-benzendiol could be transferred to three different kinds of intermediates, i.e., 2-chloro-5-methyl-phenol, 2-chloro-6-methyl-phenol, and 4-chloro-3-methyl-phenol [Figure 7(a)].

If the OH radical attaches to the para- position, 2-chlorophenol could be connected with an OH radical and produce 2-chloro-*p*-hydroquinone. The OH radicals further attack the OH functional group sites and transfer 2-chloro-*p*-hydroquinone to 2-chloro-p-benzoquinone. Furthermore, 2-chloro-p-benzoquinone could react with the free chloride radical to form 2,6-dichloro-2,5-cyclohexadiene-1,4-dione production (Figure 7(b)). The relative concentrations of the aforementioned

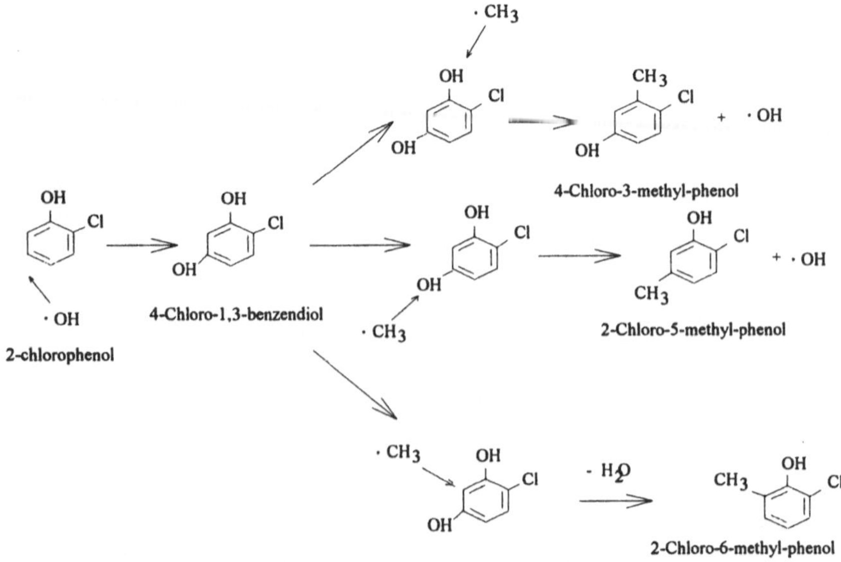


**Figure 7(a). Chlorinated methyl-phenol formation pathway for the ultrasound/Fenton degradation of 2-chlorophenol.**

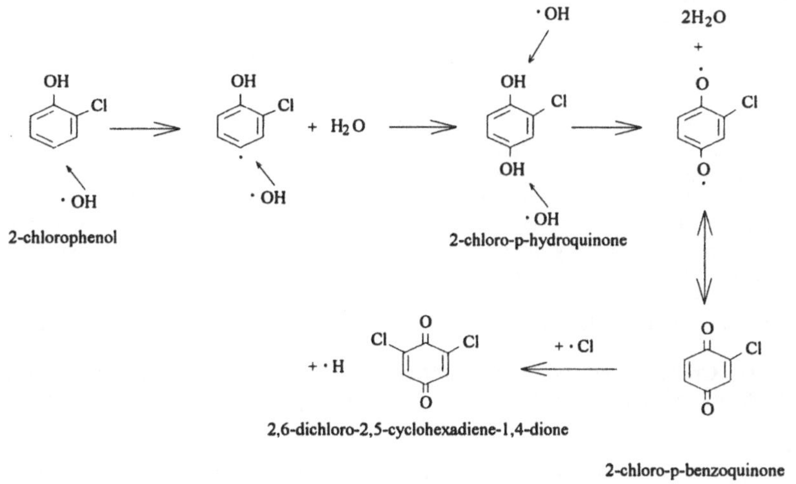


**Figure 7(b). Chlorinated benzoquinone formation pathway for the ultrasound /Fenton degradation of 2-chlorophenol.**

intermediates can be used to evaluate which mechanism, *para-* or *meso-* attack mechanism is predominant in the destruction of 2-cp. For example, at $Fe^{2+}$ 2 mg/L and $H_2O_2$ 10 mg/L, the total concentration of 2-chloro-*p*-benzoquinone and 2,6-dichloro-2,5-cyclohexadiene-1,4-dione (40.6 mg/L) are greater than the total concentration of three different kinds of methyl chlorophenol (7.9 mg/L). At higher concentration levels of $H_2O_2$, methyl chlorophenol is not detected. Therefore, the *para*-position reaction between OH radicals and 2-cp is suggested as the predominant destruction mechanism.

## CONCLUSIONS

This study demonstrates that the ultrasound/Fenton process is a promising and effective method for the destruction of 100 mg/L of 2-chlorophenol (2-cp) in water or wastewater. With addition of 10 mg/L $Fe^{2+}$ and 500 mg/L $H_2O_2$, nearly 100% of 2-cp is destroyed and more than 80% is mineralized as indicated by TOC removal. The $H_2O_2$ utilization ratio reaches 70%. Thereby, this combination of 10 mg/L $Fe^{2+}$ and 100 mg/l $H_2O_2$ is a good choice for the treatment of wastewater containing 100 mg/L 2-cp. The decomposition of 2-cp and utilization of $H_2O_2$ can be well fitted by a pseudo first-order kinetic model. During the oxidation of 2-cp, seven different intermediates are detected with 2-chloro-*p*-benzoquinone being the major product. Based on the distribution of these intermediates, the *para-* position reaction between OH radical and 2-cp is proposed as the major 2-cp destruction mechanism.

## ACKNOWLEDGMENT

This study was financially supported by the National Science Council, ROC (NSC 84-2211-E-009-052).

## REFERENCES

1. Orzechowaska, G.E., Poziomek, E.J., Hodge, V.F., and Engelmann, W.H. "Use of sonochemistry in monitoring chlorinated hydrocarbons in water." *Environ. Sci. Technol.*, 29(5), pp. 1373–1379, 1995.

2. Werderitsch, W., Lahnsteiner, J. and Frischherz, H. "Process for degradation of pesticides and VOC." *International Water Supply Association Year Book*, pp. 311–316, 1996.
3. Arnold, S. M., Hickey, W. J., and Harris, R. F. "Degradation of atrazine by Fenton's reagent: condition optimization and product quantification." *Environ. Sci. Technol.*, 29(8), pp. 2083–2089, 1995.
4. Scheck, C.K. and Frimmel, F.H. "Degradation of phenol and salicylic acid by ultraviolet radiation/hydrogen peroxide/oxygen." *Water Res.*, 29(10), pp. 2346–2352, 1995.
5. Ding, Z.Y., Aki, S.N.V.K., and Abraham, M.A. "Catalytic supercritical water oxidation: phenol conversion and product selectivity." *Environ. Sci. Technol.*, 29(11), pp. 2748–2753, 1995.
6. Canonica, S., Stemmler, U., Jans, K., and Hoigné, J. Transformation kinetics of phenols in water: photosensitization by dissolved natural organic material and aromatic ketones. *Environ. Sci. Technol.*, 29(7), pp. 1822–1831, 1995.
7. Minero, C., Pelizzetti, E., Malato, S. and Blanco, J. "Large solar plant photocatalytic water decontamination: degradation of pentachlorophenol." *Chemosphere*, 26(12), pp. 2103–2119, 1993.
8. Lin, J.-G., Chang, C.-N., Wu, J.-R. and Ma, Y.-S. "Treatment of wastewater containing 2-chlorophenol using ultrasound/$H_2O_2$ process." *Proceedings of International Conference on Environmental Technology and Business Opportunities*, November 6–7, 1995, Shah Alam (Malaysia), 1995.
9. Lin, J.-G., Chang, C.-N. and Wu, J.-R. "Decomposition of 2-chlorophenol in aqueous solution by ultrasound/$H_2O_2$ process." Water. Sci. Tech., 33(6), pp. 75–81, 1996a.
10. Lin, J.-G., Chang, C.-N., Wu, J.-R. and Ma, Y.-S. "Enhancement of decomposition of 2-chlorophenol with ultrasound/ $H_2O_2$ process. *Water Sci. Tech.*, 34(9), pp. 41–48, 1996b.
11. Minero, C., Pelizzetti, E., Pichat, P., Sega, M. and Vincenti, M. "Formation of condensation products in advanced oxidation technologies: the photocatalytic degradation of dichlorophenols on $TiO_2$." *Environ. Sci. Technol.*, 29(9), pp. 2226–2234, 1995.
12. Savage, P.E. and Smith., M.A. "Kinetics of acetic acid oxidation in supercritical water." *Environ. Sci. Technol.*, 29(1), pp. 216–211, 1995.
13. Gruebel, K.A., Davis, J.A., and Lecjie, J.O. "Kinetics of oxidation of selenite to selenate in the presence of oxygen, titania and light. *Environ. Sci. Technol.*, 29(3), pp. 586–594, 1995.
14. Martin, S.T., Lee, A.T. and Hoffmann, M.R. "Chemical mechanism of inorganic oxidants in the $TiO_2$/UV process: increased rates of degradation of chlorinated hydrocarbons." *Environ. Sci. Technol.*, 29(10), pp. 2567–2573, 1995.
15. Karlsson, S., Kaugare, S., Grimvall, A., Boérn, H. and Sävehed, R. "Formation of 2, 4, 6-trichlorophenol and 2, 4, 6-trichloroanisole during treatment and distribution drinking water." *Water Sci. Tech.*, 31(11), 99–103, 1995.
16. Muftikian, R., Fernando, Q. and Korte, N. "A method for the rapid dechlorination of low molecular weight chlorinated hydrocarbons in water." *Wat. Res.*, 29(10), pp. 2434–2439, 1995.
17. Smeds, A., Franzén, R. and Kronberg, L. "Occurrence of some chlorinated enol lactones and cyclopentene-1, 3-diones in chlorine-treated waters." *Environ. Sci. Technol.*, 29(7), pp. 1839–1844, 1995.
18. Ollis, D.F., Pelizzetti, E., and Serpone, N. "Photocatalytzed destruction of water contaminants." *Environ. Sci. Technol.*, 25(9), pp. 1522–1529, 1991.
19. Glaze, W.H. "Drinking-water treatment with ozone." *Environ. Sci. Technol.*, 21(3), pp. 224–230, 1987.
20. Mason, T.J., Lorimer, J.P. and Bates, D.M. "Quantifying sonochemistry: casting some light on a black art." *Ultrasonics*, 30(1), pp. 40–42, 1992.
21. Petrier, C., Micolle, M., Merlin, G., Luche, J.-L. and Reverdy, G. "Characteristics of pentachlorophenate degradation in aqueous solution by means of ultrasound." *Environ. Sci. Technol.*, 26(8), pp. 1639–1642, 1992.
22. Okouchi, S., Nojima, O. and Arai, T. "Cavitation-induced degradation of phenol by ultrasound." *Water Sci. Tech.*, 26(9), pp. 2053–2056, 1992.
23. Kotronarou, A., Mills, G., and Hoffmann, M.R. "Ultrasonic irradiation of *p*-nitrophenol in aqueous solution." *J. Phys. Chem.*, 95, pp. 3630–3638, 1991.
24. Ma, Y.-S. and Lin, J.-G. "Kinetics of pretreatment of wastewater containing 2-chlorophenol with ultrasonic/$H_2O_2$ process." *Proceeding in the 2nd Conference on Pretreatment of Industrial Wastewaters*, October 16–18, 1996, pp. 750–753.
25. Elizardo, K. "Fighting pollution with hydrogen peroxide." *Pollut. Eng.*, pp. 106–109, 1991.
26. Huang, C.-P., Shu, C.-S., Chen, C.-W. and Dong, C. "Electrochemically generated Fenton's reagent for the removal of chlorophenols in water." *Proceedings in the First Workshop on*

*Drinking Water Management and Treatment Technologies* (Principles and Applications of Advanced Oxidation Process), April 24–27, Taipei, Taiwan ROC, 11-1–11–39, 1995.

27. Kochany, J. and Boltom, J.R. "Mechanism of photodegradation of aqueous organic pollutant. 2. Measurement of primary rate constants for reaction of OH radicals with benzene and some halobenzenes using and EPR Spin-Trapping method following the photolysis of $H_2O_2$." *Environ. Sci. Technol.*, 26(2), pp. 262–265, 1992.
28. Sun, Y. and Pignatello, J.J. "Photochemical reactions involved in the total mineralization of 2,4-D by $Fe^{3+}/H_2O_2$/UV. *Environ. Sci. Technol.*, 27(2), pp. 304–310, 1993.
29. Takemura, Y., Kengo, S., Tatsuo, M. and Motoyuki, S. "Decomposition organic chlorine compounds in dry cleaning wastewater by Fenton's reaction on reticulated iron." *Water Sci. Tech.*, 30(3), pp. 129–137, 1994.
30. Mascolo, G., A. Földényi, Lopez, R., Passino, R. and Tiravanti, G. "Prometryne oxidation by sodium hypochlorite in aqueous solution kinetics and mechanism." *Environ. Sci. Technol.*, 29(12), pp. 2987–2991, 1995.
31. Sedlak, D.L. and Andren, A.W. "Oxidation of chlorobenzene with Fenton's reagent." *Environ. Sci. Technol.*, 25(4), pp. 777–782, 1991.
32. Pelizzetti, E. and Minero, C. "Photocatalyric activity and selectivity of titania colloids and particles prepared by the sol-gel technique: photooxidation of phenol and atrazine." *Langmuir*, 9, pp. 2995–3001, 1993.

# 37 REMOVAL AND REUSE OF AMMONIA AND PHOSPHATE BY PRECIPITATION OF STRUVITE

**Brian A. Dempsey**
Department of Civil & Environmental Engineering
The Pennsylvania State University
University Park, Pennsylvania 16802-1408 USA

## ABSTRACT

The objective of this research was to efficiently remove ammonia from industrial wastewaters by precipitation of struvite. Struvite ($MgNH_4PO_4$) causes problems when it precipitates as a scale, which often occurs in anaerobic digesters and in nutrient-rich industrial wastewaters. Several processes have been patented for the removal of ammonia by precipitation of struvite. The emphasis in this research was on the engineering aspects of the process; i.e. minimizing cost and difficulty, minimizing the use of hazardous materials, and maximizing potential reuse of the residual struvite. The precipitation environment was optimized in terms of thermodynamic equilibria and kinetic parameters that controlled the granular structure of the precipitate. The resulting solid was a relatively pure struvite residual, easily separable and dewatered, and appropriate for reuse.

## INTRODUCTION

Removal of ammonia and phosphate by precipitation of struvite ($MgNH_4PO_4$) is attractive because the two nutrients can be simultaneously removed, because the process is particularly well-suited to wastewaters that are very concentrated in the nutrients, and because the residual may be valuable as a fertilizer. Other nutrient removal processes (such as biological systems for nutrient removal, stripping into a gas phase, use of alum or ferric salts to precipitate phosphates, ion exchange, or reverse osmosis) may require multiple process steps, remove only one of the nutrients, result in a nuisance residual, or may be too expensive to use on concentrated wastewaters.

The equilibrium chemistry of struvite has been well-documented. Pertinent formation constants are listed in Table I. Precipitation occurs rapidly when conditions are appropriate, and the product can be identified by X-ray diffraction techniques. One must take precautions during drying operations, however, because the precipitate decomposes (losing ammonia) at temperatures greater than 40°C.

Struvite commonly precipitates in anaerobic digesters that are used to treat wastewater sludges. In such cases, struvite is a nuisance because the precipitates form as scales in zones

**Table I. Formation constants for the system $Mg^{2+}$-$PO_4^{3-}$-$NH_3$-$H^+$**

| Species | Common Name | Log $*K_f$ |
|---|---|---|
| $MgNH_4PO_4 \cdot 6H_2O$ (s) | Struvite | 13.15 |
| $MgHPO_4$ $H_2O$ (s) | Newberyite | 5.82 |
| $Mg_3(PO_4)_2 \cdot 8H_2O$ (s) | Bobierrite | 25.2 |
| $Mg_3(PO_4)_2 \cdot 22H_2O$ (s) | Bobierrite (am) | 23.1 |
| $Mg(OH)_2$(s) | Brucite | −16.792 |
| $MgHPO_4^{\circ}$ | Soluble complex | 15.22 |

*52nd Purdue Industrial Waste Conference Proceedings*, 1997, Ann Arbor Press, Chelsea, Michigan 48118. Printed in U.S.A.

where the pH rises, often due to degassing of $CO_2$ on the suction side of pumps or in effluent pipes that are at lower pressure than in the digester itself. Accordingly, steps have sometimes been taken to inhibit precipitation; e.g, manipulation of temperature, pressure, pH, or constituent concentrations. Thus Pitt et al.[1] added Fe salts to precipitate vivianite ($Fe_3(PO_4)_2$) which forms as a suspension rather than a scale; then the precipitation of struvite is limited by the availability of phosphate. Several investigators have added meta-phosphates or other scale inhibitors. Borgerding[2] and others have diluted digester supernatant and avoided feeds rich in nutrients. Perhaps the most common solution is to acidify the waste stream prior to arrival at locations where struvite formation had been detected, typically where $CO_2$ was off-gassing.

Recently, several patents have been granted for processes in which struvite is precipitated for removal of nutrients; e.g. according to processes described by Schulze-Rettmer[3] or Zdybiewska and Kula.[4] The major differences among these techniques lie in the reagents that are used. Zdybiewska and Kula[4] rapidly adjusted pH to 9.5 and likely formed bobierrite. They also used relatively expensive reagents. Schulze-Rettmer[3] used primarily $H_3PO_4$ and MgO but also used NaOH and brought the pH rapidly to pH between 8 and 9. Other processes often use HCl to prepare reagents and NaOH to adjust pH. As described below, these techniques do not adequately address kinetic and thermodynamic issues that affect removals and physical characteristics of the precipitates. Although the physical characteristics of the struvite precipitates from these processes are poorly documented, it is desirable that the residuals be easily separated, cleaned, and dewatered.

The objective of this research was to improve the existing technologies for removal of ammonia from industrial wastewaters by precipitation of struvite. The precipitation environment was manipulated in terms of thermodynamic equilibria and kinetic parameters that might control the purity and structure of the precipitate. Engineering aspects were emphasized; i.e. minimizing cost and difficulty of the process, minimizing the use of hazardous materials, and maximizing potential reuse of the residual struvite.

## MATERIALS AND METHODS

All chemicals were reagent grade unless otherwise noted. In some experiments, commercial or technical grade MgO and $H_3PO_4$ were used as reagents. Industrial wastewaters (especially condensate from an animal food processing industry) were used in some of the experiments.

$Mg^{2+}$, $NH_4^+$, and $H_2PO_4^-/HPO_4^{2-}$ were analyzed by ion chromatography (IC), after filtration through 0.22 m pore-size filters. In some cases, whole samples were diluted and acidified. Nonlinear IC responses for ammonium and phosphate were corrected. Meter and electrode for pH were calibrated at least daily using a two-point calibration. Samples were air-dried, and the mass of such samples was stable after two days. Specific resistance to filtration was measured using standard techniques.

In order to precipitate struvite, stoichiometric amounts of $Mg^{2+}$, $NH_4^+$, and $PO_4^{3-}$ must be present and the pH must be within the range where struvite dominates as the solid phase. These needs are met by adding a *stoichiometric reagent* and an *alkaline reagent*. The *stoichiometric reagent* includes salts or acids of phosphate and Mg(II), especially $H_3PO_4$ and MgO which are relatively inexpensive. MgO or NaOH are the best candidates for the *alkaline reagent*, which is necessary to counteract the acidity of the wastewater and any acidic reagents. Alkaline chemicals that contain Ca(II) should be avoided, because of precipitation of calcium phosphates which inhibits the precipitation of struvite. The alkalinity in MgO is slow-acting, especially relative to NaOH, but MgO contributes a stoichiometric reagent and, as detailed below, the slow provision of alkalinity is helpful for production of a clean struvite precipitate.

## THERMODYNAMIC CONSIDERATIONS

Pertinent formation constants are listed in Table I. Thermodynamic constants for all of the Mg(II) phosphate precipitates (including struvite) or complexes are taken from Taylor et al.[5,6] and these values are also reported in current critical reviews of stability constants. In addition, sequential $pK_a$ values are 2.2, 7.2, and 12.2 for $H_3PO_4$ and 9.3 for $NH_4^+$. Several Mg-phosphate soluble complexes, in addition to $MgHPO_{4¡}$, have been reported. For compatibility with Taylor's calculations of solid formation constants, we have considered only $MgHPO_{4¡}$ among these complexes.

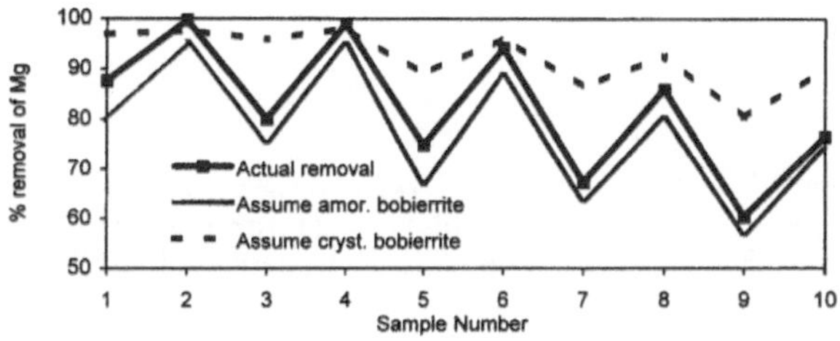


**Figure 1. Results from 10 experimental conditions are shown in this plot. Experimental data (filled squares) are consistent with possible presence of crystalline struvite, crystalline newberyite, and amorphous bobierrite.**

Taylor also reported that time and certain conditions are necessary for the conversion from amorphous to crystalline bobierrite. Both kinetic and thermodynamic analyses of the precipitation of struvite are strongly dependent on which of these isomorphs predominates over the experimental time-frame. Our experimental results were always consistent with the existence (when saturated) of amorphous bobierrite rather than crystalline bobierrite. For example, results from a series of experiments are shown in Figure 1 (heavy line with data points) along with predicted solubility of Mg(II) in equilibrium with newberyite, struvite, and crystalline bobierrite (top dashed line) or with newberyite, struvite, and amorphous bobierrite (bottom line). The data were best represented by formation of amorphous bobierrite plus struvite or newberyite. Results for soluble $NH_4^+$ and $PO_4^{3-}$ also were also consistent with formation of amorphous bobierrite instead of the crystalline phase.

## Reagent Containing Mg and P

When $NH_4^+$ is in stoichiometric excess and both phosphate and Mg(II) must be added, then it is advantageous to combine these reagents prior to addition to the wastewater flow, since MgO is sparingly soluble at alkaline pH values and the rate of dissolution of MgO is quite slow at neutral or alkaline pH.

However, newberyite ($MgHPO_4$) precipitates for many combinations of MgO and $H_3PO_4$. Both equilibrium calculations and empirical observations demonstrated that a soluble reagent containing Mg and P at concentrations approaching 1 mol/kg can be produced only if the molar ratio MgO/(HCl + $H_3PO_4$) is less than or equal to 0.5. For example, the experimentally determined ratios of filterable to total Mg(II) are shown in Figure 2 for various Mg to P ratios. Overall, the best results were obtained using a reagent containing 1 mol/kg $H_3PO_4$, 0.5 mol/kg MgO and no HCl. This reagent does not provide all of the Mg nor all of the alkalinity that is necessary for the precipitation of struvite, so that additional MgO or other reagent must be added later. As demonstrated later in this paper, sequential addition of alkalinity reagents is useful for production of the cleanest and most filterable struvite phase. Therefore, the Mg/P reagent that is described above is also the most effective initial reagent for production of a struvite precipitate that is easy to handle. It is also the cheapest among alternative reagents and does not require the use of hazardous NaOH or HCl.

## Precipitation of Struvite

The first experiments involved rapid adjustment of pH to obtain desired final conditions. Some examples of removals of N and P are shown in Figure 3, where all experiments were performed

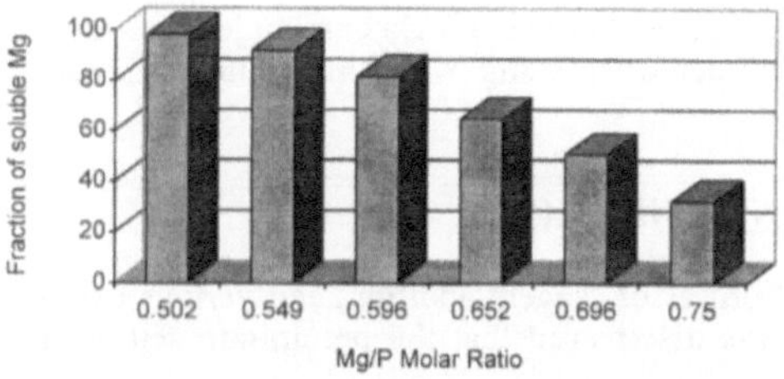


**Figure 2. Fraction of total Mg that is soluble after preparation of reagents containing no HCl and various molar ratios of MgO to $H_3PO_4$.**

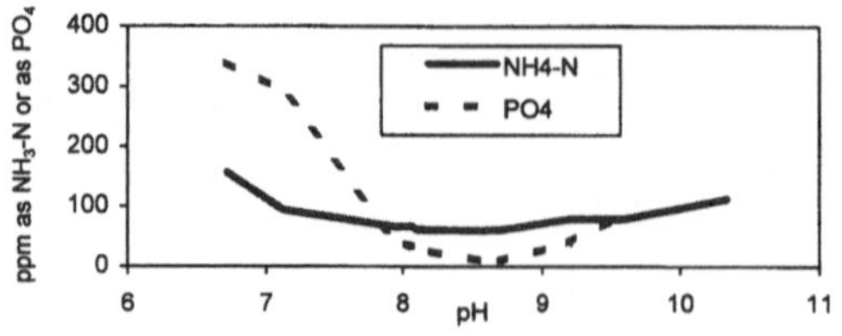


**Figure 3. Removals of NH4+ and PO43- for initial reagent with 1.5/1 Mg/P and using 2/1/1 molar ratios of Mg/P/N**

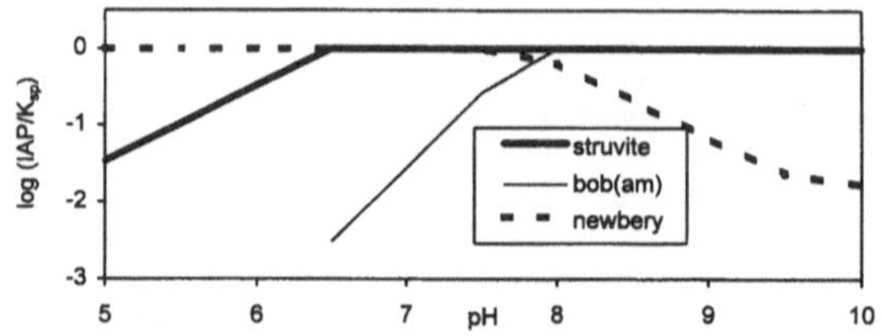


**Figure 4. Calculated stable phases for the condition 0.06M Mg, 0.05 M $PO^{3-}$, and 0.05 M $NH_4^+$. log $IAP/K_{sp}$ of 0 means that the phase is thermodynamically stable. Values less than 0 mean undersaturation for that phase.**

for the same initial concentrations of Mg, P, and N. Since the fractional removals of P and N changed as a function of pH, therefore the ratios of the elements in the precipitated phase also changed.

The formation of multiple phases can also be shown based on theoretical calculations using stability constants reported in Table I. Some results of these calculations are shown in Figure 4 for the initial condition 0.06 M $Mg^{2+}$, 0.05 M $PO_4^{3-}$, and 0.05 M $NH_4^+$. As the pH is increased from slightly acid to slightly basic values there is a change in predominant solid species from newberyite to struvite to bobierrite (am). Although MgO is undersaturated for all conditions shown in Figure 3, increasing pH also results in a decreased driving force for the dissolution of MgO.

Two solid phases exist simultaneously for certain conditions. For pH between 6.4 and 7.7, as illustrated in Figure 4, both newberyite and struvite are themodynamically stable. Both struvite and bobierrite (am) are stable phases for pH above 8.1. As described below, precipitation of more than one phase affects the granular structure of struvite, the purity of the precipitated phase, and the extent of removal of ammonia and phosphate.

### Alkaline Reagent

NaOH and MgO are quite different in terms of the operational benefits from their use. MgO is inert and of limited solubility compared to NaOH. The relatively slow dissolution and limited solubility of MgO have been reported by numerous investigators. In fact, these characteristics are perceived as benefits since they impart pH buffer capacity due to the addition of MgO. Lime cannot impart this benefit until much higher pH values. Limitations on the solubility of MgO are exacerbated by the common-ion effect, since $Mg^{2+}$ is also added in the *stoichiometric reagent.*

The following advantages are associated with the use of MgO. First, it is possible to eliminate the use of NaOH entirely, thus reducing the cost of treatment. Second, MgO ultimately provides $Mg^{2+}$ for the precipitation of struvite. Third, we have discovered (see below) that the purity and dewaterability of the precipitate is improved when the pH rise takes several minutes (as with MgO) rather than immediate (as with NaOH).

Experiments were run using several alkaline reagents, but the experimental objectives (dealing with cost and quality of the residual) were best met when using MgO. Accordingly, the use of MgO is emphasized in this report.

## NUCLEATION CONDITIONS

Experiments were conducted using different orders of reagent addition, and various pH environments as a function of reaction time. It was discovered that the precipitate settled most rapidly and the settled-sludge was densest when the *stoichiometric reagents* containing Mg and P were added to the ammonia wastewater in a fashion that ensured an acidic environment (i.e., pH from 5 to 7, where newberyite is predominant) for at least a few seconds prior to raising the pH to values most appropriate for formation of struvite (pH 7.5 to 9). Precipitates that were eventually predominantly struvite settled much faster after passing through a lower pH regime. Microscopic observation of the precipitates revealed irregularly-shaped and clustered precipitates when the pH passed through the lower values, whereas needle-shaped precipitates formed when the pH was initially within the region where struvite is predicted to be the most stable solid phase.

These results were repeated several times. For most of the experiments that are reported in this paper, the pH passed through slightly acidic pH regions after the addition of the *stoichiometric reagents*. Use of MgO as the *alkaline reagent* ensured that the pH rise occurred over a few minutes rather than immediately.

These results are interpreted as initial homo-nucleation of newberyite or a mixed precipitate of newberyite and struvite. Due to changing pH conditions, the newberyite became unstable, leading to hetero-nucleation of struvite at multiple sites and thus bunched and irregular crystal shape. These conditions also prevented formation of bobierrite, since the pH was not high enough for precipitation of amorphous bobierrite until the struvite precipitation was complete. In fact, appropriate dosage of MgO prevented bobierrite from becoming the dominant or even a stable phase.

## RATE OF PRECIPITATION

When stoichiometric reagents are present and the pH is raised to about 8.5, the rate of precipitation is extremely fast, with most of removable $NH_4^+$ filterable within seconds. However, the quality of the precipitate improved when the pH rise was slower. It is preferable to add a reagent containing Mg/P of 0.5 to 1, wait a few minutes for formation of newberyite with some coprecipitated $NH_4^+$, and then add additional slurry of MgO to accomplish dosing and pH for best removals and optimal precipitate characteristics.

The progress of a reaction is illustrated in Figure 5. The initial reagent was added at 0 minutes, with additional slurry of MgO added at 5 minutes. The initial pH of the wastewater was above 9.5, dropped rapidly with addition of the first reagent, and then rose slowly after addition of MgO. $Mg^{2+}$ was removed immediately due to precipitation of newberyite. Less than 10% of $NH_4^+$ was removed within 5 minutes, since conditions were not yet appropriate for precipitation of struvite (see Figure 4). Even though the pH eventually approached 9, the relatively low Mg/P/N ratio ensured no precipitation of bobierrite. The rate of pH rise can be controlled by manipulating the ratio of MgO to $H_3PO_4$ to $NH_4^+$, as shown in Figure 6 for a wastewater that already contained some alkalinity and some Mg(II) and phosphate. The effects of reagent doses on pH history and removal of nutrients must be evaluated for every wastewater.

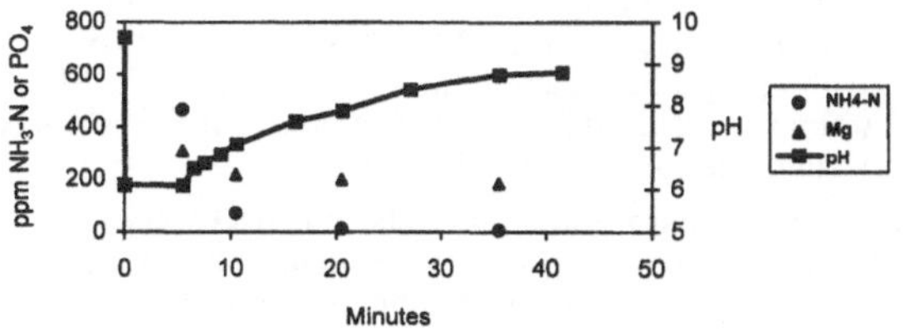


**Figure 5. Progress of a reaction involving addition of MgO/H3PO4 reagent at zero minutes and slurry of MgO at five minutes.**

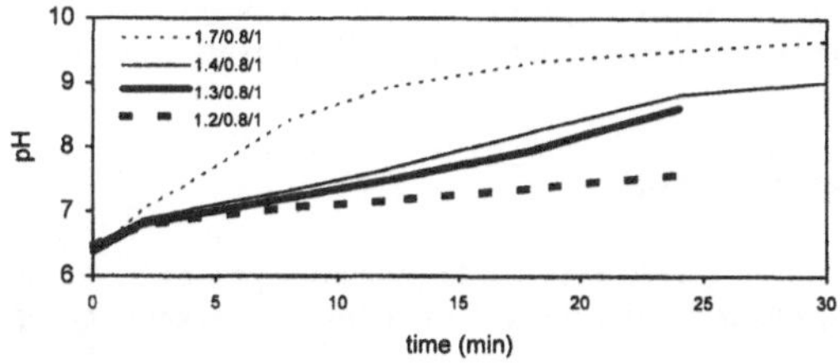


**Figure 6. Rate of change of pH for various amounts of Mg/P/N, where the N represents amount in the wastewater and Mg/P represents added reagents.**

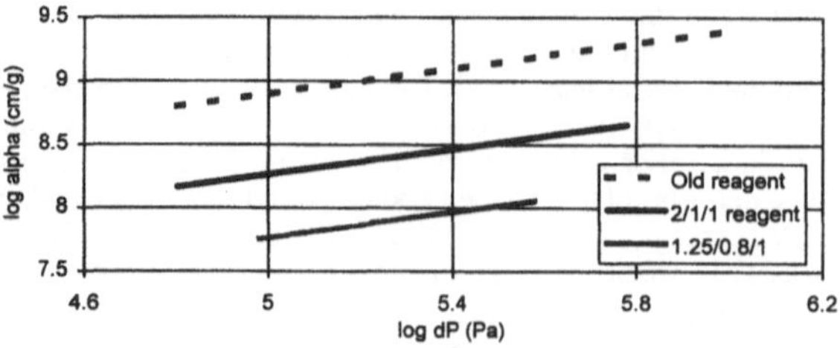


**Figure 7. Specific resistance to filtration (cm/g) as a function of differential pressure (Pa) for three reaction conditions.**

## REMOVAL OF NUTRIENTS

It is possible to eliminate more than 95% of the initial $NH_4^+$ by precipitation of struvite, but this can only be accomplished with stoichiometric surpluses of $Mg^{2+}$ and perhaps phosphate. There are two problems associated with this condition; surplus $Mg^{2+}$ and phosphate leads to precipitation of bobierrite(am) and in some cases higher soluble phosphate in the process effluent.

Also, it is operationally undesirable to require an extremely accurate and proportional dosing of phosphate (i.e., 100% molar ratio of P/N), especially if the ammonia loading is variable. A better alternative, from both operational and theoretical viewpoints, is to slightly underdose with phosphate. In our research, this provided several advantages: (1) operational control was easier; (2) underdosing prevented the formation of bobierrite(am); (3) residual $NH_4^+$ buffered the pH, permitting faster dissolution of MgO; (4) the treatment cost was decreased, and (5) the residual phosphorus was always less than 3 mg/L when P/N ratio was 0.95 or less. As long as the initial pH rise was not too rapid, then the removal of $NH_4^+$ for a given P/N depended only on the pH.

Various constituents of the wastewater may affect these reactions. For example, Ca(II) forms precipitates with phosphate. The extent of this competing reactions is dependent on pH as well as $Ca^{2+}$ concentration.

## DEWATERABILITY

Control of the nucleation environment; i.e., passing through a pH zone where newberyite predominates, results in solids that settle and dewater very well. Interfacial settling velocities of 3 m/hr were achieved when treating industrial wastewaters.

Specific resistance to filtration for some precipitates is reported in Figure 7, which shows the effects of immediate adjustment of pH to the target zone (top line), slow adjustment of pH (middle line), and slow adjustment plus stoichiometric control to keep phosphate limiting (bottom line). There is more than an order of magnitude decrease in resistance to filtration going from the top to the bottom condition.

Improvements in specific resistance to filtration were due to improved crystal habit for dewatering (slow pH change) and avoiding precipitation of bobierrite (P/N molar ratio less than 1/1). The middle line is for solids formed with the proper nucleation environment but with P/N of 1.05/1. The bottom line is for similar conditions, except P/N was 0.8. The latter solid was easily dewatered and washed, friable, and a good candidate for reuse as a fertilizer.

The recommended treatment process is based on many additional tests. It is necessary to have at least a 30 second delay between additions of the *stoichiometric reagent* and the *alkaline reagent.* Based on limited experimental data, no negative effects are obtained from recirculation or back-mixing of solids. Mg/N and P/N ratios should be kept as low as feasible. Mg/N ratio should be determined by evaluating the pH rise rate compared to the operational detention time and by the final pH for the particular wastewater. P/N ratio should usually be kept below 1.00 and may be kept as low as 0.8/1, especially for wastewaters with variable flux of $NH_4^+$. Chemical costs are minimized by this reagent strategy. In addition, decreased P/N leads to lower solids handling costs and greater residual value.

## CONCLUSIONS

Nutrient removal by precipitation of struvite is a very reasonable industrial process, and empirical results are very compatible with theoretical predictions. The reaction is fast and equilibrium with struvite is achieved within a few minutes. Although the precipitation of crystalline bobierrite is very slow, the formation of amorphous bobierrite can result in decreased removal of ammonia and in deteriorated quality of the residual solids.

Manipulation of the nucleation environment leads to significant improvements in settleability and dewaterability of the precipitate. It is important that the nucleation phase pass through a zone where newberyite is the predominant solid phase.

The elimination of HCl and NaOH, while retaining a mixed *stoichiometric reagent* results in decreased cost and process difficulty as well as a reduction of hazard.

Decreasing the P/N molar ratio to less than one increases the purity and dewaterability of the residual solids, decreases the soluble phosphorus concentration, and decreases the cost of the process per removal of ammonia.

## REFERENCES

1. Pitt, P., Mamais, D., Cheng, Y.W., Loaicono, J., and Jenkins, D. Struvite control using ferric chloride at the Southeast Water Pollution Control Plant, San Francisco. *Proceedings of the 65th Annual Conference, WEF*, New Orleans, LA, pp. 111–122, 1992.
2. Borgerding, J. Phosphate deposits in digestion systems. *J. WPCF* 44(5), pp. 813–819, 1972.
3. Schulze-Rettmer, R. The simultaneous chemical precipitation of ammonium and phosphate in the form of magnesium-ammonium-phosphate. *Water Sci. Tech.* 23, pp. 659–667, 1991.
4. Zdybiewska, M.W. and Kula, B. Removal of ammonia nitrogen by the precipitation method, on the example of some selected waste waters. *Water Sci. Tech.* 24(7), pp. 229–234, 1991.
5. Taylor, A.W., Frazier, A.W., and Gurney, E.L. Solubility products of magnesium ammonium and magnesium potassium phosphates. *Trans. Faraday Society* 59, pp. 1580–1584, 1963a.
6. Taylor, A.W., Frazier, A.W., Gurney, E.L., and Smith, L.P. 1963b. Solubility products of di- and trimagnesium phosphates and the dissociation of magnesium phosphate solutions. *Trans. Faraday Society* 59, pp. 1585–1589, 1963b.

# Section Ten
# PHYSICAL-CHEMICAL: METALS PRESENCE, FATE, AND IMPACT

## 38 LEACHING PROCESS FOR REMOVAL OF HEAVY METALS IN SOIL

**Alton Day Stone**, Senior Environmental Scientist
Pennoni Associates Inc.
Hopkinton, Massachusetts 01748

**James C. O'Shaughnessy**, Professor
Department of Civil & Environmental Engineering
Worcester Polytechnic Institute
Worcester, Massachusetts 01609

### ABSTRACT

This paper presents the results of bench-scale experiments conducted to evaluate heavy metals sorption and desorption phenomena for soils common to New England, and the suitability of soil leaching (flushing) for in-situ remediation of soils contaminated with heavy metals. The results of the investigations indicated that the soil physical parameters mass fines/unit mass soil, bulk density, coefficient of uniformity, and coefficient of concavity were not good predictors of soil sorptive or desorption capacity.

### INTRODUCTION

Environmental contamination by heavy metals is relatively common. Lead, often in high concentrations is ubiquitous in urban soils due to the past use of leaded gasolines and paints. Formerly, arsenic was widely used as a general pesticide. The majority of sites contaminated with concentrated heavy metals are associated with industrial processes such as mining, smelting, plating, coating, machining and manufacturing, and locations where disposal of associated wastes has occurred. The Environmental Protection Agency (EPA)[1] estimates that approximately 10% of the Superfund sites involve heavy metal contamination. Concentration ranges of metals reported in soil at EPA sites are included in Table I. In recognition of the hazard posed by exposures to heavy metals, federal and state agencies have established regulations to limit environmental contamination and reduce human and environmental exposures. The EPA currently lists 13 heavy metals as "priority pollutants" under the Comprehensive Environmental Recovery, Compensation and Liability Act (CERCLA) commonly known as "Superfund": antimony, arsenic, beryllium, cadmium, chromium, copper, lead, mercury, nickel, selenium, silver, thallium, zinc. Each of these metals, except barium and copper, is listed as a "hazardous material" in the Massachusetts Contingency Plan (MCP, 310 CMR 40.0000), promulgated under M.G.L. c. 21e the "Oil and Hazardous Materials Act."

Industrial solid wastes which contain sufficient leachable amounts of certain metals (Table I), as determined by the Toxicity Characteristic Leaching Procedure (TCLP) chemical analytical procedure, are listed "hazardous wastes" which are regulated under the Resource Conservation and Recovery Act (RCRA). The TCLP extraction (leaching) analysis is intended to duplicate leaching of the waste by rainwater in the open environment. If the extract contains a metal concentration above an established criteria, the waste exhibits the characteristic of "toxicity," and is classified as hazardous. These wastes are considered to be potential sources of environmental

*52nd Purdue Industrial Waste Conference Proceedings*, 1997, Ann Arbor Press, Chelsea, Michigan 48118. Printed in U.S.A.

**Table I. Background Concentrations and Regulatory Standards for Heavy Metals**

| | Eastern US Surface Soil Concentrations[2] | | Trace Element Content of Soils [3] | | | | |
|---|---|---|---|---|---|---|---|
| Metal | Range (ppm) | Average (ppm) | Range (ppm) | Average (ppm) | MCL[4] | TCLP[5] | Maximum Concentration at FIT Sites[6] |
| Antimony | 1–8.8 | 0.76 | 2–10.0 | — | 0.006 | — | 15 |
| Arsenic | 0.7–73 | 7.4 | 1–50.0 | 5 | 0.05 | 5 | 11,700 |
| Barium | 10–1,500 | 420 | 100–3,000 | 430 | 2 | 100 | 0.2 |
| Beryllium | 1–7.0 | 0.85 | 0.1–40 | 6 | 0.004 | — | 38 |
| Cadmium | — | — | 0.01–0.7 | 0.06 | 0.005 | 1 | 118,300 |
| Chromium | 1–1,000 | 52 | 1–1,000 | 100 | 0.1 | 5 | 136,000 |
| Copper | <1–700 | 22 | 2–100 | 30 | 1.3 | — | 186,300 |
| Lead | <10–300 | 17 | 2–200 | 10 | 0.015 | 5 | 466,000 |
| Mercury | 0.01–3.4 | 0.12 | 0.01–0.3 | 0.03 | 0.002 | 0.2 | 83,200 |
| Nickel | 15–700 | 18 | 5–500 | 40 | 0.1 | — | 8.8 |
| Selenium | <0.1–3.9 | 0.45 | 0.1–2 | 0.3 | 0.05 | 1 | 1.3 |
| Silver | — | — | 0.01–5 | 0.05 | — | 5 | 18 |
| Thallium | — | — | — | — | 0.002 | — | — |
| Zinc | <5–2,900 | 52 | 10–300 | 50 | — | — | 110,000 |

contamination, and must be disposed of in secure landfills or rendered nontoxic by stabilization or other processes prior to disposal. Under the Safe Drinking Water Act the EPA has established "maximum contaminant levels" (MCLs) for concentrations of certain metals (Table I) in drinking water supplies. These metals are considered particularly toxic when chronically ingested at even very low concentrations.

Cleanup of heavy metals-contaminated soils has proved difficult. Unlike organic substances, metals cannot be mineralized or detoxified in situ by biological or other processes, or easily incinerated. The most common procedure for soil remediation is excavation and disposal at a hazardous waste landfill, with or without stabilization. Costs for excavation, transportation, tipping, and taxes typically are in the range of $200–$300 per ton. Off-site disposal is practical for small soil volumes, but becomes prohibitively expensive for large sites. The major drawback to off-site disposal of metal-contaminated soil is that the method merely transfers a large volume of relatively unconcentrated waste from one location to an other. Regulatory agencies, in particular the EPA, industry, and parties responsible for cleanup have been investigating methods for on-site and in-situ cleanup of contaminated soils. In-situ control technologies include stabilization of the soil-metal complex. Stabilization procedures involve the addition of an inert substance such as grout, or electrically induced vitrification of the soil. Soil leaching, in which removal of metal mass from the soil column may be achieved through application and recovery of a solvent solution, is being recognized as a viable in-situ remedial technology.

## STUDY OVERVIEW AND OBJECTIVES

The fate of heavy metals in the soil is a function of the complex physical and chemical interactions between the soil, water, metal, and biota. In particular, the leaching and transport of metals through the soil column are controlled by two interdependent but separate categories regarding the chemistry of metals in soil:[7] (1) solution chemistry and (2) interfacial chemistry. The overall objectives of the research were to investigate sorption and desorption of heavy metals from soils common to Superfund sites and the suitability of various solvents to enhance the extraction of metals from contaminated soil using a leaching (flushing) process. Research results have been presented by Canniff,[8] St. Germain,[9] Macaulay,[10] Venkataramanappa,[11] and O'Shaughnessy and D'Andrea et al.[12]

The subject research data used in this paper were collected by Macaulay,[10] and Venkataramanappa.[11] The objectives of the research were to investigate:

1. The sorption of cadmium, chromium, copper, nickel, and zinc by soils common to New England, under laboratory conditions which approximated those of an actual chemical spill.
2. Determine if physical soil properties such as unit mass of fines and bulk density could be used to predict metals sorption in soil.
3. Determine if the presence of plastic (clay mineral) fines affected the sorption of heavy metals.
4. Evaluate the effectiveness of water and selected chemical solvents to desorb and leach heavy metals from the soil column.
5. Evaluate the kinetics of the leaching process, for the purpose of predicting rates of heavy metal removal from the soil column.

## MATERIALS AND METHODS

In the experiments, 16 study soil columns were prepared by filling Plexiglas cylinders with specific soil types. Eight different soil types were prepared and two columns (total of 16 columns) were packed with each soil type. Physical and chemical parameters of each soil type and column were determined prior to initiating the contamination/decontamination sequence. The soil columns were then contaminated by pouring a solution containing known masses of metals into the columns and allowing the solution to stand for a four-day period. The columns were drained and the volume of effluent and concentrations of metals determined. Each column was then subjected to a series of solvent leaching treatments using tap water, tap water-acid solvent, and organic-chelating agent-acid-tap water solvent. Each treatment consisted of a number of rinses of one soil column pore volume each. The volume of effluent from each rinse and metals concentrations in the effluent were determined. The resulting data were used to quantitatively describe sorption and desorption-leaching processes.

Five heavy metals, cadmium (Cd), chromium (Cr), copper (Cu), nickel (Ni), and zinc (Zn), were chosen for leaching studies. Each of the metals is an EPA Priority Pollutant and each metal except Ni is present in high concentrations at sites investigated by the EPA Field Investigation Team (Table I).

### Soil Characterization

Soils used in this column study were natural soils collected from locations in Massachusetts and New Hampshire (the parent soil), or were manufactured soils produced by combining material from two or three of the parent soils. Complete descriptions of the soil characteristics and soil classification methods are reported by Stone.[13] These soils were comprised of an outwash sand, a fine sand, Fuller till, a Moulton silt, and a Cambridge clay. Eight soils or soil types, numbered 1 through 8, were used in the column studies. Three of the test soils were parent soils and five were manufactured by combining two or three of the parent soils. Each of the eight soil types was placed into two columns. Column soils composition and other parameters are listed in Table II.

### Study Soil Columns

Soil columns were constructed and prepared using the same procedures as were employed during earlier column studies conducted at WPI.[8–11] Columns were constructed of clear Plexiglas cylinders 81.3 cm long by 7.0 cm in diameter. Both ends of the cylinders were fitted with water tight Plexiglas caps having a 2.5 cm hole in the center. To prevent soil from being washed from the columns, the bottoms were fitted with three layers of precision woven mesh having size openings of 1,000, 250, and 74 microns. The mesh retained the soil but permitted passage of the countermeasure solutions. The bottom cap of each column was fitted with a valve to control the flow of soil leachate from the cylinder. Soil leachate drained from the columns was collected. The soil column design is shown in Figure 1.

Two columns were packed with each of the eight soil types, for a total of 16 soil columns. Two empty columns were used as equipment blanks. To prepare each column, approximately 3,316 grams of soil were moistened to achieve a water content of approximately 10%. Soil was then packed into a column in approximately 5 cm thick lifts, to achieve a total depth of 45.7 cm and a total soil column volume ($V_T$) of approximately 1,752 cm$^3$. Soil was packed using a custom-

**Table II. Physical Characteristics of Soil Columns**

| Column # | Soil Number and Type | Unit Weight $\nabla$ g/cm$^3$ | Bulk (Dry) Density $D_b$ g/cm$^3$ | Dry Column Soil Mass $dM_{soil}$ g | Water Content $\theta$ % | Void Ratio e | Porosity n | Void Volume $V_v$ cm$^3$ |
|---|---|---|---|---|---|---|---|---|
| 1 | 1 OWS + 10% NPF | 1.89 | 1.72 | 3013 | 0.9.80 | 0.60 | 0.38 | 666 |
| 2 | 1 OWS + 10% NPF | 1.89 | 1.74 | 3048 | 8.90 | 0.58 | 0.37 | 648 |
| 3 | 2 Fuller Till | 1.58 | 1.44 | 2523 | 10.10 | 0.90 | 0.47 | 823 |
| 4 | 2 Fuller Till | 1.58 | 1.43 | 2505 | 10.20 | 0.92 | 0.48 | 841 |
| 5 | 3 OWS + 4% NPF | 1.76 | 1.63 | 2856 | 8.30 | 0.69 | 0.41 | 718 |
| 6 | 3 OWS + 4% NPF | 1.76 | 1.59 | 2786 | 10.40 | 0.73 | 0.42 | 736 |
| 7 | 4 OWS + 16% NPF | 1.89 | 1.73 | 3031 | 9.20 | 0.59 | 0.37 | 648 |
| 8 | 4 OWS + 16% NPF | 1.89 | 1.70 | 2978 | 10.10 | 0.62 | 0.38 | 666 |
| 9 | 5 OWS + 21% NPF | 1.89 | 1.72 | 3013 | 9.40 | 0.60 | 0.37 | 648 |
| 10 | 5 OWS + 21% NPF | 1.91 | 1.74 | 3048 | 9.90 | 0.58 | 0.37 | 648 |
| 11 | 6 Fine Sand | 1.60 | 1.47 | 2575 | 9.00 | 0.87 | 0.47 | 823 |
| 12 | 6 Fine Sand | 1.55 | 1.42 | 2487 | 9.20 | 0.94 | 0.49 | 858 |
| 13 | 7 OWS + 5% PF | 1.95 | 1.65 | 2891 | 9.30 | 0.67 | 0.40 | 701 |
| 14 | 7 OWS + 5% PF | 1.58 | 1.46 | 3084 | 10.50 | 0.56 | 0.36 | 631 |
| 15 | 8 OWS + 5% PF + 5% NPF | 1.58 | 1.45 | 2540 | 9.00 | 0.90 | 0.47 | 823 |
| 16 | 8 OWS + 5% PF + 5% NPF | 1.79 | 1.64 | 2873 | 9.40 | 0.68 | 0.40 | 710 |

OWS = Outwash Sand
NPF = Non Plastic Fines
PF = Plastic Fines

$\nabla$ = Wet column sil mass/column soil volume
$D_b$ = Dry column soil mass/column soil volume
$\theta$ = Volume water in wet soil column/column soil volume

e = Volume voids/volume solids
n = Volume voids in soil column/column soil volume
$V_v$ = n * soil column volume

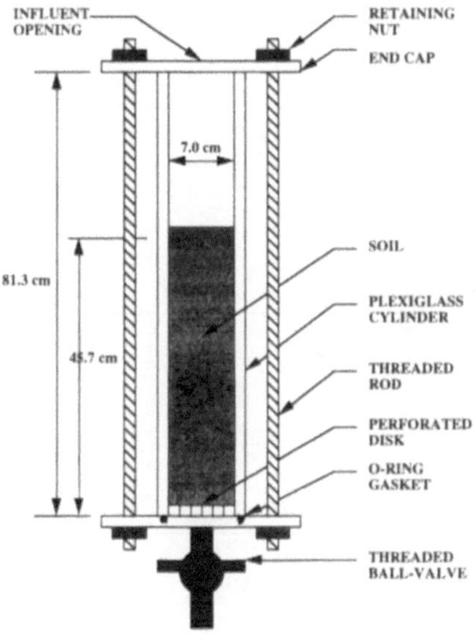


**Figure 1. Soil column apparatus.**

made, controlled drop hammer which fit inside the column. Each lift was tamped five times by dropping the hammer from a height of 30.5 cm above the soil surface to achieve a desired compaction density of 105 pcf. Upon completion of packing, each column was weighed to determine the soil column weight $W_T$.

The hydraulic conductivity or coefficient of permeability (K) of each soil column was determined several times over the course of the soil column studies, for the purpose of evaluating whether K changed with the number of rinses. In general, K was calculated after the 20, 22, 46, and 77 rinses. With few exceptions, there was little change in *K* during the column studies.

## Soil Column Contamination

Soil columns were contaminated by saturating the soil column with a solution containing established concentrations of each of the study metals (Cd, Cr, Cu, Ni, Zn). The mixed metal solution was prepared by dissolving known weights of cadmium sulfate, chromic acid, cupric sulfate, nickelous nitrate, and zinc sulfate in a known volume of distilled water. The concentrations of metals in the solutions used to contaminate each column are other laboratory preparation steps covered in detail in Stone.[13] Each column was contaminated by adding one liter of mixed metal solution to the column, and allowing the solution to stand in the column for a four-day contact period. Each column was then drained by gravity for a two-day period, the volume of effluent solution measured, and the effluent analyzed to determine the concentration of each metal.

Each column was then subjected to a series of four countermeasure treatments (i.e., leaching solvents) having a combined total of 30 to 77 rinses of approximately one soil column pore volume (void volume $V_v$) per rinse. The effluent from each rinse was collected, volume measured, and the effluent analyzed for metals concentrations.

Countermeasures consisted of tap water, or solutions consisting of several combinations of tap water with ethylenediaminetetraacetic acid (EDTA), acetic acid (AA), hydrochloric acid (HCl), and/or citric acid (CA). Countermeasures, treatments, and rinse sequences are summarized in Table III. The tap water treatment was intended to duplicate leaching of contaminated soil in the field by the downward percolation of storm water. EDTA is an organic sequestering (chelating) agent. Previous studies 8,9,10,11 demonstrated that EDTA had a strong capability to desorb metals from soil and was not intrinsically toxic. The acids were added to enhance metals leaching by reducing the pH of the soil column water.

The samples were prepared for analysis following Test Methods for Evaluating Solid Waste (SW 846); Method 3010 Acid Digestion of an Aqueous Solution (EPA, 1987), and the concentration of metals were determined using a Atomic Absorption Spectrophotometer (FLAA).

**Table III. Chemical Countermeasures, Soil Column Treatments, and Rinse Sequences**

| Rinse # | Countermeasure | pH | Treatment # |
|---|---|---|---|
| | Columes #: 1, 5, 7, 9, 11 | | |
| 1 to 20 | Tap Water | 5.4 | 1 |
| 21 to 45 | 1 g / EDTA / L + AA | 4.0 | 2 |
| 46 | Tap Water | — | — |
| 47 to 63 | 2 g CA/L | 2.6 | 3 |
| 64 to 77 | 0.5 g CA/L | 3.0 | 4 |
| | Columns #: 2, 6, 8, 10, 12 | | |
| 1 to 20 | Tap Water | 5.4 | 1 |
| 21 to 45 | 1 g/EDTA/L + HCl | 4.0 | 2 |
| 46 | Tap Water | — | — |
| 47 to 63 | 1 g/EDTA/L + 2 g CA/L | 2.9 | 3 |
| 64 to 77 | 1 g/EDTA/L + 0.5 g CA/L | 3.3 | 4 |

| Treatments | #1 | #2 | | #3 | #4 |
|---|---|---|---|---|---|
| Column # | Tap Water | 1 g EDTA/L + AA | Tap Water | 2 g CA/L | 0.5 g CA/L |
| 3 | 1–20 | 21–34 | 35 | 36–53 | — |
| 13 | 1–14 | 15–39 | 40 | 41–56 | 57–71 |
| 15 | 1–10 | 11–35 | 36 | 37–53 | 54–68 |
| | Tap Water | 1 g EDTA/L + AA | Tap Water | 1 g EDTA/L + 2 g CA/L | 1 g EDTA/L + 0.5 g CA/L |
| 14 | 1–16 | 17–31 | — | — | — |
| 16 | 1–10 | 11–35 | 36 | 37–53 | 54–65 |

EDTA = ethylenediaminetetraacetic acid; AA = acetic acid; CA = citric acid.

## Heavy Metal Sorption Studies

Sorption is proportional to surface area. The specific surface $S_s$ (surface area/unit mass or surface area/unit volume) of a soil is inversely proportional to grain size of a soil.[14] The $S_s$ of mineral soil particles increases dramatically with decreasing grain size and range in orders of magnitude from $10^3$ for silt, $10^4$ for clay, to $10^5$ for colloidal clay.[14] Cation exchange capacity CEC also increases with $S_s$ as particle size decreases. CEC range from 3 meq/100 g soil for silt, to 22 meq/100 g for clay, to 63 meq/100 g for very fine colloidal clay.[2] In particular, clay colloids are negatively charged and have extensive external and internal surfaces, making the particles ideal site for both physical, chemical, and ion replacement sorption.

Because sorption and CEC are proportional to $S_s$, soil physical properties which express or are a function of the soil particle size distribution, were considered as possible parameters which could be used to predict the sorptive capacity of soils. The total sorbed moles concentration $T[Mol_s]$ was used as the dependent variable for each soil column and was plotted as a function of soil column mass fines $M_{fines}$. The total sorbed moles concentration $T[Mol_s]$ was also plotted against the following; the coefficient of concavity $C_c$, the coefficient of uniformity $C_u$, and bulk density $D_b$ acting as the independent variables. Regression analysis was performed to evaluate the correlation between variables.

Because only Soils 2, 7, and 8 contained plastic fines (i.e., clay minerals), and the actual mass of fines in each column was not determined, a specific quantitative evaluation between individual metal species $[Mol_s]$ or total sorbed moles of metals $T[Mol_s]$ and mass of plastic fines could not be performed. However, soil columns were ranked by the total metals adsorbed $T[Mol_s]$, to evaluate the importance of the mass of plastic fines in sorption of heavy metals.

Under the assumption that one metallic ion occupies one sorption site, converting mass to mole concentrations would provide for a better comparison of the sorptive capacities of the vari-

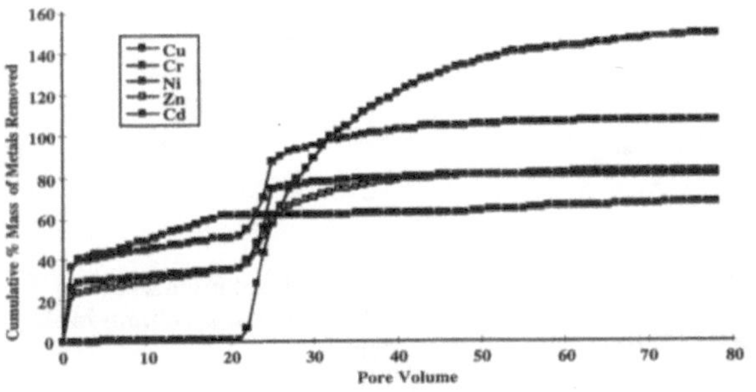


**Figure 2. Cumulative mass of metals removed from a contaminated outwash sand with 4% plastic fines.**

ous soils for the different metals. Sorbed moles concentrations were calculated using the approach that sorbed metals represented those remaining in the column after the first pore volume had been flushed. At the completion of the two-day drainage period, water content in the columns was at field capacity, and water was retained under capillary pressure in the smaller soil pores. The first pore volume water rinse typically removed 30% to 60% of the initial mass retained ($iM_R$) of Cd, Cr, Ni, and Zn. An example of the large percentage of metals removed from the column with the first pore volume flush is shown in Figure 2. Mass removed in first pore volume was considered to primarily be dissolved phase mass present in the soil capillary pore water which was flushed from the column by the initial rinse. Thus, the remaining metals ($M_s$) was considered to be representative of the actual sorbed mass of each metal.

## Sorption and Soil Physical Parameters

The total initial mass retained (total $iM_r$), mass of metals in the first pore volume ($tM_s$), mass of the five sorbed metals ($tMol_s$), and unit metals sorbed / unit mass of soil ($T[Mol_s]$) values for each soil column were calculated. The moles sorbed per kg of soil $[Mol_s]$ values for each metal and soil column are presented in Table IV. The total moles sorbed concentrations $T[Mol_s]$ for each study soil column were considered representative of the total capacities of the soil columns for sorption of heavy metals. $T[Mol_s]$ values were plotted as a function of several soil column physical parameters, and regression analysis was performed using the least squares method to evaluate whether a correlation existed which could be used to predict sorption. An example of these plots is shown in Figure 3. Examples of the regression coefficients are as follows

**Tab le IV. Sorbed Moles Concentrations by Column and Soil Type**

| Column # | Soil # and Description | Cd | Cu | Cr | Ni | Zn |
|---|---|---|---|---|---|---|
| | | Sorbed Moles Concentration (millimoles metal/kg soil) | | | | |
| 1 | 1: OW Sand + 10% NPF | 2.10 | 0.28 | 1.15 | 0.60 | 1.75 |
| 2 | 1: OW Sand + 10% NPF | 1.93 | 0.27 | 1.52 | 0.78 | 1.65 |
| 3 | 2: Fuller Till | 3.86 | 0.33 | 4.75 | 1.52 | 2.01 |
| 5 | 3: OW Sand + 4% NPF | 1.66 | 0.29 | 1.73 | 1.39 | 5.68 |
| 6 | 3: OW Sand + 4% NPF | 1.99 | 0.30 | 1.85 | 0.82 | 1.64 |
| 7 | 4: OW Sand + 16% NPF | 5.96 | 0.83 | 3.83 | 2.36 | 5.25 |
| 8 | 4: OW Sand + 16% NPF | 7.37 | 0.83 | 5.08 | 2.76 | 5.79 |
| 9 | 5: OW Sand + 21% NPF | 6.49 | 0.83 | 3.66 | 2.39 | 5.71 |
| 10 | 5: OW Sand + 21% NPF | 7.42 | 0.83 | 3.87 | 2.68 | 3.80 |
| 11 | 6: Fine Sand | 3.25 | 0.25 | 5.83 | 0.16 | 0.69 |
| 12 | 6: Fine Sand | 3.42 | 0.27 | 5.84 | 0.17 | 0.68 |
| 13 | 7: OW Sand + 5% PF | 10.77 | 0.72 | 7.27 | 8.81 | 9.43 |
| 14 | 7: OW Sand + 5% PF | 8.18 | 0.83 | 4.49 | 3.26 | 4.42 |
| 15 | 8: OW Sand + 9% NPF/PF | 10.19 | 1.09 | 6.37 | 3.34 | 10.06 |
| 16 | 8: OW Sand + 9% NPF/PF | 7.86 | 0.72 | 7.43 | 8.71 | 9.90 |

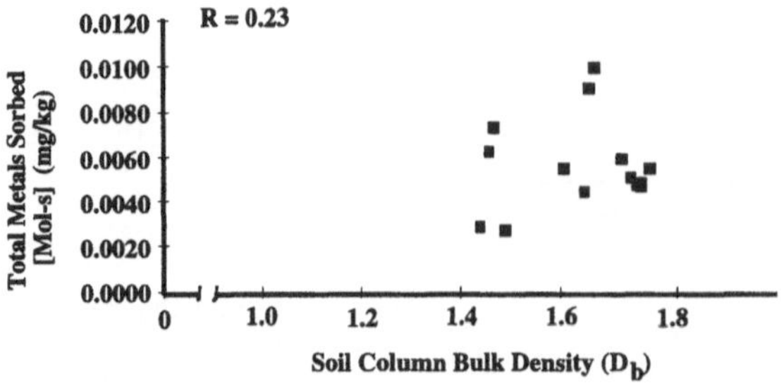


**Figure 3. Total metals sorbed [Mol – s] vs soil column bulk density.**

| | |
|---|---|
| Mass of Fines | R of 0.0018 and 0.1966, |
| Coefficient of Concavity | R of 0.0859 and 0.1107, |
| Coefficient of Uniformity | Regression analysis was not performed because the observed data similar to that of $C_c$ vs. T[$Mol_s$] and a poor correlation was observed. |
| Bulk Density | R of 0.2288 indicates there is not a strong correlation. |

The results of the investigations indicate that soil column mass fines and other general soil physical parameters are not good predictors of soil sorption. This probably is because the physical parameters do not distinguish between mass of silt and mass of clay minerals. As previously shown, the specific surface $S_s$ increases by orders of magnitude as particle size decreases from silt to mineral clays.

## Sorption and Soil Type

In Table V soil types for odd and even numbered columns are ranked in order of total sorbed moles concentrations T[$Mol_s$]. Histograms for sorption and soil type are presented in Figure 4. The Fine Sand, which did not contain fines, exhibited the least sorption; 0.004 moles/kg soil. Sorption increases with the addition of nonplastic fines; however, the increase appears independent of the percentage (mass) of nonplastic fines. Sorption again increases with the addition of plastic fines. Soils 2, 7, and 8, which contain plastic fines generally exhibited the largest sorption, with maximum concentrations of about 0.011 moles/kg soil. These results suggest that fines (silt- and clay-sized particles) and in particular plastic fines (clay minerals), are important sites

**Table V. Summary of the Total Metals Sorbed by Soil Type**

| Soil # | and Description | Moles Metals / kg Soil | Column # |
|---|---|---|---|
| 7 | Outwash Sand with 5% PF | 0.0112 | 13 |
| 8 | Outwash Sand with 5% NPF + 5% PF | 0.0103 | 16 |
| 8 | Outwash Sand with 5% NPF + 5% PF | 0.0086 | 15 |
| 2 | Fuller Till | 0.0075 | 3 |
| 4 | Outwash Sand with 16% NPF | 0.0072 | 8 |
| 7 | Outwash Sand with 5% PF | 0.0068 | 14 |
| 3 | Outwash Sand with 4% PF | 0.0068 | 6 |
| 1 | Outwash Sand with 10% NPF | 0.0064 | 1 |
| 5 | Outwash Sand with 21% NPF | 0.0063 | 9 |
| 5 | Outwash Sand with 21% NPF | 0.0061 | 10 |
| 4 | Outwash Sand with 16% NPF | 0.0060 | 7 |
| 1 | Outwash Sand with 10% NPF | 0.0059 | 2 |
| 3 | Outwash Sand with 4% PF | 0.0057 | 5 |
| 6 | Fine Sand | 0.0042 | 12 |
| 6 | Fine Sand | 0.0040 | 11 |

PF = plastic fines; NPF = nonplastic fines.

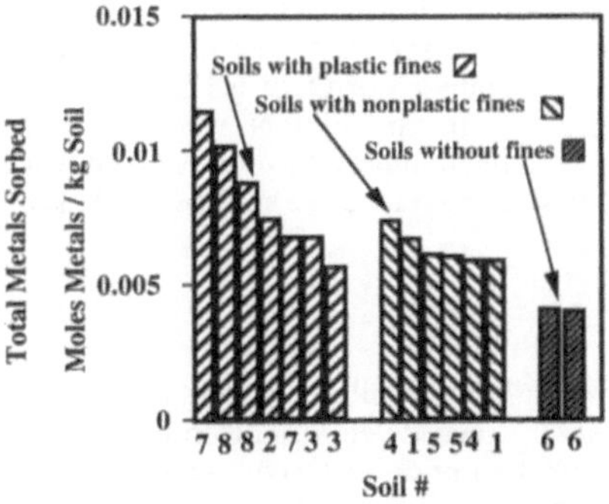


**Figure 4. Amount of total metals sorbed vs soil fines.**

for sorption of heavy metals. These results are not unexpected. In the absence of organic humus, mineral clay particles generally are the major sites for cation sorption.

Data for sorption of individual metals on each soil type was organized both by metal and soil type. This data array highlights the affinity of a metal for a certain soil type. In Table VI data are organized by soil type. This data analysis highlights the sorptive capacity of a soil for a specific metal.

Results for $[Mol_s]$ of individual metals were in general similar to the total metal results. Soil 6, the Fine Sand generally exhibited the least sorption of the eight soil types. Soils 2, 7, and 8 which contained plastic fines generally exhibited the highest sorption.

Cadmium (atomic weight = 112.40) was the most readily sorbed of the metals, and was relatively well sorbed by all soils. Cadmium was followed by Cr (AW = 51.996) and Zn (AW = 65.37), which were approximately equally sorbed. Nickel (AW = 58.71) was the least well sorbed. Sorption data for Cu was difficult to interpret due to the very low concentration of Cu available for sorption in the original contaminating solution.

**Table VI. Summary of the Individual Metals Sorbed by Soil Type**

| | Soil 1 | | Soil 2 | | Soil 3 | | Soil 4 | |
|---|---|---|---|---|---|---|---|---|
| Col. # | 1 | 2 | 3 | 4 | 5 | 6 | 7 | 8 |
| Metal | | | | | | | | |
| Cd | 0.0021 | 0.0020 | 0.0021 | — | 0.0017 | 0.0020 | 0.0019 | 0.0024 |
| Cr | 0.0016 | 0.0012 | 0.0026 | — | 0.0017 | 0.0021 | 0.0013 | 0.0017 |
| Cu | 0.0003 | 0.0003 | 0.0003 | | 0.0003 | 0.0003 | 0.0003 | 0.0003 |
| Ni | 0.0006 | 0.0007 | 0.0008 | | 0.0006 | 0.0008 | 0.0008 | 0.0009 |
| Zn | 0.0017 | 0.0017 | 0.0017 | — | 0.0014 | 0.0016 | 0.0013 | 0.0019 |

| | Soil 5 | | Soil 6 | | Soil 7 | | Soil 8 | |
|---|---|---|---|---|---|---|---|---|
| Col. # | 9 | 10 | 11 | 12 | 13 | 14 | 15 | 16 |
| Metal | | | | | | | | |
| Cd | 0.0021 | 0.0024 | 0.0013 | 0.0014 | 0.0033 | 0.0026 | 0.0029 | 0.0024 |
| Cr | 0.0012 | 0.0013 | 0.0023 | 0.0023 | 0.0021 | 0.0014 | 0.0016 | 0.0017 |
| Cu | 0.0003 | 0.0003 | 0.0001 | 0.0001 | 0.0002 | 0.0003 | 0.0003 | 0.0020 |
| Ni | 0.0008 | 0.0009 | 0.0001 | 0.0001 | 0.0027 | 0.0011 | 0.0009 | 0.0027 |
| Zn | 0.0019 | 0.0012 | 0.0003 | 0.0003 | 0.0029 | 0.0014 | 0.0029 | 0.0030 |

1. Outwash Sand with 10% NPF
2. Fuller Till
3. Outwash Sand with 4% PF
4. Outwash Sand with 16% NPF
5. Outwash Sand with 21% NPF
6. Fine Sand
7. Outwash Sand with 5% PF
8. Outwash Sand with 5% NPF + 5% PF

Both Cr and Cd were relatively well sorbed to the Fine Sand. The highest Cr [$Mol_s$] occurred with Fine Sand, and the second highest Cd [$Mol_s$] was observed for the Fine Sand. Nickel was relatively poorly sorbed to all soils except Soils 7 and 8, comprised of Outwash Sand with the addition of 5% plastic fines derived from the Cambridge Clay. In particular, Ni was not strongly sorbed to the Fuller Till which exhibited plasticity and contained approximately 5% clay which is native to the soil. Overcash 15 notes that in soils, Ni is sorbed on reactive surfaces such as hydrous oxides of Mn and Fe.

## Summary of Metals Adsorption

- For mineral soils with very low organic content, the physical soil parameters mass or percentage of fines $M_{fines}$, coefficient of concavity $C_c$, coefficient of uniformity $C_u$, and bulk density $D_b$ are not good predictors for soil sorption of heavy metals. The cause of this result probably is the fact that the physical parameters which were investigated do not express the content of clay mineral (plastic) fines in the soil.
- In general, the sorptive capacity of soils for heavy metals may be ranked in the following order of increasing capacity T[$Mol_s$]: Soils containing plastic fines > Soils containing nonplastic fines > Soils devoid of fines.
- Total sorbed moles concentrations T[$Mol_s$] ranged from 0.004 moles/kg soil for the Fine Sand (Column 11) to 0.0112 moles/kg soil for the Outwash Sand + 5% plastic fines (Column 13).
- The affinity of metals for sorption may be ranked in the following order of increasing affinity: Cd.> Cr > > Ni > Zn.
- Cr and to a lesser extent Cd, were relatively well sorbed by the Fine Sand which did not contain fines.
- The following ranges for metals sorption [$Mol_s$] of individual metals were observed and are represented as (mols /kg – soil):
  - Cd: 0.0013 (Fine Sand) to 0.0033 (Outwash Sand + 5% PF)
  - Cr: 0.0012 (Outwash Sand + 10% NPF, Outwash Sand + 21% NPF) to 0.0026 (Fuller Till)
  - Ni: 0.0001 (Fine Sand) to 0.0029 (Outwash Sand + 5% NPF + 5% PF)
  - Zn: 0.0003 (Fine Sand) to 0.0030 (Outwash Sand + 5% NPF + 5% PF)

## Desorption and Leaching Aspects

Under field conditions, liquid would drain to deeper soil until the gas-liquid-soil system reached hydraulic equilibrium. Field drainage would produce a contamination gradient downward through the soil column d[$Mol_s$]/dz. In the column studies, the contaminating solution was drained from the columns. This produced (assumed) uniform contamination of the study column; d[$Mol_s$]/dz = 0. In effect, the study soil column duplicated the top section of the field soil column (uniform soil profile) exposed to the spill, where sorbed mass would be maximum.

The study soil columns were then subjected to a leaching process having n stages, where n equaled the number of pore volume rinses. Four countermeasure treatments (T1 through T4) were used as leaching solvent; T1-tap water was followed by three chemical treatments consisting of EDTA and acid, or acid. The treatment schedule is presented in Table III. At the conclusion of each rinse the volume of effluent was recorded and the concentration of each metal was determined. Mass balance calculations were conducted to determine the mass removed in each rinse, and the cumulative percent of the original mass removed at the conclusion of each rinse.

The tap water treatment was intended to duplicate the leaching of metals from the soil column by percolation of storm water runoff, which would naturally occur in the field if remediation of a spill was not completed within a short time period. Evaluation of the data provided baseline information on the ability of water to leach sorbed heavy metals from the soil. However, the limitations of Treatment 1 in approximating field leaching are noted. In the laboratory, application of the water rinse was relatively steady compared to the natural environment, and each water rinse was drained from the soil column to permit analysis and quantification of the leaching process. In the field, both the temporal and physical aspects of storm water application are random events. Due to the randomness of application, heterogeneity of soil and variability in hydraulic boundary conditions (e.g., depth to water table), actual field conditions could not be duplicated in the laboratory.

Upon completion of the tap water leaching, the subsequent chemical treatments were intended to evaluate the effectiveness of a chemical countermeasure to enhance desorption and speed contaminant removal. A plot of one of the column metals test data is shown in Figure 2. Table VII summarizes the treatment removal effectiveness for Treatments 2, 3, and 4. The table presents the percent of metals in units of moles ( $rM_s$) removed by each treatment for each soil type and metal. Although the table is based on only one set of empirical results, the table presents a matrix which provides guidance on which countermeasure would be most suitable for desorbing metals from various soils. For example, to desorb Ni from a silty sand (e.g. Soils 1, 3, or 4) either 1g EDTA/L + AA or HCl is appropriate. However, for desorbing Ni from a soil with plastic fines (e.g., Soil 8), 1g EDTA/L + AA appears to be preferable. Other findings from this portion of the study include the following.

- The first pore volume rinse of Treatment 1 (water) was moderately to highly effective in removing dissolved-phase Cd, Cr, Ni, and Zn which remained in the undrained pore solution at the completion of the contamination period, or mass that was very weakly sorbed.
- In general, following the initial rinse, water had a low effectiveness for desorbing Cd, Cr, Ni, and Zn. However, metal mass was more or less continually removed by water rinses which indicates that water will gradually leach sorbed metal mass from the soil column.
- Water rinses were not effective in desorbing Cu from any of the soils except Soil 6, a Fine Sand.
- Water was the most effective countermeasure for desorbing Cr from all soils. However, the majority of mass removed was removed in the first few rinses, and probably was dissolved-phase mass or weakly sorbed mass.
- Treatment 2 countermeasure, EDTA, was effective in desorbing Cd, Cu, Ni, and Zn from all soils except Soil 6, Fine Sand. Treatment 2 was ineffective in desorbing Cr. The general order of effectiveness of Treatment 2 countermeasure solution in desorbing the metals was: Cu > Ni >> Zn > Cd >> Cr.
- Treatment 3 countermeasures, CA or EDTA + CA, were ineffective to slightly effective in desorbing Cd, Cr, and Zn, and had slight to low effectiveness in desorbing Cu and Ni. This may be because substantial mass had already been removed by Treatments 1 and 2.
- Treatment 4 countermeasures, CA or EDTA + CA, were ineffective in desorbing all metals. This may be because substantial mass had already been removed by Treatments 1 and 2.
- Countermeasure treatments were ineffective in desorbing Cd, Cr and Zn from the Fine Sand (Soil 6). Treatment 2 countermeasure, EDTA, was low-moderately to moderately effective in desorbing Cu and Ni from the Fine Sand. The general ineffectiveness of the countermeasure solutions may be a function of the initial low sorbed mass.

## Kinetics Studies

Development of process kinetics of metals desorption used the Treatment 2 countermeasures (1g EDTA/L + AA) for odd numbered columns and (1g EDTA/L + HCl) for even numbered columns. Apparent, overall rates of desorption are functions of sorbed metals concentrations, the chemical and physical properties of the metal, soil type and countermeasure solution, and the chemical and physical properties of the soil column at large (e.g., pH and hydraulic conductivity). Data from the soil column leaching studies for the given treatments usually were from rinse 21 through rinse 45. Cumulative pore volumes of rinse were converted to cumulative time, by calculating the hydraulic residence time (HRT), for each pore volume of rinse.

Cumulative mass removal and related data transformations were plotted as functions of elapsed process time for the purpose of evaluating whether desorption could be described by zero, first, second order kinetic expressions. Rate expressions were formulated for each metal and soil column, and rate coefficients k were calculated. Rate coefficients were plotted as a function of $H_t$ to evaluate whether the desorption process was strongly transport controlled. Rate coefficients were also plotted as a function of soil column mass fines to evaluate whether specific soil surface was a rate controlling factor.

**Table VII. Summary of Metals Desorbed Erfficiency in Percent Removal**

| | Soil #1 | | | | | Soil #2 | | | | | Soil #3 | | | | | Soil #4 | | | | |
|---|---|---|---|---|---|---|---|---|---|---|---|---|---|---|---|---|---|---|---|---|
| Treatment | Cd | Cr | Cu | Ni | Zn | Cd | Cr | Cu | Ni | Zn | Cd | Cr | Cu | Ni | Zn | Cd | Cr | Cu | Ni | Zn |
| 2: 1g EDTA/L + AA | 54 | 3 | 114 | 106 | 74 | 45 | 1 | 93 | 83 | 46 | 55 | 2 | 109 | 112 | 57 | 59 | 2 | 106 | 95 | 43 |
| 2: 1g EDTA/L + HCL | 61 | 3 | 98 | 95 | 68 | — | — | — | — | — | 69 | 3 | 130 | 110 | 68 | 54 | 2 | 112 | 89 | 61 |
| 3: 2g CA/L | 1 | 14 | 6 | 20 | 5 | 3 | 3 | 15 | 14 | 6 | 2 | 7 | 5 | 7 | 2 | — | — | — | 10 | — |
| 3: 1g EDTA/L + 1g CA/L | 2 | 11 | 6 | 14 | 5 | — | — | — | — | — | 1 | 9 | 13 | 5 | 3 | 1 | 9 | 6 | 15 | 7 |
| 4: 0.5g CA/L | 0 | 2 | 0 | 1 | 1 | — | — | — | — | — | 0 | 2 | 0 | 1 | 1 | — | — | — | — | — |
| 4: 1g EDTA/ + 0.5g CA/L | 0 | 5 | 4 | 2 | 1 | 0 | — | — | — | — | 0 | 4 | 5 | 1 | 1 | 0 | 4 | 0 | 1 | 1 |

| | Soil #5 | | | | | Soil #6 | | | | | Soil #7 | | | | | Soil #8 | | | | |
|---|---|---|---|---|---|---|---|---|---|---|---|---|---|---|---|---|---|---|---|---|
| Treatment | Cd | Cr | Cu | Ni | Zn | Cd | Cr | Cu | Ni | Zn | Cd | Cr | Cu | Ni | Zn | Cd | Cr | Cu | Ni | Zn |
| 2: 1g EDTA/L + AA | 75 | 4 | 117 | 63 | 63 | 0 | 2 | 45 | 6 | 5 | 47 | 6 | 158 | 36 | 42 | 75 | 5 | 116 | 105 | 61 |
| 2: 1g EDTA/L + HCL | 47 | 14 | 109 | 71 | 84 | 0 | 1 | 39 | 19 | 9 | 95 | 3 | 147 | 104 | 82 | 80 | 2 | 161 | 31 | 47 |
| 3: 2g CA/L | 2 | 12 | 5 | 8 | 6 | 0 | 2 | 11 | 0 | 3 | 1 | 7 | 9 | 3 | 1 | 2 | 12 | 7 | 14 | 3 |
| 3: 1g EDTA/L + 1g CA/L | 1 | 13 | 11 | 16 | 9 | 0 | 1 | 0 | 0 | 6 | — | — | — | — | — | 2 | 10 | 17 | 6 | 5 |
| 4: 0.5g CA/L | 0 | 4 | 0 | 2 | 1 | 0 | 1 | 8 | 0 | 2 | 0 | 2 | 0 | 0 | 1 | 0 | 3 | 1 | 2 | 1 |
| 4: 1g EDTA/L | 0 | 4 | 0 | 2 | 2 | 0 | 0 | 1 | 0 | 1 | — | — | — | — | — | 0 | 3 | 2 | 1 | 0 |

1. Outwash Sand + 10% NPF
2. Fuller Till
3. Outwash Sand + 4% NPF
4. Outwash Sand + 16% NPF
5. Outwash Sand + 21% NPF
6. Fine Sand
7. Outwash Sand + 5% PF
8. Outwash Sand + 5% NPF + 5% PF

All reported values are a % of the sorbed metals that were removed by each treatment. These values are following the first treatment, which uses tap water only.

NPF = nonplastic fines PF = plastic fines

Reaction rates were assumed to be either zero, first, or second order. Differential and integrated forms of rate expressions for the three cases, expressed for the simplest case in which only one reactant is involved, were used. Although during the soil column studies, desorption of several metals was occurring simultaneously, desorption of each metal was modeled as an independent separation process. Therefore, desorption of an individual metal was modeled as a simple case: $A_{(reactant)} \gg A_{(product)}$. The half-life $t_{1/2}$ of a reaction process is the time required for a reactant concentration at time t, to be reduced by 50%.[16]

*Zero Order*

$$-d[C_i]/dt = k$$
$$[C_i]_{(t)} = [C_i]_{(o)} - kt$$
$$t_{1/2} = [C_i]_{(o)}/2k$$

For a zero order reaction, a plot of $[C_i]$ vs. t is linear, the rate constant k is the plot slope, and the Y intercept equals $[C_i]_{(o)}$.

*First Order*

$$-d[C_i]/dt = k[C_i]$$
$$2.303 \log [C_{i(t)}/[C_i]_{(o)} = -kt$$
$$\log [C_i]_{(t)} = -kt/2.303 + \log [C_i]_{(o)}$$
$$t_{1/2} = (2.303/k) \log 2 = 0.693/k$$

For a first order reaction, a plot of log $[C_i]$ vs. $t$ is linear, the plot slope equals $-k/2.303$, and the Y intercept equals log $[C_i]_{(o)}$.

*Second Order*

$$-d[Ci]/dt = k[C_i]^2$$
$$1/[C_i]_{(t)} - 1/[C_i]_{(o)} = kt$$
$$1/[C_i]_{(t)} = 1/[C_i]_{(o)} + kt$$
$$t_{1/2} = 1/(k\,[C_i]_{(o)})$$

For a second order reaction, a plot of $1/[C_i]_{(t)}$ vs. $t$ is linear, the plot slope equals $k$, and the Y intercept equals $1/[C_i]_{(o)}$.

Desorption rates were evaluated using the released metals as $[Mol_{s]}$, log $[Mol_s]$ and $1/ [Mol_s]$ were plotted as a function of time for each metal and soil column, to evaluate whether zero, first or second order kinetic models were applicable as rate expressions for the desorption process. Plots were performed for the entire sequence of Treatment 2 rinses in each column, and also for other rinse sequences. Regression analysis using the least squares method was performed for each plot. Regression coefficients R were tabulated to evaluate the suitability of each rate expression model. Rate expressions were then formulated for each metal and soil column, and process rate coefficients k obtained from the slopes of the regression lines. Typical Plots are shown in Figures 5, 6, and 7. Over 225 plots were completed and correlation coefficients R for each plot were calculated by metal. The resulting rate expression formulas and process rate coefficients k for each metal and soil column were evaluated and the best order expression was determined The resulting rate coefficients k for each metal and soil column using the rate expression formulas of best fit are summarized in Table VIII.

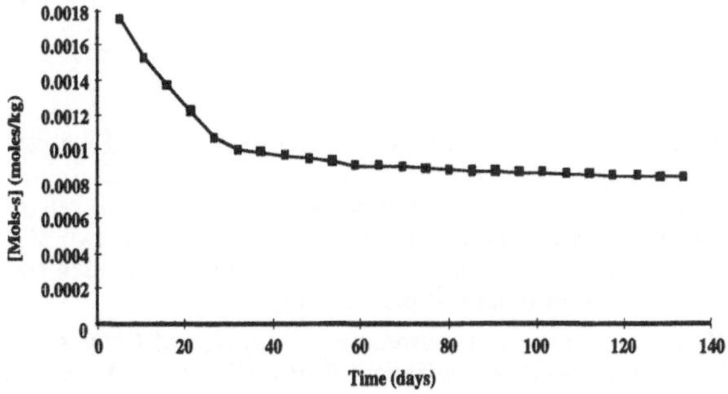


**Figure 5. Zero order rate data for cadmium (time vs moles – s).**

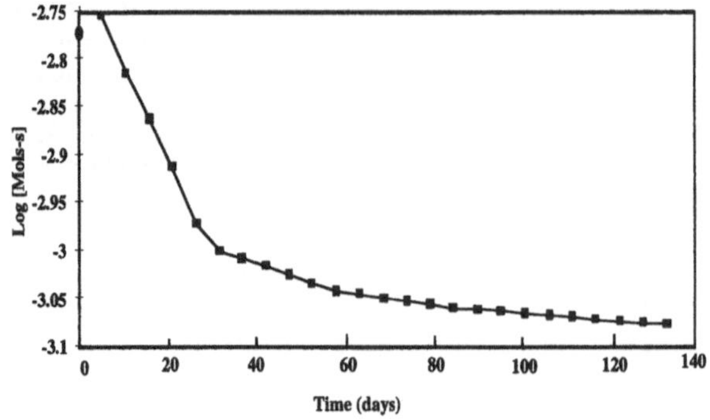


**Figure 6. First order rate data for cadmium (time vs log moles – s).**

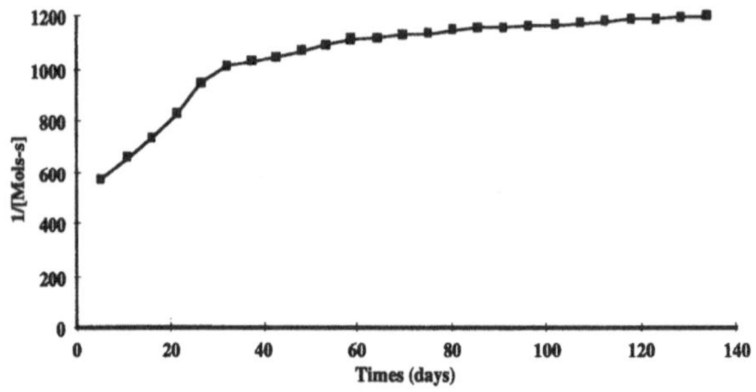


**Figure 7. Second order rate data for cadmium (time vs 1/moles – s).**

**Table VIII. Process Rate Coefficient k and Hydraulic Residence Times (HRT)**

| | | | Process Rate Coefficient k | | | | |
|---|---|---|---|---|---|---|---|
| Column # | Soil # | HRT (days) | Cadmium (2nd Order) | Chromium (2nd Order) | Copper (0 Order) | Nickel (1st Order) | Zinc (1st Order) |
| 11 | 6 | 0.43 | 0.34 | 0.68 | 1.61 E$^{-06}$ | 0.02 | 13.8 |
| 12 | 6 | 0.53 | 0.28 | 0.51 | 1.56 E$^{-06}$ | 0.05 | 34.9 |
| 5 | 3 | 1.04 | 29.9 | 0.76 | 1.68 E$^{-06}$ | 0.29 | 42.8 |
| 6 | 3 | 1.36 | 33.4 | 0.70 | 1.98 E$^{-06}$ | 0.31 | 49.8 |
| 13 | 7 | 2.02 | 4.4 | 0.75 | 1.60 E$^{-06}$ | 0.01 | 5.1 |
| 16 | 8 | 2.07 | 31.2 | 0.20 | 1.34 E$^{-06}$ | 0.01 | 6.0 |
| 15 | 8 | 2.13 | 17.2 | 0.55 | 9.71 E$^{-06}$ | 0.09 | 10.6 |
| 2 | 1 | 4.36 | 4.1 | 0.30 | 2.45 E$^{-06}$ | 0.03 | 13.9 |
| 8 | 4 | 5.05 | 3.9 | 0.14 | 3.74 E$^{-06}$ | 0.02 | 8.1 |
| 1 | 1 | 5.36 | 3.8 | 0.18 | 2.97 E$^{-06}$ | 0.03 | 14.4 |
| 10 | 5 | 6.22 | 1.67 | 0.33 | 2.18 E$^{-06}$ | 0.03 | 24.7 |
| 14 | 7 | 6.63 | 92.1 | 0.31 | 5.68 E$^{-06}$ | 0.04 | 36.4 |
| 7 | 4 | 7.65 | 21.6 | 0.20 | 2.87 E$^{-06}$ | 0.05 | 10.8 |
| 9 | 5 | 11.69 | 5.9 | 0.18 | 1.75 E$^{-06}$ | 0.02 | 4.1 |
| 3 | 2 | 14.36 | 2.4 | 0.02 | 1.65 E$^{-06}$ | 0.01 | 3.1 |

*Cadmium:* Overall, for the full 25 pore volumes of Treatment 2, desorption of Cd was expressed as second order kinetics. The mean correlation coefficient R of 0.8944 indicates a moderately good straight line fit for the plot of 1/ [$Mol_s$] vs. time data. Rate coefficients k ranged from 1.761 to 33.404. However, desorption of Cd was best expressed in terms of a two step process. Although each step was best expressed as a second order process, expressing each step as a zero order process apparently would result in little loss of predictability.

*Chromium:* Overall, for the full 25 pore volumes of Treatment 2, desorption of Cr was best expressed as second order kinetics. The mean correlation coefficient R of 0.9944 indicates a very good fit for the plot of 1/ [$Mol_s$] vs. time data. Rate coefficients k ranged from 0.0241 to 0.7638. The relatively low k values indicate that desorption of Cr was relatively slow. In general, desorption was depicted on plots as a smooth straight or nearly straight line without pronounced deceleration or acceleration Based on the mean R values, Cr desorption could be modeled as a zero order (R = 0.9891) or first order (R = 0.9890) process with little apparent loss of predictability.

*Copper:* Overall, desorption of Cu was best expressed as zero order kinetics. The mean correlation coefficient R was 0.9498 Rate coefficients k ranged from $1.56 \times 10^{-06}$ to $2.452 \times 10^{-04}$. Almost all plots of [$Mol_{s]}$ vs. time depicted a smooth deceleration curve.

*Nickel:* Overall, for the full pore volume sequence of Treatment 2, desorption of Ni was best expressed as first order kinetics. The mean correlation coefficient R was 0.9630. Rate coefficients k ranged from 0.007 to 0.314.

*Zinc:* Overall, for the full 25 pore volumes of Treatment 2, desorption of Zn was best expressed as second order kinetics. The mean correlation coefficient R of 0.9685 indicates a good curve fit . Rate coefficients k ranged from 3.1 to 49.8. Based on the mean, R values Zn desorption could be modeled as a first order (R = 0.9509) process with little apparent loss of predictability.

## Hydraulic Residence Time

Table VIII also summarizes process rate coefficient k and hydraulic residence time HRT data for each metal and soil column. Figure 8 is a typical plot of k as function of HRT. Correlation coefficients R for each data plot were as follows:

| | |
|---|---|
| Cadmium, second order | 0.04 |
| Chromium, second order | 0.77 |
| Copper, zero order | 0.51 |
| Nickel, first order | 0.36 |
| Zinc, second order | 0.44 |

The R values do not suggest a generally strong relationship between k and HRT. However, regression of the k vs. $H_t$ plot for Cr suggests the process rate coefficient for Cr desorption may be at least in part controlled by HRT.

## Soil Column Mass Fines—Specific Soil Surface

Stone[13] summarizes process rate coefficient k and soil column mass fines $M_{fines}$ data for each metal and soil column. Mass of fines for each soil column was used as an indicator of the relative specific soil surface $S_s$ for the column. A typical plot of k as a function of $M_{fines}$ is presented in Figure 9. Correlation coefficients R for each data plot were as follows:

| | |
|---|---|
| Cadmium, second order | 0.25 |
| Chromium, second order | 0.72 |
| Copper, zero order | 0.43 |
| Nickel, first order | 0.40 |
| Zinc, second order | 0.48 |

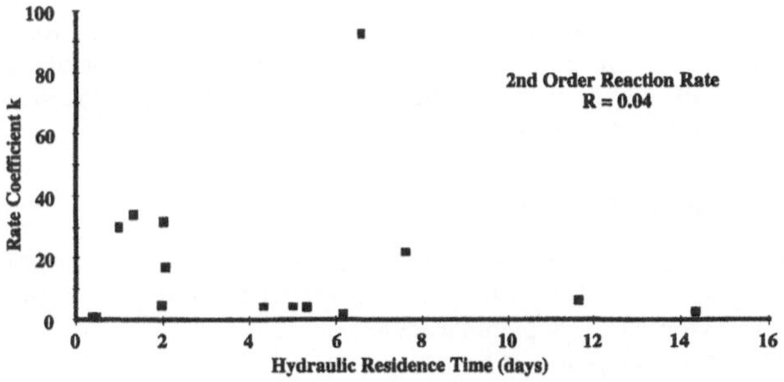


**Figure 8. Process rate coefficient vs hydraulic residence time.**

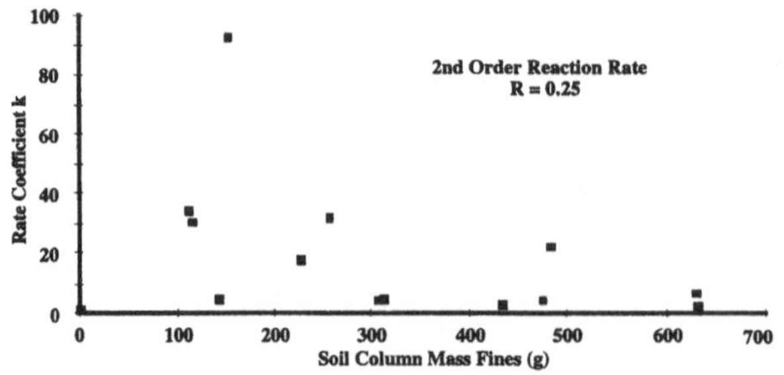


**Figure 9. Process rate coefficient vs. soil mass column fines.**

The R values do not suggest a general strong relationship between k and $M_{fines}$. However, regression of the k vs. $M_{fines}$ plot for Cr suggests the process rate coefficient for Cr desorption may at least in part be controlled by $M_{fines}$. The relationship between $M_{fines}$ and desorption of Cr may relate to hydraulic retention time $H_t$, which also is a function of $M_{fines}$. The apparent lack of a strong relationship between $M_{fines}$ and k probably is because mass of fines (i.e. – 0.075 mm particle size), does not make a distinction between silt and clay sized particles, or identify the presence of clay mineral (plastic) fines. The $S_s$ increases by orders of magnitude between silt particles and mineral clay particles.

## Process Rate Conclusions

- Desorption-leaching process rates for each metal were best expressed using the following kinetic rate orders:

| | |
|---|---|
| Cadmium | Second order |
| Chromium | Second order |
| Copper | Zero order |
| Nickel | First order |
| Zinc | Second order |

- In general, a strong correlation was not observed between desorption-leaching process rate coefficients and hydraulic residence time. This result suggests that the observed processes rates were not primarily transport controlled. This result also suggests that hydraulic residence times achieved during the column studies were sufficient to achieve equilibrium between sorbed- and dissolved-phases, and hydraulic residence times which might be achieved during field application would be sufficient to achieve equilibrium.
- In general, a strong correlation was not observed between desorption-leaching process rates and mass of fines in the soil column. This result suggests, by inference, that the observed processes rates were not primarily controlled by the specific surface of the soil.

## ACKNOWLEDGMENTS

A portion of the work performed in this study was sponsored by a grant from the United States Environmental Protection Agency, Office of Exploratory Research. The research described has not been subjected to agency review (USEPA) and therefore does not necessarily reflect the view of the agency. No official endorsement should be inferred.

## REFERENCES

1. USEPA. The RPM Primer: An Introductory Guide to the Role and Responsibilities of the Superfund Remedial Project Manager (RPM). OSWER Directive 9344.1-02, Office of Emergency Remedial Response, Washington, D.C., September, 1987.
2. SW-874 Hazardous Waste Land Treatment, April, 1983.

3. U.S. Geological Survey Professional Paper 1270. Element Concentrations in Soils and Other Surface Materials.
4. Maximum Contaminant Level. Drinking Water Regulations and Health Advisories. U.S. EPA Office of Water, 11/1992.
5. Toxicity Characteristic Leaching Procedure. U.S. EPA, Solid Waste Amendments, 1986.
6. Maximum concentration in soil recorded at uncontrolled waste sites investigated by the EPA Field Investigation Team.
7. Sims, R. C. and Wagner, K. In Situ Treatment Technologies Applicable to Large Quantities of Hazardous Waste Contaminated Soils. Proceedings of the National Conference on Management of Uncontrolled Hazardous Waste Sites, October 31–November 2, 1983, Washington, D.C.
8. Canniff, M. R. Heavy Metals Release From Contaminated Soil. M.S. Thesis, Worcester Polytechnic Institute, Worcester, MA, 1993.
9. St. Germain, T. J. Sorption Phenomena in Heavy Metal Contaminated Soils. M.S. Thesis, Worcester Polytechnic Institute, Worcester, MA, 1993.
10. Macaulay, M. J. Enhancement of Heavy Metals Removal From Contaminated Soils. M.S. Thesis, Worcester Polytechnic Institute, Worcester, MA, 1994.
11. Venkataramanappa, S. Sorption and Desorption of Selected Heavy Metals in Soil. M.S. Thesis, Worcester Polytechnic Institute, Worcester, MA, 1994.
12. O'Shaughnessy, J.C. and D'Andrea, R.A., et al., "Evaluation of In Situ Soil Flushing Techniques for Heavy Metals Removals from Contaminated Soils," *Proc. of the 48th Industrial Waste Conference*, Purdue University, West Layfette, IN, May, 1993, pp. 123–139.
13. Stone, A.D., Leaching Process for Removal; of Heavy Metals in Soils, M.S. Thesis, Worcester Polytechnic Institute, Worcester, MA, 1995.
14. Bowles, J. E. *Physical and Geotechnical Properties of Soils.* McGraw-Hill Book Company, New York, 1979.
15. Overcash, M. R., and Pal, D. *Design of Land Treatment Systems for Industrial Wastes.* Ann Arbor Science Publishers Inc., Ann Arbor, MI, 1979.
16. O'Connor, R. *Fundamentals of Chemistry.* Harper & Row, Publishers, New York, 1974.

# 39 HEAVY METAL UPTAKE DURING THE GROWTH OF DURUM WHEAT FROM APPLICATION OF SEWAGE SLUDGE AND COMMERCIAL FERTILIZER

**Heather L. Frost**, Graduate Research Assistant
**Lloyd H. Ketchum, Jr.**, Associate Professor
**Jinesh C. Jain**, Assistant Professional Specialist
**Clive R. Neal**, Associate Professor
Department of Civil Engineering and Geological Sciences
University of Notre Dame, Notre Dame, Indiana 46556

## ABSTRACT

Repetitive application of commercial fertilizer continually introduces toxic heavy metals to the soil. Such metallic toxins are subsequently taken up by food crops. Sewage sludge, usually disposed of in landfills, contains high nutrient and organic contents and is now being recycled and applied to agricultural land worldwide in increasing amounts. Metals in sewage sludge are generally organically bound and generally less available for plant uptake than the more mobile metal salt impurities found in commercial fertilizers. However, it is predicted that a slow mineralization of organic matter in sludge releases metals into more soluble forms over time.[1] This research is a preliminary project which gives indication as to metal accumulation in durum wheat roots and the difference between metal salt accumulation and the risk of a 20-year-old sludge with organically bound metals. Future studies will determine metal accumulation in different regions of the durum wheat plant. This information is especially important to phytoremediation issues, giving an indication as to which part of the wheat plant provides maximum uptake of a specific metal and as such, could be used to remove metal from the soils.

Initially, this project focused on screening of U.S. durum wheat lines, common commercial fertilizers, and sludge for 15 potentially threatening elements using inductively coupled plasma mass spectrometry (ICP-MS). Metal uptake studies in a high Cd durum wheat were carried out in greenhouse grown plants using applications of a 20-year-old sewage sludge from the Goshen, Indiana Waste Water Treatment Plant and a commonly used 13-13-13 commercial phosphate fertilizer. Results of this first-phase of research include: (1) an assessment of the effect of sludge application on Cd uptake in durum wheat roots; (2) a comparison of accumulation of Cd uptake in durum wheat using commercial fertilizers; (3) an evaluation of differing metal uptake following use of single or multiple application strategy; and (4) an identification of the metals in addition to Cd that may serve as a threat to the export of U.S. durum wheat.

## INTRODUCTION

Sewage sludge is the solid portion which remains after wastewater treatment and which is frequently disposed of in landfills. Because of its high content of nutrients and organic matter, sludge has been applied worldwide in increasing amounts to agricultural lands for the past several decades. The total cost of land applying sludge is estimated to be 25% of the cost of alternative disposal methods such as elutriation, filtration, or incineration.[2] Effective industrial pretreatment and monitoring by wastewater treatment facilities can ensure the sludge produced is low in toxic contaminants, both organic and inorganic, and high in nutrients and humus, thus rendering sewage sludge as an effective and cheap alternative to commercial fertilizers. However, the numerous heavy metal contaminants in sewage sludge and phosphate fertilizers are either useless in

*52nd Purdue Industrial Waste Conference Proceedings*, 1997, Ann Arbor Press, Chelsea, Michigan 48118. Printed in U.S.A.

plant metabolism such as Hg, Cd, Cr, Ni, Pb, or essential in only trace quantities such as Cu, Fe, and Zn.[3] Heavy metals are the most significant restraint relative to land application of sludges and can often negate the benefits of land application. Even so, metals in sewage sludge are generally organically bound and therefore less available for plant uptake than the more mobile metal salt impurities found in commercial fertilizers.[4]

Concern as to the environmental effects of sludge disposal methods has led to detailed studies of metals in the sludge (e.g., Patoine et al.[5]). The concern over recycling or disposal of sludge exists because toxic metals entering a wastewater facility are usually associated with the solids and remain bound in this form following treatment of the liquid phase.[6] Furthermore, anaerobic digestion of wastewater greatly enhances the partitioning of the toxic metals into the solid phase.[7] Contributing to the concern over the fate of metals in the land applied sludge is an ongoing controversy regarding the fate of metals in land applied sewage sludge. The sludge time bomb hypothesis states that there is a slow mineralization of organic matter in sludge that releases metals into more soluble forms.[1] As discovered in some studies, although organic matter may be significant in the soil 10 years after addition, the concentration of organic carbon may be only one-half of the initial concentration.[8] The sludge protection hypothesis supports the idea that plant uptake reaches a maximum as sludge application to soils increases, with application over sufficient time, converting the surface soil nearly completely into sludge residue.[9] This would result, at worst, in a linear relationship between cumulative sludge application and metal bioavailability because the sludge is adding adsorptive properties to the soil. Furthermore, the number of sludge applications, which have the same total input over time, have been shown to generate differing response from plants. Use of a 20-year-old weathered sludge in this experiment assists in determining the capacity of an aged sludge to immobilize trace metals in comparison to a soil with commercial fertilizer, and the effect of single versus multiple sludge application strategy in the ultimate fate of the trace metals.

The heavy metal content of commercial phosphate fertilizers is also highly variable, depending mainly on the phosphate rock source and the fertilizer production process. Phosphate rock contains various metals as constituents in the ore, and varying amounts of these elements are transferred to the fertilizers in production processes.[10] Superphosphate is produced by reacting the phosphate rock with $H_2SO_4$, with all of the constituent heavy metals resultant in the fertilizer, but triple superphosphate and diammonium phosphate are produced through wet-process $H_3PO_4$, with only some of the heavy metals resultant in the fertilizer.[10]

One of the more toxic trace metals introduced to food crops, either by use of commercial fertilizers or sewage sludge, is cadmium. The concerns about the dietary effect of Cd on human health has prompted consideration of a maximum tolerance level of 100 ppb Cd for cereal grains. The concentration of Cd in U.S. durum wheat has been found in excess of 200 ppb, mainly due to uptake from commercial fertilizers. If the maximum tolerable level of 100 ppb Cd is adopted internationally, U.S. export of durum wheat will be threatened. Until now, no comprehensive study which evaluates the uptake of metals from varying sources and their region of accumulation within the durum wheat has been conducted.

This chapter describes preliminary laboratory experiment assessing the effect of sludge application specifically on Cd as well as other heavy metal uptake in durum wheat roots. Heavy metal uptake in durum wheat roots using commercial fertilizer is compared with metal uptake after single or multiple sewage sludge application. This research focuses on screening of U.S. durum wheat lines, common commercial fertilizers, and a sewage sludge for potentially toxic metals using inductively coupled plasma mass spectrometry (ICP-MS). Metal uptake studies in durum wheat were conducted in triplicate consisting of greenhouse grown plants using either application of sewage sludge, a commonly used commercial fertilizer (13-13-13), or a nitrogen control (NH4Cl).

## METHODS

This greenhouse study was carried out in a temperature controlled greenhouse at the University of Notre Dame in Notre Dame, Indiana. Luvisol soil was collected from a site in Southern Indiana which was formed due to glacial outwash. This soil was selected because of its similarity to soils where future full-scale studies are planned. Dried sewage sludge was obtained from the Goshen, Indiana Waste Water Treatment Plant and was approximately 20 years old. This sludge was selected because of its high metal content and its age. Propell pelleted homogenized plant food (13-13-13) was selected as the commercial fertilizer source, after screening several fertiliz-

**Table I. Sludge and Soil Inventory Report**

| Characteristic | Sludge | Soil |
|---|---|---|
| total exchange capacity (ME/100 g) | 19.7 | 17.5 |
| pH 1:1 (solids/$H_20$) | 5.9 | 6.4 |
| organic matter (humus)% | 23.8 | 3.1 |
| total nitrogen (ppm) | 1400 | 234 |

ers. Durum wheat seeds of durox seed line were obtained from the United States Department of Agriculture in Fargo, North Dakota. Durum wheat was chosen as the crop of choice in this experiment, as changes in regulation of the maximum tolerance level of Cd concentration in cereal grains are under consideration.

Prior to ICP-MS analysis, both the soil and the sludge were evaluated on the basis of organic matter content, nitrogen content, phosphorus content, and pH. Nitrogen, phosphorus, and other chemical analyses on the fertilizer were obtained from the distributor. These preliminary data were collected to ensure a proper loading of sludge or fertilizer to the soils. Those characteristics tested which are relevant to loading are displayed in Table I.

Plastic pots, with a 30 cm surface diameter and approximately 7 inches of working depth, were washed with 2M $HNO_3$ prior to filling with soils. Sludge (0.14% N), fertilizer (13% N), and the $NH_4Cl$ chemical control (26% N) were all loaded to achieve a 100 lb N acre-1 equivalent concentration. This application rate was chosen based on commercial practice in the study area.[11] The sludge, fertilizer, and nitrogen chemical additions were each incorporated into the top 3 inches of soil in the respective pots. Each addition was incorporated using a one-time application, with the exception of one sludge replicate which received two applications due to a nitrogen loading calculation error. This replicate received one light application (26 grams) of sludge with a subsequent corrective heavy application (159 grams) 20 days later. Seeds were planted approximately within the top one inch of soil. Eight seeds were planted in each pot. The greenhouse was kept at approximately 20 degrees Celsius and plants were watered with tap water every two days. All plants received approximately the same amount of water which was poured from a plastic watering can. Plants were pulled from the soil after 82 days in order to analyze the root system. Soils from each of the plants were shaken off and the plant roots rinsed free from remaining soil using 18 $M\Omega$ nanopure water (i.e., water from which dissolved ions had been removed by a reverse osmosis procedure, resulting in high resistivity). Next, roots were cut from the plants using plastic scissors and dried in a drying oven for 24 hours at 60 degrees Celsius. All samples were prepared and digested in a clean lab using 16 M ultrapure doubly distilled nitric acid, ultrapure 60% doubly distilled hydrofluoric acid, 18 $M\Omega$ nanopure water, and a Teflon-covered hot plate set at 100 degrees Celsius. Any samples which still contained organic matter were digested using 50% $H_2O_2$. Each digested sample batch included procedural acid blanks, internal standards, and standard reference materials (durum wheat, palm leaves, and a Hawaiian basalt) and a three-step nitric acid wash procedure using 5% doubly distilled nitric acid.

Quantification of trace element (V, Cr, Co, Ni, Cu, Zn, As, Mo, Cd, Sn, Sb, Tl, Pb, Th, U) content was performed by injecting digested samples into a Fisons VG PlasmaQuad II Inductively Coupled Plasma Mass Spectrometer (ICP-MS). Samples were introduced, using a liquid aspiration rate of 1mL/min, into a radio frequency induced plasma with a temperature of approximately 6000K. The high thermal energy and electron-rich environment of the ICP-MS results in conversion of sample atoms to ions. The quadruple mass spectrometer permits rapid sequential detection of the sample ions at each pre-selected mass using peak-jumping mode of detection. Table II shows the ICP-MS operating conditions used in this experiment.

## RESULTS AND DISCUSSION

After the 82-day growth period, roots of the control replicate, the fertilizer applied replicate, and the two sludge applied replicates were analyzed. Table III shows trace metal concentrations for the roots, in parts per million (ppm) expressed to two standard deviations. Two standard deviations were used to express errors in order to prevent outlying values from contributing to a skewed range. In most cases, the concentration in the root is the same regardless of the nitrogen application source. With the only exception being Cu, the uptake into the plant was never consid-

**Table II. Operating Conditions for VG PlasmaQuad II Inductively Coupled Plasma Mass Spectrometer**

| Condition | Setting |
|---|---|
| Forward Power | 1350 W |
| Reflected Power | 2 W |
| Cooling Gas Flow | 15.0 L/min |
| Auxiliary Gas Flow | 2.0 L/min |
| Nebulizer Gas Flow | 0.850 L/min |
| Sample Flow Rate | 1.0 mL/min |
| Spray Chamber | double-pass scott-type water cooled to 4 degrees Celsius |
| Sampler Cone | nickel 1mm orifice |
| Skimmer Cone | nickel 0.75mm orifice |
| Vacuum Expansion Stage | 2.2 mbar |
| Vacuum Intermediate Stage | $10^{-4}$ mbar |
| Vacuum Analyzer Stage | $3.0 \times 10^{-6}$ mbar |
| Measurement Mode | PC, Peak jump |
| Dwell Time | 10.4 ms |
| Points/Peak | 3 |
| Measuring Time | 1 min |

**Table III. ICP-MS Analysis of Trace Element Root Concentration**

| Trace Elements | Control Root Concentration (ppm) | Fertilizer Root Concentration (ppm) | Sludge Root Concentration (ppm) | Adjusted Sludge* Root Concentration (ppm) |
|---|---|---|---|---|
| V | 7.47 ± 3.93 | 9.14 ± 4.63 | 4.16 ± 2.53 | 3.23 ± 3.22 |
| Cr | 5.19 ± 2.84 | 6.03 ± 3.14 | 12.20 ± 12.47 | 2.98 ± 1.84 |
| Co | 1.86 ± 1.09 | 2.34 ± 1.55 | 1.54 ± 0.96 | 0.83 ± 0.53 |
| Ni | 4.64 ± 2.48 | 6.16 ± 4.07 | 11.01 ± 6.39 | 5.88 ± 2.81 |
| Cu | 23.07 ± 27.19 | 17.55 ± 9.11 | 77.30 ± 37.43 | 51.83 ± 10.41 |
| Zn | 466.92 ± 225.21 | 493.28 ± 458.58 | 441.74 ± 155.47 | 547.31 ± 320.89 |
| As | 1.55 ± 0.84 | 1.78 ± 0.82 | 2.46 ± 2.12 | 1.11 ± 0.45 |
| Mo | 0.88 ± 0.43 | 0.81 ± 0.87 | 1.95 ± 1.04 | 2.47 ± 1.25 |
| Cd | 2.23 ± 1.28 | 1.23 ± 0.69 | 6.66 ± 1.77 | 3.45 ± 0.83 |
| Sn | 0.35 ± 0.20 | 0.42 ± 0.41 | 3.10 ± 2.69 | 0.57 ± 0.27 |
| Sb | 0.10 ± 0.05 | 0.14 ± 0.10 | 0.31 ± 2.69 | 0.08 ± 0.09 |
| Tl | 0.17 ± 0.18 | 0.16 ± 0.08 | 0.08 ± 0.01 | 0.03 ± 0.05 |
| Pb | 3.27 ± 1.33 | 4.32 ± 2.24 | 12.00 ± 10.95 | 2.99 ± 1.65 |
| Th | 1.03 ± 0.64 | 1.42 ± 0.87 | 0.56 ± 0.53 | 0.35 ± 0.45 |
| U | 0.38 ± 0.22 | 0.57 ± 0.37 | 0.23 ± 0.18 | 0.12 ± 0.16 |

All concentrations measurements are the average of five samples and are recorded in parts per million (ppm) expressed to two standard deviations ($\pm 2\sigma$) (i.e., including 99.7% of all values).
*The adjusted sludge replicate received the total sludge application of 185 grams in two applications. Total sludge application consisted of an initial dose of 26 grams and a subsequent dose of 159 grams 20 days later.

erably greater from the sludge than in the case of the control. However, there is a slight increase in the root concentration of Cr, Ni, Cu, Sn, and Pb, depending on the number of sludge applications used. The replicate which received the total sludge within one application had metal concentrations which were greater in the case of these metals. This leads to the conclusion that considerable metal uptake occurs within the first 20 days of growth.

The accumulation of trace metals in the roots expressed with regard to load of metal to the plant indicates that although there was not a considerable difference in uptake to the plants with respect to source, there was a considerable difference in metal loading to the soil. Data displayed

**Table IV. Selected Trade Element Root Concentrations Based on 100 lb N acre$^{-1}$ Dosage**

| Trace Elements | Metal Dosage From Fertilizer (mg) | Resulting Metal Concentration in Roots (ppm) | Metal Dosage From Sludge (mg) | Resulting Metal Concentration in Roots (ppm) |
|---|---|---|---|---|
| Cr | 0.06 | 6.03 ± 3.14 | 28.36 | 12.20 ± 12.47 |
| Co | 0.004 | 2.34 ± 1.55 | 4.42 | 1.54 ± 0.96 |
| Cu | 0.20 | 17.55 ± 9.11 | 109.75 | 77.30 ± 37.43 |
| As | 0.004 | 1.78 ± 0.82 | 6.40 | 2.46 ± 2.12 |
| Cd | 0.01 | 1.23 ± 0.69 | 7.32 | 6.66 ± 1.77 |
| Pb | 0.19 | 4.32 ± 2.24 | 13.87 | 12.00 ± 10.95 |

All concentration measurements are the average of five samples and are recorded in parts per million (ppm) expressed to two standard deviations (±2σ) (i.e., including 99.7% of all values).

in Table IV show the dosage of important metals from fertilizer and the one-time sludge application, with the resultant concentration of these selected metals in the root system. This loading difference was more exaggerated than in a typical situation in which sludge is used as an alternative fertilizer. In this laboratory experiment, the high metal sludge was loaded to achieve the nitrogen concentration of 100 lb N acre$^{-1}$ used with commercial fertilizers. In order to achieve this nitrogen concentration, the load of metals to the soil through the application of sludge was considerably higher than is permitted under federal regulation. In field conditions, Federal 503 Guidelines establish sludge quality and loading restrictions with respect to heavy metal concentrations.[12] The high sludge dosage was used in this experiment as the overall research concern focused on metal uptake when sludge is used to fulfill the nitrogen requirements.

Background element concentrations of trace elements for the soils, water, seeds, sludge, fertilizer, and the NH4Cl are displayed in Appendix A (Table A.I). The average machine drift for this trace element background concentration run was 3%. A soil sample from each pot in which a plant was pulled, as well as a rhizosphere soil sample from each analyzed plant, were also analyzed for trace metal content. There was no difference in metal concentration between the soil sample and the rhizosphere soil sample within each application group (i.e., metal concentration of the control rhizosphere soil was equivalent to the metal concentration in the control soil).

## APPENDIX

**Table A.I. ICP-MS Analysis of Trace Element Background Concentration**

| Trace Elements | Soil Concentration ppm ± 2σ | Durox Seed Concentration ppm ± 2σ | Water Concentration ppb ± 2σ | $NH_4Cl$ Concentration ppm ± 2σ | Sludge Concentration ppm ± 2σ | Fertilizer Concentration ppm ± 2σ |
|---|---|---|---|---|---|---|
| V | 39.82 | 0.08 | 0.73 | 6.74 | 11.30 | 33.49 |
| Cr | 31.54 | 0.51 | 1.78 | 20.85 | 153.28 | 29.94 |
| Co | 7.38 | 0.08 | 0.28 | 0.00 | 23.89 | 1.96 |
| Ni | 11.42 | 0.46 | 6.17 | 0.22 | 59.99 | 6.04 |
| Cu | 14.70 | 6.43 | 178.36 | 0.06 | 593.23 | 99.68 |
| Zn | 112.59 | 19.10 | 300.80 | 0.93 | 1056.03 | 920.12 |
| As | 6.56 | 0.19 | 0.42 | 1.33 | 34.58 | 1.98 |
| Mo | 1.32 | 1.33 | 0.98 | 0.00 | 15.07 | 8.56 |
| Cd | 0.39 | 0.41 | 0.89 | 0.00 | 39.55 | 3.72 |
| Sn | 1.50 | 0.39 | 0.14 | 0.01 | 56.38 | 7.99 |
| Sb | 0.69 | 0.05 | 0.12 | 0.00 | 2.39 | 1.05 |
| Tl | 0.44 | 0.01 | 0.04 | 0.00 | 0.08 | 0.40 |
| Pb | 22.21 | 0.07 | 6.24 | 0.09 | 74.98 | 96.26 |
| Th | 8.05 | 0.04 | 0.00 | 0.00 | 0.37 | 2.79 |
| U | 2.77 | 0.04 | 1.51 | 0.00 | 0.36 | 28.14 |

All concentration measurements are the average of five samples and are recorded in parts per million (ppm) expressed to two standard deviations (±2σ ) (i.e., including 99.7% of all values).

## CONCLUSIONS

These data show that despite a tremendous metal load through the application of a contaminated sewage sludge, metal uptake was nearly equivalent, in most cases, to the uptake from chemical fertilizer and a nitrogen control. This leads to the conclusion that organic matter has not mineralized over time, leading to a tremendous concentration of bioavailable metals, subsequently taken up by the wheat. However, there was a difference noted in root metal concentration depending on the number of sludge applications used to dispense the total 185 grams of sludge. When one application was used, the metal concentration of the roots was greater for many metals than when two applications were used. This indicates that considerable metal uptake is occurring within the first 20 days of growth (before the second application of sludge occurred in the adjusted sludge replicate).

This research not only contributes to greater understanding of the metal accumulation but also identifies the growth period in which metal uptake occurs in durum wheat roots. This research will also eventually lead to greater understanding of the technology of phytoremediation. By understanding the region of accumulation for a particular metal and the period of growth in which metals are taken up, the metal removal from contaminated soils can be maximized.

## ACKNOWLEDGMENT

This research was supported, in part, by Sakti Jana of the Department of Crop Science at the University of Saskatchewan, Canada. Research was executed at the University of Notre Dame with assistance of the Center for Bioengineering and Pollution Control. Clive R. Neal was supported by NSF grant ECS92-14596. Jinesh C. Jain was supported by NSF grant EAR 93-15930.

## REFERENCES

1. McBride, M.B. "Toxic metal accumulation from agricultural use of sludge: Are USEPA regulations protective?" *J. Environ. Qual.*, 24, pp. 5–18, 1995.
2. Wang, M.-J., S.P. Mc Grath, and K.C. Jones. "Chlorobenzenes in field soils with a history of multiple sewage sludge applications." *Environ. Sci. Technol.* 29, pp. 356–362, 1995.
3. Weier, T.E., C.R. Stocking, and M.G. Barbour. *Botany, an Introduction to Plant Biology*. 5th edition. Wiley, New York, 1974.
4. Hendrickson, L.L. and R.B. Corey. "Effect of equilibrium metal concentrations on apparent selectivity coefficients of soil complexes." *Soil Sci.*, 131, pp. 163–171, 1981.
5. Patoine, A., R. Cote, and M. Paquet. "Anaerobic detoxification of a chemi-thermomechanical pulp effluent by two acclimating sludges." *Water Pollut. Res. J. Can.*, 29, 4, pp. 471-486, 1994.
6. Page, A.L., T.J. Logan, and J.A. Ryan. *Land Application of Sludge: Food Chain Implications*. Lewis Publishers, Inc., Chelsea, MI, 1988.
7. Wagner, G.J. Accumulation of Cd in Crop Plants and Its Consequences to Human Health. *Advances in Agronomy*. Academic Press, 1993, pp. 173–213.
8. Bell, P.F., B.R. James, and R.L. Chaney. "Heavy metal extractability in long-term sewage sludge and metal salt-amended soils." *J. Environ. Qual.* 20, pp. 481–486, 1991.
9. Ryan, J.A. and R.L. Chaney. Regulation of Municipal Sewage Sludge Under the Clean Water Act Section 503: A Model for Exposure and Risk Assessment for MSW-Compost. *Science and Engineering of Composting: Design, Environmental, Microbiological, and Utilization Aspects*. Renaissance Publishers, Worthington, OH, pp. 422–450, 1993.
10. Mortvedt, J.J. "Heavy metal contaminants in inorganic and organic fertilizer." *Fert. Res.*, 43, pp. 55–61, 1996.
11. Mansfield, C. and S Hawkins. Wheat Production and Fertilization in Indiana. *Agronomy Guide*. Purdue University Cooperative Extension Service. West Lafayette, IN, 1992.
12. Bastian, R. "25 Page Summary of the Standards for the Use or Disposal of Sewage Sludge 40 CFR Part 503 (58 Fr 32:9248–9415)." Environmental Protection Agency. Office of Wastewater Enforcement and Compliance Municipal Technology Branch.1–25, 1993.

# 40 OPTIMIZATION OF TREATMENT PARAMETERS DURING BIOREDUCTIVE PRECIPITATION OF URANIUM USING A BENCH SCALE FERMENTER

**Zabin F. Rahman, Kevin G. Robinson,** and **Gregory D. Reed**
Department of Civil and Environmental Engineering
219A Perkins Hall
The University of Tennessee
Knoxville, Tennessee 37996

## ABSTRACT

Evaluation of a bench-top anaerobic bioreactor for microbially-mediated reductive precipitation of hexavalent uranium was addressed in this work. Selected treatment parameters were studied to determine the impact of each on treatment performance. Under the conditions tested, soluble uranium was effectively removed from solution.

## INTRODUCTION

Traditional treatment practices for uranium-bearing waste streams at many Department of Energy (DOE) facilities generate significant quantities of secondary sludges which are classified as mixed wastes. Currently no economical technology exists for the permanent disposal of these mixed-waste sludges. Selective removal of uranium, the principal radioactive component in many mixed-waste sludges, would allow the remaining waste stream to be treated as a Resource Conservation and Recovery Act (RCRA) waste by conventional treatment. In addition, volume reduction would make the storage or disposal of radioactive uranium sludge more economical. Conceivably, uranium could be recycled with minimal additional processing.

The recent discovery that certain microorganisms can enzymatically catalyze reduction of aqueous uranium provided a potential alternative treatment option for uranium removal. In a nutrient limiting environment, anaerobic sulfate reducing bacteria use uranium as an electron sink for respiratory metabolism. The reduced form of uranium is sparingly soluble in water and tends to precipitate quickly under favorable conditions. Since microbial reactions are induced by target specific enzymes, this bioreduction method can selectively remove uranium from aqueous waste streams. Batch experiments conducted to evaluate the scientific basis of this technology revealed that a sulfate reducing bacteria, *Desulfovibrio desulfuricans*, successfully reduced hexavalent uranium to the tetravalent state in 25-mL pressurized test tubes under anaerobic conditions. To effectively use this bioreduction process as a uranium removal technique, additional information concerning the technological aspects of treatment is required.

The focus of this research was to scale-up the uranium bioreductive precipitation process using a bench top anaerobic bioreactor and to evaluate biological performance under nongrowth conditions. The influence of treatment parameters such as the extent of anaerobic gas input, reactor temperature, the rate of mixing during uranium reduction, and uranium loading were evaluated using a synthetic test solution containing 0.03 M bicarbonate mixed with depleted uranium (aqueous) and a bacterial inoculum.

---

*52nd Purdue Industrial Waste Conference Proceedings*, 1997, Ann Arbor Press, Chelsea, Michigan 48118. Printed in U.S.A.

# MATERIALS AND METHODS

## Culture Organisms

A pure culture of sulfate reducing bacteria, *Desulfovibrio desulfuricans* (ATCC #29577), was used as a mediator for the bioreduction process. The culture conditions and procedures established by Ganesh[1] were followed for the growth and extraction of microorganisms. The culture was grown in a medium[2] containing 2.5 gm $NaHCO_3$, 1.5 gm $NH_4Cl$, 0.6 gm $NaH_2PO_4$, 1 gm KCl, 1.5 gm $MgSO_4$, 1.5 gm $Na_2SO4$, 2.0 gm yeast extract, 5.0 gm polypeptone, 20 mL (60%) D-L lactate syrup, 10 mL of vitamin mix, and 10 mL of mineral mix in one liter of deionized water. In this medium, lactate was the carbon source and sulfate was the terminal electron acceptor. After dissolving the nutrients in deionized water, the growth medium was heated to near boiling on a hot plate while purging the solution with a hydrogen-rich anaerobic gas mixture (80% $H_2$ and 20% $CO_2$) for 15 minutes, and then a nitrogen-rich oxygen-free gas mixture (85% $N_2$, 10% $CO_2$, and 5% $H_2$) for another 15 minutes. To minimize trace oxygen levels, the gas mixtures were passed through a hot copper column at 320°C before contacting the test solutions.[3] While purging with the anaerobic gas mixture, 98 mL of medium was transferred to 100-mL serum bottles which were capped with 20 mm butyl rubber stoppers. Aluminum seals were then crimped over the rubber stoppers before autoclaving the media at 250°C and 150 psi to ensure sterility. One milliliter of L-cystene (a reducing agent) was added (stock concentration 10 gm/200 mL) to each of the serum bottles after cooling the media to room temperature. The samples were inoculated with 1 mL of *D. desulfuricans* ($10^6$ cells/mL) using sterile disposable syringes and needles.

Cells were incubated at 32°C for 60 hours, then centrifuged at 4°C at a rotational speed of 11,000 rpm for 30 minutes. After decanting the supernatant, the biomass pellet was resuspended in a 0.03 M bicarbonate solution and centrifuged again for 20 minutes to remove any impurities associated with cell growth. The transfer of cells was made as quickly as possible under a fume hood and exposure to ambient air containing oxygen was minimized. The washed and consolidated cell mass was resuspended in 50 mL of bicarbonate solution before inoculating the sterile anaerobic reactor containing liquid phase depleted uranium.

## Experimental Setup

A bench-scale fermenter (Omniculture Fermenter, Virtis Co, NY) with a working volume of 2 liters was used as the anaerobic bioreactor (Figure 1). The reactor vessel was made of autoclavable, nontoxic borosilicate glass. The top of the reactor consisted of a noncorrosive stainless

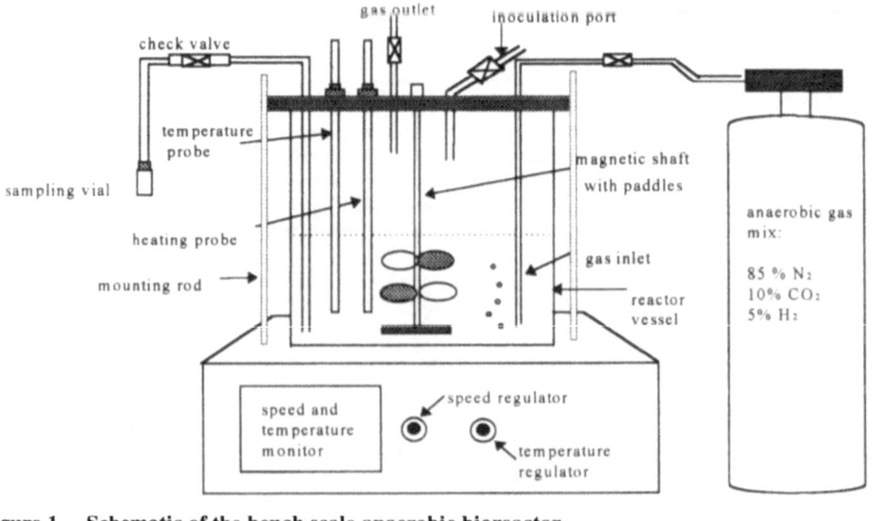


**Figure 1. Schematic of the bench scale anaerobic bioreactor.**

steel plate containing a temperature port, a heating port, an inoculation port, a sampling port, a magnetic shaft with paddles, a heat exchanger, a gas inlet and a gas outlet. The reactor, along with its accessories, was mounted on a control box with monitors for speed and temperature regulation. A rubber washer, placed between the steel plate and the 2 liter jar, prevented any diffusion of ambient air into the reactor when properly clamped. The gas inlet was connected by stainless steel tubing to a tank containing 85% $N_2$, 10% $CO_2$, and 5% $H_2$. The outlet port was connected by rubber tubing to a fume hood for disposal of flammable $H_2$ gas, a component of the anaerobic gas mixture, and any noxious fumes that might have been produced. A heater inserted into the heating port helped set the temperature inside the reactor to the desired level. The temperature monitor indicated the actual temperature of the solution sensed by the temperature probe at all times. The mixing speed was regulated by a toggle switch on the control box. The agitation system was driven by an induced magnetic couple that rotated the disk attached to the bottom of the paddle-shaft.

The sampling port consisted of two stainless steel tubes; one inserted into the vessel at an adjustable depth, and the other outside the reactor with a sampling vial screwed into it. The two steel tubes were connected by a small piece of rubber tubing and a check valve. When there was a sufficient amount of gas pressure inside the reactor and all other openings were sealed, release in the check valve at the sampling port would allow the sample to be drawn from solution to the sampling vial. The vial would be quickly removed and replaced by another vial for the next sampling.

## Preparing Test Solution

The test solution was a synthetic waste containing bicarbonate and depleted uranium. Two and one-half grams of $NaHCO_3$ were dissolved in one liter of deionized water containing 10 mM lactate syrup (2 mL of 60% lactate in one liter solution ). The bicarbonate was used to buffer the test solution and alter the chemical speciation of uranium such that the uranyl-bicarbonate complex predominates. Uranium in this form is readily available as an electron acceptor in the respiratory metabolic pathway of *D. desulfuricans*. Lactate served as the major electron donor.[4]

The reactor, containing the uranium-free solution, was sealed and then autoclaved to ensure sterility. The autoclaved solution, while still hot (~ 85°C), was purged with the anaerobic gas mixture (85% $N_2$, 10% $CO_2$, 5% $H_2$) for at least 30 minutes. Elevated temperature enhanced oxygen degassing and promoted the anaerobic environment necessary for subsequent metabolic process in the reactor. After initial degassing, the solution temperature was stabilized at 32°C, then uranium from the deoxygenated stock solution was added. Five milliliters of 20 gm/L uranium acetate were injected through the inoculation port to give an initial concentration of 100 mg/L. Other volumes of the stock solution were used to vary the initial uranium concentration. Bacteria, isolated from the growth medium, were inoculated into the reactor and samples were drawn periodically to measure soluble uranium over time.

To ensure proper conditions for the sulfate reducers, the head space was filled with anaerobic gases at all times. Initially this was done by providing a continuous supply of anaerobic gas through the gas inlet port with subsequent automatic discharge of the gases through the outlet port during the experiment. The perforated end of the gas inlet was submerged into the liquid to purge the solution and maintain a constant positive pressure inside the reactor at all times. However, for the intermittent gas input experiment, the gas supply was shut off after the initial degassing process. Anaerobic conditions were maintained by securing all openings such that they were gas-tight. The sampling technique required the vessel to have enough gas pressure inside the reactor to force a portion of the test solution into the sampling vial. This resulted in an intermittent purge cycle during sampling such that strict anaerobic conditions were maintained throughout the experiment.

The rotating speed of the mixer (rpm) and the operational temperature (°C) were set according to experimental requirements. Most experiments were conducted at 32 ± 0.5°C, since the optimum temperature for microbial action of the *D. desulfuricans* is 32–35°C. A lower temperature was studied to evaluate temperature impacts on uranium bioreduction. The paddles and baffles in the reactor helped to create a homogeneous mixture of deoxygenated uranium-bicarbonate solution and an even temperature throughout the solution. Mixing speed was varied to evaluate operational performance.

## Sampling and Analytical Methods

Oxidized uranium (hexavalent) in solution was measured over time using a kinetic phosphorescence analyzer (KPA-11, Chemchek Instruments). The KPA-11 instrument uses a pulsed laser dye to produce a wavelength of 425 nm that excites hexavalent uranium in the sample, causing it to phosphoresce (emit green light). The intensity of the light is directly proportional to uranium concentration based on a standard calibration curve.

Samples drawn from the reactor were analyzed as quickly as possible with minimum exposure to ambient air. A series of analyses performed on the same bioreduced sample at different times indicated that the soluble hexavalent uranium concentration increased with time. This was due to reoxidation of tetravalent uranium to the hexavalent form in the presence of air. To further minimize air contact, samples were diluted with deoxygenated deionized water before analysis. An acrodisk type filter (0.2 m) was used to separate soluble from precipitated uranium ($UO_{2s}$ -uraninite) in the test mixtures. The filter isolated the solid uraninite and biomass from the aqueous solution. Portions of the filtered and unfiltered samples were then reoxidized by adding a strong oxidant ($HNO_3$). Reoxidized unfiltered samples yielded the total analytical concentration of the hexavalent uranium (both liquid and solid phase) used for the test. The reoxidized filtrate measured residual hexavalent uranium and enzymatically reduced uranium which did not precipitate from solution.

# RESULTS AND DISCUSSION

## Extent of Anaerobic Gas Input

The initial scale-up of the bioreduction process was conducted in a bench-scale reactor with continuous purging using an anaerobic gas mixture. The continuous supply of gas simulated the strict anaerobic environment maintained in the previous batch pressure tube experiments.[1] It should be noted that for the batch tests, solutions were purged with a hydrogen rich gas mixture followed by a nitrogen rich gas mixture. But for the reactor tests, uranium-bicarbonate solutions were purged with a nitrogen rich mixture only and the gas mixture was not passed through a copper column for removal of trace oxygen. Figure 2 shows hexavalent uranium removal from solution over time with continuous purging using an oxygen-free gas mixture. The uranium concentration dropped from 162 mg/L to ~ 20 mg/L in about seven hours. The rate of reduction (based on the slope of the linear portion of the curve) was constant (~ 20 mg/L-hr) through the first 5 hours of incubation. Small dark specks of solid uraninite began to appear at the bottom of the reactor as reduced uranium (IV) precipitated from the solution. Both filtered and unfiltered sam-

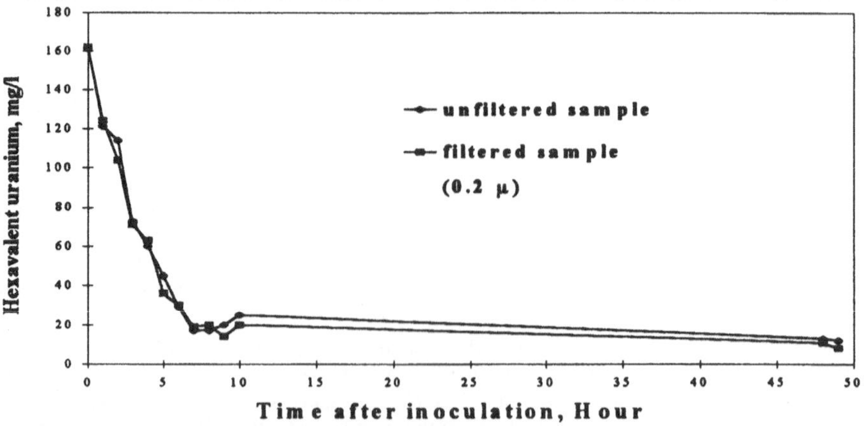


**Figure 2. Soluble hexavalent uranium remaining in solution after inoculation with *D. desulfuricans* at 32°C. The mixing rate was constant at 100 rpm, and supply of anaerobic gas was continuous.**

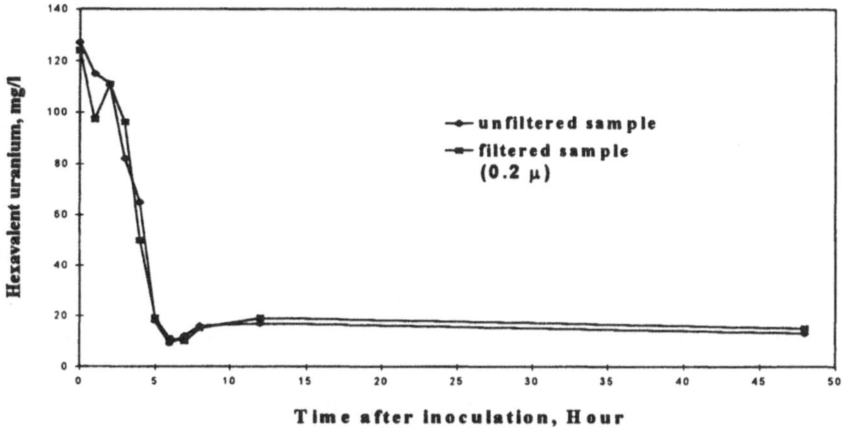


**Figure 3. Soluble hexavalent uranium remaining in solution after inoculation with *D. desulfuricans* at 32°C. The mixing rate was constant at 100 rpm, and supply of anaerobic gas was intermittent.**

ples contained the same concentration of hexavalent uranium at any given time, indicating that little of the oxidized uranium was associated with the solids or sorbed to the filter. The extent of uranium removal approached 95% over the time course of the experiment. Final hexavalent uranium levels in solution remained at ~ 8 mg/L after 48 hours of incubation.

Continuous purging with the anaerobic gas mixture resulted in rapid uranium reduction; however, the amount of gas supplied posed a hazard because of the flammable nature of $H_2$ gas (a component of the anaerobic gas mix). Additionally, the process was not economical due to the high price of the specialty gas mixture. As an alternate to the continuous supply of this anaerobic gas, test solutions were intermittently purged for five minutes during sampling to provide the headspace pressure required to draw each sample from the reactor. Once sampling was complete, purging was discontinued and all valves were closed to maintain an anaerobic environment inside the bioreactor. The trend in hexavalent uranium reduction using intermittent purging (Figure 3) was similar to that observed with continuous purging. However, the rate of reduction (in the linear region) was higher (~ 30 mg/L-hr) with intermittent gas flow. The lower rate found with continuous purging may be due to $H_2$ inhibition during lactate oxidation.[4] Since intermittent purging did not adversely impact uranium bioreduction and reduced overall cost of treatment, all subsequent experiments were conducted using intermittent gas purging.

## Temperature Effect

The optimum temperature range for *D. desulfuricans* metabolic activity is 32–35$^0$C (mesophillic). Maintaining such an elevated temperature inside a bioreactor during large-scale treatment would be costly. Therefore, enzymatic reduction experiments were performed at room temperature (24 to 27°C) to evaluate temperature fluctuations on treatment performance. Figure 4 shows that at the lower temperature, uranium reduction was slower than at the higher temperature. The extent of reduction was slightly lower; however, the rate of reduction during maximum uranium removal was significantly lower than that determined at the higher temperature. A reduction in rate at lower temperature is consistent with microbially mediated reactions. Based on these results, subsequent experiments were conducted at the higher temperature (32°C).

## Effect of Mixing Inside Bioreactor

The *D. desulfuricans* organism is typically found in quiescent environments; therefore, active mixing may adversely impact metabolic activities. However, mixing inside the reactor was necessary to minimize mass transfer limitations and to achieve a uniform temperature and uranium concentration in solution. To evaluate the effect of turbulent conditions on bioreduction, experi-

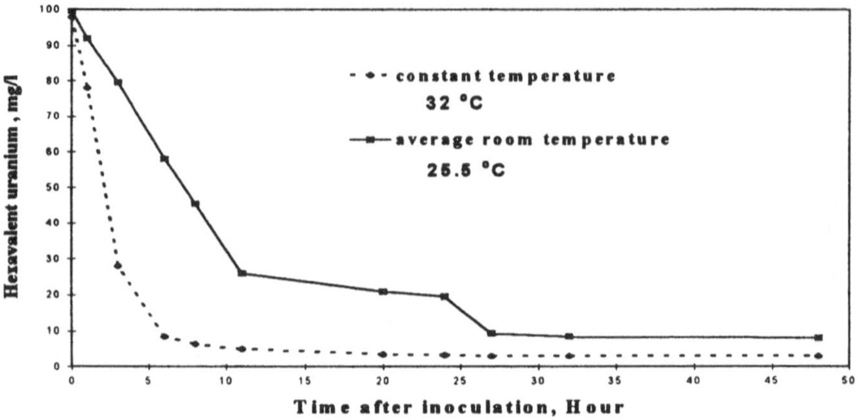


**Figure 4. Soluble hexavalent uranium remaining in solution after inoculation with *D. desulfuricans* at different incubation temperatures.**

ments were conducted in which mixing rates were varied (5, 50, 100 and 150 rpm) while all other test parameters remained constant. As can be seen from Figure 5, the lowest mixing rate (5 rpm) resulted in a lower extent as well as rate of uranium reduction. From the results it appears that a mixing rate of 100 rpm resulted in greater uranium reduction in the shortest time. Uranium reduction rates (slopes of the curves in the linear range) were determined (by linear regression) for different mixing speeds and tabulated in Table I.

**Table I. Effect on Uranium Reduction by Mixing Rates**

| Mixing Speed, rpm | Rate of Reduction, $(-dU^{+6}/dt)$, mg/L-hr |
|---|---|
| 5 | 11 |
| 50 | 12 |
| 100 | 30 |
| 150 | 12 |

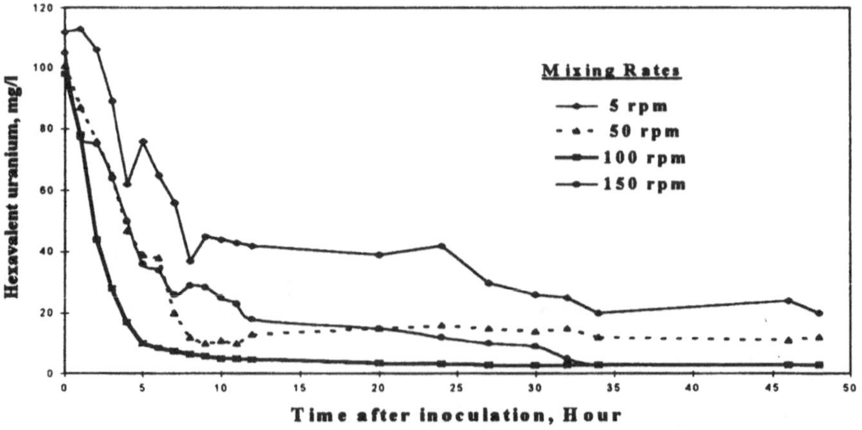


**Figure 5. Distribution of hexavalent uranium in solution after inoculation with *D. desulfuricans* at 32°C for different mixing rates.**

The utilization rate of uranium at 100 rpm was more than twice the rate at other mixing speeds. Lower reduction rates at slow mixing speeds may be due to limitations in mass transfer. Bacteria can only utilize molecules which are bound by the enzyme at an active site. Adequate mixing ensures contact between active enzyme sites and rate limiting materials such as electron acceptors. At the highest mixing speed (150 rpm), the uranium reduction rate was adversely impacted. The pure culture used in these experiments may not be accustomed to such high mixing intensities and may become less active under these conditions. Although the reduction rate was slower at 150 rpm than 100 rpm, the overall extent of uranium removal was the same for both mixing speeds. Since the criteria for optimization was set on the basis of reduction rates, the optimum mixing speed for later experiments was set at 100 rpm.

## Uranium Loading

To observe the impact of elevated loading on uranium removal, the initial uranium concentration was raised to 200 mg/L. In a full-scale treatment plant, fluctuation of waste loading is very common. It was important to determine if higher uranium concentrations would be inhibitory to the microbial population. As evident from Figure 6, the pattern of bioreduction was similar to that for 100 mg/L uranium loading.The uranium biotransformation rate in the linear range was calculated to be ~ 38 mg/L-hr which is ~ 25% higher than the rate found for the 100 mg/L uranium concentration. This is typical for biological systems where utilization rates increase with increasing concentration of the limiting compound for first order reactions.

As reported by other researchers,[1,2,5] *D. desulfuricans* does not grow when uranium is the only electron acceptor present in solution. The applicability of this process as a treatment option for uranium-containing mixed-waste streams will be closely related to the viability of microorganisms over time. An experiment was performed to determine how long the test organisms can actively utilize uranium as an electron acceptor. Removal of soluble uranium under anaerobic condition was evaluated during repeated treatment periods using a single inoculum of *D. desulfuricans* at an initial concentration of $5 \times 10^8$ cells/mL (the same as all other experiments). The uranium concentration at the beginning of each treatment period was 100 mg/L. Since most uranium reduction occurred in the first 10–12 hours, a cycle of 12 hours was set for each treatment. Uranium reduction was monitored over the 12 hour period. At the end of each cycle, uranium was injected into the reactor to bring the concentration back up to ~ 100 mg/L. The microorganisms continued to rapidly reduce uranium over five successive treatment cycles during the course of 60 hours (Figure 7), indicating that the population remained viable for repeated treatments.

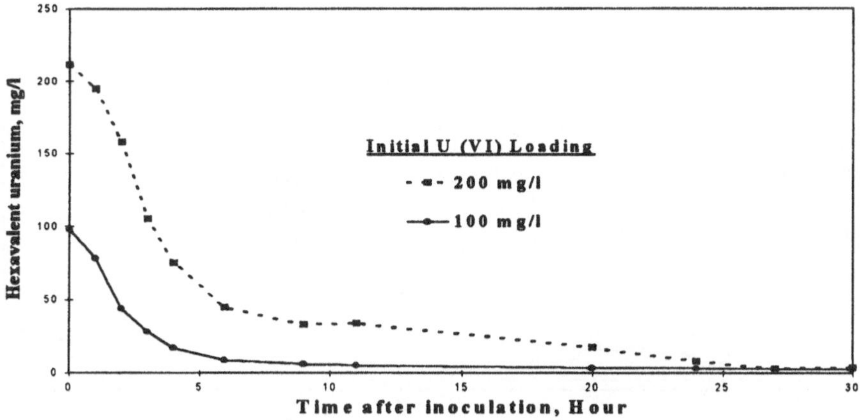


**Figure 6. Effect of uranium loading on enzymatic bioreduction and precipitation of uranium at 32°C and constant speed of 100 rpm.**

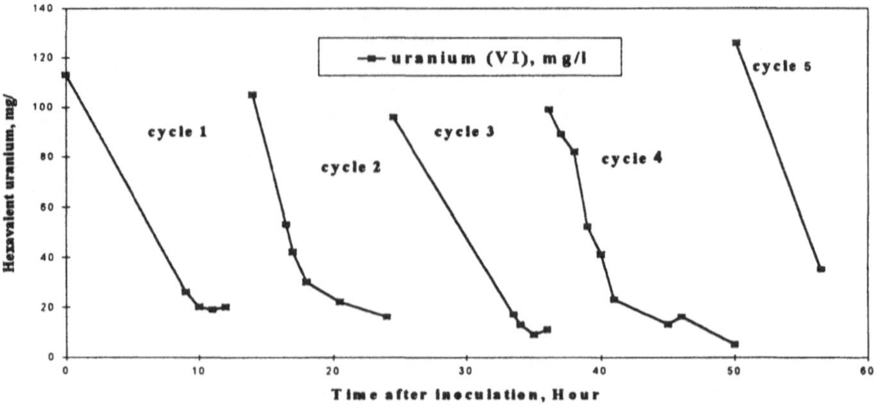


**Figure 7. Bioreduction of U(VI) after one time inoculation with *D. desulfuricans* and repeated loading of U(VI) in a cycle span of 12 hours.**

## Uranium Mass Balance

The disappearance of soluble uranium over time indicated that enzyme mediated reduction of hexavalent uranium and physical precipitation of tetravalent uranium had occurred. Although reduced uranium is highly stable in the solid phase, some amount of tetravalent uranium may remain in the solution and not precipitate immediately after reduction due to kinetic constraints. To account for nonprecipitated reduced uranium, subsamples were filtered to remove solids (precipitated uranium and biomass), then the filtrate was reoxidized by adding $HNO_3$ (2N) to yield the total aqueous uranium concentration in solution ($U^{+6}$ + $U^{+4}$). Nitric acid oxidation converts reduced uranium to the oxidized state which can be analyzed. The difference between the concentration of total uranium in solution (filtrate) and that of oxidized uranium ($U^{+6}$ ) in solution yielded the concentration of reduced uranium ($U^{+4}$) which remained in solution. The total uranium level in each sample ($U^{+6}$ + $U^{+4}$) was determined by oxidizing the unfiltered sample with nitric acid prior to uranium analysis.

A typical uranium distribution profile during bioreduction is shown in Figure 8. Uranium mass balances of > 90 % were obtained after reoxidation of unfiltered samples. Note that reduced ura-

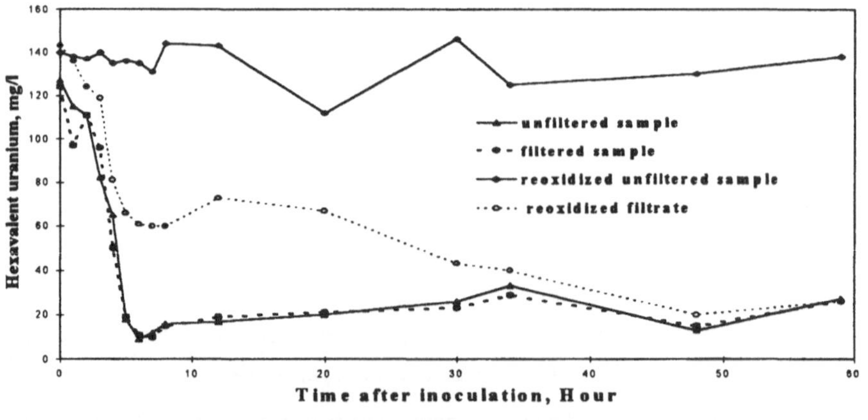


**Figure 8. Hexavalent uranium distribution after bioreduction and reoxidation.**

nium did not immediately precipitate from solution (the difference between unoxidized and oxidized filtrate samples). A significant mass of reduced uranium remained in solution between 7–30 hours. Ultimately, all of the reduced uranium did precipitate from solution after 58 hours. Since bioprecipitation consists of two distinct processes (chemical reduction followed by abiotic precipitation), competing reactions, especially ligand complexation reactions, may adversely impact successful precipitation of uranium from solution.

## CONCLUSIONS

- Bioreduction and precipitation of uranium can be achieved using a continuously stirred batch reactor.
- Continuous purging of the uranium-containing waste stream with an anaerobic gas mixture i not required for effective uranium removal. Intermittent gas flow is more cost-effective and does not adversely impact uranium reduction.
- Bioreduction rates appear to be maximized at a mixing speed of 100 rpm.
- Lower temperature (than 32 $^0$C) reduces the metabolic rate of the sulfate reducers, resulting in a slower uranium reduction rate.
- An increase in uranium loading increased the uranium utilization rate.
- *D. desulfuricans* can remain viable for successive reduction of U(VI) up to five consecutive treatment operations.

## REFERENCES


1. Ganesh, R., Impact of Organic Ligands on the Bioreductive Precipitation of Hexavalent Uranium, Ph.D. dissertation, The University of Tennessee, Knoxville, TN, 1995.
2. Lovely, D.R., and E.J.P. Phillips, "Bioremediation of uranium contamination with enzymatic uranium reduction," *Environ. Sci. Technol.*, 26, pp. 2228–2234, 1992.
3. Hungate, R.E., A Roll Tube Method for Cultivation of Strict Anaerobes, In: *Methods in Microbiology*, 3B, pp. 117–132, 1969.
4. Bryant, M.P., L.L. Campbell, C.A. Reddy, and M.R Crabill, "Growth of *Desulfovibrio* in lactate and ethanol media low in sulfate in association with H utilizing methanogenic bacteria," *Appl. Environ. Microbio.*, 33, pp. 1162–1169, 1977.
5. Lovely, D.R., E.J.P. Phillips, Y.A. Gorby, and E.R. Landa, "Microbial reduction of uranium," *Nature*, 350(40), pp. 414–416, 1991.

# Section Eleven
# OIL AND PETROCHEMICAL WASTES

## 41 CHARACTERIZATION OF HYDROCARBONS REMOVALS IN WASTEWATER FROM AN INTEGRATED OIL REFINERY

**Noah Galil, Dafna Zeira-Blum** and **Menahem Rebhun**
**Technion - Israel Institute of Technology**
**Faculty of Civil Engineering**
**Haifa 32000, Israel**

## ABSTRACT

The wastewater from an integrated oil refinery is characterized by a diversity of pollutants including hydrocarbons, in free and emulsified form, phenols, including cresols and xylenols, mercaptans, sulfides, ammonia, and cyanides. The wastewater treatment of the reported oil refinery is based on several treatment stages, consisting of gravity separation, chemical coagulation-flocculation followed by dissolved air flotation, and a combined biological treatment based on two aerated ponds in series and lime flocculation-sedimentation. The effluent is recycled and reused as makeup to the recirculated cooling water system of the refinery. The aim of this study was to identify and quantify the main hydrocarbons groups after the different treatment stages. API separators provided substantial removal of hydrocarbons from about 1,000 mg/L to less than 100 mg/L. Chemical flocculation and dissolved air flotation (DAF) enabled further removal of about 70 mg/L of the hydrocarbons, to 20 mg/L, based on average values. This removal included aliphatics, phenols, naphtalens, and benzenes. This process involved mechanisms of coagulation-flocculation accomplished by physical separation, as well as binding of hydrophobic compounds to nonpolar fractions and stripping of volatile compounds. The combined biological-chemical treatment (BCT) provided removal of hydrocarbons to average values of less than 4 mg/L, and efficient removals of over 90% of the phenols, naphtalens, and benzens.

## INTRODUCTION

An integrated oil refinery is characterized by a diversity of production processes including distillation, catalytic cracking, platforming, ethylene production, sulfur recovery, and others. The results of these various processes is the diversity of pollutants found in wastewater, which contains hydrocarbons, phenol, methyl and di-methyl phenols (cresols and xylenols) mercaptans, sulfides, and ammonia. Spent caustic streams from washeries, especially from gasoline washeries, are characterized by very high concentrations of phenols. Acid water at the bottom of stripping towers contains high concentrations of organic matter, sulfides, and cyanides.

The wastewater is usually treated in order to achieve the appropriate effluent quality for either discharge to natural recipients or to be reused by the industry. While a substantial part of the hydrocarbons can be removed from the aqueous phase by gravity separation and by dissolved air flotation, all the other organic materials, including residual hydrocarbons, phenols, and other dissolved organics, can be removed only by biological treatment processes.

The adopted solutions in order to achieve reliable cost-effective wastewater treatment and reclamation[1,2] was based on several treatment stages which include the following elements:

---

*52nd Purdue Industrial Waste Conference Proceedings*, 1997, Ann Arbor Press, Chelsea, Michigan 48118. Printed in U.S.A.

a. Specific wastewater streams, having low rates of flow, but containing very high concentrations of pollutants (such as acid tar and spent caustic) are separated from the sewerage system and transferred to a special plant, working on the treatment of toxic wastes.
b. Flow regulation of rain floods and unexpected streams resulting from accidents at the production units.
c. Primary oil separation operating before regional wastewater pumping stations.
d. Oil-water separation is performed by an API type separator equipped with bottom and surface scrapers for good collection of settled sludge and floating oil.
e. Chemical flocculation and dissolved air flotation (DAF) operate for achieving substantial improvement in the effluent quality in terms of suspended solids and oil (mainly emulsified) removal. A cationic polyelectrolite is used as flocculant.
f. Biological treatment is performed by two aerated ponds in series, with a hydraulic detention time of about 24 hours each, followed by a lime chemical flocculation-clarification. This combined biological and chemical treatment (BCT) is based on the idea of sharing the task of the bioprocess between aerated ponds and chemical clarification, for the removal of biologically synthesized suspended solids and colloids. Although aerated ponds are not considered as intensive biotreatment, the experience accumulated over more than six years of continuous operation proved that the BCT provided a reliable and cost-effective solution.[3]

The effluent is reused as makeup to the recirculated cooling water system. The cooling system has good biooxidation capacity; therefore, it could be utilized also for the completion of the bioprocess, especially for nitrification.[4]

The objectives of this study were to identify and to quantify specific contaminants after different treatment stages. The characterization included general parameters as total organic matter, expressed as chemical oxygen demand (COD), hydrocarbons, phenols as a group, and also specific contaminants such as phenol, o-cresol, p-cresol, 2,6 dimethylphenol, 2,4 dimethyl-phenol, 3,5 dimethylphenol, naphtalene, 1 methylnaphtalene, 2 methylnaphtalene, toluene, ethyl benzene, m+n xylene and o-xylene.

## MATERIALS AND METHODS

The chemical characterization was based on composite samples collected once a week, during 24 hours from sampling points located after the API, after the DAF, and after the BCT, as indicated in Figure 1, between October 1994 and July 1995.

General analysis of organic matter was performed according to *Standard Methods for the Examination of Water and Wastewater*[5] for COD, following Method 5520D, for hydrocarbons extractable by hexane following Method 5520D and for phenols following Method 5530.

GC/MS and GC analysis were used for the identification of specific compounds. The equipment included: a. Finnigan Mat GC/MS, type ITS 40 (ion trap) with capillary column DB-5 of

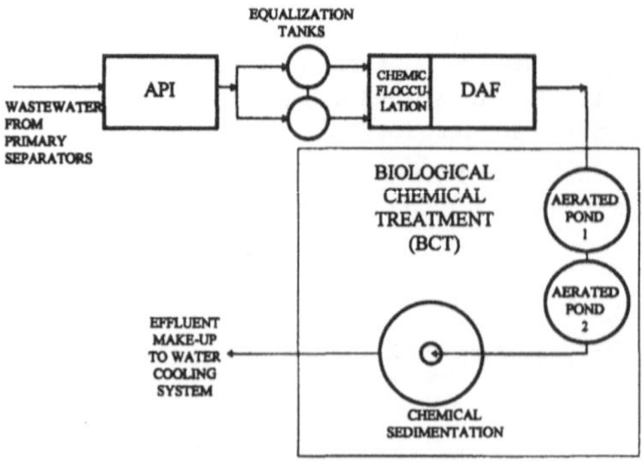


**Figure 1. General description of the wastewater treatment system.**

30m × 0.22 mm, 0.25 μm; b. Varian GC, type 3300 with FID and a DB-5 of 15 m × 0.22 mm, 0.25 μm, capillary column.

The method used for the identification of pollutants was based on double phase extraction by methylene chloride, first after rising the pH over 11, and after acidification to pH < 2. This method was reported as being used for similar purposes by Lopez-Avila et al.,[6] Burks and Wagner,[7] Reece and Burks,[8] Clark et al.,[9] Gulyas et al.,[10] and Tows et al.[11]

**Phenol, cresols and xylenols:** concentrations were analyzed after (a) acidification to pH = 4 by phosphoric acid in the presence of cuprum sulfate, for avoiding possible biodegradation;[12,13] (b) distillation; (c) additional acidification to pH = 2; and (d) extraction by methylene chloride which contained 25 mg/L of 2-nitrophenol/chlorophenol as internal standard. The extract was injected to the GC/FID following a program based on 50°C rising by 2°C/min to 98°C, then by 15°C/min up to 220°C.

**Polyaromatics:** were analyzed by extraction with methylene chloride in alkaline environment. GC/MS was programmed at 60°C for 1 minute, increasing the temperature by 20°C/min to 250°C for 16 minutes.

**Mono and alkyl aromatics:** acidification by sulfuric acid to pH<2, extraction by methylene chloride which contained 2-nitrophenol/chlorophenol as internal standard following a program based on 50°C for 4 minutes, increasing the temperature by 2°C/min to 98°C for 2 minutes, then by 15°C up to 220°C.

## RESULTS AND DISCUSSION

The results are expressed by using cumulative distributions of all the values obtained during the monitoring period. This way of presentation enabled obtaining the probability of a concentration valuing equal to or less than the stated magnitudes. The results are summarized in Table I for different probabilities of 15, 50, and 85%, as well as the average and the standard deviation values.

Gravity separation by API enables the removal of hydrocarbons, mainly aliphatic compounds, in the range of hundreds of mg/L. This relatively simple and very reliable process is of tremendous importance as the first stage of the wastewater treatment sequence, since it could remove more than 90% of the total raw wastewater pollution potential. However, the API effluent still contains residual organic matter at levels of about 900 mg/L, expressed as average concentrations of COD, including about 90 mg/L of hydrocarbons (Fig. 2 and Fig. 3). About 25% of these hydrocarbons could be considered as "soluble," since they passed a 0.45 μm filter during Buchner funnel filtration tests. The other 75% were mostly hydrocarbons in emulsified form.

Chemical flocculation followed by DAF separation removed up to 90% of the hydrocarbons, which were found initially in emulsified form, by creating oily sludge. The DAF process removed from the aqueous phase also about 50% of the soluble hydrocarbons, apparently aromatic and other volatile organic compounds (VOCs) due to stripping mechanisms. In order to avoid emissions of VOCs to the atmospheric environment, a new covered dissolved gas (nitrogen) flotation unit (DGF) has been installed and operated since summer 1996. The average total hydrocarbon concentration in the DAF effluent was 20 mg/L; 12 mg/L of it could be considered as soluble (Table I).

Removal of phenols and naphtalens by chemical flocculation and DAF was obtained at levels of about 30% (Fig. 4). This could be explained by possible binding of these compounds to hydrocarbons which have been separated by DAF.

Based on the effluent specific characterization, it seems that in the case of wastewater treatment at an integrated oil refinery, three different mechanisms are involved in the flocculation—DAF process: (a) coagulation-flocculation of the emulsified hydrocarbons, which enable physical separation by flotation; (b) stripping of VOCs; and (c) binding of hydrophobic compounds to nonpolar phases which are included in the materials separated by flotation.

The chemical flocculation—DAF can be considered as an important protection of the biological treatment, by lowering the total loading, as well as by partial removals of methyl benzens, phenols, and naphtalens, known for their relatively low biodegradability and inhibitory effects.

The combined biological-chemical treatment (BCT) achieved high removals of phenols, and naphtalenes (always over 90%); however, the general organic matter concentrations indicated residuals of average COD total 270 and COD soluble 180 mg/L. BOD5 analyses indicated al-

**Table 1. Effluent Qualities After Treatment Stages Expressed on Different Statistical Means**

| | API | | | | | DAF | | | | | BT2 | | | | |
|---|---|---|---|---|---|---|---|---|---|---|---|---|---|---|---|
| Parameter | 15% | 50% | 85% | Ave. | STD | 15% | 50% | 85% | Ave. | STD | 15% | 50% | 85% | Ave. | STD |
| COD- total | 497 | 894 | 1352 | 877 | 346 | 304 | 413 | 536 | 418 | 119 | 193 | 260 | 392 | 271 | 83 |
| COD- soluble | 261 | 330 | 502 | 360 | 100 | 230 | 313 | 479 | 326 | 109 | 128 | 171 | 245 | 181 | 65 |
| Phenols- total | 7.0 | 9.1 | 12.7 | 8.7 | 3.4 | 4.1 | 8.4 | 11.6 | 8.0 | 3.2 | 0.02 | 0.07 | 0.38 | 0.14 | 0.15 |
| Phenols- soluble | 4.5 | 8.4 | 11.9 | 8.4 | 3.3 | 4.0 | 8.2 | 11.2 | 7.5 | 3.1 | 0.00 | 0.03 | 0.25 | 0.08 | 0.13 |
| Hydrocarbons- total | 57 | 87 | 147 | 92 | 39 | 10.3 | 20.3 | 36.3 | 20 | 10 | 2.3 | 2.8 | 7.6 | 3.8 | 2.4 |
| Hydrocarbons- soluble | 15 | 21 | 36 | 22 | 11 | 5.5 | 11.3 | 24.5 | 12 | 8 | 1.0 | 1.2 | 5.8 | 2.1 | 2.4 |
| Phenol | 5.6 | 7.5 | 8.2 | 6.6 | 1.3 | 1.3 | 4.3 | 7.4 | 4.0 | 2.6 | 0.06 | 0.13 | 0.26 | 0.12 | 0.07 |
| o-Cresol | 1.4 | 2.0 | 4.5 | 2.1 | 1.1 | 0.75 | 1.5 | 2.3 | 1.4 | 0.6 | 0.02 | 0.06 | 0.11 | 0.05 | 0.03 |
| p-Cresol | 2.0 | 2.5 | 6.1 | 2.8 | 1.6 | 0.57 | 2.2 | 3.8 | 2.0 | 1.2 | 0.04 | 0.05 | 0.22 | 0.08 | 0.08 |
| 2,6 Dimethyl phenol | 0.17 | 0.33 | 0.75 | 0.32 | 0.21 | 0.04 | 0.13 | 0.63 | 0.20 | 0.21 | 0.01 | 0.01 | 0.02 | 0.01 | 0.004 |
| 2,4 Dimethyl phenol | 0.50 | 0.85 | 1.4 | 0.76 | 0.41 | 0.15 | 0.49 | 1.2 | 0.49 | 0.35 | 0.03 | 0.04 | 0.05 | 0.04 | 0.007 |
| 3,5 Dimethyl phenol | 0.51 | 0.64 | 1.1 | 0.68 | 0.26 | 0.28 | 0.52 | 0.83 | 0.48 | 0.21 | 0.02 | 0.03 | 0.72 | 0.04 | 0.03 |
| Naphtalen | 0.21 | 0.34 | 1.3 | 0.55 | 0.42 | 0.14 | 0.24 | 1.0 | 0.35 | 0.33 | 0.01 | 0.02 | 0.03 | 0.01 | 0.008 |
| 1 Methyl naphtalen | 0.22 | 0.32 | 0.80 | 0.37 | 0.20 | 0.11 | 0.25 | 0.4 | 0.23 | 0.11 | 0.02 | 0.02 | 0.03 | 0.02 | 0.006 |
| 2 Methyl naphtalen | 0.15 | 0.50 | 0.90 | 0.46 | 0.31 | 0.10 | 0.24 | 0.52 | 0.25 | 0.16 | 0.02 | 0.03 | 0.04 | 0.03 | 0.008 |
| Toluene | | | | 1.9 | 1.3 | | | | 0.92 | 0.62 | | | | 0.08 | 0.007 |
| Ethyl Benzene | | | | 4.1 | 1.9 | | | | 2.2 | 0.78 | | | | 0.03 | 0.006 |
| m + p Xylene | | | | 0.77 | 0.08 | | | | 0.41 | 0.14 | | | | 0.01 | 0.003 |
| o-Xylene | | | | 1.2 | 0.34 | | | | 0.82 | 0.26 | | | | 0.08 | 0.01 |

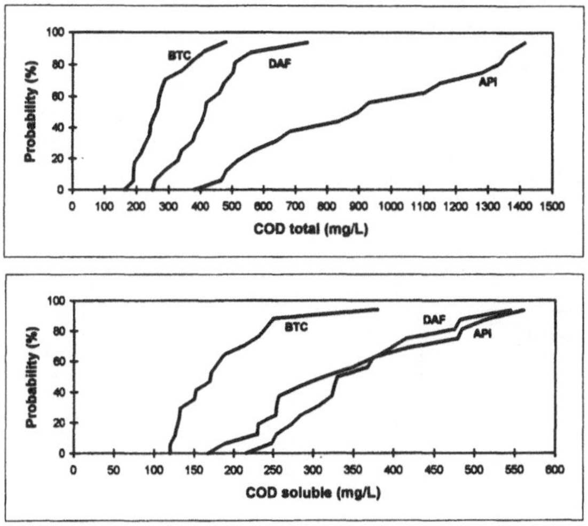


**Figure 2. Removal of organic matter, as COD.**

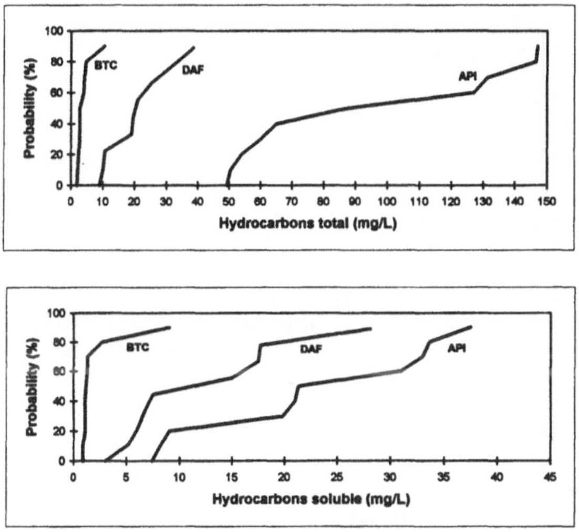


**Figure 3. Removal of hydrocarbons.**

ways values lower than 10 mg/L. It seems that most of the residual organic matter in the BCT effluent could be, apparently, metabolic products of the biological process, characterized by low biodegradability.

The results indicate that the biological treatment plays an important role in the removal of residual organic matter, found at this stage mainly in soluble form. This matter includes phenols, naphtalens which are removed at efficiencies of over 90%. The chemical flocculation - clarification following the aerated ponds enabled the achievement of a final effluent almost clear of suspended solids and colloidal materials. The fact that more than 40% of the residual hydrocarbons in the BCT effluent were connected with the solid phase, indicates that sorption of hydrophobic contaminants plays an important role in the biological treatment process of this type of wastewater (Fig. 5).

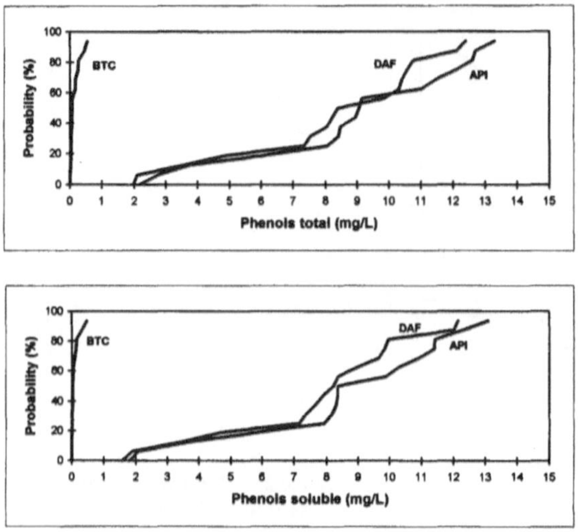


**Figure 4. Removal of phenols.**

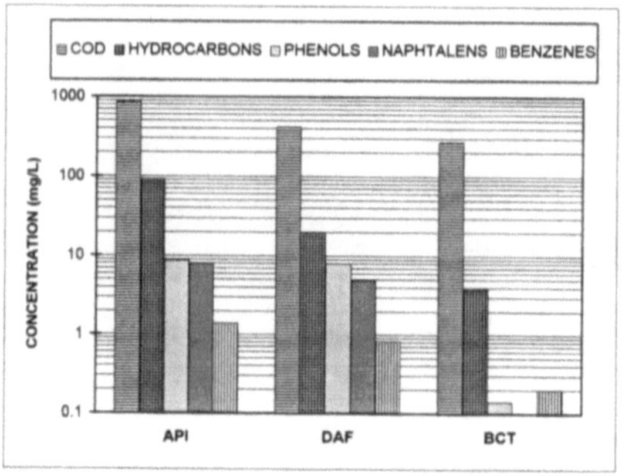


**Figure 5. Residual concentrations of hydrocarbons after different treatment stages (based on average values).**

## CONCLUSIONS

The wastewater treatment sequence at the reported integrated oil refinery included a combination of physical, chemical, and biological treatment processes.

API separators could provide substantial removal of hydrocarbons, from about 1,000 mg/L to less than 100 mg/L.

Chemical flocculation-DAF enabled additional removal of about 70 mg/L of hydrocarbons, including aliphatics, phenols, naphtalens, and benzenes, down to 20 mg/L, based on average values. This process involved mechanisms of coagulation-flocculation accomplished by physical separation, as well as binding of hydrophobics to nonpolar fractions and stripping of VOCs.

The combined biological-chemical treatment (BCT) provided removal of hydrocarbons to average values of less than 4 mg/L, including efficient removals of phenols, naphtalens, and benzenes, always over 90%.

## REFERENCES

1. Galil, N. and Rebhun, M. Environmental Solutions Based on Recycling of Water, Oil and Sludge at an Integrated Oil Refinery, *Proc. of the 46th Ind. Waste Conf., Purdue Univ.*, Lewis Pub. Inc., Chelsea, MI, 1991.

2. Rebhun, M. and Galil, N. Cost Effective Combination of Biological and Chemical Processes for Treatment and Recycling of Wastewater in a Large Petrochemical Complex, *Proc. WEFTEC, 96, 69th Annual Conf. & Exposition*, Vol. 3, Part II, 521–530, Dallas, TX, 1996.
3. Galil, N., Levinsky, Y., and Rebhun, M. Cost Effective Wastewater Treatment and Recycling in a Large Industrial Complex, *Proc. of the 2nd Specialized Conference on Pretreatment of Industrial Wastewaters*, IAWQ, Athens, Greece, 1996.
4. Rebhun, M. and Engel, G. "Reuse of wastewater for industrial cooling systems," *J. Water Pollut. Control Fed.*, 60, pp. 237–242, 1988.
5. *Standard Methods for the Examination of Water and Wastewater* 18th edition, prepared and published jointly by APHA, AWWA and WEF, 1992.
6. Lopez Avila, V., Northcutt, R., Onstot, J., Wickham, M. and Billets, S. "Determination of 51 priority organic compounds after extraction from standard reference materials," *Anal. Chem.*, 55, 6, pp. 881–889, 1983.
7. Burkes, S. L. and Wagner, J. Characterization and Treatment of Aqueous Wastes and Residue from Petroleum Refineries, Oklahoma State University, Stillwater Gov. Report Announcement and Index (GRA&I) issue 26, 1983.
8. Reece, C.H. and Burks, S.L. Isolation and Chemical Characterization of Petroleum Refinery Wastewater Fractions Acutely Lethal to *Daphnia Magna*, Oklahoma State Univ., Oklahoma City, FAD DA8325840, 1985.
9. Clark, L.B., Rosen, R.T., Hartman, T.G., Alaimo, L.H., Louis, J.B., Hertz, C. Ho, C.T., and Rosen, J.D. "Determination of non-regulated pollutants in three New Jersey publicly owned treatment works," *J. Water Pollut. Control Fed.*, 66, 2, pp. 104–113, 1991.
10. Gulyas, H., Reich, M., and Sekulov, I. Characterization of a Biologically Treated Wastewater from Oil Reclaiming: Recording of Low Molecular Weight Organics and Estimation of Humic Substances, *Proc. of the 2nd Specialized Conference on Pretreatment of Industrial Wastewater*, IAWQ, Athens, Greece, 1993.
11. Tows, I., Albers, G., Gulyas, H., Erickhoff, H.P. Reich, M., and Sekulov, I. (1993) Identification of Three Organics in a Treated Lubricating Oil Refinery Wastewater, *Proc. of the 2nd Specialized Conference on Pretreatment of Industrial Wastewater*, IAWQ, Athens, Greece, 1993.
12. Galil, N., Rebhun, M., and Brayer, Y. "Disturbances and inhibition in biological treatment of wastewater from an integrated oil refinery," *Water Sci. Technol.*, 20, 10, pp. 21–29, 1988.
13. Carron, J. M. and Afgan, B.K. (1989) "Environmental aspects and analysis of phenols in aquatic environment," in: *Analysis of Trace Organics in the Aquatic Environment*, Afgan, B.K. and Chan, A.S.Y., Eds.

# 42 BIOREMEDIATION OF A PETROLEUM—CONTAMINATED SITE: A FEASIBILITY ANALYSIS

**T. Viraraghavan**, Professor
Faculty of Engineering
University of Regina
Regina, SK, Canada S4S 0A2

**D. Mihial**, Project Engineer
Agra Earth and Environmental
Regina, SK, Canada

**R. B. Thomson and M. D. Mortin**, Students
Faculty of Engineering
University of Regina
Regina, SK, Canada.

## ABSTRACT

A study was conducted to determine if bioremediation was a feasible alternative for the remediation of a particular petroleum-contaminated site in Saskatchewan, Canada. The feasibility study included the following: site investigation, site assessment, degradation rates, bioremediation alternatives (landfarming, composting, and biopile/biofiltration integration), feasibility analysis and preliminary design of the selected alternative. It was determined that the biopile composting alternative was best suited for this project.

## INTRODUCTION

Petroleum-contaminated sites are a common occurrence in our environment today. The sites are contaminated through spills, leaking petroleum storage tanks, and earlier disposal of hazardous wastes at these sites. There are numerous alternatives that are suitable for remediating the site to acceptable standards. The selection of a suitable method for the remediation of a contaminated site depends on site and hazardous waste characteristics and regulatory guidelines.

In the mid-1960s, a company in Saskatchewan, Canada excavated a pit and used it as a dumping ground for petroleum wastes that were generated by its operations. This process continued for a number of years before the pit was finally covered. There were no environmental concerns or strict regulations to worry about and abide by at that time. The company is now planning to move its operations to a new location in a period of two to three years. The company is now well aware of its environmental obligations and is planning to remediate the site to comply with regulatory guidelines. Based on the approximate costs of various remediation alternatives shown in Figure 1, it was determined by the company that bioremediation should be examined further to remediate the site.

The main objective of this study was to determine if bioremediation was a feasible technique for the remediation of the company's contaminated pit based on short-term effectiveness (2–3 years limit), cost (limited budget), and ease of implementation.

## SITE INVESTIGATION

The exact boundaries of the contaminated pit were not known. Site investigation and analyses were performed to determine the following: types of hydrocarbons present in the contaminated

*52nd Purdue Industrial Waste Conference Proceedings*, 1997, Ann Arbor Press, Chelsea, Michigan 48118. Printed in U.S.A.

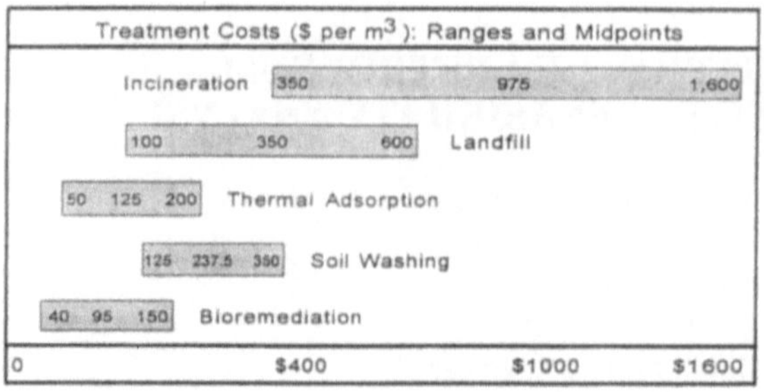


Figure 1.

pit; total petroleum hydrocarbon concentrations (TPH); benzene, toluene, ethyl-benzene, and xylene concentrations (BTEX); phenol concentrations; ethylene glycol concentrations; PCB concentrations; and trace element concentrations.

Analyses were performed on samples taken from a field drilling program. The program included drilling a number of test holes (61) in the estimated area of the pit. The test holes drilled were broken down into two categories. In category 1 there were 40 test holes drilled and logged. In category 2, there were 21 test holes drilled, logged, sampled and tested (Figure 2).

## Site Assessment

From the test holes, observations of physical soil characteristics and chemical analyses were combined to determine the pit boundary, dimensions, stratigraphy, estimated volume, and the magnitude of contamination. For site assessment, the Saskatchewan Environment and Resource Management (SERM) and Canadian Council of the Ministers for the Environment (CCME) regulatory guidelines were followed.[2, 3]

Gas chromatographic analyses determined the various types of hydrocarbon products located within the pit. Table I distinguishes the different types of hydrocarbons present based on test hole number and depth.

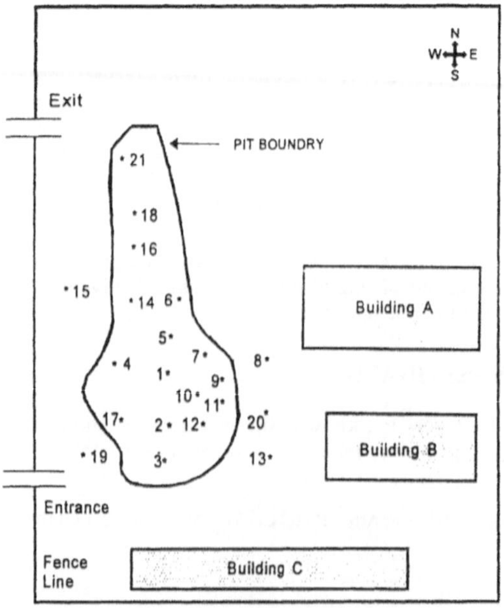


Figure 2.

Results showing the TPH values are provided in Table II. Only three test-hole samples (asterisk in the table) were below the guideline value of 1000 ppm.[2]

Results from the benzene, toluene, ethylbenzene, and xylene analyses showed that these compounds exceeded the SERM and CCME guideline values in some test holes (Table III).

The analyses results showed that phenols, ethylene glycol, and PCBs were all below regulatory guidelines. Therefore, these parameters were not of concern. From the ICP trace element analysis, the results (Table IV) showed that all the elements were below regulatory guidelines.

**Table I. Types of Hydrocarbons Present**

| Test Hole | Depth (m) | Type of Hydrocarbon |
|---|---|---|
| 1 | 3.3 | Gasoline, paint thinners, diesel fuel, lube oil |
| 2 | 2.2 | Gasoline, paint thinners, diesel fuel, lube oil |
| 3 | 1.5 | Gasoline, paint thinners, diesel fuel, lube oil |
| 6 | 3.8 | Gasoline, paint thinners, diesel fuel, lube oil |
| 7 | 3.8 | Gasoline, paint thinners, diesel fuel, lube oil |
| 9 | 3.0 | Gasoline, paint thinners, diesel fuel |
| 10 | 3.8 | Gasoline, paint thinners, diesel fuel, crude oil |
| 11 | 5.0 | Diesel fuel |
| 12 | 2.3 | Diesel fuel |
| 16 | 6.1 | Diesel fuel, lube oil |
| 17 | 3.0 | Gasoline, paint thinners, diesel fuel, lube oil |
| 18 | 5.3 | Gasoline, paint thinners, diesel fuel, lube oil |
| 20 | 3.8 | Diesel fuel |
| 21 | 5.3 | Diesel fuel, lube oil |

**Table II. TPH Concentrations[a]**

| Test Hole # | Avg TPH/Hole (ppm) | Test Hole # | Avg TPH/Hole (ppm) |
|---|---|---|---|
| 1 | 66,620 | 12 | 76,177 |
| 2 | 94,473 | 13 | 2,259 |
| 3 | 50* | 14 | 21,644 |
| 4 | 47,319 | 15 | 34* |
| 5 | 23,820 | 16 | 28,388 |
| 6 | 23,699 | 17 | 12,384 |
| 7 | 46,233 | 18 | 44,195 |
| 8 | 5,709 | 19 | 221* |
| 9 | 54,800 | 20 | 2,253 |
| 10 | 65,500 | 21 | 58,735 |
| 11 | 84,562 | | |

[a]Average TPH concentration for the site = 39, 896 ppm.

**Table III. BTEX Concentrations (mg/kg)**

| Test Hole # | Benzene | Toluene | Ethylbenzene | Xylene |
|---|---|---|---|---|
| 1 | 47 | 36 | 37 | 835 |
| 2 | 11 | – | 5.5 | 26 |
| 12 | – | 9.5 | 6.0 | 36 |
| 14 | – | – | – | 5.1 |
| 17 | 2.3 | – | – | 97 |
| 21 | 1.5 | – | – | – |
| SERM Guidelines | 0.5 | 3.0 | 5.0 | 5.0 |
| CCME Guidelines | 0.5 | 3.0 | 5.0 | 5.0 |

**Table IV. Trace Element Analysis (mg/kg)**

| Parameter | Test Hole #1 (3.8) | Test Hole #12 (3.0) | Test Hole #14 (3.0) | Test Hole #17 (2.3 m) | CCME Guideline |
|---|---|---|---|---|---|
| Arsenic | 9 | 7 | 8 | 8 | 30 |
| Barium | 138 | 167 | 133 | 186 | 500 |
| Cadmium | 0.5 | 0.5 | 0.5 | 0.5 | 5 |
| Chromium | 17 | 18 | 18 | 28 | 250 |
| Copper | 52 | 28 | 26 | 82 | 100 |
| Lead | 131 | 90 | 35 | 110 | 500 |
| Mercury | 1 | 0.1 | 0.1 | 0.1 | 2 |
| Nickel | 17 | 14 | 16 | 22 | 100 |
| Zinc | 72 | 63 | 51 | 273 | 500 |

Using TPH concentrations as an indicator, the boundary of the contaminated pit (approximately 25 m × 60 m) was determined. The boundary is shown in Figure 2. It was estimated that the topography of the naturally deposited soils in the surrounding area of the pit, prior to excavation, was flat with a vertical relief of less than 1 m. The stratigraphy of the pit and the surrounding area is shown in Figure 3.

The water table was located 15 m below the existing surface and had a relatively impermeable clay layer above; therefore, groundwater contamination was not a serious concern. It was estimated that contaminant concentrations exceeding the regulatory guidelines were confined within depths of 10 m below existing surface and within 2 m of the estimated boundaries of the pit. The sidewalls of the pit were sloped. From the test results and calculated extrapolation, the estimated volume of contaminated soil within the pit was approximately 8000 $m^3$.

## Degradation Rates and Constants

For the purpose of this study, it was necessary to obtain an approximate estimate for the degradation rates of the hydrocarbons in the contaminated pit. Experimental studies would have to be performed on samples from the site to determine the actual degradation rates. Therefore, by making some assumptions and conducting a literature search, an approximate value can be developed for the degradation rates. The rates were based on reviews of case studies that were similar in nature and information from a handbook on environmental degradation rates. In developing such an estimate, it was assumed that all volatilization had occurred due to long period of storage. Approximate ranges, high and low, of degradation times were estimated.

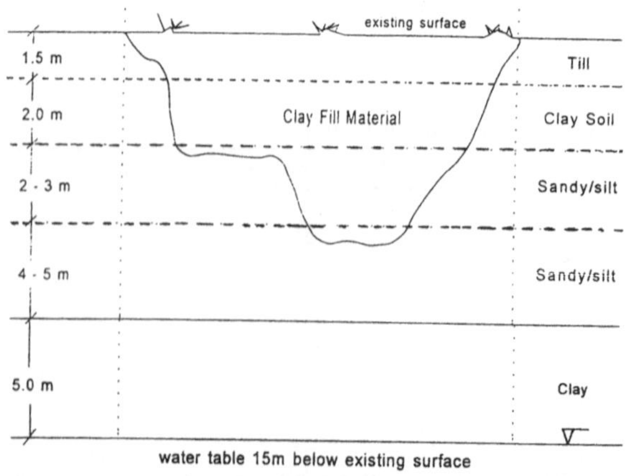


**Figure 3.**

**Table V. Estimated Degradation Time in Soil**

| Parameter | Minimum Time (days) | Average Time (days) | Maximum Time (days) |
|---|---|---|---|
| Benzene | 33 | 65 | 105 |
| Toluene | 14 | 42 | 79 |
| Ethylbenzene | 9 | 17 | 29 |
| Xylene | 52 | 136 | 206 |
| TPH | n/a | n/a | n/a |

**Table VI. Approximated Times for Our Site to Meet Guidelines**

| Parameter | Landfarming Time (days) | Composting Time (days) | Enhanced Process Time (days) |
|---|---|---|---|
| Benzene | 80–105 | 60–70 | 50–58 |
| Toluene | 55–79 | 30–40 | 20–28 |
| Ethylbenzene | 20–29 | 22–28 | 20–26 |
| Xylene | 180–206 | 70–78 | 60–68 |
| TPH | n/a | 184–230 | 130–184 |

Degradation rates were based on a first order reaction.

$$\text{In } C/C_0 = -kt$$

where $C$ = final concentration
$C_0$ = initial concentration
$k$ = degradation rate
$t$ = time

The equation can be rearranged to solve for time.

$$t = \frac{\ln(C/C_0)}{-k}$$

Initial concentrations were obtained from site investigations, the final concentrations were taken from the guidelines, and the estimated degradation rate were all obtained from literature. Therefore, the degradation time can be calculated. Using the rates from Howard et al.,[4] the degradation time ranges for BTEX were calculated, and these are shown in Table V.

The degradation times (Table V) were used to get an approximate idea as to how long it would take for the various hydrocarbons to degrade in just the soil itself (i.e., basic landfarming). As the conditions change through the use of different bioremediation alternatives, so will the degradation rates. To estimate these degradation rates for each alternative, some case studies that were similar in nature were examined. The study included: a case study on basic composting,[5] three case studies on biopile composting,[6] and one case study on biopile/biofiltration.[7] From a review of these case studies, the degradation rate for each alternative was estimated. These estimated degradation rates were used to determine the degradation time for the contaminated site. These estimated degradation times for the site are shown in Table VI.

## BIOREMEDIATION ALTERNATIVES

In-situ bioremediation, like all biological treatment methods, involves stimulating native microbes to convert organic compounds to carbon dioxide and water. The types of waste treated by in situ processes are limited to gasolines, diesel fuel, jet fuels, and other compounds $C_{20}$ and lighter. The process is best suited to permeable soils, such as sands and gravels. In silt, degradation occurs at a slower rate, but clay is normally unsuitable for treatment. It is extremely useful when the excavation of the soils would be difficult or costly.

Costs are dependent on the hydrogeology of the site, and length of treatment required to meet the objectives. Installation and equipment will cost $40,000 to $200,000 for small sites. Operation, maintenance, and monitoring costs range between $15,000 and $70, 000 per year. This translates to overall costs between $25 and $75 per tonne.

## Landfarming

### *Process Description*

A landfarm is constructed to decompose and immobilize contaminants within the zone of incorporation and the underlying layers, but without polluting the groundwater, subsoils, or nearby watercourses. Landspreading is the application of wastes to the upper soil zones for decomposition and incorporation in to the soil.[8] The soil is applied evenly over the surface of an engineered site and allowed to dry, or injected just below the surface. After the waste has achieved a similar moisture content as the native soil, it is tilled into the top 15 cm of soil. It is also tilled at future dates to promote aeration and mixing of any additives. The first application of waste should be less than the anticipated application rate to allow the microbes to acclimatize themselves to the new substrate.[8]

Landfarming has proved successful with a wide range of hydrocarbons, from gasoline to heating oils. Heavier oils are usually treated by composting because the potential for emission of noxious odors from biodegradation is eliminated. Petroleum sludges are generally applied in 2 to 4 applications per year, perhaps more in favorable climates, up to a maximum of 50 $kg/m^2$ or 500 $m^3/ha$.[9] The biological processes perform best at natural pH levels, between 6.0 and 7.5. The process is of a degradable nature; therefore, the contaminants are completely removed. Perhaps the greatest advantage of landspreading is that a minimal amount of equipment is required for its operation and maintenance. Although the effectiveness is dependent primarily on climatic conditions, it is also dependent on moisture content, temperature, oxygen level, nutrient availability, and the contaminants present. Despite these parameters, removal efficiencies should be between 68% and 99%, excluding volatilization losses.

### *Treatment Rates and Costs*

In southern Canada, treatment times should be 1–3 years of treatment, consisting of a 3–4 month treatment season, depending, of course, on the type of contaminant and climate. Average costs for the operation range from $30 to $125 per tonne, depending on the amount of enhancement provided.

## Composting

Waste decomposition is achieved at high temperatures which result from the high biological activity occurring from the degradation process. Composting has been used for centuries to degrade organic solids, and has proved effective for treatment of highly contaminated soils. The microbes are primarily of a thermophilic origin, and the process occurs above 55°C. For hazardous wastes, mesophilic microbes may suffice. This would make the process easier to implement in northern climates. The waste and soils are placed in piles 1 to 2 meters high, and aerated by mechanically turning the pile (windrows) or by forcing air through a system of perforated pipes (static piles). Bulking agents are added in either case to increase the void space which facilitates mixing requirements and oxygen transfer. A windrow operation consists of waste and organics piled in long rows. As aeration of the pile is essential, each row is mechanically turned 1 to 7 times per week. The turning aerates the pile, and releases excess heat and volatiles. Static piles are similar to windrows, but the piles are not turned. Instead, the piles are built on top of a system of perforated pipes, and air is either forced through the system, or removed using negative pressure. An enclosed reactor system is a contained operation that is initially optimized for future batches. Because the system is contained, the emissions can easily be controlled by collecting the air in the headspace above the compost. The reactor can be in a batch or continuous mode, and can be run with or without agitation.

Most hazardous wastes do not contain enough organic material to sustain efficient composting. Usually, this is overcome by adding or mixing highly degradable solids. The biodegradable solids act as both bulking agents and a thermal source. The bulking agent creates a void space for air passage. The thermal production is from exothermic biological reactions. Typically, two materials are used to accomplish this. An inert material is recommended for the bulking agent, as it can be recycled and it therefore can save costs. The thermal source must not be a preferential carbon source for the microorganisms, but it may serve as a co-metabolite. Because of the elevated temperatures, composting is a fairly quick process. Treatment goals can usually be reached in less than 100 days. Due to the operation and maintenance of a compost pile, the initial investment is higher than other options. However, the ongoing expenses are significantly less than other processes. Over the long term, composting has been shown to be a much cheaper option.

Composting will remove greater than 70% of the total petroleum hydrocarbons present. Windrow treatment will be completed in less than 90 days, while treatment in static piles can be completed in under 60 days. In an enclosed batch reactor, after optimization, the composting process can be completed in as little as three days. A windrow operation may range in costs between $20 and $40 per tonne, while a static pile process should range between $60 and $120 per tonne of soil treated.

*Accelerated Composting*

Accelerated composting occurs in an aerated heap, or biopile. This is becoming increasingly popular for large volumes of soils containing low concentrations of organic wastes. In this process, the soil is moved to a treatment area equipped with a lower liner, berms, a system of perforated pipes, and blowers and/or vacuum pumps. The soil is piled into static heaps 3 to 4 meters high. Bulking agents are often added to improve air flow, and humidified air is forced through the pile. The entire operation is enclosed within a second liner to control emissions, either by recycling the air, or by mechanisms at the vent. By recycling the humidified air, the water is conserved, and there is little need to add more water. This can be advantageous where water is expensive or unavailable .

In advanced accelerated composting, a biofilter can be added at the vent to degrade any volatiles that are emitted. The gases are passed through a reactor where the desired microorganisms are contained on a filter material. The filter material may vary from soil beds to engineering material, but often contains an organic component such as peat or compost. The filter has a high porosity, and control of water is critical. For this reason, the biofilter is almost always preceded by a humidifier/dehumidifier. The other major design criterion is the gas retention time. There are now commercially available biofilter systems.

Biopiles and biofiltration systems are compatible with sand and gravel soils, but clay and silt should be mixed with a more permeable material to facilitate air flow. Due to the time required to complete the treatment, PAHs are usually degraded. The process has no emissions and any residue is left only on the impermeable liner, so it is considered environmentally friendly. Only a small plot size is required for the soil, as it can be piled relatively high. The process can treat large amounts of soil, with an upscaling of the operation.

*Effectiveness*

Early tests have shown a reduction in total petroleum hydrocarbons of greater than 90%. Controlled tests have reduced 80% to 90% of the monocyclic aromatic hydrocarbons (essentially benzene, toluene, ethylbenzene, and xylene). Overall removal rates are achieved in 7 weeks for light hydrocarbons ($C_{20}$ and lighter), and 20 weeks for heavier hydrocarbons. Initial cost estimates are between $50 and $90 per cubic meter. This makes it one of the least expensive bioremediation technologies available.

## Integrated Biopile/Biofiltration

The combination of a biopile and biofiltration processes is a relatively new approach to bioremediation. It was developed to give an effective, innovative, and integrated approach for the bioremediation of petroleum contaminated soils. The typical integrated biopile/biofiltration system (shown in Figure 4) consists of:

- an asphalt or high-density membrane pad on which contaminated soils are stockpiled;

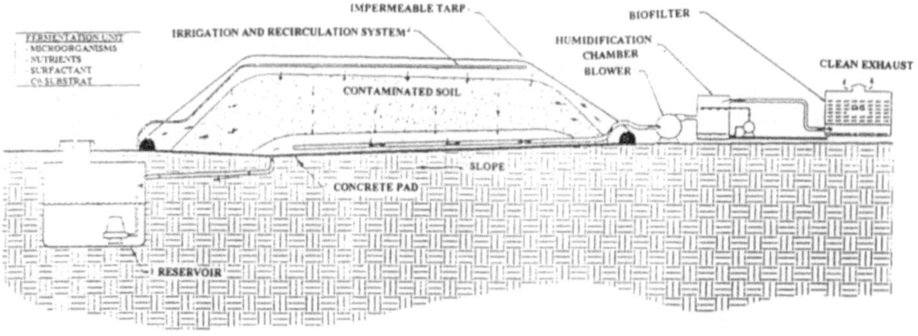


**Figure 4.**

- an overhead spray irrigation and recirculation system for optimizing soil moisture and adding nutrients to soils;
- an underdrain connected to a reservoir to collect leachate;
- a pump to draw air or push air into the soil pile and to supply the soil with oxygen;
- a fermentation unit which holds nutrients, microorganisms, and any leachate produced;
- an impermeable sheeting/tarp which covers the pile for air and water control during treatment; and a biofiltration unit which consists of a humidifier and a bed of biologically active material (biofilter) to treat the air withdrawn from the pile containing volatile hydrocarbons.

The petroleum products are removed from the soils by biodegradation and volatilization in the biopile process. The biofiltration process then takes any air emissions from the biopile containing volatile hydrocarbons and treats them through the use of a bed of biologically active material (biofilter). From the biofilter the air emissions are then released into the atmosphere. This process continues until the contaminated soil is remediated to acceptable levels.

The integrated biopile/biofiltration process is a relatively new treatment option for hydrocarbon contaminated soils. There have been a number of successful studies that have proved that this is a viable treatment option. The biopile/biofiltration process is applicable to such target contaminants as gasolines, jet fuels, heavy oils (including crude oils), lube oils, and tank bottom sludge. The biopile/biofiltration can treat a wide variety of soil types ranging from sandy soils to heavy clay and sludge.

*Effectiveness*

The effectiveness of this process depends on the type of soil, contaminant properties, and the following factors:

- total aerobic heterotrophic microorganisms: greater than $10^6$ colony forming units (CFU)- per gram
- hydrocarbon-degrading microorganisms greater than $10^4$ CFU/gram
- pH of 6 to 8
- temperature of 20°C to 40°C
- available nitrogen: greater than 50 mg/kg
- available phosphorus: greater than 30 mg/kg
- oxygen: greater than 4% by volume in gaseous phase and moisture: 15% to 25% for soils; 50% to 60% for compost filters.[10]

It has been reported that:

- various petroleum products have been remediated and biodegradation rates have been as high as 4400 $g/m^3$ of soil per day.
- removal efficiencies are greater than 80% in mineral oil and grease and more than 95% in monocyclic aromatic hydrocarbons.[10]
- data collected during several years of a biofiltration operation indicate a removal efficiency of 80% to 90% volatile hydrocarbons is achievable at surface loads of up to 100 $m^3/m^2/h$. The removal rates for volatile hydrocarbons range from 10 to 60 $g/m^3$ of filter per hour.[10]

## Treatment Rates and Costs

Data has been collected from a number of projects involving this type of treatment process. From these projects it has been determined that gasoline contaminated soils have been remediated within 5 to 7 weeks and diesel contaminated soils within 20 weeks. These numbers can vary depending on the type of soil and waste type. The costs of the biopile/biofiltration process depends on treatment time and volume of contaminated soil. The costs range from $50.00 to $90.00 per cubic meter of soil.

## CONCLUSIONS AND RECOMMENDATIONS

A study was performed to determine all possible bioremediation alternatives. From the site data and laboratory test results, a short list of alternatives was considered. There were several processes available to the company varying in both cost and performance. It was readily determined that bioremediation was a feasible alternative for the company. Due to the restraints placed on the project, and for the best performance at the least cost, biopile composting was recommended. It can be constructed on the north side of the property, with estimated dimensions of 40 m by 60 (Figure 5). The design would include an impermeable pad, an airtight cover, a water and nutrient recirculation system, and an air blower for air venting and purging (Figure 6). With proper monitoring and operation, the remediation objectives will be accomplished well within the two year time frame.

After completion of the task, the system would be dismantled. The pit was proposed to be lined with bentonite clay to minimize leaching from the remediated soil. After the soil had been placed in the pit, a separate bentonite clay liner would be installed over the top to promote drainage away from the pit, and prevent the collection of water. The estimated cost for the construction and operation of the process was between $450,000 and $650,000.

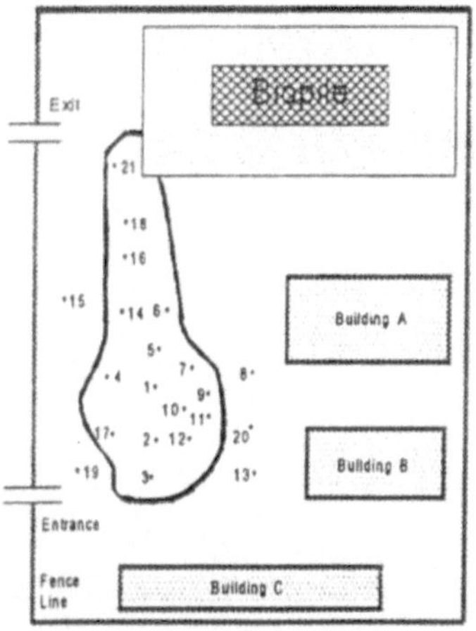


**Figure 5.**

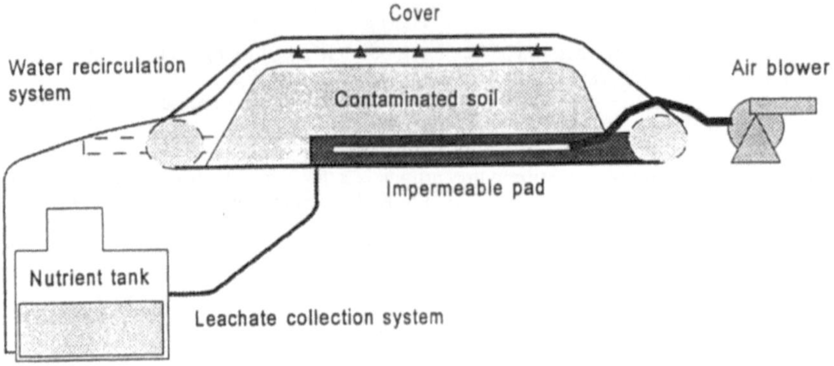


**Figure 6.**

## REFERENCES


1. Leahy, M.C., and Brown, R.A. "Bioremediation: optimizing results." *Chem. Eng.*, 101(5), pp. 108–116, 1994.
2. SERM. Risk Based Corrective Actions for Petroleum Contaminated Sites. Regina, Saskatchewan, Canada, 1995.
3. CCME. Interim Canadian Environmental Quality Criteria for Contaminated Sites. Report CCME EPC-CS34, Canadian Council of Ministers of the Environment, Winnipeg, Manitoba, 1991.
4. Howard, P.H., Boethling, R.S., Jarvis, W.F., Meylan, W.M., and Michalenko, E.M. *Handbook of Environmental Degradation Rates.* Lewis Publishers, Chelsea, MI, 1991.
5. Hardy, J.J., and Mabbula, S. "Close a site for $3000." *Soils*, 1994.
6. Hildebrandt, W.W., and Wilson, S.B. "The use of modern onsite bioremediation systems to reduce crude oil contamination on oilfield properties." *J. Petroleum Technol.* January, 1991.
7. Cyr, B., Sansregret, J. L., Pouliot, Y., Faucher, C., and Lei, J. Integrated Biopile and Biofiltration Process for the Remediation of Hydrocarbon Contaminated Soils. Biogenie Inc., Sainte-Foy, Province of Quebec, 1993.
8. PACE. Manual for Landspreading of Petroleum Industry Sludges. Report No. 81-5B, Petroleum Association for the Conservation of the Canadian Environment, Calgary, Alberta, 1981.
9. Rowell, M.J., Ashworth, J., and Qureshi, A. A. Bioremediation of Petroleum Hydrocarbons in Soils Environments. Norwest Soil Research Limited, Edmonton, Alberta, 1992.
10. Lei, J., Sangregret, J.L., and Cyr, B. "Biopiles combined for soil cleanup." *Pollut. Eng.* June, 1994.

# 43 OIL-BASED LUBRICANT/COOLANT TREATMENT BY HIGH-SHEAR ROTARY ULTRAFILTRATION

**Brian E. Reed**, Associate Professor
**Wei Lin**, Res. Assistant Professor
**Roger Viadero, Jr.**, Ph.D Candidate
Dept. of Civil and Environmental Engineering
West Virginia University
Morgantown, West Virginia 26506-6103

## ABSTRACT

The SpinTek high-shear rotary ultrafiltration system was effective in concentrating oily feeds from about 5% oil to as high as 75% while maintaining a high permeate flux. The decoupling of the turbulence promotion (used to reduce concentration polarization) from feed pressurization by rotating the membrane was the primary reason for the increase in system performance over that commonly observed with conventional cross-flow membrane systems.

## INTRODUCTION

Concentrated (>5%) oily wastes are produced by various metal-working operations (rolling mills, metal cutting and working). For many companies, the concentrated oily waste is mixed with other oil or nonoil waste streams and then treated with conventional cross-flow membrane systems (e.g., tubular, spiral wound, etc.). The residual that results from conventional membrane systems is usually treated further by physical-chemical methods (e.g., chemical addition-settling, heat cracking) prior to disposal or recycling.

The ability of conventional cross-flow membrane technology to treat a concentrated oily waste directly or the residual from conventional membrane systems is limited because of the low permeate flux observed at high oil concentrations. Conventional cross-flow systems rely on high recirculating velocities (≈ 10 ft/s) to scour or clean the membrane surface so that a satisfactory permeate flux is maintained. As the concentration of the feed increases, the maintenance of a high velocity is difficult because of the increase in feed viscosity. The decoupling of the cleaning action from feed recirculation/pressurization can be accomplished by moving the membrane surface. For example, SpinTek (Huntington Beach, CA) has developed a high-shear rotary cross-flow membrane system in which membrane disks are rotated at high speeds. Membrane rotation provides the turbulence required to clean the membrane while the pump is only required to provide transmembrane pressure and a small amount of recirculation.

The objective of this research was to assess the efficacy of using the SpinTek high-shear rotary cross-flow ultrafiltration system for the direct treatment of rolling mill coolants/lubricants. Two types of membrane materials, ceramic and polymeric, and two rolling mill coolants/lubricants were investigated. System performance was measured using the following parameters: (1) permeate flux, (2) permeate quality, (3) membrane cleaning and durability.

## BACKGROUND

Ultrafiltration is a pressure-driven membrane technique that uses porous membranes for the separation of material in the 1 nm–10 μm size range or compounds with molecular weights in excess of 5000. Colloidal material, macromolecules, and micelles are examples of items that can be

*52nd Purdue Industrial Waste Conference Proceedings*, 1997, Ann Arbor Press, Chelsea, Michigan 48118. Printed in U.S.A.

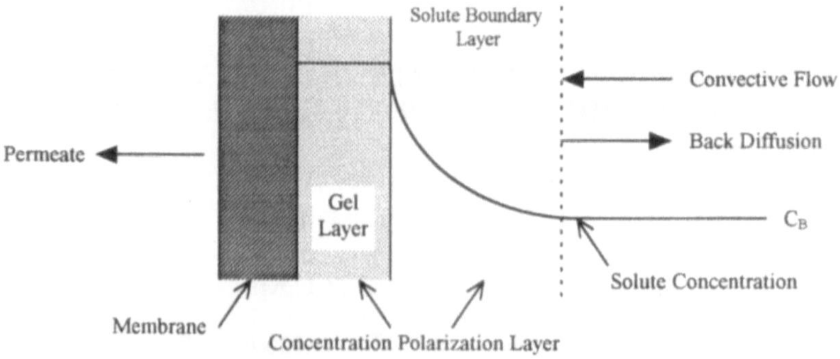


**Figure 1. Concentration polarization at the membrane surface.**

fractionated. "Clean" water (permeate) is forced through the porous membrane while the solute is retained by the membrane, concentrating the feed with time. Numerous researchers have reported on UF's effectiveness in treating oil/grease wastewaters. UF reduced the volume of a waste-cutting oil emulsion by 95%–98% and concentrated oil and solids as much as 60%.[1] Bodzek and Konieczny[2] reported oil reductions of 95 to 99% and COD reductions of 91 to 98% in the UF permeate from a metal industry emulsion. Zaidi *et al.*[3] reported that the oil content of several oilfield brines was reduced to less than 20 mg/L and the short-term permeate flux was about 80 gal/ft$^2$-d. Cheryan[4] summarized several case studies in which UF was demonstrated to be an effective treatment method for oily wastes.

In all membrane processes, a solute boundary layer (dynamic membrane) will form at the membrane surface due to convective mass transport. This phenomenon is referred to as "concentration polarization" and is shown schematically in Figure 1.[4] Concentration polarization is one reason why the permeate flux for a waste is lower than the clean water flux (CWF). The buildup of the solute at the membrane surface is reduced by back diffusion of the solute and the thickness of the boundary layer is reduced by providing turbulence at the membrane surface. If the solute concentration at the membrane is high enough, a gel layer can form. Additionally, solute can be adsorbed or be forced into the membrane pores, resulting in membrane fouling.

A schematic of the SpinTek high-shear rotary cross-flow ultrafiltration system (referred to as ST-II system hereafter) is presented in Figure 2. The ST-II system uses a series of flat, round membrane disks set on a hollow rotating shaft inside a cylindrical housing (only one disk shown in Figure 2). The fluid stream enters the membrane chamber under pressure and is distributed across the membrane surface. Permeate is forced through the membrane and is collected in the hollow shaft and is discharged. The concentrate exits at the edge of the membrane packs. To reduce concentration polarization in conventional UF systems, a large portion (≈98%) of the con-

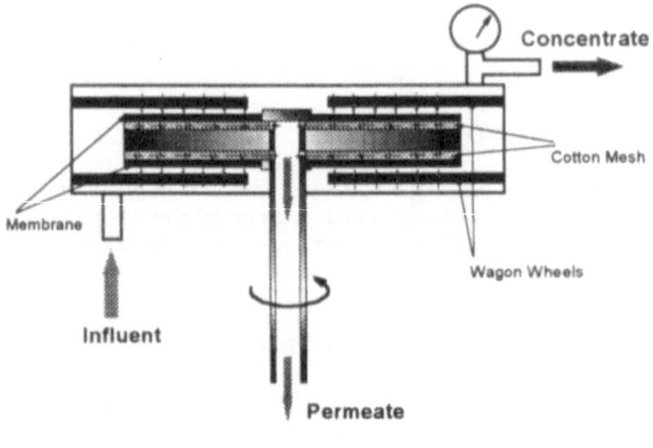


**Figure 2. Schematics of ST-II System (only one disk shown).**

centrate is recycled back to the membrane unit, producing large liquid velocities near the membrane surface. The large velocities increase turbulence which reduces the thickness of the solute boundary layer. In the ST-II system, the rotation of the membrane disk produces the required turbulence. In addition to the rotational action, turbulence promoters ("wagon wheels") are located on each side of the membrane to prevent rotational flow (vortex formation) from occurring. In conventional UF systems, maximum liquid velocities of about 15 ft/s are possible, while with the ST-II system liquid, velocities of 60 ft/s are typical. As the feed thickens with treatment time, a conventional UF system is not able to maintain the high cross-flow velocities because of the difficulty in pumping viscous material at high flow rates. Because the ST-II system does not rely on pumping to produce the required liquid velocities, extremely concentrated wastes can be treated/produced.

## MATERIALS AND METHODS

A total of eight runs were completed using two coolants/lubricants from an aluminum rolling mill operation. In the first phase, the effect of membrane type was investigated. Three runs were conducted using a polymeric membrane (PV-100K) and three were completed using a ceramic membrane. In the second phase, the two coolants/lubricants were concentrated to high oil contents using the ceramic membrane. Characteristics of the two membranes are presented in Table I. The coolants/lubricants had an initial oil content of about 5% and were taken directly from the milling process. Both coolants contained surfactants, film strength additives, and antifoaming agents but specific recipes cannot be disclosed for proprietary reasons.

A schematic of the experimental apparatus is presented in Figure 3. Prior to a run, the clean water flux (CWF) was measured using DI water under the following conditions: (1) vessel pressure = 42 psi, (2) membrane rotational speed = 1750 rpm, (3) return flow = 1.2 gal/min, and (4) feed tank temperature = 110 ± 2°F (controlled by heat exchanger). Permeate flux was measured every 5 minutes until it had stabilized for a 30-minute period.

Following the CWF determination, the system was drained and coolant was pumped from the original waste drum(s) through a 100 mesh sieve into the wastewater storage tank. A predeter-

**Table I. Characteristics of Membranes**

| Characteristic | PV-100K | Ceramic |
|---|---|---|
| Manufacturer | Advanced Membrane Technology | TRUMEN |
| Material | Polyvinyliden-fluoride | $TiO_2$ - $Al_2O_3$ |
| Area | 0.53 $ft^2$ | 0.53 $ft^2$ |
| Pore Size | 100,000 MWCO[a] | 0.11 m |
| Max. Operating Pres. | 150 psi | >150 psi |
| Max. Operating Temp. | 120°F | >160°F |

[a]Molecular weight cutoff. A MWCO of 100,000 ≈ 0.01 m.

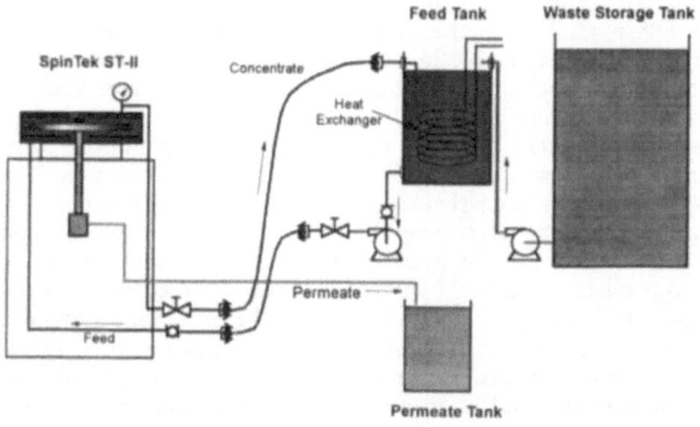


**Figure 3. Schematic of experimental set-up.**

mined volume of coolant ($V_{system}$) was then pumped from the storage tank to the feed tank. The system was operated in semibatch mode (feed tank level kept constant at $V_{system}$ by pumping waste from the storage tank) until all coolant in the storage tank had been treated. At this point, batchdown operation began (volume in the feed tank reduced). Batchdown was stopped when either the lowest feed tank volume had been reached, or the flux or permeate quality had decreased to an unacceptable level. The operational conditions for waste treatment were the same as those presented earlier for the CWF determination. Permeate flow rate and turbidity, feed tank temperature, and vessel pressure were measured frequently. The volumetrically-based concentration factor was calculated by measuring the volume of permeate produced ($V_{perm}$) and the following equations:

$$CF_{SB} = 1 + V_{perm}/V_{system} \quad (1)$$

$$CF_{BD} = CF_{SB} \times [V_{system}/(V_{system} - V_{perm})] \quad (2)$$

Where $CF_{SB}$ = concentration factor during semibatch operation, $V_{perm}$ = volume of permeate produced (gal), $V_{system}$ = volume of waste in the feed tank (gal), and $CF_{BD}$ = concentration factor during batchdown operation. At predetermined values of the CF, the permeate was sampled for O/G, TSS, COD (PV-100K runs 1 and 2 only) and pH, and analyzed according to *Standard Methods*.[5]

At the conclusion of batchdown operation, all concentrate was drained from the system and its volume measured. The system was first flushed using tap water (≈135°F) until the liquid from the return hose was clear and then was washed. For all the polymeric runs and ceramic run 1, the washing solution consisted of 181 grams of a commercial laundry detergent and ≈30 mL of a commercial dishwashing liquid in 5 gallons of hot tap water (≈135°F). The pH of the washing solution was 10.3. For ceramic runs 2 and 3, the washing solution was changed because a white substance, believed to be the detergent, had been deposited on the inner portion of the membrane. A new washing solution, consisting of 30 mL of the dishwashing liquid, pH = 10 (adjusted with NaOH) was used. Following washing, the system was flushed with tap water (110°F) and then the CWF was determined. After determining the CWF, the membrane was visually inspected. If after visual observation it was determined that the membrane was permanently fouled or the CWF was not close to the CWF taken prior to the run, the membrane was replaced.

Deviations from the above-mentioned procedure are as follows: (1) for ceramic runs 2 through 5, the system was operated in the recycle mode (permeate routed back to feed tank) at whole number CFs for an extended period of time to assess long-term performance of the system and (2) for ceramic run 3, temperature and rotational speed excursions were conducted at CF = 4X. Temperature was increased to 140°F for about 20 hours and the rotational speeds of 1000, 1250, and 1500 rpm were employed for about 5 hours each.

## RESULTS AND DISCUSSION

### Effect of Membrane Type

#### *Permeate Flux*

*Polymeric Membrane:* In Figure 4, the permeate flux and concentration factor (CF) versus time are presented for polymeric runs 1 through 3. After the temperature had stabilized at 110°F, the flux during semibatch operation was relatively stable in runs 1 and 3 but declined in run 2. In conventional cross-flow UF systems, flux decline at the start of a run is much larger than what was observed in Figure 4. For example, Lin and Reed[6] and Reed *et al.*[7] reported *that* for a 0.5% oil waste, the flux declined by as much as 200% from its initial value to the flux observed several hours into the run. The decline in flux was attributed to concentration polarization. While concentration polarization is also operative in high-shear systems, the rotational induced turbulence keeps the flux decline to a minimum (by decreasing the thickness of the solute boundary layer) despite the waste having an oil content that was an order of magnitude larger (5% compared to 0.5%). The highest flux was observed in run 2 and could be the cause of the most obvious flux decline. The higher the flux, the more solute is transported to the membrane surface and the larger the concentration polarization effect. The more obvious flux decline could also be due to variations in the coolants. As expected, the flux decreased sharply during batchdown operation. During batchdown, the oil content of the feed (as measured by the CF) increased rapidly.

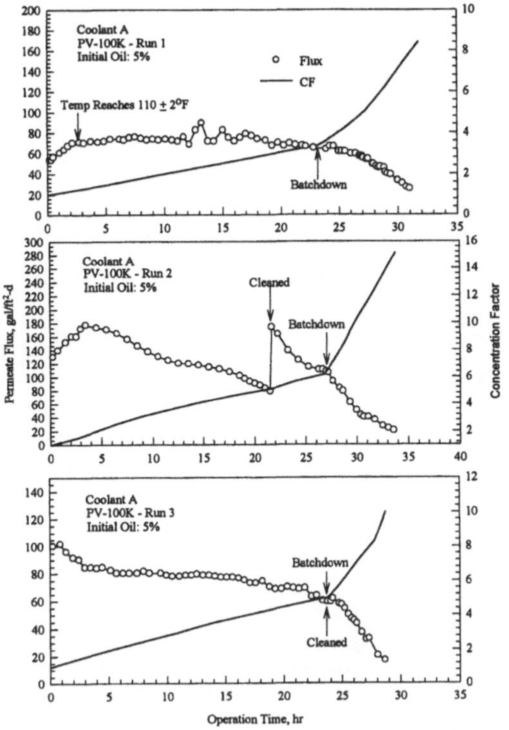


**Figure 4. Permeate flux and concentration factor (CF) vs. operation time for the PV-100K membrane.**

For run 2, the system was cleaned at a CF of 5.14 (21.5 hr) because of high turbidity levels in the permeate. The flux increased after cleaning (but at the same CF) to approximately the value that was observed at the beginning of the run (≈ 170 gal/ft$^2$-d) but then decreased sharply. This indicates that the flux decline was due in part to membrane fouling in addition to concentration polarization. In contrast, for PV-100K run 3, the system was cleaned at a CF = 5 (just before batchdown) but the flux before and after cleaning was the same. A constant flux before/after cleaning indicates that concentration polarization is primarily responsible for the flux decline. The fouling of the membrane during run 2 may have been bacterial in nature. The company has reported the presence of a large population of bacteria that degrade the emulsifying additives. As the emulsifier concentration is lowered, the oil droplet size increases and the emulsion is more likely to "break" (free oil formation). Free oil formation on the PV-100K membrane surface may be the cause of the decrease in permeate quality at high CFs. The degradation of the emulsifying package could also explain the higher flux observed in PV-100K run 2. As the oil droplet size increases, a larger flux can result because the oil droplets are packed less densely at the membrane surface (resistance of the dynamic membrane is reduced). This phenomenon has been reported by other researchers.[8]

*Ceramic Membrane:* In Figure 5, the permeate flux and concentration factor (CF) versus operation time are presented for ceramic runs 1 through 3. For runs 2 and 3, the system was operated in the recycle mode at whole number values of CF to determine if the flux was temporal dependent at a given CF. In run 2 (coolant A), the flux decreased during changes in CF but then increased during the subsequent recycle operation to about one-half or more of the value observed at the previous CF. For run 3 (coolant B) this phenomenon was not as apparent and could be due to the differences between the coolants. Membrane cleaning during the run was not required for any of the runs.

In a full-scale system, $V_{system}$ is large (10,000–40,000 gal) and the system is operated for long periods of time before there is a large change in the CF. For coolant A, if the predesign treatability study was conducted without recycling the permeate at predetermined CFs (as was done in

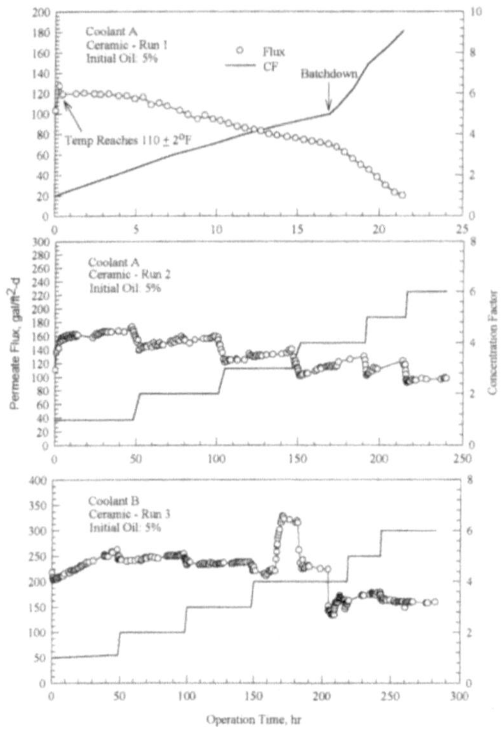


**Figure 5. Permeate flux and concentration factor (CF) vs. operation time for the ceramic membrane.**

ceramic run 1), the design flux would be underestimated. Cheryan[4] has suggested that the flux versus CF relationship be determined using discrete membrane tests (i.e., make up samples of the waste already at the desired CF and then run the membrane system in the recycle mode). After each discrete test, the system would be cleaned and the clean water flux measured to ascertain if fouling has occurred. While this approach is useful when dealing with synthetic wastes, it is not feasible when actual wastes are studied.

The effect of temperature and rotational speed on permeate flux was assessed during ceramic run 3. In Figure 6, permeate flux versus operational time for the temperature and rotational excursion portion of Run 3 is presented. At 166 hours and CF = 4X, the feed tank temperature was increased from 110°F to 140°F resulting in a flux increase from 220 to 319 gal/ft$^2$-d. The viscosity of water, which decreases by 24% over the temperature range investigated, accounts for only a portion of the flux increase. The remaining increase in flux is due to changes in the viscosity, density, and diffusivity of the oil droplets. Coolant B's viscosity-temperature relationship was determined at a CF = 4X (data cannot be presented for proprietary reasons). The coolant viscosity decreased more sharply with temperature compared to what is predicted for water and this is believed to be the primary reason for the 45% increase in flux with temperature. Because the flux-temperature relationship is so strong, a full-scale system should be operated at the highest temperature possible. Thus, the ceramic membrane, which has an maximum operating temperature of 160°F, has a distinct advantage over the polymeric membrane (max. temp. of 120°F).

At 204 hr, rotational speed (ω) excursions were started. The flux decreased from 230 to 140 gal/ft$^2$-d as the rotational speed was decreased from 1750 to 1000 rpm. When the rotational speed was increased from 1000 to 1500 rpm and then back to 1750 rpm, the flux did not fully recover to the prerotational excursion value. Lipp et al.[8] reported the gel layer, once formed, may be sufficiently stable to withstand increases in turbulence. It is hypothesized that this phenomenon was operative in the later portion of ceramic run 3. Until additional research is conducted on the J-ω relationship, the highest possible rotational speed should be used in actual operations. Permeate water quality (turbidity, O/G and TSS) was not affected by changes in temperature or rotational speed.

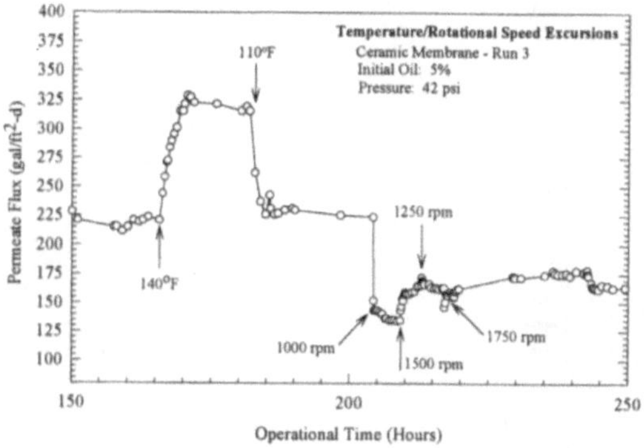


**Figure 6. Permeate flux vs. operation time for temperature/rotational excursions.**

*Permeate Quality*

In Table II, a summary of the permeate quality results for both membranes during semibatch and batchdown operation and for the entire run are presented. Results from each membrane will be discussed separately.

*Polymeric Membrane:* In Figure 7, permeate turbidity and O/G versus operation time are presented for PV-100K runs 1 through 3. The permeate TSS followed the same trend as the turbidity and O/G thus, TSS data will not be presented in graphical form. For all runs, the permeate tur-

**Table II. Summary of Permeate Water Quality Results**

| Parameter | Average ± std | | | Removal, % |
|---|---|---|---|---|
| | Entire Run | Semibatch | Batchdown | |
| ***Turbidity, NTU*** | | | | |
| PV-100K | | | | |
| Run 1 | 81 ± 182 | 20 ± 9.3 | 175 ± 267 | — |
| Run 2 | 210 ± 313 | 90 ± 80 | 597 ± 459 | |
| Run 3 | 147 ± 355 | 4.3 ± 6.1 | 595 ± 518 | |
| Ceramic | | | | |
| Run 1 | 81 ± 181 | 8.0 ± 6.1 | 355 ± 255 | — |
| Run 2 | 1.21 ± 0.9 | 1.25 ± 0.96 | 0.99 ± 0.38 | |
| Run 3 | 0.23 ± 0.09 | 0.19 ± 0.04 | 0.27 ± 0.07 | |
| ***O/G, mg/L*** | | | | |
| PV-100K | | | | |
| Run 1 | 221 ± 300 | 98 ± 10 | 289 ± 363 | 98.8 |
| Run 2 | 832 ± 1,172 | 426 ± 494 | 1,481 ± 1,684 | 94.8 |
| Run 3 | 945 ± 2,632 | 89 ± 11 | 2,310 ± 4,132 | 99.6 |
| Ceramic | | | | |
| Run 1 | 187 ± 107 | 121 ± 13 | 252 ± 121 | 99.4 |
| Run 2 | 37 ± 5 | 40 ± 3 | 32 ± 0 | 99.9 |
| Run 3 | 20 ± 15 | 11 ± 10 | 33 ± 2.6 | 99.9 |
| ***TSS, mg/L*** | | | | |
| PV-100K | | | | |
| Run 1 | 56 ± 63 | 53 ± 28 | 57 ± 74 | 99.3 |
| Run 2 | 120 ± 111 | 82 ± 107 | 161 ± 107 | 99.0 |
| Run 3 | 59 ± 114 | 3.4 ± 3.4 | 147 ± 151 | 99.5 |
| Ceramic | | | | |
| Run 1 | 12 ± 15 | 3 ± 2.5 | 23 ± 17 | 99.5 |
| Run 2 | 8.8 ± 8.3 | 7.7 ± 6.9 | 15 ± 11 | 99.7 |
| Run 3 | < 1 | 1.2 ± 1.4 | < 1 | 99.7 |

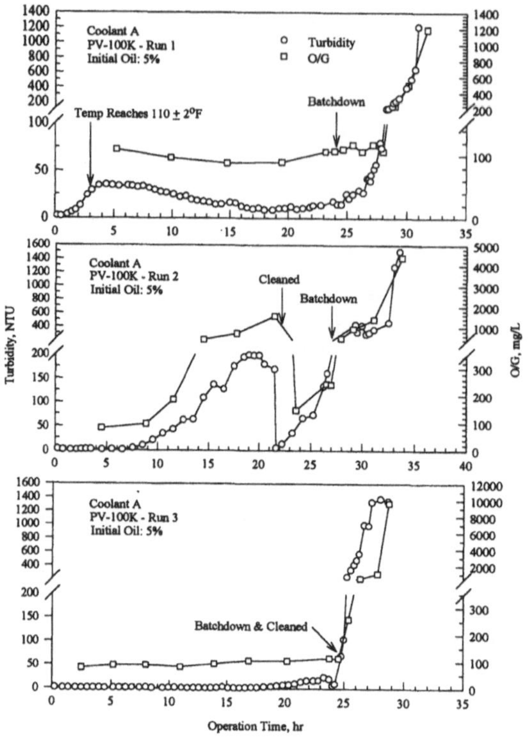


**Figure 7. Permeate turbidity and O/G vs. operation time for the PV-100K membrane runs 1, 2 and 3.**

bidity, O/G, and TSS increased with operation time with the largest increase taking place during batchdown operation. In run 1, the turbidity decreased from about 5 hr to 20 hours and could be due to the formation of the solute boundary layer that acted as a second membrane.

In run 2, the system was cleaned at 21.5 hr; however, turbidity and O/G were lowered only for a short time, after which both rose sharply. Permeate COD was measured for runs 1 and 2 and followed the same trends as permeate turbidity, O/G and TSS. Initially, the permeate COD was approximately 3,000 to 3,500 mg/L and was a measure of the low molecular weight additives (free surfactant, antifoaming agents, biocides, etc.) that were not retained by the membrane. As the feed thickened, the COD increased to as high as ≈ 15,000 mg/L. The high COD is attributed to oil droplets that were "squeezed" through the membrane pores. For all polymeric runs, the membrane was not effectively cleaned at the end of the run and a new membrane was used for each run. This could be due to the ineffectiveness of the membrane cleaning solution/process or irreversible fouling of the membrane (i.e., no cleaning solution/process would be effective). For all runs there was a good correlation between permeate turbidity, O/G, and TSS; therefore, turbidity may be used as a real-time indicator of permeate quality.

*Ceramic Membrane:* In Figure 8, permeate turbidity and O/G versus operational time are presented for ceramic membrane runs 1 through 3. In run 1, the turbidity and O/G followed the same trend as observed with the polymeric membranes. However, in Runs 2 and 3, the turbidity was very low throughout the run and the O/G was always below 45 mg/L. In over nine additional runs on a variety of oily wastes, the same ceramic membrane always produced permeate that was similar in quality to that observed in runs 2 and 3. It is not clear why the membrane performed better in runs 2 and 3 (and in other subsequent runs) compared to run 1. The virgin ceramic membrane may have been compressed in run 1, producing a tighter membrane for later runs. The physical/chemical nature of the membrane may also have been altered, for the better, by additives found in the coolant or in the cleaning solution. In future research, the physical/chemical nature (e.g., oil and water contact angles and surface tensions) of the membrane should be determined to examine the effect of the waste on membrane behavior. Differences in the coolants

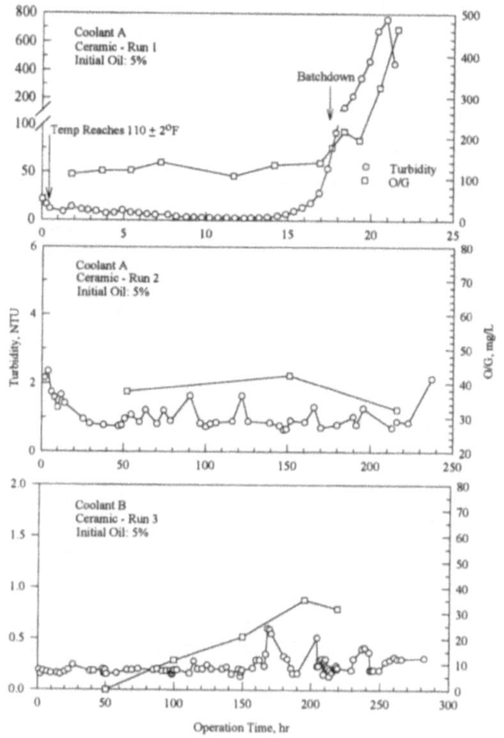


**Figure 8. Permeate turbidity and O/G vs. operation time for the ceramic membrane, runs 1, 2 and 3.**

could also account for the difference in permeate quality. However, as was mentioned previously, in numerous subsequent runs, the permeate quality (at comparable CFs) was of similar quality to that observed in runs 2 and 3. For runs 2 and 3 the system was operated for an extended period of time (between 250 and 300 hours). For the entire duration of these runs, the permeate quality was excellent and long-term operation of the system should not significantly affect permeate quality.

## Comparison of Polymeric and Ceramic Membrane Fluxes

In Table III, a summary of permeate flux results are presented for the six runs. A similar table was presented earlier for permeate quality parameters (Table II). The ceramic membrane had a larger flux compared to the PV-100K membrane. This was most likely due to the larger pore size of the ceramic membrane [0.1 μm compared to 100,000 MWCO (about 0.01 μm)]. However, the use of a membrane with a larger pore diameter does not necessarily guarantee a higher flux. If the solute particle is similar in size to the membrane pore, the solute can plug the pore opening. In some cases, the flux can be controlled by the resistance of the dynamic membrane (i.e., solute boundary layer) and not by the resistance of the membrane.

For the polymeric membrane, a new membrane was required for each run. There was a noticeable layer of oil on the polymeric membranes in all runs, while for runs 2 and 3, the clean water flux was never restored to near its original value. In contrast, the same ceramic membrane was used for all runs; however, the washing solution was altered (no detergent) between runs 1 and 2. During ceramic run 1 cleaning, the CWF was high but eventually decreased to zero. The membrane vessel was opened and a whitish deposit, believed to be the commercial detergent, was observed on the membrane. After removing the deposit and sponging the membrane with ≈ 1% sulfuric acid solution, the clean water flux was restored above its virgin value.

**Table III. Summary of Permeate Flux Results**

| | Average Flux[a] ± std, gal/ft²-d | | | CWF,[b] gal/ft²-d | |
|---|---|---|---|---|---|
| Run ID | Entire Run | Semibatch | Batchdown | Before | After |
| *PV-100K* | | | | | |
| Run 1 | 64 ± 14 | 73 ± 4.8 | 50 ± 12 | 118 | 118 |
| Run 2 | 104 ± 44 | 128 ± 31 | 61 ± 31 | 589 | 172 |
| Run 3 | 69 ± 18 | 78 ± 8.5 | 45 ± 14 | 718 | 115 |
| *Ceramic* | | | | | |
| Run 1 | 87 ± 30 | 99 ± 18 | 61 ± 17 | 320 | 159 |
| Run 2 | 135 ± 23 | 142 ± 19 | 103 ± 9 | 513 | 431 |
| Run 3 | 205 ± 11 | 235 ± 13 | 165 ± 8 | 431 | 481 |

[a]At 1102°F.
[b]Clean Water Flux.

## High Concentration Factor Experiments

In Figure 9, permeate flux and concentration factor versus operational time are presented for coolants A (run 4) and B (run 5). Concentration factors of 14 and 15 were reached for coolants A and B, respectively. Based on an initial oil content of 5%, this is equivalent to a final concentrate oil content of 70 and 75%. During batchdown the transmembrane pressure was difficult to control due to the high viscosity of the concentrate. The decrease in transmembrane pressure with increasing CF is not critical, as the permeate flux is independent of pressure at high feed oil concentrations.[4] For coolant A, the average flux decreased from 179 gal/ft²-d at CF = 1X to 31

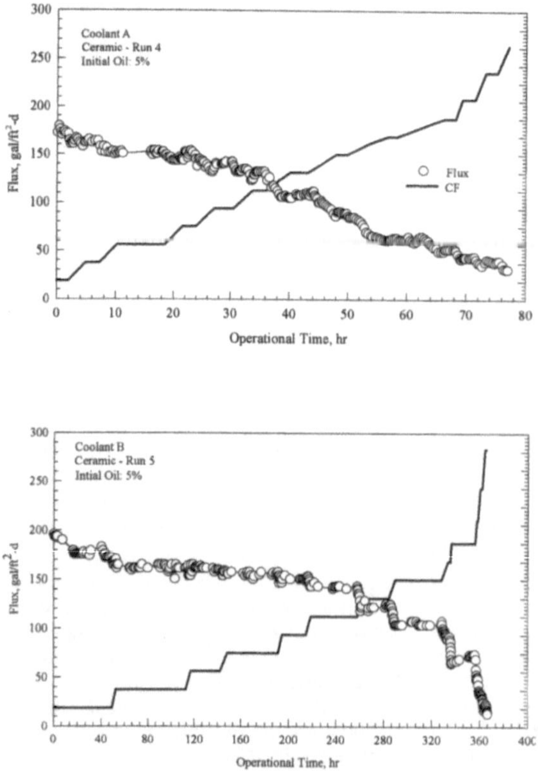


**Figure 9. Permeate flux and CF vs. operation time for the high CF experiments (ceramic runs 4 and 5).**

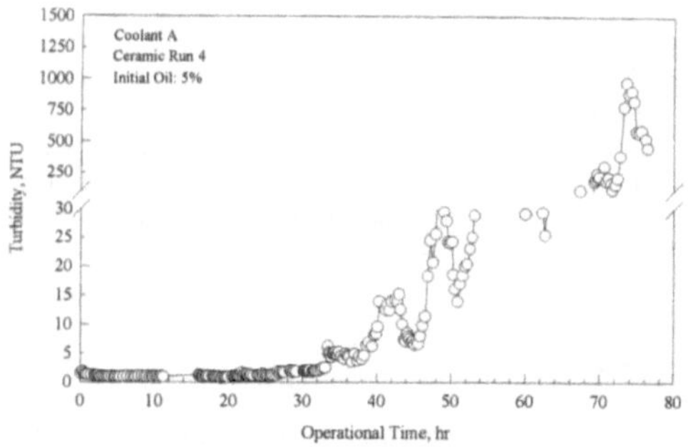


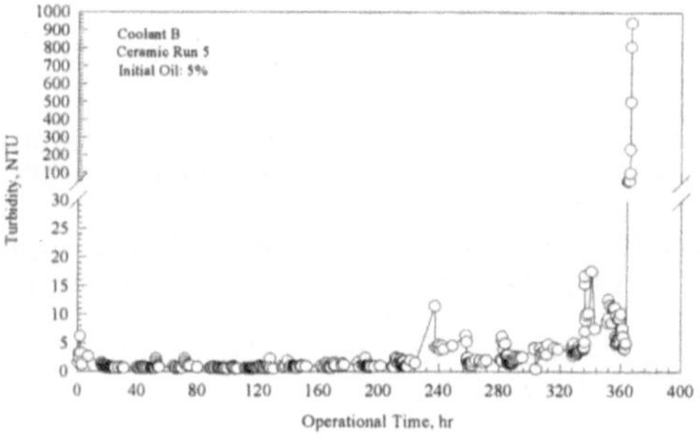


**Figure 10. Permeate turbidity vs. operation time for the high CF experiments.**

gal/ft$^2$-d at CF = 14X. The average flux over the duration of the run was 110 gal/ft$^2$-d. For coolant B, the average flux decreased from 196 gal/ft$^2$-d at CF = 1X to 19 gal/ft$^2$-d at CF = 15X ($\approx$ 27 gal/ft$^2$-d at CF = 14X). The average flux over the duration of the run was 130 gal/ft$^2$-d.

In Figure 10, permeate turbidity versus operation time are presented for ceramic runs 4 and 5. Turbidity was low during the initial phases of experiment when the oil content in the feed tank was relatively low. As the feed was concentrated, turbidity increased, especially during the latter portion of batchdown. For coolant A, the feed tank temperature was difficult to control because of the low volume of liquid in the feed tank. The increase in feed temperature (as high as 138°F) correlated directly with the large increase in turbidity. The higher temperatures, coupled with the large oil content, allowed the passage of emulsified oil droplets through the membrane. As demonstrated in run 5, with better feed tank temperature control, the permeate turbidity can be kept at low levels until very high oil contents are reached. It should be noted that permeate discharge without additional treatment was not expected, given the high oil content of the wastes. Despite these difficulties, results from runs 4 and 5 demonstrate the feasibility of using the high-shear rotary UF system to treat coolants/lubricants directly or to treat residuals from conventional cross flow membrane systems.

## CONCLUSION

The SpinTek ST-II system was effective in concentrating oily wastes from about 5% to as high as 75% while maintaining a high permeate flux. The decoupling of turbulence promotion from feed pressurization/recirculation by rotating the membrane is the primary reason for the increase in system performance over that observed with conventional cross-flow ultrafiltration systems. The following specific conclusions can be forwarded: (1) The ceramic membrane was superior to the PV-100K polymeric membrane in regard to flux quantity and quality as well as membrane cleaning/durability. In addition, because of its higher operating temperature (160°F compared with 120°F for the PV-100K membrane), its use in full-scale operations is recommended; (2) Permeate flux was dependent on the feed temperature and rotational speed. Flux increased by about 45% when the temperature was increased from 110 to 140°F. A larger decrease in waste viscosity, over that predicted for water alone, was the primary reason for the stronger than expected flux-temperature relationship. The flux-rotational speed (ω) relationship was described by $J = f(\omega)^{0.90}$ and the gel layer exhibited some sustainable stability with increases in ω. Until additional research is conducted on the J-ω relationship, the highest possible rotational speed should be used in actual operations. Permeate water quality was not affected by temperature or rotational speed excursions.

## REFERENCES

1. Pinto, S.D. Ultrafiltration for Dewatering of Waste Emulsified Oils. *Proceedings, 1st International Conference,* IIT Research Institute, Chicago, Ill., 1978, pp. 1–4.
2. Bodzek, M. and K. Konieczny, K. *Waste Management*, 12, p. 75, 1992.
3. Zaidi, A., et al. "The use of micro/ultrafiltration for the removal of oil and suspended solids from oilfield brines." *Water Sci. Technol.,* 25(10), pp. 163–173, 1992.
4. Cheryan M., *Ultrafiltration Handbook,* Technomic Publishing Comp, Inc., Lancaster, PA, 1986.
5. *Standard Methods for the Examination of Water and Wastewater.* American Public Health Association, Washington, D.C., 1992.
6. Lin, W. and Reed, B.E. Comparison of Cross–Flow Ultrafiltration and Centrifugal Membrane Technology in Treating an Oil/Grease Wastewater, *Proceedings 28th Mid-Atlantic Industrial and Hazardous Waste Conference*, Buffalo, NY, 1996.
7. Reed, B.E., W. Lin, C. Dunn, P. Carriere, and G. Roark. "Treatment of an Oil/Grease Wastewater Using Ultrafiltration: Pilot-Scale Results," *Separation Sci. Technol.*, 32(9), pp. 1493–1511, 1997.
8. Lipp, P., Lee, C.H., Fane, A.G., and Fell, C.J.D. "A fundamental study of the ultrafiltration of oil-water emulsions." *J. Membrane Sci.*, 36, pp. 161–177, 1984.

# 44 A COMPARISON OF SPECTROPHOTOMETRIC AND GAS CHROMATOGRAPHIC MEASUREMENTS OF HEAVY PETROLEUM PRODUCTS IN SOIL AND AQUEOUS SAMPLES

**Farhad Nadim, Shili Liu, George E. Hoag, Jiang Ping Chen,** and **Robert J. Carley**
Environmental Research Institute, Box U-210
Storrs, Connecticut 06269

**Orest P. Polishchuk**
Dept. of Chemistry
I. Franko State University of Lviv
6 Kyrylo and Methodiy St.
290005, Lviv, Ukraine

**Ludmila V. Sidorova**
State University "Lvivska Polytechnica"
POB 7346, 11/55 Dovzendo St.
290070, Lviv, Ukraine

## INTRODUCTION

Petroleum hydrocarbons originate from liquid fossil fuels called oil, or more properly, *petroleum.* Petroleum is a naturally occurring liquid with widely different composition and complexity. Except for a few surface seepages, the vast portion of petroleum is located well below the surface of the earth and it is extracted only by drilling. Crude oil is a complex mixture of hydrocarbons. Its elemental composition is carbon-hydrogen, with variable quantities of oxygen and sulfur, and trace amounts of nitrogen, metals, and other elements. Refining of crude oil usually begins with washing with water to remove salt and other inorganic impurities, followed by distillation to separate the various fractions ranging from methane to asphalt. Fuel oil such as gasoline and diesel are separated during this procedure. Modern refineries use a procedure called "cracking" to fractionate unused cuts. In this process the material is heated to temperatures above 230°C at different pressures with or without a catalyst. Cracking converts heavy and complex hydrocarbons to more simple molecules with lower boiling points with the loss of hydrogen (J. A. Plambeck, 1995).

Being a major industrial nation, the United States uses over 250 billion gallons of oil and petroleum products each year on average. At every point in the oil production, distribution, and consumption process, oil is invariably stored in storage tanks. The potential for an oil spill is significant, and the effects of spilled oil many times pose serious threats to the environment.

Protecting the ecological system and water resources from petroleum contamination are of main concern.

## PURPOSE OF THIS STUDY

EPA method-418.1, known as TPH-IR is based upon a methodology developed to determine the oil and grease content of industrial wastewaters, sewage, and sludges. The source methodology is presently documented as Standard Method 5520 in *Standard Methods for the Examination of Water and Wastewater*, Seventeenth Edition (1989), jointly published by the American Public

*52nd Purdue Industrial Waste Conference Proceedings*, 1997, Ann Arbor Press, Chelsea, Michigan 48118. Printed in U.S.A.

Health Association, the American Water Works Association, and the Water Pollution Control Federation.

The TPH-IR method extracts "oil and grease" from a sample with 1,1,2-trichloro-1,2,2-trifluoroethane (Freon-113). The first signs of ozone depletion in the stratosphere layer that surrounds the Earth from 10–50 km above the Earth's surface were observed by scientists in the early 1970s, the reduction was not significant and many researchers thought of it as computer errors. In the early 1980s a 60% reduction of ozone was observed over the Antarctic region and raised major concern in the scientific community. A comprehensive study revealed that chlorofluorocarbons (CFCs) and halons (a chemical used to extinguish fire), were the major cause of ozone destruction in stratosphere. Finally, in the spring of 1987 a Protocol was signed in Montreal by major industrial nations that was aimed to reduce the production of CFCs by 50% by 1998. After the signing of the original Protocol, new measurements showed worse damage to the ozone layer than was originally expected. During the World Environment Conference in Rio de Janeiro, Brazil in 1992, the developed countries decided to completely end the production of halons by the beginning of 1994 and of CFCs by the beginning of 1996.

A halt on the production of CFCs would mean substituting Freon-113 with another solvent in EPA Method 418.1 or seeking another analytical method for the extraction and quantitation of total petroleum hydrocarbon in soil and water samples. Researchers have studied new Freon alternatives for environmental laboratories. These alternatives include but are not limited to: hexane-gravimetric analysis, supercritical CO2 extraction, tetrachloroethene(PCE)-IR, tetrachlorohexafluorobutane-IR, accelerated solvent extraction, and methylene chloride extraction followed by GC-FID analysis. Each of the methods mentioned above require certain solvents and instrumentation that could make the method cost-limiting in real application.

A gas chromatograph equipped with a flame ionization detector (GC-FID) is an instrument that most environmental labs use for their daily sample analysis. This study was aimed to compare the TPH-IR and GC-FID applications for the quantitation of total petroleum hydrocarbons in different types of oil samples.

The Department of Environmental Protection in the State of Connecticut, CT-DEP, is seeking cost-effective methods to be substituted for the TPH-IR method. This method would be based on the use of a GC-FID for quantitation of total petroleum hydrocarbon in soil and water samples. The detection limit (action level) of the method would be 500 PPM in soil, 0.5 PPM in groundwater, and 50 PPB in potable water.[20]

The study should provide a comparison between the data obtained from TPH-IR, and GC-FID-TPH for widely used petroleum based oils, such as #2 to #6 fuel oils, mechanical lubricating oil, diesel, kerosene, jet fuel and weathered gasoline. The solvents used for this study would be methylene chloride and trichlorotrifluoroethane (Freon-113).

## OVERVIEW OF STATES METHODS FOR TPH ANALYSIS

Regulators of different states have determined certain cleanup standards for soils contaminated with petroleum hydrocarbon (TPH). These levels range from as low as 10 ppm to 1000 ppm. Each state may use a certain procedure to achieve the cleanup criteria. Some states set their cleanup numbers based on human health risk, and some calculate it using transport equations and models that estimate the quantity of TPH that can be transported from soil to surface and groundwater under varying climatological conditions. Use of standard equations and assumptions to estimate exposure and cancer risk for workers and residents have revealed that state-derived soil cleanup levels for TPH is a very conservative ranking tool for soils that are contaminated with petroleum compounds.[13] For instance, toxicological studies show that soil cleanup targets in the range of 1000 to 11000 ppm for #2 diesel oil would be protective of human health.

Benzene is a potential exception among the petroleum hydrocarbons because it is highly mobile in soil due to its high water solubility. On the other hand, benzene levels are much lower than those assumed in the evaluations because benzene is highly volatile, and low in concentration in weathered products. Benzene is also a biodegradable compound and it biodegrades under aerobic conditions in the vadose zone.

In an effort to review the methods that other states use for the detection and quantitation of total petroleum hydrocarbon in soil and water samples, a comprehensive study was carried out and a table was established; however, this table is available in the technical report submitted to the Environmental Research Institute at the University of Connecticut.[11, 17, 19, 20]

## SUMMARY OF EPA METHOD 418.1

Method 418.1 defines "oil and grease" as any compound that can be extracted from a sample by Freon-113. Petroleum hydrocarbon, fatty matter, and oil from animal and vegetable sources, waxes, organic dyes, sulfur compounds, chlorophyll, and other compounds are included as extractable materials within this method. Freon-113 does not fully dissolve hydrocarbons that are heavier than C28. Gasoline, diesel, and kerosene range hydrocarbon molecules are between C6 and C28, but #4 and #6 fuel oil and coal tar contain hydrocarbon that are heavier than C28 and cannot be totally extracted with Freon-113.

After the extraction of oil and grease from a sample the "Total Petroleum Hydrocarbons" are derived by adding silica gel to the "oil and grease" extracted from a sample. The role of silica gel is to adsorb the polar material such as vegetable oils and animal fats. The method considers all "oil and grease" materials that are not eliminated by silica gel adsorption as "petroleum hydrocarbons."

An Infrared Spectrophotometer (IR) quantifies the extracted petroleum hydrocarbon. The absorbance of the extracted sample is measured at 2930 cm-1, and compared to the absorbance of a known standard. According to Beer's law, within a certain limit there is a linear relationship between absorption and concentration. All compounds with carbon-hydrogen linkages have absorption bands between 3.0 and 3.65 µm wavelength due to C-H stretching vibrations. Benzene's C-H absorption band at 3.3 µm is strong and it is a unique band.[16]

Method 418.1 is used for the detection of total petroleum hydrocarbon (TPH) in water samples. By substituting a Soxhlet extraction procedure for the separatory funnel procedure, this method may be modified and used for the analysis of total petroleum hydrocarbon in solid matrices. For extraction of solid matrices, sonication with a Fisher Sonic Dismembrator, Model 300, or a similar sonicator at 50% power for 3 minutes is recommended. Sodium sulfate is added to the sample during the extraction procedure and in the filtration process to eliminate any existing moisture in the sample.

## LIMITATIONS OF METHOD 418.1

There are several limitations associated with method TPH-IR that can be outlined in the three following categories. First, compounds with high degree of volatility may be lost during analysis. Second, the method solvent is not strong at extracting heavy hydrocarbon molecules (such as hydrocarbon molecules found in #4 and # 6 fuel oils). And third the method does not have an analytical specification.

The IR instrument measures the amount of petroleum hydrocarbon in the Freon-113 extracts by the light absorbance of carbon to hydrogen bonds at 2930 $cm^{-1}$. Freon-113 does not contain any carbon-hydrogen bonds, therefore it has a very low light absorbance at 2930 cm-1 and does not interfere with measurements. The method lacks analytical specificity because its "Petroleum Hydrocarbon" measured by the TPH-IR method are any compound that is extractable with Freon-113, unabsorbed with silica gel and has an IR absorbance at 2930 $cm^{-1}$. Researchers have conducted a number of experiments to determine the common errors associated with the TPH-IR method. The results of their studies have shown that analytical errors may result from three problems. First, errors resulting from differences between the sample and the calibration standard. Second, suppression of relatively polar petroleum hydrocarbon such as aromatics and alkenes with silica gel, and third, errors resulting from the inability of silica gel to eliminate the interferences of organic matter and vegetative interferences.

It is recommended that the TPH-IR be used for soil samples for soil samples that are contaminated with #2 fuel or diesel oil.

Some studies have shown that the TPH-IR method for soil samples as limestone, clays, and silts produce results that may be as much as four times the actual TPH concentration in the samples.[21]

Petroleum hydrocarbon may be divided into two groups; the *Volatile Petroleum Hydrocarbon* (VPH) and the *Extractable Petroleum Hydrocarbons* (EPH). VPH is basically defined as volatile aliphatic and aromatic hydrocarbon ranges: $C_5$ through $C_{12}$, that correspond to a boiling point range between approximately 36°C and 220°C. EPH are those aliphatic and aromatic hydrocarbons that range from $C_9$ through $C_{36}$ with a boiling point range between approximately 170°C

**Table I. Analytical Techniques for Determination of TPH in Soil and Aqueous Samples**

| Method | Instrument | Analyte |
|---|---|---|
| EPA-8015 (Modified) | GC/FID | TPH |
| EPA-8020 | GC/PID | BTEX |
| EPA-9071 | Gravimetric | TRPH |
| EPA-418.1 | Infrared | TRPH |
| LUFT | GC/FID | TPH |

TPH: Total Petroleum Hydrocarbon.
BTEX: Benzene, Toluene, Ethylbenzene, and Xylene.
TRPH: Total Recoverable Petroleum Hydrocarbons.
LUFT: Leaking Underground Storage Tank.

**Table II. New Analytical Techniques for Determination of TPH in Soil and Aqueous Samples**

| Method | Instrument | Analyte |
|---|---|---|
| API-GRO | GC/FID | GRO |
| API-DRO | GC/FID | DRO |
| API-PHC | GC/FID | PHC |
| TDHL-GRO | GC/FID | GRO |
| TDHL-DRO | GC/FID | DRO |
| WDNR-GRO | GC/FID | DRO |
| WDNR-DRO | GC/FID | DRO |

and 430°C. Current analytical techniques for determination of total petroleum hydrocarbon in soil are summarized in Table I.

All the methods mentioned above have as a basis a determination by gas chromatography-flame ionization detection (GC-FID).

New analytical techniques for determination of total petroleum hydrocarbon in soil are summarized in Table II.

New developmental methods are:

- High Resolution Gas Chromatograph Method with GC/FID
- Purge and Trap technique with an elevated purge temperature of 85°C
- Supercritical Fluid Extraction (SFE)
- Synchronous Scanning Fluorescence Spectroscopy.

## OBJECTIVES

The Department of Environmental Protection (DEP) of the State of Connecticut has set action levels for total petroleum hydrocarbon (TPH) in soil and water. For drinking water this level is 50 μg/L, for groundwater it is 500 μg/L, and for soil it is 500 μg/Kg.

Through this study the extraction efficiencies of Freon-113 and methylene chloride for different grades of fuel oil and their subsequent quantitation with an Infrared Analyzer (IR) and a Gas Chromatograph will be determined. Soils contaminated with petroleum oil will be extracted with Freon and methylene chloride and will be analyzed with both instruments.

## EXPERIMENTAL PROCEDURE

### Instrumentation

A HP-gas chromatograph Model-5890 equipped with HP-ChemStation version 4.00, 1988, was used for analysis of hydrocarbon compounds. The column used was a 30-m-long, 0.53mm ID, 0.50 μm (film thickness) DB-1, J&W Scientific. Using a standard mix of straight chain alkanes

Table III. Physical and Chemical Properties of the Petroleum Products

| Product Grade of Fuel | Carbon Range | Specific Gravity | Boiling Point Range |
|---|---|---|---|
| Gasoline | C6–C10 | 0.72 - 0.78 | 60 °C–170 °C |
| Diesel | C8–C21 | 0.84 at 24°C | 260 °C–360 °C |
| Kerosene | C8–C16 | 0.83 at 15°C | 190 °C–260 °C |
| Jet Fuel | C8–C16 | 0.82 at 15°C | 190 °C–260 °C |
| No. 2 | C8–C21 | 0.84 at 24°C | 260 °C–360 °C |
| No. 4 | C8–C30 | 0.93 at 15°C | Oil Specific |
| No. 6 | C20–C78 | 0.95 at 15°C | Oil Specific |
| Crankcase oil | C22–C50 | Oil Specific | Oil Specific |
| Lubricating oil | C20–C40 | 0.87 at 24°C | Oil Specific |
| Light crude oil | Oil Specific | 0.80 at 24 °C | Oil Specific |

Table IV. Straight Chain Alkanes

| Alkane | Concentration (μg/mL) |
|---|---|
| Decane(10) | 0.0145 |
| n-Undecane(11) | 0.0236 |
| n-Pentadecane(15) | 0.0301 |
| Hexadecane(16) | 0.0151 |
| Octadecane(18) | 0.0153 |
| Eicosane(20) | 0.0168 |
| Pentacosane(25) | 0.0163 |
| Triacontane(30) | 0.0171 |
| Detriacontane(32) | 0.0171 |
| Hexatriacontane(36) | 0.0184 |
| n-Tetracontane(40) | 0.0169 |
| Tetratotracontane(44) | 0..0173 |

and different temperature programs, the optimum program was chosen. The oven temperature was set at 45°C for 2 min and raised to 290°C at 15°C/min, then held at 290°C for 10 min. The total run time was 28 minutes. For spectral analysis of the Freon extracted fuel, a MIRAN-1A infrared analyzer was used.

## Material

For this study, high grade #2 fuel oil, regular #2 fuel oil, commercial diesel, heavy lubricating oil, kerosene, and #6 fuel oil were used.

## Standards

For the performance check of the GC-FID instrument a standard mixture of straight chain alkanes dissolved in isooctane was used. The alkanes and their concentrations are listed in Table IV.

## Extraction Procedure

The samples were extracted with Freon-113 and with methylene chloride. Stock solutions of 1000 mg/L were made by weighing 0.25 gram of the target oil and dissolving it in the solvent in a 250 mL grade A volumetric flask, with concentrations of 500 mg/L, 250 mg/L, 100 mg/L Due to the presence of heavy molecular weight hydrocarbon, Freon did not have the strength for complete extraction of #6 fuel oil. Methylene chloride seemed to completely solubilize #6 fuel oil. This phenomenon was clearly visible in the volumetric flasks used for extraction. A stock solution of 1000 mg/L of each fuel oil was prepared in a 250 mL grade A volumetric flasks (Fisher

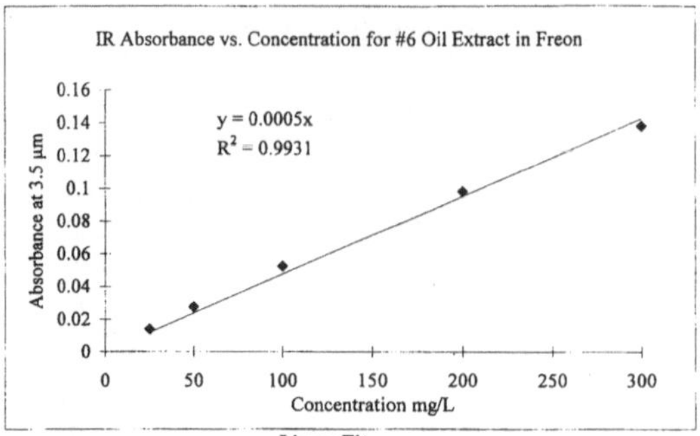


**Linear Fit**

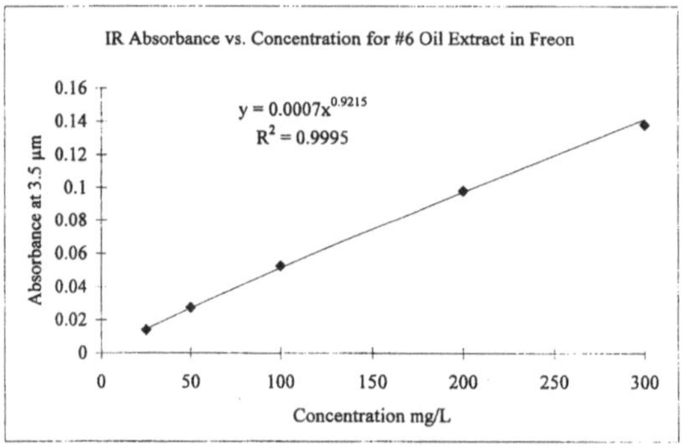


**Power Fit**

**Figure 1. Comparison of linear and power fits to absorbance data for #6 fuel oil.**

Scientific). Concentrations of 500 mg/L, 250 mg/L, 100 mg/L,50 mg/L, and 25 mg/L were prepared in 50 ml volumetric flasks by diluting the 1000 mg/L stock solution.

## Discussion

The extracted oil samples were simultaneously analyzed with GC/FID and IR. Calibration of the IR was done by preparing five standards over the concentration range of 25 mg/L to 500 mg/L. Concentration of 1000 mg/L was beyond the range of instrument. During the calibration of the IR instrument it was observed that for heavier fuel oils (#6 and lubricating oil) power fit into the data points yields a better fit than linear regression (Figure 1).

The sensitivity of the gas chromatography was low for hydrocarbon concentrations of 50 mg/L and less in the extracted solvent.

The GC/FID peak areas were normalized by dividing all the readings to the highest concentration area and the same procedure was done for the data obtained with the IR. A typical normalized graph(lubricating oil) is shown in Figure 2.

For IR measurements, the measure of angle between the line connecting the data points and the horizontal concentration axis is larger than the corresponding angle for GC/FID measurements. The graph indicates that IR is a more sensitive tool in the range of concentrations that it measures. The IR absorbances for different fuel oil were measured and are shown in Figure 3. Figure 3 shows that as the concentration of the oil in Freon-113 increases, the difference in absorbance units for different fuel grades becomes larger (#6 oil and kerosene have the lowest and the highest absornbances, respectively).

During the extraction of #6 oil with Freon-113 and methylene chloride it was observed that a portion of this oil does not dissolve in Freon-113 and the oil was completely miscible in methylene chloride. GC-FID analysis of 250 mg/L and 500 mg/L of #6 oil in Freon-113 and methylene chloride did not show any significant difference (Figure 4). Therefore, it may be concluded that the portion of petroleum hydrocarbons that do not dissolve in Freon-113 and are miscible in

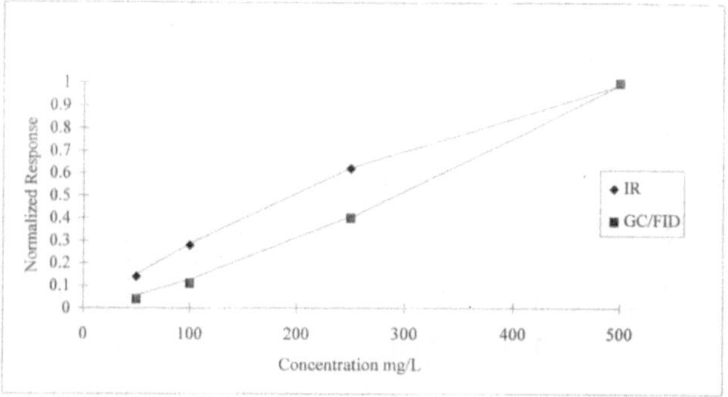


**Figure 2. Comparison of the GC-FID and IR normalized responses for lubricating oil.**

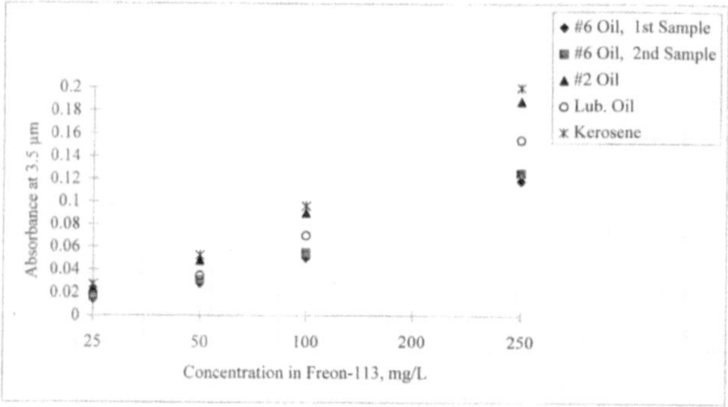


**Figure 3. Comparison of IR absorbance for different fuel oils.**

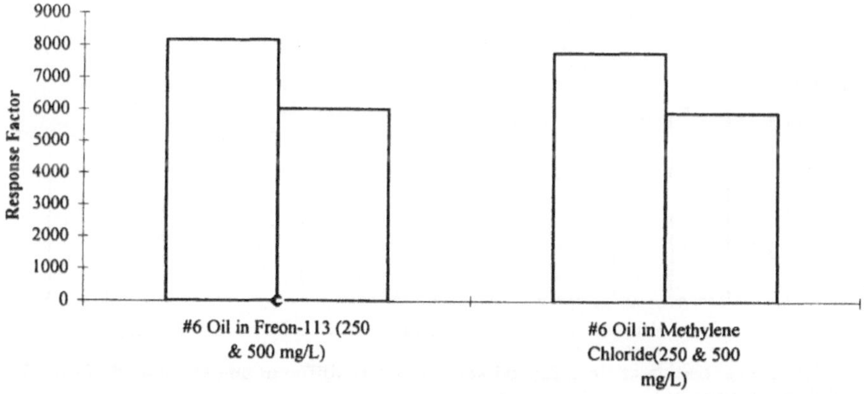


**Figure 4. Comparison of GC-FID response factors for 250 and 500 mg/L of #6 fuel oil in Freon-113 and methylene chloride.**

methylene chloride do not produce significant signals(peaks) when they are analyzed with flame ionization detectors, and their contribution to the total peak area is insignificant.

Three soil samples contaminated with #2 fuel oil were extracted with Freon-113 and methylene chloride. The extractants were analyzed with TPH-IR and GC-FID. For all three soil samples the IR instrument showed higher concentrations than the GC-FID quantitational analysis of the Freon-113 and the methylene chloride extractants. The GC-FID response factors of methylene

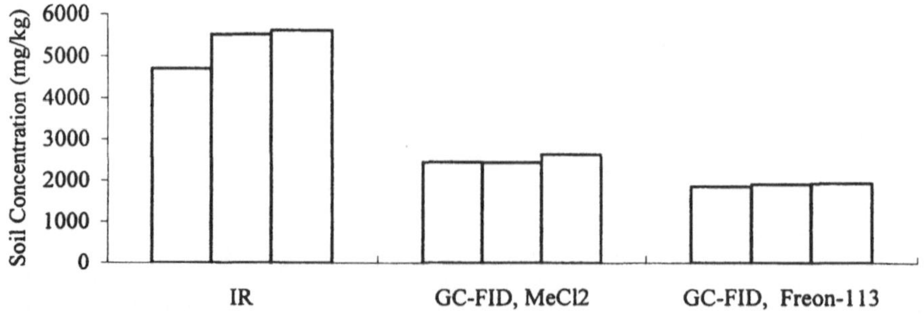


**Figure 5. Concentration of No. 2 fuel oil measured with different analytical methods for Soil-1 (triplicate extraction).**

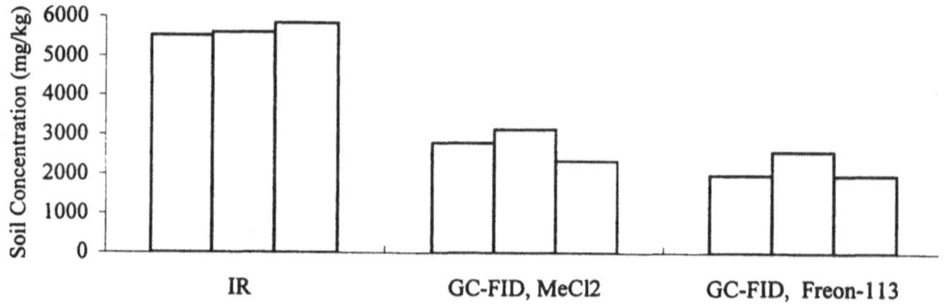


**Figure 6. Concentration of No. 2 fuel oil measured with different analytical methods for Soil-2 (triplicate cxtraction).**

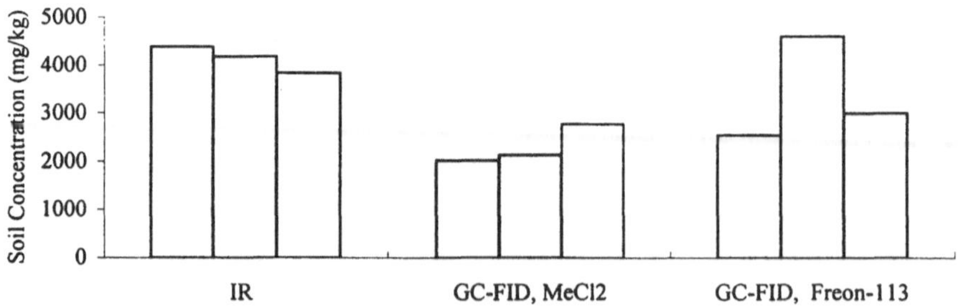


**Figure 7. Concentration of No. 2 fuel oil measured with different analytical methods for Soil-3 (triplicate extraction).**

chloride and Freon-113 extraction of the contaminated soils did not show any significant difference (Figures 5–7).

## CONCLUSIONS

The results of this study have indicated that EPA Method 418.1 (TPH-IR) is a relatively fast method for determination of total petroleum hydrocarbons in soil and water samples. However, qualitative determination of TPH in solid and aqueous samples is not feasible with this method.

EPA Method 418.1 for TPH analysis should be used with extreme care because any outside interference may lead into erroneous results. Due to the ban on the production of Freon-113 in the United States, EPA Method 418.1 using Freon-113 as the extraction solvent may not be practical anymore in the near future and if other solvents such as high-grade tetrachloroethylene are going to be substituted for Freon-113, the associated cost with this method due to purchasing and disposal should be considered when comparing it to a GC/FID based TPH analysis.

Solvent extraction and use of GC/FID is more practical than other techniques developed for the analysis of petroleum compounds because most analytical laboratories have access to this instrument and its cost of maintenance and operation is usually less than many other available techniques. GC/FID may also be used to identify the nature and source of the petroleum contamination. When analyzing samples that are contaminated with light hydrocarbon such as diesel oil, and kerosene, concentrating the sample is not recommended because substantial loss of samples may occur during the concentration procedure.

## ACKNOWLEDGMENTS

This study was funded by the State of Connecticut's Department of Environmental Management.

The participation of Drs. Polishchuk and Sidorova in this research project was made possible under the auspices of the U.S. Agency for International Development through the Institutional Partnership Program for the NIS administered by the International Research and Exchanges Board.

We would like to send our special thanks to Dr. Mark L, Bruce at Quanterra Environmental Services, North Canton, Ohio, and Mr. Peter Zack at the Bureau of Waste Management, Leaking Underground Storage Tank Program at the State of Connecticut Department of Environmental Protection.

## BIBLIOGRAPHY

1. Department of Natural Resources, Emergency & Remedial Response Section, Guidance for Landspreading Petroleum Contaminated Soil, (PUBL-SW-512-95), (12/95), Madison, WI.
2. Department of Natural Resources, Emergency & Remedial Response Section, A Guide to Wisconsin's LUST Analytical Guidance, (PUBL-SW-130REV), (7/93), Madison, WI.
3. Department of Natural Resources, Emergency & Remedial Response Section, A Guide to Wisconsin's LUST Program-Owner's Fact Sheet, (PUBL-SW-179), (4/92), Madison, WI.
4. Department of Natural Resources, Emergency & Remedial Response Section, Comparison of Field Screening Techniques with Fuel-Contaminated Soils , (PUBL-SW514-95), (10/95), Madison, WI.
5. Department of Natural Resources, Emergency & Remedial Response Section, Solid Waste Analytical Requirements for Landfilling of Petroleum Contaminated Soil (4/92), Madison, WI.
6. Department of Natural Resources, Emergency & Remedial Response Section, Site Assessments for Underground Storage Tanks (UST Closure Guidance and Documentation Checklist) (PUBL-SW-175-93), (9/92), Madison, WI.
7. Department of Natural Resources, Emergency & Remedial Response Section, Underground Tanks, Property Transfers and Environmental Cleanups, (PUBL-SW-500), (4/94), Madison, WI.
8. EPA Method 418.1, Total Recoverable Petroleum Hydrocarbon by IR, Groundwater Analytical Inc., Sanford, Jonathan R., (Vice President), Buzzard Bay, Massachusetts, Groundwater Technical Analytical Bulletin, Winter 91/92, pp. 1–4.
9. Fan, C.-Y., S. Krishnamurthy, and C.T. Chen, *A Critical Review of Analytical Approaches for Petroleum Contaminated Soil*, Soils Contaminated with Petroleum Constituent, pp. 61-74.
10. Fetter, C.W., *Contaminant Hydrogeology*, Macmillan Publishing Company, New York, 1993.
11. Grasso, D., *Hazardous Waste Site Remediation: Source Control*, Lewis Publishers, Chelsea, MI, 1993.
12. Hounslow, A.W. *Water Quality Data, Analysis and Interpretation*, Lewis Publishers, Chelsea, MI, 1995.
13. Heath, J. and S. Atwood, *State Standards Reach for the Moon*, Soils, Analysis, Monitoring and Remediation, December, 1992, pp. 12–13.

14. Ott, L. *An Introduction to Statistical Methods and Data Analysis*, Third edition, PWS-Kent Publishers, Boston, MA, 1988.
15. Mac Berthouex, P., and L.C. Brown, *Statistics for Environmental Engineers*, Lewis Publishers, Chelsea, MI, 1994.
16. Miran-1A, General Purpose Infrared Analyzers, Operation and Maintenance Manual, Miran, Foxboro, MA, 1986.
17. Kostecki, P., Association for the Environmental Health of Soils (AEHS), Amherst, MA. *State-by State Summary of Cleanup Standards*, Soils, Analysis, Monitoring and Remediation, December, 1992, pp. 14–24.
18. Parr, L.J., R.E. Claff, D.S. Kocurek, and J.C. Lowry, *Interlaboratory Study of Analytical Methods for Petroleum Hydrocarbons*, Soils Contaminated with Petroleum Constituent, Quanterra Environmental Services, Arvada, CO, April 1996, pp. 52–60.
19. Smith, R.K. Apichemical Consultants, *Handbook of Environmental Analysis*, Second edition, Genium publishing Corporation, New York, 1994.
20. State of Connecticut Department of Environmental Protection, Proposal for the Connecticut Clean-up Standard Regulations, Hartford, Connecticut, December, 1994.
21. Thomey, N., D. Brathberg, Applied Earth Sciences, Inc. Houston, TX, and C. Kalisz, Star Enterprise, Irving, TX, *A Comparison of Methods for Measuring Total Petroleum Hydrocarbon in Soil* (Report).
22. Woodruff, E.B., H.B. Lammers, and T.F. Lammers, *Steam-Plant Operation*, Fifth Edition, McGraw-Hill Book Company, New York, 1995.

# Section Twelve
# MANAGEMENT AND REUSE STRATEGIES

## 45 BENEFICIAL USE OF FOUNDRY EXCESS SYSTEMS SANDS FOR CONSTRUCTION PRODUCTS

**Raymond W. Regan, Sr.**, Professor
Dept. of Civil and Environmental Engineering
University Park, Pennsylvania 16802

**Elaine S. Batchev**, Graduate Research Assistant
Environmental Resources Research Institute
University Park, Pennsylvania 16802

**Robert C. Voigt**, Professor
Dept. of Industrial and Manufacturing Engineering
Penn State University
University Park, Pennsylvania 16802

### ABSTRACT

This chapter will explain how foundry excess systems sand (FESS) can be successfully incorporated into many standard construction products, such as asphalt concrete, portland cement, portland cement concrete, and flowable fill material, along with some illustrative experiences. Strength and durability problems of some construction products using FESS with a higher clay content are considered. Other difficulties reported with substituting FESS into the products will be discussed, along with viable solution approaches.

### INTRODUCTION

Foundry excess systems sand (FESS) is a high-volume residual material that can be reused by substituting it for traditional materials in typical construction products. Reusing FESS for this purpose is legally recognized as a beneficial use in many states. In the spring of 1996, for example, a general permit was issued by the Pennsylvania Department of Environmental Protection (PaDEP) to 37 foundries represented as sources of the beneficially reused materials. In order to qualify for beneficial use, the materials must be classified as nonhazardous and meet the same industry performance criteria as the material being replaced.

The objectives of the chapter include the following:

- Show by chemical leaching tests results that, to a large degree, foundry residual sands are nonhazardous. Their substitution into construction products will not adversely affect the environment any more than products made with traditional construction material.
- Review the physical characterizations of the foundry excess systems sands (FESS) to demonstrate that this material has similar gradations as typical construction fine aggregate material, except that they generally contain too large a fraction of fine material; that is, the fraction less than the 200 mesh sieve size. Scalping this fine fraction brings the foundry sand closer to required gradation.
- Review the experiences with FESS in the construction industry, including their use in asphalt cement and cement concrete products, including flowable fill, a.k.a. controlled low-strength material (CLSM).

*52nd Purdue Industrial Waste Conference Proceedings*, 1997, Ann Arbor Press, Chelsea, Michigan 48118. Printed in U.S.A.

## BACKGROUND

The environmental impact of the solid residuals derived from the Pennsylvania metal casting industry have been presented previously.[1]

### Rationale

There is a lack of understanding of how FESS can be incorporated into typical construction products. Both the physical and chemical characteristics of the proposed uses are outlined, and how well "typical" FESS corresponds with these guidelines. It is proposed that many of the problems involving the incorporation of FESS into a beneficial use product may be solved by relatively simple techniques.

In the case of asphalt concrete, particles less than the No. 200 size sieve (including clays and dusts) fill in the voids between the interstices of the aggregate material. The result is that the percentage of voids is decreased, which *decreases* the asphalt concrete's strength. For cement concrete, the clay not only consumes moisture, but also uses up a great deal of the cement surrounding the aggregate particles as well, leaving less cement available to bind the larger aggregate particles together. With less cement to coat the aggregates available, the compressive strength of the concrete product can be decreased dramatically.

Scalping the fine particles (including clay and organics) is an approach to some of these deleterious effects, since these particles can reduce the strength of the product dramatically.

### Legal Specifications for the General Beneficial Use Permit in Pennsylvania

In order for foundry residual to be considered for beneficial use in the state of Pennsylvania and other states as well, certain conditions apply. The residual must have substantially similar physical and chemical characteristics of the material it is replacing. In a physical sense, the material must have similar performance characteristics as the aggregate material it is replacing. For example, makers of concrete specify a mix of fine aggregate material that must have a certain percentage of very fine particles in the mix, along with a certain percentage of medium sized particles in the mix, and so on. Hence, in order for FESS to be substituted for the fine aggregate fraction, it must be similarly sized and have a similar range of sizes for the particles.

In addition to the size distribution of the particles, the *shape* of the particles can sometimes play a major role. Some producers desire rounded particles, such as cement concrete producers. Rounded particles need less water and cement to maintain the same water/cement ratio and overall strength. However, producers of asphalt concrete desire angular particles for their aggregate. This is because angular particles increase the frictional resistance of the mix, thereby increasing its shear strength. Shear strength is vital for asphalt so that the pavement does not flow under load. If FESS is substituted into concrete, this product should now have the same compressive strength, tensile strength, durability, and resistance to freeze-thaw cycles as normally-made concrete. There cannot be any deterioration in the performance of the product.

### Sources of Solid Wastes

In addition to FESS, other materials fractions from the metal casting industry include core sand, slag, and baghouse dust.

Amounts of these wastes generated at an individual foundry vary depending on the processes used. The combination of process waste streams from 54 Pennsylvania foundries is shown in Table I. The table indicates that on average, excess systems sand comprises 67% of the materials

**Table I. Solid Waste Generation Rates From 54 Pennsylvania Foundries for 1990–1991**

| Category | Solid Waste (tons/year) | Percentage of Solid Waste |
|---|---|---|
| Excess systems sand | 94,408 | 67% |
| Baghouse dust | 23,901 | 17% |
| Slag | 23,232 | 16% |

generated from a typical foundry. Baghouse dust and slag generation rates comprise 17% and 16%, respectively, of the solid waste generated.[3]

Other lesser sources of spent materials include cupola ash (ashes produced from coal burning in the cupola) and refractories (insulating material lining the bottom of the cupola), which are commonly combined with the residual slag materials.

In the past, all of these wastes were combined in a common stockpile at the foundry prior to disposal. However, segregation of the waste piles is important to keep a more uniform stock of material if the residual is to be managed as a beneficially used material.

## RESULTS

### Chemical Characterization of Pennsylvania Foundry Excess Systems Sands

Excess systems sands from 33 foundries were analyzed for their leaching characteristics. The eight primary metals and four secondary metals listed in the drinking water standards in Table II[4] were analyzed for their leachability by the Toxic Characteristic Leaching Procedure (TCLP).

In order for a waste to be considered nonhazardous, it must be found to leach by the TCLP test less than 100 times its drinking water standard. For each metal, this will be referred to as the Toxicity Standard. Additionally, interim guidance from the PaDEP indicated that any material leaching at less than 25 times the drinking water standard for any constituent was considered beneficially usable. Therefore, for each metal, this value was referred to as the Beneficial Use Standard.

The TCLP test results typically indicated that all of the eight primary metals leached at concentrations less than their respective Toxicity Standards, as shown in Table III. In addition, all of the primary metals leached below the Beneficial Use Standards, with one exception where lead had a higher concentration from one foundry (Cluster 4). In addition, the four secondary metals were also analyzed. Copper and zinc both leached below the Beneficial Use Standard. Manganese always leached below the Toxicity Standard, and except for the three foundries in Cluster 3, usually leached below Beneficial Use Standards as well. Except for the three foundries in Cluster 2, iron leached below its Toxicity Standard of 30 mg/L, but only one of 33 of the FESS leached below the Beneficial Use Standard (Cluster 3).

Generally speaking, Pennsylvania FESS were classified as nonhazardous. In addition, a factor that may preclude the use of these sands beneficially was the high iron content. Iron leached at concentration above the beneficial use standard 32 out of 33 samples.

**Table II. Federal Drinking Water Standards for Common Metals in Foundry Wastes**

| Metal | Drinking Water Standard (mg/L) | Toxic Standard (mg/L) | Beneficial Use Standard (mg/L) |
|---|---|---|---|
| PRIMARY | | | |
| Arsenic | 0.05 | 5 | 1.25 |
| Barium | 1.00 | 100 | 25 |
| Cadmium | 0.01 | 1.00 | 0.25 |
| Chromium | 0.05 | 5.00 | 1.25 |
| Lead | 0.05 | 5.00 | 1.25 |
| Mercury | 0.002 | 0.2 | 0.05 |
| elenium | 0.01 | 1.0 | 0.25 |
| Silver | 0.05 | 5.0 | 1.25 |
| SECONDARY | | | |
| Copper | 1.00 | 100 | 25 |
| Zinc | 5.0 | 500 | 125 |
| Iron | 0.3 | 30 | 7.5 |
| Manganese | 0.05 | 5.0 | 1.25 |

**Table III. TCLP Results (mg/L) from 33 Pennsylvania Foundries**

| Chemical | Cluster 1 (28 Foundries) | Cluster 2 (3 Foundries) | Cluster 3 (1 Foundry) | Cluster 4 (1 Foundry) |
|---|---|---|---|---|
| Arsenic | 0.006 | 0.002 | 0.002 | 0.002 |
| Barium | 0.33 | 0.3 | 0.3 | 0.3 |
| Cadmium | 0.031 | 0.087 | 0.03 | 0.1 |
| Chromium | 0.1 | 0.1 | 0.1 | 0.16 |
| Lead | 0.25 | 0.99 | 0.27 | 1.55 |
| Mercury | 0.0005 | 0.0005 | 0.0005 | 0.005 |
| Selenium | 0.029 | 0.032 | 0.83 | 0.002 |
| Silver | 0.06 | 0.06 | 0.6 | 0.06 |
| Copper | 0.19 | 0.09 | 2.64 | 0.85 |
| Zinc | 0.39 | 4.17 | 1.21 | 70.3 |
| Iron | 27.6 | 490.3 | 1.9 | 207 |
| Manganese | 0.35 | 4.79 | 0.14 | 1.03 |

## Physical Characterization of Pennsylvania Foundry Excess Systems Sands

A study was conducted to determine the physical characteristics of typical Pennsylvania foundry excess systems sand.[5] All of the samples were obtained from either iron or steel foundries, since these represent the majority of the foundries and thus the majority of waste residuals generated in the state. Residual green sand was tested since it is the most common type of binding system used.

The results from this analysis provided the following observations:

1. The FESS were generally considered to be fine-grained when compared to Type B, No. 2 bituminous fine aggregate material. Furthermore, these sands were too fine when compared to the gradation required for Cement Concrete Sand, Type A, used in the production of concrete.
2. The FESS had a narrower distribution of particle sizes than do most fine aggregates used in construction. Therefore, typical FESS do not completely fulfill the full spectrum of gradation requirements of typical construction aggregate material. The authors suggested, that to meet the specification for fine aggregate, a larger proportion of the material between 0.25 and 0.375 inch was required. The authors suggested that slag could improve the gradation distribution.
3. FESS exhibited typical compaction behavior of construction materials. The ability to compact is important for such applications as fill material and roadbed material, where the material undergoes high compressive loads. The material must be able to compact to provide a strong support for any loads that are placed on it. The results of the test showed that FESS had higher compactibility, as demonstrated by their higher maximum bulk unit weights, than the virgin sand. In addition, the optimum moisture content (OMC) of the FESS was similar to that of the virgin sand. The OMC is the moisture content of the specimen at the point of maximum bulk unit weight. Soils that have similar maximum bulk unit weights and optimum moisture contents tend to exhibit similar compressive strengths.
4. FESS have sufficient shear strength for use in compacted fills and embankments. This, once again, is a particularly important observation, since most of the loading that will occur on fill material will need very large shear strengths to prevent the shifting of the fill material under load. Therefore, typical Pennsylvania FESS have comparable shear strengths to sand usually used in construction material.
5. The authors suggested that some processing of the waste may be needed for some beneficial uses. For example, large core butts, etc. would be removed. Also, the FESS may have to be washed to remove some debris. The removal of core butts and other sources of organic material would increase the strength characteristics of the waste sand, since the primary reason for a lowered strength potential is due to the organic material, especially clay, in the FESS.

Overall, the authors suggested that, based on physical properties alone, the FESS may be used as a partial replacement for fine aggregate used in construction material. Complete gradation for construction aggregate may be met by blending FESS with coarser material.

## EXPERIENCES WITH BENEFICIALLY USABLE MATERIALS

### Substituting Foundry Excess Systems Sand as a Fine Aggregate for Asphalt Concrete

Asphalt concrete is the most popular paving material used on all highways and roadways. Typically, asphalt concrete consists of approximately 95% graded aggregate and 5% bitumen (by weight). The graded aggregate includes particles ranging from fine sand to 1/2- to 1-inch gravel. Depending on the mix design and the ultimate strength requirements of the product, the fine-size particle fraction may comprise as much as 36% of the asphalt concrete. This fine aggregate material may be possibly replaced by FESS.

### Parameters Considered in the Design of Asphalt Concrete

The most critical factors in designing asphalt concrete are its (1) stability; (2) durability; (3) skid resistance; and (4) stripping resistance.

Stability is the property that requires that the asphalt concrete not flow under loading conditions. The stability of asphalt concrete is determined by the amount of asphalt in the hot mix (HMA), the gradation of the aggregate and the particle size of the aggregate. Too many fines in the aggregate causes the stability to decrease. In addition, producers of HMA require angular aggregate particles which will increase the concrete's shear strength. If the aggregate particles are rounded, the shear strength of the mix would be reduced, and the asphalt pavement would typically undergo a process called rutting, where the asphalt concrete moves so much under load that it forms bumps on the surface of the pavement.[6]

Typically, foundry sand contains too many fines and too few angular particles to fulfill the fine aggregate requirement of HMA concrete. However, foundry sand can still be used as a partial replacement of the fine aggregate by being mixed with a coarser and more angular material. Studies have shown that substituting 15% of the total fine aggregate with a typical FESS fulfills the concrete's stability requirements.[7,8]

An important factor in asphalt concrete production is that it must be durable. Durability refers to the cohesion of the pavement. Excess fines in the aggregate leads to a loss of cohesion in the mix by reducing the air voids. Reduction in the air voids leads to increased viscosity and thermal cracking.[6] Once again, FESS contains too many fine particles to meet this requirement. However, by mixing the foundry sand with another fine aggregate, durability requirements can be met.

Another important requirement of asphalt concrete is skid resistance. This property requires that the asphaltic surface retains a high-friction surface that does not become smooth and polished after repeated loading. Once again, angular particles provide greater skid resistance by increasing the shear strength of the asphalt concrete.

The final important requirement is that the asphalt concrete is strip resistant. This property is particularly important when using FESS that is contained mostly of silica. Silica sand, due to its chemical composition, is hydrophilic, meaning that it tends to draw water to itself. Organic material and clay are also hydrophilic. This hydrophilic property can strip the asphalt cement coating the aggregate grains, resulting in the degradation of the asphalt surface. If strip potential exists, it can be corrected by adding an antistripping additive such as hydrated lime.[8] However, the costs of adding such an additive must be taken into consideration.

### Specification for Using Foundry Excess Systems Sand as Kiln Feed

Cement requires sand as a source of silica, alumina and iron oxides. Although sand is also used as an aggregate material, this need is secondary to its use as a source of these chemicals. Hence, no specifications for the gradation of the sand used as kiln feed are written. However, what is important is that the sand has a high silica content. In general, a silica content of over 80% is required.[9] In addition, slag from a foundry can serve as a raw material for the aluminum oxide and magnesia content. Furthermore, for green sand which has a high percentage of clay (4–10%), the clay in the foundry sand can be a source of iron and aluminum oxides. The percentages of lime, iron, silica, alumina, gypsum, and magnesia depend upon what type of cement is specified for the job-site conditions.

## Requirements for Cement Concrete Products

Portland cement concrete (PCC) is a mixture of sand and gravel or crushed stone bonded by a hardened paste of portland cement and water (additives for workability, air-entrainment, etc. are typically used). PCC mixes are designed by the ready mix supplier/contractor to DOT requirements. The characteristics of PCC which influence its performance are described below.

- The concrete must develop enough strength to withstand loads. PCC pavements normally are specified to have a nominal design compressive strength of 25 Mpa (3600 psi)[8]. Traffic can be allowed on the road when the concrete has a strength of 15 Mpa (2180 psi).[8]
- The concrete must be durable enough to withstand freeze/thaw cycles, salt penetration, traffic abrasion, and sulfate attack.
- The concrete mix must be able to cure and set completely within a several-day time period.

Significant physical parameters of concrete include particle size distribution and fineness modulus, dust content, density, organics content, deleterious materials content and grain shape.

The particle size distribution affects cement and water requirements, workability, economy, porosity, shrinkage and durability of concrete. Too many fine particles can lower the strength of concrete. One way to express the fineness of a mix is by calculating the fineness modulus. The fineness modulus is calculated by adding the cumulative percentage of materials that is coarser than sieves 100, 50, 30, 16, 8, 4 and 3/8″ and dividing by 100, as defined by ASTM standards. The greater the fineness modulus, the coarser the mixture.

The fine particle fraction (–200 mesh sieve) must be limited. These particles include clays and other dusts. A large percentage of fine particles can interfere with the bonding of the aggregate material, since fines attract much of the binder to itself, leaving less to bind the aggregates. In addition, fine particles also absorb moisture in the mix, once again leaving less to coat the aggregate material. Due to all of these results, ASTM C33 allows a maximum of 3% particles finer than the No. 200 sieve size for fine aggregate material.

Density must be a minimum of 75–110 $lbs/ft^3$, according to ASTM C330 for general fine aggregate. High density is required since the concrete will be subjected to high compressive loads.

The organics content and deleterious materials content is specified by ASTM C33. This specification requires that the maximum amount of clay lumps and friable particles allowed is 3%. The organic content is restricted since a high degree of organics causes drying problems and a deterioration in strength.

Finally, the aggregate particles must be rounded. Rounded particles need less water and cement to coat their surfaces, and promote more workable mixtures than one made with angular particles.

## Flowable Fill Material

Flowable fill is a common product used in highway and structural application. Flowable fill is a low-strength concrete product commonly referred to by the concrete industry as Controllable Low-Strength material (CLSM). The uses of flowable fill are many and can include *backfills* for sewer trenches, utility trenches, bridge abutments, pile excavations, retaining walls, and conduit trenches; and *structural fills* for foundation sub-base, subfooting, and floor slab base. In addition, flowable fill can be used for filling abandoned underground storage tanks, wells, and manholes.

Flowable fill consists of fly ash, sand, water and cement. Typical concentrations of these constituents are given as:

- 82% fine aggregate material (sand)
- 14% fly ash (pozzolan)
- 4% portland cement

The fine aggregate fraction can consist of natural sand, manufactured sand, and/or reused sand. Fly ash is used as a pozzolan, where pozzolan is defined from ASTM standards as a siliceous material which behaves like a cement only in the presence of water. Specifically, in the presence of moisture, it reacts with calcium hydroxide at ordinary temperatures to form compounds "possessing cementitious properties" (ASTM C125). Additionally, the amount of portland cement content is adjusted on a case-by-case basis. If the fill needs to be stronger, adding more cement would increase the strength of the fill material. Finally, the percentage of water is also adjusted on a case-by-case basis.

Flowable fill is transported on-site in a revolving drum mixer, just as is ready mix concrete. Flowable fill is placed by pouring it into the cavity to be filled, and is self-leveling and self-compacting, due to this natural settling action. Its strength increases over time, achieving its maximum strength in 21–28 days, just as concrete does.

## FINAL SUMMARY

### Technical Issues

Foundry excess systems sand may be used as a partial replacement of fine aggregate material in the production of asphalt concrete, portland cement concrete, and flowable fill. Typical FESS nearly meets the fine aggregate gradations specified for these products, except that it contains too many fine particles less than the 200 mesh sieve size. Scalping the fines and clay content usually allows the FESS to fulfill gradation requirements of these products.

### Nontechnical Issues

FESS may only be used if there is a market demand for it. Producers of construction material must have an incentive to integrate the foundry material into their products. The FESS must not only meet physical and chemical requirements, but also be an inexpensive source of material. New machinery may be needed by the construction user to process the FESS to specifications, which increases costs. Therefore, the construction company needs to weigh the benefits vs. the costs involved with incorporating any such residual materials. Furthermore, foundries are not the only industry that has approached construction companies to use their residual products. For example, producers of residual glass, rubber, and roofing shingles are currently trying to get their products recycled into asphalt concrete as well.[7] In order for FESS to become incorporated into a beneficial use product, it must be less expensive than other available materials, including other recyclable residuals, while maintaining construction specifications.

## ACKNOWLEDGMENTS

The authors are grateful for the metal casting industry's longstanding support of this beneficial use effort.

Funding for this project was provided by the members of the PFA, the Advanced Technology Center of Central and Northern Pennsylvania, and the University. Details of the research presented are available.[10]

## REFERENCES

1. Fahnline, D. E. and R. W. Regan. Leaching of Metals from Beneficially Used Foundry Residuals into Soils. *Proceedings, 50th Purdue/Industrial Waste Conference.* May, 1995.
2. Pennsylvania Department of Environmental Resources. *Pennsylvania Bulletin*, 22, 7, p. 3461, 1992.
3. Voigt, R. C. and R. W. Regan. Beneficial Reuse and Reclamation of Foundry Residual Wastes. Project No. 92C.111OC. The Metal Casting Center and the Environmental Resources Research Institute. The Pennsylvania State University. October 1993.
4. Regan, R.. W., R. C. Voigt, W. C. Paletski and R. Massell. Chemical Characterizations of Spent Molding Sands: Environmental Issues. AFS Technical Paper 94-126. The Metal Casting Center and the Environmental Resources Research Institute. The Pennsylvania State University. May 1994.

5. Hitunen, D., R., R. Roque and X. Wu. Beneficial Reuse of Foundry Residual Wastes in Pennsylvania: Physical Characterization of Materials. The Metal Casting Center and the Environmental Resources Research Institute. The Pennsylvania State University. October 1992.
6. American Foundrymen's Society, Inc. Final (Phase I) Report on Alternate Utilization of Foundry Waste Sand. Illinois Department of Commercial and Community Affairs. Grant No. 90-82109, 1990.
7. Leidel, D. S. "Sand reuse: User requirements." *Modern Casting*, pp. 38–39. July 1993.
8. John Emery Geotechnical Engineering Unlimited (JEGEL). Spent Foundry Sand Alternative Uses Study. Etobicoke, Ontario. July 1993.
9. Leidel, D. S., M. Novakowski, and D. Pohlman. "Beneficial sand reuse: Making it work." *Modern Casting*, pp. 28-31. August 1994.
10. Batchev, E. A.. The Beneficial Use of Foundry Excess Systems Sands for the Production of Construction Products, unpublished ME report in Environmental Engineering, The Pennsylvania State University, University Park, PA 16802. August, 1995.

# 46 IMPLEMENTING AN ENVIRONMENTAL MANAGEMENT PROGRAM: WHOSE DECISION IS IT? THE NATIONAL PARK SERVICE/ENVIRONMENTAL PROTECTION AGENCY PARTNERSHIP PROJECT

**Marie Bushway Zanowick**
EPA Region VIII Office of Pollution Prevention
999 18th Street, Suite 500
Denver, Colorado 80202

**Michael Schene**
NPS Intermountain Region
Rocky Mountain and Colorado Plateau Support Office
12795 West Alameda Parkway
Denver, Colorado 80225

## ABSTRACT

The mission of the National Park Service, Intermountain Regional (IMR) is to promote sustainable park environments. This involves blending environmental preservation with visitor services and uses. In order to support this mission, the Intermountain Region, Rocky Mountain and Colorado Plateau System Support Office is committed to the goal of empowering parks to develop and implement an effective environmental management program that will meet the goals outlined in their initiative, *Park Environments 2000: Investing in Safe, Unimpaired Park Environments.* The strategy of the *Park Environments 2000 Initiative* is to create a coalition of System Support office, NPS Headquarters and park operations to protect and repair environmental damage from human use through development and implementation of an integrated environmental management program within each park that centers around sustainable development.

In order to implement this strategy of developing sustainable parks, the NPS/EPA Partnership Project has focused its efforts on incorporating pollution prevention concepts into park practices. To achieve these goals, the NPS IMR looked for support and expertise from the Environmental Protection Agency Region VIII because of their commitment to a similar mission: "To restore and protect the ecological integrity of the Rocky Mountains and the plains and the health of their inhabitants."

The partnership was formalized through a federal Interagency Agreement (IAG) between EPA Region VIII and the NPS IMR. Under this agreement, NPS provides financial and management support to the project and coordinates activities with the cluster parks. The EPA provides technical expertise in pollution prevention and regulatory compliance and oversees the field work for pollution prevention assessments of the parks.

This project was underway during a structural reorganization of the National Park Service. Questions concerning the role of the NPS central office in environmental decision making and how to empower and support environmental managers at the local park level were addressed.

The partnership between the NPS IMR and EPA Region VIII has resulted in quantifiable environmental results that support the goals of both agencies. The cost savings to the Park Service to date is estimated at over $500,000. EPA Region VIII has been able to gain valuable field experience and insight by working directly with a regulatory "customer."

---

*52nd Purdue Industrial Waste Conference Proceedings*, 1997, Ann Arbor Press, Chelsea, Michigan 48118. Printed in U.S.A.

The NPS IMR will collaborate with parks to determine the next steps to meet the project goal: To empower parks to develop and implement an effective environmental management program that will meet the goals of the *Park Environments 2000 Initiative*.

## BACKGROUND

The National Park Service (NPS) was established in 1916 through congressional legislation. The putative agency was mandated to both preserve and protect significant natural and cultural resources while providing for their use and enjoyment. This tension in the agency's mandate has been difficult to resolve and remains a fundamental issue today. The NPS, like other federal agencies, remained a highly centralized agency even though President Richard Nixon tried to decentralize the federal government in the 1970s. In the 1990s, the imperative to reduce the federal deficit coupled with American society's growing distrust of public institutions has resulted in actions that have been directed at "reinventing government." This forced federal agencies, including the NPS, to reorganize. The NPS adopted many of the ideas popularized by David Osborne in his best-selling book *Reinventing Government: How the Entrepreneurial Spirit is Transforming the Public Sector.* Among the most important of these adopted ideas was the decision to decentralize the agency and emphasize more decision-making at the local (park) level. As a result, the traditional roles for a central and regional office of policy formulation and implementation have been largely eliminated and have been replaced by functions that are defined as "value-added" customer services—-technical assistance and training. The Government Performance and Results Act (GPRA) has added another element, and the NPS is now trying to measure the value of services provided to the public.

President William Clinton has long recognized the value of federal partnerships and has encouraged agencies to share resources. The NPS has had a long and successful history of partnership work and recently stated that it is one of the agency's primary mission goals.[1] The National Park Service/Environmental Protection Agency (EPA) partnership is so effective because it is based on the shared value of environmental protection. Both agencies have a mission to preserve and protect the environment as well the people, animals, and plants that inhabit it. The partnership is also predicated upon shared resources. The NPS has managed to leverage a minimal investment of funds and management oversight into more than a $500,000 savings! The EPA, for its part, has been able to derive case studies and lessons learned from a land-managing agency that has direct application to other public institutions. And the EPA has been able to develop these data from a collaboration rather than an adversarial relationship.

## PARK ENVIRONMENTS 2000

In 1993, former NPS Rocky Mountain Regional Director Robert M. Baker became committed to the *Park Environments 2000*, an initiative that would produce safe, unimpaired park environments by the year 2000, then some seven and now two years away. The initiative defined the following goals that would be met by parks in the Rocky Mountain Region (now the Intermountain Region; see Figure 1).

- All parks and concessionaires are administered with the greatest sensitivity and consideration toward environmental protection.
- No park or concession operation produces hazardous wastes; all waters and beaches are safe for public use without fear of human waste contamination.
- All parks and concessionaires are in compliance with state and federal environmental law.
- Solid wastes are minimized through recycling.
- Land and water pollution from human waste is eliminated.
- All parks and concessionaires are prepared to respond effectively to hazardous materials spills.
- All old dump sites are identified and rendered environmentally safe.
- Employees are protected against harmful working environments.
- Environmentally safe practices and products are integrated into park and concessioner operations.
- Emissions of air pollutants from park and concession operations are minimized to the extent technologically possible.
- Energy efficiency is achieved as technologically possible.

# National Park Service

## Intermountain Field Area

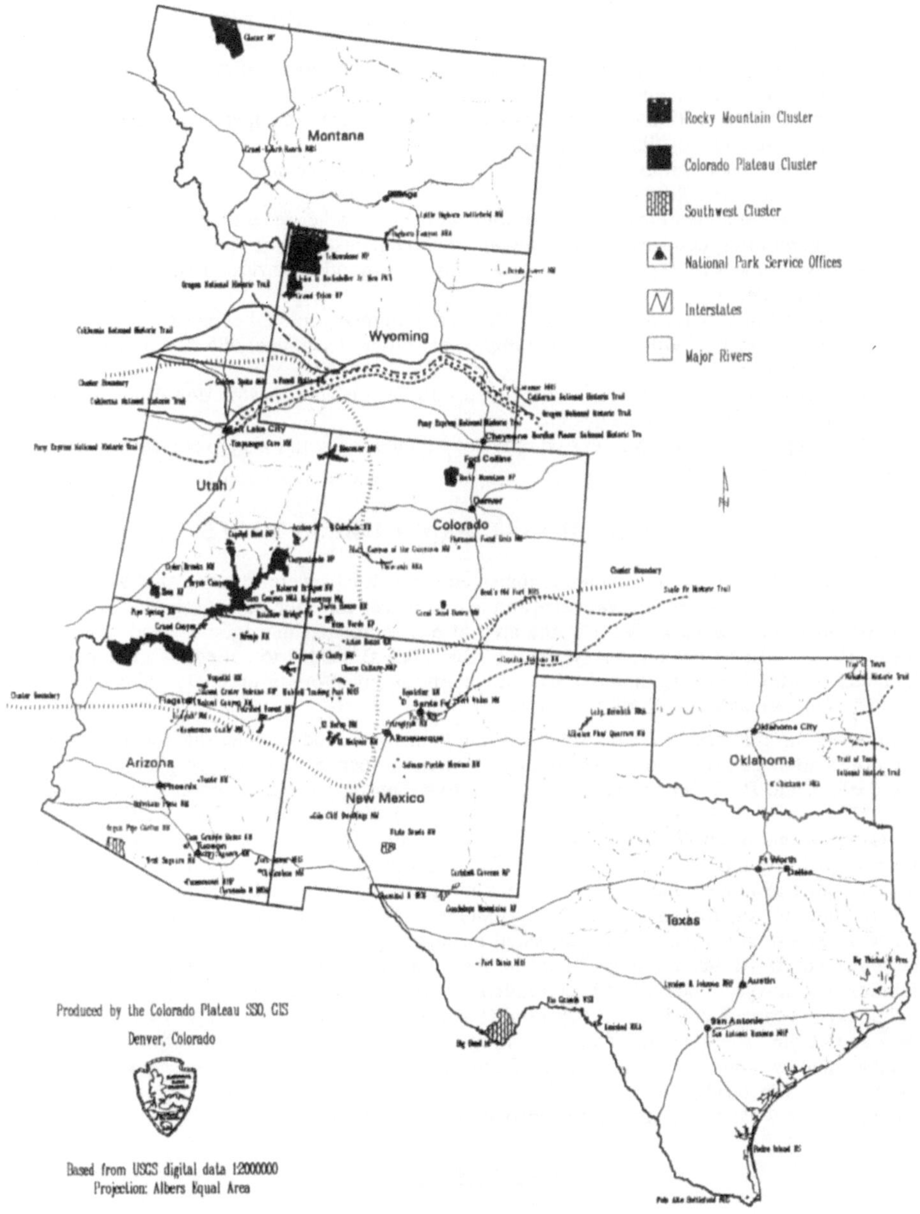


MAP 1

**Figure 1. National Park Service, Intermountain Field Area.**

The strategy for accomplishing the *Park Environments 2000* is based on a web of relations between the NPS headquarters office, support office or technical center, and park staff to protect and repair environmental damage from human use through development and implementation of an integrated environmental management program within each park that centers around sustainable development, focused on the following elements.

1. Environmental assessments of all NPS units and concession operations within the Intermountain Region;
2. Development of comprehensive, "user-friendly" environmental compliance guidelines for managers;
3. Pollution Prevention integrated into all park and concession operations; source reduction and recycling emphasized;
4. Environmental training for all employees; total employee participation in *Park Environments 2000*;
5. Open communications and cooperation with states, other federal agencies, and local communities; cooperative and mutual aid agreements in place;
6. Development of expertise and agreements to respond to hazardous materials spills and incidents;
7. Integration of environmental compliance requirements into all concession contracts; concessionaires will develop and implement programs comparable to *Park Environments 2000*;
8. Environmental protection is integrated into design and construction projects with emphasis on sustainable design;
9. Occupational health surveys completed; engineering solutions developed to protect employee health.

## ENVIRONMENTAL MANAGEMENT TOOLS

In order to assure that environmental management practices were implemented within the National Parks, a basic approach was reviewed and selected. Baseline data were needed in order to understand where the parks were in the area of environmental management and how they got there. The Pollution Prevention Assessment Team (Team) needed to collect baseline data which included information on product purchasing records, an inventory of products and wastes on-site, as well as historical waste disposal records.

In collecting this baseline data, several assumptions were made. One, the Team assumed that the information gathered would be used by the parks to better manage their programs and meet the goals of the *Park Environments 2000* initiative. During the initial scoping for the pollution prevention and regulatory compliance assessments, the Team met with the park's Hazardous Materials Coordinators and Facility Managers to determine what kind of data were needed and what format would be most useful. Second, the Team assumed that the IMR office would use these data to direct resources, including programmatic guidance, technical support, training, and project funding, for the region-wide area. By looking at all the parks in the region. the Team could identify "holes" in the parks' environmental management programs and deficiencies that should be remedied. Third, the Team assumed that these data could be used to validate the success, or failure, of the project once it had been implemented. The Team developed a measurement system that looked at the effectiveness of the training, the quality of the pollution prevention assessment data and at the effectiveness of incorporating the recommendations into parks' environmental management programs. And finally, the Team assumed that the information gathered would be used by the NPS headquarters office to assess the environmental status of the parks in the IMR and determine what next steps were needed at the national level.

## FIELD APPROACH

### Project Implementation

To achieve these goals, the NPS Intermountain Region looked for support and expertise from the Environmental Protection Agency Region VIII because of their commitment to a similar mission: "To restore and protect the ecological integrity of the Rocky Mountains and the plains and the health of their inhabitants."

The partnership was formalized through a federal Interagency Agreement (IAG) between EPA Region VIII and the NPS IMR for work to be done with the Rocky Mountain and Colorado Plateau clusters. Under this agreement, the NPS provides financial and management support to the project and coordinates activities with the cluster parks. The EPA provides technical expertise in pollution prevention and regulatory compliance and oversees the field work.

The objective of this IAG included:

a. Develop and present regulatory compliance and pollution prevention training for park personnel.
b. Conduct assessments of the parks in the Rocky Mountain and Colorado Plateau Cluster. (This effort has been expanded to include the Southwest Cluster, IMR.)
c. Provide environmental baseline information for these parks, along with recommendations to eliminate and/or reduce waste in these parks.
d. Design and conduct an implementation pilot project to determine what barriers, if any, exist in carrying out these recommendation.
e. Evaluate the prevention assessments to verify the quality and usefulness of the recommendations.
f. Develop a comprehensive, EPA/NPS "user-friendly" pollution prevention and environmental management tool kit for parks.

## Regulatory Compliance and Pollution Prevention Training

The EPA developed a training course for Park Service employees. The training team included EPA specialists in Resource Conservation and Recovery Act; Comprehensive Environmental Response, Compensation, Liability Act; Emergency Response; and Community Right-to-Know regulations, as well as pollution prevention/waste minimization practices and technologies. The goal of the training was to educate the Park Service on applicable environmental regulations and to give them the environmental tools they need to implement an environmental management program that would include regulatory compliance, pollution prevention, and integrated solid waste management practices.

As part of the training course, participants visited a vehicle maintenance facility in order to get "hands-on" training on how to use information presented in the course. During the tour, the park personnel identified the types and volumes of wastes (oil, glycol, solvents, etc.) generated by the facility, identified options to reduce and/or eliminate these wastes through better management practices, and assessed whether these options were "workable" (i.e., will it work at *this* facility, is there room for new equipment, is there a better way to purchase and/or store these products?).

Interaction during the training class and the field trip provided the EPA with valuable information concerning the specific needs of the Park Service in developing an environmental management program. These lessons can be applied to other federal land management agencies and used to assist them in developing their own programs.

## Pollution Prevention Assessments

Pollution prevention assessments of parks in the Rocky Mountain and Colorado Plateau Clusters were conducted during the spring, summer, and fall of 1995 (and will be completed for the parks in the Southwest Cluster by the fall of 1997). The specific objectives of this task included: (1) Identify current practices utilized by the parks for hazardous waste management and establish a baseline profile of waste types generated by a typical NPS facility (park); (2) Identify pollution prevention techniques that are being used, or could be utilized, to eliminate or reduce the amount of waste generated by the parks; (3) Establish solid waste baseline data on the amounts and current practices of solid waste generation, recycling, and disposal; (4) Identify any abandoned hazardous waste site(s) requiring more detailed investigation; and (5) Make recommendations for pollution prevention options to be adopted by each park in support of the goals of the *Park Environments 2000* initiative.

The EPA worked with a community college in the Denver metropolitan area to hire and train five environmental management students to conduct the assessments. The students were given an orientation to National Park Service operations and general guidelines for how they would interact with park personnel. They recognized that they would be "ambassadors" for the region's hazardous materials program and were there to help the parks solve problems. This work included helping parks obtain Material Safety Data Sheets, labeling hazardous waste containers, and providing product information. The students were supervised by a faculty member with extensive field experience. Additionally, the EPA project manager was on-site for several of the assessments. The assessments were discussed with park personnel during the training and were used as an incentive for each park to "rate high" during the assessments by putting into place the information received during the training. The assessments also served as a follow-up for the park Hazardous Materials Coordinators to clarify issues and questions raised during the training and to identify barriers to implementing pollution prevention ideas.

Each Pollution Prevention Assessment ranged in length from two days to two weeks, depending on the size and geographical constraints of the park operations. The pollution prevention survey staff, consisting of the students and the faculty advisor, initially met with NPS personnel to detail the objectives and outcomes of the visit, request introduction to appropriate park personnel, identify sources of information for the assessment, and to answer other questions regarding the project. The survey staff was particularly concerned about their liability if they found threats to human health and the environment and did not report these possible violations to the appropriate regulatory agency. It was agreed that if the survey personnel found such a threat that they would contact the IMR hazardous materials coordinator and this individual would decide on the proper course of action with park staff. The confidentiality of data was maintained through a procedure whereby the raw assessment forms were sent directly to the IMR and were never seen by EPA staff.

After the pollution prevention assessment was completed, the survey staff would meet with the park staff for an exit interview to discuss the observations and outline the pollution prevention recommendations for the park. These were identified as:

- *Immediate:* Recommendations that should be implemented immediately (0 to 6 months) due to potential for large environmental compliance penalties and fines, impact on the public and/or employee health and safety, or potential impacts on the environment.
- *Short-Term:* Recommendations that could be implemented within 1 to 2 years with additional funding and/or training of park employees.
- *Long-Term:* Recommendations that would require additional time, significant capital investments or planning to be implemented.

The park superintendent, or someone designated by the superintendent, was requested to sign the list of recommendations after agreeing or disagreeing with them. The pollution prevention assessments provide a valuable baseline data report for the parks.

## Implementation Pilot

A majority of the pollution prevention assessments demonstrate the need for additional environmental management training as well as a need to conduct a park-wide hazardous materials cleanup in order to achieve some of the goals of the *Park Environments 2000 Initiative.* To support this work, the IMR selected one park to serve as a pilot for implementing these recommendations of the pollution prevention assessment. The Implementation Pilot was designed to determine what next steps would work well to implement the recommendation, and what barriers exist that would hinder the effort. Mesa Verde National Park, located in southwest Colorado and internationally famous for its cultural resources, was selected because of their interest and motivation in meeting the goals of the *Park Environments 2000 Initiative*. The project included:

- Hazardous materials and spill prevention training updates for park personnel involved in the project.
- Selection of the short-term recommendations to be tested.
- Identification of the type and cost of equipment needed for the project.
- Development of the implementation plan.
- Technical assistance during the implementation activity.
- A final report detailing next steps and barriers.

The Implementation Pilot provided both the EPA and NPS with valuable information about the pathways to success, as well as the barriers, in implementing an environmental management program. This information is being used to plan the next steps for the program.

### Evaluating the Program Effectiveness - Measurement

Follow-up is needed to determine if the pollution prevention surveys and subsequent reports were valuable to the parks and to validate the information that was gathered in the field. A measurement project is currently underway which is designed to determine the quality of field data, to evaluate and quantify to what extent pollution prevention has been implemented in the parks following the pollution prevention assessments, and to document present barriers to implement pollution prevention options and collect suggestions from the parks on how the NPS regional office can provide additional support to the parks in achieving the goals of the *Park Environments 2000 Initiative*.

The NPS IMR will collaborate with parks to determine the next steps to meet the project goals: To empower parks to develop and implement an effective environmental management program that will meet the goals of the *Park Environments 2000 Initiative*; to find out what motivates the parks to change, what would it take for a park to make a commitment to environmental management, and where the IMR SuppOffice can best assist in reaching those goals.

### NPS Guidance Document and Resource Guide

The EPA used the information gathered during this project to design a "user-friendly" tool kit for environmental management. This tool kit includes a guidance document containing pollution prevention and regulatory compliance information, along with a resource manual containing practical information on how to obtain equipment and supplies.

## CONCLUSION

After investing some resources, both human and financial capital, we are faced with the question: So What? Have the partnership efforts really made a difference? Are Intermountain parks now generating less hazardous waste? Are they reducing their solid waste? Are employees committed to pollution prevention practices and processes? Right now, we are unable to answer these questions definitively. Indications from the budget process and feedback to the Support Office suggest that we have made a difference in some parks. In others, we would find it more difficult to see changes in practices or attitudes. So, what are some of the barriers?

### Top Down and Bottom Up

This is a top-down and bottom-up process, and will only work with committed people on both ends. We have heard regularly from staff people that they want to do the right things, but do not have management support. James Champy, chairman of Perot Systems Corporation, argues that "major change programs must be top-down and vision-driven" and, "at least initially, . . . must be actively led by senior management."[2] Likewise, managers complain about their overwhelming burdens, all done with shrinking budgets. Budget prospects are not favorable and the solution lies, at least to some extent, in creative ideas that have been developed by people like David Osborne. Work processes also must be subjected to a rigorous analysis, as suggested by the reengineering gurus, Michael Hammer and James Champy.[3]

### The Fallacy of the Big Stick

It is always tempting to suggest that the best way to gain a manager's attention is to use legal or regulatory leverage. This has proved to have short-term success, but long-term failure. As soon as managers feel that the pressure is no longer present, they will return to "business as usual." Rather, there has to be a process that will engage the manager so that he or she decides that there is some value to be derived from investment of their time or their staff.

## Data is Not the Answer

As noted above, we have completed comprehensive pollution prevention assessments of the 50 parks in the Rocky Mountain and Colorado Plateau clusters. Park staffs participated in the generation of these data, and park management approved all recommendations. Yet we know that in at least some parks, these assessments are not being used at all. When we mention them in training courses, for example, some people will ask how they can get copies. The budget statements submitted yearly often fail to reference the assessments, or the projects include requests that were never recommended. It is apparent that more has to be done to encourage the use of the assessments. The answer is not only knowledge, but empowerment of these people.

The NPS is committed to a decentralized structure, with much of the decision-making taking place at the local (park) level, combined with an evolving system of monitoring and oversight. Regulatory review and recent fines by the Environmental Protection Agency in one region of NPS have seemingly not generated much concern or alarm in other regions. And this would only produce some short-term changes in behavior. Given this framework, in order for an environmental management program to succeed, there have to be sophisticated initiatives that involve the stakeholder in the development, measurement, and implementation of the program. Outside interests provide technical assistance in facilitation, social science research, assessment, and the design of the outcome measures. But "ownership" of the program still rests with the park and its staff. They decide what they want to accomplish and how to measure the accomplishment, both in the short-term and long-term. For public sector agencies, the measurement piece will be increasingly more important as they struggle to comply with the Government Performance and Results Act. One such pilot program will be started next year in the Intermountain Region and will use the theory of outcomes evaluation and outcome measures to develop an environmental management program that will be driven by park interests. This innovative effort will challenge park management to select not only their own environmental goals, but also the means to evaluate success.[4]

## Think No Small Thoughts Nor Be Afraid to Implement Them

The world-renowned management consultant Peter F. Drucker some two years ago was so disturbed by public sector reinvention efforts that he wrote a long essay on the subject for *The Atlantic Monthly*. In this piece he asserts that "able people [in the federal government] are getting nowhere fast because their basic approach is wrong. They are trying to patch and to spot-weld here, there, and yonder . . . . There will be no results unless there is a radical change in the way the federal government and its agencies are managed and paid. The habit of continuous improvement has to be built into all government agencies, and has to be made self-sustaining."

## REFERENCES

1. *National Park Service Strategic Plan*, 1996, p. 26.
2. Champy, J.A., "Preparing for organizational change." In: *The Organization of the Future*, Hesselbein, F., Goldsmith, M., and Beckhard, R., Eds. Jossey-Bass Publishers, San Francisco, CA, 1997, p. 11.
3. Hammer, M. and Champy, J. *Reengineering the Corporation: A Manifesto for Business Revolution*, Harper Collins Publishers, New York, 1993.
4. Salmon, J. and Schene, M. "Outcomes measures and their implementation in environmental management programs," *Total Quality Environmental Management*, forthcoming.
5. Drucker, P.F. "Really reinventing government," *The Atlantic Monthly*, 275(2), February 1995, p. 52.

# 47 THE EFFECTS OF THE PRETREATMENT PROGRAM ON HAULED WASTE IN WISCONSIN

**Ken Denow** and **Vanessa Thompson**
Wisconsin Department of Natural Resources
Milwaukee, Wisconsin 53212

## ABSTRACT

In 1983, the Wisconsin Department of Natural Resources (WDNR) began delegating the pretreatment program* to Publicly Owned Treatment Works (POTWs) with design flows of greater than or equal to (≥) 5 million gallons per day (MGD). The delegation required these POTWs to establish procedures to identify and characterize hauled waste. POTWs permitted haulers, analyzed the waste, verified compliance, and enforced their sewer use ordinance (SUO).

POTWs without pretreatment programs are not required to establish the same procedures and/or compliance measures. This is causing an inequity in the control of hauled wastes between POTWs with pretreatment delegation and those without. The result is a shift in the disposal of hauled waste to POTWs without pretreatment programs. There is evidence that smaller POTWs were accepting noncompliant categorical wastes, some POTWs were experiencing negative plant impacts, and some POTWs with pretreatment programs were no longer accepting industrial hauled waste.

These trends caused us to study hauled waste programs at POTWs. Our research involved collecting information from POTWs through surveys, on-site inspections, personal contacts, informational reporting forms, and background literature review.

Information was obtained on the volume and types of waste being discharged, facility size and treatment capability, how the waste was being identified and characterized, frequency of negative plant impacts, and guidance and training needed by POTW operators.

The results confirmed:

- The volume of waste hauled in Wisconsin has increased dramatically since 1985.
- POTWs <5 MGD are accepting considerably more waste since the inception of the pretreatment program.
- POTWs <5 MGD are experiencing negative plant impacts due to hauled waste.
- POTW operators want and need additional information on hauled waste.

The results support the premise that the pretreatment program affects where and how hauled wastes are disposed. How hauled waste is managed needs to be recognized and understood by the regulatory, environmental engineering, and operations community in meeting the needs of POTWs and the environment.

The United States Environmental Protection Agency (U.S. EPA) and states need to spend more effort on educational and informational programs to provide the needed tools to all POTWs. The U.S. EPA needs to evaluate existing rules and regulations to learn if they are causing inequity between the acceptance and regulation of hauled waste at POTWs.

---

*The pretreatment program is delegated to those POTWs with design flow >5 million gallons per day and is a permit requirement. The term POTW ≥5 will represent those POWTs with pretreatment programs, while POTW <5 will represent those without.

*52nd Purdue Industrial Waste Conference Proceedings*, 1997, Ann Arbor Press, Chelsea, Michigan 48118. Printed in U.S.A.

## INTRODUCTION

The Wisconsin Department of Natural Resources (WDNR) delegated the pretreatment program to Publicly Owned Treatment Works (POTWs) with design flows of greater than or equal to (≥) 5 million gallons per day (MGD) in 1983. The delegation required these POTWs to establish procedures to identify and characterize hauled waste types, control waste discharged, and verify waste discharged. The pretreatment program established procedures and controls for POTWs ≥ 5 MGD to follow when accepting hauled waste. POTWs <5 MGD do not have to establish the same level of procedures and controls.

Since 1983, the WDNR has had the authority and requirement to conduct an audit of a POTW's pretreatment program. The audit authority is limited to POTWs ≥ 5 MGD. The information obtained from these audits since 1983 showed a trend of less waste taken to POTWs ≥ 5 MGD.

This study determined why this was happening and whether the volume of hauled waste discharged at POTWs <5 MGD had increased. A comparison of the volume of waste discharged at the two types of POTWs was performed to determine if the shift to POTWs <5 MGD occurred statewide. Finally, this study compared the procedures and controls used at the POTWs.

POTWs ≥ 5 MGD who accept hauled waste must develop and implement procedures to accept hauled waste. These procedures must allow them to control and verify the waste discharged including identification of waste types to determine if they are subject to hazardous waste regulations, categorical discharge limits, and local limits. This means they must:

- identify and locate all industrial users who might be subject to the pretreatment program
- identify the character and volume of pollutants contributed to the POTW by industrial users
- notify industrial users of applicable pretreatment standards and requirements including those relating to user charges and solid or hazardous waste disposal
- randomly sample and analyze the effluent from industrial users and conduct surveillance and inspection activities in order to identify problems, independently from the information supplied by industrial users
- evaluate the need for local limits and adopt and apply local limits to industrial users. It is important to note that these local limits are site-specific and typically contain discharge limits for metals and other pollutants.

These procedures and controls are not required to be adopted and implemented by POTWs <5 MGD design flow. It is at the discretion of POTWs <5 MGD to implement a hauled waste program. In addition, these POTWs are not required to conduct a local limit evaluation, and most of these POTWs' sewer use ordinances typically contain discharge limits for conventional pollutants only.

The research that resulted in this paper was triggered by the following observations in southeastern Wisconsin:

- **Failure of some POTWs ≥5 MGD with pretreatment programs to implement a hauled waste program.** This included not identifying and characterizing hauled waste and not ensuring the waste complied with local limits.
- **A shift of hauled waste to POTWs <5 MGD without pretreatment programs.** When WDNR scrutinized the hauled waste programs at POTWs ≥5 MGD, the disposal of waste frequently shifted to POTWs <5 MGD.
- **The occurrence of some plant upsets and operational problems due to hauled waste.** In most of these cases, no violations of permit limits occurred, and therefore, interference or pass-through could not be demonstrated.
- **Noncompliant categorical wastewater hauled to POTWs <5 MGD without pretreatment programs.** In most of these cases, the operator did not know the waste was subject to categorical discharge limits and had no procedures in place to make this determination.
- **POTWs with uncontrolled discharge sites.** WDNR staff observed remote discharge sites with no controls, no sampling, no recording log for incoming waste, etc.

- **POTWs soliciting haulers and generators to bring waste to their plants.** By doing this, plant operators expected to generate revenue to help pay debt from plant construction and to avoid increasing the local user fees.
- **Wisconsin Wastewater Operators Association, Inc. (WWOA) members' concerns about hauled waste were also increasing.** The organization contacted WDNR staff and requested additional information about hauled waste programs.

## METHODOLOGY

The research consisted of collecting data from three sources: a qualitative questionnaire, a quantitative questionnaire, and on-site inspections. WDNR personnel conducted on-site inspections and interviews of POTW operators, superintendents, and administrators at 36 POTWs in Wisconsin from April through December 1996. These inspections were conducted at 28 POTWs <5 MGD and at 8 POTWs ≥ 5 MGD. These inspections were conducted to learn about the discharge point locations, the access and control at the discharge points, and other information the operator could provide. The information gathered revealed several problem areas, with additional data needed from the POTWs. Later in 1996, WDNR and WWOA worked together to develop and distribute a qualitative hauled waste questionnaire to obtain information on hauled waste programs POTWs had in place to effectively handle hauled waste. The questionnaire was published in WWOA's June 1996 trade journal, the *Clarifier*, and was mailed to all WWOA members and distributed at regional WWOA meetings. WWOA is a statewide association. One hundred twelve responses were received, of which 102 POTWs indicated they accepted hauled waste. Wisconsin has approximately 700 POTWs.

The purpose of the questionnaire was to determine if:

- The POTW accepts hauled waste, and if so, what types of waste are accepted.
- The POTW operator needs additional information and training on hauled waste.
- The POTW has procedures to approve, control, and audit hauled waste received.

The information requested was to determine if the POTW had procedures for approving, controlling, and auditing hauled waste and to learn the effectiveness of them. These questions are the same types of questions WDNR or the U.S. EPA asks when conducting an audit of a POTW's pretreatment program.

The questionnaire asked POTWs for the parameters analyzed on hauled waste samples. No information was requested about what parameters are analyzed for different waste types, nor for the analytical methods used.

Information was requested to learn what POTWs asked when accepting waste: Does the hauler or generator sign a form, do you use a form to identify the type of waste, do you obtain the name/address of the generator, and do you use a form which lists the waste characteristics?

Another set of controls listed in the questionnaire were: Does the POTW employee monitor the dumping of waste, do you control the hours of discharge, do you control access to the discharge point, and do you screen haulers or inspect generators?

Finally, POTWs were asked: Do you differentiate between domestic and nondomestic wastes, do you identify waste that could cause operational problems, do you identify Hazardous Waste, and do you identify waste subject to Categorical standards?

POTWs were asked what additional training and guidance they need to effectively implement a hauled waste program. Eighty-one percent of the POTWs stated additional guidance is needed. The specific areas of concern varied, depending on the individual POTWs. Some POTWs wanted additional information and training on special waste types such as manufacturing waste, LUST, leachate, and remediation wastewater.

A quantitative questionnaire was mailed to the 367 Wisconsin POTWs with design flow ≥ 0.1 MGD (including 25 POTWs with pretreatment programs). There was an overall response rate of about 30% for POTWs <5 MGD and representing all regions of the state. POTWs ≥ 5 MGD had a 100% response rate.

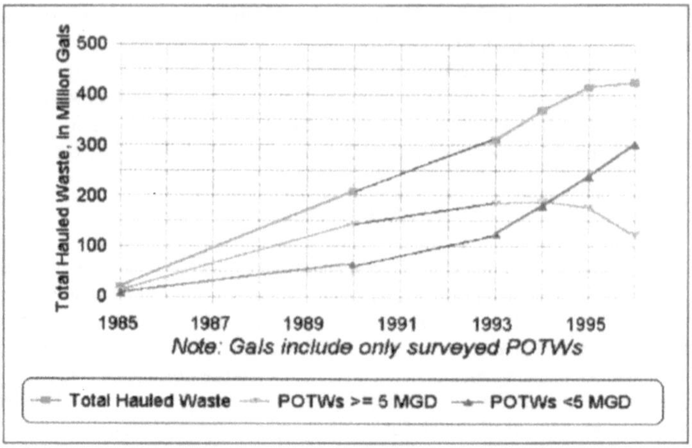


**Figure 1. All hauled wastes to Wisconsin POTWs. Destination of waste, by plant size.**

## RESULTS

The volume of waste hauled to POTWs in Wisconsin has changed dramatically since 1985 (Figure 1 and Table I). The data from the quantitative questionnaire show a steady increase in the total volume of hauled waste from 19 million gallons (Mgals) in 1985 to 415 Mgal in 1995. The percentage of hauled waste discharged to POTWs with pretreatment programs (POTWS ≥ 5 MGD design flow) has decreased. In 1990, POTWS ≥ 5 MGD took 69% of all hauled waste. By October 1996, they received only 29% of the total hauled waste, according to a recent questionnaire (Figure 2). The total volume of hauled waste has increased by 207 millions gallons since 1990. Eighty-five percent or 175 million gallons of the increased volume is going to POTWs <5 MGD (POTWs with design flows less than 5 MGD).

The second questionnaire requested the design flow of the POTW, limits for BOD (biological oxygen demand), TSS (total suspended solids), ammonia and phosphorus; volumes of waste accepted in 1985, 1990, 1993, 1994, 1995, and 1996, and for the following categories: domestic septage, domestic holding tank, nondomestic holding tank, LUST (Leaking Underground Storage Tanks), other (which was sorted and renamed leachate), and manufacturing waste. The survey also asked the respondents why the volume of hauled waste increased or decreased. Tables I–VII show the raw data from the second survey, sorted by hauled waste type and by POTW size.

**Table I. Total of All Hauled Waste Types**

| | All POTWs | POTWS ≥ 5 MGD | | | POTWS < 5 MGD | | |
|---|---|---|---|---|---|---|---|
| Year | Gallons of Waste Hauled | Gallons of Waste Hauled | % of Total Waste Hauled | Number of POTWs Accepting Waste | Gallons of Waste Hauled | % of Total Waste Hauled | Number of POTWs Accepting Waste |
| 1985 | 19,926,776 | 11,393,859 | 57% | 4 | 8,532,917 | 43% | 26 |
| 1990 | 208,524,295 | 143,783,552 | 69% | 16 | 64,740,743 | 31% | 42 |
| 1993 | 309,668,978 | 184,436,508 | 60% | 22 | 125,232,470 | 40% | 68 |
| 1994 | 369,044,253 | 188,345,180 | 51% | 25 | 180,699,073 | 49% | 89 |
| 1995 | 415,660,351 | 175,869,112 | 42% | 24 | 239,791,239 | 58% | 104 |
| 1996[a] (10 mo) | 424,450,331 | 121,824,941 | 29% | 23 | 302,625,390 | 71% | 102 |

[a]1996 data represents January through October.

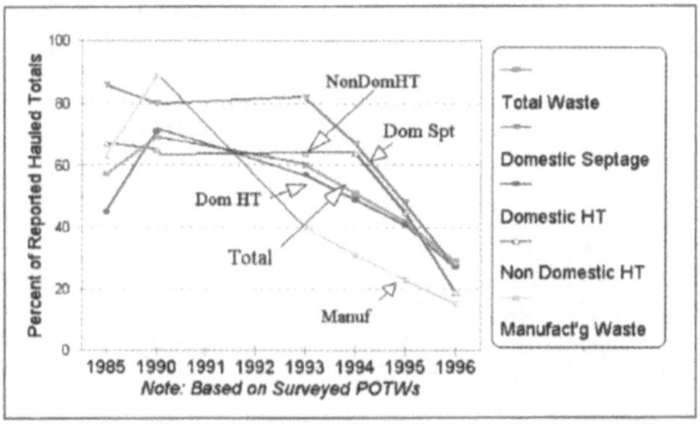


**Figure 2. POTWs > 5 MGD & select waste types (percentage of reported hauled totals).**

**Table II. Domestic Septage**

| | All POTWs | POTWS ≥ 5 MGD | | | POTWS < 5 MGD | | |
|---|---|---|---|---|---|---|---|
| Year | Gallons of Waste Hauled | Gallons of Waste Hauled | % of Total Waste Hauled | Number of POTWs Accepting Waste | Gallons of Waste Hauled | % of Total Waste Hauled | Number of POTWs Accepting Waste |
| 1985 | 2,248,050 | 1,941,700 | 86% | 2 | 306,350 | 14% | 13 |
| 1990 | 18,513,257 | 14,854,830 | 80% | 8 | 3,648,427 | 20% | 24 |
| 1993 | 21,901,716 | 17,934,915 | 82% | 11 | 3,966,801 | 18% | 35 |
| 1994 | 28,836,983 | 19,217,091 | 67% | 11 | 9,619,892 | 33% | 35 |
| 1995 | 24,686,313 | 11,955,681 | 48% | 14 | 12,730,632 | 52% | 40 |
| 1996[a] (10 mo) | 13,134,204 | 3,533,394 | 27% | 11 | 9,600,810 | 73% | 40 |

[a]1996 data represents January through October.

**Table III. Domestic Holding Tank**

| | All POTWs | POTWS ≥ 5 MGD | | | POTWS < 5 MGD | | |
|---|---|---|---|---|---|---|---|
| Year | Gallons of Waste Hauled | Gallons of Waste Hauled | % of Total Waste Hauled | Number of POTWs Accepting Waste | Gallons of Waste Hauled | % of Total Waste Hauled | Number of POTWs Accepting Waste |
| 1985 | 9,422,950 | 4,210,800 | 45% | 2 | 5,223,150 | 55% | 16 |
| 1990 | 138,515,293 | 98,467,718 | 71% | 12 | 40,047,575 | 29% | 37 |
| 1993 | 169,037,571 | 95,754,434 | 57% | 17 | 73,283,137 | 43% | 56 |
| 1994 | 248,526,232 | 121,535,439 | 49% | 18 | 126,990,793 | 51% | 63 |
| 1995 | 261,773,792 | 106,451,818 | 41% | 17 | 155,321,974 | 59% | 70 |
| 1996[a] (10 mo) | 254,060,150 | 69,763,969 | 27% | 17 | 184,296,181 | 73% | 70 |

[a]1996 data represents January through October.

In 1985, the domestic holding tank waste only made up 47% (9 Mgals) of total volume of hauled waste and by 1995 domestic holding tank waste made up 63% (261 Mgals) of the total volume of hauled waste. The remaining 1995 percentages of waste were 15%, leachate; 8%, non-domestic holding tank; 6%, domestic septage; 4%, LUST; and 4%, manufacturing waste.

POTWs ≥ 5 MGD reported the volumes of hauled waste discharged have decreased. In 1990, POTWs ≥ 5 MGD received 69% of the total volume of waste. The percentage of waste discharged at POTWs ≥ 5 MGD continued to decline through 1996 by approximately 9% per year.

**Table IV. Nondomestic Holding Tank**

| | All POTWs | POTWS ≥ 5 MGD | | | POTWS < 5 MGD | | |
|---|---|---|---|---|---|---|---|
| Year | Gallons of Waste Hauled | Gallons of Waste Hauled | % of Total Waste Hauled | Number of POTWs Accepting Waste | Gallons of Waste Hauled | % of Total Waste Hauled | Number of POTWs Accepting Waste |
| 1985 | 6,209,559 | 4,134,659 | 67% | 1 | 2,074,900 | 33% | 3 |
| 1990 | 12,250,430 | 7,929,180 | 65% | 4 | 4,321,250 | 35% | 5 |
| 1993 | 29,458,506 | 18,724,539 | 64% | 8 | 10,733,967 | 36% | 12 |
| 1994 | 24,126,882 | 15,419,200 | 64% | 9 | 8,707,682 | 36% | 20 |
| 1995 | 34,021,132 | 15,325,825 | 45% | 9 | 18,695,307 | 55% | 25 |
| 1996[a] (10 mo) | 54,615,973 | 10,144,518 | 19% | 9 | 44,471,455 | 81% | 29 |

[a]1996 data represents January through October.

**Table V. Manufacturing Waste**

| | All POTWs | POTWS ≥ 5 MGD | | | POTWS < 5 MGD | | |
|---|---|---|---|---|---|---|---|
| Year | Gallons of Waste Hauled | Gallons of Waste Hauled | % of Total Waste Hauled | Number of POTWs Accepting Waste | Gallons of Waste Hauled | % of Total Waste Hauled | Number of POTWs Accepting Waste |
| 1985 | 162,650 | 101,700 | 63% | 1 | 60,950 | 37% | 2 |
| 1990 | 1,044,618 | 928,978 | 89% | 4 | 115,640 | 11% | 4 |
| 1993 | 8,992,770 | 3,557,420 | 40% | 7 | 5,435,350 | 60% | 11 |
| 1994 | 12,342,112 | 3,669,926 | 31% | 7 | 8,672,196 | 70% | 28 |
| 1995 | 18,256,405 | 4,065,200 | 23% | 7 | 14,191,205 | 77% | 38 |
| 1996[a] (10 mo) | 24,007,758 | 3,569,200 | 15% | 8 | 20,438,558 | 85% | 45 |

[a]1996 data represents January through October.

**Table VI. Leaking Underground Storage Tanks (LUST)**

| | All POTWs | POTWS ≥ 5 MGD | | | POTWS < 5 MGD | | |
|---|---|---|---|---|---|---|---|
| Year | Gallons of Waste Hauled | Gallons of Waste Hauled | % of Total Waste Hauled | Number of POTWs Accepting Waste | Gallons of Waste Hauled | % of Total Waste Hauled | Number of POTWs Accepting Waste |
| 1985 | 1,000 | 0 | | 0 | 1,000 | 100% | 1 |
| 1990 | 239,000 | 2,000 | 1% | 1 | 237,000 | 99% | 4 |
| 1993 | 15,305,459 | 11,808,854 | 77% | 6 | 3,496,605 | 23% | 9 |
| 1994 | 11,309,219 | 6,950,698 | 61% | 10 | 4,358,521 | 39% | 17 |
| 1995 | 14,552,804 | 11,932,201 | 82% | 9 | 2,620,603 | 18% | 30 |
| 1996[a] (10 mo) | 21,236,982 | 13,432,063 | 63% | 10 | 7,804,919 | 37% | 31 |

[a]1996 data represents January through October.

By 1994, the POTWs <5 MGD that responded to the quantitative questionnaire reported an increase in the volume of waste discharged and the number of POTWs <5 MGD accepting hauled waste. In 1993, 68 POTWs accepted hauled waste and by 1995, this number had increased to 104. POTWs <5 MGD reported that a major reason for the increase of hauled waste at their plant was due to other POTWs having the following in place: more expensive disposal costs, more restrictions on how and what waste types are accepted, and enforced discharge limits.

**Table VII. Leachate**

| | All POTWs | POTWS ≥ 5 MGD | | | POTWS < 5 MGD | | |
|---|---|---|---|---|---|---|---|
| Year | Gallons of Waste Hauled | Gallons of Waste Hauled | % of Total Waste Hauled | Number of POTWs Accepting Waste | Gallons of Waste Hauled | % of Total Waste Hauled | Number of POTWs Accepting Waste |
| 1985 | 1,871,567 | 1,005,000 | 54% | 2 | 866,567 | 46% | 3 |
| 1990 | 37,961,697 | 21,600,846 | 57% | 6 | 16,360,851 | 43% | 11 |
| 1993 | 64,972,956 | 36,656,346 | 56% | 11 | 28,316,610 | 44% | 18 |
| 1994 | 43,902,825 | 21,552,836 | 49% | 10 | 22,349,989 | 51% | 23 |
| 1995 | 62,369,905 | 26,138,387 | 42% | 11 | 36,231,518 | 58% | 24 |
| 1996[a] (10 mo) | 57,395,264 | 21,381,797 | 37% | 13 | 36,013,467 | 63% | 25 |

[a]1996 data represents January through October.

Most of the operators at POTWs <5 MGD responding to the quantitative questionnaire did not distinguish between domestic and nondomestic holding tank waste in their reported volumes. In such cases, all holding tank waste was listed under "Domestic HT."

The volume of manufacturing waste has increased from 0.16 Mgals in 1985 to 18.26 Mgals in 1995. (Manufacturing waste included any type ranging from metal finishing to food processing). During 1996, the POTWs ≥5 MGD began to report an increase in the number of manufacturing facilities that transport and discharge their wastewater to other POTWs, rather than discharge wastewater via the sanitary sewer.

The qualitative questionnaire asked POTWs to report the frequency of negative plant impacts due to hauled waste. The specific question was: How often, if ever, has the hauled waste been incompatible with your POTW? Forty-six POTWs <5 MGD (57%) reported impacts at least once a year, referred to as Group A and 35 POTWs < 5 MGD (43%) reported impacts as less than once per year or never, referred to as Group B. The physical and administrative controls these two groups employed were compared to each other to determine, if possible, which controls are the most important to consider when implementing an effective hauled waste program.

A higher percentage of POTWs in Group B reported they obtain more information (Table VIII), they carry out more controls (Table IX) and they differentiate waste types (Table X) more frequently than the POTWs in Group A.

**Table VIII. Forms Used by POTWs <5 to Gather Information**

| Group | Signature | Type of Waste | Generator Information | Waste Characteristics |
|---|---|---|---|---|
| Group A | 26% | 41% | 17% | 13% |
| Group B | 45% | 54% | 48% | 22% |

**Table IX. Physical Controls Used by POTWs <5**

| Group | Monitor Discharge | Control Access | Control Hours | Screen Haulers/ Inspect Generators |
|---|---|---|---|---|
| Group A | 26% | 28% | 36% | 6% |
| Group B | 65% | 74% | 71% | 51% |

**Table X. Criteria Used by POTWs <5**

| Group | Domestic vs. Non | Problem Wastes | Hazardous Wastes | Categorical Waste |
|---|---|---|---|---|
| Group A | 67% | 34% | 17% | 6% |
| Group B | 57% | 57% | 42% | 28% |

The data show there is an inverse relationship between the number and frequency of negative plant impacts and the degree of control exerted over the actual discharge of the waste at these POTWs < 5. POTWs said they ask for analytical data on the waste, review the data received, and determine if the waste is compatible prior to accepting the waste.

Many of the POTWs reporting a decrease in the volume of waste received, stated the decline occurred when they started to monitor and control the waste discharged.

Most of the POTWs, regardless of size, stated a general handbook (88%) and manual (53%) would best fit their needs for written guidance, while about one in four felt statewide pricing, state and federal regulations, and receiving station design would be beneficial.

Sixty-four percent of the POTWs ≥ 5 MGD and 88% of the POTWs < 5 MGD stated they need additional training. The types of training most preferred were seminars (52%) and presentations at annual and regional Wastewater Operator meetings, 60% and 80%, respectively. Eighteen percent of the respondents indicated on-site training would be useful.

The receiving stations ranged from a manhole at a remote location to a fully automated system where the entrance and the discharge are recorded electronically.

Many POTWs did not control access to the discharge location and did not verify that the waste discharged was the same waste that had been approved. For example, one POTW used a mailbox to collect manifests from haulers. The mailbox was open and overflowing with manifests from the prior four months. In this situation, it was clear there were very few or no controls in place over the discharge of hauled waste.

One POTW ≥ 5 MGD requires the hauler to take a sample from the truck while at the plant, then take the sample into the POTW's laboratory and personally hand the sample to lab personnel. This POTW automatically samples every tenth load received from each hauler and periodically analyzes all samples on a randomly selected day. This POTW reports no problems with wastes discharged or with haulers.

Another POTW ≥ 5 MGD requires haulers to collect a sample, record discharge in a log book, and use a three-part manifest. The samples collected are compared daily to the physical characteristics recorded on the waste profile. In addition, a POTW employee randomly appears at the discharge station to collect the samples from the discharging trucks. These same samples are then analyzed to verify that the waste received was the waste approved. Again, this POTW reports few or no problems with discharges or with the haulers.

In both of these cases, the POTWs are ≥ 5 MGD. The haulers are permitted, the waste must be approved prior to discharge, and the discharge location is only open during the day when staff is at the POTW. Both of these POTWs reported that when these controls were put in place, the volume of waste received decreased.

## CONCLUSIONS

The volume of hauled waste has increased dramtically since 1985, as has the number of POTWs <5 MGD accepting hauled waste. The volume of hauled waste discharged at POTWs < 5 MGD increased from 8 Mgals in 1985 to 302 Mgals as of October 1996. This accounted for 71% of the total waste discharged at POTWs. This is significant because a larger volume of waste is going to POTWs <5 MGD while the focus of the regulatory agencies is on the POTWs ≥ 5 MGD. The number of POTWs < 5 MGD who accepted hauled waste increased to 104 in 1995 from 26 in 1985. Forty-six percent of the POTWs < 5 MGD reported negative plant impacts at least once a year due to hauled waste. The data show that negative plant impacts directly correlate to the administrative and physical controls exerted over hauled waste. Those POTWs experiencing negative plant impacts had fewer administrative and physical controls in place.

This study shows POTWs < 5 MGD receive a higher percentage of all hauled waste in Wisconsin, but are not mandated to implement hauled waste program. Only POTWs ≥ 5 MGD have the requirement to implement a hauled waste and accept larger volumes of waste.

The results support the premise that the pretreatment program affects the disposal of hauled waste. The regulations and restrictions imposed by federal, state, and local agencies brought about much of the change. The management of hauled waste needs to be recognized and understood by the regulatory, environmental engineering, and operations community. This will allow them to meet the needs of all POTWs and the environment.

The data obtained demonstrate that the frequency of negative plant impacts is directly related to controls exerted over the waste discharged and a "presence" by the POTW. The comparison of the POTWs in Group A and Group B (Tables VIII–X) demonstrate that the Group B POTWs experienced far fewer negative plant impacts due to hauled waste. A greater percentage of POTWs in Group B required enough information about a waste to enable them to make informed decisions about accepting a waste, differentiating waste types, requiring signatures from the haulers, and requiring generator information.

The POTWs obtain analytical data on the wastes but no determination was made to see if the parameters analyzed were proper for the waste type and if the analytical methods used were appropriate (40 CFR Part 136).

POTWs reporting a decrease in waste provided the following reasons for the decrease: regulatory obligations to comply with the pretreatment regulations, lower disposal costs at other POTWs, and lack of controls over the disposal of hauled waste at other POTWs.

The data show that to implement an effective hauled waste program, a POTW needs to have a three-step program in place. It should contain an approval process, controls over the actual discharge of the waste, and audit procedures. The data do not suggest that an extensive analyzing program is needed, but that a "presence" is effective.

At POTWs where an established hauled waste program was in place, the operators reported that a "presence" over the program is necessary. They explained that while every hauled waste program may be unique to each POTW, it is important to verify what the hauler is discharging. The verification frequency may vary depending on the number of haulers and number of different waste types. Without this, they stated, theyw ould begin to receive waste that did not comply with local discharge limits and/or waste that had not been approved (i.e., the hauler would try to sneak something in). They also reported the reliability of the hauler varied from hauler to hauler and this is why controls are important.

POTWs stated additional training and guidance is needed to effectively implement a hauled waste program. A guidancd manual would be the most useful. The U.S. EPA should develop and distribute this manual to ensure national consistency. The U.S. EPA and the states need to emphasize education and information programs to all POTWs so they all can implement effective hauled waste programs.

The U. S. EPA and WDNR need to evaluate this inquity between POTWs further to determine the impact this has on the POTWs and the environment. This study determined that the pretreatment program does have an impact on hauled waste.

More importantly, further work is needed to design a stepped-control program which is appropriate for the time and resources for operators at smaller plants while ensuring that all wastes discharged at all POTWs are controlled equally.

## RECOMMENDATIONS

Based on the information gathered for this paper and our experience, it is recommended that the following items be considered by all POTWs who receive hauled waste:

1. At a minimum, POTWs should determine the following prior to accepting waste:
   - if the waste is domestic
   - what process generated the waste
   - what pollutants and concentration are present
   - if the waste is subject to categorical pretreatment standards
   - if the waste is a hazardous waste
   - why the generator wants to haul this waste
2. All POTWs must have procedures in place to approve, control, and audit wastes discharged.
3. The U.S. EPA needs to develop guidance and training on how to implement a hauled waste program. This should be designed specifically for POTWs without pretreament programs.

## BIBLIOGRAPHY

WDNR. 1992. Wisconsin Administrative Code, Chapter NR 211, General Pretreatment Regulations.

U.S. Environmental Protection Agency, Office of Water Enforcement and Permits, Washington, DC. 1994. Guidance Manual for the Identification of Hazardous Wastes Delivered to Publicly Owned Treatment Works by Truck, Rail, or Dedicated Pipe.

Paul A. Blakeslee. Surface Water Quality Division Field Operations Guidance. State of Michigan. Acceptance of Trucked Wastewater at Publicly Owned Treatment Works, 1994.

Wisconsin Wastewater Operators Association, Inc. (WWOA) and WDNR. Surveys of Hauled Waste Received at Wastewater Treatment Plants, 1996.

WDNR. Volume of Hauled Waste Questionnaire Results from Publicly Owned Treatment Works. Wastewater Section Southeast District, 1996.

U. S. Environmental Protection Agency. Federal Register 40 CFR Part 136, Guidelines Establishing Test Procedures for the Analysis of Pollutants under the Clean Water Act, 1984.

U.S. Environmental Protection Agency. Federal Register 40 CFR 40 Part 403, General Pretreatment Regulations for Existing and New Sources of Pollution. Water Management Division, 1983.

United States Environmental Protection Agency, Office of Water, Industrial User Permitting Manual, September 1989.

# 48 RISK-BASED CORRECTIVE ACTION AT A CHEMICAL MANUFACTURING FACILITY

**Greg A. Gilliland**, Principal Engineer
Texas Eastman Division, Eastman Chemical Company
Longview, Texas 75607

**Ram D. Ramaswami**, Senior Project Manager
Parsons Engineering Science, Inc.
Denver, Colorado 80290

## ABSTRACT

This chapter summarizes a risk-based approach to support bioremediation as a technology for site closure. The approach to closure was presented to the Texas Natural Resource Conservation Commission (TNRCC). TNRCC has approved the permit modification, and full-scale bioremediation of the lagoon sludge and soil mixture is currently underway. This risk-based corrective approach by Texas Eastman Division will result in substantial cost savings and is protective of human health and environment.

## INTRODUCTION

Risk assessment is used as a risk management tool to integrate remedial action alternatives and to establish cleanup criteria. The risk assessment process is a detailed scientific evaluation of the site including hazard identification, toxicity assessment, and exposure assessment, resulting in characterization of site-specific risks.

As a remedial alternative, bioremediation has the advantages of being cost-effective as well as environmentally sound. The risk assessment is complementary to bioremediation because it ensures cleanup to acceptable levels based on health and environmental protective criteria without excessive costs. The risk-based approach to closure using bioremediation as a remedial alternative is being implemented at the Texas Eastman Division (TEX) of Eastman Chemical Company.

TEX manufactures organic chemicals at the Longview, Texas Facility. Past operations in the wastewater treatment lagoons have generated sludges that are defined as hazardous under the Resource Conservation and Recovery Act (RCRA). Wastewater discharges in Lagoon 1 ceased in 1983. Under the original closure plan,[1] the pumpable sludge in the lagoon was removed and treated in an on-site incinerator. Approximately 15,000 cubic yards of sludge and soil mixture remain in Lagoon 1. TEX found it impractical to continue burning at the high rate anticipated in the closure plan. High silica content of the sludge and soil mixture has caused operational problems including kiln shutdown. The generated ash consumes a large amount of scarce hazardous landfill space. In the revised Closure Plan for Lagoon 1, TEX proposed to bioremediate the remaining sludge and soil mixture.

This chapter is organized into six sections. The introduction includes a brief description of the project. Section 2 addresses the bioremediation demonstration study and the results of the study. Section 3 presents the site-specific risk assessment approach and the results of this approach. Section 4 presents the current status of the project. Section 5 acknowledges the various people and organizations who made significant contributions to this project, and Section 6 presents the references used in this study.

*52nd Purdue Industrial Waste Conference Proceedings*, 1997, Ann Arbor Press, Chelsea, Michigan 48118. Printed in U.S.A.

## BIOREMEDIATION DEMONSTRATION STUDY

Bioremediation is a managed treatment process that uses microorganisms to degrade and transform organic chemicals in contaminated soils, sludges, and residues. The final metabolic products in an aerobic environment usually are cell mass, carbon dioxide, and water. Generally, degradation rates of organic chemicals are faster in an aerobic environment and supply of oxygen becomes one of the critical factors.

### Treatability Evaluation

Initial bioremediation feasibility studies were conducted in laboratory-scale reactors and pilot-scale solid-phase reactors in the field.[2] The objective of these studies was to develop site-specific information on the bioremediation of Lagoon 1 sludge.

Based on the results of the treatability evaluation, a field demonstration was performed. This demonstration was conducted to obtain engineering data on the operation of a large-scale land treatment unit. The Direct Contact Unit (DCU) design, experimental protocol followed during the DCU study, and the results are presented in subsequent sections.

### Field Demonstration Study

*Objective*

The primary objective of the field demonstration was to evaluate bioremediation to determine the time required for treatment. Secondary objectives were to gain field experience in operating full-size bioremediation units; determine toxic effects on the natural microbes, if any; and generate analytical data to prove to the regulatory agency that bioremediation is an effective method for closure of the lagoon.

*Waste Matrix*

The waste matrix used for the field demonstration study came from a Lagoon 1 that was used to store sludge during plant operations. Sludge samples from several locations in the lagoon were characterized, and a master composite sample from these locations was analyzed for the various parameters and chemical constituents. Results of the master composite are presented in Table I.

The master composite sample had an ash content of 13.3%, a solids content of 30.1% by weight, and a specific gravity of 1.02. Total Organic Carbon (TOC) value of the composite was 35,100 parts per million (ppm). Total nitrogen (N) content was 323 parts per million (ppm) as N. Total N consisted of an Organic-N content of 251 ppm and an inorganic N content of 72 ppm as ammonia-N. Total phosphorus content in the composite was high at 32,400 ppm as P. Redox potential of 354 millivolts (mV) indicate that the sludge is moderately reduced and will require aeration. The sludge pH was found to be 7.15 standard units (S.U.). The total concentrations of polycyclic aromatic hydrocarbons (PAHs) and bis-(2ethyl-hexyl) phthalate ester (BEHP) in the composite were 15,588 ppm and less than 500 ppm, respectively.

*DCU Design*

The Direct Contact Unit (DCU) was constructed with two cells, each 50 ft by 100 ft. These cells were used to bioremediate the sludge from the lagoon. During the test, an area outside the active cells was constructed to serve as a staging area. The DCU was built about 4 feet above grade to control drainage. A leachate collection system, comprised of a network of perforated pipes, leachate collection sump, and pump was installed. Additionally, glass block lysimeters were installed below the treatment zone to assess the potential of vertical migration. A sprinkler system consisting of sprinkler heads, piping, and pump was installed to facilitate application of water.

*Experimental Protocol*

The two cells were labeled east and west cells. The effects of high loading rates were evaluated in the east cell, and the effects of using acclimated bacteria were evaluated under lower

**Table I. Master Composite of Sludge Sample Characteristics from Several Locations in the Lagoon**

| SLUDGE CHARACTERISTICS from Master Composite | | CHEMICAL CONSTITUENTS from Master Composite in ppm | |
|---|---|---|---|
| BOD-5 day (ppm) | 118,000 | Bis(2-ethylhexyl)phthalate | <500 |
| Chemical oxygen demand | 877,000 | | |
| Total organic carbon (ppm) | 35,100 | Naphthalene | 6,500 |
| Redox potential (mv) | 354 | 2-Methylnaphthalene | 2,500 |
| Ammonia nitrogen as N (ppm) | 72 | Acenaphthylene | 1,300 |
| Kjeldahl nitrogen as N (ppm) | 323 | Acenaphthene | 800 |
| Nitrate+nitrite (ppm) | <1 | Fluorene | 1,600 |
| Organic nitrogen (ppm) | 251 | Phenanthrene | 2,300 |
| Total nitrogen as N (ppm) | 323 | Anthracene | <500 |
| Ortho phosphate as P (ppm) | 1.31 | Fluoranthene | <500 |
| Organic phosphorus as P(ppm) | 32,400 | Pyrene | 588 |
| Total phosphorus as P (ppm) | 3 | | |
| | 2,400 | Chrysene | <500 |
| Ash content (%) | 13.0 | Benzo(a) anthracene | <500 |
| Bulk density (gm/cc) | 1.013 | Benzo(b) fluoranthene | <500 |
| Specific gravity (gm/cc) | 1.020 | Benzo(k) fluoranthene | <500 |
| Moisture content (%) | 69.99 | Benzo(a) pyrene | <500 |
| pH (S.U.) | 7.15 | Indeno(1,2,3) pyrene | <500 |
| Solids content | 30.1 | | |
| Total volatile solids (%) | 16.8 | TOTAL PAHs | 15,588 |
| | | Benzene | 230 |
| | | Ethylbenzene | 1,300 |
| | | Styrene | 19 |
| | | Toluene | 500 |
| | | Xylenes | 700 |
| | | TOTAL VOLATILES | 2,749 |

loading rates in the west cell. Sludge for the cells was dewatered and applied with a front end loader in layers 12-inches thick. About 20% of the partially bioremediated sludge from the east cell was used to provide acclimated microbes for the west cell. Both cells were routinely tilled to maintain an oxygenated environment within the treatment zone. No chemical fertilizer was added. Chicken manure was analyzed for its physical and chemical properties and was added in the appropriate ratios to provide adequate nitrogen. When feasible, the moisture content of the sludge was maintained between 30 and 70% of the sludge water holding capacity. No chemicals were added to control pH, which fluctuated between 6 and 7 S.U. during the course of the study.

Each cell was operated for a period of four months. East cell was operational during the summer/fall months and west cell was operational during the fall/winter months to address ambient temperature changes.

*Effectiveness Monitoring*

The effectiveness of bioremediation during this study was determined by measuring operational parameters (pH, temperature, microbial count, organic carbon, phosphorus, and moisture content) and environmental parameters (PAH, BEHP, VOCs, Microtox assay, and TCLP). Temperature, pH, and moisture were measured in the field on a routine basis. The analytical plan was based on the use of EPA-approved methods or standards used by the water and wastewater industry.

*Results*

The results of the field study are summarized below:

- Volatile organic compounds (VOCs) quickly degraded aerobically (i.e., in the presence of oxygen) in soils.[3] The bioremediation demonstration studies involving benzene, ethylbenzene, styrene, and toluene in sludge indicated extensive biotransformation (measured as loss of compound).[4] In sludge, biodegradation (and volatilization) can be very rapid, provided adequate amounts of oxygen, moisture, and nutrients (e.g., nitrogen, phosphorus) are available. The reduction in concentration of these constituents was so rapid that site-specific half-lives could not be calculated.[4] Howard et al.[5] reported that the half-life of these VOC constituents range from 3 to 28 days.
- It is unlikely that BEHP released to soil will either evaporate or leach into groundwater.[5] Data from the bioremediation demonstration shows that it biodegrades under aerobic conditions with half-life of 19 to 48 days. This rate is slightly longer than the half-life of 5 to 23 days reported by Howard et al.[5]
- The PAHs adsorbed onto soil can be biodegraded. Biodegradation rates vary widely depending on oxygen availability, pH, nutrient level, moisture content, temperature, and microbial community. In general, PAHs with 2 or 3 rings (e.g., naphthalene and 2-methylnaphthalene) were more readily degraded than PAHs with 4 or more rings.[6] Zero and first-order rate constants and half-life values were computed for selected PAHs. Computed half-life values for the east cell study are presented in Table II. The site-specific half-lives for selected PAHs ranged from 19 to 71 days.[4] These values generally were faster than the literature-based values.[5]

**Table II. Computed Half-Life Values**

| | East Cell Study[a] | | | | |
|---|---|---|---|---|---|
| | Zero Order Reaction | | First Order Reaction | | |
| Chemical Constituent | Half-Life (days) | Coefficient $qR^2$ | Half-Life (days) | Correlation Coefficient $R^2$ | Literature-Based Half-Life Values (days) |
| Benzene | — | — | — | — | 5 to 16 |
| Toluene | — | — | — | — | 4 to 22 |
| Ethylbenzene | — | — | — | — | 3 to 10 |
| Xylene | — | — | — | — | 7 to 26 |
| Styrene | — | — | — | — | 14 to 25 |
| Bis(2-ethylhexyl)phthalate | 65 | 0.9235 | 45 | 0.9403 | 5 to 23 |
| Naphthalene | 47 | 0.7298 | 19 | 0.9501 | 17 to 48 |
| 2-Methylnaphthalene | 53 | 0.8075 | 24 | 0.8888 | 4 |
| Acenaphthylene | 60 | 0.8606 | 35 | 0.8485 | 43 to 60 |
| Acenaphthene | 62 | 0.9614 | 39 | 0.9643 | 12 to 102 |
| Fluorene | 50 | 0.9214 | 26 | 0.8494 | 32 to 60 |
| Phenanthrene | 55 | 0.8712 | 29 | 0.8242 | 16 to 200 |
| Anthracene | 70 | 0.9409 | 60 | 0.9188 | 50 to 460 |
| Fluoranthene | — | — | — | — | 140 to 440 |
| Pyrene | 96 | 0.4139 | 71 | 0.4755 | 210 to 1900 |
| Chrysene | — | — | — | — | 372 to 993 |
| Benzo(a) anthracene | — | — | — | — | 102 to 679 |
| Benzo(b) fluoranthene | — | — | — | — | 360 to 610 |
| Benzo(k) fluoranthene | — | — | — | — | 908 to 2139 |
| Benzo(a) pyrene | — | — | — | — | 57 to 530 |
| Indeno(1,2,3) pyrene | — | — | — | — | 500 to 730 |
| AVERAGE OF PAHs | 53 | 0.8869 | 23 | 0.9518 | — |

[a]Note: $R^2$ of less than 0.7 is probably not significant.

- In general, the results of the pilot study and field-scale performance study compared well. Relatively faster degradation rates were observed in the field-scale system. Treated residuals from the bioremediation of sludge had low to nondetectable levels of organic compounds in the leachate extract of the sludge. Toxicity tests and microbe counts showed that high levels of organics present in the sludge were not toxic to microbes. This demonstration study showed that bioremediation of organics by direct contact land treatment is an effective technology for closing of the lagoon.

## RISK ASSESSMENT

A risk assessment was completed for Lagoon 1 at TEX in Longview, Texas, under Risk Reduction Standards 2 and 3 pursuant to Section 31 Texas Administrative Code (TAC) Chapter 335, Subchapters A and S.[7] The mechanics of the risk assessment process in Texas conforms to EPA as well as TNRCC rules and was patterned after Superfund risk assessment.[8-10] The assessment includes the components: Site Characterization; Constituent Characterization; Exposure Assessment, Toxicity Assessment, Preliminary Remediation Goals; and sections on groundwater protection, uncertainty, conclusions, and references.

The purpose of the risk assessment was to develop preliminary remediation goals (PRGs) for the upper 2 feet of residual contaminated soil using Risk Reduction Standard 3. The PRGs represent a level of constituents that can be left on-site without posing an unacceptable risk to human health. The risk assessment incorporated the industrial nature of the TEX facility. As required under Risk Reduction Standard 3, this risk assessment report evaluated the potential adverse effects under both current and future conditions in the absence of any actions to control the hazard.

Results of the risk assessment (completed according to Risk Reduction Standard 3) show that after treatment, some constituents may remain in the soil above the PRGs. Because these constituents could pose a threat to future site workers, TEX has proposed to close Lagoon 1 by placing a cap over the bioremediated sludge and soil mixture, and by maintaining the existing groundwater containment system. The threat to such workers (in the absence of any control measures) is that of incidental ingestion of particulates from Lagoon 1 soil. This cap would consist of 2 feet of clean soil with a permanent vegetative cover. This configuration would allow oxygen and moisture to migrate into the residual sludge and soil mixture. This modified cap would allow biodegradation of the constituents in the sludge and soil mixture after closure.

The potential for constituents to migrate from the lagoon to the underlying groundwater was assessed using a leaching test according to Risk Reduction Standard 2. The test showed that before remediation, a few constituents lighter than pyrene could leach from the sludge and potentially impact groundwater. However, none of these constituents have been reported in a series of downgradient monitoring wells. In addition, a downgradient slurry wall and series of recovery wells would prevent any future off-site migration if these constituents did reach the groundwater.

In summary, the continuation of access restrictions and routine use of personal protective equipment during the closure period will eliminate site worker exposure to constituents in the sludge and soil mixture. Placement of a modified cap over the bioremediated sludge and soil mixture will permanently eliminate future exposure.

## CURRENT STATUS

Subsequent to completion of the bioremediation study and the risk assessment process, this approach to closure was presented to the TNRCC. The TNRCC has approved the permit modification, and TEX is currently in the process of reducing the toxicity, mobility, and volume of the lagoon sludge and soil mixture (residual material) through bioremediation. A detailed sampling and analysis plan (SAP)[11] was prepared to be used in collection and analyzing samples in support of remediation of the wastes placed in the Corrective Action Management Unit (CAMU) at the TEX facility. As a part of the remedial strategy, TEX is placing residual material in Lagoon 1 into two treatment units approximately 4-acres each in size, with 12-inch high lifts, tilling and watering the material for 90 days and collecting confirmation samples to indicate when the material is successfully remediated. If the samples report benzene below 1 ppm and naphthalene concentrations below 408 ppm, and a composite of the confirmation samples has a synthetic potential leaching procedure (SPLP) leachate concentration of naphthalene of less than 1.46 milligrams per liter (mg/L), then the next lift is emplaced. At the present time, TEX has successfully remediated approximately 50% of the Lagoon 1 sludge/soil mixture.

In conclusion, the bioremediation demonstration study that is discussed in this chapter showed that most of the constituents in the sludge and soil mixture were amenable to bioremediation and that bioremediation is the preferred technology for the residual sludge and soil mixture in Lagoon 1. Results of the risk assessment show that by providing engineering controls, TEX plans to eliminate any potential threat to future site workers.

This risk-based corrective approach by TEX will result in substantial cost savings and is protective of human health and environment.

## ACKNOWLEDGMENTS

The contributions of the following colleagues are gratefully acknowledged:

- Dr. James L. Grant and Mr. Dipak Patel of Grant Environmental,Inc., Denver, Colorado, for their support on the bioremediation study.
- Dr. Gary McGinnis of Michigan Technological University, Houghton, Michigan, for the treatability evaluation study.
- Dr. Tom McAninch, Dr. Jerry Caplinger, Dr. Roger Kenton, and Dr. Karolyn Hardway of TEX for the performance of analytical work.
- Mr. Robert Chapin, Dr. Stephen M. Ellingson, and Mr. Daniel McCarthy of Geraghty & Miller for risk assessment study and confirmation sampling and analysis plan.

## REFERENCES

1. Texas Eastman (TEX). Original Lagoon 1 Closure Plan, 1985.
2. McGinnis, G.M. "Bioremediation of Lagoon 1 Sludge at the Texas Eastman Plant in Longview, Texas." September, 1991.
3. Kostecki, P.T. and E.J. Calabrese. *Petroleum Contaminated Soils.* Volume I. Remediation Techniques, Environmental Fate, Risk Assessment. Lewis Publishers, Inc. Chelsea, MI, 1989.
4. James L. Grant and Associates, Inc. (JLGA). Bioremediation Demonstration - Lagoon 1. September, 1993.
5. Howard, P.H. et al. *Handbook of Environmental Degradation Rates.* Lewis Publishers, Inc., Chelsea, MI, 1991.
6. McKenna, E.J. and R.D. Heath. Biodegradation of Polynuclear Aromatic Hydrocarbon Pollutants by Soil and Water Microorganisms. Water Resources Center, University of Illinois, Research Report No. 113. UILV-WEC-76-0113, 1976.
7. Geraghty & Miller, Inc. Risk Assessment for Lagoon 1. Texas Eastman Division, Eastman Chemical Company, Longview, TX. September, 1995.
8. Texas Administrative Code (TAC). 1993. Subchapter S, Risk Reduction Standards, 335.551 - §§335.599. Texas Natural Resources Conservation Commission (TNRCC), Austin, TX.
9. U.S. Environmental Protection Agency (USEPA). Risk Assessment Guidance for Superfund, Human Health Evaluation Manual. Interim Final. Office of Emergency and Remedial Response, Washington, D.C. EPA/540/1-89/002. December, 1989a.
10. U.S. Environmental Protection Agency (USEPA). Risk assessment Guidance for Superfund, Human Health Evaluation Manual, Volume 1, Part A. Interim Final. Office of Emergency and Remedial Response, Washington, D.C. EPA/540/1-89/002. December, 1989b.
11. Geraghty & Miller, Inc. Sampling and Analysis Plan for Closure Plan, Lagoon 1 Corrective Action Management Unit. Texas Eastman Division, Eastman Chemical Company, Longview, TX. July, 1995.

# Section Thirteen
# INTERNATIONAL ACTIVITIES

## 49 WASTEWATER IMPROVEMENT PROJECT IN CHINA

**X. Ching Wu**, Senior Engineer
**Gerald E. Seaburn**, Vice President
Law Engineering and Environmental Services, Inc.
Kennesaw, Georgia 30144

### ABSTRACT

Recently, Law and Engineering Environmental Services, Inc. (LAW) conducted a wastewater improvement project for the City of Tianjin, the third largest industrial city in China. The project included the analysis and recommendations for financial, institutional, and technical issues to improve the city's wastewater management practices. This chapter will focus on the overview of the project and a general discussion on how an environmental project was conducted in that country.

### INTRODUCTION

During 1994 through 1995, a wastewater improvement project covering technical, institutional, and financial issues was conducted in China by GibbAnglian Ltd., a joint venture of Gibb and Anglian Water. Gibb is the international division of Law Engineering and Environmental Services, Inc., located in Reading, England. Anglian Water is an UK-based multinational water company. An eight-volume study report covering the findings, evaluations, and recommendations was completed, and three wastewater treatment/recovery facilities were designed. This paper is presented to provide information on how this multitask project was managed and executed by this foreign consulting firm in China. Background information on current environmental conditions which is thought to be helpful for conducting environmental projects in the country is presented as well.

### BACKGROUND INFORMATION

#### General Background and Objectives

Tianjin is the third biggest industrial city in China, located in north China, close to the capital city, Beijing. Since the implementation of economic reform started in China in 1980, there has been considerable development in Tianjin. Nevertheless, for many years the construction of urban drainage, sewerage, and wastewater treatment (domestic and industrial) has lagged far behind the development and environmental requirements for the city.

The Tianjin Wastewater Improvement Project is part of the Tianjin Urban Development and Environmental Project (TUDEP) co-financed by the Tianjin Municipal Government (TMG) and the World Bank. This project focuses on sewage collection and treatment, as well as measures used to improve the institutional controls on sewage and pollution control. To ensure that the TMG achieves its overall objectives of optimizing water consumption and reducing pollution loads, it has placed particular emphasis on the reduction of industrial water consumption and the improvement of effluent discharge quality.

*52nd Purdue Industrial Waste Conference Proceedings*, 1997, Ann Arbor Press, Chelsea, Michigan 48118. Printed in U.S.A.

The Government of the People's Republic of China and the United Kingdom, through the Overseas Development Administration (ODA) agreed to collaborate on a study (Tianjin Wastewater Improvement Project) to complement the overall TUDEP program, to identify the most cost-effective strategies for reducing and treating industrial wastewater and minimizing the contribution made by industry to the overall pollution load. GibbAnglian Ltd. was awarded the contract by the ODA to undertake the study. The study was divided into three tasks according to the Terms of Reference (TOR) issued by ODA:

1. **Technical**—To assist the Environmental Protection Bureau (EPB) of Tianjin in identifying the most cost-effective solutions for reducing and treating industrial wastewater.
2. **Institutional**—To develop the most appropriate division of responsibilities between drainage and sewerage departments of the Tianjin Municipal Engineering Design Bureau (MEB) and to strengthen the capacity of the local design bureaus to identify, technically appraise, and design a limited range of wastewater minimization and reuse technologies.
3. **Financial**—To strengthen the capacity of People's Construction Bank of China (PCBC) to assist the Fund for industrial Pollution Control (FIPC) to undertake economic and financial evaluations of loan applications for industrial pollution control.

The GibbAnglian Project Team started the study in Tianjin on April 18, 1994. The final eight-volume report covering above listed fields was submitted to the ODA and TMG in December 1996.

## Organizational Structure and Collaborating Institutions

To coordinate the TUDEP, the TMG established a Project Leading Group (steering committee) chaired by the Director of Urban and Rural Construction Commission (URCC), and a Project management office. Several local government agencies, including the Tianjin Science and Technology Commission (TSTC) and the Tianjin Environmental Protection Bureau (TEPB) are represented in the steering committee. The TEPB is responsible for pollution research and monitoring, supported by two research organizations: the Environmental Protection Science Institute (EPSI) for wastewater treatment and the Environmental Protection Technology Development Center (EPTDC) for industrial waste minimization.

The ODA project falls under the overall umbrella of the Project Leading Group formed by URCC but is administered and supervised by the TSTC, with the TEPB as the local partner organization. Figure 1 illustrates the project management structure.

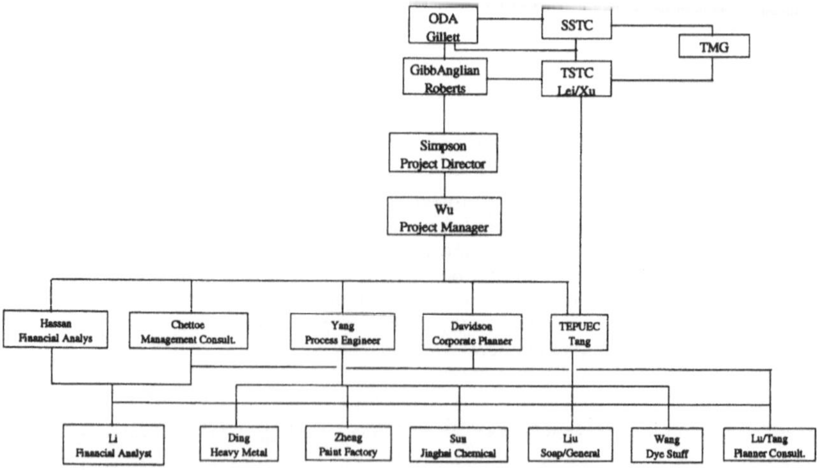


**Figure 1. Wastewater improvement project organizational chart.**

## PROJECT EXECUTION

### Project Team

A project team, consisting of technical, management, and economic experts, was established by GibbAnglian to conduct the three tasks. A local work force was selected and recommended to the project team by the TSTC. The individuals selected are from the Tianjin Municipal Engineering Bureau (MEB) and the Tianjin Environmental Protection Bureau (EPB). It was realized that local involvement and support is one of the key elements to the success of the project. The project team was divided into three groups to carry out the three tasks. Each group had local experts and was led by GibbAnglian team member. A biweekly meeting was held to provide a routine communication channel for the whole project team.

### Information Collection

*Technical*

In order to carry out the technical assignment, the following information were considered necessary:

- Legislation governing industrial pollution control;
- Current industrial pollution control activities;
- Wastewater conditions in the specific industry sectors.

Investigation was conducted through site visits, document review, interviews, and sampling. The findings are summarized as follows.

*Legislation.* The legal framework for environmental management in China is provided by the Environmental Protection Law (EPL) of 1989, and the associated regulations. In the EPL there are three main concepts: prevention, the polluter pays principle, and management enforcement.

The EPL states that prevention means (1) the incorporation of environmental protection into the design, construction, and operation of facilities, and (2) mandatory environmental impact assessments (EIAs) on all new expansion or renovation projects. The EPL, in implementing the above concepts, also states that:

- It is prohibited to place new polluting facilities in or near reserved or populated areas.
- Old polluting facilities should be relocated to comply with the EPL.
- Recycling will be promoted by adjustments to prices or tax for goods made from recycled materials, any additional profits being returned to the enterprise for investment in antipollution control.
- More emphasis will be placed on and encouragement given to the use of low polluting energy sources and district heating systems.

In addition, EIAs are now required in the early stage of the project.

The EPL adopted the "polluter pays" principle which has already been adopted by a number of developed countries, and states that to implement this principle the following actions will be taken:

- a percentage of the capital cost of new projects must be set aside for environmental protection;
- a similar set-aside for renovating existing enterprise;
- deadlines for pollution reductions by industry;
- a penalty for noncompliance, known as a levy system, based on the amount of emissions exceeding standards, the funds collected to be redistributed to finance.

The penalties for noncompliance are far-reaching. In addition to fines, compensation may be assessed for damage caused by polluters to crops or animals, plants may be closed, and unit leaders may be held personally responsible for pollution. In late 1992, more than 200 national standards had been issued for environmental quality (air, surface water, noise), emission standards for industry and vehicles, and measurement methods. Standards were issued by local governments as well. However, local standards can only be approved by the National Environmental Protection Bureau if they are more stringent than national standards.

A wastewater discharge permit system is being tested in 230 cities in China over the period 1992–1995. The discharge limits are mainly based on concentrations levels in China; however, China is also experimenting with mass-based effluent standards for pollution control.

Environmental management enforcement is provided in the EPL by:

- mandating the creation of a national environmental protection agency to draft policies, regulations, and standards, and to supervise their enforcement, monitor pollution, and promote research and development;
- setting up environmental protection bureau at the provincial level, county, and township level, with specific responsibility for implementing the legislation;
- setting up environmental protection sections in large and medium-sized enterprises.

Environmental protection sections have been set up in most commissions and bureau since then.

From these EPL provisions, a large workforce has been created for studying, supervising, monitoring, and enforcing environmental protection work. It is estimated at about 200,000 people in government agencies at national, provincial, county, and district levels, and industries are involved in environmental protection work. Each industry has its environmental protection department/division in charge of preparation of implementation plans which will monitor, examine, and report on environmental issues.

The National Environmental Protection Bureau (NEPB) has issued a set of regulations for the discharge of industrial wastewater. These regulations are industry-specific and have been adopted nationwide. The regulations were published nationwide and are available in local EPBs.

*Industrial Pollution Control Activities in Tianjin.* Industrial Wastewater in Tianjin. There are some 5,000 separate industrial units in Tianjin City. The primary industries in Tianjin are chemical, metallurgical, textiles, and electronics. Most of the primary industries (60%) have limited wastewater treatment on-site. Over the past 20 years there has been a severe shortage of water, and farmers have been using untreated wastewater (municipal combined with industrial) from Tianjin for agricultural purposes. Heavy metals, particularly cadmium, zinc, and some organic-metallic compounds, have caused a serious threat to public health and the environment. Since the late 1970s China has focused on reducing these pollutants in industrial wastewater discharges.

Discharge Quality Control of Industrial Wastewater. In Tianjin, industry is subjected to tight controls for discharge of wastewater to the city sewer system, which consists of open channel system (canals), limited sewer collection system, and two municipal wastewater treatment plants. These controls are established by the Tianjin Environmental Protection Bureau (Tianjin EPB) in the form of regulations. The Tianjin EPB has been most active in improving the quality of water in the rivers and canals in Tianjin. The Tianjin EPB monitors and sets targets for the city's overall water quality and enforces pollution control regulations such as mandatory screening of industrial investments. It collects relatively large amounts of pollution levies assessed on discharges above effluent standards, and uses the revenues for various pollution control measures. The Tianjin EPB applies a rigorous approach to the enforcement of the regulations issued by the NEPA throughout industry in Tianjin. It is recognized that the pollution control activities of the Tianjin EPB are among the most sophisticated in China. However, its pollution control measures and regulation enforcement need to be improved.

Current Compliance Situation in Tianjin. To show the impact of pollution control legislation on industry in Tianjin, the capital and operating costs of effluent treatment facilities of the industrial complexes in one of the city sewer systems (Xian Yang Lu) have been analyzed.

The survey shows that among the 154 industries in Xian Yang Lu system, there are:

- 49 industries that have a treatment facility which meets standards,
- 4 industries that have remove suspended solids only, and
- 71 industries that have additional treatment equipment.

It appears from this data that industry is having a great deal of difficulty in meeting current discharge consents.

*Wastewater Conditions in the Specific Industries.* Investigations were made in the following industries:

- Jinghai Chemical Factory
- Changcheng Dye Factory
- Tianjin "Beacon" Paint Factory
- Tianjin Soap Factory
- Tianjin Pulp and Paper Factory

Wastewater treatment systems were evaluated and recommended for Jinghai and Toyo Ink factories, and a chemical recovery system was recommended for Changcheng Factory. To give an example, only information on Jinghai and Changcheng is included in this paper.

Jinghai Chemical Complex is in the Jinghai County and about 40 km from downtown Tianjin. It was put in operation in 1990. There are ten manufacturers on-site. This group manufactures organic intermediate products for perfume and toiletry industry in China and abroad. Products include p-nitro-chlorobenzene, benzoic alcohol, guaiacol, cumarin, musk ketone, musk zylene, musk jiale, crystal rose, etc. The total production output is about 14,350 ton/yr. About 90% of the product is exported to foreign markets. The group's turnover is 350 million Yuan. The raw materials used in the plant are 38 compounds including inorganic acids, alkalis, salt, and organic chemicals. The current annual consumption of raw materials is approximately 40,000 ton.

Process wastewater including heavily polluted waste liquors from the reaction vessels are discharged in batches into a central wastewater holding canal. Another source of wastewater is rectification process. In addition, floor washing and vessel washing produce large wastewater volumes. Certain waste liquors containing high acid concentration are stored in containers and sold to other factories for reuse. Distillation residuals and solids wastes also sold to other companies, either for material recovery or used as fuel. A small amount of waste acid is disposed of at a landfill site.

There are no wastewater treatment facilities on-site and all the effluent discharged is of an unacceptable quality to the environment. The effluent contains chlorinated hydrocarbons and other toxic organic compounds and presents a chronic environmental health hazard. It is vital that an acceptable solution be found to treat this wastewater before the groundwater and surrounding environment are severely polluted.

Changcheng Chemical Factory produces 2-naphthol, 2-amino-5-napthol-7-sulfonic acid known as J-acid, J-acid urea, and 2-napthylamine-1-sulfonic acid known as Tobias acid. The Tobias acid is reacted with concentrated sulfuric acid to produce the J acid. The J acid is an interim product in the manufacture of reactive and acid dyes. These compounds are used for production of azo dyes. The plant also produces a small range of inorganic compounds for the dye industry.

The Changcheng chemical factory produces approximately 300 $m^3$/d of wastewater, of which 78 $m^3$/d process wastewater consists of four streams containing various aromatic compounds which are toxic and carcinogenic. These wastewater streams represent a COD load greater than 2.332 t/d. The untreated wastewater is discharged into a pond and is seeping into the groundwater. When the retention pond overflows, the wastewater has to be discharged either to the southern sewage canal or to Duliujian River. The discharge of untreated process wastewater streams is causing grave pollution problems in the local river and groundwater. This high organic toxic load discharged to the environment makes the Changcheng factory probably one of the worst polluting industries in Tianjin. The pollution is severely affecting the economic viability of the local farming community. The farmers are apparently having difficulty in marketing their produce because of pollution contamination. Should this gross pollution not be eliminated, the health of the local inhabitants will be at risk.

*Institutional*

The project team had numerous meetings and interviews with key personnel to collect and understand the existing management structure, main facilities, and operations of the Municipal Engineering Bureau (MEB) and the Drainage Management Department (DMD). Although the structure of these two organizations is set to serve their needs, its principle is commonly applied to most government agencies in China.

*Main Purpose and Facilities.* The Tianjin Municipal Engineering Bureau (MEB) reports to the Tianjin Urban and Rural Construction Commission (URCC). It is a large organization by any standard, in range of activities, number of organizational units and number of staff, which amounts to about 34,000 people. The overall aim and scope of MEB is to build and maintain to state standards, roads, highways, bridges, sewage treatment plants, wastewater and stormwater discharge infrastructure in the city area; also to build and maintain main roads over the suburbs, outer districts, and counties in the metropolitan area. The two functions of construction and management of facilities are two key features of MEB and are reflected in its departmental structure. The MEB also provides technical and sometimes financial support to the counties, suburbs, and outer districts where there are municipal engineering units.

The Drainage Management Department (DMD), also known as Tianjin Sewerage Management, is responsible to the MEB for the operation and maintenance TMG facilities including discharge operations over an area of 240 $km^2$, 1,788 kms of interceptors and main sewers within the city boundary, 131 pumping stations, 171 kms of river embankments, and 2 sewage treatment plants, according to the current TMG regulations. These regulations include the sewage tariff chargeable to industrial and commercial users of the sewerage system in the city area. The DMD is also charged with recommending extensions and improvements to the above system to keep up with the latest development in the industry.

*Top Management Structure.* The MEB, responsible to the Urban and Rural Construction Commission, is headed by its Director and assisted by seven members of the MEB Standing Committee of the Communist Party of China (CPC), who are equivalent to deputy directors in status, and five other deputy directors. Each of these officials is responsible for part of the MEB's activities. It has been made clear that CPC matters take precedence over MEB management matters at this level. Less important decisions which are beyond the authority of unit heads (although different unit heads have different powers) are made by a management committee which includes the CPC Standing Committee and the remaining five deputy directors mentioned above. This management structure emphasizes the central importance of the CPC to the running of MEB and, it is presumed, TMG also.

Operationally, the Director is assisted by five deputy directors responsible for, respectively: foreign projects, foreign-funded (e.g., World Bank) projects in Tianjin, and overseas training; China-based projects outside Tianjin and "profit-generating" enterprises; projects within Tianjin municipality; MEB facilities management subsidiaries such as DMD, regulations and labor and wages; and planning and finance, and preproject development. The extent of the authority delegated to these deputy directors is apparently difficult to specify: "routine" decisions are said to be taken by deputy directors, while bigger issues require the Director's input. The appointment and dismissal of the Director and his deputy directors is recommended by the MEB CPC Standing Committee and confirmed by the URCC and, in the case of the Director, the TMG mayors.

The Drainage Management Department (DMD) is headed by the DMD Director, who reports administratively to the Director of the Municipal Engineering Bureau (MEB), but operationally to three Vice Directors responsible for, respectively, projects within Tianjin Municipality, infrastructure management, and investment planning and financial management. He also reports operationally to the MEB Chief Engineer in charge of Science and Technology and the MEB Deputy Secretary General of the CPC in charge of CPC affairs. There is a significant CPC presence which is separate from the administrative system, yet an integral part of the DMD management system. The DMD CPC Standing Committee has the greatest power within the DMD and it makes all major decisions which are within DMD's authority and not referred to the MEB. Figure 2 gives the organizational chart of DMD.

*Collection of Pollution Levies and Penalties from Industry.* The Municipal Engineering Bureau (MEB) charges industry at a fixed rate of ¥0.209/$m^3$ ($0.025/$m^3$ or $0.10/1,000gal) based on 90% of their water consumption. This is collected quarterly. The levies are a major source of funding for the maintenance of the sewerage network and sewage treatment plants.

The MEB has an Overall Monitoring Organization which monitors pollution load, specifically biological oxygen demand and suspended solids, throughout the Tianjin City. This monitoring organization is divided into six submonitoring units to cover the six sewerage and drainage re-

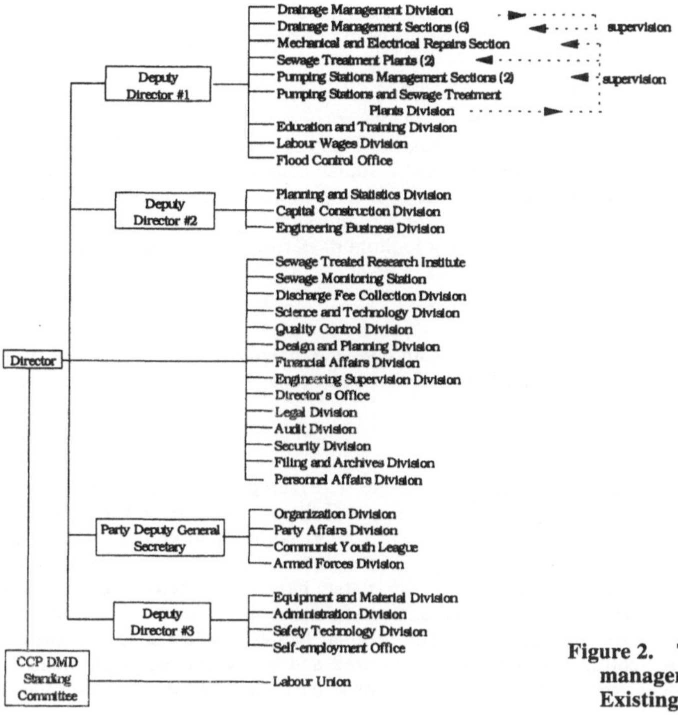


**Figure 2. Tiajin drainage management department: Existing organization.**

gions of Tianjin. Each sub-unit is equipped with a laboratory. The subunits report to the Overall Monitoring Organization.

The EPB collects the levies covering penalties when the industries exceed the consent limits. The levies are listed in the Tianjin Environmental Protection Standards Regulations. The levies are to be used as a major source of financing for pollution control activities in industry, especially the installation of wastewater treatment plants and training programs. The EPB allocates about 20% of the revenues to its own monitoring and research activities, and allocates the rest to the industries for their pollution control activities. As a policing agent, the EPB monitors through the Environmental Monitoring Center (EMC), discharges from industries. The EPB does not monitor the final effluent quality from the MEB's sewage treatment works and outflows to the environment. The EPB operates a database which contains information (process type, quality, and quantity of waste) on the industrial wastewater and air pollution activities in Tianjin.

### *Financial*

*Background.* Since the mid-1980s, the Tianjin Environmental Protection Bureau (EPB) has been using a proportion of the pollution levies to experiment with a small-scale loan program to assist the industrial polluters. In addition, it has also provided assistance through the bureau of each sector to give small grants to industries to construct cheap and simple facilities for reduction in pollution. In 1992, under the Tianjin Urban Development and Environment Project (TUDEP), the World Bank boosted this program by providing assistance with setting up the Fund for Industrial Pollution Control (FIPC). The main objective is to establish a sustainable mechanism for financing pollution control measures by industrial enterprises and increasing the benefits of the pollution levy revenues. In July 1993, a permanent office of FIPC was established.

*Management Structure.* FIPC is an autonomous entity with its own charter, organization, and operating procedures. It is managed by a board of directors consisting of seven members: two are nominated by the EPB; one by the Sanitation Administration Committee; and two each by the

Tianjin Economic Council and the Tianjin Finance Bureau. The directors are appointed for a period of four years and they can be reappointed. The board elects the chairman and the executive director.

The FIPC office is also supported by other institutions in project appraisal and in providing technical assistance to the borrowers. The People's Construction Bank of China (PCBC) is a leading capital financing institution in the country and its Tianjin branch, which helped with the earlier experimental loan program, is handling actual disbursement and loan collection. It also assists FIPC in financial and economic appraisal of proposed projects and management of reserve funds. The Tianjin Urban and Rural Construction Commission (URCC) provides assistance during the implementation phase of the project and monitors its performance in operation.

## Study and Analysis

Based on the information gathered, the project team conducted analysis and comparison studies. Alternatives were developed and evaluated for recommendations.

### *Technical*

During the study, one big challenge the technical group was facing is the adaptation of technologies for a specific application. Shortage of capital and lack of environmental management experience were observed to be common phenomena in China. With no differences from other developing countries, China has limited funds to spend on pollution control. Industries normally focus on their products and try not to spend any more than they have to on environmental issues. Although the Chinese government has emphasized environmental issues in recent years, the majority of the industries still need to be educated. The original factories selected for the project team to study and develop treatment options were reviewed and a few changes were made in order to better serve the local needs. The changes were made based on the following considerations:

- implementation feasibility, depending on the fund availability and technical feasibility;
- environmental urgency, or seriousness of environmental damage caused by the waste generated from the factory.

The project team worked closely with the local engineers and made efforts to understand the local situation so that the most cost-effective and acceptable measures can be selected for the recommendations. Every effort was made to make sure that the treatment facilities designed by the project team will be actually installed and properly operated.

During the second phase of the project, it was requested by the local government that a new industry, Toyo Ink Factory, be added to the project list. This factory was a new facility installed in the Xiqing Development Zone in Tianjin by the Tianjin Toyo Ink Limited, a Chinese-Japanese joint venture. It has the capital, and needs treatment for its effluent.

According to the revised list, a wastewater treatment system employing a biological reactor for Jinghai chemical plant and Toyo Ink Factory, wastewater treatment/chemical recovery using solvent extraction, and biological processes for Changcheng Chemical plant were designed. It is reported that the installation of the treatment system was started in 1996 and will be completed in 1997.

In addition to providing a high quality technical product, the project team also provided on-job training to the local people who worked on the project. By the end of the project, a CADD department and a treatability study team equipped with a full-scale biological treatment facility were established.

### *Institutional*

It was realized that the management system in China is very different from those in the UK or US. Although privatization and management structure reform have been an ongoing movement since the '90s in China, little has been done in environmental, water/wastewater operations. In order to be able to make the recommendations easy to understand and, moreover, to be acceptable, data collected from interviews and meetings were studied with a positive attitude. Preliminary conclusions and recommendations derived from the comparison of the Chinese system and the western system were presented to local government officials for their comments. Meetings were held for questions and answers, and, more importantly, discussions. The final recommendations were well accepted. To date, the reform in the EMB and EPB are in process. However, to see a real change, there is still a long way to go.

*Financial*

Case study and a work shop were used to help the FIPC familiarize itself with the western system. The project team member spent half a day giving seminars and half a day working with a FIPC member on case studies for weeks, to educate and train the local people. The FIPC members, most of whom are recent university graduates, showed great enthusiasm and learning skills in absorbing those new concepts, methods, and procedures. Compared to the local technical and institutional groups, the FIPC group is younger and like all younger generation, they are more open and eager to learn about the outside world. The project team was always amazed by the progress of their knowledge and language skills. By the end of the project, the FIPC management team, equipped with a well-defined procedure for environmental project appraisal, was ready to take off for their new assignments.

## CONCLUSIONS

During the project, LAW established good working relationships with local consultants and government agencies in various fields and have become aware of environmental legislation and environmental health and safety (EH&S) compliance issues affecting local industries, joint ventures, and foreign companies in China. We realized, based on our international project experience, that just like other developing countries, China is a large market for environmental technology, equipment, management practices, etc. Although there are differences that need to be overcome, by working together with the local people, the global environmental market, including China, is reachable and manageable.

# 50 BIOLOGICAL POST-TREATMENT OF ANAEROBICALLY-TREATED BLACK LIQUOR SPILLS FROM KRAFT PULP-MILL

**C. Estrada-Vázquez, R. Hernández-Vera, I. Magaña-Plaza, and A. Hernández-González**
Centre for Advanced Studies and Research (CINVESTAV)
Environmental Biotechnology and Anaerobic Processes R&D Group-DBB
Mexico City, Mexico

**H.M. Poggi-Varaldo**
P3 Consulting Engineers and Ecatepec Institute of Technology-TESE
P.O. Box 75-202
Mexico D.F., 07300, Mexico

**G. Fernández-Villag Mex**
Centre for Disaster and Hazards Prevention (CENAPRED)
Mexico City, Mexico

## ABSTRACT

Previous research has shown that anaerobic treatment in biological fluidized bed reactors of kraft black liquor spills effluents (DL) is an effective alternative. However, important concentrations of recalcitrant organic matter and color still remain in the anaerobic effluent (AnE). Aerobic bacterial post-treatment was ineffective, while adsorption onto activated carbon removed both color and organic matter, but at potentially unfeasible costs.

The purpose of this work was to evaluate the post-treatment of the AnE by the application of selected ligninolytic fungi strains under a variety of experimental conditions. A factorial experiment $2^3$x3 with repetition was planned, the factors being: fungal strain (2 levels, *Lentinus edodes*, *L.e.*, and *Trametes versicolor*, and *T.v.*); effluent concentration (2 levels, 50% and 100% v/v); carbon source supplementation (2 levels, 0.0 and 0.5 g/L glucose), and fungus immobilization (none, alginate, and activated carbon). The experimental responses were the net specific removals of color (1/mg biomass) and ligninoid substances (g lignin/mg biomass) discounted for adsorptive removal.

It was found that *L.e.* effected significantly higher color and ligninoid removals (0.008 1/mg and 0.012 g lg/mg biomass) than *T.v.* Interestingly, whole AnE (100% concentration) resulted in higher responses (approximately 0.011 1/mg and 0.011 g lg/mg biomass) than dilute effluent (50%). Glucose supplementation at 0.5 g did not have a conclusive effect: supplemented flasks showed a slightly higher ligninoid reduction than the nonsupplemented, but the color removal was significantly lower than the corresponding to nonsupplemented flasks. Fungi immobilization on alginate showed the highest pollutant removals (0.012 1/mg and 0.012 g lg/mg biomass). Free mycelium was second, while immobilization on activated carbon was last. Overall, the best color and recalcitrant organic matter removals were obtained for *Lentinus edodes*, whole AnE, no carbon-source supplementation, and fungus immobilizaton on alginate.

## INTRODUCTION

Several works have shown that aerobic post-treatment (using bacterial consortia) of anaerobically-pretreated pulp mill effluents cannot remove more than 10 to 15% (as chemical oxygen demand, COD) of the remaining organic matter load (activated sludge for bleaching effluents,[1] ro-

*52nd Purdue Industrial Waste Conference Proceedings*, 1997, Ann Arbor Press, Chelsea, Michigan 48118. Printed in U.S.A.

tating biodisc reactor for black liquor spill effluents[2–5]) and no apparent further removal of the anaerobic effluent toxicity.[3,4] This suggests that the anaerobic stage effectively reduces the biodegradable organic matter, while the remaining organic load is recalcitrant to both anaerobic and aerobic bacteria.[2,5–7]

Some physicochemical post-treatments of the anaerobically-pretreated effluents from a kraft liquor spill wastewater (AnE) have been tested by Camacho-Martínez.[5] He found that coagulation/flocculation removed up to 80% COD; however, the acute toxicity to *Daphnia magna* significantly increased. Camacho-Matínez[5] and Estrada-Vázquez[8] reported on activated carbon (GAC) adsorption post-treatment of AnE at batch and column level, respectively. They found up to 90% removal of COD, color, and toxicity, but concerns about the high costs of such post-treatment prompted the need for exploring other alternatives.

Since the recalcitrant organic load of the AnE from pulp wastewaters mainly consist of ligninoids (lignin-related substances), the ligninolytic fungi pertaining to the Basidiomycete group show promise, due to their ability for attacking the lignin and lignin-derivatives.[7,9–11] Enzymes from some Basidiomycete fungi and whole fungi were shown to be very effective for color removal of raw bleaching kraft wastewaters.[12–18] Moreover, this type of fungi has been applied with varying success to the pulp-making process (bio-delignification of wood chips, bio-bleaching of brown stock).[19–30] In general, it was reported that savings of bleaching chemicals and less water pollution could be achieved by the use of ligninolytic fungi or their enzymes.

However, emphasis on ligninolytic fungi application has been made on raw bleaching wastewaters. Little is known about its use for the treatment of kraft black liquor spills wastewaters (diluted liquor, DL) or to anaerobically-treated DL (AnE). Thus, the purpose of this work was to evaluate the post-treatment of the dilute liquor treated in an anaerobic fluidized bed reactor (RANLEF) by the application of selected ligninolytic fungi strains under a variety of experimental conditions.

## MATERIALS AND METHODS

### Experimental Design

A factorial experiment $2^3$x3 with repetition was planned (see Tables I and II). The factors considered were: fungal strain (factor A at 2 levels, *Lentinus edodes*, *L.e.*, and *Trametes versicolor*, and *T.v.*); AnE concentration (factor B at 2 levels, 50% and 100% v/v); carbon source supplementation (factor C at 2 levels, 0.0 and 0.5 g/L glucose), and fungus immobilization (factor D at 3 levels: none, calcium alginate, and activated carbon). The experimental responses were the net unit removals of color (CUR, 1/mg biomass) and ligninoid substances (LUR, g lignin/mg biomass) discounted for adsorptive removal.

**Table I. Description of Factors and Levels**

| Factor | Levels |
|---|---|
| Fungal strain | (–) *Lentinus edodes*<br>(+) *Trametes versicolor* |
| Anaerobic effluent concentration | (–) 50%<br>(+) 100% |
| Carbon source concentration | (–) 0.0 g/L glucose<br>(+) 0.5 g/L glucose |
| Immobilization type | (–) calcium alginate<br>(0) free mycelium<br>(+) activated carbon |

Responses:
Color Unit Removal Efficiency (*CUR*)
Lignin Unit Removal Efficiency (*LUR*)
(removal per unit biomass, discounted for abiotic removal).

**Table II. Experimental Design Lay-Out**

| Exp. Unit | A | B | C | D |
|---|---|---|---|---|
| 1 | – | – | – | – |
| 2 | + | – | – | – |
| 3 | – | + | – | – |
| 4 | + | + | – | – |
| 5 | – | – | + | – |
| 6 | + | – | + | – |
| 7 | – | + | + | – |
| 8 | + | + | + | – |
| 9 | – | – | – | 0 |
| 10 | + | – | – | 0 |
| 11 | – | + | – | 0 |
| 12 | + | + | – | 0 |
| 13 | – | – | + | 0 |
| 14 | + | – | + | 0 |
| 15 | – | + | + | 0 |
| 16 | + | + | + | 0 |
| 17 | – | – | – | + |
| 18 | + | – | – | + |
| 19 | – | + | – | + |
| 20 | + | + | – | + |
| 21 | – | – | + | + |
| 22 | + | – | + | + |
| 23 | – | + | + | + |
| 24 | + | + | + | + |

A: Fungal strain factor
B: AnE concentration
C: Carbon source concentration (glucose)
D: Immobilization type

**Table III. Main Characteristics of the Anaerobic Effluent**

| Parameter | |
|---|---|
| COD | 576.18 mg/L |
| Color | 0.510 |
| Ligninoids | 0.069 g/L |

Dark brown color
moderately toxic to *Daphnia magna*.

## Methods and Analysis

The induction of the ligninolytic enzymatic systems of the fungi was carried out using MK[31–36] medium for *L.e.* and MB medium[17,18,22] for *T.v.*, both media being nitrogen-deficient.

The AnE was the effluent from a mesophilic RANLEF treating DL at 2-day HRT[6]. The main characteristics of the AnE are shown in Table III; more details can be found elsewhere.[6,8] It was a highly colored effluent (dark brown), with a moderate-to-high organic load (recalcitrant to bacterial degradation), an important content of ligninoids, and a considerable toxicity to *Daphnia magna*.

The post-treatment tests were run in 500 mL flasks containing AnE and fungal biomass (total working volume 200 mL), according to experimental design, under aseptic conditions at 25°C/100 rpm and a 7-day incubation period, batch mode. The initial pH was adjusted to 5.0 ± 0.1 (with $H_2SO_4$ or NaOH 1 N). Units with GAC contained 200 mg vegetal activated carbon/flask[8].

Several controls were included (by duplicate): units with no fungal biomass (no free mycelium, GAC without fungus, and calcium alginate beads with no fungus), and units with sterile mycelium. Controls were handled identically to the active units.

The *L.e.* strain was provided by the Microbial Collection of the CINVESTAV (CDBB-H-985, optimum growth temperature 25°C). The *T.v.* was kindly supplied by Dr. Ian Reid (PAPRICAN, Montreal, Canada) and preserved in the Microbial Collection-CINVESTAV (catalog number CDBB-H-1051, corresponding to PAPRICAN #52 and ATCC-20869, optimum growth temperature 28°C).

Color and ligninoid concentration were determined spectrophotometrically at wavelengths of 465 (visible) and 280 nm (UV), respectively. A 1-cm width quartz cuvette was used; absorbance was measured on previous dilution down to an absorbance value nearly 0.3. Sample pH was adjusted to 7.0 with 1 mL of a phosphate buffer; this buffer contained 305 mL of $Na_2HPO_4$ 0.2 M and 195 mL of $NaH_2PO_4$ 0.2 M in a 1000 mL total volume.

## RESULTS AND DISCUSSION

The results of this work were evaluated according to the standard techniques of statistical experimental design analysis[37] and using the SAS software package.[38] The results of the ANOVAs for the CUR and the LUR are displayed in Table IV and Table V, respectively.

First, the main effects will be examined. Regarding the factor fungal strain (A), the overall best results were obtained for *L.e.* for both color and ligninoid removals (Fig. 1 and Fig. 2). On the average, *L.e.* color and ligninoid reductions were 146% and 105% higher than those achieved by *T.v.*. This factor was highly significant for both responses (Prob. 0.003 for CUR, Prob. 0.0001 for LUR; see Table IV and Table V).

The AnE concentration factor (B) results were somewhat unexpected role. It was thought that units at low concentration (50%) would give better removals than those at high concentration (100% or whole), because of dilute toxicity in the first case. However, the CUR was outstandingly higher (5-fold) for units with 100% AnE than for units with 50% AnE (see Fig. 3). The same trend was observed for LUR, but the difference was only 1.8- to 2-fold (Fig. 4). This AnE

**Table IV.** ***Anova* for Net Unit Color Removal Efficiency (*CUR*)**

| Source | DF | Sum of | Mean | F Value | Pr>F |
|---|---|---|---|---|---|
| Model | 23 | 0.00244 | 0.00011 | 3.39 | 0.0021 |
| Error | 24 | 0.00075 | 0.00003 | | |
| Corr. Total To | 47 | 0.00319 | | | |

| R-Square | C. V. | Root MSE | CUR Mean |
|---|---|---|---|
| 0.764 | 88.87 | 0.00559 | 0.00629 |

| Source | DF | Mean Square | F Value | Pr>F |
|---|---|---|---|---|
| ***D*** | 2 | 0.00025 | 8.06 | ***0.0021*** |
| C | 1 | 0.00004 | 1.27 | 0.2706 |
| ***B*** | 1 | 0.00055 | 17.54 | ***0.0003*** |
| ***A*** | 1 | 0.00034 | 10.86 | ***0.0030*** |
| B*A | 1 | 0.00001 | 0.38 | 0.5432 |
| C*A | 1 | 0.00011 | 3.42 | 0.0766 |
| D*A | 2 | 0.00008 | 2.51 | 0.1024 |
| C*B | 1 | 0.000001 | 0.04 | 0.8482 |
| ***D*B*** | 2 | 0.00011 | 3.48 | ***0.0471*** |
| D*C | 2 | 0.00003 | 0.97 | 0.3941 |
| C*B*A | 1 | 0.000005 | 0.14 | 0.7078 |
| D*B*A | 2 | 0.00010 | 3.10 | 0.0637 |
| D*C*B | 2 | 0.00002 | 0.75 | 0.4828 |
| D*C*A | 2 | 0.00003 | 0.90 | 0.4194 |
| D*C*B*A | 2 | 0.00008 | 2.40 | 0.1121 |

**Table V. *Anova* for Net Unit Ligninoid Removal Efficiency (*LUR*)**

| Source | DF | Sum of | Mean | F | Pr>F |
|---|---|---|---|---|---|
| Model | 23 | 0.00272 | 0.00012 | 6.04 | 0.0001 |
| Error | 24 | 0.00047 | 0.00002 | | |
| Corr. Total | 47 | 0.00319 | | | |

| R-Square | C. V. | Root MSE | LUR Mean |
|---|---|---|---|
| 0.853 | 51.44 | 0.00442 | 0.00860 |

| Source | DF | Mean Square | F Value | Pr>F |
|---|---|---|---|---|
| ***D*** | 2 | 0.00055 | 28.20 | ***0.0001*** |
| C | 1 | 0.000007 | 0.34 | 0.5647 |
| ***B*** | 1 | 0.00013 | 6.49 | ***0.0177*** |
| ***A*** | 1 | 0.00049 | 24.81 | ***0.0001*** |
| B*A | 1 | 0.000009 | 0.48 | 0.4939 |
| C*A | 1 | 0.00007 | 3.82 | 0.0625 |
| ***D*A*** | 2 | 0.00012 | 6.12 | ***0.0071*** |
| C*B | 1 | 0.00002 | 1.02 | 0.3525 |
| D*B | 2 | 0.00002 | 1.25 | 0.3047 |
| *C | 2 | 0.00002 | 0.83 | 0.4462 |
| C*B*A | 1 | 0.00004 | 2.07 | 0.1632 |
| ***D*B*A*** | 2 | 0.00007 | 3.59 | ***0.0433*** |
| D*C*B | 2 | 0.00003 | 1.43 | 0.2593 |
| ***D*C*A*** | 2 | 0.00010 | 5.30 | ***0.0124*** |
| ***D*C*B*A*** | 2 | 0.00006 | 3.18 | ***0.0598*** |

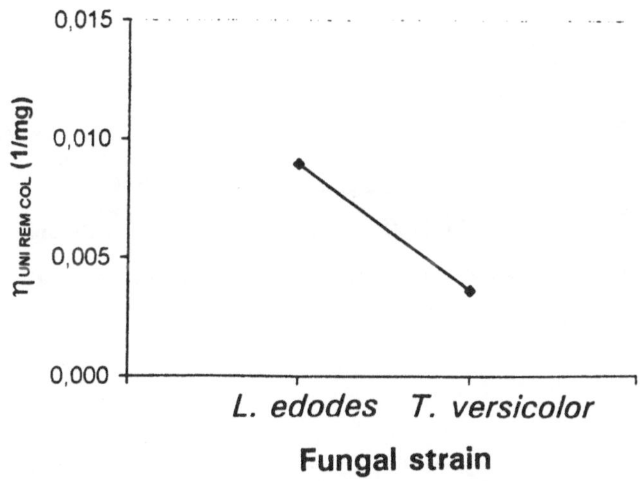


**Figure 1. Main effect of the fungal strain factor on the unit color removal efficiency (CUR).**

concentration factor was highly significant for CUR (Prob. 0.0003) and moderately significant for LUR (Prob. approx. 0.02). These results have an important and positive implication from the practical standpoint: whole AnE can be fed to the post-treatment stage; no dilution is required. Then, small reactors would be needed. Furthermore, with caution due to the exploration of only two levels of the factor in this experiment, it can be said that no toxic effects of the AnE on the fungi should be expected.

The main effect of the carbon source concentration factor (C) was not significant (see Table IV and Table V), although it seems to be a 15% decrease in CUR and a 8% increase in LUR with the addition of glucose (Fig. 5 and Fig. 6, respectively). Most research on treatment of bleaching effluents with ligninolytic fungi found that glucose or other carbohydrate supplementation was beneficial for color removal. However, our results are in the line of the findings of Esposito et

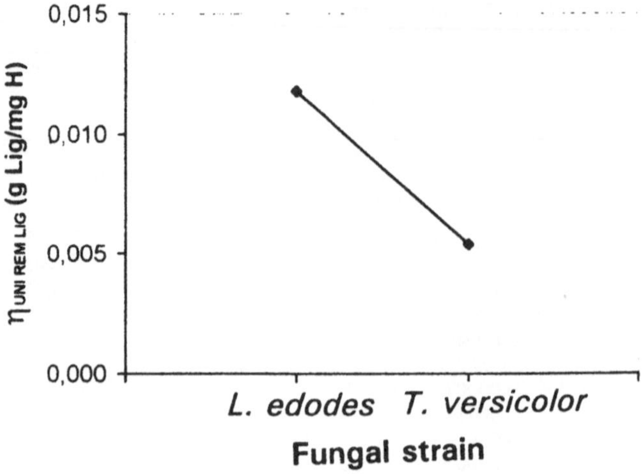


**Figure 2. Main effect of the fungal strain factor on the unit ligninoid removal efficiency (LUR).**

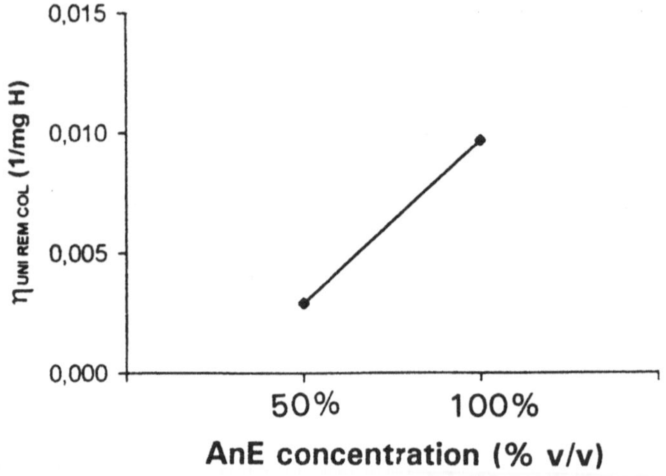


**Figure 3. Main effect of the anaerobic effluent concentration factor on the unit color removal efficiency (CUR).**

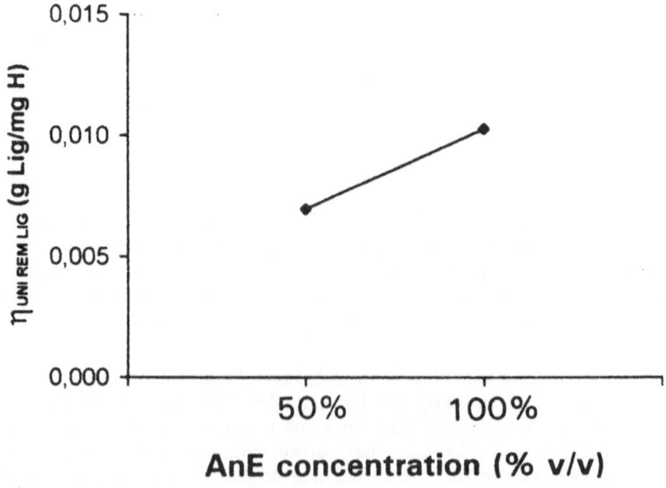


**Figure 4. Main effect of the anaerobic effluent concentration factor on the unit ligninoid removal efficiency (LUR).**

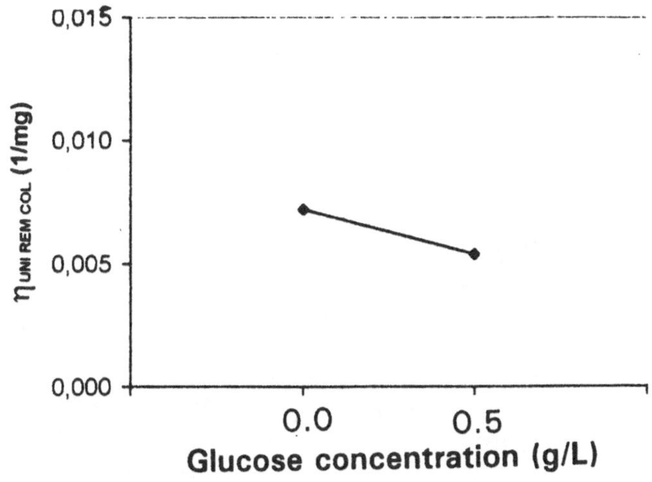


**Figure 5. Main effect of the concentration of the carbon source (glucose) factor on the unit color removal efficiency (CUR).**

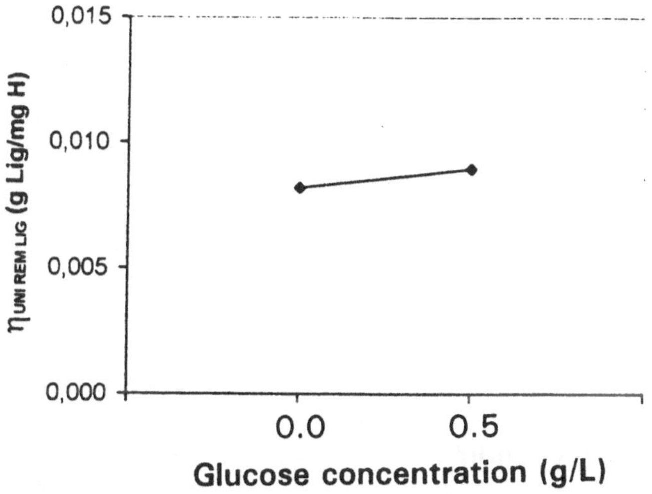


**Figure 6. Main effect of the concentration of the carbon source (glucose) factor on the unit ligninoid removal efficiency (LUR).**

al.,[14] who screened lignin-degrading fungi for color removal from pulp mill effluents, with no carbohydrate supplementation. Again, from the practical point of view, it is an advantage that no further additions of foreign carbon sources are needed to enhance the CUR and LUR of the AnE.

The factor immobilization (D) was highly significant for both CUR and LUR (Table IV and Table V). The best results were obtained for fungi immobilized on calcium alginate (see Fig. 7 and Fig. 8). Surprisingly, the poorest results were for fungi immobilized on GAC. This could be due to the fact that the *net* removals were considered in this evaluation; that is, removals discounted for abiotic (adsorption) reductions. An analysis based on gross removals (biological plus abiotic) showed that fungi immobilized on GAC gave the best results (data not shown).

Some factor interactions were very significant, particularly the second order interactions Immobilization x AnE concentration (DxB) for CUR and Immobilization x fungal strain (DxA) for LUR (Table IV and Table V, respectively). This is quite reasonable from the statistical standpoint, since D and B are highly significant factors for CUR and they would "drag" their weight into the interactions they contribute to.[37] A similar statistical reasoning is valid for factors D, A, and their interaction DxA for LUR. However, the biological or physical interpretation might be

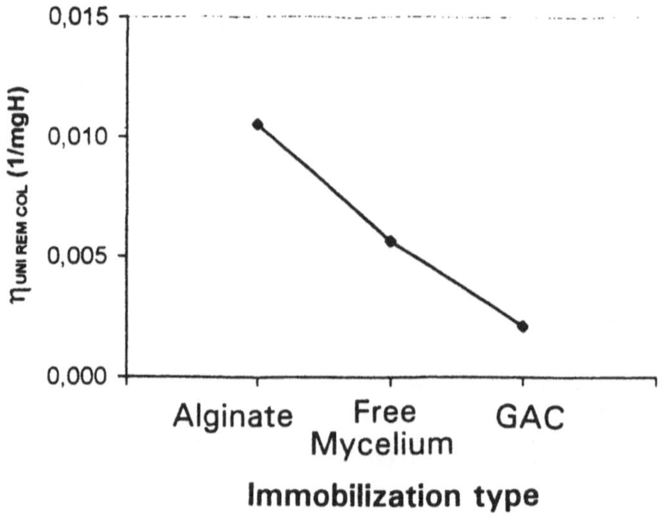


**Figure 7. Main effect of the immobilization type factor on the unit color removal efficiency (CUR).**

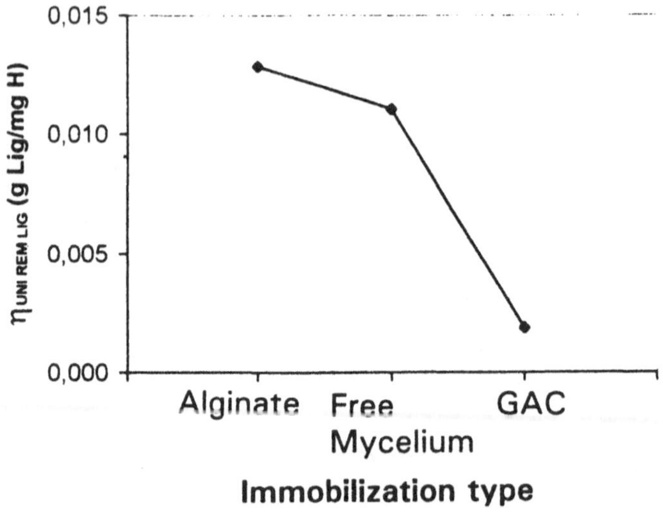


**Figure 8. Main effect of the immobilization type factor on the unit ligninoid removal efficiency (LUR).**

more complex and will not be attempted in this chapter. A few selected higher order interactions were also significant (see Table IV and Table V).

## CONCLUSION

The following conclusions can be drawn from this work:

- *L.e.* color and ligninoid reductions were 146% and 105% higher than those achieved by *T.v.*. The factor fungal strain was highly significant for both responses CUR and LUR;
- Units containing whole AnE (100% concentration) gave better removals than those with diluted (50%) AnE. The CUR was five-fold higher for units with 100% AnE than for units with 50% AnE. The same trend was observed for LUR, but the differ-

ence was only 1.8- to 2-fold. The AnE concentration factor was highly significant for CUR and moderately significant for LUR;
- Glucose supplementation did not significantly improve CUR and LUR of the AnE;
- The factor immobilization was highly significant for both CUR and LUR. The best results were obtained for fungi immobilized on calcium alginate. Interestingly, the poorest results were for fungi immobilized on GAC.
- With an eye to a future practical post-treatment based on ligninolytic fungi process, this experiment suggested at least two advantages:
  1. Whole AnE can be fed to the post-treatment stage (highest CUR and LUR) with no apparent toxicity to the fungal biomass. As there is no need to dilute the AnE, more compact reactors could be used and savings made on investment costs;
  2. No benefits from supplementing the AnE with a carbohydrate were found, so savings in O&M can be achieved because no external source of carbon is necessary.

Overall, the best color and recalcitrant organic matter removals were obtained for *Lentinus edodes*, whole AnE, no carbon-source supplementation, and fungus immobilizaton on alginate. This combination deserves further studies for the post-treatment of anaerobically-treated kraft black liquor effluent at lab reactor level.

## ACKNOWLEDGMENTS

The authors wish to thank Dr. I. Reid (PAPRICAN, Montreal, Canada) for kindly supplying the strain of *Trametes versicolor*; Prof. Jovita Martínez-Cruz (Microbial Collection Curator, CINVESTAV) for her help with training and strain preservation; and Dr. Fernando Esparza-García (DBB-CINVESTAV) for his comments during the development of the experiments. Support from the Mexican Council of Science and Technology (CONACYT) as a graduate scholarship to CE-V is gratefully acknowledged. This work was partially funded by the ICAITI (Guatemala)-AID, INE-Ministery of Environment, Natural Resources and Fisheries (Mex.) and CINVESTAV (Mex). Critical reading of the manuscript by Prof. Noemí Rinderknecht-Seijas (ESIQIE del IPN) is greatly appreciated.

## REFERENCES

1. Rintala, J. and R. Lepistö, "Thermophilic anaerobic-aerobic and aerobic treatment of Kraft bleaching effluents," *Water Sci. Technol.*, 28(2), pp. 11–16, 1993.
2. Poggi-Varaldo, H.M. "Anaerobic methanogenic consortia for pretreatment and detoxification of pulp and paper mill wastewaters." *Proc. 1st Intl. Mtg. Microb. Ecol.*, 8–12 May, 1995, CINVESTAV-IPN, México, p. 53.
3. Hernández-Vera, R. and Poggi-Varaldo, H.M. "Aplicacion de pruebas de bioensayos con *Daphnia magna* en el tratamiento de un efluente derivado de la industria ee la celulosa y papel en tratamiento biol and paper mill wa*Intl. Congress of Biotechnol. Havana'95*. Nov. 13–18, 1995, Havana, Cuba.
4. Hernández-Vera, R., Poggi-Varaldo, H.M., and Loa-Ramos, A. "Detoxification of an effluent from a pulp mill by a combined biological treatment." *Proc. 1st Intl. Mtg. Microb. Ecol.*, 8–12 May, 1995, CINVESTAV-IPN, México, p. 53.
5. Camacho M.E. Depuration of a Kraft Pulp Mill Wastewater by Anaerobic Treatment and Three Post-Treatment Alternatives: Aerobic, Coagulation-Flocculation and Adsorption on Activated Carbon. B.S. Chem. Eng. Thesis. Instituto TecnolMex.) and CINVESTAV (Mex)co, 1994. In Spanish.
6. Estrada-Vázquez, C., Poggi-Varaldo, H.M., Fernández-Villag Anaerobic Treatment and Three Post-Treatment Alternatives: Aerobic, Coagulation-*Proc. 51st. Purdue Industrial Waste Conference,* Ann Arbor Press Inc., Chelsea, MI, 1997, pp. 651–662.
7. Martín, C. and Manzanares, P. "A study of the decolourization of straw soda-pulping effluents by *Trametes versicolor.*" *Bioresource Technol.* 47, pp. 209–214, 1994.
8. Estrada-Vázquez, C. Post-Treatment of Anaerobically-Treated Industrial Wastewaters. M.S.Thesis. Dept. Biotechnol. and Bioeng., CINVESTAV, Mexico DF, Mexico, 1997. In Spanish.
9. Hatakka, A.I. and Uusi-Rauva, A.K. "Degradation of 14C-labelled poplar wood lignin by selected white-rot fungi." *Eur. J. Appl. Microbiol. Biotechnol.* 17, pp. 235–242, 1983.

10. Kirkpatrick, N., Reid, I.D., Ziomek, E., Ho, C., and Paice, M.G. "Relationship between fungal biomass production and the brightening of hardwood Kraft pulp by *Coriolus versicolor.*" *Appl. Environ. Microbiol.* 55(5), pp. 1147–1152, 1989.
11. Kirkpatrick, N., Reid, I.D., Ziomek, E., and Paice, M.G. "Biological bleaching of hardwood Kraft pulp using *Trametes (Coriolus) Versicolor* inmobilized in polyurethane foam." *Appl. Microbiol. Biotechnol.* 33, pp. 105–108, 1990.
12. Ferrer, I., Dezotti, M., and Durán, N. "Decolorization of kraft effluent by free and inmobilized lignin peroxidases and horseradish peroxidase." *Biotechnol. Lett.* 13(3), pp. 577–582, 1991.
13. Galeno D.G. and Agosin, T.E. "Screening of white-rot fungi for efficient decolourization of bleach pulp effluents." *Biotechnol. Lett.* 12(11), pp. 869–872, 1990.
14. Esposito, E., Canhos, V.P., and Durán, N. "Screening of lignin-degrading fungi for removal of color from Kraft mill wastewater with no additional extra carbon-source." *Biotechnol. Lett.* 13(8), pp. 571–576, 1991.
15. Royer, G., Livernoche, D., Desrochers, M., Jurasek, L., Rouleau, D., and Mayer, R.C. "Decolorization of Kraft mill effluent: kinetics of a continuous process using inmobilized *Coriolus versicolor.*" *Biotechnol. Lett.* 5(5), pp. 321–326, 1983.
16. Royer, G., Desrochers, M., Jurasek, L., Rouleau, D., and Mayer, R.C. "Batch and continuous decolorization of bleached kraft effluent by a withe-rot fungus," *J. Chem. Technol. Biotechnol.* 35B, pp. 14–22, 1985.
17. Archibald, F., Paice, M.G., and Jurasek, L. Decolorization of Kraft Bleached Effluent Lignins by *Coriolus versicolor.* In: *Biotechnology in Pulp and Paper Manufacture, Applications and Fundamental Investigations.* (Proceedings of the 4th ICBPPI). Kirk, T.K. and Chang, H-m, Eds. Butterworth-Heinemann, 1990, pp. 1–13.
18. Archibald, F., Paice, M.G., and Jurasek, L. "Decolorization of Kraft bleachery effluent chromophoros by *Coriolus (Trametes) versicolor.*" *Enzyme Microb. Technol.* 12(11), pp. 846–853, 1990.
19. Trotter, P.C. "Biotechnology in the pulp and paper industry: a review. Part 2: Upgrading pulp properties, papermaking, effluent treatment, and converting biomass to fuels and chemicals." *Tappi J.* 73(5), pp. 201–205, 1990.
20. Livernoche, D., Jurasek, L., Desrochers, M., and Dorica, J. "Removal of color from Kraft mill wastewaters with cultures of white-rot fungi and with immobilized mycelium of *Coriolus versicolor.*" *Biotechnol. Bioeng.* 25, pp. 2055–2065, 1983.
21. Livernoche, D., Jurasek, L., Desrochers, M., and Velisky, I.A.. "Decolorization of a Kraft mill effluent with fungal mycelium inmobilized in calcium alginate gel." *Biotechnol. Lett.* 3(12), pp. 701–706, 1981.
22. Paice, M.G., Ho, C., Jurasek, L., Bourbonnais, R., and Archibald, F. Optimization of Kraft Pulp Bleaching by *Coriolus (Trametes) versicolor.* In: *Biotechnology in Pulp and Paper Manufacture. Applications and Fundamental Investigations.* (Proceedings of the 4th ICBPPI). Kirk, T.K. and Chang, H.-m., Eds.. Butterworth-Heinemann, 1990, pp 1–13.
23. Reid, I.D., and Paice, M.G. "Biological bleaching of Kraft pulps by white-rot fungi and their enzymes." *FEMS Micob. Rev.* 13, pp. 369–376, 1994.
24. Reid, I.D., and Paice, M.G. "Effect of residual lignin type and amount on bleaching of Kraft pulp by *Trametes versicolor.*" *Appl. Environ. Microbiol.* 60(5), pp. 1395–1400, 1994.
25. Reid, I.D., Paice, M.G., Ho, C., and Jurasek, L. "Biological bleaching of softwood Kraft pulp with the fungus *Trametes (Coriolus) versicolor.*" *Tappi J.* 7(8), pp. 149–153, 1990.
26. Valmaseda, M., Almendros, G., and Martínez, A.T. "Substrate-dependent degradation patterns in the decay of wheat straw and beech wood by ligninolytic fungi." *Appl. Microbiol. Biotechnol.* 33, pp. 481–484, 1990.
27. Setliff, E.C., Marton, R., Gransow, S.G., and Eriksson, K.L. "Biomechanical pulping with white-rot fungi." *Tappi J.* 7(8), pp. 141–147, 1990.
28. Leatham, G.F., Myers, G.C., and Wegner T.H. "Biomechanical pulping of aspen chips: energy savings resulting from different fungal treatments." *Tappi J.* 7(5), pp. 197–200, 1990.
29. Sachs, I.B., Leatham, G.F., Myers, G.C., and Wegner, T.H. "Distinguishing characteristics of biomechanical pulp." *Tappi J.* 7(9), pp. 249–254, 1990.
30. Trotter, P.C. "Biotechnology in the pulp and paper industry: a review. Part 1: Tree improvement, pulping and bleaching, and dissolving pulp applications." *Tappi J.* 73(4), pp. 198–204, 1994.
31. Agosin, E., Daudin, J.-J., and Odier, E. "Screening of white-rot fungi on (14C)lignin-labelled and (14C)whole-labelled wheat straw." *Appl. Microbiol. Biotechnol.* 22, pp. 132–138, 1985.

32. Bonnarme, P., and Jeffries, T.W. "Mn (II) regulation of lignin peroxidases and manganese-dependent peroxidases from lLignin dDegrading white rot fungi." *Appl. Environ. Microbiol.* 56(1), pp. 210–217, 1990.
33. Leatham, G.F., and Griffin, T.J. "Adapting liquid spawn *Lentinus edodes* to oak wood." *Appl. Microbiol. Biotechnol.* 20, pp. 360–363, 1984.
34. Leatham, G.F. "The ligninolytic activities of *Lentinus edodes* and *Phanerochaete chrysosporium*." *Appl. Microbiol. Biotechnol.* 24, pp. 51–58, 1986.
35. Wang, S.H., Ferguson, J.F., and McCarthy, J.L. "The decolorization and dechlorination of Kraft bleach plant effluent solutes by use of three fungi: *Ganoderma lacidum, Coriolus versicolor* and *Hericiumerinaceum*." *Holzforschung* 46(3), pp. 219–223, 1992.
36. Sermanni, G., D'Annibale, A., Di Lena, G., S. Vitale, N., Di Mattia, E., and Minelli, V. "The production of exo-enzymes by *Lentinus edodes* and *Pleurotus ostreatus* and their use for upgrading corn straw." *Bioresource Technol.* 48, pp. 173–178, 1994.
37. Montgomery, D.C. Diseño y analisis de experimentos. Grupo Editorial Iberoamericano, S.A. de C.V. México D.F., México, 1991.
38. SAS version 6.04. SAS Institute, NC, USA.

## LIST OF ABBREVIATIONS

| | |
|---|---|
| AnE | anaerobic effluent from a mesophilic anaerobic fluidized bed reactor |
| ANOVA | analysis of variance |
| avg | average value |
| $BOD_7$ | biochemical oxygen demand at 7 days incubation time |
| COD | chemical oxygen demand |
| CUR | net color unit removal efficiency |
| DF | degrees of freedom |
| DL | diluted liquor, black liquor-contaminated wastewater |
| GAC | granular activated carbon |
| HRT | hydraulic retention time |
| *L.e.* | *Lentinus edodes* |
| LUR | net ligninoid unit removal efficiency |
| lg | ligninoids |
| RANLEF | Spanish acronym for the anaerobic fluidized bed reactor |
| *T.v.* | *Trametes versicolor* |
| UV | ultraviolet range |

Greek letters:

| | |
|---|---|
| $\eta$ | removal efficiency (COD, color, ligninoids) |

# 51 FEASIBILITY STUDY FOR REMEDIATION OF AN OILY WASTE POND IN VENEZUELA

**Eric J. Dawson**
Parsons Engineering Science Inc.
Houston, Texas 77034

**Kevin A. Tynes**
Parsons Engineering Science International, Inc..
Lima, Perú

**Rocco M. Palazzolo**
Parsons Engineering Science, Inc.
Caracas, Venezuela

**Francisco Brito** and **Diana Gilbert4**
Corpoven, S.A.
Puerto La Cruz, Venezuela

## ABSTRACT

This chapter focuses on the development of a closure plan for an oily waste pond based on characterization data, treatability studies, and assessment of effectiveness, implementability, and cost of closure alternatives. Emphasis is given to the most promising closure alternatives, consisting of hydrocarbon recovery, centrifugation, stabilization/solidification, and landfilling.

## INTRODUCTION

Corpoven, S.A., an affiliate of Petróleos de Venezuela, S.A. (PDVSA), owns and operates a 200,000 barrel per day refinery in Puerto La Cruz, Venezuela. The refinery has been in operation since the early 1950s. Since the early 1980s, waste sludges, tank bottom residues, and other semiliquid wastes produced at the refinery have been discharged into a natural depression, known as Los Nisperos pond, located inside the refinery boundary. A site investigation and feasibility study were conducted for the Los Nisperos site, with the objective of developing a closure strategy for the pond.

The study of the Los Nisperos site, performed in response to Corpoven's efforts to comply with Venezuelan national environmental regulations promulgated in 1992, is the first large-scale site investigation/feasibility study completed in Venezuela. The study was conducted in four phases:

- site investigation
- identification and evaluation of treatment technologies
- development and evaluation of remediation alternatives
- closure plan development.

The site investigation was comprised of reviewing existing information, obtaining additional data through sampling and analysis of the pond contents, soil, and groundwater, reviewing relevant regulations to determine requirements and limitations for closure of the pond, and a discussion of overall refinery plans and operations that could be affected by closure of the pond or could impact use of the site. Following the site investigation, potentially applicable treatment technologies were identified and evaluated through review of technical literature, consultation

---

*52nd Purdue Industrial Waste Conference Proceedings*, 1997, Ann Arbor Press, Chelsea, Michigan 48118. Printed in U.S.A.

with equipment suppliers and environmental contractors, and laboratory-scale treatability studies. Once effective treatment technologies were identified, complete alternatives, comprised of one or more technologies, were developed and evaluated. All alternatives were compared relative to each other on the basis of effectiveness, implementability, and cost. The most appropriate alternatives were then developed in further detail, including conceptual cost estimates, for final evaluation. In addition to cost, effectiveness, and implementability, the final evaluation of the best alternatives considered other more intangible aspects such as the overall strategy for development of the refinery and a facility-wide waste management plan. The results of the site investigation/feasibility study conducted for the Los Nisperos pond are the focus of this paper.

## SITE DESCRIPTION

The Los Nisperos pond is located within the boundaries of the Corpoven refinery at Puerto La Cruz, Venezuela. Puerto La Cruz is a resort city of approximately 400,000 inhabitants situated along the coast of northeastern Venezuela in the State of Anzoátegui. The city and refinery are situated in a narrow coastal plain, surrounded by mountains to the east and south, and the Caribbean Sea to the north and west, as shown in Figure 1.

The Los Nisperos pond is approximately one kilometer from the Caribbean Sea. The pond occupies the location of the former confluence of two streams which flowed out of nearby hills. When the refinery was constructed, the flows in the streams were diverted around the refinery, leaving dry streambeds. Subsequent site grading for the construction of a tank farm filled in the streambeds above and below the confluence, forming a closed depression. Grading and construction activities have also resulted in fill being placed around the pond, thus forming higher dikes around the pond. The pond is irregular in shape and occupies an area of approximately two hectares. Runoff into the pond is limited to the rainfall that falls on the interior slopes of the depression.

The principal materials discharged to the Los Nisperos pond are tank bottom residuals. The refinery facilities include a large marine terminal for crude oil and refined products; therefore, there are a large number of crude oil tanks, as well as refined product tanks. As a result, the tank bottom residuals that have been placed in the pond result from a variety of sources, including

**Figure 1. Site location map.**

crude oil tanks, heavy fuel oil tanks, diesel tanks, gasoline tanks, including leaded gasoline tanks, and other petroleum product storage tanks. Other sources of materials that have been, or may have been, placed in the pond include ballast basin sludge, API separator sludge, dissolved air flotation (DAF) unit sludge, miscellaneous sludges from refining processes, and catalysts. The pond was selected for temporary storage of the sludges, tank bottoms, etc. based on a study of potential sites. The study of potential sites included an evaluation of soil characteristics, location, and other parameters. Oily materials were first placed in the pond in the early 1980s.

The pond surface is covered with floating oil and floating matted vegetation, and there are dead trees standing in some areas. Additionally, there are other types of assorted rubbish present in the pond such as lumber, paper, and plastic.

Prior to the site investigation/feasibility study, little characterization work had been performed at the site. The volume and characteristics of the pond contents were not known, and the hydrogeological conditions of the pond site had not been investigated.

## SITE INVESTIGATION

The site investigation consisted of collection and analysis of existing site information, a limited hydrogeologic investigation, characterization of the pond contents, and a topographical survey of the pond area. Field investigation activities were conducted during August and September 1994, with subsequent measurements of water levels in monitoring wells occurring after the initial investigation activities.

### Characterization of Pond Contents

In order to develop and evaluate closure alternatives, it was necessary to know the characteristics and volume of the pond contents. The characteristics of the pond contents were determined by analysis of samples taken from the pond. A topographic and bathymetric survey was conducted to determine the volume of the pond.

Depth probing revealed that the pond contained approximately 17,000 $m^3$ of material, had a relatively uniform bottom, and three distinct layers. The layers consisted of a bottom sludge layer and a two floating liquid layers. The top floating layer was comprised of a layer of water directly over the bottom sludge, with a layer of floating oil over the water layer. Of the total pond contents, the bottom sludge was estimated to comprise approximately 30%, the free water layer approximately 20%, and the floating oil layer approximately 50%.

Analysis of the samples of the pond contents showed that the bottom sludge layer was an emulsion of oil, water, and fine solids, and contained approximately 40% fixed solids by weight and approximately 15% water by volume. The floating oil layer consisted mainly of hydrocarbons, and contained approximately 3% fixed solids by weight and approximately 5% water by volume.

The bottom sludge and floating oil were both characterized by:

- high viscosities,
- high heavy hydrocarbon contents,
- low light hydrocarbon contents,
- lack of distillable fractions,
- high heat values,
- absence of PCBs and other chlorinated organic compounds,
- relatively high concentrations of iron, most likely due to corrosion and wear of pipelines, pumps, and tanks, and
- relatively low concentrations of sulfur, vanadium, lead, and nickel.

### Characteristics of Surrounding Area

Hydrogeologic conditions in the area of the pond were investigated through the installation and sampling of four groundwater monitoring wells around the perimeter of the pond, as shown in Figure 2. The groundwater level was found to be below the bottom of the pond, so there was no direct communication between the pond and the groundwater. Soil and groundwater samples collected from the monitoring wells indicated that there had not been migration of contaminants

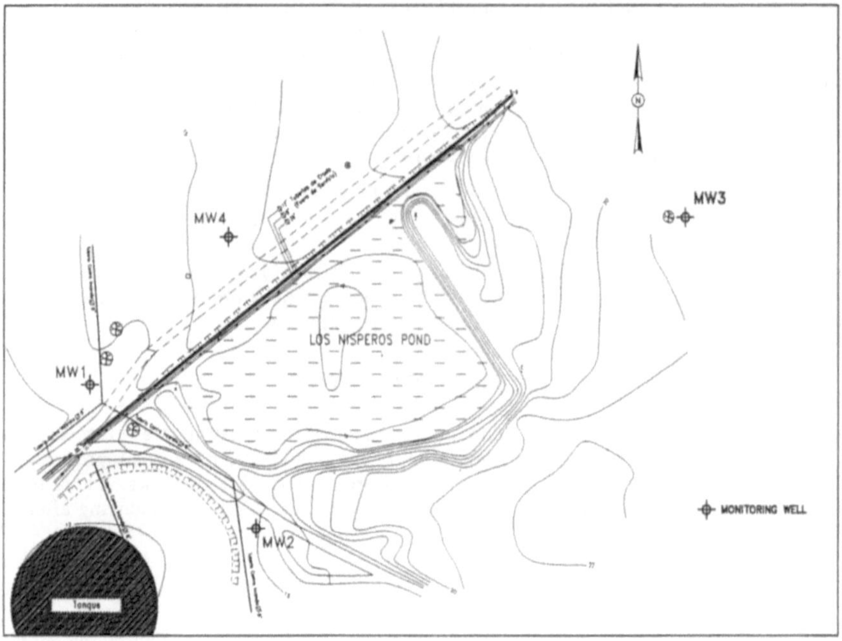


**Figure 2. Monitoring well locations.**

from the pond. As a result of the contaminants being contained within the pond, remediation of groundwater and soil outside of the pond would not be necessary.

## TREATMENT PROCESS IDENTIFICATION AND EVALUATION

Several treatment processes were identified and evaluated on the basis of effectiveness, implementability, and cost. These processes included technologies for recovery of hydrocarbons for fuel blending (including centrifugation, solvent recovery, and filtration), reuse technologies (including distillation/catalytic cracking, delayed coking, production of asphalt cement, and use of sludge as cement kiln fuel), and sludge/solids management technologies (including landfilling, landfarming, stabilization/solidification, thermal desorption, and bioslurry treatment). Laboratory studies were performed in support of this evaluation. Although the detailed evaluation is not provided here, this section summarizes the findings of the treatment process identification and evaluation.

Recovery of hydrocarbons from the sludge was an attractive alternative for various reasons. Recovery would reduce the amount of waste that would have to be managed, and the recovered hydrocarbons could potentially be used as a source of revenue. In addition, since the sludge does not contain high concentrations of contaminants like chlorinated organic compounds and heavy metals, hydrocarbon recovery became more feasible. The sludge did, however, contain significant percentages of water and solids, so a separation process would be necessary to obtain high quality hydrocarbons suitable for reuse. Laboratory-scale testing of separation processes showed that the viscosity of the sludge prevented effective separation of solids and water without the use of a dilutant, such as diesel or fluid catalytic cracking oil, to reduce the viscosity. Centrifugation was found to be the most effective separation process, and produced a hydrocarbon with low solids and water content. The levels of sodium, aluminum, and silica were, however, too high for the recovered hydrocarbon to be used directly as a fuel oil, but it would be possible to blend the recovered hydrocarbon with fuel oils produced at the refinery and achieve a product of acceptable quality. There was no benefit to combining the hydrocarbons recovered from the sludge with the crude oil processed at the refinery since the hydrocarbons in the sludge did not contain a significant distillable fraction. Filtration and solvent extraction did not prove to be feasible or cost-effective. Hydrocarbon recovery by centrifugation was retained for final evaluation.

Several reuse technologies were considered, but were eliminated for various reasons. For example, reuse technologies such as distillation/catalytic cracking, delayed coking, production of asphalt cement, and the use of sludge as cement kiln fuel, were not attractive options because appropriate facilities were not available locally, and the sludge would have to be transported over large distances, making these alternatives economically unfeasible.

Landfilling, landfarming, and stabilization/solidification are all well proven and demonstrated techniques for managing petroleum sludges that contain emulsions of water and solids. Although landfarming was considered, this alternative was not considered for final evaluation because of space restrictions within the refinery. Since liquids cannot be landfilled, landfilling could not be implemented without solidifying the semiliquid sludge. Laboratory-scale stabilization/solidification testing demonstrated that the sludge could be effectively solidified using cement kiln dust, which is available locally and economically. Leaching tests were performed on stabilized/solidified samples, and the results indicated that the stabilized/solidified sludge was appropriate for disposal in a sanitary, rather than hazardous or secure, landfill. Other processes such as, thermal desorption, and bioslurry treatment did not prove to be technically possible or feasible. Landfilling and stabilization/solidification were retained for final evaluation.

Three treatment processes remained for final evaluation, recovery of hydrocarbons by centrifugation, stabilization/solidification, and landfilling. These treatment processes were combined into two potential alternatives: (1) stabilization/solidification and landfilling the entire contents of the pond (no hydrocarbon recovery), and (2) recovering hydrocarbons from the sludge combined with stabilization/solidification and landfilling the remaining solids and contaminated soil. A more in-depth analysis of these two alternatives was then performed, including development of conceptual cost estimates, to differentiate between these two best alternatives.

Alternative 2, recovery of product by centrifugation and stabilization/solidification of residual solids which would then be landfilled, was selected as the best alternative for the following reasons:

- slightly lower cost,
- potential for recovery of some of the closure cost through sale of recovered product, and
- expectation that the product recovery process could be integrated with a refinery waste minimization program for oily wastes.

In Alternative 2, sludge from the pond will be centrifuged to recover as much hydrocarbon as possible. Recovered hydrocarbons will be used for fuel blending. The water separated from the sludge will be treated in the refinery wastewater treatment system. The remaining pond contents, including contaminated sediments, and solids separated during centrifugation will be stabilized/solidified and landfilled in the pond. It was decided to use the pond area for landfilling because the bottom of the pond is above the groundwater table, and there was no appropriate landfill facility located convenient to the refinery. The remaining volume in the pond will be backfilled with clean fill to raise the grade to match the surrounding terrain. A low permeability cap will be constructed over the area to minimize the infiltration of rainwater through the stabilized/solidified mass. Because of limited land area within the refinery, the stabilized/solidified mass, clean backfill, and low permeability cap will be constructed to allow the construction of a storage tank over the land area currently occupied by the pond. Long-term monitoring of the site to maintain the integrity of the low permeability cap and to verify that contaminants do not migrate from the site will be required.

A closure plan for the pond was developed for this selected alternative. The closure plan outlined the steps and equipment necessary to implement closure of the pond. The main elements of the closure plan included:

- outlining responsibilities of the parties involved,
- summarizing a schedule for the closure activities,
- describing the closure process with process flow diagrams for the product recovery and stabilization/solidification activities,
- describing the steps to implement closure of the pond,

- listing major equipment and materials, including process flow diagrams,
- defining preliminary design criteria and conceptual layout for the low permeability cap, including a construction quality assurance plan,
- summarizing post-closure care and monitoring programs, and
- summarizing health and safety issues.

## PROJECT SPECIFIC EXPERIENCE

Specific environmental legislation addressing the remediation of sites such as Los Nisperos has only recently been promulgated. As a result of the relatively recent promulgation of national environmental regulations in Venezuela, an environmental investigation infrastructure is in the early stages of development. For example, at the time of the study very few firms had experience with environmental drilling, and analytical laboratories did not have programs in place to accommodate extensive analysis of environmental samples. The number of environmental drillers and laboratories with environmental capabilities has increased significantly, but specialized environmental remediation equipment is not yet available locally on a wide basis, and there are not many local environmental remediation contractors with extensive experience.

Since Venezuelan environmental regulations are relatively new, they are largely untested, and standard interpretations had not been established. The movement seems to be, however, toward a site-specific, risk-based approach to environmental cleanups which mirrors the trend in the U.S. The Los Nisperos site will likely become an example of that approach.

Venezuela has been a major producer of petroleum for over 50 years. Consistent with historic petroleum operations, there are other oily waste ponds in Venezuela similar to the Los Nisperos pond. The Venezuelan government and the PDVSA companies are committed to the minimization of the sources of these impoundments, and to the remediation of those sites that pose significant threats to the Venezuelan populace and environment. As more environmental projects of this type are conducted in Venezuela, there will be an increase in the resources available and the continued development of an infrastructure to support such efforts.

## ACKNOWLEDGMENT

This study was funded in part by a grant from the U.S. Trade and Development Agency.

# Section Fourteen
# POLLUTION PREVENTION

## 52 MODELING ATOMIZATION PROCESSES

**S. D. Heister**, Associate Professor
School of Aeronautics and Astronautics
Purdue University
West Lafayette, Indiana, 47907

### ABSTRACT

Recent nonlinear simulations of various atomization processes are have been generated using computational schemes based on Boundary Element Methods. Both single fluid and two-phase computational models are discussed. Example simulations of finite-length liquid jets and droplets under forced excitation are provided to highlight current capabilities. A brief summary is included for future prospects using these tools, and the application to problems of interest in waste management.

### INTRODUCTION

The stringency of current and near-term federal regulations regarding industrial waste has prompted a renewed emphasis on increasing our understanding of basic processes germane to this field. In particular, injection and atomization of liquids is a process which has relevance to many industrial waste problems. While we currently cannot provide complete physical description of a complex spray emanating at high-speed from a nozzle, great strides have been made in recent years to successfully model lower speed flows. In this paper, we shall inform the reader as to the current state-of-the-art in atomization modeling, and provide some specific examples from this author's particular research program.

There are two primary obstacles to successfully modeling the nonlinear surface deformations present in any atomization problem. The presence of surface tension as an active force in these flows implies that local surface pressures depend on surface shape. However, the instantaneous surface shape cannot be determined without a knowledge of the pressure field. This situation implies a basic nonlinearity in the free-surface boundary conditions which must be addressed to successfully model flows of this nature. The second obstacle involves topological changes in the computational mesh which occur during "atomization events," i.e., when droplets are pinched from the parent surface.

Classical works due to Rayleigh,[1] Weber,[2] and others[3–6] ignore both obstacles by performing linear analyses under the assumption of infinitesimal deformations. While this body of knowledge is still drawn upon for many of today's practical injection problems, none of these analyses address surface shapes under large deformations. While there have been a variety of "weakly nonlinear" analyses performed,[7,8] a complete description of the atomization process has eluded us until the past couple of decades. Computational methods based on finite element methods (FEMs) and Boundary Element Methods (BEMs), and the Volume of Fluid (VOF) methodology have been developed to provide high-resolution simulations of nonlinear free-surface problems. Most FEM simulations[9] include the effect of viscosity; the main difficulty in using these methods for atomization lies in the grid distortion which occurs near pinch points. The VOF schemes[10] interpolate surface location from a fixed computational mesh (thereby eliminating the grid distortion issue). However, the interpolation process leads to poor resolution of the surface tension force since this force depends on surface curvature. Most recently, two new schemes

---

*52nd Purdue Industrial Waste Conference Proceedings*, 1997, Ann Arbor Press, Chelsea, Michigan 48118. Printed in U.S.A.

have been developed to improve upon the VOF methodology by use of front-tracking[11] and Level Sets.[12] While these techniques show much promise, they are currently limited to low Reynolds number (i.e., low-speed) flows due to computational limitations.

A large number of atomization problems have been treated using BEMs.[13–22] The main advantage of the BEM is that this technique requires a computational mesh be placed *only on the boundary* of the domain. This unique mesh permits high resolution of surface curvature (and hence surface tension forces), and also eliminates mesh-distortion issues near pinching events. In fact, BEM solutions have been run *beyond* droplet pinching events, as will be illustrated in subsequent sections of this chapter. The main disadvantage of the BEM approach is that it is most easily applied to inviscid problems; very few viscous solutions have been obtained, with the exception of weakly viscous[23] and creeping[24] (Stokes) flows.

In this chapter, we will focus on results obtained via the BEM approach. The following section summarizes the computational methodology for a variety of physical flow situations. The subsequent section provides a sample of results for finite-length liquid jets and droplets under forced excitation. We conclude the paper with a brief description of future efforts.

## MODEL DESCRIPTION

Most of the models developed to date were generated for inviscid flows using Boundary Element Methods (BEMs) as a basic mechanism to describe fluid dynamics. A variety of problems have been considered, including coupled simulations involving liquid and exterior gaseous phases. Calculations have been performed which consider body forces such as gravity, although these forces have been neglected in many problems.

Our interest lies in developing models which can address capillary (surface tension) forces at the interface. By choosing the liquid density ($\rho$), the size/height of the liquid body ($\alpha$), and a farfield velocity ($U$) as dimensions, the gas/liquid density ratio:

$$\epsilon = \rho_g/\rho \tag{1}$$

and the Weber number based on gas density:

$$We = \frac{\rho_g U^2 a}{\sigma} \tag{2}$$

become the two dimensionless parameters characterizing the flowfield. Here, the Weber number measures the ratio of inertial forces imposed by the gas phase to surface tension ($\sigma$) forces. In the following development, we presume that the nondimensionalization described above has been applied.

In the bulk of the modeling performed to date, we assume that both phases can be represented as incompressible, inviscid fluids. In this case, a velocity potential (whose gradient is simply the velocity) exists. Let $\phi$ and $\phi_g$ represent the velocity potential in liquid and gaseous phases, respectively. Continuity requires that both velocity potentials satisfy Laplace's equation:

$$\nabla^2\phi = \nabla^2\sigma_g = 0 \tag{3}$$

Proceeding with a standard BEM formulation, the integral form of Eq. 3 for the liquid domain becomes:

$$\alpha\phi(\vec{r_i}) + \int_\Gamma \left[\phi\frac{\partial G}{\partial n} - qG\right] d\Gamma = 0 \tag{4}$$

where $\phi(\vec{r_i})$ is the value of the potential at a point $\vec{r_i}$, $\Gamma$ denotes the boundary of the domain, and $G$ is the free-space Green's function corresponding to Laplace's equation. An analogous form of Eq. 4 can also be derived for the gas phase potential. For a well-posed problem, either $\phi$ or $q = \partial\phi/\partial n$ must be specified at each "node" on the boundary. Here $n$ is the outward normal to the boundary so that $q$ represents the velocity normal to the boundary. The quantity $\alpha$ in Eq. 4 results from singularities introduced as the integration passes over the boundary point, $\vec{r_i}$.

Models have been developed for both two-dimensional and axisymmetric flowfields. If we let $r$ and $z$ denote radial and axial coordinates, respectively, and denote the base point with subscript "i," the Green's function solution to the axisymmetric Laplacian can be written:

$$G = \frac{4rK(p)}{\sqrt{(r+r_i)^2 + (z-z_i)^2}} \tag{5}$$

where

$$p = \frac{(r-r_i)^2 + (z-z_i)^2}{(r+r_i)^2 + (z-z_i)^2} \tag{6}$$

and $K(p)$ is the complete elliptic integral of the first kind. For computational efficiency, this quantity is calculated using a curve fit[18] which has an accuracy to $10^{-8}$.

In the case of a 2-D flow (letting $x$ and $y$ represent the coordinates), we have:

$$G = \frac{1}{2\pi}\ln|\vec{r}-\vec{r}_i| = \frac{1}{4\pi}\ln\left[(x-x_i)^2 + (y-y_i)^2\right] \tag{7}$$

In all cases, we presume that the quantities $\phi$, $\phi_g$, $q$, and $q_g$ vary linearly along a given segment of the boundary. For the 2-D flows, integration across a segment can be carried out analytically. Singularities resulting from integration across a segment containing the base point are also integrable. In the case of axisymmetric flow, the integrations must be carried out numerically. In this case, we choose a four-point Gaussian quadrature for evaluation of integrals. Logarithmic singularities which arise in the elliptic integral when the segment contains the base point are treated with a special Gaussian integration designed to accurately treat this condition. Additional details regarding the numerical implementation can be found in References 18 and 21 for the axisymmetric and 2-D cases, respectively.

## Free Surface Treatment

The main challenge in developing models capable of tracking large deformations of an interface lies in the treatment of the free surface itself. Since capillary forces are important, it is crucial to develop a treatment capable of accurately determining surface curvature at all times during the simulation. For this reason, all models employ fourth-order centered differencing (on a generalized, variable spacing mesh) to determine surface curvature.[18]

In current models, we track surface nodes along the local *liquid* velocity vector. Under this assumption, for an axisymmetric situation, flow kinematics require:

$$\frac{Dz}{Dt} = \frac{\partial\phi}{\partial z} \qquad \frac{Dr}{Dt} = \frac{\partial\phi}{\partial r} \tag{8}$$

where the notation $D()/Dt$ implies a Lagrangian derivative for points on the surface moving with the local liquid velocity.

Recognizing that our BEM solver will return velocities normal to the surface, we employ the velocity transformations:

$$\frac{\partial\phi}{\partial r} = \frac{\partial\phi}{\partial s}\sin(\beta) + q\cos(\beta) \qquad \frac{\partial\phi}{\partial z} = \frac{\partial\phi}{\partial s}\cos(\beta) + q\sin(\beta) \tag{9}$$

where $\beta$ is the local wave slope and $\partial\phi/\partial s$ is the velocity tangential to the local surface. This tangential velocity is calculated using 5-point centered differences on $\phi$, except for nodes adjacent to ends of the free surface, where a 3-point formula is employed. The local wave slope, $\beta$, is calculated by determining the slope of a parabola fit through three points in the region of interest.[18] A completely analogous treatment can be employed for 2-D flows by replacing $r$ with $y$ and $z$ with $x$ in Eqs. 8 and 9.

Dynamics of the interface are addressed through the unsteady Bernoulli equation. In an Eulerian system where time derivatives are assumed to occur at a fixed spatial location, the dimensionless form of this relation for the liquid surface is:

$$\frac{\partial \phi}{\partial t} + \frac{1}{2}(\nabla \phi)^2 + P_g + \frac{\kappa}{We} = 0 \tag{10}$$

where $P_g$ is the gas pressure at the interface, and $\kappa$ is the surface curvature. The Eulerian-Lagrangian transformation for nodes on the interface moving with the *liquid velocity* is:

$$\frac{D(\,)}{Dt} = \frac{\partial(\,)}{\partial t} + \nabla\phi \cdot \nabla(\,) \tag{11}$$

Using this transformation, the Bernoulli equation in the liquid and gas phases are written:

$$\frac{D\phi}{Dt} = \frac{1}{2}(\nabla\phi)^2 - Pg - \frac{\kappa}{We} \qquad \epsilon\frac{D\phi_g}{Dt} = \epsilon\nabla\phi \cdot \nabla\phi_g - \frac{\epsilon}{2}(\nabla\phi_g)^2 - Pg \tag{12}$$

Equations 8 and 12 are integrated in time using a fourth-order Runge-Kutta scheme. A stable, consistent procedure[21] has been developed to handle the coupled, nonlinear boundary conditions at the interface (Eq. 12). Since the nodes on the interface are allowed to move with their local velocity, over time they tend to group themselves in regions of high curvature. This phenomenon leaves regions of lower curvature poorly defined. To alleviate this problem, the surface mesh is regridded using a series of cubic splines (for surface coordinates, $\phi$, and $\phi_g$) at each time step to keep the spacing between the nodes constant along the surface. The regridding process does provide a natural "smoothing" of the surface. Many previous authors[14–16] have been forced to implement smoothing procedures to alleviate "zig-zag" instabilities which develop on the surface after a large number of time steps.

## Example Computational Domains

A typical computational domain denoting various boundary conditions used in the liquid jet models is shown in Figure 2. Here, $q$ and $q_g$ are velocities normal to the boundary in the liquid and and gas domains respectively. Gas nodes are denoted with an "x," and liquid boundary nodes with a "o" in this figure. In addition, a capability exists to determine velocity profiles within the fluid at a given station (such as the orifice exit plane). These interior nodes are denoted with a "+" in Fig. 2. Here, it is important to note that interior nodes are not required to obtain arbitrary unsteady solutions using the methodology employed in this research. Note that the actual orifice internal geometry can be modeled, although we have also performed simulations which begin with prescribed orifice exit planes conditions as will be discussed in the following section.

The computational domain and the boundary conditions used in droplet models are displayed in Figure 2. In this particular domain, which was used to study a droplet's response to a forced acoustic excitation, symmetry planes exist for both r and z axes. Other calculations have been performed which utilize the upper half plane in order to study secondary atomization (droplet breakup) in a quiescent gas. As in Figure 1, gas nodes are denoted with a "x," while the liquid nodes are labeled with an "o."

Along the radial line of symmetry for both domains, $q = 0$ in the liquid and $\partial\phi_g/\partial r = 0$ in the gas. Bernoulli equations provide the boundary condition along the interface. For the forced excitation problem, the outer boundary of the gas domain is modeled as a sinusoidal flow in the axial direction:

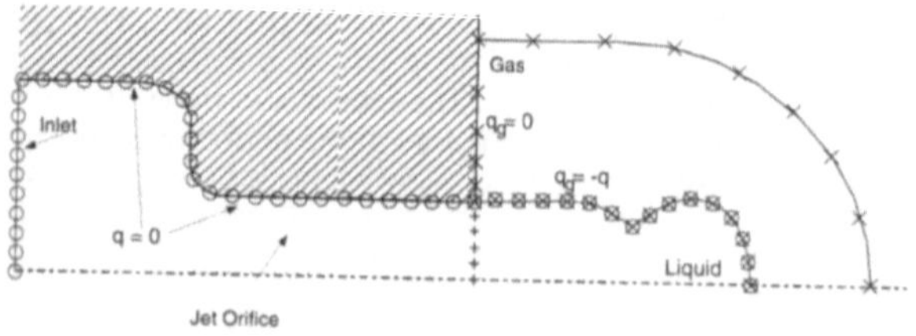


**Figure 1. Schematic of computational domain and boundary conditions for liquid jet simulations.**

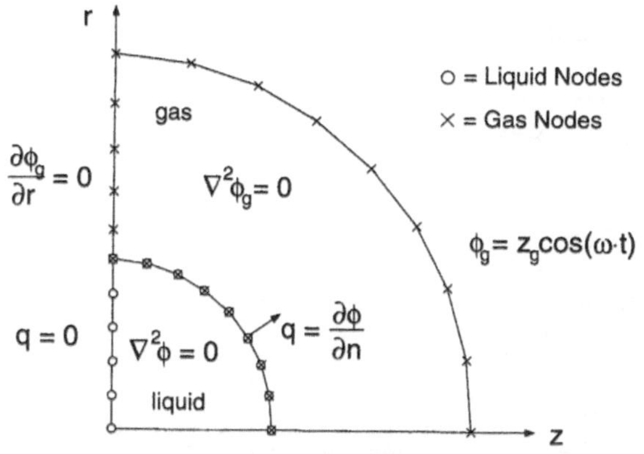


**Figure 2. Schematic of computational domain and boundary conditions for droplet simulations.**

$$\phi_g = z_g \cos(\omega t) \tag{13}$$

Nodes are not required along the line $r = 0$ since the Green's functions vanish along this boundary and no net contributions to the integrals in Eq. 4 are generated.

## RESULTS

Figures 3–10 present results from liquid jet simulations. Results in Figs. 3–9 are obtained for a finite-length jet assuming a specified oscillatory inflow at the orifice exit plane under negligible gas-phase interactions. The oscillatory inflow was utilized to simulate experimental conditions in which a speaker or piezoelectric driver is used to force disturbances in the jet of a given wavenumber, $k$. The assumed velocity for nodes traversing the orifice exit plane is:

$$q = [1 + q' \sin(kt)] \tag{14}$$

where $q'$ is the amplitude of the oscillatory disturbance.

All calculations begin with a long column of fluid, usually about 15 to 20 orifice radii, outside the orifice in order to ensure that perturbations from the assumed spherical end cap do not influ-

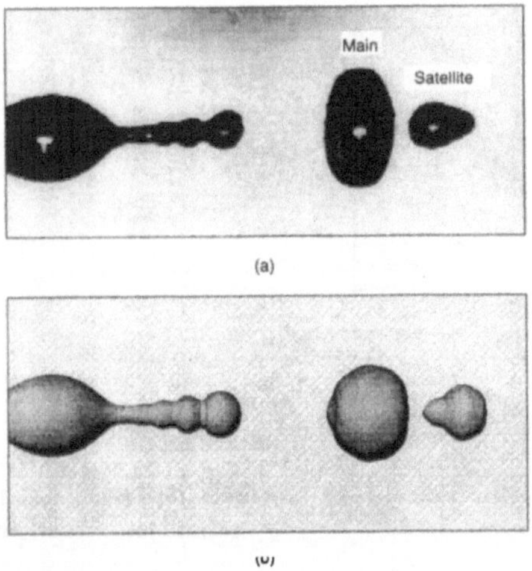


**Figure 3. Jet profile comparison near the pinch location, (a) experiment of moses, (b) BEM result, We = 17.6, Bo = 0.0109, $k$ = 0.447.**

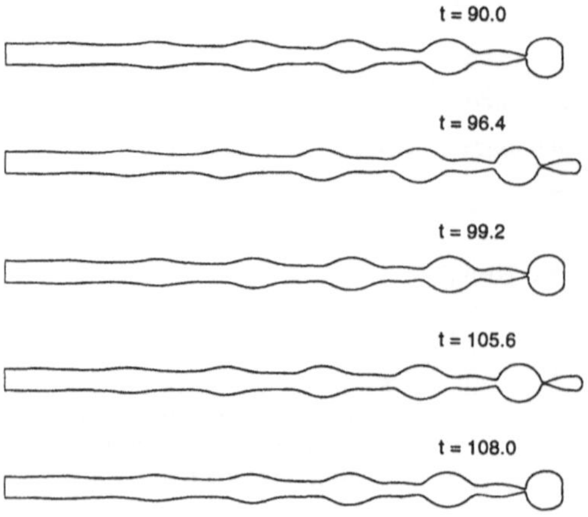


**Figure 4. Typical jet evolution, We = 100, $k$ = 0.70, $q'$ = 2%.**

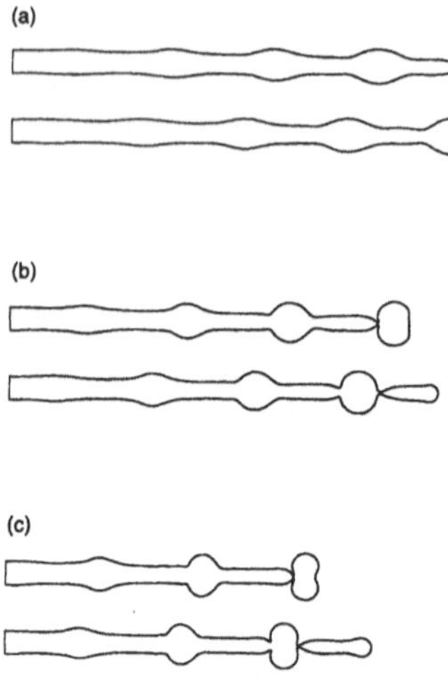


**Figure 5. Effect of inflow velocity disturbance magnitude, We = 100, $k$ = 0.70, (a) $q'$ = 2%, (b) $q'$ = 4%, (c) $q'$ = 6%.**

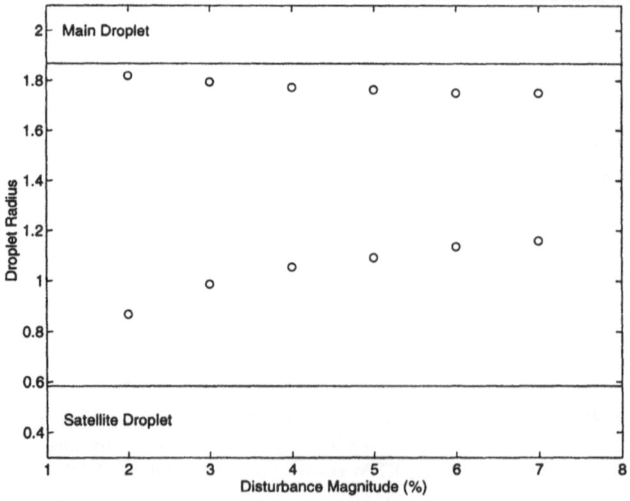


**Figure 6. Droplet radii vs. inflow velocity disturbance magnitude for We = 100, $k$ = 0.7 (o finite-length calculation; __ infinite jet calculation.**

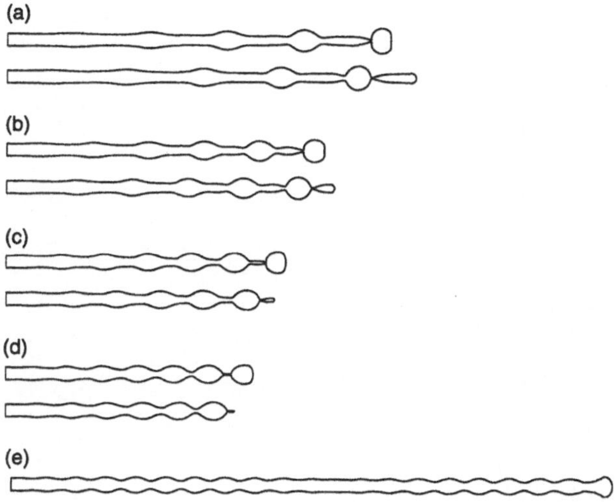


**Figure 7. Effect of disturbance wave number, We = 100, $k$ = 0.70, $q'$ = 2%, (a) $k$ = 0.5, (b) $k$ = 0.7, (c) $k$ = 0.9, (d) $k$ = 1.1, (e) $k$ = 1.3.**

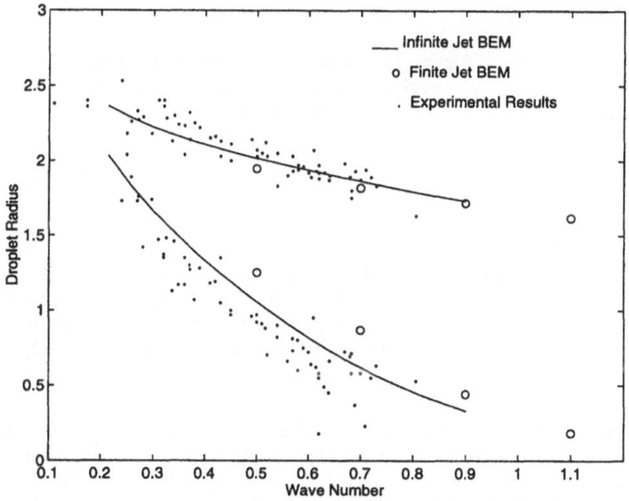


**Figure 8. Droplet radii vs. disturbance wave number.**

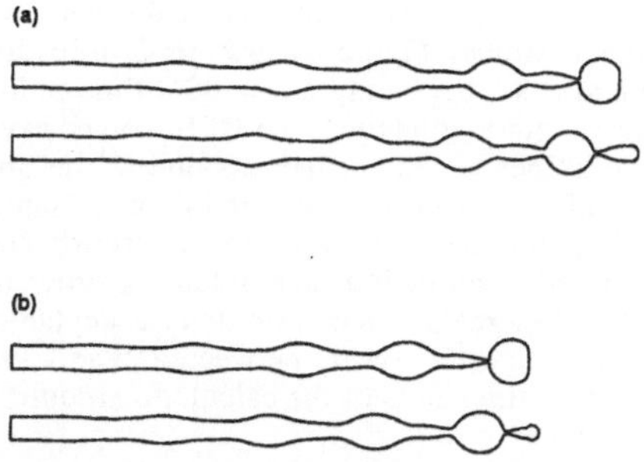


**Figure 9. Effect of Weber number on jet profile, $k$ = 0.7, $q'$ = 2%, (a) We = 100, (b) We = 50.**

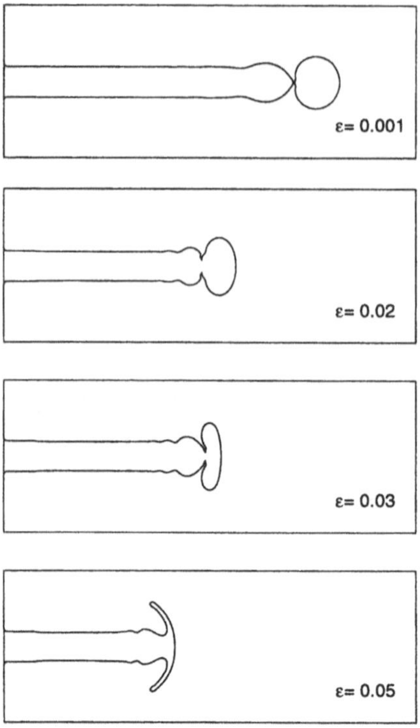


**Figure 10. Effect of density ratio on the evolution of a jet under steady inflow conditions, *We* = 17.6.**

ence the development of waves due to the imposed perturbation. The calculations presented here use a grid spacing of 20% of orifice radius, a time step of $\Delta t$ = 0.0025 time units (unless noted otherwise), and separate droplets from the calculation when the radial coordinate at the pinch location is within 5% of the orifice radius. Numerical experiments indicate that results are insensitive to this "pinch criteria," as well as the assumed initial geometry, as long as simulations are run through a few pinch events.

A BEM calculation was performed in order to match the results from experiments conducted by Moses et. al.[25] Figure 3 shows a comparison between the experimental image (a) and a three-dimensional rendering of the calculated jet profile (b) at time $t$ = 92.4 time units. Here, the BEM result was aligned axially with the experimental image such that the breakup point occurs at the same axial location; the simulation tends to underpredict jet breakup length due to the neglect of viscous effects. This figure shows the excellent qualitative agreement between the experimental and numerical results. Note how the tip of the jet shows a series of three or four ripples formed by the recoil of the surface after the pinching event.

A standard case is chosen as We = 100, $k$ = 0.7, and $q'$ = 2%. Linear stability analysis shows that this wave number is the most unstable for this Weber number. Figure 4 shows jet profiles for five successive droplet shedding events for the standard case, beginning at $t$ = 90.0 time units. By this time, the calculated drop sizes and jet breakup length predicted by the BEM solver have a repeating pattern with very little variation. Periodic bulges appear in the jets due to the unsteady inflow. Nonlinear effects lead to the formation of two drops (termed "main" and "satellite" droplets) for each wave generated by the assumed disturbance. A moderate 2% disturbance magnitude was chosen over typical acoustically-generated perturbation magnitudes in order to keep the jet breakup length to a reasonable size, and therefore make calculation times acceptable. After the jet sheds the fluid from the initial assumed geometry, there were between 255 and 290 nodes on the jet surface during this calculation. The first 77 time units of the calculation required about 500,000 CPU seconds on an IBM RISC/6000 computer.

### Effect of Oscillation Magnitude

The effect of the oscillation magnitude on the jet breakup is investigated by six simulations of $q'$=2% through 7%. Figure 5 shows jet profiles for three of these disturbance magnitudes. Each pair of profiles shows the jet at times just before a main droplet and a satellite droplet are shed from the calculation. As this figure shows, increasing the size of the perturbation decreases the breakup length, and changes the shape of the droplets. At larger disturbance magnitudes, the main droplets take on a "squashed" shape as a result of the high velocity fluid encountering lower velocity fluid which has already exited the nozzle. In the other half of the imposed oscillation the jet is "stretched" since the fluid immediately outside the orifice is at higher velocity than the inflow. Breakup times are reduced (as compared to the infinitesimal case) primarily due to the fact that the troughs begin at lower height.

The effect of a finite-amplitude disturbance on the predicted droplet sizes is presented in Figure 6, where solid lines denote predictions assuming an infinitesimal disturbance.[17] The satellite size is shown to increase with disturbance magnitude at the wave number ratio selected for these simulations. Necking and pinching of the jet under finite-amplitude disturbance occur at locations much closer to the peak of the wave. Extrapolation to higher $q'$ values would presumably lead to a monodisperse case; droplet trains of this type have been created experimentally by Dressler and co-workers.[26]

### Effect of Oscillation Wave Number

Since the jet radius and mean orifice exit velocity are chosen as characteristic dimensions, nondimensional wave numbers are equivalent to a dimensionless disturbance frequency. Figure 7 shows the effect of wave number on the character of the jet for We = 100 and $q'$ = 2%. Each of the first four pairs of profiles show the jet at times just before a main droplet and just before a satellite droplet are shed from the calculation. Increasing the wave number from $k$ = 0.5 to $k$ = 1.1 tends to decrease the breakup length even beyond the most unstable wave number, $k_{max}$ = 0.7; a trend not predicted by linear theory. In addition, the size of satellite drops tends to decrease with increasing frequency in this range. At a wave number of $k$ = 1.1, satellite drops have nearly vanished, indicating that a wave number near this value can produce monodisperse atomization. At $k$ = 1.3, the breakup length of the liquid jet increases drastically from the previous calculation, and the model no longer predicts the formation of satellite droplets. Figure 7(e) shows the jet profile at time $t$ = 107.0 time units; the jet length length is still increasing at this point in the calculation.

Quantitative predictions of droplet sizes are shown in Figure 8. As in Figure 6, the solid lines are results for infinitesimal disturbances and experimental results are from References 27 and 28. The finite-amplitude results are for $q'$ = 2%, and they show a tendency to drive conditions toward a monodisperse result by increasing satellite drop size and decreasing main drop size over the entire range of wave numbers investigated. At higher disturbance amplitudes, local curvature in the transition region between peaks and troughs can be driven to large enough values to promote nonlinear instability at wavelengths predicted to be stable using linear analysis.

### Effect of Weber Number and Gas Density

For fixed orifice radius and liquid density, decreasing the Weber number corresponds to either increasing the surface tension or decreasing the inflow velocity. As the surface tension is increased, the wave formed on the surface of the jet from the unsteady inflow should tend to grow more quickly, leading to shorter breakup lengths. This effect is shown in Figure 9 for We = 50 and 100, using $k$ = 0.7 and $q'$ = 2%. Each profile shows the jet just before a main droplet and satellite droplet are shed from the calculation. Increasing the Weber number also tends to decrease the size of the satellite droplets for the chosen conditions.

Figure 10 depicts the influence of gas density on a low-speed jet under steady inflow conditions. Simulations in this figure were conducted using the fully-coupled gas/liquid solution procedure applied to the domain shown in Figure 1. Increases in $\epsilon$ have the effect of decreasing jet breakup length, and flattening the drop produced when the jet pinches. This flattening becomes increasingly violent such that at $\epsilon = 0.05$, the droplet has flattened into a mushroom cap. It is interesting to note that $\epsilon$ must be increased by an entire order of magnitude (relative to the air/water value of $\epsilon \approx 0.001$) before significant differences in jet shape are observed.

## Results from Droplet Simulations

Figures 11–13 highlight some of the results obtained from droplet simulations. In Figure 11, the results from fully-coupled gas/liquid calculations (using the domain shown in Figure 2) are summarized for the liquid drop subjected to acoustic excitation from the gas phase. Here, the ratio of the excitation frequency ($\omega_g$) to the droplets natural frequency ($\omega_n$) is an additional parameter characterizing the disturbance. Using the nondimensionalization discussed above, Lamb's result for $\omega_n$ can be written:

$$\omega_2^2 = \omega_n^2 = \frac{24}{(3+2\epsilon)We} \tag{15}$$

In Figure 11, the Weber number was held at 0.5779 and the density ratio was 0.00123. These conditions correspond to acoustic excitation of a 100 micron water droplet with a 160 db distur-

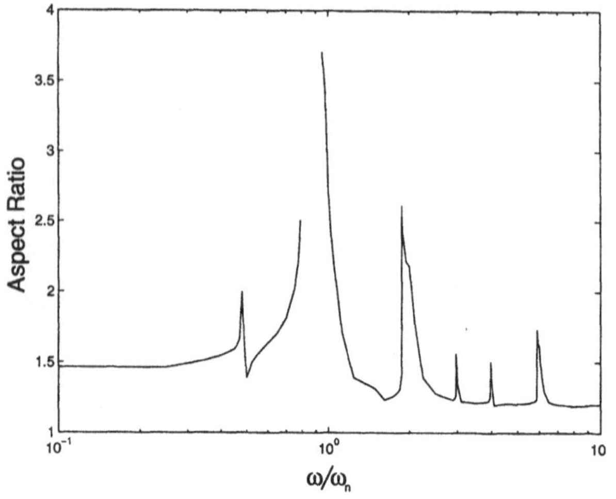


**Figure 11. Droplet frequency response as a function of gas-phase acoustic excitation frequency.**

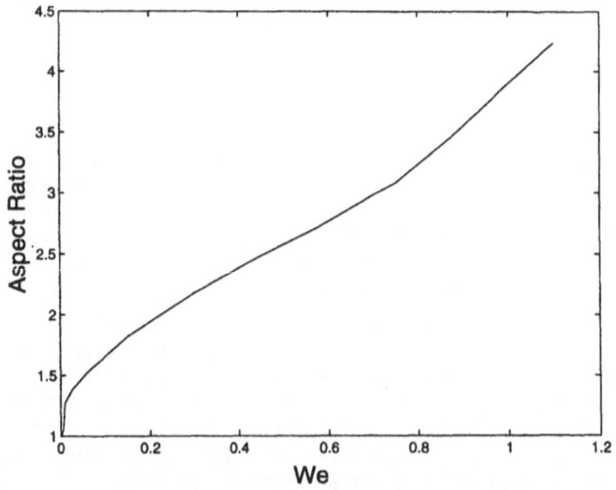


**Figure 12. Effects of the Weber number upon droplet behavior ($\epsilon$ = 0.00123, $\omega = \omega_n$).**

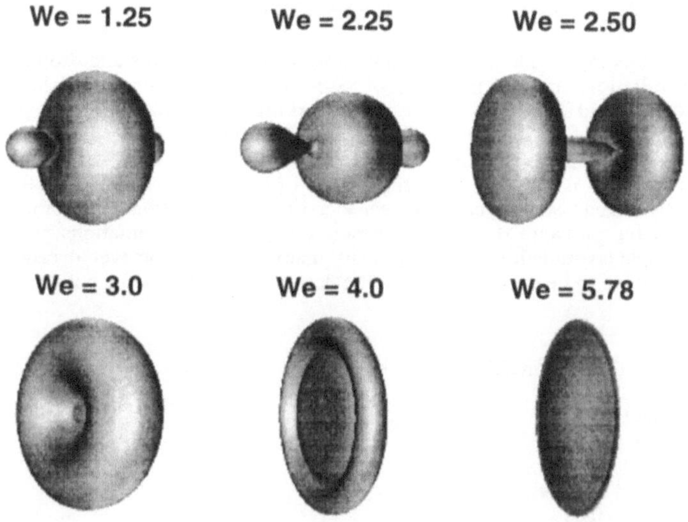


**Figure 13. Droplet atomization shapes at various Weber numbers (e = 0.00123, $\omega = \omega_n$).**

bance in ambient air. The maximum aspect ratio obtained during the unsteady simulation provides a measure of the overall droplet response depicted in this figure.

Areas of significant response occur for $\omega/\omega_n$ = near 0.5, 1.0, 2.0, 3.0, 4.0, and 6.0. The peaks near $\omega/\omega_n$ = 0.5, 1.0, and 2.0 are primarily driven by second mode oscillations, while the peaks near $\omega$ = 3.0, 4.0, and 6.0 contain significant fourth-mode effects. It is important to note that the fundamental frequency of the fourth mode is three times that of the fundamental frequency of the second mode, $\omega_4 = 3\omega_n$, for a drop in a low density gas ($\epsilon \ll 1$). It is worth mentioning that a frequency ratio near 5.0 does not produce a notable response. A noteworthy area is the breakup region that exists for frequency ratios between 0.80 and 0.90. In this band, the disturbance is tuned to the natural frequency of the droplet to an extent such that atomization occurs. All peaks are shifted to frequencies slightly less than the linear result due to the nonlinear frequency shifts which are known to occur under finite deformation of the drop. In fact, the continual shifting of droplet natural frequency with deformation level leads to a bounded response over much of the frequency range (for this Weber number). In these regions, the acoustic perturbation continues to excite the droplet until the oscillations of the drop become out of phase with the disturbance, thereby tending to reduce the amplitude of the oscillation. This process repeats indefinitely or until the drop breaks up.

The influence of the intensity of the acoustic wave was investigated by conducting a series of simulations at fixed density and frequency ratios, but with varying Weber number. For these simulations, the time step was set to be approximately 1/1000 of the period of a droplet oscillation for the second mode. Once again, the aspect ratio was used as an overall measure of droplet response; Figure 12 provides results for the case $\epsilon = 0.00123$, $\omega/\omega_n = 1.0$. During these simulations, it was observed that droplet atomization occurred at Weber numbers above 1.10. Here it is important to note that this "critical Weber number" is dependent on both density and frequency ratios; the 1.10 value is for harmonic excitation of a water droplet in air (e = 0.00123).

Droplet breakup was studied by conducting trials using Weber numbers greater than the critical value. From this analysis, three regimes of breakup were identified. Examples of these modes are shown in Figure 13. In the "nipple" breakup regime (1.1 < We < 2.5), two small satellite droplets are formed as a result of the nonlinear motion of the drop. Since the acoustic intensity is barely above the threshold value, breakup in this region takes a substantial amount of time. The nipple regime is similar to the mode of breakup experienced in the breakup band of frequencies, as shown in Figure 11. As Weber number is increased to 2.5, atomization occurs with the formation of two satellite droplets which are larger than the central drop. The "kidney" regime exists for 2.5 < We < 3.0 and is characterized by breakups which occur in 1–2 periods of the acoustic wave.

## CONCLUSIONS/FUTURE EFFORTS

Current technology enables us to provide accurate simulations of low-speed jets and droplets. However, computational limitations and the influence of viscosity limit capabilities for high-speed jet simulations at the present time. We are presently working on a zonal BEM/integral method model to address early stages of deformation of the high-speed viscous jet.[29] In addition, simulations of droplet breakup in a quiescent, viscous gas are currently underway. Finally, a full three-dimensional BEM capability is very nearly complete, and preliminary results from this model should be available by the end of this year. The ever-increasing computational capabilities brought about by new computer hardware should enable more complex, 3-D simulations in the coming years. While the complete simulation of a spray is still many years (maybe even decades) away, future analyses could prove to be quite valuable in the design of waste systems with minimal environmental impacts.

## NOMENCLATURE

$a$ = orifice or column radius
$G$ = free space Greens function
$k$ = wave number
$K(p)$ = complete elliptic integral of the first kind
$n$ = coordinate normal to local surface
$P$ = pressure
$q$ = velocity normal to local boundary
$r$ = radial coordinate (axisymmetric flow)
$s$ = coordinate aligned with local surface
$t$ = time
$We$ = Weber number, $We = \rho_g U^2 a/\sigma$
$x$ = axial coordinate (2-D flow)
$y$ = transverse coordinate (2-D flow)
$z$ = axial coordinate (axisymmetric flow)
$\alpha$ = boundary point singularity contribution
$\beta$ = surface slope
e = gas/liquid density ratio
$\Gamma$ = domain boundary
$\kappa$ = surface curvature
$\phi$ = velocity potential
$\rho$ = density
$\sigma$ = surface tension
$\omega$ = frequency

Subscript
$(\ )_g$ = gas phase

## REFERENCES

1. Rayleigh, L., "On the instability of jets," *Proc. London Math. Soc.* 10, p. 4, 1878.
2. Weber, C., "On the breakdown of a fluid jet," *Z. Agnew. Math. Mech.* 11, p. 136, 1931.
3. Leib, S.J., and Goldstein, M.E., "The Generation of capillary instabilities on a liquid jet," *J. Fluid Mech.*, V 168, pp. 479–500, 1986.
4. Li, H.-S. and Kelly, R. E., "The instability of a liquid jet in a compressible airstream," *Phys. Fluids A*, V4, pp. 2162–2168, 1992.
5. Sterling, A.M. and Sleicher, C.A. "The instability of capillary jets," *J. Fluid Mech.*, 68, pp. 477–495, 1975.
6. Lin, S.P. and Lian, Z.W., "Mechanisms of the breakup of liquid jets," *AIAA J.*, V28, pp. 120–126, 1990.
7. Yuen, M.-C., "Non-linear capillary instability of a liquid jet," *J. Fluid Mech.*, V33, pp. 151–163, 1968.
8. Chaudhary, K.C. and Redekopp, L.G. "The nonlinear capillary instability of a liquid jet." Part 1. Theory," *J. Fluid Mech.*, 96, 257–274, 1980.
9. Foote, G.B., "A numerical method for studying simple drop behavior: Simple oscillation," *J. Computational Phys.*, 11, pp. 507–530, 1973.

10. Hirt, L.W. and Nichols, B.D. Volume of fluid (VOF) method for the dynamics of free boundaries, *J. Computational Phys.*, 30, pp. 201, 1981.
11. Unverdi, S.O., and Tryggvason G., "A front tracking method for viscous, incompressible, multi-fluid flows," *J. Computational Phys.*, V100, pp. 25–37, 1992.
12. Osher, S. and Sethian, J.A., "Fronts Propagating with Curvature Dependent Speed: Algorithms based on Hamilton-Jacobi Formulations," UCLA Computational and Mathematics Report 87-12, 1987.
13. Liggett, J A., and P.L.-F. Liu, *The Boundary Integral Equation Method for Porous Media Flow*, Allen and Unwin, London, 1983.
14. Mansour, N.N., and Lundgren, T.S., "Satellite formation in capillary jet breakup," *Phys. Fluids*, V2, pp. 1141–1144, 1990.
15. Longuet-Higgins, M.S., and Cokelet, E.D. The deformation of steep surface waves on water. I. A numerical method of computation, *Proc. R. Soc. Lond. A*, 1976, 350, pp. 1–26.
16. Medina, D.E., Liggett, J.A., Birchwood, R.A., and Torrance, K.E. "A consistent boundary element method for free surface hydrodynamic calculations," *Internat. J. Numerical Methods in Fluids*, 12, pp. 835–57, 1991.
17. Spangler, C.A., Hilbing, J.H., and Heister, S.D. "Nonlinear modeling of jet atomization in the wind-induced regime," *Phys. Fluids*, 7, pp. 964–71, 1995.
18. Hilbing, J.H., Heister, S. D., and Spangler, C. A., "A boundary element method for atomization of a finite liquid jet," *Atomization and Sprays*, 5, (6), pp. 621–638, 1995.
19. Hilbing, J.H. and Heister,, S.D., "Droplet size control in liquid jet breakup," *Phys. Fluids*, 8, (6) pp. 1574–1581, 1996.
20. Heister, S.D., Rutz, M., and Hilbing, J., "Effect of acoustic perturbations on liquid jet atomization," *J. Propulsion and Power*, 13, (1), pp. 82–88, 1997.
21. Heister, S.D., "Boundary element methods for two-fluid free surface flows," to appear, *Engineering Analysis with Boundary Elements*, 1997.
22. Murray, I.F. and Heister, S.D., "On a droplet's response to acoustic cxcitation," *Internatl. J. Multiphase Flow*, to appear, 1997.
23. Lundgren, T.S. and Mansour, N.N. Oscillations of drops in zero gravity with weak viscous effects, *J. Fluid Mech.*, 194, pp. 479–510, 1988.
24. Kelmanson, M.A. Boundary integral equation solution of viscous flows with free surfaces, *J. Eng. Math.*, 17, pp. 329–43, 1983.
25. Moses, M.P., Collicott, S H., and Heister, S.D. Visualization of Liquid Jet Breakup and Droplet Formation, 7th International Symposium on Flow Visualization, Seattle, WA, 1995.
26. Dressler, J.L. Two-Dimensional, High Flow, Precisely Controlled Monodisperse Drop Source, Wright Laboratory WL-TR-93-2049, 1993.
27. Lafrance, P. "Nonlinear breakup of a laminar liquid jet," *Phys. Fluids*, 18, pp. 428–432, 1975.
28. Rutland, D.F. and Jameson, G.J. "Theoretical prediction of the sizes of drops formed in the breakup of capillary jets," *Chem. Eng. Sci.*, 25, pp. 1689–1698, 1970.
29. Hilbing, J.H., and Heister, S.D., "Nonlinear simulation of a high-speed, viscous, liquid jet," in review, *Atomization and Sprays*, 1997.

# 53 ENTRAINMENT BY LIGAMENT-CONTROLLED EFFERVESCENT ATOMIZER-PRODUCED SPRAYS

**J.J. Sutherland, P.E. Sojka,** and **M.W. Plesniak**
Thermal Sciences and Propulsion Center
School of Mechanical Engineering
Purdue University
West Lafayette, Indiana 47907-1003

## ABSTRACT

Entrainment of ambient air into sprays produced by a new type of effervescent atomizer is reported. Entrainment data were obtained using a device similar to that described by Ricou & Spalding.[1] Entrainment data were analyzed using the model of Bush & Sojka,[2] in concert with measured momentum rate data that were also acquired as part of this study. The analysis shows that entrainment by sprays produced using this type of atomizer is predicted to within about 35% by the expression

$$E = \frac{\dot{m}_e}{x\sqrt{\rho_e M_o}} \tag{1}$$

where $\dot{m}_e$ is the entrained gas mass flow rate, x is the distance along the spray axis measured from the dispenser exit orifice, $\rho_e$ is the density of the entrained air, $M_o$ is the spray momentum rate at the exit orifice, and E is the experimentally determined entrainment number whose value is $0.15 \pm 0.056$ ($2\sigma$).

## INTRODUCTION

Effervescent atomization is characterized by actively introducing gas bubbles into a liquid stream immediately upstream of an exit orifice, thereby forming a two-phase flow. This allows efficient transfer of energy between the atomizing gas and the liquid. The result is a high quality spray (i.e., small mean drop size) produced at air-to-liquid mass flow rate ratios substantially lower than most atomizers.

A number of investigators have studied sprays produced by effervescent atomizers. They have been concerned almost exclusively with atomizer performance; i.e., mean drop size, and have reported their results in terms of the spray Sauter mean diameter (SMD or, equivalently, $D_{32}$).

One topic which has not been studied in detail is entrainment. Entrainment, for the purposes of this discussion, is defined as the quantity of ambient gas which is drawn in through the interface of a spray as it expands downstream of the nozzle. It is the result of momentum transfer, from both liquid drops and any atomizing gas used, to the ambient air surrounding the spray.

Entrainment has important implications in many engineering applications. In consumer product sprays, the entrained mass flow rate has a cause and effect relationship on carrier liquid evaporation. In combustion applications such as gas turbines and diesel engines, entrainment has a significant effect on the local equivalence ratio and, therefore, a direct impact on $NO_x$ formation. In furnaces, entrainment has a large impact on droplet residence time, since higher entrainment rates lead to greater rates of spray deceleration and, therefore, slower moving drops. Entrainment is also important in spray drying, where the goal is to remove as much liquid as possible from a droplet during the drying process. Finally, entrainment can influence the finish quality of painted or coated surfaces through both transfer efficiency and solvent evaporation.

---

*52nd Purdue Industrial Waste Conference Proceedings*, 1997, Ann Arbor Press, Chelsea, Michigan 48118. Printed in U.S.A.

A key to understanding entrainment is the ability to measure it. Three methods are common: experimentally measuring the velocity profile of the entrained air and then integrating it over a control volume that encompasses the spray, measuring the global entrainment rate, and numerically simulating the two-phase spray field. Studies that have used these techniques, and that are relevant to the current work, are summarized below.

Binark & Ranz[3] measured velocities outside the spray using a constant-resistance hot wire anemometer, and inside the spray, using an impact-static probe. Data were used to develop an expression that yielded induced air velocities for a set of homologous nozzles. Unfortunately, their expression is of limited use for estimating entrainment into sprays because atomizer-specific experimental data are necessary in order to apply their model to other types of atomizers. Binark & Ranz[3] also developed a theoretical model that yielded results within an order of magnitude of their experimental data. However, it required the assumption that the velocity of the drops was constant and equal to the injection velocity. If this assumption were correct, no transfer of momentum would occur between the drops and the surrounding air and, thus, no air would be entrained.

Rasbash & Stark[4] correlated spray entrainment in terms of "reaction at the nozzle," or the force exerted on a flat obstacle oriented normal to the spray axis. The fact that Rasbash & Stark[4] present their results as an empirical correlation suggests that their expression is applicable only to sprays formed using their type of atomizer.

Briffa & Dombrowski[5] measured entrained air velocity by seeding the air in and around a flat spray with lycopodium powder and taking double-flash (microsecond duration) photographs. Air velocities inside the spray were measured assuming the air velocity that was tracked by the 15 μm diameter droplets. A model was developed which demonstrated a linear relationship between entrained mass flow rate and axial distance. However, a nonlinear relationship was observed experimentally

Benatt & Eisenklam[6] used the Ricou & Spalding[1] method to measure global entrainment and compared results obtained to predictions from a model based on their knowledge of the liquid sheet breakup process, spray dynamics, momentum loss calculations, and induced gas flow observations. The expected linear scaling of entrainment with axial distance was observed, and experimental results confirmed their theory. However, the theoretical proportionality constant was lower than the experimental value. Benatt & Eisenklam's[6] model is restricted to pressure-swirl atomizers, since it relies on calculations based on a particular breakup process.

Tishkoff[7] studied air entrainment into two different pressure-swirl atomizers by visualizing the entrained air flow patterns using helium jet seeding and measuring velocities using a constant temperature hot-wire anemometer. He developed a correlation for the entrainment number, which is limited to pressure-swirl atomizer produced sprays.

MacGregor[8] developed a simple model based on jet momentum rate and the drag on spherical droplets and then used a Ricou & Spalding[1] device to test it. Experimental results obtained from two nozzles show a linear relationship with inlet mass flow rate. However, the gradient of the experimental data implies that there exists a critical mass flow rate below which little or no entrainment occurs, suggesting little or no jet breakup is occurring under these conditions, and predictions did not agree with experimental results. Nonetheless, MacGregor[8] used his model to demonstrate the effects of drop size distribution parameters and noted that entrainment rates were a linear function of axial distance for drops having diameters greater than 1000 μm, that entrainment rates fell off at large axial distances for smaller drops, and that entrainment rates were relatively insensitive to the width of the drop size distribution.

Boysan & Binark[9] and Rothe & Block[10] performed numerical simulations in order to predict induced air flows into sprays. Boysan & Binark[9] assumed a Rosin-Rammler drop size distribution and numerically solved the partial differential equations for stream function and vorticity transport coupled with the ordinary differential equations of spray motion. Good agreement was observed upon comparison of predicted and measured air velocities. However, this agreement is misleading due to the incorporation of adjustable parameters into their model. These parameters would have to be determined for each new type of spray.

Rothe & Block[10] were interested in the cone angle contraction due to entrained gas flow that was noted earlier by Binark & Ranz.[3] They numerically solved equations for drag, axial momentum, and drop trajectory in order to calculate the induced air velocity and amount of spray contraction. Comparisons of model predictions with experimental data of previous researchers were presented. Again, the level of agreement between model predictions and experimental results is misleading due to the use of fitting parameters in their analysis.

Ruff *et al.*[11] were among the first to consider entrainment in the dense region of the spray; i.e., where the liquid is breaking up. Their analysis was based on the assumption of a locally homogenous flow (LHF). Mean streamwise and radial entrainment velocities were obtained using laser Doppler anemometry near the edge of the spray. These velocities were then integrated to provide experimental entrainment rates. Predicted values were greater than measured values in all cases. This is because entrainment rates are strongly related to flow properties near the edge of the spray. Unfortunately, this is also a dilute region of the spray where the LHF model overpredicts quantities such as the entrainment rate.

Until recently, no studies addressing entrainment into effervescent sprays had appeared in the literature. Bush & Sojka[2] were the first to study entrainment in such sprays. As such, they were the first to study entrainment in two-phase jets characterized by velocity slip between the two phases at the nozzle exit. Bush & Sojka[2] developed a two-phase model, analogous to that of Ricou & Spalding,[1] that was based on dimensional analysis and the conservation of momentum. A Ricou & Spalding[1] device was then used to measure entrainment rates. They found that "conventional" effervescent sprays entrain air similar to other sprays and single-phase gas jets, in that the normalized entrainment rate scales linearly with the dimensionless axial distance. However, their results also show that "conventional" effervescent sprays are not characterized by a single entrainment number, as are their single-phase gas jet counterparts, instead requiring scaling by liquid density and nozzle diameter. Recent work by Luong[12] suggests this lack of a single value for entrainment number is due to unsteadiness inherent in conventional effervescent sprays.

This review of the spray entrainment literature demonstrates that previous work cannot be directly applied to ligament-controlled effervescent atomizer produced sprays. Consequently, the current study extends previous work to this important class of atomizer by answering the practical question "At what rate does a ligament-controlled effervescent atomizer produced spray entrain surrounding air?"

## EXPERIMENTAL APPARATUS

Previous work by Lund et al.[13] and Santangelo & Sojka[14] indicated that atomizer performance could be improved (i.e., $D_{32}$ reduced) by producing smaller ligaments at the injector exit. The current atomizer, shown schematically in Figure 1, achieves that goal through use of a porous disk located at the exit orifice. As can be seen in Figure 1, the atomizer consists of a top plate, a

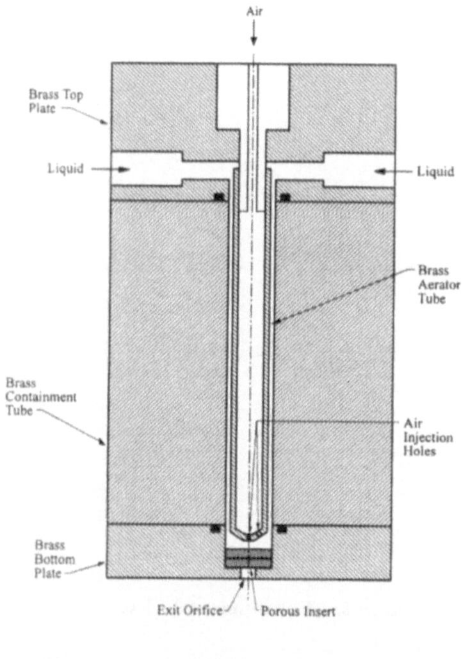


**Figure 1. Ligament-controlled effervescent atomizer.**

containment tube, an aerator tube, and an exit orifice plate. Liquid is fed into the side of the atomizer and flows through an annular gap between the containment and aerator tubes. Air is injected into the liquid through two holes located at the tip of the aerator tube. The resulting two-phase flow passes through a porous medium before leaving through the exit orifice. The choice of pore size is important. This particular medium was chosen by trial-and-error as one that would provide acceptable spray performance (a mean drop size less than 70 μm) while requiring a minimal operating pressure (nozzle supply pressure less than 800 kPa) for ALRs below 1%.

The nozzle operates using supplies of metered atomizing air and liquid. The air flow rate was monitored using a rotameter and regulated using a needle metering valve. The liquids were supplied from a reservoir, with mass flow rates measured using rotameters. Flow rates were regulated using a needle valve.

The atomizer can deliver liquid mass flow rates of between 0.5 and 1.0 g/s. It can operate at air-liquid ratios by mass (ALRs) of 0.5 % and above. Data presented here are restricted to 0.5 and 0.6 g/s and $0.75 \leq ALR \leq 3.75$ %.

Global entrainment rates were measured using a device similar to the one developed by Ricou & Spalding.[1] Their device consisted of a cylindrical housing that enclosed a nozzle mounted to a back plate. Entrained gas was injected into the cylinder and forced to pass through a porous inner cylinder before mixing with the jet. The porous boundary served to create a uniform radial velocity profile for the entrained gas. A mask of the same diameter as the jet was placed at the cylinder exit under zero pressure drop conditions. This mask helped to prevent any inflow or outflow of gases that might affect the entrainment measurement. By supplying entrained gas such that the pressure differential across the exit mask is zero, the entrained mass flow rate may be measured. Since there are no pressure gradients in an ambient environment, the flow conditions of a jet spraying into ambient air are duplicated when the pressure differential across the mask is zero.

The entrainment device discussed above was successfully adapted by Bush[15] to measure entrainment rates of effervescent sprays. A schematic of the device is shown in Figure 2, while exact design specifications may be found in Bush.[15]

A key to adapting the device to sprays was proper sizing of the exit mask. To accomplish this, 12 square Plexiglas plates were constructed, each with a hole bored in its center. Hole diameters range from 13 to 152 mm, which allowed rapid fitting of the proper mask size at each spray operating condition. A proper size is achieved when the exit mask is large enough to allow the spray to pass through (i.e., no droplets striking the mask), but not so large as to leave a gap be-

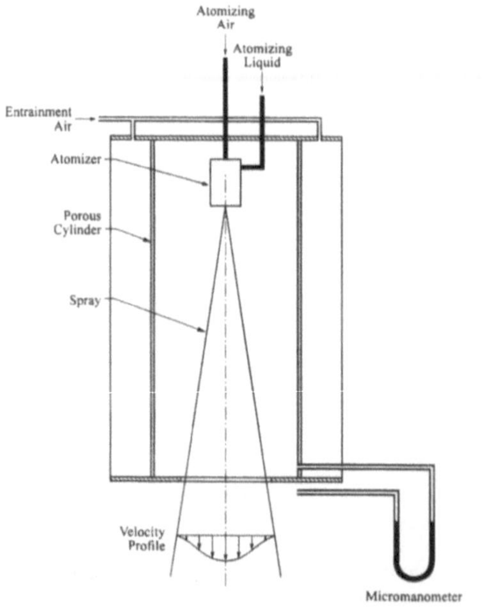


**Figure 2. Entrainment device.**

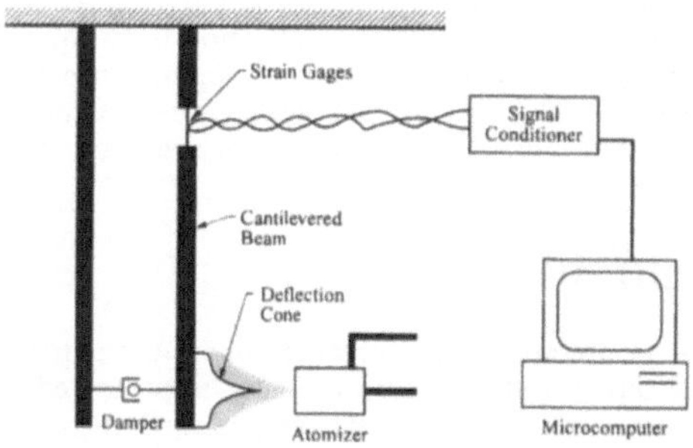


**Figure 3. Momentum rate probe of Bush et al.[16]**

tween the edge of the spray and the mask. Once the proper mask is installed, data acquisition consists of adjusting the entrained air mass flow rate such that there is no pressure gradient across the exit mask and then recording the corresponding rotameter setting. The pressure differentials across the exit mask were monitored using a differential pressure transducer with a 3.5 digit process indicator. The entrained air mass flow rates were measured using a rotameter and controlled by a needle valve.

A number of previous spray entrainment studies have demonstrated the importance of jet momentum rate on entrained gas flow rates. Bush et al.[16] constructed a device for determining the axial momentum of two-phase jets based on the design of Deichsel & Winter.[17] That probe was used in this study. A summary of its principles and operation is provided below.

The momentum rate of a spray is measured by converting the axial flow to a radial flow by spraying against a deflection cone whose contour is obtained from the equation of streamlines for an incompressible, axisymmetric, stagnation point flow.[18] It is easily shown that the amount of reaction force needed to hold the cone in place is equal to the momentum rate of the spray.

For this study, the deflection cone was mounted on a cantilevered beam. When the spray is directed at the cone, the beam is deflected, resulting in a strain at the base. This strain can be measured with strain gauges and appropriate signal conditioning hardware. Bush et al.[16] describe the design details of the deflection cone, the strain gauge beam, and the signal conditioner. A schematic of the momentum rate probe is shown in Figure 3.

Sprays were formed using a total of six separate fluids. They were formed from either mixtures of water and commercially available glycerine, or commercially available refined hydrocarbons (Texaco solvent neutral oils, SNO-100 and SNO-320, or Bennzoil Universal Calibration Fluid, UCF-1). Their compositions and physical properties are presented in Table I. Note that five of the fluids differ in viscosity (0.020, 0.040, or 0.080 Pa-s) and surface tension (either 0.030 Pa-m for 0.020 and 0.040 Pa-s or 0.067 Pa-m for all three viscosities). The sixth fluid was water. These viscosities and surface tensions were chosen to span the range of current alcohol-based consumer products and their (projected) water-based counterparts.

Fluid viscosities were measured using a falling ball viscometer. Surface tensions were measured using a du-Nuoy ring tensiometer. Fluid densities were calculated from the quotient of a known volume of fluid and its measured weight.

**Table I. Composition and Physical Properties of Spray Fluids (at room conditions)**

| Fluid Number | Composition, Weight-% | Viscosity, Pa-s | Surface Tension, Pa-m | Density, $kg/m^3$ |
|---|---|---|---|---|
| 1 | 63/37 Glycerine/Water | 0.020 | 0.067 | 1170 |
| 2 | 72/28 Glycerine/Water | 0.040 | 0.067 | 1197 |
| 3 | 80/20 Glycerine/Water | 0.080 | 0.067 | 1217 |
| 4 | 75/35 SNO 100/Bennzoil | 0.020 | 0.030 | 840 |
| 5 | 90/10 SNO 100/SNO 320 | 0.040 | 0.030 | 847 |
| 6 | Water | 0.001 | 0.072 | 998 |

The uncertainty in determining the entrainment number, E, was calculated using the expression developed by Bush.[15] The mass flow rate uncertainty calculated using this equation is as high as 30%. The only other significant source of uncertainty is associated with the momentum rate probe. This is approximately 7%, as noted by Bush et al.[16] Assuming uncorrelated uncertainties, a value of 31% was calculated as the maximum overall uncertainty for the entrained mass flow rates.

## RESULTS AND DISCUSSION

Normalized entrainment and momentum rate measurements were obtained using the entrainment device and momentum rate probe described above. Entrainment rate data are presented as normalized entrainment rate versus dimensionless axial distance. Normalized entrainment rate is defined as $\dot{m}_e$, $\dot{m}_l$, where $\dot{m}_e$ is the entrained air mass flow rate and $\dot{m}_l$is the liquid mass flow rate. Dimensionless axial distance is defined as $x/d_o$, where x is the axial distance and $d_o$ is the atomizer exit orifice diameter. Momentum rate results are plotted as momentum rate versus ALR. Six fluids, including water, were sprayed in order to determine the influence of operating conditions and fluid physical properties on the entrainment behavior and momentum rates of ligament-controlled effervescent atomizer produced sprays.

Bush[15] reported normalized entrainment results for the Lund et al.[13] "conventional" effervescent atomizer. His data showed that normalized entrainment measurements scaled linearly with dimensionless axial position, as was predicted by the dimensional analysis of Ricou & Spalding.[1] Normalized entrainment rates were also reported to increase with increasing air/liquid ratio (ALR). Normalized entrainment rates, plotted against dimensionless axial distance for a 1.0 g/s water spray produced using a Lund et al.[13] style atomizer, are shown in Figure 4.

Figure 5 shows normalized entrainment rates versus dimensionless axial distance for a ligament-controlled effervescent atomizer spraying water under the same conditions as for Figure 4. Again, normalized entrainment rates are found to scale linearly with dimensionless axial distance and to increase with increasing ALR ($r^2$0.984 for all cases). These features are common to all normalized entrainment measurements for all liquids sprayed in this investigation. The ALR scaling, in particular, is not surprising since increasing ALR increases the exit momentum rate of the spray. In addition, increasing ALR results in a small, but noticeable,[19] decrease in mean drop size, which should result in more effective momentum transfer from the spray to the surrounding air. The more effective momentum transfer is expected to increase entrainment.

When comparing the magnitudes of the normalized entrainment rates, values decreased upon changing from the Lund et al.[13] atomizer to the ligament-controlled effervescent atomizer. This is due to the dependence of entrainment behavior on the exit momentum rate of the spray. Unlike the Lund et al.[13] atomizer, the ligament-controlled effervescent atomizer requires the two-phase flow to pass through a porous insert before exiting. This alters the near nozzle breakup structure and thus the exit momentum rate of the spray, thereby causing a change in the entrainment behavior. As will be demonstrated, the end result is a decrease in the dimensionless entrainment number, E.

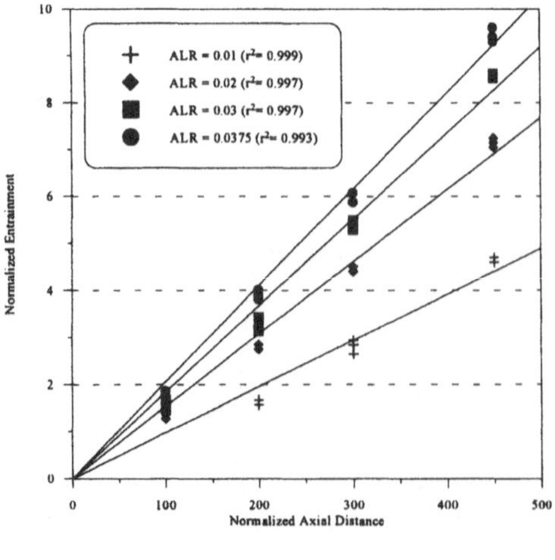


**Figure 4. Normalized entrainment versus dimensionless distance for water being sprayed at a mass flow rate of 1.0 g/s using a conventional effervescent atomizer.**

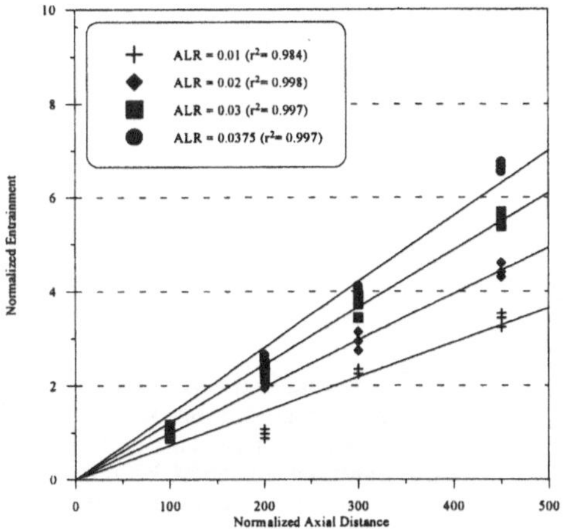


**Figure 5. Normalized entrainment versus dimensionless distance for water being sprayed at a mass flow rate of 1.0 g/s using a ligament-controlled effervescent atomizer.**

Entrainment data for ligament-controlled effervescent atomizer produced sprays using working fluids other than water are presented in Table II. A representative plot is included as Figure 6. Plots for the remaining fluids are reported by Sutherland.[19] The five fluids sprayed have surface tensions of 0.030 or 0.067 Pa-m and viscosities varying from 0.020 to 0.080 Pa-s. In each case, the data exhibit a linear relationship between normalized entrainment and normalized axial distance; the correlation coefficients range from 0.989 to 0.998. In addition, the slope of the normalized entrainment versus normalized axial distance lines increases with ALR in all cases. This behavior is anticipated since an increase in ALR leads directly to an increase in exit orifice momentum rate.

The effect of viscosity on normalized entrainment rate is observed through comparison of data for fluids having similar surface tensions. Comparison of data from Figure 5 with that of Fluids 1 through 3 (at equal ALRs) indicates an increase in normalized entrainment of 25 to 35% as viscosity is increased from 0.001 to 0.020 Pa-s, but a smaller effect on the normalized entrainment rates when the viscosity is 0.020 Pa-s or higher (normalized entrainment rates were within 16% of the mean for each ALR for viscosities between 0.020 and 0.080 Pa-s). Similar viscosity scaling is obtained from sprays having a common surface tension of 0.030 Pa-m; i.e., Fluids 4 and 5—the data show that viscosity has little effect on the normalized entrainment rates for viscosities of 0.020 and 0.040 Pa-s, with variations being less 7% for each ALR.

The effects of surface tension can also be obtained from comparisons of Table II data. The Fluid 1 and 4 entries contain data from sprays with surface tensions of 0.067 and 0.030 Pa-m, respectively, and a common viscosity of 0.020 Pa-s. Comparison of these data indicates a slight de-

**Table II. Slopes and Coefficients of Determination for Least Squares Linear Fits to Normalized Entrainment Versus Normalized Axial Distance Data for Liquids Sprayed at 0.6 g/s**

| ↓ Fluid / ALR → | 0.075 | 0.01 | 0.015 | 0.02 |
|---|---|---|---|---|
| 1 | 0.0082, $r^2 = 0.998$ | 0.0097, $r^2 = 0.994$ | 0.012, $r^2 = 0.998$ | 0.015, $r^2 = 0.998$ |
| 2 | 0.0083, $r^2 = 0.998$ | 0.0091, $r^2 = 0.998$ | 0.011, $r^2 = 0.997$ | 0.014 $r^2 = 0.998$ |
| 3 | 0.0084, $r^2 = 0.989$ | 0.011, $r^2 = 0.996$ | 0.014, $r^2 = 0.997$ | 0.017, $r^2 = 0.998$ |
| 4 | 0.0081, $r^2 = 0.994$ | 0.0095, $r^2 = 0.993$ | 0.011, $r^2 = 0.998$ | 0.014, $r^2 = 0.997$ |
| 5 | 0.0081, $r^2 = 0.997$ | 0.0098, $r^2 = 0.996$ | 0.011, $r^2 = 0.996$ | 0.015, $r^2 = 0.998$ |

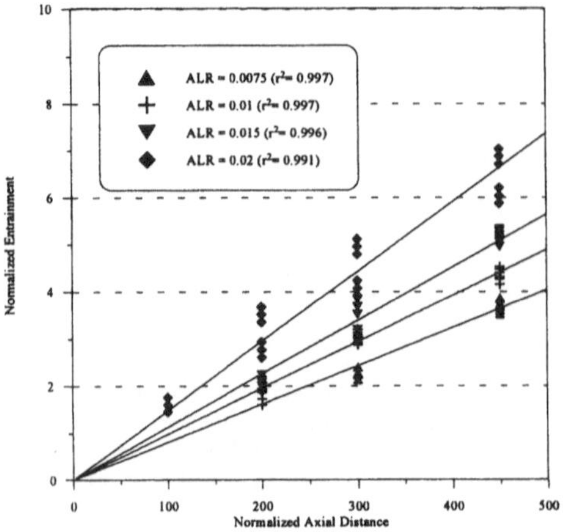


**Figure 6. Normalized entrainment versus normalized axial distance for a fluid (Fluid 5) having a viscosity of 0.040 Pa-s, a surface tension of 0.030 Pa-m, and operating at liquid mass flow rates of 0.5 and 0.6 g/s.**

crease in normalized entrainment (<12% in all cases) with a decrease in surface tension. The Fluid 2 and 5 entries contain data from sprays with surface tensions of 0.067 and 0.030 Pa-m, respectively, and a common viscosity of 0.040 Pa-s. Comparison of these data show little surface tension scaling; normalized entrainment rates are within 7% of each other. Since any scaling observed due to surface tension is small (well within experimental uncertainty), it can be concluded that surface tension has little effect on normalized entrainment rates for fluids with viscosities above 0.020 Pa-s.

Figure 6 presents entrainment data obtained at two different liquid mass flow rates (0.5 and 0.6 g/s). The data exhibit a linear dependence of normalized entrainment on dimensionless axial distance ($r^2 \geq 0.990$ for all cases), as well as an increase in the slope of these lines with an increase in ALR. Most importantly, the data for both mass flow rates collapse to a single line for any particular ALR value.

Momentum rate data corresponding to the entrainment rate data of Figures 5 and 6 and Table II are presented in Figures 7 and 8 and Table III, respectively. Figures 7 and 8 are representative of all momentum rate data obtained during this study and show that momentum rate is linearly proportional to ALR. Coefficients of determination ($r^2$) are above 0.962 in all cases.

Momentum rate for a water spray is plotted versus air/liquid ratio and shown in Figure 7. When compared to data obtained by Bush[15] at the same liquid mass flow rate; i.e., 1.0 g/s, the

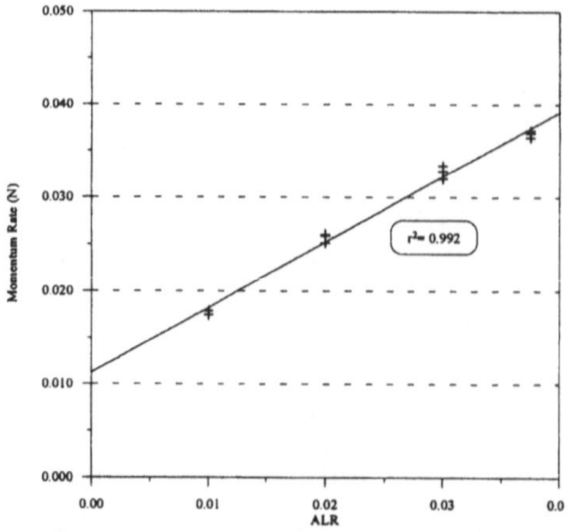


**Figure 7. Momentum rate versus air-liquid ratio (ALR) for water.**

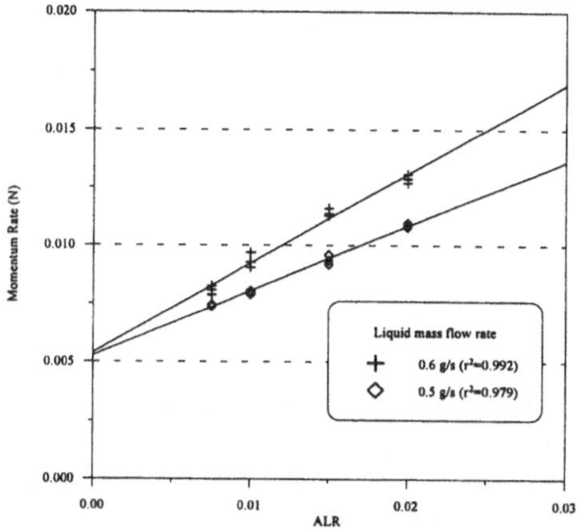


**Figure 8. Momentum rate versus air-liquid ratio (ALR), at two mass flow rates, for a fluid (Fluid 5) having a viscosity of 0.040 Pa-s and a surface tension of 0.030 Pa-m.**

momentum rates for ligament-controlled effervescent atomizer produced sprays are lower over the entire range of air-to-liquid mass flow rate ratios. This supports the hypothesis that the addition of a porous insert does indeed alter the exit momentum of the spray and, ultimately, the entrainment behavior.

The influence of surface tension on momentum rate can be determined by comparing the Fluid 1 and 4 or 2 and 5 entries in Table III. In both cases, momentum rates decrease when the surface tension is increased. This same trend was reported by Bush[15] and was attributed to differences in liquid density, not surface tension. Lower liquid density results in a lower void fraction at the atomizer exit and, thus, higher interphase velocity slip. Higher slip results in a higher momentum rate.

The influence of viscosity on momentum rate is unclear. The Fluid 1 through 3 entries in Table III show momentum rates for liquids having viscosities of 0.020, 0.040, and 0.080 Pa-s, respectively, and a common surface tension of 0.067 Pa-m. An increase in viscosity from 0.020 to 0.040 Pa-s results in a decrease in momentum rate (slope), while a further increase in viscosity from 0.040 to 0.080 Pa-s results in an increase in momentum rate (slope). The relative changes in momentum rate are small (~20%) and could be due to experimental uncertainty.

When examining the influence of viscosity on momentum rates for fluids with a common surface tension of 0.030 Pa-m, the effects are less obvious. The Fluid 4 and 5 entries in Table III contain data for mixtures having viscosities of 0.020 and 0.040 Pa-s, respectively, and operating at two liquid mass flow rates. These data show that momentum rate (slope) increases with viscosity at the low mass flow rate, yet exhibits only a minor variation with viscosity for the higher mass flow rate.

The effects of liquid mass flow rate on momentum rate may also be inferred from Table III. In each case, the momentum rate (slope) increases with an increase in liquid mass flow rate, from

**Table III. Slopes, Intercepts, and Coefficients of Determination for Least Squares Linear Fits to Momentum Rate versus Air-liquid Ratio (ALR) Data.**

| Fluid @ Liquid Mass Flow Rate | Slope, Intercept; $r^2$ |
|---|---|
| 1 @ 0.6 g/s | 0.39, 0.0060; $r^2 = 0.988$ |
| 2 @ 0.6 g/s | 0.32, 0.0057; $r^2 = 0.993$ |
| 3 @ 0.6 g/s | 0.37, 0.0065; $r^2 = 0.962$ |
| 4 @ 0.5 g/s | 0.17, 0.0057; $r^2 = 0.992$ |
| 4 @ 0.6 g/s | 0.37, 0.0086; $r^2 = 0.989$ |
| 5 @ 0.5 g/s | 0.28, 0.0053; $r^2 = 0.992$ |
| 5 @ 0.6 g/s | 0.38, 0.0054; $r^2 = 0.979$ |

**Table IV. Average Entrainment Numbers**

| Fluid @ mass flow rate | E 2 σ |
|---|---|
| 1 @ 0.6 g/s | 0.158 ± 0.058 |
| 2 @ 0.6 g/s | 0.153 ± 0.053 |
| 3 @ 0.6 g/s | 0.175 ± 0.065 |
| 4 @ 0.5 g/s | 0.124 ± 0.040 |
| 4 @ 0.6 g/s | 0.153 ± 0.056 |
| 5 @ 0.5 g/s | 0.150 ± 0.046 |
| 5 @ 0.6 g/s | 0.148 ± 0.063 |

0.17 to 0.37 for Fluid 4 and from 0.28 to 0.38 for Fluid 5. This is expected, since the momentum rate is expected to increase proportional to the liquid mass flow rate.

The entrainment rate data were combined with the momentum rate to calculate entrainment numbers using the model developed by Bush.[15] Bush's[15] model is based on the work of Ricou & Spalding,[1] who employed a Buckingham Pi analysis using the exit momentum rate, entrained gas density, and axial distance as the normalizing parameters to obtain [1].

Entrainment numbers were calculated for five different fluids with varying viscosities and surface tensions by correlating experimental data via [1]. They are listed in Table IV. A representative sample of the data is included as Figure 9, while Figure 10 is a composite of all the data acquired in this study (Fluids 1 through 5, plus water).

Figure 9 shows entrainment number results for a water spray. A noticeable increase in entrainment number with air-to-liquid mass flow rate ratio is observed. This behavior occurred for all fluids considered during this investigation.

Inspection of Figure 10 leads to the conclusion that air-to-liquid mass flow rate ratio has some influence on the value of the entrainment number. In all cases, the value of the entrainment number increases with increasing ALR. It might be argued that this variation is due to experimental uncertainty, since the entrainment number values differ from the mean by less than 40% in all cases, which is approximately the range of uncertainty involved in the calculation. However, consideration of the limiting cases of pure gas and pure liquid jets demonstrates some ALR scaling should be expected. For the pure gas jet, the entrainment number has been shown by Ricou and Spalding[1] to be approximately 0.28. This is nearly twice the value reported here for ligament-controlled effervescent atomizer-produced sprays. For the pure liquid jet, the entrainment number would be zero, if defined to be equal to the amount of surrounding gas crossing the spray boundary, as it is in this study. If the entrainment number is instead defined to be proportional to the amount of surrounding gas set in motion by the liquid jet, a simple boundary layer analysis indicates that E would be approximately 0.02. This is less than 8% of the Ricou & Spalding[1]

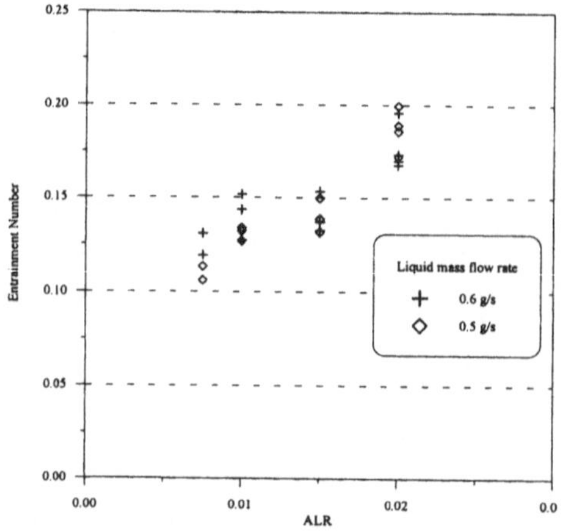


**Figure 9. Entrainment number versus air/liquid ratio (ALR) for a fluid (Fluid 5) having a viscosity of 0.040 Pa-s and a surface tension of 0.030 Pa-m, operating at two liquid mass flow rates.**

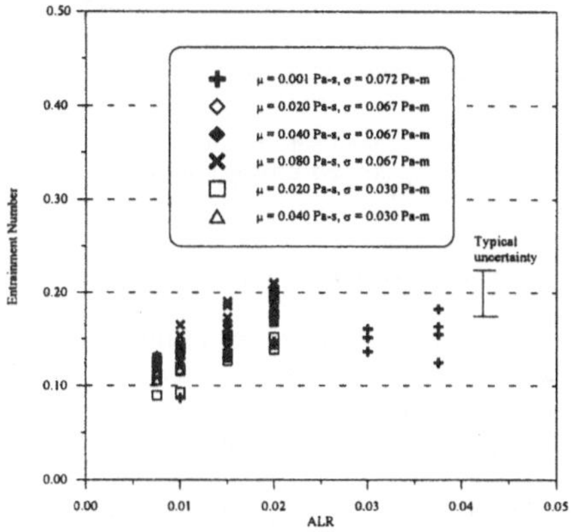


**Figure 10. Entrainment number versus air/liquid ratio (ALR) for all fluids.**

value and only 14% of the value reported here. Regardless of the approach taken to estimate the zero ALR limit for E, the conclusion is that E should increase with ALR.

Table IV may be used to demonstrate the influence of the fluid physical properties on the entrainment number. A comparison of data from the Fluid 1 through 3 entries, characteristic of water-based mixtures, indicates that entrainment number is relatively insensitive to fluid viscosity—average entrainment numbers differ by less than 15%. The lack of an influence of viscosity is also observed when comparing data from equal mass flow rate Fluid 4 and 5 entries, which contain results characteristic of hydrocarbon-based mixtures—average entrainment numbers differ by less than 21%. This leads to the conclusion that fluid viscosity has a negligible impact on entrainment number throughout the range of conditions considered in this study.

The effects of surface tension on the entrainment number are demonstrated when comparing data from Table IV entries for Fluid 1 with Fluid 4 and for Fluid 2 with Fluid 5, all at equal mass flow rates. As can be observed, variations in mean entrainment number due to changes in surface tension are less than 4% and therefore well within experimental uncertainty of the mean value. This indicates that surface tension has little effect on entrainment number.

Figure 10 reinforces the conclusion that the entrainment number is insensitive to fluid physical properties, and also demonstrates that ALR does have an effect. Considering the data as a whole, the resulting value for E is $0.15 \pm 0.056$ ($2\sigma$). Note that $2\sigma$ corresponds closely to the uncertainty in these estimates (±30%). Also note that E values for all fluids sprayed under all combinations of liquid mass flow rates and ALR are within 40% of the mean value. Finally, since normalized entrainment scales linearly with normalized axial distance (to within our experimental accuracy), we conclude that the model accurately estimates entrainment in low air-to-liquid mass flow rate ratio, ligament-controlled effervescent atomizer-produced sprays.

The single E value reported here, for a wide range of conditions, is in contrast to the earlier results of Bush,[15] who noted that entrainment number depends on the liquid density and the diameter of the exit orifice for sprays produced by conventional effervescent atomizers. Bush[15] attributed the observed scaling to a possible shift in the flow structure at the nozzle exit or a significant shift in mean drop size. These phenomena were not observed in this investigation and their absence may explain the similarity in the entrainment number for all fluids sprayed. The single value of E may also indicate that the current atomizer provides relatively steady sprays, in contrast to those produced by conventional effervescent atomizers.[12]

## SUMMARY AND CONCLUSIONS

The results of this study are summarized as follows:

- Normalized entrainment by ligament-controlled effervescent atomizer produced sprays is linearly proportional to normalized axial distance, as has been reported for gas jets and for some other types of sprays.

- Normalized entrainment by these sprays increases with atomizing air-to liquid mass flow rate ratio (ALR), as would be expected since an increase in ALR increases the initial momentum rate of the spray; entrainment data obtained at two liquid mass flowrates collapsed on to a common line.
- Liquid surface tension has a negligible effect on normalized entrainment; increasing liquid viscosity from 0.001 to 0.020 Pa-s increased normalized entrainment by approximately 30 to 50%, with further increases in liquid viscosity having a much smaller impact (less than 12%).
- The initial spray momentum rate is proportional to ALR and to liquid mass flow rate, as expected.
- Liquid surface tension had a negligible effect on momentum rate; the influence of viscosity was mixed with momentum, increasing with an increase in viscosity in some cases and decreasing with an increase in viscosity in others.
- Entrainment number is relatively insensitive to liquid physical properties—all variations were within experimental uncertainty.
- Entrainment number increases with ALR, as would be expected by considering the limiting cases of pure liquid (ALR = 0) and pure gas (ALR = $\infty$) jets; the value of entrainment number determined here, $0.15 \pm 0.056$ ($2\sigma$), lies between the gas jet value, 0.282, reported by Ricou & Spalding[1] and the liquid jet (issuing into air) value of 0.

The following conclusions can be drawn, based on the results of this study:

- Entrainment into steady two-phase jets where the gas and liquid streams exhibit inter-phase velocity slip can be modeled using the momentum rate approach of Ricou & Spalding,[1] although their entrainment number value is no longer applicable. The experimentally determined entrainment number for the two-phase jets studied here increases with ALR.
- The appropriate experimentally determined entrainment number, E, for ligament-controlled effervescent atomizer produced sprays is $0.15 \pm 0.056$ ($2\sigma$). This value of E predicts entrainment to within 40% for sprays considered in this study.
- Entrainment depends on the structure of the spray present at the atomizer exit, as shown by a comparison of results from this study with those of Bush.[15]

## ACKNOWLEDGMENTS

We thank our colleague Professor J.P. Gore for advice given during many discussions about entrainment. This work was performed under EPA Cooperative Agreement No. CR822618. The technical monitor is Kelly W. Leovic.

## REFERENCES

1. Ricou, F. P and Spalding, D. B. "Measurements of entrainment by axisymmetric turbulent jets." *J. Fluid Mech.*, 11, pp. 21–32, 1961.
2. Bush, S.G. and Sojka, P.E. Entrainment by Effervescent Sprays at Low Mass Flow Rates, presented at the 6th ICLASS, Rouen, France, 1994.
3. Binark, H. and Ranz, W.E. Induced Air Flows in Fuel Sprays. ASME Paper No. 58-A-284 (1958 Annual Meeting of the ASME), 1958.
4. Rasbash, D.J. and Stark, G. "Some Aerodynamic Properties of Sprays." *Chemical Eng.*, December, A83–A88, 1962.
5. Briffa, F.E.J. and Dombrowski, N. "Entrainment of air into a liquid spray." *AIChE J.*, 12(4), pp. 708–717, 1966.
6. Benatt, F.G.S. and Eisenklam, P. "Gaseous entrainment into axisymmetric liquid sprays. *J. Inst. Fuel*, August, pp. 309–315, 1969.
7. Tishkoff, J. Air Entrainment into Sprays from Swirl Chamber Atomizers. Proceedings 3rd ICLASS, London, UK, 1985.
8. MacGregor, S. A. "Air entrainment in spray jets." *Internatl. J. Heat Fluid Flow*, 12(3), pp. 279–283, 1991.
9. Boysan, F. and Binark, H. "Predictions of induced air flows in hollow cone sprays." *J. Fluids Eng.*, 101, pp. 313–318, 1979.

10. Rothe, P.H. and Block, J.A. "Aerodynamic Behavior of Liquid Sprays." *Internatl. J. Multiphase Flow*, 3, pp. 263–272, 1977.
11. Ruff, G.A., Sagar, A.D., and Faeth, G.M. "Structure and Mixing Properties of Pressure-Atomized Sprays." *AIAA J.*, 27(7), pp. 901–908, 1989.
12. Luong, J.T.K., Unsteadiness in Low Mass Flow Rate Effervescent Atomizer Produced Sprays, MSME Thesis, Purdue University, 1996.
13. Lund, M.T., Sojka, P.E., Lefebvre, A.H., and Gosselin, P.G. "Effervescent atomization at low mass flow rates. Part I: The influence of surface tension." *Atomization and Sprays*, 3(1), pp. 77–89, 1993.
14. Santangelo, P.J. and Sojka, P.E. "A Holographic Investigation of the Near-Nozzle Structure of an Effervescent Atomizer-Produced Spray." *Atomization and Sprays*, 5(2), pp. 137–155, 1995.
15. Bush, S.G. Entrainment by Effervescent Sprays at Low Mass Flow Rates. MSME Thesis, Purdue University, 1994.
16. Bush, S.G., Bennett, J.B., Sojka, P.E., Panchagnula, M.V., and Plesniak, M.W. "A momentum rate probe for use with two-phase flows." *Rev. Sci. Instruments*, 67(5), pp. 1878–1885, 1996.
17. Deichsel, M. and Winter, E.R.F. "Adiobatic two-phase pipe flow of air-water mixtures under critical flow conditions." *Internatl. J. Multiphase Flow*, 16(3), pp. 391–406, 1990.
18. White, F.M. *Viscous Fluid Flow, Second Edition*. McGraw-Hill, New York, 1991.
19. Sutherland, J.J., Ligament-Controlled Effervescent Atomization. MSME Thesis, Purdue University, 1996.

# 54 CALCULATING TRANSFER EFFICIENCY OF A LIQUID FINISHING SYSTEM

**John P. Owed**
ITW Ransburg Electrostatic Systems

## AUTOMATIC SYSTEM CONSIDERATIONS ECONOMIC JUSTIFICATIONS

Essential to all manufacturing processes is a good understanding of the cost related to a specific product. Within the paint shop there are many things that contribute to the coating cost of a finished product. These included pretreatment chemicals, labor, utilities, waste disposal, paint filters, coating materials, and many other things. The objective of this paper is to help the manufacturer identify the per unit cost of their coating material. Typically, this number will be one of the larger percentages of the total finishing cost. This number can easily be calculated by dividing the number of parts coated per minute, hour, day, or year into the number of gallons of coating material used in the same time frame. Once this number is identified the manufacturer can calculate the per unit cost of the coating material. It has been said that you pay for coating materials four times:

1. You pay someone to buy it originally.
2. You pay someone to apply it.
3. You pay someone to clean it up.
4. You pay someone to dispose of it.

With this in mind, any opportunity to identify a potential savings in the above-mentioned areas should be pursued. Once the manufacturer identifies the per piece cost of their coating material, they have also identified how much material is being applied per piece. This number or volume of material can then be used to evaluate the performance of the finishing process. The manufacturer then knows how their existing system is performing and can compare it to known performance characteristics of other systems.

The following items will be covered in detail:

- Fluid Flow Measurement and Methods
- Calculating "Parts per Gallon"
- Calculating Actual Transfer Efficiency
- Theoretical Transfer Efficiency
- Calculating Return on Investment

## FLUID FLOW MEASUREMENT AND METHODS

Fluid flow is a measurement of how much material is flowing per applicator in a given amount of time. Typically, fluid flow rate is identified in cubic centimeters per minute (cc/min) or ounces per minute (oz/min). In an effort to have a repeatable process, the manufacturer must monitor and/or control the fluid delivery system.

Some manufacturers do not know where their fluid flow rates are set. They have operators that adjust fluid deliveries by "I": "I think it looks good." This can be a very costly method; note the following example. Example #1 - Assume that the target fluid delivery for a particular part is 200 cc/min. And at this rate the desired film thickness of 1.0 mils dry is obtained. What if 1.1 mils is actually being applied because the fluid flow rate is at 220 cc.min (10% high)?

---

*52nd Purdue Industrial Waste Conference Proceedings*, 1997, Ann Arbor Press, Chelsea, Michigan 48118. Printed in U.S.A.

220cc = actual fluid flow rate
200cc = target fluid flow rate
20cc = excess material applied

$$\frac{20\text{ cc}}{\text{min}} \times \frac{60\text{ min}}{\text{hr}} \times \frac{2080\text{ hr}}{\text{yr}} \times \frac{1\text{ gal}}{3785\text{ cc}} = \frac{659.44\text{ gal}}{\text{yr}}$$

$$\frac{659.44\text{ gal}}{\text{yr}} \times \frac{\$25}{\text{gal}} = \$16,486 \text{ per year for a single applicator.}$$

The above example illustrates why it is important to know where fluid flow rates are set. Fluid flow can be measured using one of the following processes:

*Beaker Method*

This is the most basic way to measure fluid delivery. A beaker or a graduated cylinder (which is preferred) is used to collect the coating material for 15, 30, or 60 seconds. For low flow rates, a longer collection time should be used. The size or graduation of the beaker should be selected considering the amount of flow anticipated. You would not want to use a 1000 cc beaker to measure a flow rate of 60 cc/min. Greater accuracy can be obtained with a graduated cylinder. Normally a 100 cc capacity graduated cylinder will work well.

*Equipment Required*

Graduated cylinder, 100 cc capacity
Stop watch
Assistant

*Procedure*

Disable voltage, atomization, fan, turbine or shaping air.
Remove the air cap, bell cup, or disk from the applicator to be tested.
Purge paint through the applicator into a grounded collection receptacle to ensure tha there is no air in the lines.
Hold graduated cylinder to fluid tip or tube.
Simultaneously start stop watch and trigger applicator for 15, 30, or 60 seconds.
Shut off trigger after designated time and place cylinder on level surface.
Record measurement in cylinder. Multiply 15 second test by four, or 30 second test by two in effort to obtain fluid flow rate in cc/min.

**Table I. Fluid Flow Calculations[a]**

| Qty. of Appl. | Flow cc/ min | Total Flow | % Over Target | Over/ yr Gallons | $10 per gallon | $20 per gallon | $30 per gallon | $40 per gallon | $50 per gallon | $75 per gallon | $100 per gallon |
|---|---|---|---|---|---|---|---|---|---|---|---|
| 1 | 300 | 300 | 10 | 989 | 9,892 | 19,783 | 29,675 | 39,567 | 49,458 | 74,188 | 98,917 |
| 2 | 300 | 600 | 10 | 1,978 | 19,783 | 39,567 | 59,350 | 79,133 | 98,917 | 148,375 | 197,834 |
| 3 | 300 | 900 | 10 | 2,968 | 29,675 | 59,350 | 89,025 | 118,700 | 148,375 | 222,563 | 296,750 |
| 4 | 300 | 1,200 | 10 | 3,957 | 39,567 | 79,133 | 118,700 | 158,267 | 197,834 | 296,750 | 395,667 |
| 5 | 300 | 1,500 | 10 | 4,946 | 49,458 | 98,917 | 148,375 | 197,834 | 247,292 | 370,938 | 494,584 |
| 6 | 300 | 1,800 | 10 | 5,935 | 59,350 | 118,700 | 178,050 | 237,400 | 296,750 | 445,125 | 593,501 |

[a]Above calculations based on 2080 hours per year (260, 8 hour days).

*Flowmeter Method*

There are various units available which allow a user to measure fluid flow rate without using beakers. These methods utilize a fluid flow meter. The flow meter is inserted into the fluid stream at the outlet of a pressure pot or color valve stack. The fluid as it is flowing through to the applicator drives gears located inside the flow meter. A pick-up located on the flow meter generates a "pulse" every time a gear tooth passes it. The flow meter is connected to a control unit via a cable which transmit the pulses from the flow meter. Using a calibration value preprogrammed, the control unit converts the pulses to a flow rate. The calibration value is based on a volume of coating material that is in the cog of the gear. Once the applicator is "triggered," the fluid flow rate in cc/min will be displayed at the control unit. If multiple applicators are fed from a common supply, one applicator at a time can be triggered to measure individual flow.

An added benefit to most flow meter systems is that they also have a "totalization" feature. This allows the operator to accumulate a volume of material over any desired time frame or production run. For example, if the operator wanted to know how much material was used to coat a specific part or run of parts, they would simple reset the Totalizer and record the setting at the desired time.

*Equipment Required*

Flow meter
Flow meter cable
Flow control unit

*Procedure*

Install flow meter into fluid stream. Flow meter should be grounded. For waterborne electrostatic system, a special fiber optic flow meter and cable is required.
Install flow control unit: connect to 120v ac power and ground.
Connect flow meter cable between flow meter and control unit.
Trigger applicator(s) and view flow rate on display at control unit.

*Weight Method*

This method, just as it sounds, involves weighing the amount of coating material applied to a specific part or for a specified amount of time. This method works best if the material is supplied to the applicators from a pressure pot which can be easily handled. Normally, a pressure pot with a capacity of two gallons or less is ideal. The resolution of the scale used is also very critical to the final accuracy obtained. A scale with a minimum accuracy of one-tenth of a gram should be used to weigh the pressure pot.

*Equipment Required*

Pressure pot with quick disconnect for fluid and air
Scale with resolution of 0.1 grams
Stop watch

*Procedure*

Fill equipment to be tested with material from pressure pot to make sure all air is purged from fluid circuit.
Shut fluid valve at pressure pot and release air pressure.
Place pressure pot on scale and record weight. Identify with a "B" for before.
Replace pressure pot; restore air pressure prior to opening fluid valve.
Simultaneously start stop watch and trigger applicator for 60 seconds or spray desired parts for designated time.
Shut fluid valve at pressure pot and release air pressure.
Place pressure pot on scale and record weight. Identify with "A" for after.
Subtract "A" from "B" and record gram weight applied.

Converting "gram" Weight: Some manufacturer track their usage using grams per square foot or part. In most cases, the gram weight will need to be converted to "cc's". This can be accomplished by using the specific gravity. Specific gravity is a measure of grams per cc. The following method should be used to calculate the specific gravity and then to convert to cc's.

*Equipment Required*

Graduated cylinder, 100 cc capacity
Scale with resolution of 0.1 grams

*Procedure*

Place the graduated cylinder on the scale and zero the scale (the scale should display "0" with the graduated cylinder on it).
Remove the graduated cylinder and fill it from the pressure pot to a mark between 50 and 100 cc.
Place the graduated cylinder on the scale and record the weight. Note the volume in the graduated cylinder.
Divide the gram weight by the number of cc's in the cylinders. The resulting number is the specific gravity (SPGR).

$$\frac{58 \text{ g} \quad \text{(scale weight)}}{50 \text{ cc (beaker volume)}} = 1.16$$

## Calculating "Parts per Gallon"

In an effort to calculate the number of parts that can be coated with one gallon of paint, you must first determine how much paint is used per part. This can be easily done using the flow meter/totalizer or weight method. The advantage of these two methods is that they take into account trigger on/off time. Either method could be used to measure one individual part or several minutes' worth of parts. The longer the run checked, the more representative the numbers will be. Once the volume of material used for a given number of parts is identified, parts per gallon can be quickly calculated using the following formulas:

$$\frac{\text{Volume of material applied (cc's)}}{\text{Number of parts coated}} = \text{"x"}\frac{\text{cc's}}{\text{parts}}$$

*Example A: Calculating Parts Per Gallon*

Material applied = 240 cc
Number of parts coated = 24

$$\frac{240 \text{ cc}}{24 \text{ parts}} = \frac{10 \text{ cc}}{\text{part}}$$

$$\frac{1 \text{ part}}{10 \text{ cc}} \times \frac{3785 \text{ cc}}{1 \text{ gal}} = \frac{378.5 \text{ parts}}{\text{gallon}}$$

*Example B: Calculating Parts Per Gallon*

Material applied = 273.6 g
Number of parts coated = 24
Specific gravity of coating = 1.14 cc

$$273.6 \text{ g} \times \frac{1.0 \text{ cc}}{1.14 \text{ g}} = 240 \text{ cc}$$

$$\frac{240\ \text{cc}}{24\ \text{parts}} = \frac{10\ \text{cc}}{\text{part}}$$

$$\frac{1\ \text{part}}{10\ \text{cc}} \times \frac{3785\ \text{cc}}{1\ \text{gal}} = \frac{378.5\ \text{parts}}{\text{gallon}}$$

## CALCULATING ACTUAL TRANSFER EFFICIENCY

Transfer efficiency is a measurement that quantifies how much of the coating material applied actually goes on the part coated. In an effort to measure transfer efficiency, great care, accuracy, and attention must be given to detail and procedure. Figure 1 identifies the "advertised" transfer efficiency of various pieces of equipment.

The numbers that appear in Figure 1 are typically generated in a lab situation using an ASME approved test standard. These numbers work well when comparing the different pieces of equipment.

Many things can influence transfer efficiency, they include:

- equipment used
- temperature
- humidity
- substrate coated
- coating material
- application method
- equipment maintenance
- booth air velocity
- operating technique
- part grounding

The following information is required to calculate transfer efficiency:

- amount of coating material sprayed by weighing the pressure pot before and after.
- amount of coating material applied by weighing the part coated before and after.
- solids by weight of the coating material.

*Equipment Required*

scale with resolution of .1 gram
pressure pot with quick disconnect for air and fluid
foil (if part weight is greater than scale can accurately weigh).

*Procedure*

Preweigh parts to be coated, or foil which will be used to wrap parts.
Fill spray system with coating to be tested, shut off fluid valve, relieve air pressure, and weigh pressure pot.

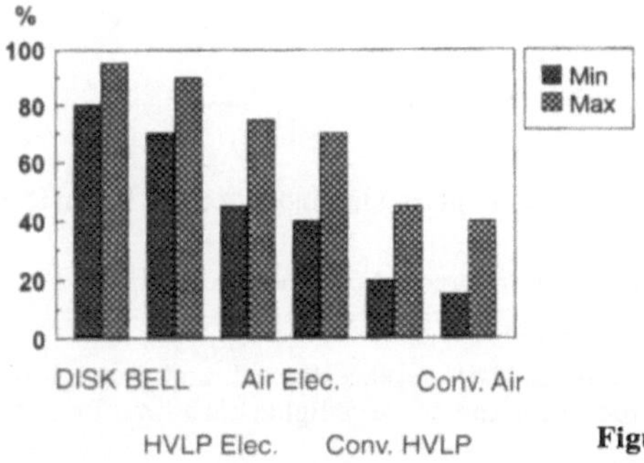


**Figure 1. Equipment type.**

**THEORETICAL TRANSFER EFFICIENCY**

| | | |
|---|---|---|
| I. FIND SOLIDS APPLIED | | |
| | SQUARE INCHES COATED = | 1728.000 SQ.INCHES |
| | AVG DRY FILM THICKNESS = | 1.000 MILS |
| | SOLIDS APPLIED = | 28.310 CC DRY |
| II. SOLIDS SPRAYED | | |
| | CC PER PART SPRAYED = | 51.060 CC/PART SPRAYED |
| III. SOLIDS BY VOLUME | | |
| | SOLIDS BY VOLUME = | 65.000 % SOLIDS BY VOLUME |
| IV. % TRANSFER EFFICIENCY= | | 85.299 % TRANSFER EFFICIENCY |

**ACTUAL TRANSFER EFFICIENCY**

| | | |
|---|---|---|
| I. FIND SOLIDS APPLIED | | |
| | WEIGHT OF PART AFTER = | 11537.000 GRAMS |
| | WEIGHT OF PART BEFORE = | 11465.500 GRAMS |
| | TOTAL WEIGHT APPLIED = | 71.500 GRAMS |
| II. SOLIDS SPRAYED | | |
| | WEIGHT OF P.POT BEFORE = | 25757.000 GRAMS |
| | WEIGHT OF P.POT AFTER = | 25638.000 GRAMS |
| | TOTAL WEIGHT SPRAYED = | 119.000 GRAMS |
| III. SOLIDS BY WEIGHT | | |
| | SOLIDS BY WEIGHT = | 78.509 % SOLIDS BY WEIGHT |
| IV. % TRANSFER EFFICIENCY= | | 76.531 % TRANSFER EFFICIENCY |

**Figure 2. Transfer efficiency worksheet.**

Reconnect pressure pot.
Spray the parts being tested.
Reweigh the pressure pot (shut off fluid valve, relieve air pressure, and weigh pressure pot).
Cure parts according to recommended parameters.
Reweigh the part(s) coated (allow to cool to ambient temperature).
(Reference Figure 2)

*Calculating Solids by Weight*

The solids by weight for coating material can typically be found on the MSDS or material data sheet. However, if the material has been reduced, this number will not be accurate. The solids by weight of the coating material should always be checked using the following procedure:

*Equipment Required*

scale with resolution of .001 gram
glass vial with lid
foil dishes (3)

*Procedure (see Figure 3)*

Collect a sample of the coating material being tested in the glass vial and place the lid on it to prevent evaporation.

Number the foil dishes (on the bottom) and weigh each one; record the "before" weight.

Weigh the vial of material and record the "before" weight.

Remove the lid and pour a quarter size sample into dish #1, replace the lid, and reweigh the bottle. Repeat this procedure for dish #2 & #3, recording the bottle weights each time in an effort to document the "wet" weight.

| | SAMPLE #1 | SAMPLE #2 | SAMPLE #3 |
|---|---|---|---|
| BOTTLE WEIGHT BEFORE | 17.136 | 16.213 | 15.227 |
| BOTTLE WEIGHT AFTER | 16.213 | 15.227 | 14.303 |
| WET SAMPLE TOTAL = | 0.923 | 0.986 | 0.924 |
| FOIL DISH WEIGHT AFTER | 2.508 | 2.514 | 2.510 |
| FOIL DISH WEIGHT BEFORE | 2.158 | 2.139 | 2.162 |
| DRY SAMPLE TOTAL = | 0.350 | 0.375 | 0.348 |
| SOLIDS FRACTION = | 37.920 | 38.032 | 37.662 |
| AVERAGE SOLIDS FRACTION = | 37.872 % SOLIDS BY WEIGHT | | |

| | SAMPLE #1 | SAMPLE #2 | SAMPLE #3 |
|---|---|---|---|
| BOTTLE WEIGHT BEFORE | | | |
| BOTTLE WEIGHT AFTER | | | |
| WET SAMPLE TOTAL = | 0.000 | 0.000 | 0.000 |
| FOIL DISH WEIGHT AFTER | | | |
| FOIL DISH WEIGHT BEFORE | | | |
| DRY SAMPLE TOTAL = | 0.000 | 0.000 | 0.000 |
| SOLIDS FRACTION = | | | |
| AVERAGE SOLIDS FRACTION = | 0.000 % SOLIDS BY WEIGHT | | |

| | SAMPLE #1 | SAMPLE #2 | SAMPLE #3 |
|---|---|---|---|
| BOTTLE WEIGHT BEFORE | | | |
| BOTTLE WEIGHT AFTER | | | |
| WET SAMPLE TOTAL = | 0.000 | 0.000 | 0.000 |
| FOIL DISH WEIGHT AFTER | | | |
| FOIL DISH WEIGHT BEFORE | | | |
| DRY SAMPLE TOTAL = | 0.000 | 0.000 | 0.000 |
| SOLIDS FRACTION = | | | |
| AVERAGE SOLIDS FRACTION = | 0.000 % SOLIDS BY WEIGHT | | |

**Figure 3. Solids fraction worksheet.**

Place the three foil dishes into a metal container (one gallon can) and cure the coating material according to the manufacturer's specifications.

Once the coating material is cured and cooled to ambient temperature, reweigh the three foil dishes and record the "after" weight.

Use the following formula to calculate solids by weight.

Bottle weight "before" - bottle weight "after" = wet sample.

Foil dish weight "after" - foil dish weight "before" = dry sample.

$$\frac{\text{dry sample weight}}{\text{wet sample weight}} = \%\ \text{solids by weight}$$

Average the three weights collected.

*Transfer Efficiency Calculation*

Once all the neccessary information has been obtained, the actual transfer efficiency can be calculated using the following formula:

$$\frac{\text{material applied}}{\text{material sprayed} \times \%\ \text{solids by weight}} = \text{Actual Transfer Efficiency}$$

## Theoretical Transfer Efficiency

When it is not possible to measure the actual transfer efficiency by weight, it can be calculated theoretically using volumes. The following information is required to theoretically calculate transfer efficiency:

Surface area of part coated
    Measure surface area of the part or reference drawings
Average coating thickness
    Use coating thickness gauge to measure average coating thickness over surfaces coated.
Solids by volume of material applied
    Reference manufacturer data sheet. If the material has been reduced, an estimate based on the amount of reduction should be made.
Amount of coating material sprayed
    Use one of the methods mentioned earlier in this chapter.

Once this information is gathered, use the following method to calculate Theoretical Transfer Efficiency:

$$\text{Surface area of part (in}^2\text{)} \times \text{coating thickness (in)} \times \frac{1\text{ gal}}{231\text{ in}^3} \times \frac{3785\text{ cc}}{\text{gal}} = \text{Amount* applied}$$

$$\frac{\text{Amount Applied}}{\text{Amount sprayed} \times \%\text{ Solids by volume}} = \text{Theoretical Transfer Efficiency}$$

Example:

Assume surface area of part = 1728 in$^2$

Average dry film thickness = .001 mils

$$1728\text{ in}^2 \times .001\text{ in} = 1.728^3\text{ in} \times \frac{1\text{ gal}}{231\text{ in}^3} \times \frac{3785\text{ cc}}{\text{gal}} = 28.31\text{ cc}$$

$$\frac{28.31\text{ cc}}{50\text{ cc} \times 65\%} = 87.1\%\text{ Theoretical Transfer Efficiency}$$

## CALCULATING RETURN ON INVESTMENT

When a manufacturer wants to upgrade or replace a finishing system, they need to have an idea of what the payback will be in an effort to write a justification. As noted previously, the cost of the coating material is typically one of the largest percentages of the entire finishing cost. In the previous text of this paper, various methods have been given for quantifying the amount of coating material that is being used for a specific process. The manufacturer should use this information to develop a baseline for their existing system performance. This information should then be used to compare the various pieces of competitive equipment available in the industry. Of the previous measurement methods listed, one of the easiest and most practical to work with is the amount of coating material used per part. Although transfer efficiency is a very good method for measuring system performance, it is often difficult for a manufacturer to generate this informa-

*Assume 100% transfer efficiency, 100% Solids by Volume

tion in-house. When conducting trials at competitive labs, the manufacturer can attend these trials armed with their material usage per part information. During the trials this information is then generated and compared with the existing information. In addition to this, the customer can also visually compare coated samples and film thickness consistency.

*Example*

Existing System: Air Atomized Electrostatic Guns on Longstroke Reciprocator

| | | |
|---|---|---|
| Material usage | – | 50cc/part |
| Production hours | – | 8000 parts/day |
| Production days | – | 260 days/year |
| Price per gallon | – | $25/gallon |

$$\frac{50\text{ cc}}{\text{part}} \times \frac{8000\text{ parts}}{\text{day}} \times \frac{260\text{ days}}{\text{year}} \times \frac{1\text{ gal}}{3785\text{ cc}} = 27,477\text{ gallons}$$

$$27,477\text{ gal} \times 25\text{ dol gal} = 686,925\text{ dollars year}$$

Proposed System: Electrostatic Rotary Atomizers on Shortstroke Reciprocator

$$\frac{30\text{ cc}}{\text{part}} \times \frac{8000\text{ parts}}{\text{day}} \times \frac{260\text{ days}}{\text{year}} \times \frac{1\text{ gal}}{3785\text{ cc}} = 16,486\text{ gallons}$$

$$16,486\text{ gal} \times 25\text{ dol gal} = 412,150\text{ dollars year}$$

| | |
|---|---|
| Existing system: | 686,925 dollars per year |
| Proposed system: | 412,150 dollars per year |
| Projected savings: | 274,775 dollars per year |

In the above example, the manufacturer can realize a savings of $274,775 per year in coating material cost alone. The price of any potential investment for new equipment can be divided into this number to determine the length of payback.

# 55 EXPERIMENTS IN SMALL-SCALE CAVITATION IN ATOMIZING ORIFICE FLOWS

**Steven H. Collicott**, Associate Professor
**Mark Henry** and **Haiyun Li**, Graduate Students
School of Aeronautics & Astronautics
Purdue University
West Lafayette, Indiana 47907-1282

## ABSTRACT

Cavitation inside of spray orifices is an inherently unsteady process which, in many cases, appears to have an effect on some properties of the spray. Experimental data on small-scale cavitation is scarce, thereby hindering development of practical numerical models. Lack of empirical and numerical models limits spray-system design to cut-and-try exercises. Experiments in a small-scale cavitating slot orifice are described, and image data presented to illustrate the complex flow structures present in low-pressure spray orifices.

## INTRODUCTION

Cavitation within an atomizing orifice has been shown to affect the resulting spray, in some cases in a favorable manner.[1,6,9] Better understanding of the characteristics of the cavitation (inception, extent, conditions for hydraulic flip, effect of asymmetric flow, . . .) is desired to improve the performance of some spray equipment, and also to improve the design methods for such systems. Experimental research can lead to better understanding of the physics of the phenomena and specialized experiments can produce an empirical database for specific design needs. Numerical modeling research can produce nonempirical design tools for a range of applications. Experimental and numerical research efforts, conducted in parallel by a research group with daily interaction, is an effective methodology for advancing both areas. Previous reports from this group describe the numerical modeling[3,4,7] and experiment geometry.[8] This chapter reports on experiments to visualize and study the cavitation at the entrance to a spray orifice in a two-dimensional flow, and examples from the numerical modeling efforts are included for comparison.

Earlier experimental efforts with optical methods applied to studying cavitation within small orifices, regardless of the pressures involved, are few. Chaves et al.[11] presents a number of images in the range of 200 μm diameter and $L/D \approx 5$, accompanied by a good discussion of the image data, discharge coefficient, and a two-point correlative velocimetry technique. Tamaki, et al.[9] present image data from small orifices, and an brief investigation into the effects of turbulence and cavitation on the spray. Images of the cavitation within the slender circular orifices are presented in both these works. Because of the aspect ratio (L/D) of the orifices, and that the cavitation occurs along the walls of the orifice where the refractive power of the curved glass-liquid or glass-vapor interfaces is strongest, these are inherently difficult images to acquire and interpret. Several more papers are found for larger-sized orifices,[2,5] though the scaling of those results down to diesel-injector sizes is impractical. In contrast, because the present experiment seeks to compare with the pseudo-density numerical model of Chen and Heister,[3,4,7] rather than with an existing injector or spray nozzle, the two-dimensional slot flow with its superior optical access is used.

The cavitating slot flow has previously been shown to be a practical geometry for generating data for code development.[8] The 2-D mean flow, the single cavitation region, and easy optical

*52nd Purdue Industrial Waste Conference Proceedings*, 1997, Ann Arbor Press, Chelsea, Michigan 48118. Printed in U.S.A.

access combine to make this a useful tool. A number of questions in code development can be answered in this geometry, and done so efficiently since the 2-D code is much quicker to run than a 3-D code. The pressures used in the present work, while low for diesel injectors, are typical of spray-coating applications.

## EXPERIMENTAL SETUP

The purpose of the experiments is two-fold. The primary motivation is to provide empirical data for comparison to the predictions of the Chen and Heister[3,4,7] two-dimensional slot orifice model. A secondary goal is to explore the fluid dynamic flow structures in a slot orifice. These insights into the fundamental physics of the flow will potentially provide valuable feedback to modeling efforts. A flow rig is constructed capable of forcing the fluid through an accurately formed slot orifice with a pressure of 10 atmospheres. A unique feature of this flow rig is that optical access is provided to the flow inside the slot orifice. Two separate optical systems are constructed to probe the flowfield inside the orifice. A third optical system is used to map the velocity profile in the slot and thus validate that the experiments indeed provide a reasonable approximation of two-dimensional flow in a slot orifice. The data of primary interest is the transient pressure response and flow rate of the system and images of the cavitation inside the orifice. These data can be compared to the predictions of Chen's model. Additionally, images are acquired of the flow inside the orifice which, although not useful for quantitative measurement, are useful for examining the fundamental flow structures present in cavitating flow in a slot orifice.

Figure 1 displays a picture of the flow rig facility constructed for the experiments. Water purified through a reverse osmosis process is used as the working fluid. The flow rig consists of several main components: an accumulator, upstream reservoir, test section, and a downstream reservoir system. Each of these systems is represented symbolically on the schematic in Figure 2, and discussed in detail in later sections. The test section contains the slot orifice, which is formed by two BK7 glass blocks.

**Figure 1. Picture of flow rig facility for low-pressure cavitation studies.**

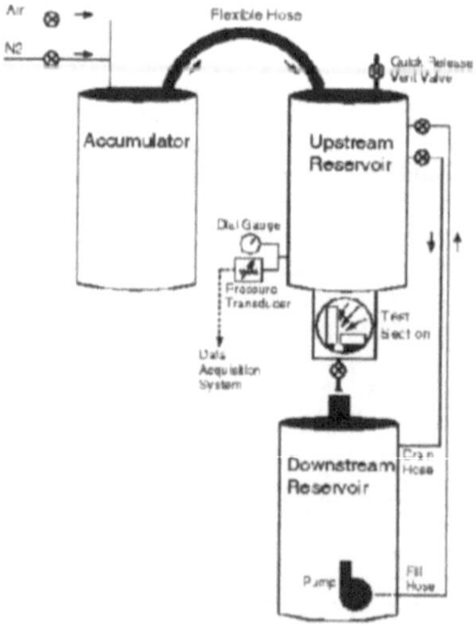


**Figure 2. Overall schematic of flow rig.**

Before the start of each experiment, the upstream reservoir is filled with water from the downstream reservoir through low-pressure lines. When the water reaches the fill line, the pump in the downstream reservoir is turned off, and the low-pressure lines are isolated from the upstream reservoir. The accumulator is then pressurized, and it in turn pressurizes the water in the upstream reservoir. A ball valve located downstream of the test section is opened to initiate the flow from the upstream reservoir, through the slot orifice contained in the test section, and finally into the downstream reservoir system. The entire flow rig is capable of operating at pressures of up to 10 atmospheres. Run times of 30–60 seconds are typical.

The working fluid is forced through the flow rig by head pressure supplied by the accumulator. The accumulator is recharged before each experiment with either nitrogen or air. The shop air supply provides up to 115 psig, and bottled nitrogen can then be switched on to boost to 150 psig. The accumulator is constructed of Schedule 80 PVC pipe. The PVC pipe is approximately 40 inches long and 10 inches inner diameter, and flanges are mounted at each end. An endcap, rubber gasket, and 3/4″ steel bolts seal each flange assembly. All-thread is used in four of the eight bolt holes in the endcap to connect the two flanges and prevent separation from the pipe, thus providing an additional safety margin in the event of an accidental over-pressure. The pipe is nominally rated at a working pressure of 260 psig, while the flanges are limited to 150 psig. A 1-1/4″ brass ball valve is mounted in the top endcap for rapid release of pressure after each experiment. The accumulator provides head pressure to the upstream reservoir through a 1″ hose mounted in the top flange.

The working fluid is pressurized in the upstream reservoir before it enters the test section. Construction of the upstream reservoir is similar to the accumulator. A Schedule 80 pipe is sealed at each end with a flange, rubber gasket, and endcap assembly. The pipe is approximately 30 inches long, and 10 inches inner diameter. The top endcap is fitted with a 1" hose that feeds from the accumulator, and a small speaker that hangs suspended inside the tank. The bottom endcap has a square section removed from the center to allow mating to the test section. All-thread is used to connect the two flanges for safety. Two clear, plastic hoses are attached via 1/2″ ball valves to the side of the PVC pipe. These are low pressure hoses that are used for filling the tank from the pump in the downstream reservoir and draining the tank to a calibrated fill level. During experiments the ball valves are closed to isolate the hoses from the pressurized tank. A pressure transducer is fitted in the side of the tank to measure gauge pressure during experiments. The pressure transducer is a Druck PTX 620 which exhibits typical total errors due to hysteresis, linearity, and thermal drift of 0.03%. The transducer outputs a proportional current which is calibrated with a dead weight tester. A large two inch dial gauge shares the same port as the PTX 620. This gauge provides a relatively crude estimate of system pressure to the flow rig operator as a safety precaution. The pressurized water in the upstream reservoir flows into the test section.

The test section is attached to the bottom endcap of the upstream reservoir with stainless steel 1/4″-20 screws. An o-ring is placed in the interface area to prevent leakage. A picture of this stainless steel vessel is displayed in Figure 3. Attached to the bottom of the test section is a two inch ball valve for initiating the flow once the upstream reservoir is full of water and the system is pressurized. The test section contains three optical windows and houses the slot orifice on a mounting block.

**Figure 3. Picture of exterior of test section.**

**Figure 4. Picture of mounting block and slot orifice.**

The circular windows, constructed of BK7 glass, provide optical access to the slot orifice. The windows are positioned in the center of three of the test section walls. Two of the windows lie on an axis opposing each other, while the third is normal to this axis. Each window is 3/8 inches thick and 2.5 inches in diameter. The surface is polished to a $\lambda/10$ finish. The windows are held in place by slip rings made of stainless steel. The rings are held in place by #8-32 screws, and o-rings seal each window assembly.

The slot orifice is formed by glass blocks mounted on a stainless steel block in the test section. Figure 4 contains a picture of this mounting block. The mount plays several roles in the experiments. First, it positions the glass blocks, and hence the slot orifice, in the field of view of the optical windows. Second, the mount successfully constrains the fragile glass blocks during the experiments. One of the greatest challenges in designing the experiments is the problem of tightly securing the fragile blocks without chipping or cracking them or obscuring optical access. The third and final design objective for the mount is to allow accurate micropositioning of the glass blocks while maintaining ease and speed of changing the slot size.

The single-sided cavitation region in the slot orifice is depicted in Figure 5. The primary advantage of the single-sided cavitation region is better optical access. A convention is defined for discussing the flow in the slot and the slot geometry itself. The direction from orifice entrance to exit planes is referred to as the stream-wise direction, across the thin dimension of the slot is cross-stream, and across the wide dimension of the slot is called length-wise. Likewise, the stream-wise length of the slot orifice is given as L, while the cross-stream thickness is D and

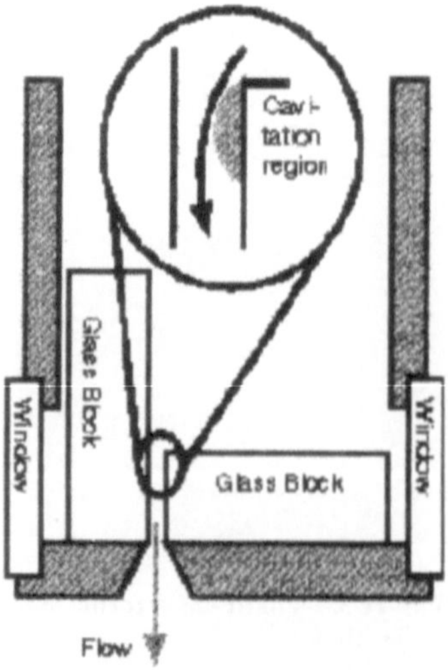


**Figure 5. Diagram of single-sided cavitation region inside slot orifice.**

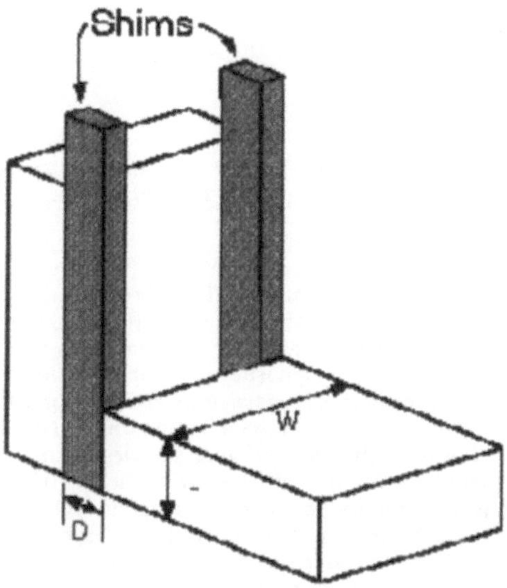


**Figure 6. Slot orifice and naming convention.**

length-wise width is W. These conventions are depicted in Figure 6. All slots formed in the experiments have a value of W of one inch. Two different slot heights are tested, one with L = 0.375 inches, and the other having L = 0.0787 inches. The chamfer on each block is at a 45° angle and approximately 0.13 inches long. The gap between the blocks, D, is initially set and maintained during the experiments by stainless steel shim strips placed at each end of the slot. The position of the shims can be seen in Figure 6. Due to the small size of the slots formed in the experiments, it is found to be imperative that the cross-stream dimension of the slot be uniform to effectively model the two-dimensional flow in Chen's pseudo-density model. Each shim is cut to approximately 0.060 inches wide by 2.5 inches in length. The flow in the slot orifice is found to be relatively unaffected by the presence of the shims as long as the shims are not wider than approximately 0.075 inches. Larger shims cause considerable cross-flow across the wide dimension of the slot.

After the flow leaves the test section it reaches the downstream reservoir system. This consists of a drainage pipe, a holding tank, a submersible pump, and a filter. The drainage pipe is made of 6 inch PVC, and its purpose is to contain the spray exiting the slot orifice, and transport the water back to the holding tank. The large diameter of the pipe ensures that the exit pressure for the slot orifice is atmospheric. The holding tank is a 10 gallon plastic tank with a cover to prevent contaminants from entering. A submersible pump is placed in the bottom of the holding tank. The pump forces the water from the holding tank through an inline filter. The water then continues into the upstream reservoir through the low pressure lines to recharge the system for the next experiment. The refilling process takes approximately three minutes.

## Optical Systems

Two separate imaging systems are used for the experiments. The first is referred to as the front-view imaging system, while the second is called the side-view imaging system. Although both of these systems are used to collect images for analysis of flow properties, the front-view images are the only ones useful for quantitative measurements. The side-view images are useful for visualization of flow structures found in the cavitating slot orifice. A third optical system, the Laser Doppler Velocimeter, is used for establishing the velocity profile across the wide dimension of the slot.

A diagram of the front-view imaging system is shown in Figure 7. Backlit images taken with the front-view imaging system are used to observe growth of cavitation in the slot orifice as a function of $\Delta P$. The images are also used for observing length scales in the liquid/vapor interface.

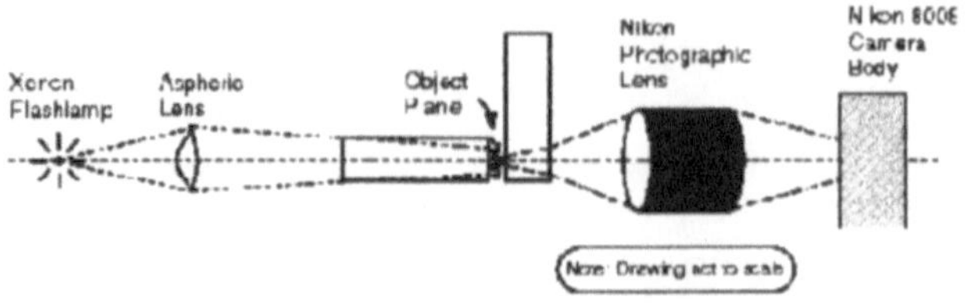


**Figure 7. Diagram of front-view imaging system.**

The illumination source for the front-view imaging system is a short-pulsed, broadband, arc Xenon flashlamp. The manufacturer specifies the flash duration to be one microsecond. The light emitted from the flashlamp is approximately collimated with an aspheric lens having a focal length of approximately 50 mm and a diameter of 30 mm. On the opposite side of the slot orifice from the illumination optics is a 35 mm format Nikon lens. This lens has a focal length of 105 mm and a maximum aperture of f/4. Also on this side of the slot orifice is a 35 mm format Nikon 8008 camera body. The lens and camera are placed in two different configurations depending on the glass blocks used to form the slot. Although some distortion is present due to the use of the Nikon lens at a much greater magnification than its optimal design point, the image is still quite acceptable. An iris diaphragm is installed between the Nikon lens and camera to increase the f-number of the system and provide more depth of field. The shutter speed on the camera is set at 1/30th of a second and the flashlamp pulsed at 29 Hz to provide at most a single burst of light on each frame. The images are recorded on Kodak color transparency slides and processed at 1600 ASA. The slides are digitized using a Nikon CoolScan scanner.

Figure 8 displays a diagram of the side-view imaging system. Scattered light images taken with the side-view imaging system are used to visualize the cavitating slot flow. The illumination source is a Q-switched frequency-doubled Nd:YAG laser with a pulse length of approximately 10 nanoseconds. A single lens is used to focus the beam to a small spot in the slot orifice. On an axis perpendicular to the laser path is a scope manufactured by Infinity Optics. This microscope provides a magnification of 16.4 when combined with a special adapter that allows a Nikon camera body to be fitted to the eyepiece. The field of view of the camera and scope combination is approximately 2.2 mm × 1.47 mm. This is sufficient to image cavitation around the 0.375 inch glass block onto 35 mm film in the camera body, as shown in Figure 9. TMAX black and white print film is used for recording the images and is push-processed to a speed of 6400 ASA with Kodak Dektol developer. The laser is pulsed at 8 Hz and the camera shutter set to 1/8th of a second to provide a single illumination pulse per frame.

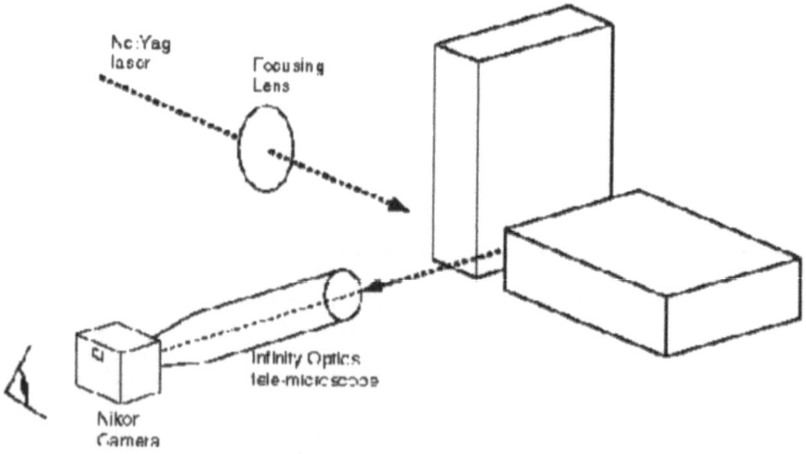


**Figure 8. Diagram of side-view imaging system.**

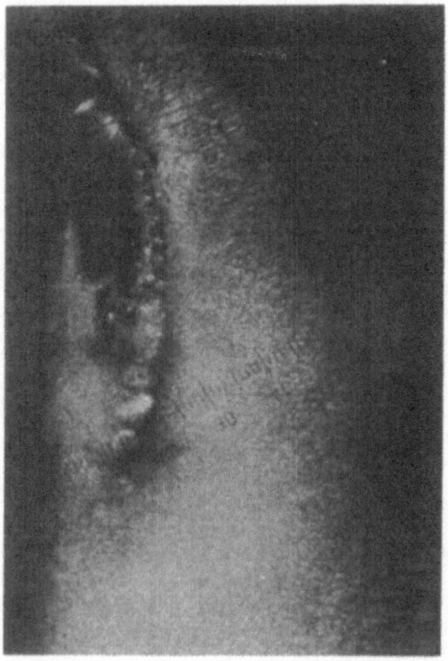

**Figure 9. Example side-view of cavitation with presumed vapor pocket.**

Ideally, these images could be used for quantitative measurements of the cavitation region inside the slot. However, in practice the slot gap cannot be maintained without the stainless steel shims, and the shims obscure the side view of the slot interior. Thus only one shim is used, and it must be placed in the center of the slot to balance the fluid forces on the glass block. This doesn't affect the nature of the cavitation, but requires a greater pressure to be applied to the upstream side of the slot to achieve the same amount of cavitation. Therefore, the images provide qualitative insight into the cavitation formation process, but are not valid for quantitative measurements.

## Data Acquisition and Reduction Methodology

A variety of slot sizes are tested in the experiments. For each new slot, D and W are measured (since L is known a priori). Then the slot is installed in the test section, and the upstream reservoir is filled with water. The accumulator is pressurized to a particular value, and the flow is started. The transient pressure is recorded and quantitative images are taken of the flow during the test. The images are registered with the corresponding values of the pressure at which they are acquired. This process is repeated for a variety of pressure values. In this way, flow in the slot orifice is tested over a range of operating conditions. After these tests are completed, the same slot geometry is tested over a range of pressures, and qualitative images are acquired with the side view imaging system.

## Observations and Results

Discharge measurements taken from the slot orifice confirm several trends reported by other authors about circular orifices. As reported by Fox and Stark,[10] the discharge coefficient is low for small Reynolds numbers, presumably because the orifice is dominated by viscous effects. The flow through the orifice is laminar in this regime and the vena contracta should be of negligible size. As the supply pressure is increased, and Reynolds number increases, the discharge coefficient reaches a steady value until the orifice reaches the point of hydraulic flip. Discharge coefficient now drops to a lower value and remains constant for further increases in supply pressure. The incipient cavitation pressure is found to increase with the L/D ratio.

Images taken with the front view imaging system confirm that the stream-wise extent of cavitation is uniform across the slot. The liquid/vapor interface changes shape dramatically for varying values of L and D. In all cases where partial cavitation is present, a break-up region is observed that is characterized by unsteady shedding of both single bubbles and "bubble clouds" that convect downstream and then collapse under the adverse pressure gradient. Figures 12 and 13 show typical front view images. Figure 12 shows a slot of D=0.889 mm and L=9.5 mm. Cavitation is just beginning to develop on the entrance lip, whereas Figure 13 shows a more fully-developed cavitation.

The side view images seem to indicate different cavitation mechanisms for different slot sizes. Some slots are observed to contain a vapor pocket at the upstream corner, followed immediately by a recirculating region that generates bubbles. Figure 9 shows an example of this. This is a side view of a slot where D = 1.0 mm and L = 9.5 mm. The flow is down in this image. Other slots seem to cavitate immediately at the upstream corner, causing bubbles and clouds of bubbles to emanate from local surface irregularities and convect downstream. The side view images are also useful for comparison with the code development effort, since the size of the two-phase bubbly region at the upstream corner should roughly resemble pseudo-density contours produced by the 2-D model. Figure 12 is a side view of the same slot depicted in Figure 9 at about the same value of (P. Figures 10 and 11 show this same slot for increased values of $\Delta$P.

## CONCLUSIONS

The slot orifice behaves similarly to circular orifices, as discharge coefficient measurements exhibit the same trends for changes in L/D and Reynolds number. The slot provides the additional advantage of superior optical access, since imaging is accomplished through two plane surfaces instead of curved ones. The formation of cavitation at the upstream corner is observed to be complex, full of small-scale structure, and with other features yet identified. Cavitation produces a vapor pocket in some cases, followed by a breakup region immediately downstream, and in other cases, it is not clear whether a vapor pocket exists or not. Front views agree with side views for extent of cavitation, so there is confidence that they are viewing the same manifestations of cavitation. For other L/D ratios, bubbles are formed at the corner and convect

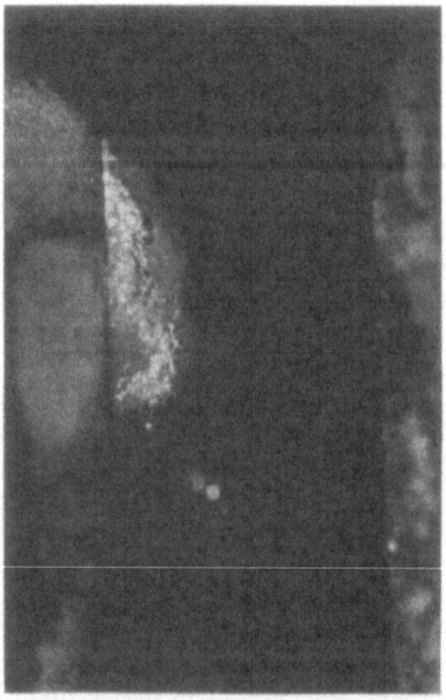

**Figure 10. Cavitation at pressure slightly greater than the minimum necessary to create cavitation.**

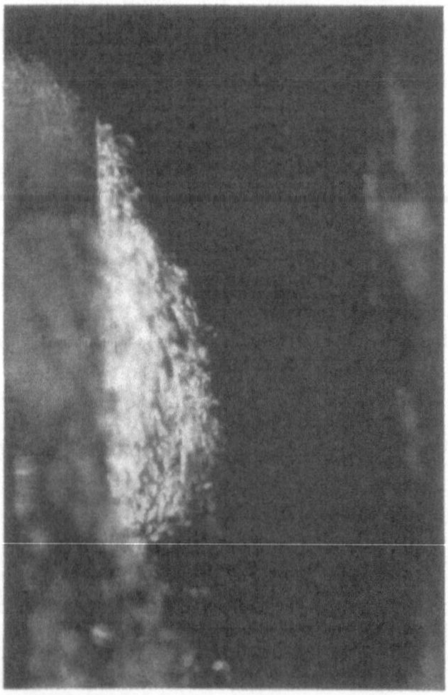

**Figure 11. Highher pressure cavitation.**

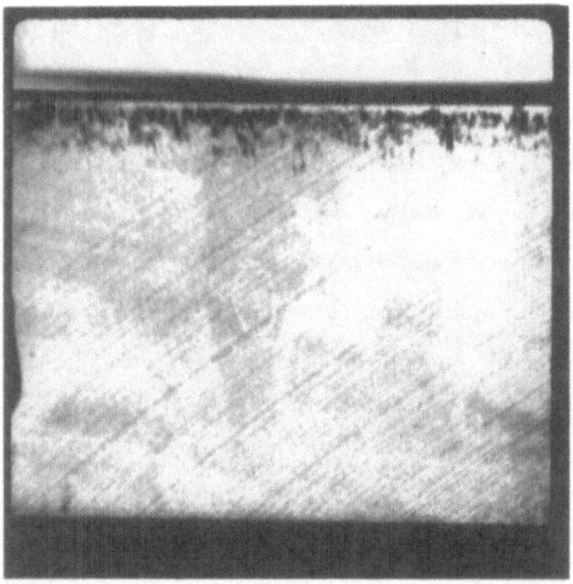

**Figure 12. Front view of cavitation near inception, as in the side view shown in Figure 10.**

**Figure 13. Front view of extended cavitation.**

downstream where they collapse under the adverse pressure gradient, and no vapor pocket is observed. In the cases where a vapor pocket is present, the collapse region is characterized by bubbles tearing off the vapor region in clumps. The shape of the liquid-vapor interface changes dramatically within the orifice for varying values of L and D. Not only does the slot orifice find utility for comparison with the current two-dimensional code development work, but it also has the capacity to answer fundamental questions about the flow field in an orifice operating in the cavitation regime.

The experiments demonstrate that low-pressure spray systems, even with substantial bluntness at the entry to the spray orifice, are susceptible to cavitation. Experiments such as those described here surpass present numerical models in identifying flow structures and length and time scales. Exploitation of cavitation to affect spray properties must be explored experimentally at this time. Effects of dissolved gas content, particulate content, non-Newtonian fluid properties, and other issues can also be explored.

## REFERENCES


1. Reitz, R.D. and Bracco, F.V., "Mechanism of atomization of liquid jet." *Phys. Fl.,* 25, pp. 1730–1742, 1982.
2. Nurick, E.H., "Orifice cavitation and its effect on spray mixing." *J. Fluids Eng.*, pp. 681–687, Dec. 1976.
3. Chen, Y. and Heister, S.D., "Modeling cavitating flows in diesel injectors," *Atomization and Sprays*, 6, pp. 709–726, 1996.
4. Chen, Y. and Heister, S. D., "Modeling hydrodynamic nonequilibrium in cavitating flows," *J. Fluids Engineering-Transactions of the ASME*, 118 (1), pp. 172–178, Mar., 1996.
5. He, L. and Ruiz, F., "Effect of cavitation on flow and turbulence in plain orifices for high-speed atomization," *Atomization and Sprays*, 5, pp. 569–584, 1995.
6. Bergwerk, W., "Flow pattern in diesel nozzle spray holes," *Proc. Inst. Mech. Eng.*, 173, (25), pp. 655–660, 1959.
7. Chen, Y., Numerical Approaches for Hydrodynamic Cavitated Flows, Ph.D. Thesis, Purdue University, School of Aeronautics and Astronautics, 1995.
8. Collicott, S. H., Chen, Y., and Heister, S. D., Cavitating Slot Flow: 2-D Modeling and Experiments, *ILASS-Americas 96*, San Francisco, CA, May, 1996, pp. 155–159.
9. Tamaki, N., Nishida, K., Hiroyasu, H. and Shimizu, M., Effect of the Internal Flow in a Nozzle Hole on the Breakup Processes of a Liquid Jet, *ILASS-Americas 96*, San Francisco, CA, May, 1996, pp. 155–159.
10. Fox, T. A., and Stark, J., "Characteristics of miniature short-tube orifice flows," *Proceedings of Institution of Mechanical Engineers—Part C*, 203 (C5), pp. 351–358, 1989.
11. Chaves, H., Knapp, M., Kubitzek, A., Obermeier, F., and Schneider, T., Experimental Study of Cavitation in the Nozzle Hole of Diesel Injectors Using Transparent Nozzles, #950290, Society of Automotive Engineers, pp. 199–211.

# 56 POLYMER FILTRATION: A NEW TECHNOLOGY FOR SELECTIVE METALS RECOVERY

**Barbara F. Smith, Thomas W. Robison, Michael E. Cournoyer, Kennard Wilson, Nancy N. Sauer,** and **Man T. Lu**

A new technology is under development to selectively recover regulated metal ions from electroplating rinse waters. The electroplating metal ions are recovered in a concentrated form with the appropriate counter ions ready for return to the original electroplating bath. The technology is based on the use of specially designed water-soluble polymers that selectively bind with the metal ions in the rinse bath. The polymers have such a large molecular weight that they can be physically separated using available ultrafiltration technology. The advantages of this technology are high metal selectivity with no sludge formation, rapid processing, low energy, low capital costs, and small size. We have tested and demonstrated the recovery of zinc and nickel (using a new alloy electroplating bath designed to replace cadmium) from rinse waters. The metal-ion concentrate was returned to the original electroplating bath. Impurity metals such as iron and copper were removed from the zinc/nickel concentrate and were not returned to the electroplating bath. Test panels were electroplated as a baseline and compared with new test panels electroplated after the addition of the recovered zinc and nickel. No adverse effects on the bath integrity were observed. The rinse water was depleted of the electroplating metals to less than 0.1 ppm zinc and nickel, and the water was satisfactory for discharge to the sewer system.

## INTRODUCTION

Closed-loop surface finishing systems and zero-discharge goals require different approaches to wastewater handling. Traditional precipitation technologies or even improving current precipitation technology to meet metal ion discharge limits is no longer acceptable.[1] New selective metal-ion recovery systems are required to recover metal value, to reduce cost and liability of toxic sludge formation and burial, and to reduce reagent, water, and sewer costs.

The concept of smaller point source treatment units in conjunction with plating line segregation becomes important because less energy is expended than in treating metal-ion mixtures which occur with end-of-pipe treatment systems. By focusing on the point source, the chemistry of metal separations can be optimized for the particular plating line, thereby optimizing the removal and recovery of the metal ions of interest. In contrast, the shotgun approach of precipitation tries to find one set of conditions that is appropriate for precipitation of most metal ions in a combined waste stream. What normally occurs is that the optimum conditions for one suite of metals is not optimum for others, and the metal discharge concentrations are variable and often times exceed EPA limits. The added problem of soluble chelators that occur in electroplating waste waters often interferes with the precipitation process, adding to the variability of metal concentration in the discharge waters.

The newer approaches involve multiple or tandem technologies to reach the near-zero discharge requirements. Some of the proposed approaches are proposed are (1) ion exchange with electro-recovery followed by evaporation, (2) evaporation with electro-recovery, or (3) reverse osmosis with electro-recovery. Other technologies that can be put together include carbon treatment, recrystallization, electrodialysis, and chelating ion exchange. The tandem-technologies chosen depend upon the ultimate goal; for example, if water is to be recovered, if the metal value is to be recovered, if sludge is is to be eliminated, if discharge limits are to be met, or all of the above.

*52nd Purdue Industrial Waste Conference Proceedings*, 1997, Ann Arbor Press, Chelsea, Michigan 48118. Printed in U.S.A.

We have been evaluating a new technology to add to the possible tandem technologies that is nonsludge-forming and can selectively recover metal-ion value for direct reuse in the original electroplating bath.

## POLYMER FILTRATION (PF)

The use of water-soluble chelating polymers in combination with ultrafiltration has been proposed for the recovery of metal ions from dilute plating bath rinse waters.[2] Metal ion concentration/separation, as shown in Figure 1, is based on size exclusion with retention (or membrane rejection) of metal ions chelated to the polymer, while smaller unbound species in solution are able to freely pass through the pores of the ultrafiltration membrane.[3] The homogeneity of PF separation processes have the advantage of rapid kinetics not ordinarily found in heterogeneous systems such as ion exchange chromatography. Mechanical stability of the polymer is not required as it is with resin-based separations systems. Since the process takes place in the aqueous phase, the use of potentially hazardous organic solvents found in liquid-liquid extraction separations are also avoided.

The relative efficiency with which a membrane retains or rejects a species can be determined experimentally with each species being assigned a numerical value between 0 and 1—called the rejection coefficient (s). A rejection coefficient of 0 means that the species freely passes through the membrane while species with a rejection coefficient of 1 are retained completely. The observation is that the small metal ions will freely pass through the membrane (s=1) unless the effective size is temporarily increased (s=1). In the case of a polymer/metal ion complex in which the polymer (P) is physically too large to pass through the ultrafiltration membrane, the rejection coefficient of the metal ions ($M^{n+}$) in the presence of a chelating polymer is a reflection of the equilibrium or stability constant of the complex, which is a measurement of the affinity the polymer has for a metal ion.

$$P + M^{n+} \rightleftharpoons PM^{n+}$$
$$K_s = \frac{[PM^{n+}]}{[P][PM^{n+}]} \qquad (1)$$

Generally there are two modes of operation in ultrafiltration.[4] The first is the concentration mode, schematically shown in Figure 2, where the volume in the retentate is reduced by simple ultrafiltration. The final concentration of any species in solution can be determined by:

$$C_f = C_o \cdot \left( \frac{V_o}{V_f} \right)^{\sigma} \qquad (2)$$

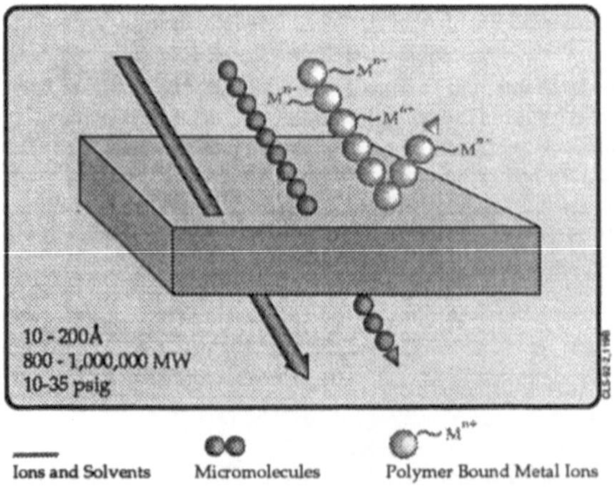


**Figure 1. Polymer Filtration (PF).**

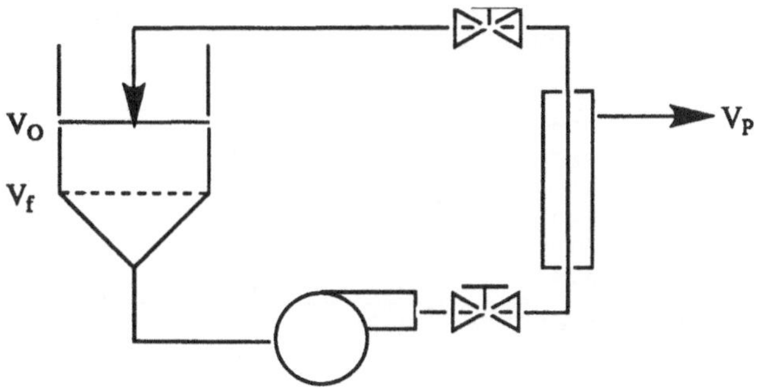


**Figure 2. Schematic diagram of concentration by ultrafiltration.**

where $C_f$ is the final concentration of the species, $C_o$ the initial concentration, $V_o$ the initial volume of solution, and $V_f$ the final volume. If the rejection coefficient of the species is 1, as would be the case for the water-soluble chelating polymers, then Eq. 2 simplifies to:

$$C_f = C_o \cdot \frac{V_o}{V_f} \tag{3}$$

For two species in solution, a polymeric/metal ion species (P) and a micromolecular impurity (A), where $s_P \gg s_A$, the ultrafiltration of the solution should result in the concentration and enrichment of P based on:

$$\left(\frac{C_A}{C_P}\right)_f = \left(\frac{C_A}{C_P}\right)_o \cdot \left(\frac{V_o}{V_f}\right)^{-(\sigma_A - \sigma_P)} \tag{4}$$

Ultrafiltration is the basis for a significant degree of purification during concentration of polymer/metal ion complexes in solution by PF.

The second mode of an ultrafiltration operation is diafiltration (Fig. 3). As permeate is generated, solute-free liquid is added to the retentate at the same rate so as to maintain a constant volume. In the diafiltration mode, the lower molecular weight species in solution are removed at a maximum rate when the rejection coefficient equals 0. The retentate is, in effect, washed free of smaller solute. Theoretically, the percent solute remaining in the retentate can be calculated by using Eq. 5:

$$C_f = V_o \cdot e^{-\frac{V_w}{V_e}(q - \sigma)} \tag{5}$$

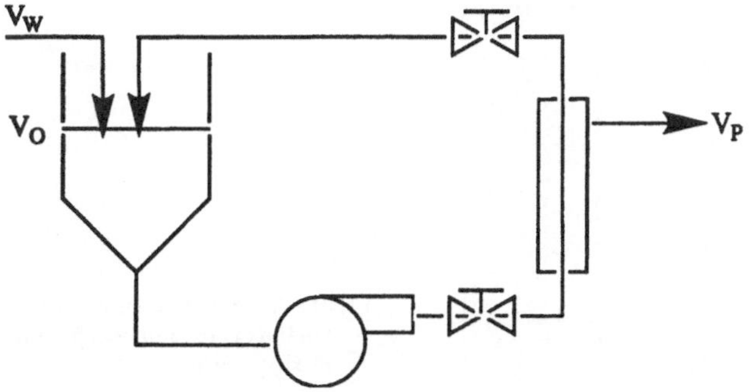


**Figure 3. Schematic of the diafiltration process.**

**Table I. Percent Solute Removal and Retained as a Function of Volume Equivalents During Diafiltration[a]**

| $V_w/V_o$ | % Solute Removed | % Solute Retained |
|---|---|---|
| 1 | 63.2 | 36.8 |
| 2 | 86.5 | 13.5 |
| 3 | 95.0 | 5.0 |
| 4 | 98.2 | 1.8 |
| 5 | 99.3 | 0.7 |
| 10 | 99.9+ | <0.1 |

[a]Sigma = 0.0 with continuous water addition assumed.

where $V_w$ is the volume of solute free liquid (volume equivalents) added, which also equals the amount of permeate processed.

Table I gives the correlation between solute removed and the percent solute remaining during a diafiltration process, assuming s = 0.0, as a function of the dilution ratio. From the table it can be seen that theoretically, after 5 volume equivalents of processed solution, that 99.3% of the lower molecular weight species should be removed. Experimentally, however, rejection coefficients of 0.0 are not ordinarily observed in PF. Even the smallest of interactions between the solute and the water-soluble polymer or the ultrafiltration membrane will give a small retention value. The effects of various rejection coefficients on the percent solute retained during a diafiltration process is shown graphically in Figure 4. The curves for low retention coefficients follow an exponential decay with each additional volume equivalent giving diminishing returns in percent solute removed, while higher rejection coefficients approach a straight line.

In the recovery of metal ions from plating bath rinse waters, both the concentration and diafiltration modes of operation are in effect. In a batch treatment of rinse waters, the polymer would be added to the rinse waters. Within a few minutes of mixing, the complexation of metal ions in solution would be complete and the volume of solution then reduced by operating in the concentration mode and the final metal ion concentration determined by Equation 3. The metal ions are released from the polymer by some chemical process, possibly lowering the pH. When the metal ions are no longer bound to the polymer, they freely pass through the membrane and are collected as permeate during diafiltration, while the regenerated polymer is retained for recycle.

When the ultrafiltration processes of concentration and diafiltration are combined, an effective method for the recovery, concentration, and purification of the metal ions in solution can be achieved. A number of industries successfully use ultrafiltration processes for various applications, including water purification, waste treatment, pharmaceuticals, life sciences, and the food and beverage industry.

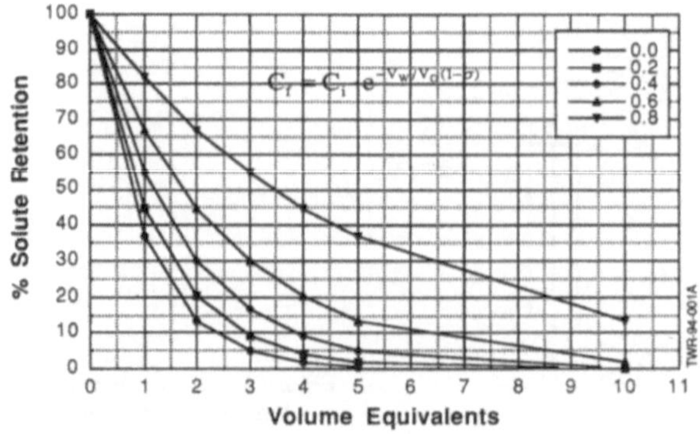


**Figure 4. Solute retained at various rejection coefficients during diafiltration.**

**Table II. Composition of Simulated Ni/Zn Plating Bath Solution**[a]

| Formula | MW | Grams/Liter | Oz/Gal |
|---|---|---|---|
| $NiCl_2 \cdot 6H_2O$ | 237.7 | 73.2 | 9.8 |
| ZnO | 81.4 | 11.2 | 1.5 |
| $NH_4Cl$ | 53.5 | 170.4 | 22.8 |
| $H_3BO_4$ | 61.8 | 20.4 | 2.7 |
| BOE-NIZ(porosity control) | | 50 mL | |

[a]pH adjusted to 6.5.

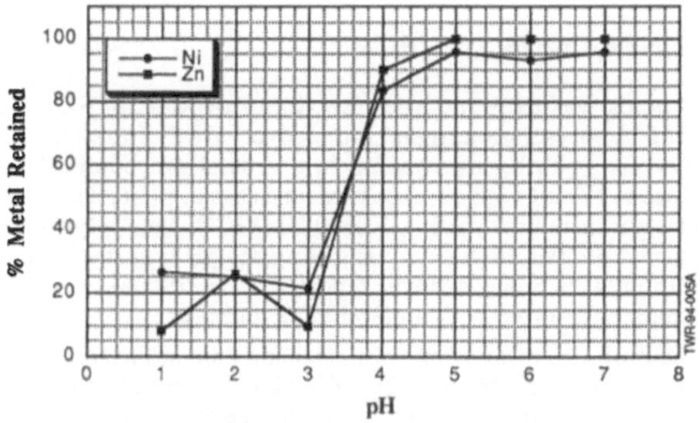


**Figure 5. Rentention of nickel and zinc by MetalSet-Z.**

## EVALUATION OF POLYMER FILTRATION FOR ELECTROPLATING APPLICATIONS

New alloy electroplating baths are under development to replace cadmium baths. The zinc/nickel alloy bath has shown the most promise at this time and was chosen for development of a metal-ion recovery system from rinse waters such that a near closed-loop system could be developed simultaneously with the bath develpment and testing. The composition of the new electroplating bath is shown in Table II.

Retention studies were performed on diluted electroplating bath solutions for evaluating the PF technology's effectiveness in retaining the plating metals in the presence of other bath constituents. BON-NIZ was deliberately omitted from the test solutions since it will be removed by carbon treatment prior to ultrafiltration in the demonstration (vida infra).

To simulate the electroplating rinse waters, the above solution was diluted 1:100, resulting in nickel and zinc metal ion concentrations in the range of 180 ppm and 90 ppm, respectively. The concentrations of the other species in solution were diluted in the same proportions. The general procedure for retention studies involved the addition of an excess of polymer (MetalSet-Z)©* to the simulated dilute rinse waters. The solutions were adjusted to the desired pH (1-7) and ultrafiltered through a 10,000 MWCO, polysulfone, flat, membrane, using an Amicon stirred cell. The volumes of retentate and permeate were measured and both analyzed for nickel and zinc by inductively coupled plasma-atomic emission spectroscopy (ICP-AES) or ICP-mass spectrometry (ICP-MS). The detection limit for these metal ions for this analytical technique is 0.01 ppm (μg/mL).

The results for the retention of nickel and zinc by MetalSet-Z are shown in Figure 5. The MetalSet-Z has good retention for both metals at pH 5 and above. Below pH 5, the polymer begins to release the metals and by pH 3, most metals are free in solution. The observation that there is

*©PolySet-Z is a proprietary polymer developed for metal removal from waste streams and is now available from Micro Set, Landisville, N.J.

very little change in the percent metal retained below pH 3 suggests that there is still some polymer/metal ion interaction at low pH (i.e., sigma ≠ 0 for the metal ions)

$$P + M^{n+} \rightleftharpoons PM^{n+}$$

While these results indicate a significant percentage of the metal ions remain in the retentate due to this polymer/metal ion interaction, during a diafiltration process the metal ions are constantly being removed from the retentate side of the membrane. Therefore, the above equilibrium is continuously being shifted toward the left, freeing more metal ions and allowing them to permeate the membrane. Within experimental error, there were no differences in the selectivity of MetalSet-Z for one metal over the other in these studies.

## REGENERATION STUDIES

The purpose of the regeneration studies was to investigate the ability of MetalSet-Z to go through a full cycle of metal ion retention/concentration, metal ion recovery by diafiltration, and polymer reloading using the simulated plating-bath rinse waters. In a general procedure, dilute electroplating bath solutions (50 mL) were treated with an excess MetalSet-Z at pH 6. These solutions were concentrated to 10 mL using a stirred cell with a 10,000 NMWCO, flat, polysulfone membrane. The metal ions were released from the polymer by adjusting the pH with hydrochloric acid and were collected by diafiltration. This procedure was repeated an additional two times after raising the pH of the polymer retentate back to 6 with ammonium hydroxide.

For polymer regeneration studies, the initial concentrations of nickel and zinc in the solution at pH 6 were 45 ppm and 24 ppm, respectively, as determined by ICP-AES. During the first concentration to 10 mL, retention of the nickel and zinc was >99% with no detectable metals in the permeate.

After adjusting the pH to 2, the solution was diafiltered and six volume equivalents collected. The percent of zinc and nickel ions retained after each volume equivalent is shown in Figure 6. With both of the curves falling above the theoretical values calculated from Equation 5, some weak polymer/metal ion interactions are still are present at pH 2.

The second and third regeneration results were similar to those of the first (Fig. 7). In both concentration steps no metals were detected in the permeate, indicating >99% retention. The two diafiltrations followed a similar trend, with all curves falling slightly above the theoretical values for sigma=0. The conclusions of the study indicate that MetalSet-Z can successfully retain electroplating bath metals from dilute solutions, that the metals can be concentrated, recovered, and the polymer recycled without the loss of performance.

## METAL ION RETENTION USING HOLLOW-FIBER MEMBRANE CARTRIDGES

An actual process could not use flat membranes as the flux rates would not be high enough. Thus, further studies with simulated electroplating rinse water (5.5L) solution using a hollow-

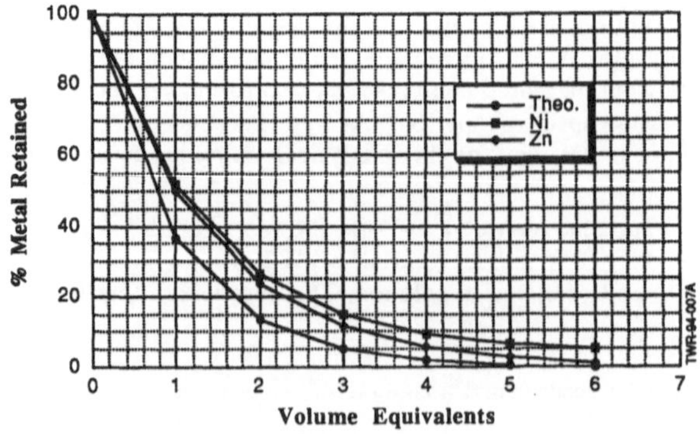


**Figure 6. The first diafiltration with MetalSet-Z for zinc and nickel.**

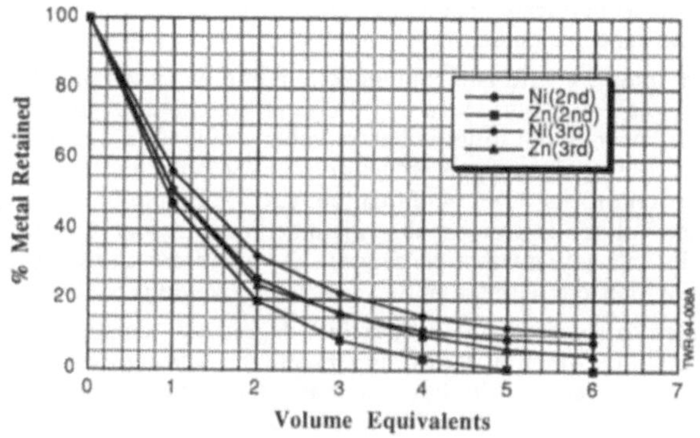


**Figure 7. The second and third diafiltrations with MetalSet-Z for zinc and nickel.**

fiber ultrafiltration cartridge were performed to determine the effectiveness of MetalSet-Z in retaining, concentrating, and recovering the plating bath metals from the bulk of the solution and to evaluate flux properties of the system. The initial concentrations of nickel and zinc in the solution as determined by ICP-MS were 164.5 ppm and 91.4 ppm, respectively. The boron concentration from boric acid was 31.7 ppm. The pH of the solution was adjusted to 6.3 by the addition of concentrated hydrochloric acid.

Conditions for the process involved the use of a 10,000 NMWCO, polysulfone, hollow-fiber cartridge (1.5 sq. ft, 0.14 sq m). Recirculation rates of ~4 L/min were maintained using a peristaltic pump. The operating back-pressure was 15 psig (pure water flux at this back pressure was 147 mL/min). The initial concentration of MetalSet-Z was 0.12% by weight. Based on 0.25 g metal/g polymer loading capacity for MetalSet-Z, quantitative zinc and nickel metal ion retention would utilize approximately 70% of MetalSet-Z's capacity. The final volume of the solution after the concentration step was ~ 350 mL or a concentration factor of about 10-fold. Permeate rates, taken every 1000 mL, are displayed in Figure 8 and indicated an interesting result. After the first 3000 mL of permeate was collected, there was no decrease in the flux rates, which were 65% that of pure water. This lack of decrease occurred despite an increase in MetalSet-Z's concentration from 0.12% to 0.21%. Only after 4000 mL of permeate was collected did the flux rates begin to drop. During this volume reduction, the average concentration of metal ions in the permeate was ~ 1 ppm for nickel and 2.3 ppm for zinc, indicating reasonable metal-ion retention by the polymer.

After the solution was concentrated by a factor of 10, the metals were released from the polymer by the addition of concentrated hydrochloric acid to a final pH of 2.0. At this point, the initial permeate from the diafiltration process was estimated to have a metal ion concentration approximately ten times that of the initial metal ion concentrations (164.5 ppm nickel, 91.4 ppm zinc). The results from the ICP-MS are in close agreement, showing 1617 ppm nickel and 1009

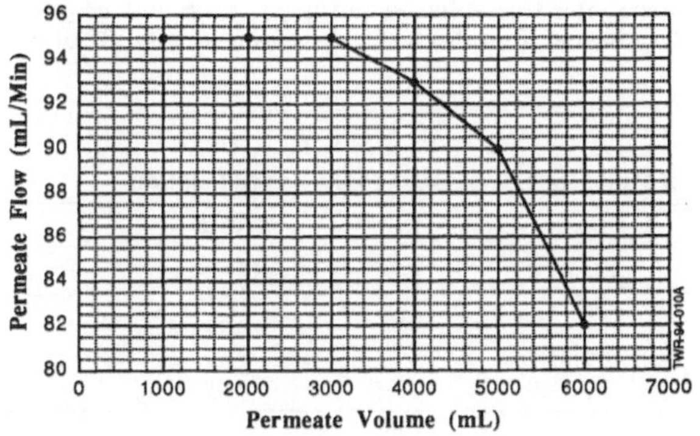


**Figure 8. The effect of increasing polymer concentration on permeate flux.**

ppm zinc in the first volume equivalent of permeate. The concentrations of nickel and zinc in successive permeate fractions follow the expected decrease.

It was determined that boron did not interfere with zinc and nickel binding. With the possibility of hydrogen bonding between the two species, boric acid might also be concentrated to a certain degree. The initial concentration of boron in the solution before volume reduction was 31.69 ppm. The concentration in the permeate during the volume reduction was ~34 ppm. and the concentration of boron during the diafiltration process after acidification was 31.53 ppm in the first permeate fraction, with each subsequent volume-equivalent showing a decrease in concentration. This is strong evidence that there is little, if any, interaction between the polymer and the boric acid, and therefore boric acid should not interfere with the polymer/metal ion binding.

A few additional tests were performed to determine if any polymer degradation occurred and resulted in loss of polymer through the hollow-fiber, ultrafiltration membrane. During 6 hours of continuous processing of a 1% MetalSet-Z solution (with both the retentate and the permeate recirculated back into the feed reservoir), no increase of polymer loss was observed through a 10,000 MWCO hollow-fiber, ultrafiltration cartridge. The temperature of the recycling solution increased from 24 to 40.5°C during the continuous cycling of the polymer through the pump (magnetic drive) and system (see demonstration details). Permeate samples taken every two hours over the six hour period were analyzed for polymer breakthrough using a spectroscopic detection method with copper which was sensitive to 1 ppm of polymer.

## METAL-ION SELECTIVITY STUDIES

Metal ion impurities often enter an electroplating bath from reagents, the materials being plated, and/or the electroplating equipment. Iron and copper are two metals that often contaminate baths. For an electroplating metal recycle system to function properly it is necessary for it to be able to selectively recover nickel and zinc and to separate them from metal impurities such that the impurities are not added back to the bath. To test the selectivity of MetalSet-Z, a solution of the electroplating rinse water (40 gallons) prepared according to Table I and diluted approximately 1:100 was spiked with low levels (~5 ppm) of copper and iron chloride. The polymer had an initial concentration of 0.15% by weight. Based on results obtained from previous loading capacity studies, calculations indicated that the retention of all metal ions in solution would give an 80% loading of the polymer. The analysis of the initial feed solution by ICP-MS is shown in Table III.

During the concentration stage of the run, the expected decrease in the permeate flow rate as well as the total volume of permeate collected as a function of time (Fig. 9) was observed. The total time to process the 40 gallons of solution was approximately one hour. The permeate collected during the concentration stage was analyzed for metal ion content and gave a nickel concentration of 1.94 ppm and a zinc concentration of 1.35 ppm.

Once the solution was concentrated to approximately 1.5 gallons, the pH of the solution was lowered to 3.5 with hydrochloric acid and diafiltered to release the metal ions. Five volume-equivalents were collected and each fraction analyzed for metal ion concentration (Table IV). It is clear from the results that pH 3.5 is not the optimum pH for zinc and nickel ion release. The polymer solution appears to have a slightly greater retention for Ni than for Zn; however, after five volume-equivalents, significant amounts of both nickel and zinc still remain in the retentate. This is contrary to theoretical models which predict that greater than 95% of the metal ions should have passed through the membrane after five volume-equivalents, if no polymer/metal ion interactions are taking place. Analysis results for other metal ions indicate at a pH of 3.5,

**Table III. Concentration of Metal Ion Species in the Initial Feed Solution**

| Metal | ppm | %RSD |
|---|---|---|
| Ni | 125.47 | 13.03 |
| Zn | 69.82 | 9.61 |
| B | 19.901 | 26.17 |
| Cu | 2.353 | 4.19 |
| Fe | 3.775 | 4.47 |

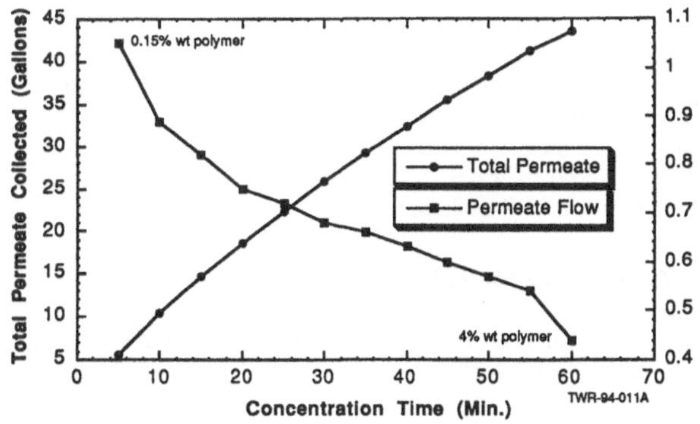


**Figure 9. Permeate flow and rate of collection during the 40 gallon concentration.**

**Table IV. Nickel and Zinc Concentrations in Various Samples**

| Sample | Ni | Zn | Cu | Fe (ppm) |
|---|---|---|---|---|
| Permeate from concentration | 1.94 | 1.35 | 0.024 | 0.25 |
| Volume equivalent 1-pH 3.5 | 533.52 | 698.99 | 0.79 | 1.52 |
| Volume equivalent 2-pH 3.5 | 422.63 | 376.51 | 0.44 | 1.37 |
| Volume equivalent 3-pH 3.5 | 398.49 | 234.46 | 0.24 | 1.25 |
| Volume equivalent 4-pH 3.5 | 331.41 | 155.99 | 0.15 | 1.25 |
| Volume equivalent 5-pH 3.5 | 325.39 | 153.15 | 0.10 | 0.52 |
| Volume equivalent 1-pH 3.0 | 472.59 | 30.37 | 2.84 | 1.24 |
| Volume equivalent 4-pH 3.0 | 44.80 | <4 | 1.41 | 0.60 |
| Volume equivalent 1-pH 1.0 | 80.62 | <4 | 14.1 | 1.76 |
| Volume equivalent 3-pH 1.0 | 22.47 | 2.97 | 2.33 | 0.74 |
| Retentate after Diafiltration | 42.06 | 1.94 | 17.46 | 26.3 |

copper and iron are still substantially retained. The pH of the retentate was then lowered a second time to 3.0. The first-volume equivalent of permeate during the second diafiltration process gave a nickel and zinc ion concentration of 472.59 ppm and 153.15 ppm, respectively. These results confirm that significant polymer/metal ion interactions were still taking place at pH 3.5. The pH of the retentate was lowered a third time to pH 1. The nickel concentration in the permeate was 80.62 ppm in the first volume-equivalent and down to 22.47 ppm after the third volume-equivalent.

Analysis of the retentate after the diafiltration process indicates that more of the nickel (42 ppm) was retained than the zinc (1.9 ppm), which suggests that there is a slightly stronger interaction between the MetalSet-Z with nickel than with zinc. Copper and iron were found to be significantly concentrated in the retentate at 17.5 ppm and 26.3 ppm, respectively. While iron showed no indication of being released from the polymer at low pH, the copper concentration in the first volume equivalent collected at pH=1 was 14.0 ppm.

Several conclusions can be drawn from the results of the experiment. All the metal ions were removed from the rinse water stream and concentrated in the first step, indicating that the process can easily meet EPA discharge limits. The optimum release of the nickel and zinc occurred at or slightly below a pH of 3.0. Diafiltration at this pH will reduce the number of volume-equivalents needed to recover >95 % of the nickel and zinc in the concentrated rinse waters to about 3. Copper and iron were retained and separated from zinc and nickel at pH 3.

## DEMONSTRATION AT BOEING ELECTROPLATING FACILITY

The focus of the demonstration at an actual plating facility at the Boeing Facility in Seattle was on the treatment of 100 gallons of electroplating rinse waters from the nickel/zinc alloy baths. Objectives for success were set:

1. Nickel and zinc had to be removed from the process rinse water to less than 1 ppm, meeting EPA discharge levels.
2. The plating metals had to be recovered in a concentrated, contaminant-free form.
3. The recovered metals had to be returned to the original plating baths without adverse effects on the plating process (the whole polymer filtration process would be performed within the chemistry of the nickel/zinc plating bath).
4. The polymer solution would be reused for further metal ion recovery.
5. The process had to be performed within an 8-hour period.

## ULTRAFILTRATION UNIT

The demonstration at Boeing required building a pilot-scale ultrafiltration unit large enough to process the 100 gallons of the plating bath rinse waters within an 8-hour period. Certain commercially available ultrafiltration units were capable of doing this; however, these units were expensive, with the smaller units costing $35,000-$40,000, and they were constructed from stainless steel, an undesirable material for metal recovery in chloride solutions. The goal in building the ultrafiltration unit was to construct a lightweight, inert, mobile, and relatively inexpensive system which could easily fit between the rows of electroplating baths at Boeing, which are spaced with approximately three feet between each row. Unlike the commercial systems, this system used PVC plumbing in place of stainless steel to reduce the cost and weight. The completed unit, which was mounted on a cart for mobility, was less than 100 pounds and cost approximately $7,000. Over a third of the cost came from the two hollow-fiber, ultrafiltration cartridges, which had 10,000 molecular weight cut-off (MWCO) membranes.

## PROCEDURE

A schematic of the polymer filtration process used at Boeing is illustrated in Figure 10. For the demonstration, one liter of the zinc and nickel alloy electroplating bath solution with nickel and zinc concentrations of approximately 17,800 ppm and 8,000 ppm, respectively (based on analyses performed at Boeing's analytical laboratory), was added to the 100 gallons of tank rinse water. The diluted solution was mixed thoroughly using compressed air, resulting in a cloudy blue-green solution at pH 7.7. The cloudiness was the result of ametal hydroxide precipitation of zinc. One-half of this solution was transferred to a 55-gallon plastic drum. The pH was adjusted to 7.3 using concentrated hydrochloric acid (12 mL), redissolving the suspended solids and clearing the solution.

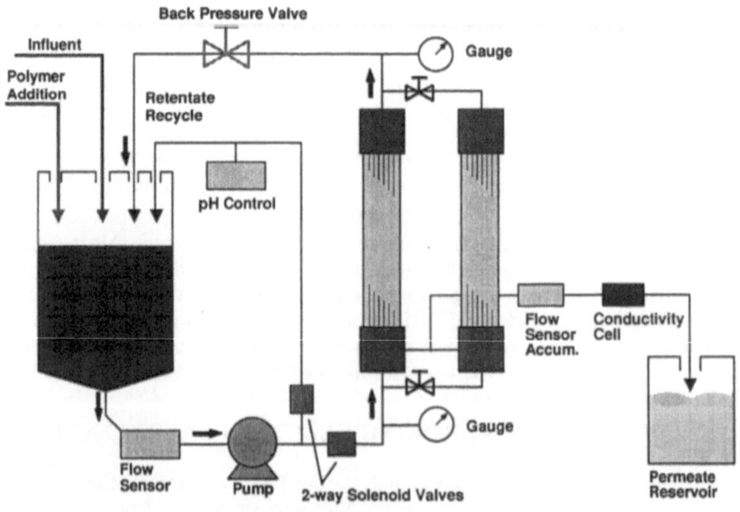


**Figure 10. Schematic of polymer filtration process using chelating polymers.**

Carbon (2 pounds) was added to the 50 gallons of solution to remove organics, including the BOE-NIZ, which could potentially foul the ultrafiltration membrane. After stirring for 20 minutes, the carbon was allowed to settle and the solution was pumped through a 5-micon course filter into a second 55-gallon plastic drum. PolySet-Z (165 g) was added to the tank, stirred for 5 minutes, resulting in a light purple solution.

The solution was concentrated to approximately 1.5 gallons using the ultrafiltration process. A pH of 7.0 to 7.1 was maintained in the feed reservoir during the concentration phase by using the pH controller to automatically add small volumes of a 15% hydrochloric acid or ammonium hydroxide solution when pH adjustment was necessary. Permeate samples were taken periodically at 5, 20, 35, and 50 gallons of permeate volume processed. The above process was repeated with the second 50 gallons of rinse water solution. After carbon filtration, the 1.5 gallons of concentrated polymer/nickel/zinc retentate from the first 50 gallons was used to treat the second 50 gallons, since only 30% of the polymer's full capacity had been reached. The second concentration process was run at the lower pH of 6.0, with permeate samples again taken for every 15 gallons of permeate generated.

Diafiltration was performed on the concentrated 1.5 gallon solution by first lowering the pH of the retentate solution to pH 2.9 using 15% hydrochloric acid. Deionized water was pumped into the feed reservoir at the same rate as the permeate was generated. Permeate samples were taken in approximately one-half volume-equivalents, or 0.8 gallons, for nickel and zinc analysis.

A second polymer filtration process was performed on a third 50 gallons of plating rinse-water solutions. For this solution, 250 mL of the plating bath and permeate volumes 4 and 5 from the first 100 gallon run were added to the 50 gallons of water. The solution was treated with 1 pound of carbon, filtered, and the pH raised from 5.5 to 7.0 using ammonium hydroxide. After filtering through a coarse filter and transferring to the second holding tank, the polymer used in the first polymer filtration was added and the solution processed under similar conditions. The pH for the process was held at 7.0. Diafiltration of the concentrated 1.5 gallons of solution was again performed at a pH of 2.9, with samples being collected at 0.5 volume equivalents for metal analysis.

A complete summary of the concentration and diafiltration processes including the results from the analysis for nickel and zinc are given in Tables V and VI.

Prior to starting, the metal recovery experiments test panels (4" × 6") of 4131steel were plated to determine the baseline behavior of the bath. Further panels were plated after each addition of recovered metal ion concentrate was added to the bath. All panels were submitted to salt-spray corrosion testing.

## RESULTS AND DISCUSSION

The 100 gallons of rinse water were analyzed on site for nickel using the Hoch test kit and were determined to be 44 ppm nickel. This amount was in close agreement with the theoretical value of 47 ppm (17,800 ppm/378 L = 47 ppm) predicted by Boeing for the nickel concentration in the actual plating bath. Subsequent analysis of the rinse water solution at Los Alamos National Laboratory by ICP-MS gave the nickel and zinc concentrations at 43 ppm and 12 ppm, respectively. The lower than expected zinc concentration may be a result of some zinc lost as a hydroxide precipitate during the original dilution step described in the previous section.

During the concentration process for the first 50 gallons, the pH of the feed solution was maintained at 7.0 . On-site analysis of the permeate for nickel using the Hoch kit indicated the concentration to be less than 1 ppm. The subsequent analysis of samples taken during the process by ICP-MS gave concentration values of <0.05 ppm for nickel and 0.097 ppm for zinc. These are well below the discharge limits for both metals. The second 50 gallons was processed at the lower pH of 6.0. At this pH, nickel was not efficiently retained, with 10–15 ppm coming through the membrane with the permeate. Zinc, at <0.05 ppm, was still well below discharge levels. These results indicate the importance of performing the concentration stage at the optimum pH to ensure all metals are retained during the process. The time for concentrating 100 gallons to 1.5 gallons was 2 hours.

After the volume reduction of 100 gallons to approximately 1.5 gallons, the pH of the solution was adjusted to 2.9 by the addition of dilute hydrochloric acid. Deionized water was introduced during the diafiltration process and volume equivalents of permeate were taken on 0.5 intervals, with 2.5 volume-equivalents total collected. The approximate concentration of nickel and zinc in the 2.5 volume-equivalents (3.75 gallons) collected was 946 ppm and 431 ppm, respectively. The

**Table V. Results of the First 100 Gallon Boeing Demonstration (Concentration)**

*Date: 4/25/94* *Time: 4:15 pm* Concentration I

| Perm Vol. | Total Time | Time | Back Press. | Perm. Flow | pH | Temp. | LANL Analysis | | Boeing Analysis | |
|---|---|---|---|---|---|---|---|---|---|---|
| | | | | | | | [Ni] | [Zn] | [Ni] | [Zn] |
| Gal | Min | Diff. | PSIG | Gal/Min | | °C | ppm | ppm | ppm | ppm |
| 0 | 0 | | | | | | | | | |
| 5* | 5:44 | 5:44 | 2 | 0.69 | 7.1 | 26 | <0.05 | 0.097 | | |
| 10 | 12:00 | 6:16 | 3 | 0.82 | 7.1 | 26 | | | | |
| 15 | 18:09 | 6:09 | 3 | 0.78 | 7 | 26 | | | | |
| 20* | 24:43 | 6:34 | 2.8 | 0.74 | 7 | 25 | | | | |
| 25 | 31:18 | 6:35 | 2.8 | 0.74 | 7 | 25 | | | | |
| 30 | 38:16 | 6:58 | 2.8 | 0.72 | 7 | 25 | | | | |
| 35* | 45:20 | 7:04 | 2.7 | 0.68 | 7.1 | 25 | | | | |
| 40 | 52:48 | 7:28 | 2.7 | 0.64 | 7.1 | 25 | | | | |
| 45 | 1:00:50 | 8:02 | 2.6 | 0.51 | 7 | 25 | | | | |
| 50* | 4:22 | 4:22 | 2.9 | 0.86 | 6.1 | 22 | 15.02 | <0.05 | 10.3 | <0.1 |
| 55 | 10:00 | 5:38 | 3 | 0.86 | 5.9 | 22 | | | | |
| 60 | 15:50 | 5:50 | 3 | 0.84 | 6.2 | 22 | | | | |
| 65* | 22:00 | 6:10 | 2.9 | 0.8 | 6.1 | 22 | | | | |
| 70 | 28:07 | 6:07 | 2.9 | 0.79 | 6.4 | 22 | | | | |
| 75 | 34:29 | 6:22 | 2.8 | 0.77 | 6.1 | 22 | | | | |
| 80* | 40:58 | 6:38 | 2.7 | 0.69 | 6.2 | 22 | | | | |
| 85 | 47:46 | 6:48 | 2.7 | 0.72 | 6.4 | 22 | | | | |
| 90 | 54:50 | 7:04 | 2.6 | 0.69 | 6.1 | 22 | | | | |
| 95* | 1:02:00 | 7:10 | 2.5 | 0.63 | 6.3 | 22 | 13.73 | 0.129 | | |
| 100 | | | | | | | | | | |
| | 96.88 total gallons processed | | | | | | | | | |

Procedure:
1. Approximately 100 gallons of rinse water were spiked with 1 liter of plating bath solution.
2. Half of this was pumped into a 50 gallon container and 12 mL of concentrated hydrochloric acid added (7.8 to 7.3)
3. Two pounds of charcoal were added and the mixture stirred for 20 minutes.
4. The solution was filtered through a 5 micron filter into a second container (pH 8.4); polymer added (1000 mL)
5. Processed maintaining the pH at 7.0 using pump and pH controller with 15% HCl.
6. Concentrated down to approximately 1.5 gallons; sat overnight.
6. Repeated the second 50 gallons using the same polymer solution; however, the pH was 6.0.

retentate of 1.5 gallons contains nickel and zinc at 166 ppm and 72 ppm. The second diafiltration process for the 50 gallons of rinse water concentrated to approximately 1.5 gallons was performed under similar conditions. The final concentration of nickel and zinc in 2.5 volume-equivalents was 449 ppm and 191 ppm, respectively. Total time for each diafiltration was approximately 15 minutes.

Corrosion protection is a requirement for zinc-nickel coatings. The plating must pass 96 hours of neutral salt spray testing per ASTM B117, "Salt Spray (Fog) Testing," with neither white corrosion products of zinc nor base metal corrosion products visible to the unaided eye. For our demonstration, corrosion protection was used as the benchmark for success. If the salt spray results were positive, we assumed proper maintenance of the electroplating bath using recycled metals. Six 4 × 6 inch 4130 alloy steel panels were zinc-nickel plated and chromate conversion coated (per BAC 5637, Type II, Class 1) to an approximate thickness of 0.0005 inch as control specimens. Two more sets of panels were plated following each addition of recovered metal solution. All three sets were then placed into salt spray testing (ASTM B117) for 96 hours. Upon removal, no corrosion products were observed on any of the panels. From these results it was concluded that the use of recovered zinc-nickel from the polymer filtration process was not dele-

**Table VI. Results of the First 100 Gallon Boeing Demonstration (Diafiltration)**

*Date: 4/26/94* *Time: 10:49 am* Diafiltration I

| Volume Eq. | Total Time | Back Press. | Per. Flow | pH | Temp. | LANL Analysis | | Boeing Analysis | |
|---|---|---|---|---|---|---|---|---|---|
| | | | | | | [Ni] | [Zn] | [Ni] | [Zn] |
| Gal | Min | PSIG | Gal/Min | | °C | ppm | ppm | g/L | g/L |
| 0.5 | | 2.3 | 0.19 | 2.9 | 25 | 1422.811 | 674.509 | 1.3 | 0.6 |
| 1 | | 2.5 | 0.17 | 2.8 | 26 | 1451.215 | 656.203 | | |
| 1.5 | | 2.6 | 0.15 | 2.9 | 26 | 934.15 | 413.994 | 0.8 | 0.4 |
| 2 | | 2.6 | 0.15 | 2.9 | 26 | 578.572 | 255.329 | | |
| 2.50[a] | 13 | 2.6 | 0.14 | 2.9 | 26 | 344.736 | 156.298 | | |
| Retentate | | | | | | 166.34 | 72.172 | | |

[a]Total volume processed was appoximately 3.75 gallons (14.2 Liters)

**Table VII. Results of the Second 50 Gallon Boeing Demonstration (Concentration)**

*Date: 4/26/94* *Time: 10:49 am* Concentration II

| Perm Vol. | Total Time | Time | Back Press. | Perm. Flow | pH | Temp. | LANL Analysis | | Boeing Analysis | |
|---|---|---|---|---|---|---|---|---|---|---|
| | | | | | | | [Ni] | [Zn] | [Ni] | [Zn] |
| Gal | Min | Diff. | PSIG | Gal/Min | | °C | ppm | ppm | ppm | ppm |
| 0 | 0 | | | | 7.2 | 21 | | | | |
| 5* | 5:44 | 5:44 | 3 | 0.87 | 7.2 | 22 | 0.197 | 0.035 | | |
| 10 | 11:00 | 5:16 | 3.1 | 0.91 | 7.2 | 22 | | | | |
| 15 | 16:27 | 5:27 | 3.2 | 0.94 | 7 | 22 | | | | |
| 20* | 21:55 | 5:28 | 3.2 | 0.9 | 7.1 | 22 | | | | |
| 25 | 27:30 | 5:35 | 3.2 | 0.85 | 7.1 | 22 | | | | |
| 30 | 33:34 | 6:04 | 3.1 | 0.82 | 7.1 | 22 | | | | |
| 35[a] | 39:34 | 6:00 | 3.1 | 0.79 | 7.1 | 22 | 0.085 | 0.014 | | |
| 40 | 48:10 | 8:36 | 1.6 | 0.54 | 7.1 | 23 | | | | |
| 45 | 57:32 | 9:22 | 1.5 | 0.52 | 7.2 | 23 | | | 0.4 | <0.1 |
| 47.6[b] | 1:01:24 | 3:52 | 1.5 | 0.51 | 7.2 | 22 | | | | |

Back pressure on this run was not held constant

Procedure:
1. Approximately 50 gallons of rinse water was spiked with 250 mL of Ni/Zn plating solution.
2. To this solution was added 12 mL of concentrated hydrochloric acid, followed by 1 pound of charcoal.
3. The mixture was stirred for 20 minutes.
4. The test for surfactant was negative.
5. The solution was pumped through a 5 micron filter to remove suspended charcoal.
6. Permeate volumes 4 and 5 were added to the solution (approx. 6 liters).
7. pH of this solution was 5.5 and was adjusted to 7.1 using ammonium hydroxide.
8. Ni concentration at this point was measured at 34 ppm according to Ken Mullen.
9. Concentration procedure was started and the volume taken down to approximately 1.5 gallons.

[a] Permeate samples taken.

[b]Total volume processed.

**Table VIII. Results of the Second 50 Gallon Boeing Demonstration (Diafiltration)**

*Date: 4/27/94* *Time: 10:30 am* Diafiltration II

| Volume Eq. | Total Time | Back Press. | Per. Flow | pH | Temp. | LANL Analysis | | Boeing Analysis | |
|---|---|---|---|---|---|---|---|---|---|
| | | | | | | [Ni] | [Zn] | [Ni] | [Zn] |
| Gal. | Min | PSIG | Gal/Min | | °C | ppm | ppm | g/L | g/L |
| 0.5 | | 2.2 | | 2.8 | 29 | 654.841 | 251.433 | 0.6 | 0.3 |
| 1 | | 2.6 | | 2.8 | 28 | 767.661 | 337.317 | | |
| 1.5 | | 2.6 | | 2.8 | 27 | 428.606 | 189.727 | | |
| 2 | | 2.6 | | 2.8 | 27 | 253.891 | 113.084 | | |
| 2.50[a] | 15 | 2.6 | 0.17[b] | 2.8 | 27 | 138.354 | 61.594 | | |
| Retentate | | | | | | 53.182 | 16.322 | | |

Procedure:
1. The plating rinse waters were concentrated to approximately 1.5 gallons.
2. The pH of the solution was adjusted to a pH of approximately 2.9 using 15% hydrochloric acid.
3. Deionized water was added at the same rate as the permeate discharge.

[a]Total volume processed was approximately 3.75 gallons (14.2 liters).
[b]Average permeate flow rate.

terious to the process and that our overall objective had been accomplished.

## SUMMARY

Based on the objectives set for a successful demonstration, the results indicate a very successful demonstration with all goals being met. The nickel and zinc were retained and concentrated from 100 gallons of plating rinse waters down to 1.5 gallons, a volume concentration factor of 67, and the permeate met discharge limits. The metals from this concentrate were successfully recovered from the polymer in a concentrated form (approximately 22-fold concentration factor) and returned to the original plating bath with subsequent electroplatings indicating no adverse effects on the quality of the plating. The regenerated polymer was successfully recycled to treat another 50 gallons of plating rinse waters and concentrate additional plating metals. The entire process was performed within a 4-5 hour period, well below the allotted 8 hours. Finally, the process was conducted within the chemistry of the actual electroplating baths as only ammonium hydroxide and hydrochloric acid were used. No additional chemicals which would be incompatible with the electroplating bath solutions were introduced into the polymer filtration process.

## CONCLUSION

The technology evaluated for the recovery of plating metal ions from rinse waters involves the process of polymer filtration, which combines the use of water-soluble chelating polymers and ultrafiltration. Metal ions in dilute rinse water solutions, typically less than 1000 ppm in total plating metal ion concentration, are retained by the polymers, while the smaller, unbound species freely pass through the ultrafiltration membrane. This process allows for the concentration of the plating metals by retention while the ultrafiltered permeate water meets EPA discharge limits for the electroplating metals. The metals were recovered from the concentrated polymer solution by pH adjustment using diafiltration and recycled to the original plating baths.

## ACKNOWLEDGMENTS

The authors would like to acknowledge the Department of Energy, Office of Industrial Technology, Industrial Waste Reduction Program for support of this development effort. We want to particularly thank Earl C. Groshart (Technical Fellow; Materials, Processing, and Physics) and Mary J. Nelson, both of Boeing Defense and Space Group, Seattle, Washington, for their help, guidance, and support, and for the use of their facilities for the demonstration phase of this project. We wish to thank Dr. Jose Olivaris for the ICP analyses.

## REFERENCES

1. Cherry, K.F. *Plating Waste Treatment*, Ann Arbor Sciences Publishers Inc., Ann Arbor, MI, 1982.
2. The principle of polymer filtration of metal ions was first proposed by A. S. Michaels, "Ultrafiltration," in E.S. Perry, Ed., *Advances in Separation and Purification,* John Wiley, New York, 1968.
3. Other references on PF include: H. Strathman, *Sep. Sci. Tech., 15(4)*, 1135–1152, 1980. K. E. Geckler, E. Bayer, B. Y.A. Spivakov, V. M. Shkinev, G. A. Vorobeva, *Anal. Chim. Acta*, 285–292, 1986. Q. T. Nguyen, P. Aptel, J. Neel, *J. Mem. Sci.*, 71–72, 1980. M. Rumeau, F. Persin, V. Sciers, M. Persin, J. Sarrazin, *J. Mem. Sci*, *73*, 313–322, 1992.
4. For a general discussion on the ultrafiltration process, see M. Cheryan, *Ultrafiltration Handbook*, Technomic, Lancaster,United Kingdom, 1986.
5. Winston, W. S. and Sirkar, K. K., Eds., *Membrane Handbook*, Van Nostrand Reinhold, New York, 1992,

# 57 ZEOLITE: A NATURALLY OCCURRING MINERAL FOR THE TREATMENT OF METALS CONTAMINATED WATER AND WASTE STREAMS

**Trudy J. Olin**, Research Civil Engineer
**R. Mark Bricka**, Research Chemical Engineer
U.S. Army Engineer Waterways Experiment Station
Vicksburg, Mississippi 39180

## ABSTRACT

This chapter presents the results of a research effort to identify and evaluate an economical, single-use sorbent for treatment of metals contaminated water and waste streams.

## INTRODUCTION

Heavy metals contamination is an environmental problem at U.S. Army installations engaged in firearms training and munitions production. Weathering and corrosion of expended munitions and leaching from wastewater lagoons, landfills, and burn pits, has resulted in heavy metals contamination of the soil at these facilities. Transport of metals to the groundwater has been confirmed in some locations, requiring treatment of the groundwater at the site. Certain treatment processes for contaminated soil produce metals-laden extracts, which also require treatment before reuse or disposal.

The principal metals encountered in firing range soils are lead, copper, and zinc. Cadmium and other metals, such as antimony, incorporated in the munitions are sometimes seen in lesser concentrations. Chromium is primarily associated with plating operations. Mercury is associated with various propellants and, while present in much smaller concentrations, is of concern because of its acute toxicity.

Conventional treatment technologies for metals contaminated groundwater and waste streams include ion exchange and activated carbon. While ion exchange is generally quite effective for the removal of metals from aqueous streams, resins are expensive and must be regenerated at additional expense. Activated carbon is not as effective for most metals, and also requires regeneration. Alternative sorbents are therefore needed that are effective and economical. In Phase I, an extensive literature search was conducted to identify low cost sorbents with potential for treatment of metals contaminated water and waste streams. Most sorbents requiring modification were eliminated from further consideration because of the additional cost introduced by the required chemical modifications. Twelve sorbents were screened in batch testing, including: activated carbon, bark, chitosan, crown ether, corn cob, xanthate, clay (kaolinite and montmorillonite), peat moss, seaweed, and reagent grade zeolite. Of these, zeolite demonstrated the highest capacity for Pb, Cr, and Cd, and was selected for Phase II testing in batch, kinetic, and column studies.

## MATERIALS AND METHODS

### Characterization

The zeolite used in the batch and column studies was provided to the Waterways Experiment Station (WES) by the United States Geological Service (USGS). The zeolite was obtained from a natural deposit of clinoptilolite-rich rock located in South Dakota (Rocky Ford SDH). Large

*52nd Purdue Industrial Waste Conference Proceedings*, 1997, Ann Arbor Press, Chelsea, Michigan 48118. Printed in U.S.A.

**Table I. Batch Studies Conducted with South Dakota Zeolite**

| | Conditions | | | |
|---|---|---|---|---|
| Test Type | AW/UW | Solution | mg°/l | Contact Time |
| Equil. | UW | Synthetic | Pb: 30K | 2 hr |
| Equil. | AW&UW | Synthetic | Pb: 20K | 2 hr |
| Equil. | AW | P soil extract | Pb: 13<br>Zn: 45<br>Cu: 44 | 2 hr |
| Kinetic | AW | P soil extract | Pb: 13<br>Zn: 45<br>Cu: 44 | 0.25 - 30 hr |
| Kinetic | UW | Synthetic | Pb: 1K<br>Zn: 50<br>Cu: 50 | 0.25 - 30 hr |
| Select. | AW | Synthetic | A: Pb: 500<br>Ca: 100<br>B: Pb: 50<br>Ca: 100 | 2 hr & 24 hr |
| Equilib | AW | FBH soil extract | Pb: 500<br>Zn: 2<br>Cu: 7 | 24 hr |

blocks of the material were crushed and the material sieved into three particle size ranges: 0.5 to 1.0 mm, 1.0 to 4.0 mm, and 2.0 to 4.7 mm. According to X-ray diffraction analysis performed on this material,[1] the dilutant minerals in the zeolite, listed in order of abundance, are plagioclase, calcite, quartz, opal, K-feldspar, and trace clay. The weight percent water capacity of this zeolite is 58. The material demonstrates high structural stability in acidic solutions. In this study the CEC and surface area of the material were also measured.

## Batch Studies

Seven batch studies were conducted: four equilibrium tests, two kinetic tests, and one selectivity test (Table I).

### *pH Control*

Control of pH in sorption studies is typically addressed by the use of a buffer. In the screening studies, acetic acid and sodium acetate were used to maintain the pH near 4.5. However, because zeolite has a high affinity for $Na^{1+}$, use of sodium acetate introduces competing ions into the metal solution. Zeolite releases carbonates when in contact with solution, which are responsible for the rise in pH over time. Acid washing removes most of the carbonates, eliminating the need for a buffer. Batch studies were conducted using both acid-washed zeolite (AW) and unwashed zeolite (UW) for performance comparison.

### *Kinetic Studies*

Kinetic studies were conducted on unwashed zeolite to determine how rapidly a pH increase occurred in the solution and to evaluate how much of the removal demonstrated by unwashed zeolite was attributable to sorption and how much to precipitation. The kinetic study was also conducted to determine the contact time necessary to achieve equilibrium. Three solutions were prepared using metal salts and DDI water, with the following initial concentrations: Pb, 1000 mg/L; Cu, 50 mg/L; Zn, 50 mg/L.

An additional kinetic study was conducted using 2-4.7 mm acid-washed zeolite with extract from a metals and organics contaminated soil taken from a burn pit (P-extract). The pH of this solution is approximately 1.1 to 1.2. The extract was prepared by sequential surfactant extraction of organics contamination, followed by acid extraction of metals. A number of metals and organics compounds were present in the soil. Analysis was restricted to Pb, Zn, and Cu concentrations for this study.

*Equilibrium Batch Tests*

Equilibrium batch studies were conducted on both acid-washed and unwashed zeolite, using a single solution concentration of Pb for varying sorbent weights to maintain uniform solution activity. Additional batch studies were conducted with the P-extract and FBH extract. The P extract was described previously. The FBH extract was produced from a firing range soil. The soil was first oxidized with 0.01 M CaO. Following oxidation, the supernatant was pumped off and the soil was extract with 0.1 M acetic acid. This was filtered through 0.5 micron Whatman 5 filter paper and stored in 55-gallon drums at room temperature prior to use.

*Selectivity Tests*

A selectivity test was performed using acid-washed zeolite and two solutions to determine how much the sorption of Pb was affected by the presence of high concentrations of competing ions. Solution A contained 500 mg/L Pb and 100 mg/L Ca. Solution B contained 50 mg/L Pb and 100 mg/L Ca. Samples were tumbled for 2 hr and 24 hr.

## Column Studies

Ten column studies were conducted (Table II). The largest particle size (2.0 to 4.7 mm) was selected for nine of the studies because it was expected to have the best hydraulic properties. The last study utilized the smallest particle size range (0.5 to 1.0 mm) for performance comparison. Both acid-washed and unwashed zeolite were tested in the columns.

Three flow rates below 40.7 L/min · $m^2$ (1 gal/min · $ft^2$) were selected to minimize side wall effects and demonstrate the relationship between capacity and flow rate. In terms of bed volumes, the flow rates were 3 BV/hr (60 mL/min), 1 BV/hr (20 mL/min), and 0.3 BV/hr (6.0 mL/min). Column tests were conducted in custom made Plexiglas columns. Inside diameter was 5.08 cm (2 inches). Bed length was 60.96 cm (24 inches). One hundred micron stainless steel porous plates were used at either end of the column bed to minimize particulate transport. Influent and effluent tubing was Tygon. Valves and fittings in contact with the extract were stainless steel. Extract was delivered to the columns with Masterflex digital peristaltic pumps.

*Influent*

Columns were challenged with the FBH soil extract utilized in the equilibrium batch studies, described previously.

**Table II. Column Studies Conducted with South Dakota Zeolite Column**

| Column Number | Packing Material | Particle Size (mm) | Flow Rate | Duration |
|---|---|---|---|---|
| 1 | Empty-Control | N/A | | 12 hr |
| 2 | Unwashed zeolite (UW) | 2.0–4.7 | 3 BV/hr | 12 hr |
| 3 | Acid washed zeolite (AW) | 2.0–4.7 | 3 BV/hr | 12 hr |
| 4 | UW | 2.0–4.7 | 1 BV/hr | 12 hr |
| 5 | AW | 2.0–4.7 | 1 BV/hr (stop flow) | 14 hr |
| 6 | UW | 2.0–4.7 | 0.3 BV/hr | 12 hr |
| 7 | AW | 2.0–4.7 | 0.3 BV/hr | 12 hr |
| 8 | AW | 2.0–4.7 | 1 BV/hr | 15 hr |
| 9 | UW | 2.0–4.7 | 0.3 BV/hr | 7 days |
| 10 | UW | 0.5–1.0 | 0.3 BV/hr | 12 hr |

*Column Packing*

Columns were loose-packed with zero headspace using oven dried zeolite. The columns were tapped during loading to consolidate the material.

*Run and Sampling Procedures*

After loading, the columns were flushed with DDI water until the effluent ran clear and the zeolite appeared to be fully saturated (approximately 5-15 minutes). The influent tubing was purged of DDI water with extract up to a three-way valve at the bottom of the column. The first influent sample was then taken from the valve, and flow was then directed through the DDI saturated column. This was taken to be time zero.

Influent was sampled manually at the three-way valve at three hour intervals throughout the duration of flow. Effluent samples were taken automatically using an ISCO Model 3700 portable sampler. Effluent was allowed to flow into a collection reservoir, which was emptied by the sampler at five to ten minute intervals, depending upon the rate of flow. The sampler was set to advance to a new sample bottle after each sampling event. Sample concentrations represent a composite of the five or ten minute interval over which they were collected.

*Analysis*

Elemental analysis was conducted using a Perkin Elmer 5100 model flame atomic absorption spectrophotometer (AA) with single element hollow cathode lamps. Liquid samples were filtered and analyzed for dissolved metals.

## RESULTS

*Cation Exchange Capacity*

The selectivity of zeolites for ammonium ions ($NH_4^+$) over sodium ions ($Na^+$) is used in the determination of CEC.[2,3] The exchange sites of the zeolite are first saturated with $Na^+$. Total cation exchange capacity (TCEC) is determined using a solution of $NH_4OAc$, which replaces the $Na^+$ in both internal and external exchange sites. Average TCEC for the three particle size ranges of zeolite ranged from 7.5 to 10.7 meq/100g. Average charge density ranged from 1.47E25 to 2.08E25. The measured TCEC is well below that indicated for commercially available zeolite, which is estimated at about 180 -220 meq/100g. This may indicate that the material used in this study was less homogeneous than commercially available materials, or may reflect the effect of the relatively large particle sizes utilized.

*Surface Area*

Specific surface area for all three particle size ranges was measured using a Gemini 2360 Surface Area Analyzer using $N_2$ gas as the standard adsorbate. Specific surface area was calculated using a Brunauer-Emmett-Teller (BET) multipoint procedure. Average specific surface for all particle sizes ranged from 29.8 to 30.8 $m^2/g$.

*Kinetic Studies: Synthetic Solutions*

The kinetic study conducted with the unwashed zeolite demonstrated that, for Pb, Cu, and Zn, sorption kinetics are quite rapid. For Pb, percent sorption was 99.8% after 15 minutes (Figure 1); for Cu, percent sorption was 99.3% and for Zn, percent sorption was 96.5%. The total pH change recorded for the Pb solution was approximately 2.5 units, increasing from 5.65 to 8.2 (Figure 2). The pH increase recorded in the first 15 minutes was only 0.6 units, however. The pH did not rise above 7 until after four hours of contact time. The percent removal of Pb is therefore thought to be attributable predominantly to sorption. The pH change for the Zn solution was greater and more rapid, measuring 7.96 at 15 minutes, and 9.17 to 9.22 at 30 hours. The pH change for the Cu solution was also rapid, increasing from an initial value of 5.33 to 7.96 at 15 minutes.

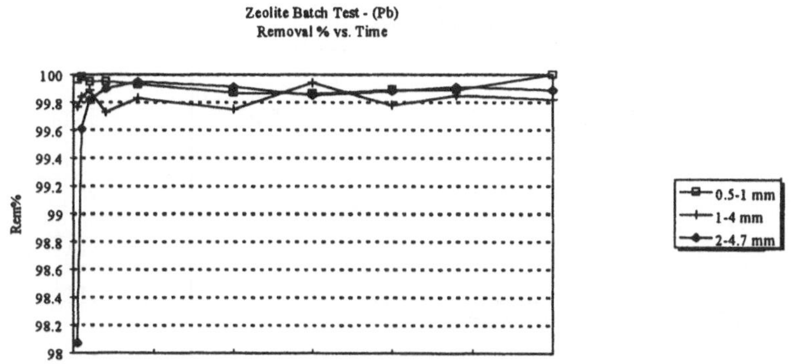


**Figure 1. Kinetic study. Pb sorption as a function of time.**

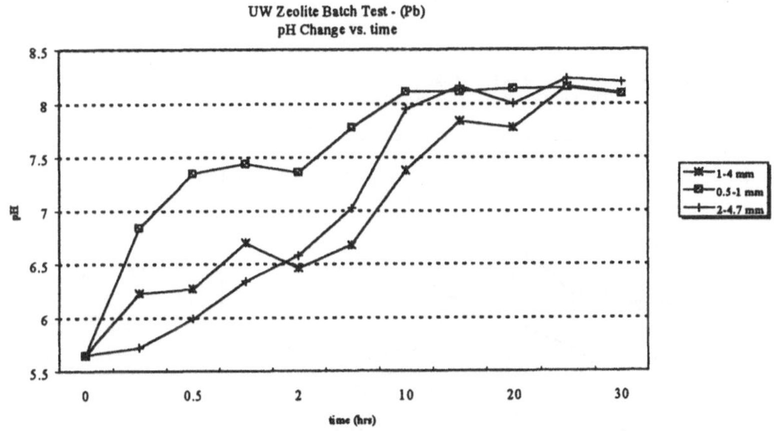


**Figure 2. Kinetic study: pH change with time for UW zeolite.**

In the kinetic study using acid washed zeolite and the P-extract, less Pb and Zn was sorbed from solution and the kinetics were slower than for the synthetic solutions and unwashed zeolite. Lower sorption is primarily attributed to hydrogen ion competition, as the pH of the P solution remained low for the duration of the study, but may also have been due to competition from other metals present in the extract. Pb and Zn appeared to have been preferentially absorbed over Cu. Initially, Zn sorbed more rapidly than Pb. However, it appears that the zeolite has a higher affinity for Pb because, as solution concentration of Pb decreased, the solution concentration of Zn increased.

*Equilibrium Batch Tests: Synthetic Pb Solution*

General data trends indicate approximately 20% higher removal for the finest material over the coarsest, for both washed and unwashed zeolite, with some exceptions attributed to the heterogeneity of the material. Lead capacity ranged from 54.4 mg/g to 366.2 mg/g for acid-washed material at an initial concentration of 17,700 mg/L Pb (Figure 3), and from 21.3 to 149.6 mg/g for unwashed material at 14,100 mg/L Pb (Figure 4). At an initial concentration of approximately 30,000 mg/L Pb, the unwashed material demonstrated a maximum capacity of 574.7 mg/g. This result and the overall data trends appear to indicate concentration dependence of capacity. At

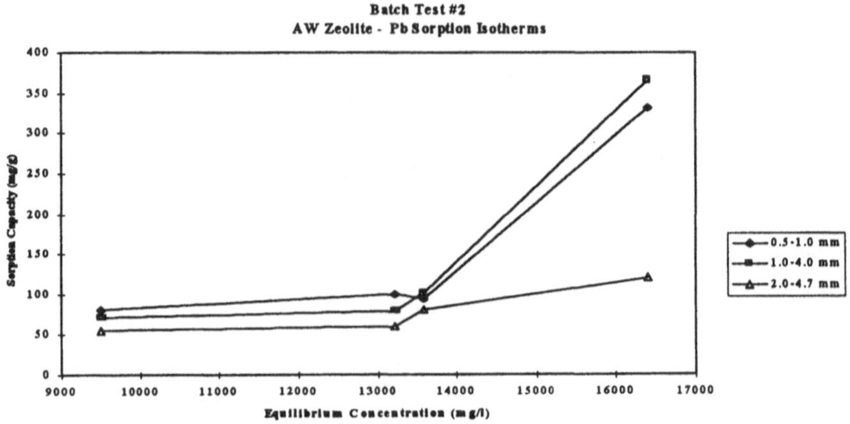


**Figure 3. Equilibrium study. AW zeolite Pb sorption isotherms.**

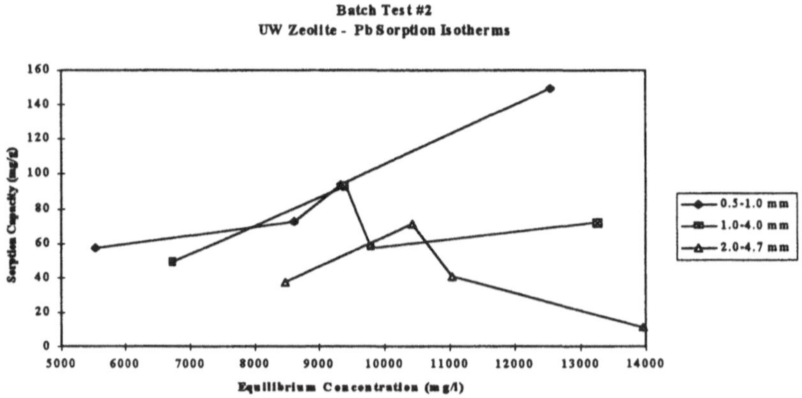


**Figure 4. Equilibrium study. UW zeolite Pb sorption isotherms.**

comparable initial concentrations, capacity of the acid-washed material was equal to or greater than the capacity of unwashed material over all particle sizes and sorbent masses tested.

Highest Pb capacity for the P-extract using 2.0 to 4.7 mm acid washed zeolite was 0.32 mg/g. Maximum Zn and Cu capacity was 0.06 mg/g. Low sorption capacity may be attributable to hydrogen ion competition due to the low pH, or to competitive sorption from other metals in solution. Analysis for other metals was not conducted. The initial contact time was two hours. The test was repeated with a 24-hour contact time with no significant improvement in sorption.

The percentage of lead removed from the FBH extract was higher than the percentage of zinc or copper. The order of removal was Pb>Zn>Cu. Maximum Pb capacity was 26.3 mg/g. Highest zinc capacity was 0.025 mg/g. Highest copper capacity was 0.044 mg/g.

*Selectivity Tests*

Results of the selectivity test indicate that Pb sorption was relatively unaffected by varying Ca concentrations (Figures 5 and 6). Sorption of Ca was slightly influenced by Pb concentration,

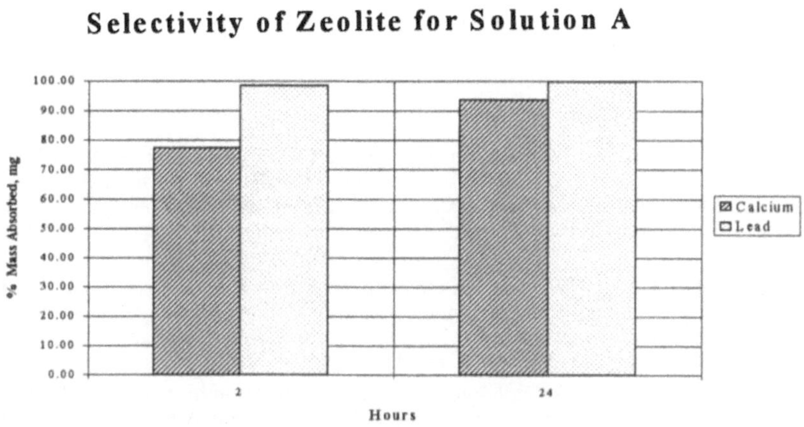


**Figure 5. Selectivity test. Solution A high Pb:low Ca.**

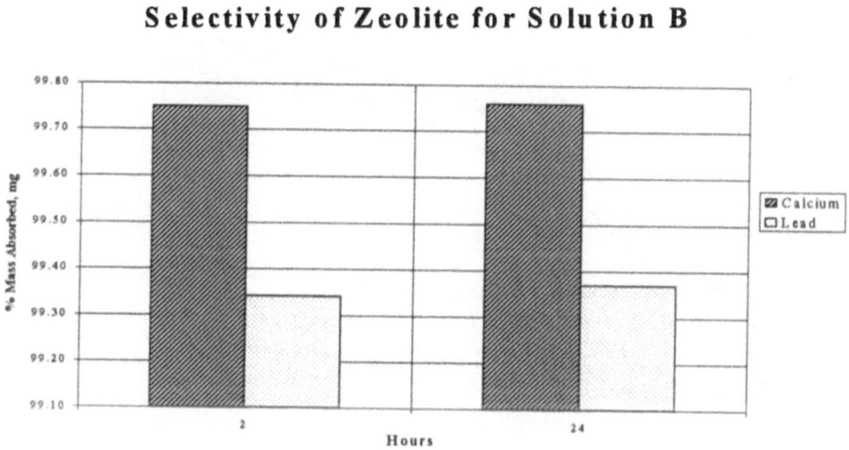


**Figure 6. Selectivity test. Solution B low Pb:high Ca.**

and was lowest for high Pb concentration and 2 hr contact time. Slightly higher Pb removal was obtained in the 24 hour contact time.

*Model Parameters*

Linear and nonlinear regression was performed on the equilibrium data for fit of Langmuir and Freundlich models. Parameter values are quite varied, and $R^2$ values indicate that neither the Freundlich nor the Langmuir are consistently good descriptors for the data. Freundlich n values greater than one may indicate a concentration dependence of demonstrated capacity. Because of the typically poor correspondence between parameters obtained from batch studies and column studies, respectively, more reliable scale-up parameters may be obtained from analysis of column performance. Kinetic data were plotted to determine order of reaction. For the synthetic Pb solution, only the data from the first 30 minutes could be satisfactorily fit, indicating a first-order reaction with $R^2$ of 0.944. Highest correlation for the Picatinny data was for a second-order reaction, with $R^2$ of 0.866.

## Column Studies

An average influent concentration was determined for all analytes from the influent samples for each column. These average values were used in the calculation of normalized effluent concentrations. Measured influent concentrations ranged from 400 to 450 mg/L Pb. Zn and Cu concentrations were approximately 2.5 and 7.0 mg/L, respectively. Influent pH was within 4.5 ± 0.06.

Performance of AW vw UW zeolite in the continuous flow studies is represented by the breakthrough curves of Columns 6 and 7 (Figures 7 and 8). Column 6 and 7 were loaded with 2.0 to 4.7 mm UW and AW zeolite, respectively, and run at 0.3 BV/hr (6 mL/min) for a period of 12 hours. Influent was sampled at 0, 3, 6, 9, and 12 hours.

The breakthrough curves for Column 5 illustrate that sorption in continuous flow was kinetically limited (Figure 9). Column 5 was loaded with 2.0 to 4.7 mm AW zeolite and run at 1 BV/hr

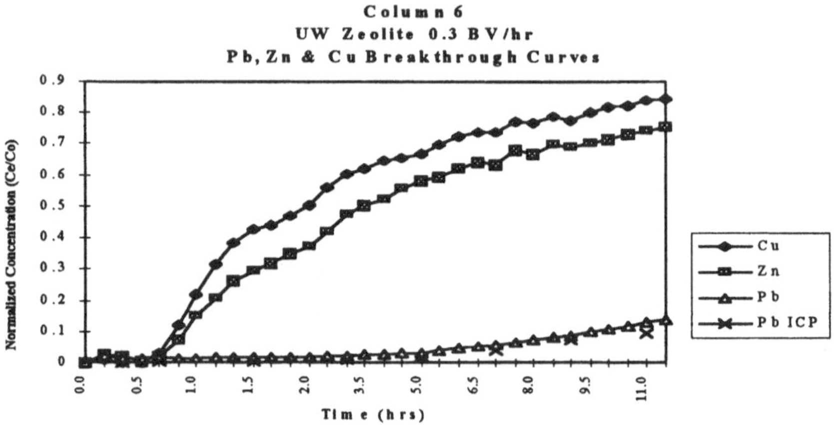


**Figure 7. Column 6 breakthrough curves.**

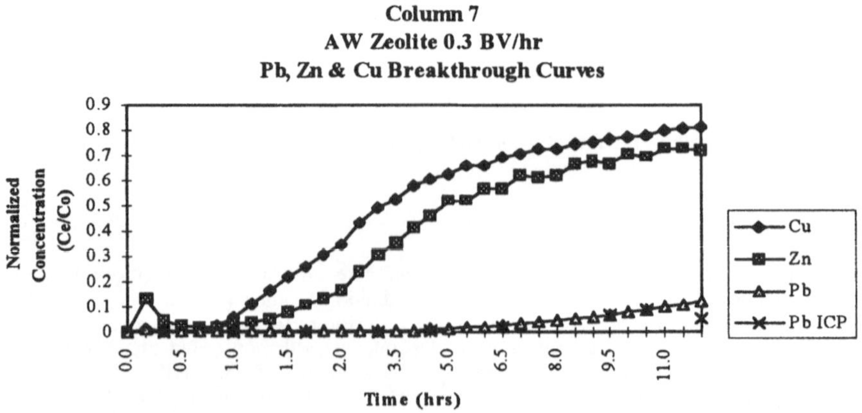


**Figure 8. Column 7 breakthrough curves.**

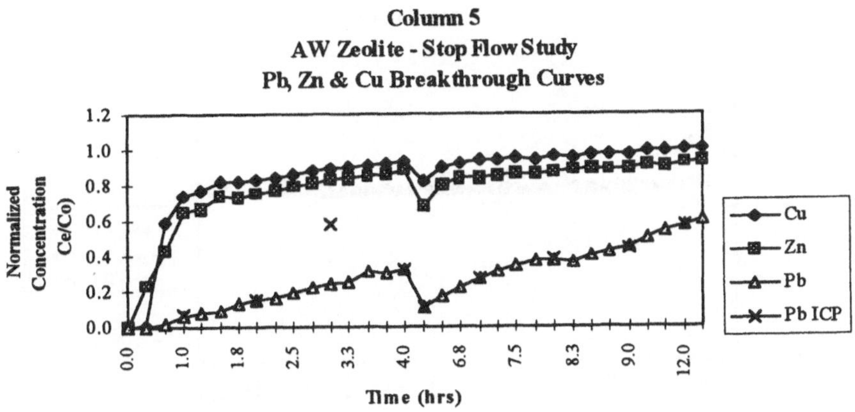


**Figure 9. Column 5 breakthrough curves.**

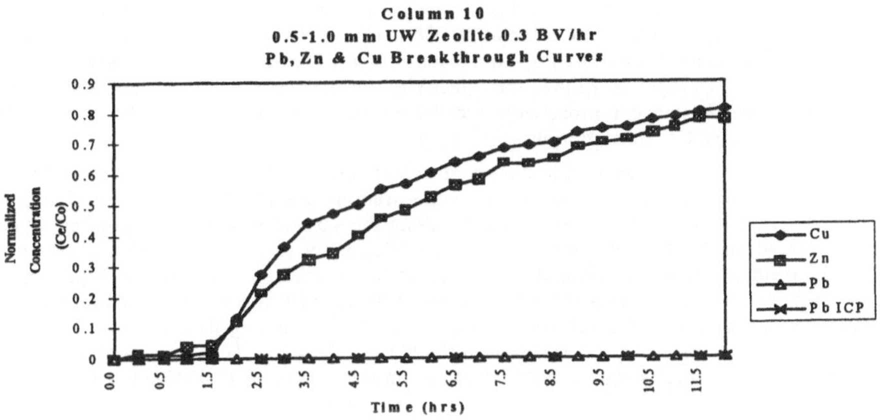


**Figure 10. Column 10 breakthrough curves.**

(20 mL/min) for a period of 14 hours, with a 2 hour flow interruption at 4 hours. Influent was sampled at 0, 3, 6, 9, and 12 hours.

The effects of particle size on the kinetics of sorption in the continuous flow system is illustrated by the breakthrough curves of Column 10 (Figure 10). Column 10 was packed with 0.5 to 1.0 mm UW zeolite and run at 0.3 BV/hr for a period of 12 hours. Influent was sampled at 0, 3, 6, and 9 hours.

The pH of effluent samples was measured to determine if metals removal was partially attributable to precipitation. Effluent pH was initially near, or slightly above, 7.0 due to pre-wetting the column with DDI water, but dropped rapidly to influent pH for both UW and AW zeolite. Change of pH with time for the columns is reflected in Figure 11.

## Discussion of Results:

Results of the batch studies indicate an apparent selectivity for Pb over Cu and Zn for both the synthetic solutions and the soil extracts, but this may have been driven by the higher relative

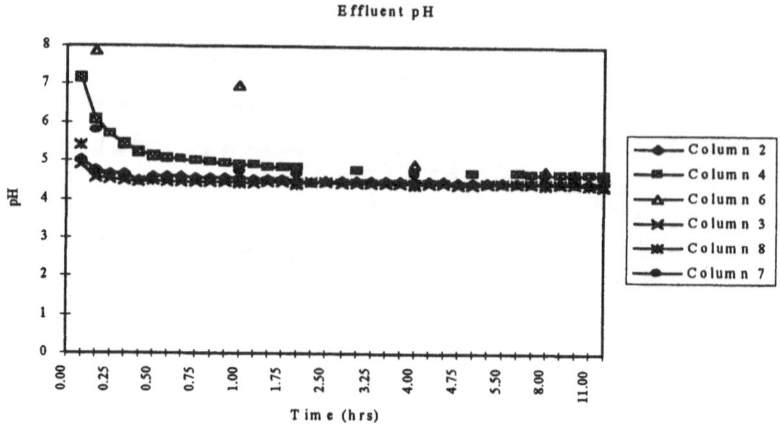


**Figure 11. Change of pH with time in column studies.**

concentration of Pb. Highest demonstrated capacity for Pb was 574.7 mg/g obtained for the unwashed zeolite using a synthetic solution at 30,000 mg/L Pb. Sorption using the FBH extract and the P extract was much lower. Highest Pb capacity for the P extract was 0.32 mg/g and for the FBH extract was 26.3 mg/g. Results of the selectivity test indicate that Pb sorption was relatively unaffected by the Ca concentration, with over 99% removal of the lead over 24 hours at both high and low relative Ca concentrations.

For all the columns, Pb transport was markedly more retarded than for Cu and Zn. At the highest flow rate (3.0 BV/hr), all analyte effluent concentrations rose rapidly to near the influent concentrations. This suggests that significant flow along the side walls may have occurred, bypassing the sorbent, or that the retention time was insufficient for significant sorption to take place. Hydraulic residence time was estimated for each of the flow rates, ranging from approximately 18 minutes to over 3 hours. This is well in excess of the contact time required for complete metals removal indicated by the batch studies, suggesting that bypassing was the most likely cause for the rapid rise in effluent concentrations at the highest flow rate. However, results of the stop flow study indicate that sorption is kinetically limited (Column 5). Effluent concentrations dropped from 138.3 mg/L Pb ($C_e/C_0$=0.32) to 48.6 mg/L Pb ($C_e/C_0$=0.11) after a two-hour period of flow cessation. Cu and Zn effluent reductions were 12% and 22%, respectively. For the UW zeolite, at 1.0 BV/hr and 0.3 BV/hr, Pb removal was 25% and 78% higher than for the high flow rate, respectively. For the AW zeolite, Pb removal was equal for the 3.0 BV/hr and 1.0 BV/hr flow rates, and 81% higher for the 0.3 BV/hr flow rate. Differences in Cu and Zn removal were less dramatic. For the UW material, 12-hour effluent Cu concentrations were 8% and 16% lower for the 1.0 BV/hr and 0.3 BV/hr flow rates, respectively. Similarly, 12-hour Zn effluent concentrations were 11% and 27% lower. For the AW material, 12-hour effluent Cu and Zn concentrations were virtually unchanged as compared to high flow rate concentrations, and were in some cases slightly higher. There was a 26% higher removal of Zn for the lowest flow rate, however, as compared to the 3.0 BV/hr flow rate. These results indicate that for the AW material, flow rate did not significantly affect Cu and Zn removal. Results of the seven day study indicate that, at 0.3 BV/hr, the zeolite was exhausted ($C_e/C_0$=0.95) at approximately 166 hours, based on the Pb breakthrough curve.

Differences in Cu and Pb removal apparently attributable to acid-washing the material were highest for the 3.0 BV/hr flow rate. Pb removal was 28% higher for AW over UW at 3.0 BV/hr, 5% higher at 1.0 BV/hr, and 18% higher at 0.3BV/hr. Cu removal was 28% higher for AW over UW material at 3.0 BV/hr, 18% higher at 1.0 BV/hr, and 4% higher at 0.3 BV/hr. Zn removal was 2% higher for AW over UW at 3.0 BV/hr, 20% higher at 1.0 BV/hr, and 8% higher at 0.3 BV/hr.

Comparison of the results of the low flow study using the 0.5 to 1.0 mm UW zeolite to the highest removal obtained with the larger material indicates a significantly higher Pb removal at

12 hours (Ce/Co=0.00 vs. 0.12). Cu removal was only slightly improved, however, and Zn removal was approximately 10% less than for the larger material at 12 hours.

*Volume Reduction Analysis*

Analysis of the results obtained for Columns 9 and 10 give some indication of the volume of liquid that could be treated to specified limits. Effluent Pb concentrations for Column 9 did not rise above 10 mg/L until after 24 hours. This represented a total volume of 8.6 liters, indicating a treatment capacity for this effluent concentration and flow rate of approximately 34.4 L/kg. Effluent Pb concentrations for Column 10 were at or below detection for the 12-hour duration of the study. This represented a total volume of 4.3 liters. Capacity based on this volume does not reflect the total capacity of the sorbent, since the duration of the study was insufficient to develop breakthrough. Capacity calculations based on this volume can be taken to be a very conservative estimate of the material capacity, and preliminary volume reduction analysis conducted using these values. Treatment capacity based on a total volume of 4.3 liters and zero Pb concentration in effluent, is 16.7 L/kg zeolite, or 7.1 mg Pb/g zeolite.

The specific gravity of clinoptilolite ranges from 1.5 g/cm$^3$ for reagent grade zeolite, to between 2 to 2.4 g/cm$^3$ for the naturally occurring mineral. Assuming a SG of 2.0, and a sorption capacity based on the total volume of extract treated in Column 10, treatment of 1000 liters of water at 425 mg/l Pb would result in a volume reduction of 97% (1,000 liters water vs. 30 liters zeolite) or a weight reduction of 94% (1000 kg water vs. 60 kg zeolite). Natural zeolite is available for approximately $94./Mg ($85.-100./short ton), or $0.18/kg ($4./50 lb) in smaller quantities, plus freight. Disposal in a hazardous waste landfill is estimated at approximately $50./200 liters. For the example developed here, and assuming a single use, zeolite cost would be approximately $5.60-$10.60/1000 liters influent. Disposal cost would be $7.50 for 30 liters of spent zeolite. This gives a total treatment cost of approximately $13.-18./1000 liter, as compared to $250./1000 liters to dispose of the extract directly, a 93 to 95% cost savings.

The cost to treat the same volume of this extract with a commercial ion exchange resin was also estimated. Because the resin was not tested with this waste, capacity was estimated based on published values for the resin, Amberlite IRC-718. Conditions at which the capacity of the resin was determined differ from those under which the capacity of the zeolite was determined, and this should be strongly considered in the final analysis. The capacity of the resin is based on a breakthrough value of 0.5 ppm Pb, for a waste stream containing 50 ppm Pb, as compared to zero breakthrough for a waste stream containing 425 ppm Pb for the zeolite. Flow rate was 8 bed volumes per hour for the resin, and 0.3 bed volumes per hour for the zeolite. A definitive cost comparison would require determining the ultimate capacity of the fine zeolite under low flow conditions (which would give a lower unit cost than that estimated here), and the actual capacity of the resin for the same extract and flow rates.

Capacity for the Amberlite resin is given as 3.6 lb/ft$^3$ for a 50 ppm Pb influent, 0.5 ppm breakthrough, 8 bed volumes per hour flow rate, at pH 4, as previously stated. Regeneration cost is estimated assuming 10 lb HCL at a concentration of 5% per cycle, and a resin life of 100 cycles. Purchase cost of the resin ranges from $420. to $435./ft$^3$, with a minimum purchase requirement of $1,000.00. Estimated treatment cost for 1,000 liters of soil extract using the Amberlite resin is approximately $3.90/1,000 liters. This cost estimate is contingent upon the following assumptions: the minimum purchase requirement is not a factor, and the total volume of extract treated over the life of the resin is 101,000 liters. If smaller volumes of wastewater were treated, the unit cost would be higher. For example, if only 1,000 liters were treated, the unit cost would be approximately $115.-$1,000/1,000 liters, depending upon whether the minimum purchase requirement applies.

## CONCLUSIONS

The highest capacity demonstrated for lead sorption in the batch studies was 574.7 mg/g for the 1.0 to 4.0 mm UW material at an initial Pb concentration of 30K mg/L. Kinetics of sorption appear to be relatively rapid, with over 99% removal of Cu, Pb, and Zn metals in the first 15 minutes of contact. Zeolite demonstrates a very strong preference for Pb over Cu and Zn, and is effectively sorbed in competition with Ca, even at high relative concentrations of Ca.

Results of the column studies indicate that, for the 2.0 to 4.7 mm material, acid-washing of the zeolite does improve sorption characteristics. Comparison of the best results obtained in the column studies using the 2.0 to 4.7 mm material to the results obtained using the 0.5 to 1.0 mm material suggest that sorption is kinetically favored for the fine zeolite, which has a much higher capacity under continuous flow conditions at the low flow rate tested. Further capacity increases could potentially be developed by acid-washing the fine zeolite. However, review of the batch studies does not indicate a consistently higher capacity for the fine zeolite over the coarser zeolite. It is likely that natural variations in the material may be partly responsible for the differences in performance, and that longer contact times will also minimize performance variation between particle sizes. Further column tests would need to be conducted to make a more definitive determination in this regard.

The volume reduction analysis indicates that significant mass and volume reduction can be achieved with zeolite in the treatment of aqueous waste. Cost comparison to commercial ion exchange resins indicates that for single use treatment of small volumes, zeolite is significantly more cost-effective than ion exchange. At higher volumes, ion exchange is likely to be more cost-effective if regenerated for multiple use cycles. This determination should be made by capacity testing of both sorbents for the specific waste stream. Further analysis should be directed at determining the ultimate capacity of the fine zeolite at low flow rates, the comparative performance of commercially available zeolite with a higher indicated CEC, and comparative costs of an ion exchange resin for treatment of this waste stream.

## ACKNOWLEDGMENT

Information presented herein, unless otherwise noted, was obtained under the SERDP at the WES. Permission was granted by the Chief of Engineers to publish this information.

## REFERENCES

1. Desborough, G.A. Some Chemical and Physical Properties of Clinoptilolite-Rich Rocks, Open-File Report 96-265, U.S. Department of the Interior, U.S. Geological Survey, Denver, CO, 1996.
2. Rhodes, J.D. *Methods of Soil Analysis, Part II, Chemical and Microbiological Properties*, 2nd ed., Chapter 8: Cation Exchange Capacity, American Society of Agronomy, Madison, WI, 1982.
3. Ming, D.W. and Dixon, J.B. "Quantitative determination of clinoptilolite in soils by a cation exchange capacity method," *Clays and Clay Minerals*, 35, pp. 463–468, 1987.

# 58 CLEANING MODIFICATION STEPS THAT RESULTED IN POLLUTION PREVENTION AND WASTE MINIMIZATION

**Raymond Lizotte**, Senior Environmental Specialist
Texas Instruments Inc.
Materials & Control Group
Attleboro, Massachusetts

**Ken Hogue**, Graduate Student
Department of Civil & Environmental Engineering
Worcester Polytechnic Institute
Worcester, Massachusetts 01609

**James O'Shaughnessy**, Professor
Department of Civil & Environmental Engineering
Worcester Polytechnic Institute
Worcester, Massachusetts 01609

## ABSTRACT

This chapter reviews four metal cleaning case studies at various manufacturing operations and presents the different approaches and technology assessments that were used to change these metal finishing operations. In each of the case studies, both pollution prevention and a positive financial solution were achieved. Each case was site- and time-specific, indicating that pollution prevention is a proactive and on-going optimization problem, similar to the actual manufacturing processes. The solutions included chemical substitution, process modification, process modernization, housekeeping improvements, and recycle improvements.

## INTRODUCTION

From 1950 to 1990, cleaning with halogenated solvent was the preferred technique within industry for cleaning material when quality, speed, and versatility were required. Beginning with the passage of the Clean Air Act Amendments of 1990, a number of environmental rules have been enacted that are pressuring manufacturers to find alternatives to these processes.

The first rule to impact halogenated solvent cleaning were the provisions of the Clean Air Act that mandated phaseout of the chemicals that have been implicated as harmful to the stratospheric ozone layer. The ozone depleting halogenated solvents (primarily CFC-113 and 1,1,1-trichloroethane) were popular cleaning solutions throughout the electronics industry. In addition to a general phaseout in the production of ozone depleting solvents, the rule required manufacturers to pay annually escalated taxes, and place warning labels on these products. By 1994, most users of ozone depleting halogenated solvent had implemented nonozone depleting substituted technologies.

In December 2, 1994, the EPA established emission limitations on trichloroethylene (TCE) vapor degreasers. This regulation, called the Maximum Achievable Control Technology (MACT) Standard for Halogenated Solvent Cleaners, requires each degreaser to demonstrate compliance with the emission limits by December 2, 1997. Failure to meet the limits would require shutdown of noncompliant degreasing operation.

*52nd Purdue Industrial Waste Conference Proceedings*, 1997, Ann Arbor Press, Chelsea, Michigan 48118. Printed in U.S.A.

This chapter presents the results of a number of case studies at different manufacturing locations in both the electronics industry and the metals finishing industry to meet the new standards. The focus of these studies was evaluating a number of pollution prevention/waste minimization approaches. Technologies evaluated include:

1. aqueous washing
2. product substitution
3. process modification
4. equipment modernization using high efficiency c onventional vapor degreasing
5. equipment modernization using high vacuum vapor degreasing

The first case involved printed circuit board cleaning and ended in chemical substitution of cleaning agents. The existing saponifier solution that was in use at an existing facility would have required a capital investment of $745,000 in wastewater pretreatment cost in a planned facilities move. The major problem in this case was the high $BOD_5$ concentrations associated with the cleaner. Various alternative cleaners were evaluated. The new cleaning agent had a low BOD and COD, and is nonchelating. The product does not contain VOCs, is nonflammable, and has a moderate pH, which was important to a solder step involved in the production process. Metal hydroxide sludge production was reduced from 66,000 to 2,500 pounds per year.

The second case focused on CFC substitution for hand-cleaning circuit boards. As a replacement to the use of Freon 113 in the cleaning of soldered components, four alternatives were evaluated. These methods were:

1. the use of isopropyl alcohol
2. aqueous based cleaners
3. not cleaning the circuit boards
4. using "CFC-free" products

In this instance isopropyl alcohol cleaning was adopted, with a disposal cost of $85 per drum. The new waste streams are blended with other fuels and used in heating and incineration processes.

The third case evaluated various technologies as a substitution for Freon that was used in vapor degreasers. In this case study, aluminum racks were being cleaned in vapor degreasers using Freon TE. Two aqueous base process substitution solutions were employed. The first modification was the installation of a conveyor in-line aqueous cleaning system. This system was used for post- wave- solder flux removal, and preconformal coat cleaning. The second system was an aqueous dip-tank cleaning system used to clean aluminum racks prior to painting. These modifications resulted in the reduction of Freon usage by more than 20,000 lb/year and a three year cost savings of $375,000. These modifications also reduced cooling water usage requirements more than 15,000 gal/day.

The fourth case study involved vapor degreasing modification and/or elimination. In this case study, a diverse metal finishing company with a historical reliance on trichloroethylene (TCE) vapor degreasing has conducted a major project to find substitute cleaning techniques. Four technologies were chosen for in-depth investigation in this case study. They were:

1. aqueous washing
2. high efficiency conventional vapor degreasing
3. high vacuum vapor degreasing
4. nonaqueous solvent washing.

The process by which the company assessed each of these substitute technologies has been presented. The outcome of the evaluation resulted in the implementation of each of the investigated options within the company. However, due to wastewater discharge constraints and aqueous washers' inherent lower versatility (average loss of 10% on yield rates), aqueous washing has not been widely implemented. Due to various reasons, high vacuum vapor degreasing and nonaqueous solvent washing have been the new technologies of choice.

## CASE STUDY #1 - AQUEOUS CLEANING OF PRINTED CIRCUIT BOARDS

Cleaning Circuit Boards—After wave soldering operations, PC boards need to be cleaned to remove flux and other contaminants such as oil and dirt. This is usually done in a conveyorized in-line cleaning machine called an aqueous cleaner. The principal steps are washing, rinsing, and drying. Cleaning performance depends on factors such as: the level of contamination, pH, temperature, agitation, and rinsing capability. PC board design also is an important factor. Cleaning chemicals tend to have a high surface tension, and may not penetrate holes readily.

The cleaning solvent is selected based upon the following criteria:

- ability to remove grease, oil, dirt, flux
- safety
- cost
- ability to remove both ionic and nonionic contamination
- ability not to damage the materials/parts being cleaned
- waste treatability.

During the previous 16 years, over a million circuit cards were produced using a saponifier solution that was essentially a 50:50 mix of ethanolamine and ethylene glycol monobutyl ether. The cleaning process took place after wave soldering operations, and/or before urethane coating of printed circuit boards, to remove flux, dirt, and grease. It was an in-line automated process taking place in either of two conveyorized "aqueous cleaning machines" about 12 feet long. The process consisted of a recirculating wash section, blow-off stations, recirculating rinse, cascading deionized water rinse, and a drying section. The process operated at 150°F. Annual usage of the saponifier solution is shown in Figure 1.

The existing wastewater treatment process produced wastewater from the two aqueous cleaners (maximum flow of 6,000 gal/day) and was pumped to one of two 1,500 gallon concentrated waste surge tanks. This concentrated wastewater had high $BOD_5$ concentrations as well as copper, chrome, and zinc content. The wastewater was metered into the main effluent stream (about 125,000 gal/day). The combined effluent stream was then pH adjusted with lime. Sodium sulfide was added to aid metals precipitation. Polymer was added to improve flocculation. Sludge was settled out in a plate separator. The filtrate overflowed to a weir box, where it received a final pH adjustment ($H_2SO_4$) before release to a POTW. The sludge was pumped by diaphragm pumps to a filter press and disposed of offsite as a hazardous waste. The existing treatment process is depicted in Figure 2.

In 1994 a proposed relocation of the aqueous cleaner production line to a different municipality would find the effluent from the cleaning process unable to meet the new POTW's sewer discharge Chemical Oxygen Demand (COD) limits without pretreatment. The new discharge limits COD = 2,000 mg/L, BOD = 1,000 mg/L, and oil/grease = 100 mg/L. The saponifier solution has

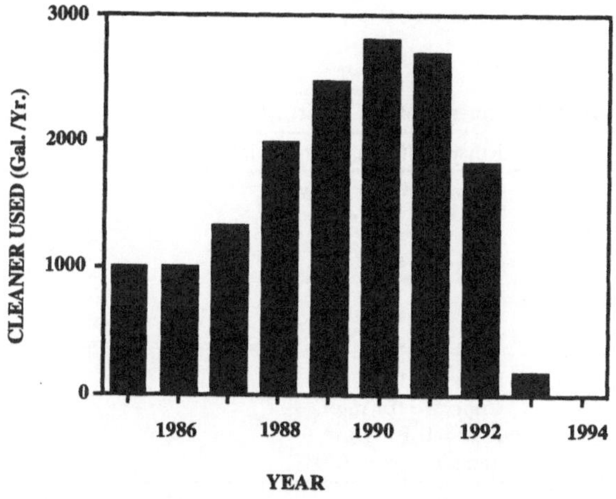


**Figure 1. Amount of saponifier solution used annually.**

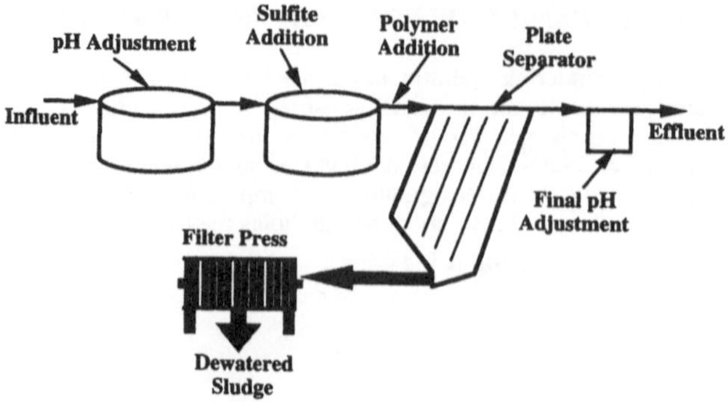


**Figure 2. Existing wastewater pretreatment system.**

a COD as high as 106,000 mg/L. At the original location, sufficient dilution from other processes reduced COD discharge levels. Wastewater flows at the original location averaged 125,000 gal/day, and were projected to total only 6,000 gal/day at the new site. Also, new VOC provisions of the Clean Air Act would affect the use of the saponifier cleaner solution.

Other drawbacks to the saponifier cleaner were its moderate toxicity and that its amine-glycol chemistry attacked the cleaning machine plumbing and pump seals, necessitating repairs and downtime. In addition, it was difficult to recycle in a closed-loop system, and required extensive treatment before discharge to the town sewer. Wastewater treatment would have to be considered when evaluating alternatives.

Further analyses of the problem material balanced around the processes and quantified the waste streams. The components of the waste streams from the aqueous cleaner machines is listed in Table I. A treatability study was performed on the cleaning waste, in preparation for moving to the new site. Results indicated that the annual sludge volume would be 136 drums. The high chelating properties of the saponifier solution increased the solid waste generated. In addition, ferric chloride would have to be added as an additional co-precipitator. The estimated cost of disposal for this solid waste was $300 per drum.

## Criteria for Selecting a Substitute Process

The following key attributes were required for any replacement process.

- chemistry that is nontoxic and nonflammable.
- airborne emissions that would not be polluting, and not regulated.
- chemistry that could be recycled or treated for sewer discharge.
- ideally, a "drop-in" replacement into existing process equipment.
- cost (last on the list, but near the top consideration).

**Table I. Composition of the Aqueous Wash Waste Streams**

| Wash Solution Characteristics | Rinse Solution Characteristics |
|---|---|
| 5% Cleaning solution | Flow = 24,000 gal/wk |
| Flow = 280 gal/wk | |
| Cleaning solution is | |
| 50% ethanolamine & | |
| 50% ethylene glycol monobutyl ether | |
| COD = 93,800 mg/LCOD = 4,400 mg/L | |
| pH =10.9 | |
| Lead = 3.0 mg/L | Lead = 0.1 mg/L |
| Copper = 2.7 mg/L | Copper = 0.1 mg/L |
| Zinc = 6.7 mg/L | Zinc = 0.1 mg/L |
| Temperature = 150°F | Temperature = 150°F |

The following treatment technologies were considered to replace the existing process;

- granular activated carbon,
- oxidation,
- biological treatment, and
- hauling the untreated wastewater away.

*Granular Activated Carbon (GAC)*

The COD concentration in the combined wastewater was 7,730 mg/L. Activated carbon is typically cost-effective for COD concentrations below 500 mg/L. Experience has shown that carbon will hold about 10% of its weight before breakthrough occurs. Based on this analysis, carbon consumption was estimated to be more than 2,000 pounds per 8 hours of operation. The price of carbon was $1.50, o that daily cost per eight hour shift would be $3,000.

*Oxidation with Ozone*

This alternative was only briefly considered. An electric discharge ozone plant would have high electricity consumption. To reduce the electrical consumption, oxygen could be purchased, but that would require an external tank, as well as truck unloading. Final cost estimates, in addition to the storage tanks, and truck unloading were unattractive to management.

*Hauling Untreated Wastewater Offsite*

This option didn't receive much consideration. The visual impact of tanker trucks hauling 6,000 gallons of wastewater away per day didn't sit well with upper management. Management wanted to maintain a clean image in the community, and within the rest of the corporation. Other considerations were the cost and liability involved. Esthetics, more than any other reason, negated this option.

*Initial Recommendations*

A continuous wastewater treatment system which contained metals removal and sludge dewatering processes would be required due to lead concentrations. An atomic absorption spectrophotometer would be required to analyze the waste stream. The equipment and engineering cost for a "basic" water and air pollution control system was estimated to be $1,009,000. This cost does include a building. Beyond that, an additional $373,000 would purchase a new aluminum processing line, $190,000 to purchase chromate and passivate treatment, and $55,000 for a 18 MEG, 10 gpm process water treatment system. This cost estimate caused plans for moving to the new site to be revaluated. An alternative cross flow metal precipitation system, followed by a sand filter, and final pH adjustment would cost $745,000. This cost was based on a 35 gpm wastewater flowrate. It would require 11,000 square feet, and increased ceiling height, which were very significant drawbacks. Simultaneously, another study identified a precipitation system and sand filter that cost $180,000. Annual operating costs were estimated at $187,000, and hazardous waste disposal costs at $26,000. As a result of these cost estimates, another potential solution was given consideration: chemical substitution.

*Chemical Substitution*

A search for equipment-compatible cleaning agents yielded very few non amine-glycol based products. This list of potential products was then evaluated based upon health, safety, environmental considerations, cost of material, and shelf-life. Semiaqueous cleaning solutions have emerged as alternatives to Freon degreasing and aqueous cleaning systems. They are effective for grease, tar, waxes, and oils, due to their strong solvency in hydrocarbons.[4] Semiaqueous systems use terpenes, dibasic esters, glycol ethers, and other hydrocarbons combined with surfactants. Alkaline based products are also on the market. They are CFC-free, are nonflammable, and do not pose any organic respiratory hazards. However, due to their high pH, they require special handling and can affect the pH of the wastewater stream. Betz Laboratories has introduced their Betz-Kleen™ line of alkaline cleaners, and claim that they are less expensive than solvent based systems.[5] Another chemical substitute considered was an alcohol based product; however, flammability and BOD considerations were a liability for this product.

The chemical product was Armakleen 2000 manufactured by Church and Dwight Company. This baking-soda based product had low BOD, COD, and was nonchelating. Chelating compounds tend to foul waste water treatment processes.[6] The product contained no VOCs, was nonflammable, and a moderate pH, and would not affect the shine of the solder. This product was evaluated using the in-house materials laboratory. This evaluation measured basic physical properties in the vendor-recommended operating concentration and temperature range. Compatibility testing with the machine plumbing and pump seals was also conducted.

A pilot study was devised where the new product would be tested in one of the two aqueous cleaning machines. The second machine would serve as a baseline for comparison. All testing was performed in split lots. Specific criteria were developed for evaluating the new product. The test criteria included:

- visual inspection for flux residue at 10X
- MIL-P-28809 cleanliness/temperature/humidity testing
- surface insulation resistance electrical testing
- high density designs were taken apart to visually inspect for flux residues in tight places
- unsealed and sensitive components were electrically tested after four passes through the system.

The results of these pilot tests indicated that the new cleaning product performed as well as the original cleaner, with no damage to the boards. Prior to performing large-scale tests on saleable product, a proposal was made to both management and the customer. After approval was received, production testing was performed in a manner similar to the above-described test procedure. No cleanliness level failures or abnormal electrical test results were observed.

Treatability study of the new cleaning process predicted the generation of 4-5 drums of metal hydroxide sludge per year. In addition, the new process also reduced the amount of chemicals used in the wastewater treatment plant to treat the wastewater. Historically, the old process generated 132 drums (66,000 pounds) of sludge per year. However, production rates were higher.

### Case Conclusions

The new product has been in use for over a year (over 20,000 circuit cards), and no deficiencies/problems have developed. The process development group is now assessing the cost benefits of recycling the water used in the cleaning process. Capital investment for this project was near zero, and material costs for the old and new cleaning products were nearly equal. Therefore, the economics boiled down to waste disposal costs. Waste disposal costs were not easily compared because production volumes had also decreased dramatically. However, the annual costs are relatively small. The sludge disposal cost is $205 per drum. Recent sludge generation rates have been:

| | | | |
|---|---|---|---|
| 1994 | 1600 lb | 4 drums | $820 |
| 1995 | 2000 lb | 5 drums | $1025 |

The major cost consideration for this project was the savings of approximately $180,000 of capital spending for a BOD treatment facility, which would have been necessary if the switch to a low BOD cleaning solvent was not undertaken.

## CASE STUDY #2 - CFC SUBSTITUTION FOR HAND CLEANING CIRCUIT BOARDS

Hand cleaning of circuit boards typically occurs after an operator inserts components that are not installed mechanically. Freon was the cleaning solvent of choice. Freon was nonflammable, not toxic, dried quickly, was inexpensive, and cleaned very well. Currently Freon use is heavily regulated, and many manufactures have switched to aqueous products, CFC-free products, and/or isopropyl alcohol. The major drawback associated with isopropyl alcohol is its flammability.

## Existing Process

Freon 113 and various CFC based aerosol products were used to clean circuit boards after hand soldering processes. Hand soldering would be performed to insert MSRPs, or to make repairs. Freon 113 was the predominant product used and was dispensed in various plunger cans, then applied to the boards by either dipping or wiping. Approximately 30 drums/year of Freon were used in this manner, generating about 6 drums of waste Freon annually. In addition, a CFC aerosol product was used. This aerosol product contained 80% trichlorotrifluoroethane and 20% chlorodifluoromethane. This product would be sprayed onto the board to remove the flux. CFC aerosols worked very well, dried fast, were nontoxic and nonflammable, and for a while, were inexpensive. Use was reported as 300 twenty ounce cans of this material per year. Other than air emissions, there was no waste generated.

The annual material costs (1990) for the existing process were; bulk Freon = $60,000; CFC = aerosol products = $1,500; waste disposal = $0. The total annual cost was = $61,500. There was no charge for the Freon waste, since the waste disposal company could recycle and resell the product.

## The Reason for Change

CFCs were identified as ozone depleting chemicals, and the Montreal Protocol was signed. This required the labeling of products made with ozone depleting chemicals, and federal regulations drove up the prices. A substitute chemical or cleaning process needed to be found. In 1990, a "Freon Task Force" was formed with the mission of eliminating the use of Freon, while minimizing cost and impact to operations. In this case a "CFC" replacement team" comprised of operations personnel was formed. The following substitutes were identified by the team:

- not cleaning the boards
- the use of a solvent (isopropyl alcohol)
- aqueous based cleaners
- the use of "CFC-free" products

### *No Clean Option*

This option was dismissed as most parts were covered by military specifications and required cleaning.

### *Solvent-Based Systems*

Offer good cleaning of oil, grease, and waxes. They dry quickly, are effective in tight areas, and are reclaimable. Their drawbacks are that they may be combustible, moderately toxic, produce odors, are VOCs, and are impacted by SARA 313.

### *Isopropyl Alcohol (IPA)*

This solvent is allowed under the military specifications. IPA has several desirable qualities. It adequately dissolves flux from PC boards, dries in a reasonable period of time, has a reasonably low toxicity, and is inexpensive. Waste IPA is also inexpensive to dispose of. In 1996, it cost $85 per drum to dispose of waste (flux laden) IPA. The waste has a high btu value and incinerators like the flammable property of this material.

The major drawback to IPA is its flammability. It does pose additional risk at the workplace, in both handling and storage of the bulk material and at the workbench. It's flammability also precludes its use in enclosed systems and in vapor degreasers. The cost to use IPA was estimated in 1994 to total $3,340 per year as follows:

| | | |
|---|---|---|
| IPA Cost | 25 drums | $1,300 per year |
| Waste Disposal | 24 drums | $2,040 per year |

*Aqueous-Based Cleaners*

Opinions vary, and often production personnel feel that aqueous-based cleaning systems still don't provide the same cleaning efficiency as was achieved when using ozone depleting systems. However, new cleaning technologies are quite effective. Several vendors use ultrasonics with aqueous technologies to improve cleaning efficiencies. One such system uses ultrasonics, two counterflow rinses, a final hot rinse, and a hot air dryer. The problems with these systems are complexity, capital cost, operating costs, process time, and the aqueous waste stream that is generated.

*CFC-Free Products*

LPS Laboratories has a non-CFC alternative which is an aerosol based mixture of heptane, acetone, methyl cyclohexane, cyclohexane, toluene, and p-amyl acetate. The propellant is carbon dioxide. This product is used in small amounts. The last cleaner evaluated was ACTREL, developed by Exxon Chemical Company. Actrel is not regulated under SARA Title 113 and is not considered a "Hazardous Air Pollutant." It is used for precision metal, plastic, and elastomer cleaning processes. The spent material can be redistilled or used as a fuel substitute.

### Case Conclusions

IPA was selected for bench-scale cleaning of PC boards. Purchases varied from 20–30 drums of IPA per year, and in 1994, 1210 gallons of waste IPA were generated. In 1995, 990 gallons of IPA waste were generated; however, the waste disposal costs are insignificant compared with the raw material cost savings, about $58,000 per year. The economics of this project were very favorable due to the escalating cost of Freon, but also because there was essentially no capital investment required. A few years prior, before Freon prices escalated, the cost reduction would have been much less. In this case taxation was the driving factor.

## CASE STUDY #3 - ELIMINATING VAPOR DEGREASERS

In order to eliminate the use of Freon in vapor degreasers, several projects were conducted concurrently. These included the following:

- Training — This task was easy to implement, and resulted in a large reduction in the use of Freon.
- Freon recycling — This project was implemented to reduce the company's overall consumption of Freon.
- Alternative technologies — Some vapor degreasers were replaced by an aqueous cleaner machine, other vapor degreasers were replaced with an aqueous dip-tank process. This project underwent further waste minimization refinement due to water consumption.

All of the alternative technology projects involved a capital outlay, but they all combined to eliminate the use of Freon, and reduce operating costs. These cost reductions were aided by the high cost of Freon, but also by other cost-saving ideas which were uncovered and implemented as part of these projects.

### Existing Process

Aluminum parts and circuit cards were being cleaned in vapor degreasers using various Freon based products. The total usage of Freon products peaked at 4,655 gallons in 1989. Freon usage is summarized in Table II. Freon waste disposal costs were negligible, as hazardous waste disposal companies credited the waste Freon, due to its reclamation value.

### Training

In 1990 the Task Force instituted a technician retraining program. This training taught vapor degreaser operators correct immersion and removal techniques, startup and shutdown proce-

**Table II. Amount and Cost of Freon**

| Year | Freon Purchased (gal/yr) | Cost ($/gal) | Total Cost ($/Year) |
|---|---|---|---|
| 1987 | 4055 | $ 15.73 | $ 63,800 |
| 1988 | 4370 | $ 15.97 | $ 69,800 |
| 1989 | 4655 | $ 22.22 | $ 103,000 |
| 1990 | 3235 | $ 39.81 | $ 129,000 |
| 1991 | 2240 | $ 43.40 | $ 97,200 |
| 1992 | 735 | $ 54.67 | $ 40,000 |

dures, and stressed keeping the degreaser cover closed when the equipment was not in use. Gasketing material was also added to the underside of the cover.

## Freon Recycling

In 1990 an expanded in-house solvent recycling program was started. This program reduced new product purchases by 35%. A total of 1,120 gallons of Freon was recycled, saving $44,500. Capital costs were $10,000. The recycling process involved redistilling waste Freon material in one of four packaged batch recovery stills. Only the cleaner material was redistilled. There were capital and operating costs associated with the equipment, but due to the cost of virgin material, it was an economical alternative.

## Aqueous Cleaner (Process Substitution)

Much of the rework and special product PC board work was cleaned in one of several vapor degreasers located throughout the plant. One aqueous cleaning machine was used primary for regular high-volume production runs. In 1991, an additional conveyorized in-line aqueous cleaning system to be used for rework and special project board cleaning was purchased and installed. The intent was to eliminate the use of many vapor degreasers. This process change saved 960 gallons of Freon ($52,000) per year. This single project reduced Freon consumption by 42%. The capital cost for the aqueous cleaner was $100,000.

## Aqueous Dip-Tank Process (Process Substitution)

In 1993, an aqueous dip-tank cleaning system used to clean aluminum parts prior to painting was purchased and installed. This process replaced several more vapor degreasers, saving an estimated 640 gallons ($35,000) of Freon annually The new equipment also replaced aging etching and chromating equipment. The capital cost for installation of the new dip tank line was $124,000. Water consumption (used for condensation in many of the vapor degreasers) was reduced by 15,000 gallons/day ($12,000/year). Water costs this facility about $2.38 per 100 cubic feet. The process diagram for the aqueous dip-tank system is shown in Figure 3.

Each of the 10 polypropylene tanks holds 250 gallons of solution. The system was run 8 hours per day. The four rinse tanks each produced 28 gpm of overflowed. Tanks #1, 3, and 10 operate at 90°F. Vapors from these heated tanks were ventilated to a rooftop packed tower water-scrubbing system. Scrubber effluent was produced at the rate of 15 gallons per hour. In 1994, the aqueous dip-tank process was modified. The four rinse tanks which were previously overflowing at 28 gpm each were modified to reduce the wastewater flow. In essence, the four tanks were converted to eight tanks, operating in pairs in a countercurrent manner, as shown in Figure 4. In countercurrent operation, 4 gpm of water is added to the second tank , which then overflows to the first tank. About 4 gpm of water then overflows the first tank to the wastewater treatment plant. The net effect of the two-stage rinsing is that the first tank has higher concentrations of wastes, and the second tank stays relatively clean. This process modification saved 12 million gallons of water ($38,000) per year. Capital costs were estimated at $20,000.

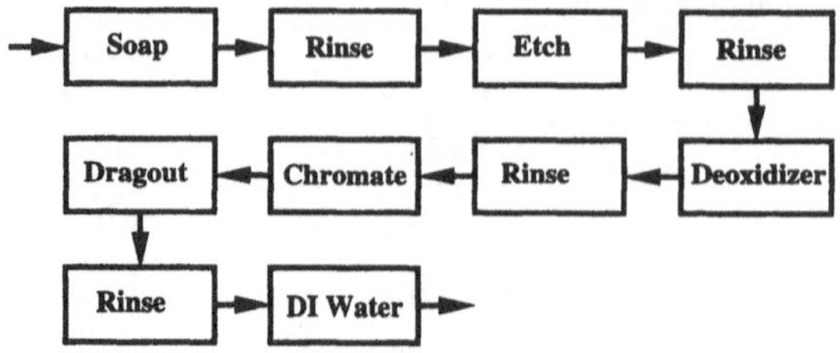


**Figure 3. Aqueous rinse tank system.**

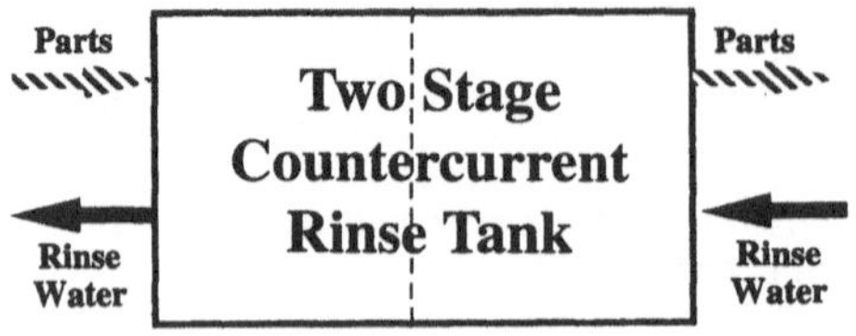


**Figure 4. Countercurrent rinse tanks.**

## Outcome

As a result of these task team efforts, by 1994 Freon usage was reduced to zero from a high of 4,655 gallons in 1989. Annual Freon cost savings for 1990 through 1993 were estimated to be as follows:

| Year | Annual Savings | Cumulative Savings |
|---|---|---|
| 1990 | $44,500 | $44,500 |
| 1991 | $96,500 | $141,000 |
| 1992 | $96,500 | $237,500 |
| 1993 | $131,500 | $369,000 |

In addition, water usage costs were reduced by $50,000 per year. The total savings was $569,000. Capital costs totaled $254,000 over the 4 years of the project. Therefore, on a total project basis, this project saved the company $315,000. These figures do not include the cost of the engineering project work, and the cost of the in-house Freon recycling project. Costs associated with the recycling project include: utilities, manhours, equipment cost, material handling, loss of cleaning efficiency, cost to dispose of still bottoms, training, etc. The total cost savings, over a four year time span, for these three cases are estimated to be $371,000

## CASE STUDY #4 - VAPOR DEGREASING MODIFICATION OR ELIMINATION

This case study occurred at the Texas Instruments, Materials and Controls Group manufacturing facility located in Attleboro, Massachusetts. Involved in the case study were the pollution prevention teams of many of the manufacturing divisions. The Environmental, Safety, and Health organization facilitated the progress of the pollution prevention teams in addition to providing specialized expertise. In all, six teams composed of 50 engineers participated.

The TI facility is classified as a major, complex source in regard to air quality regulations. In addition to being a major source for the combustion criteria pollutants (CO, NOx, SOx), the site is also classified as major for the emissions of volatile organic compounds (VOCs) The majority of the VOC emissions are trichloroethylene discharges from the site's trichloroethylene vapor degreasers. Emissions of trichloroethylene from vapor degreasers also make the site subject to the halogenated solvent cleaning MACT standards.

As stated earlier, pollution prevention teams were a major resource conducting the substitute cleaning technology assessments. TI has been utilizing the pollution prevention investigating

**Table III. Texas Instruments Pollution Prevention Teams**

| Team # | Operation | No. of Degreasers | Degreaser Type |
|---|---|---|---|
| 1 | MC & PC Disc Cleaning | 1 | Batch, loose parts |
| 2 | IMT Parts | 1 | Batch, loose parts |
| 3 | Materials Business | 7 | Continuous Strip |
| 4 | Thermostat Material | 1 | Batch, loose parts |
| 5 | PEP Parts | 1 | Batch, loose parts |
| 6 | PC Device | 1 | Batch, loose parts |

procedures described in the Massachusetts Toxics Use Reduction Act as its primary methodology for addressing new regulatory requirements. One of the basic premises of the Act is that the preferred method of complying with environmental regulations is to eliminate use of the toxic chemical that causes the entity to be subject to the regulation. Therefore, the first step taken by TI when it became aware that it was subject to the Halogenated Solvent Cleaning MACT standard was to form pollution prevention teams. Six pollution prevention teams were formed around the operations manufacturing operations listed in Table III.

Each pollution prevention team began by brainstorming potential options to investigate. Each team would then evaluate each option and determine which would most appropriately meet the needs of their operation. The option to be investigated by each team was:

- Determine if the existing vapor degreaser can be upgraded to meet the Halogenated Solvent Cleaning MACT standard..
- Investigate the applicability of halogenated-solvent-based cleaning techniques, including high efficiency conventional vapor degreasing and high vacuum vapor degreasing.
- Investigate the applicability of modifications to the manufacturing process such that the need to clean is eliminated.
- Investigate the applicability of conventional substitute cleaning technique such as aqueous, semiaqueous and nonaqueous washing technologies.
- Investigate the applicability of leading edge substitute cleaning techniques such as supercritical fluid extraction or dry ice impact cleaning technologies.

The first task of each team was to evaluate the functionality of their existing vapor degreasers. Included in this evaluation were:

- annual solvent usage and emissions
- annual and hourly product throughput
- compliance factors including previous years emissions and solvent/air interface measurements.

At this point, each team assessed the feasibility of their existing vapor degreasers being upgraded to meet the halogenated solvent cleaning MACT standards. Degreaser vendor representatives were brought in to examine the unit. These representatives concluded that, except for Team 6's (PC Device) unit, the units would not be able to meet the new emission standards. This was due primarily to the age (most about 40 years old) and subsequent condition of the units. Team 6 decided that since they anticipated going to a no-clean manufacturing option within two years, they would upgrade the PC Device Degreaser to meet the emission standards.

The remainder of the teams proceeded to evaluate the remaining four options to be investigated in accordance to plans developed by each team. For example, Team 1, MC & PC Disc Cleaning, determined immediately that if they accelerated a program to modify their production processes that they could completely eliminate the need to clean prior to the December 2, 1997 deadline.

Teams 2 and 5 narrowed their investigations to halogenated solvent vapor degreasing technologies including high efficiency conventional vapor degreasing and high vacuum vapor degreasing. Teams 3 and 4 focused their search on conventional substitute technologies such as aqueous and nonaqueous technologies.

The following provides a summary of the results of the investigations performed by these teams.

*High Efficiency Conventional Vapor Degreasing*

This technology basically consists of installing a number of vapor control methods into a conventional open-top vapor degreaser. These methods include tight airlock style covers, extended freeboard inert gas blankets, and super cooling coils. The best performing equipment using these technologies achieve emission rates just below the emission limits. Even though the equipment supplier claimed that the equipment would continuously meet the emission limit, the teams felt that operating so close to the limit would be a problem. As a major manufacturer, TI routinely demonstrates compliance with regulations such as the Halogenated Solvent Cleaner MACT standard. The cost of demonstrating compliance and the adverse impact if, as the equipment ages, it drifts out of compliance, resulted in the decision not to pursue this technology.

*High Vacuum Vapor Degreasing*

High Vacuum Vapor Degreasing eliminates one of the root causes of emissions from a conventional vapor degreaser—the mixing of solvent and air. SEREC, a company located in Providence, RI, has demonstrated that mixing air with solvent vapor is not necessary for the cleaning process to work. The technology consists of a chamber in which soiled parts are placed. Vacuum pumps remove most of the air. At this point, solvent is introduced to remove the soil. A different set of vacuum pumps remove the solvent and return it to a sealed reservoir. When the chamber is opened, only a small quantity of solvent is emitted because it is impossible to pull an absolute vacuum.

An assessment was conducted comparing conventional vapor degreasing, high efficiency convention vapor degreasing, and high vacuum vapor degreasing. These results are listed in Table IV. The emissions from high vacuum vapor degreasing are so low (5 lbs/yr in the test device) that they not only obviously meet the Halogenated Solvent Cleaner MACT standard, thus eliminating costly compliance demonstrations, but they are below the threshold in Massachusetts requiring permitting, thus eliminating the cost of a new source review permit.

High vacuum vapor degreasers do have a high initial capital investment ($150,000 to $350,000), but produce significant cost savings on chemical purchases and disposal costs. The financial assessment of purchasing this technology resulted in a cost recovery (payback) in about a 4 year time period.

*Aqueous Washers*

From 1984 to 1990, TI had installed 20 aqueous washers in its manufacturing operations. Consequently, the company has considerable experience with the technology In general, aqueous washing is not recommended as a replacement for conventional vapor degreasing for two reasons:

1. Site wastewater discharge capabilities that handle wastes from aqueous washers are almost at the maximum. Any further additions of aqueous washer waste streams

**Table IV. Conventional vs. High Vacuum Vapor Degreasing**

| Criteria | Conventional | High Vacuum |
|---|---|---|
| Weight of parts cleaned | 200 kg / hr | 200 kg / Hr |
| Solvent purchased | 22,000 lb /yr | 6,200 lb /yr |
| Solvent emissions | 10,000 lb/yr | 5,000 lb /yr |
| Solvent recycle | 10,000 lb/yr | 6,000 lb /yr |
| Solvent incinerated | 2,000 lb/yr | 195 lb/yr |
| Energy rating | 100 % | 95 % |

would require wastewater offsets or a significant capital investment to add new capacity.

2. Aqueous washers generally are less versatile than vapor degreasers and the manufacturing area would have to absorb a 10% yield rate reduction. For some areas, this is a considerable cost (over $100,000 per year).

If the two issues described above are resolved, aqueous washing does provide a number of benefits: the technology is readily available at moderate costs ($40,000 to $100,000 per unit) and customers will approve their use.

Nonaqueous Washing: This technology consists of a tank containing a nonchlorinated, nonaqueous solvent below its boiling point into which material is introduced. Pollution prevention Team 3 has been investigating the use of this technology for the past several years. In order for nonaqueous washing to be suitable, the loss of solvent must be sufficiently low as to meet new source review rules that require new VOC emission sources to meet BACT and LAER.

In 1995, Team 3 developed an innovative nonaqueous washing application with emission rates low enough to meet BACT and LAER. A prototype unit was installed in August 1995 and data collected during the pilot were used to demonstrate to company management, the customer base, and regulators that it would be an appropriate replacement technology. In May 1996, one of the existing vapor degreaser tanks was converted to a nonaqueous washer. By reusing part of existing equipment, the technology has become more cost-effective to implement. In addition, the technology consumes less than 100 gallons of solvent per month, thus, making TI exempt from Massachusetts permitting requirements.

## Implementation Plan

Teams 1, 2, and 5 have decided to implement high vacuum vapor degreasings to meet the Halogenated Solvent Cleaner MACT standard. Team 1 decided to purchase a small unit as backup in case their program to move to a new production process is delayed. Teams 2 and 5 decided that their business would be best served by having a single cleaning unit service both areas. The unit will be located in the IMT manufacturing area and the PEP will transport its parts to IMT for cleaning.

Team 3 decided to convert six of its seven TCE vapor degreasers to the nonaqueous washing process. The seventh is scheduled to be converted to an aqueous washer. The cost of conversion is approximately $20,000 per unit. Annual reductions in the cost to operate the nonaqueous washers over the vapor degreasers amount to nearly $100,000 per year. These figures indicate that implementing this technology will have about a two-year payback.

Team 3 also determined that if the nonaqueous washing technology use was expanded to include operation that currently do not have cleaning capability, additional cost savings could be realized. The team is currently evaluating installing the cleaning technology in another 8 operations.

The installation of 14 nonaqueous washers will result in overall emission less than the MACT standard. Three year average of emissions from the TCE vapor degreasers was 75.3 tons. Allowable emissions for the same units under the MACT standard would be 14.8 tons per year. The VOC emissions from the nonaqueous washers is estimated in the worst case as 12.5 tons per year.

The site has filed a comprehensive plan approved application with the Massachusetts Department of Environmental Protection to gain concurrence with the department that the nonaqueous washers represent BACT. Also, the application requests that the department recognize that the annual emission credits of 2.3 tons per year VOC (the difference between the MACT allowable emission rate and the potential emission rate of the project) be set-aside credits for future use or sale at TI's discretion.

Team 4 decided to implement an aqueous loose parts washing system. As part of their project, the team was required to find water savings elsewhere within the facility to offset the anticipated increase in rinsewater generated by the washer. Overall, the project investment was nearly $450,000, with a payback of a little under 4 years.

## SUMMARY

Each of these case studies resulted in both pollution prevention and a positive financial analysis. Each case was site- and time-specific, indicating that each analyses is a proactive and ongoing optimization problem, similar to the actual manufacturing processes.

## REFERENCES

1. Risotto, S., "Alternatives for Ozone-Depleting Cleaners," Center for Emissions Control, Washington DC, *Industrial Finishing Magazine*, July 1993, p. 56
2. "Parts Cleaner Skirts CFCs," Staff, *Environment Today Magazine*, April 1993.
3. Candeas, B.G. and Gould, D.Z., "Successfully Implementing a New PWA Cleaning Solvent in a DoD Manufacturing Environment."
4. "Parts Cleaners Search of Better CFC Replacements," Staff, *Environmental Management Today Magazine*, June 1995, pp. 15 and 19.
5. Linden, E., "Who Lost The Ozone?" *Time*, May 10, 1993, pp. 56–58.
6. "The Ozone Hole is Spreading," Staff, *US News & World Report* December 13, 1993, pp. 87–90.

# 59 INVESTIGATION INTO THE NATURE AND EXTENT OF MICROBIAL CONTAMINATION PRESENT IN A COMMERCIAL METALWORKING FLUID

**Keith M . Rinkus**, Ph.D. Candidate

**Wei Lin**, Res. Assistant Professor

**Alka Jha**, M.S. Candidate

**Brian E. Reed**, Associate Professor
Department of Civil and Environmental Engineering
West Virginia University
Morgantown, West Virginia 26506-6103

## ABSTRACT

A commercial metalworking fluid, 5% stock solution/95% deionized water, was found to contain multiple microbial species including bacteria, most probably *Actinobacter, Pseudomonas, Aerobacter* and *Bacillus*, and fungi. Bacterial concentrations in the metalworking fluid increased with decreasing temperature ($5.0 \times 10^7$ @ 50%C and $2.5 \times 10^8$ @ 37%C). The bacterial population was capable of using all the stock solution components as substrates. However, fatty acids and fatty acid esters were metabolized at a much greater rate than the other components due to the ease of the β-oxidation metabolism of these components. Rates of biological growth decreased with increasing temperature.

## INTRODUCTION

Metal manufacturing facilities use rolling processes to press large ingots into forms suitable for various processes and products. For metal rolling operations (MR), lubrication of the machine workpiece area is of major concern for production of high quality products. Lubrication of the machine workpiece area is achieved through the application of oil-water emulsions referred to generically as metalworking lubricants or fluids (MWFs). In addition to lubrication effects, MWFs also serve the purposes of workpiece and tool cooling and rust formation inhibition in the rolling operation area.

Establishment of an optimum metal rolling operation is dependent upon the establishment of a stable MWF and therefore, a stable oil-water emulsion. However, an oil-water MWF is inherently thermodynamically unstable. Therefore, if the composition or concentration of the emulsion ingredients, the pH, the ambient temperature or any of a number of different parameters is changed, the characteristic properties of the MWF may also change, affecting product quality. One factor which may compromise the integrity of MWF is the presence of microorganisms. The growth of microorganisms leads to the destruction, through metabolic activity, of lubricant ingredients and a production of acidic materials which leads to a weakening in emulsion stability. Emulsion breakdown results in oil/water separation, which adversely impacts the primary function of lubrication. Production of acidic materials may also lead to corrosion of equipment and workpieces, pH imbalances, and objectionable odors.

The purpose of this study was to investigate the nature of the microorganisms inhabiting a MWF used in an metal rolling operation. This investigation was composed of a microscopic examination, culture of microorganisms in both MWF and various microbiological media and determination of microbial substrate utilization rates using oxygen uptake (OxU) respirometric

*52nd Purdue Industrial Waste Conference Proceedings*, 1997, Ann Arbor Press, Chelsea, Michigan 48118. Printed in U.S.A.

studies. The study was performed to determine the nature of the microorganisms inhabiting the MWF, and to elucidate possible methods for microbial control. Presently, the MWF in the rolling process is kept at a temperature of 50°C to control microbial growth. At that temperature, it is unlikely that microorganisms are being destroyed. However, it is possible that the high temperature is retarding the proliferation of the microbes, and limiting the number of species which can remain viable and flourish.

## BACKGROUND

MWFs used in MR processes contain 3% to 5% of oil mixtures (stock solution) emulsified in water (Beyers, 1994). The oil stock solution is composed of mineral oil, emulsifier(s), fatty acids and various coupling and wetting agents used to produce a MWF with the desired properties.

The general components of the MWF stock solution are presented in Table I. The MWF investigated in this study contained a stock solution consisting of napthenic mineral oil, oleic acid, two esters (esterified fatty acids), two amines, and a few other components for emulsifying, and other purposes. The exact composition of the stock solution and the identity of some of the additives is proprietary information. For this study, the authors defined the MWF as a 5% stock solution, 95% deionized water mixture. The stock solution consisted of the six listed components (as supplied by the metalworking facility), present in the percentages listed in Table I.

Passman (1988) and Guyman and Bennett (1958) have reported that bacteria are commonly present in metalworking fluids at high concentrations ($1.0 \times 10^9$ CFU/mL) and are capable of degrading MWF components. However, because of the complex nature of the microbial population and the composite nature of the MWF, it is difficult to determine which of the MWF components is being utilized as a primary carbon source. There are four main stock solution component types in the MWF, based on chemical and biochemical characteristics. The components are base oil, fatty acid ester emulsifier and lubricants, oleic acid boundary lubricant, and amine additives. Typically, the fatty acids are the components easiest to utilize as a substrate, as the fatty acids can enter directly into the β-oxidation metabolic pathway inherent in all aerobic cells. The fatty acid esters may also be easily degraded if the bacterial culture has an esterase capacity to convert the esters into fatty acids. Foxall-Vanken et al (1986) observed this esterase capability in multiple studies of bacteria isolated from MWFs. Oil components, both mineral and synthetic, and amines are not as easily degraded as the esters or fatty acids, although in some studies these components have been observed to support minimal microbial populations (Foxall-Vanken et al., 1986). Bacteria isolated from MWF commonly include aerobic species from *Acinetobacter*, *Pseudomonas, Aerobacter,* and *Bacillus* genera.

**Table I. Composition of MWF Stock Solution**

| Percent of Stock Solution | Component | Purpose | Molecular Formula | mg ThOD/mg Compound |
|---|---|---|---|---|
| 68 | Base oil | Lubrication | $C_{16}H_{34}$ | 3.47 |
| 17 | Ester A | Emulsifier | $C_{22}H_{42}O_5$ | 2.46 |
| 5 | Ester B | Extreme pressure lubricant | $C_{22}H_{42}O_5$ | 2.46 |
| 5 | Oleic acid | Boundary lubricant | $C_{18}H_{34}O_2$ | 2.89 |
| 1.5 | Triethanolamine (TEA) | Secondary emulsifier stabilizing agent | $C_6H_{15}O_3N$ | 1.77 |
| 1.5 | Amine A | Stabilizing agent quench control | $C_{18}H_{40}O_6N_2$ | 2.1 |

## EXPERIMENTAL DESIGN

### Experimental Approach

The research was divided into the following three tasks: (1) culturing of microorganisms in MWFs, (2) an investigation of microbial characteristics, and (3) the determination of carbon source utilization for the microorganisms. In Task 1, a sterile MWF was inoculated with microbes from actual plant MWF to study the microbial growth in bench-scale reactors which were maintained at conditions approximating plant conditions. Microbial concentrations in the two bioreactors are indicative of concentrations which may exist in actual coolant systems. Reactors were maintained at two different temperatures (37°C and 50°C) so that temperature effects on microbial populations in MWF could be investigated. Also, the bioreactors were maintained so that a steady supply of microorganisms was available for other experimental procedures.

In Task 2, two sets of experiments were conducted; a microscopic examination of lubricant samples and selective culturing of microorganisms on different agar media. The microscopic examination included observation of microorganisms using hanging drop slides, and Gram staining and endospore presence staining to clarify the type and nature of the microorganisms. Bacteria can be divided into two groups, Gram negative and positive, based on cell membrane structure. In Gram positive microorganisms, the cell wall is composed of many layers of peptidoglycan, a protein/carbohydrate macromolecule forming a thick rigid structure. Gram negative microorganisms, in contrast, have cell walls that contain one or only a few peptidoglycan layers and an outer cell membrane. It is this outer membrane composed of lipoproteins which confers resistance of Gram negative cells to some antibiotics and chemical biocides. Therefore, the Gram designation of the majority of the bacterial population inherent in the MWF can dictate which physical methods, biocides, and antibiotics would be most effective at microbial control.

The identification of endospore-producing species of bacteria is important, as endospore must be destroyed in addition to vegetative cells to ensure microbial population control. Endospore are, in effect, dormant bacterial cells. When environmental conditions become unfavorable for a microbial population that produce endospore, the vegetative bacterial cells die and release endospore. These spores contain DNA, RNA, and a small variety of essential compounds in a small desiccated cell surrounded by thick peptidoglycan layers. Endospore are capable of withstanding conditions that would eradicate viable bacteria. When favorable conditions again appear, the endospore germinates, creating a viable bacterium from the spore contents.

Agar culturing experiments were designed to identify microbia type in the MWF. Media selective for bacteria and fungi were prepared to qualify the presence of different types of microbes. The presence and effect of different eukaryotes are important because if the microbial problem is caused by different types of fungi or bacteria, different fungicides, biocides, or changes in pH, temperature, or nature of the MWF must be used to control microbial growth.

In Task 3, substrate utilization rates were determined using oxygen uptake (OxU) measurements of closed batch reactors containing microbes and selected compounds. Three different types of OxU samples were prepared. Type I consisted of reactor fluid (MWF with microbes). Type I experiments were conducted at both 20°C and 50°C to determine differences in metabolic activity of the microbes in the MWF at room temperature (20°C) as opposed to metalworking operation temperature (50°C). Type II samples contained MWF and microorganisms, but the amount of nutrients added to each sample was varied. Nutrient variation was done to determine whether the MWF had all necessary nutrients, or was nutrient deficient. Type II experiments were conducted at 20°C. Type III samples consisted of a nutrient broth, a microbial inoculum, and one of the six MWF components. Type III samples were designed to determine which of the MWF components could be used as a carbon source (substrate) by the MWF microbial population. Type III experiments were conducted at 20°C and 50°C.

### Materials and Methods

#### *Task I: Bioreactor Experiments*

A diagram of the bioreactor setup is presented in Figure 1. Two bioreactors were constructed and operated in the fill and draw mode. Reactor 1 was maintained at 50°C (approximating plant conditions), and reactor 2 was maintained at 37°C (optimum microbial growth conditions). Both reactors were aerated with humidified air, maintained at a pH of 7.3 to 7.6, and maintained at a

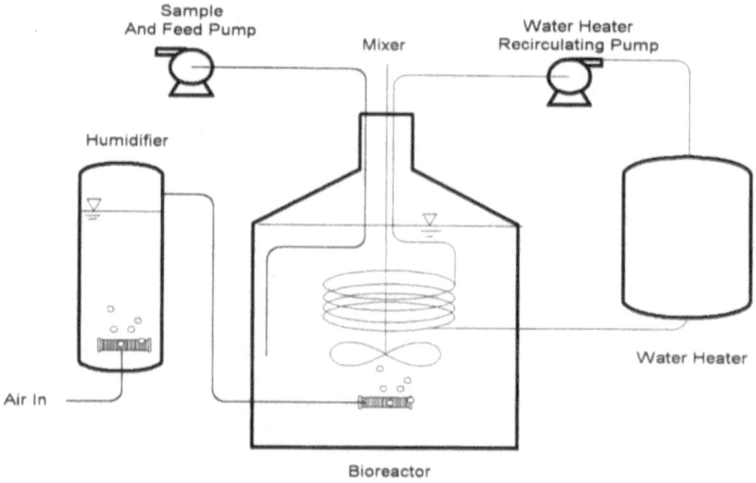


**Figure 1. Sample bioreactor schematic.**

constant volume. Evaporative losses were made up with deionized water, and samples were removed and replaced with fresh sterile MWF. The reactors had a hydraulic residence time of 100 days. During actual rolling operations, the plant MWF system maintains a MWF residence time of ~375 days. However, the entire volume of MWF is replaced over 10 to 12 weeks due to decreases in MWF quality. Therefore, the residence time of the bioreactors was calculated to correspond to this 3 month residence time. pH in the reactors was maintained with the addition of 0.01 M phosphoric acid or triethanolamine (TEA) as needed. A peristaltic pump was used for sampling and adding the fresh MWF to the reactor. Reactor 1 was started by adding 9.5 L of MWF, freshly prepared using sterile water and stock solution, to the sterilized vessel. The reactor was then inoculated by adding 500 mL of actual plant MWF, sampled from the MWF reservoir during normal rolling operations. Microorganism concentrations (as CFU/mL) were measured every two days from each reactor to monitor fluctuations in bacterial population over time. Differences in color and shape of the colonies were noted as well as microbial concentrations. Reactor 2 was begun in the same way, but had a total volume of 3 liters.

*Task II: Investigation of Microorganism Characteristics*

All microscopic observations were conducted on samples from Reactor 1 because the 50°C samples were representative of actual plant conditions. Hanging drop slides were prepared for direct coolant microorganism inspection. Gram and endospore stained slides of microorganism samples were prepared to determine the Gram nature and presence or absence of bacterial endospore. Samples of MWF used to culture bacteria and fungi agar plates were also obtained from Reactor 1. Three different types of agar plates were cultured, one bacterial agar and two fungi agar types. Bacterial agar plates were made with Difco nutrient agar. The two types of fungi plates contained Difco mold and yeast agar. One mold and yeast plate type also contained two antibiotic agents to control bacterial contamination; streptomycin sulfate (40 mg/L) and chlortetracycline (35 mg/L). Twenty mg of streptomycin and 17.5 mg of chlortetracycline were aseptically mixed with 50 mL of sterilized DI water. This solution was added to the agar (500 mL) after autoclaving and when the agar had cooled down to 40°C. Petri dishes were stored at 4°C until used.

*Task III: Oxygen Uptake Experiments*

Oxygen uptake (OxU) rates of microorganisms in various solutions were measured using a 6-bay N-Con Comput-OX WB series respirometer. Reactors were placed in the respirometer for a test duration of 240 hours, during which the reactors were automatically sampled for oxygen uptake every half hour. All respirometer samples had a volume of 300 mL. Each of the 500 mL respirometer sample bottles, caps, and associated equipment was sterilized before use. Type I OxU samples, Reactor 1 MWF, consisted of 300 mL of Reactor 1 solution. Type II samples, nutrient supply determination samples, were composed of 200 mL of sterilized Difco Bushnell Haas (BH) nutrient broth or sterilized DI water to which 5% v/v MWF stock solution was added. To this 200 mL, 100 mL of Reactor 1 solution was added. Type III samples, carbon source determination, contained 260 mL of BH nutrient broth to which was added 1 mL of the selected MWF concentrate component. Forty mL of inoculum was then added to the 260 mL of BH broth. The inoculum was prepared by centrifuging 40 mL of Reactor 1 MWF effluent at 10,000 RPM for 60 min, discarding the centrate, and resuspending the pellet in 40 mL of BH broth. Respirometer reactor pH was maintained between 6.8 to 7.8. If the sample solution did not fall into the prescribed range without additives, phosphate buffer was added to the stock solution to attain the desired pH. MWF carbon source (substrate) utilization values (ΔS) were calculated based on oxygen uptake measurements from the batch respirometer samples using the following equation.

$$OxU = \Delta S - \Delta X \tag{1}$$

where OxU = Oxygen uptake (mg)
ΔS = Change in substrate concentration (mg soluble COD)
ΔX = Growth of biomass (mg cell COD, i.e. insoluble COD)

Use of Equation 1 is valid with two assumptions: (1) substrate degradation and associated biomass growth are the only events contributing to oxygen uptake, and (2) the amount of soluble product from substrate degradation is negligible. Biomass is typically expressed in mass units (mg). However, when an oxygen mass balance is calculated for a system such as a respirometer unit, units of biomass COD (mg) is usually used. (Templeton and Grady, 1988 and Dang *et al.*, 1989). The biomass COD can be determined by direct measurement or by calculation of theoretical oxugen demand (ThOD) based on the commonly accepted formula of bacterial biomass: $C_5H_7O_2N$. The ThOD calculation results in a conversion factor of 1.42 mg biomass COD/mg biomass.

The relationship between biomass growth and substrate removal is expressed as yield, in units of mg cell COD produced/mg substrate COD removed. Dang et al. (1988) has found a typical yield (Y) of 0.30 mg $COD_{biomass}$/mg $COD_{substrate}$ for various organic compounds. Substituting ΔS*Y for Δ X in Equation 1 and rearranging yields:

$$\Delta S = \frac{OxU}{(1 - Y)} \tag{2}$$

using a Y value of 0.30 in Equation 2 yields:

$$\Delta s = 1.43(OxU) \tag{3}$$

Substrate removed can be converted from soluble COD to mass of substrate by determining the ThOD for each of the substrates (see Table I) and using the following equation:

$$\text{Mass of Carbon Source Consumed (mg)} = \frac{mg\ soluable\ COD_{Removed}}{\left(\frac{mg\ ThOD_{Carbon\ Source}}{mg\ Carbon\ Source}\right)} \tag{4}$$

## RESULTS AND DISCUSSION

### Task I: Bioreactor Study

The concentration of microbes (CFU/mL) in Reactors 1 and 2 versus time from reactor stabilization is presented in Figure 2. For the first period of operation of the bioreactors, constant conditions were not maintained because of evaporative losses and pH fluctuations. Therefore, only data recorded after constancy was established are presented here. The initial pH (before microbial inoculation) of both bioreactors was 8.15. Microbial activity leads to production of acidic compounds which reduces the solution pH. Constancy was established using the same methods

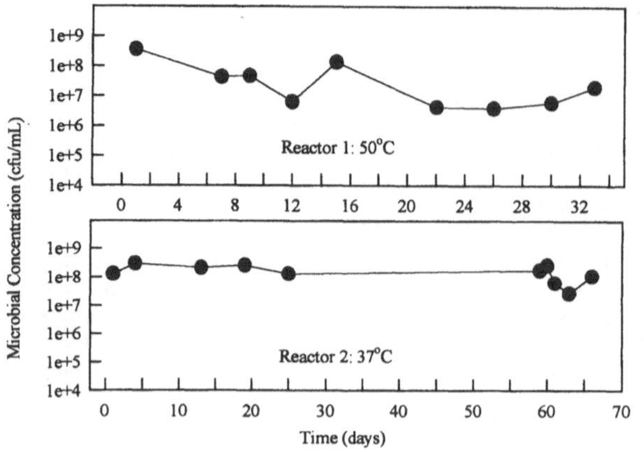


**Figure 2. Microbial concentration in Reactors 1 and 2 versus time (after constancy).**

utilized by the actual metalworking plant, the periodic addition of DI water, fresh MWF, and TEA. The average microbial concentration of Reactor 1 was $5.0 \times 10^7$ CFU/mL, and of reactor 2 was $2.5 \times 10^8$ CFU/mL. Hill (1977) reported MWF bacterial concentrations from $10^6$ to $10^9$ in actual metalworking plants. Therefore, the benchscale reactors provide an environment similar enough to plant conditions to produce analagous effluent . There were also morphological differences between the microbes from the two bioreactors.

Reactor 2 (37°C) bacterial agar plates contained greenish colonies in addition to the yellow colonies evident in both reactor bacterial agar plates. Since inoculation of the MWF was occurring continuously from work pieces, airborne dust, workers, and general production activity, the majority of microbiological species found in the immediate area are present in the MWF. However, some of these microorganisms do not flourish in the high temperature MWF environment. Therefore, when the MWF temperature decreases (between production runs or in isolated sections of the MWF circulation loop), a larger variety of organisms are able to grow, leading to an increase in microorganism diversity as well as overall concentration.

## Investigation into Microbial Characteristics

### *Microscopic Investigation*

Reactor 1 MWF samples contained a number of different types of microorganisms. Both cocci (spherical) and rod (bacillus) shaped bacteria were observed, some of which appeared to be motile. No higher organisims (protists) were observed. A large amount of noncellular material was also observed. It is unclear whether the noncellular material was extracellular products secreted by the microorganisms or came from other sources such as dirt and debris from the manufacturing process. However, hanging drop slide observations were conclusive in that the coolant did contain multiple microbial species.

Based on Gram stained slide observation, it appeared that the majority of microorganisms in the MWF were Gram negative. However, small numbers of dark red stained organisms indicate that both Gram negative and Gram positive bacteria are present in the MWF. Based on the shape, Gram designation, and literature, the majority of the Gram negative bacteria are probably from the *Acinetobacter*, *Pseudomonas*, and *Aerobacter* genera which are ubiquitously found in soils (Tortora et al, 1995), and can flourish in the MWF environment.

The Gram negative designation of the majority of microbes in the MWF imparts certain characteristics to the microbial population. Gram negative cells contain an outer membrane of lipoproteins, lipopolysaccarides, and phospholipids. This outer membrane confers resistance to detergents, and certain antibiotics such as penicillin and sulfonamide. Members of the Gram negative *Pseudomonas* genera, which are prevalent in MWFs, (Foxall-Vanaken et al., 1986) have even higher than normal resistance to chemical control methods. For instance, *Pseudomonas* bacteria are known to actually grow in many antiseptics and disinfectants. Conversely, Gram

**Table II. Fungi Plate Count Results**

| Fungi Counts (CFU/mL) | | |
|---|---|---|
| Reactor Temperature °C | Incubation Temperature °C | |
| | 37 | 50 |
| | Mold and yeast agar without antibiotics | |
| 37 | $2.4 \times 10^7$ | 333 |
| 50 | $2.7 \times 10^3$ | 0 |
| | Mold and yeast agar with antibiotics | |
| 37 | $4.0 \times 10^6$ | 0 |
| 50 | $3.3 \times 10^3$ | 0 |

negative cells are more susceptible than Gram positive cells to drying and streptomycin, chloramphenicol, and tetracycline antibiotics (Tortora et al., 1995).

Based on obervations of the endospore stained slides, some of the bacterial strains inhabiting the coolant produce endospore. These organisms were rod shaped, Gram positive and most probably from the *Bacillus* Genera. Endospore are much harder to destroy than vegetative cells. Endospore have been known to survive boiling water for up to 3 hours (Tortora et al. 1995). Endospore are usually destroyed by long-term (> 3 hours) boiling or autoclaving (121°C at 15 psig) for 20 min or more. However, these control methods are not realistically implementable in a MWF during rolling operations. Control of endospore producing species is most likely to be effective by controlling the vegetative cells themselves. If the MWF environment can be kept uninhabitable to the *Bacillus* bacteria (vegetative cells), the endospore present in the MWF will never germinate and grow, and therefore, will not disrupt MWF integrity.

*Fungi Agar Culture*

The results from the fungi culturing experiment are presented in Table II. Eight sets of agar plates were prepared to identify if appreciable amount of yeast and mold existed in the MWF. When incubated at 50°C, very little fungi were found in samples taken from either Reactor 1 or 2 (50°C and 37°C, respectively). Conversely, when identical samples were incubated at 37°C, high concentrations of fungi were observed. The higher number of CFU/mL present in the plates without antibiotics is probably due to bacterial contamination, as it was not observed in the more selective media containing antibiotics. Generally, the lower fungal concentration evident in Reactor 1 (50°C) when compared to Reactor 2 (37°C) is indicative of spore destruction in the higher temperature reactor. Fungi do not usually grow in solution, preferring surfaces. Therefore, fungi, both mold and yeast, probably exist as nonvegetative spores in solution, beginning to vegetate when attached to surfaces such as tank walls, machinery, or flooring. Overall, the environment of the MWF at 50°C was not conducive to the proliferation of fungi.

*Oxygen Uptake Experiments*

Cumulative oxygen uptake versus time for Reactor 1 (Type I, MWF) samples is presented in Figure 3. Samples 1 and 2 were duplicates and had average OxU rates of 2.56 and 3.1, mg $O_2$/L/hr, respectively. The lag phase for each of the 20°C samples was approximately 45 hours. The lag time in the 20°C samples was probably due to the change in solution temperature (bioreactor was at 50°C, and the first respirometer run was at 20°C). The microorganisims did not show appreciable activity for the first 45 hours as they were becoming acclimated to the new temperature. Oxygen uptake for the 50°C sample (Sample 3) was significantly lower than the 20°C samples, indicating a lower rate of microbial metabolic activity. Oxygen uptake versus time for the MWF samples containing Bushnell Haas (BH) broth or deionized (DI) water with 5% stock solution are presented in Figure 4. The BH and DI samples had average OxU rates of 4.85, and 4.13 mg $O_2$/L/hr respectively. Three blanks are also presented in Figure 4. The first blank contained the nutrient broth and stock solution but no microorganism seed. The second

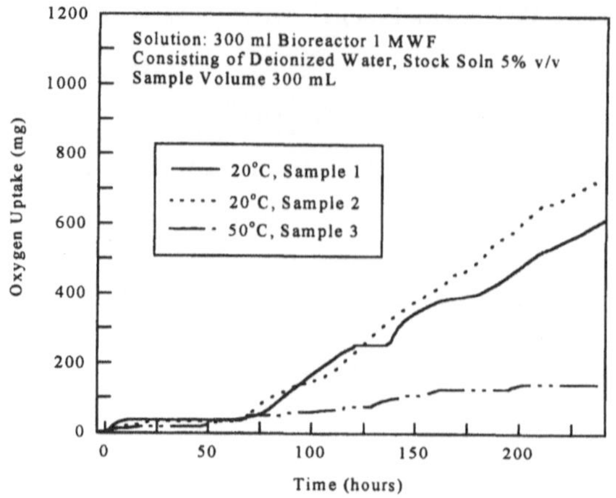


**Figure 3. Oxygen uptake versus time (Type I sample, MWF).**

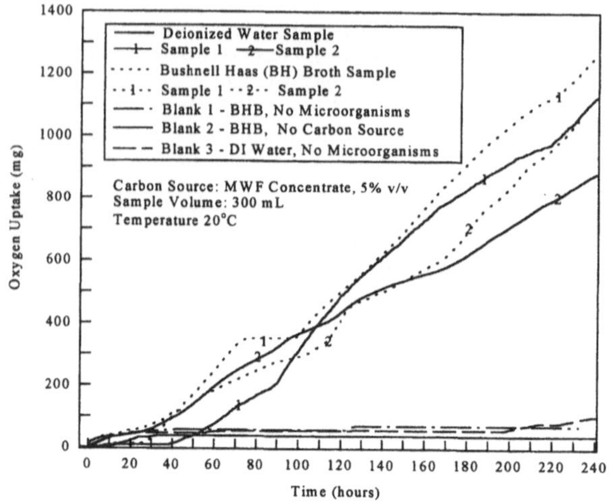


**Figure 4. Oxygen uptake versus time (Type II nutrient samples).**

blank contained the nutrient broth and the microbial seed, but no stock solution. The third blank contained DI water and microorganisms but no stock solution. None of the blanks had OxU rates over 0.6 mg $O_2$/L/hr, indicating little metabolic activity . The degree of microbial activity (as measured by OxU) in the BH and DI sample types was very similar. BH broth is a nutrient broth that contains all compounds necessary for aerobic growth except a carbon source. (0.2 g/L magnesium sulfate, 0.02 g/L calcium chloride, 1 g/L monopotassium phosphate, 1 g/L ammonium phosphate dibasic, 1 g/L potassium nitrate and 0.05 g/L of ferric chloride) In each of the two sample types, the MWF stock solution was the carbon source. Since the BH and DI samples exhibited similar OxU over time, it can be postulated that the MWF made of stock solution and DI water contains the nutrients necessary for unhindered microbial growth.

Microbiological substrate utilization versus time at 20°C for each of the six stock solution components is presented in Figure 5. Samples containing base oil and TEA exhibited little microbial activity. Amine A had a 13% decrease in concentration. The three components which had the highest utilization over the ten day batch run were the two esters and oleic acid. Ester A, the emulsifier, had a 68% decrease in concentration, and ester B, a 55% decrease. Oleic acid had a 54% decrease in concentration. Low utilization for the base oil may be due to the low solubility of oil in water. Respirometer samples were vigorously mixed, allowing oil droplets to become

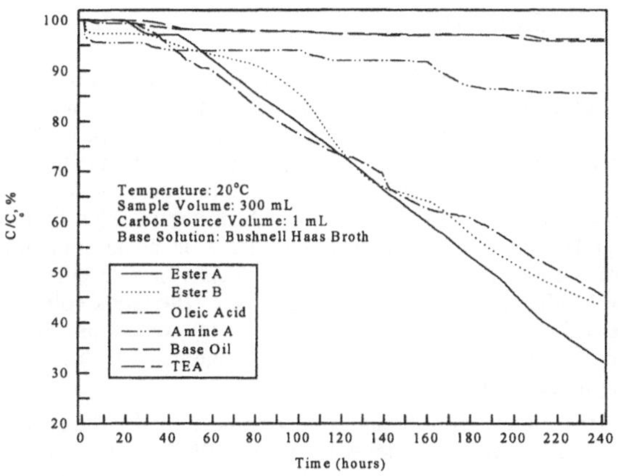


**Figure 5. Substrate utilization versus time at 20°C (Type III samples).**

dispersed in the BH solution; however, no additional emulsifier was added to any samples. Neither of the amines were substantially degraded, although both were soluble. It is likely that the bacteria do not have the proper enzymes to metabolize the amines. The degradation of oleic acid was expected because of the ease of fatty acid metabolism. However, the fatty acid esters were degraded at approximately the same rate as oleic acid. Therefore, the bacteria present in the MWF have the esterase enzyme necessary to convert fatty acid esters to fatty acids.

Substrate utilization versus time at 50°C for each of the stock solution components is presented in Figure 6. Only two of the stock solution components was removed in an appreciable fraction: oleic acid and ester A (emulsifier). Oleic acid had the highest utilization at 36% (ester A, 31%). A much lower fraction of ester B was used at 50°C than at 20°C. The differences in substrate utilization in both type and quantity are indicative of a change in microorganism metabolic processes or microbial population from temperatures of 20°C to 50°C. At 20°C, the majority of the microbial population can degrade the esters. At 50°C, the majority of the microbes, if undergoing metabolic activity, can remove oleic acid and ester A but not ester B. Another possibility is that the esterase enzymes which convert the esters to fatty acids do not function as well (or denature) at 50°C. If the esterase capability is decreased or changed, the decrease in ester to fatty acid conversion would limit the amount of substrate available for metabolic use.

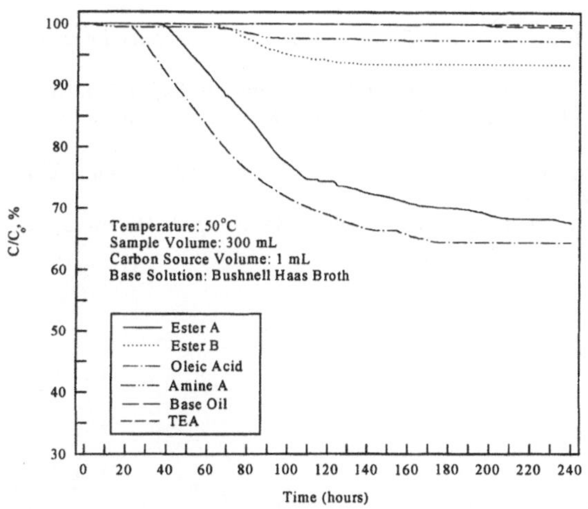


**Figure 6. Oxygen uptake versus time at 50°C (Type III samples).**

## CONCLUSIONS

The production MWF contains a variety of bacteria, both cocci and bacillus. Based on the bacterial shape, Gram designation and literature, the bacteria are probably Gram negative *Acinetobacter, Pseudomonas* and *Aerobacter*, but also containing some Gram positive bacteria such as *Bacillus* which produce endospore. The following specific conclusions can be drawn: (1) The predominant form of microorganisim inhabiting the MWF is bacterial (multispecies ~ $5 \times 10^7$ cfu/mL), along with a nonvegetative spore population of fungi, both yeasts and molds. No higher organisims (i.e., protists) were found; (2) The bacterial population can grow on basic emulsion (DI water with 5% MWF stock solution) without the need for added materials or nutrients; (3) The bacterial population contained in the MWF is capable of utilizing most of the stock solution components as a substrate; (4) Substrate utilization rates are higher for the two esters and oleic acid at 20°C and for ester A and oleic acid at 50°C than for the other components; (5) Substrate utilization rates decreased with increasing temperarure.

It appears that at least partial control of the microbial population may be possible by the reduction or replacment of fatty acid and fatty acid ester components from the MWF stock solution. Removing or replacing the fatty acid and ester components of the stock solution with less degradable compounds may result in a decrease in microbial activity due to a decrease in available substrate. Another method of bacterial growth control may be the increase in the minimum MWF temperature. Increase in temperature may inhibit the metabolic processes (through denaturation of metabolic enzymes) of the bacterial population.

## BIBLIOGRAPHY


Bennett, E.O., "The Biology of Metalworking Fluids." *J. Am. Soc. Lubr. Eng.* 28(6), pp. 237–47, 1972.

Brock, D.V., "Lubricant Base Oil." *Lubr. Eng.* March, pp. 184–185, 1987.

Byers, J.P., Ed. *Metalworking Fluids*. Marcel Dekker, Inc. New York, 1994.

Dang, J.S., D.M. Harvey, A. Jobbagy, and C.P L. Grady, Jr., "Evaluation of Biodegradation Kinetics with Resporometric Data." *Res. J., WPCF.* 61 (11/12), pp. 1711–1721, 1989.

Fabian, F.W., and Pivnick, H. "Growth of Bacteria in Soluble Oil Emulsions." *Appl. Microbiol.* April, pp. 199–203, 1953.

Foxall-Vanaken, S., Brown, JR., J.A., Young, W., Salmeen, T., McClure, T., Napier, JR., S., and Olsen, R.H., "Common Components of Industrial Metal-Working Fluids as Sources of Carbon for Bacterial Growth." *Appl. Environ. Microbiol.* June, pp. 1165–69, 1986.

Guynes, G.J., and Bennett, E.O., "Bacterial Deterioration of Emulsion Oils," *Appl. Micribiol.* August, pp. 117–121, 1958.

Hill, E.C., "Microbial Infection of Cutting Fluids." *Tribology Internatl.* February, pp. 49–54, 1977.

Hunz, R.P., "Water-Based Metalworking Lubricants." *J. Am. Soc. Lubr. Eng.* October, pp. 549–53, 1983.

Isenberg, D.L. and Bennett, E.O., "Bacterial Deterioration of Emulsion Oils." *Appl. Microbiol.* August, pp. 121–125, 1958.

Passman, F.J., "Microbial Problems in Metalworking Fluids." *J. Soc. Tribologists Lubr. Eng.* May, pp. 431–33, 1988.

Templeton, L.L. and C.P.L. Grady, Jr., "Effect of Culture History on the Determination of Biodegradation Kinetics by Batch and Fed-Batch Techniques." *Res. J., WPCF.* 60(5), pp. 651–658, 1988.

Tortora G.J., B.R. Funke, and C.L. Case. *Microbiology: An Introduction*. The Benjamin/Cummings Publishing Company Inc., Redwood City, CA, 1995.

Wheeler. H.O. and Bennett, E.O., "Bacterial Inhibitors for Cutting Oil." *Appl. Microbiol.* December, pp. 122–126, 1955.

# 60 POLLUTION PREVENTION AND WASTE MINIMIZATION: DANA'S WAY

**John Clark**
Dana Corporation
2100 West State Street
Fort Wayne, Indiana 46801

Dana Corporation, a Fortune 200 Company, supplies components to the automotive industry. The Columbia City, Indiana, Brake Products Plant activities are the basis of this presentation. This facility has completed an aggressive minimization and pollution prevention program over the last six years.

In 1990, senior management of Dana Corporation from the Weatherhead Division assigned a three-person task force to evaluate the eight plants, their processes and environmental practices. The assignment was to eliminate or minimize the cost and liability of the manufacturing processes as it pertains to environmental issues. The assignment also included reviewing the facilities to assure that the community and the worker were existing in a safe environment. The personnel assigned to this task force were the personnel responsible for facility engineering, environmental compliance, and capital planning for the three largest facilities. This cross-functional team allowed a review of all plants by personnel with strong experience in operations, environmental, and manufacturing. The team visited each site and evaluated the processes and practices being used, the wastes being generated, and the disposal activities of the wastes. The team then produced a recommended five year plan that include process changes, new capital equipment with projected costs and savings, and corrective action plans for issues in which no process change was recommended. This plan was presented to the entire Weatherhead Division Staff, including the Division General Manager, Controller, Engineering Manager, and all Plant Managers. The plan was reviewed by the Division Staff and implemented June of 1991.

The responsibility for implementing the plan fell back to the task force personnel and their counterparts at the individual plants. They had authority to plan the changes, budget them into the capital plan, and between themselves and the Plant Managers, monitor and report to the Division Staff on the status of the plan. The Columbia City Brake Plant began implementation of the plan in mid-1990 and completed it in 1996. The following are the actions taken by this facility under this plan.

At the beginning of the plan the facility maintained seven trichloroethylene vapor degreasers, utilized methyl ethyl ketone as a thinner for a painting operation, used anhydrous ammonia in a blueprint machine, and reclaimed only 70% of the oil trapped in the metal turnings.

The first process change was a modification to the paint operation that utilized methyl ethyl ketone (MEK) as a thinner and drying aid. This operation was very simple; it required placing a paint stripe around a hose at a set distance from the end of the hose as an assembly guide. The hose was rolled through a device where it contacted a spinning wheel which was coated with paint. This coated wheel placed a 1/16″-wide stripe on the circumference of the hose. Testing was conducted to modify this process; it was determined that a standard water-based paint would not dry properly and caused the stripe to smear, making the stripe ineffective as an assembly guide. Working with the local Sherwin Williams paint distributor, a special mixture of a water-based latex paint was produced. This mixture, combined with an infrared heat source on the discharge chute of the striping machine, corrected the drying issue, allowing the process change.

This change allowed the facility to eliminate 5,700 lb of MEK air emissions, as well as disposal of 880 lb of hazardous waste per year.

*52nd Purdue Industrial Waste Conference Proceedings*, 1997, Ann Arbor Press, Chelsea, Michigan 48118. Printed in U.S.A.

The next project was to replace the old print coping machine which used anhydrous ammonia as a media. Contacting different companies in the office machine industry showed the facility that alternative equipment was available which utilized a less volatile and hazardous material. The copier was purchased and installed, eliminating the need for anhydrous ammonia. This equipment change eliminated 900 lb of air emissions.

With completing the paint striping and copier replacement projects at low capital costs, but still succeeding in meeting the plans goal, strong support was fostered in the plant that other issues could be corrected as well. Plant personnel became active in the larger projects, allowing a more effective implementation of the plan.

The third phase of the plan was to eliminate the use of trichloroethylene as a cleaning fluid. To accomplish this, the facility more actively investigated water based washing systems. A cross-functional team from the plant including Engineering, Skilled Trades, and Production personnel was established to choose the equipment and oversee runoff approvals. The team evaluated the facility layout, present processes, and customer cleanliness specifications, and made the determination that several types of washers would be required. Small part-on-part tumbling or vibrating media type units would be needed for deburring of Cold Forming, Punch Press, and cutoff saw operations. Multiple stage units that wash, rinse twice, and dry would be required for final operations to assure cleanliness specs were met. The final determining factor was that floor space was an expensive and critical commodity for each facility; therefore: a small footprint for any new cleaning system was mandatory.

Two different prototype units were designed and built by facility personnel for in-house testing and concept approval. The team evaluated the prototypes, in-house test results, and testing conducted at the washer manufacturers to reach the final determination of which units to purchase. The team evaluated systems from six suppliers, and pilot units were purchased from three of them. After actual operation in the plants, one of the pilots was returned to the manufacturer and capital requests were formulated for the purchase of the balance of the required units.

The testing of water-based cleaning equipment began in 1990 and was completed in 1991. Installation of the machines purchased to replace the trichloroethylene washers was begun in 1991 and completed in 1994. The final installation included five steel shot media Roto Finish vibrating bowls, one part-on-part Ultramatic vibrating bowl, three part-on-part Hurricane rotating drum spray washers, and five Bowden rotating rack three-tank washers. The vibratory bowls are utilized as parts to remove the residual oil from the extruding process from our supplier as well as removing any burrs from the saw operation. The Ultramatic vibrating bowl is used to remove oils from our Punch Press operations, as well as any forming burrs. The three part-on-part rotating Hurricane units are used for washing of oils and for burr removal from initial machining and Cold Forming operations, prior to secondary machining operations. The Bowden rotating rack units are used to remove oils from fittings between operations, or as a final operation.

With the installation of these machines, all units using trichloroethylene were eliminated. The elimination of this material from our processes reduced our air emissions from 41 tons to 1.5 tons per year. The facility also reduced its hazardous waste disposal by 42,000 lb per year.

The change of process eliminated the hazardous material but created new waste streams, both hazardous and nonhazardous. In utilizing water as a cleaning media, the facility found that the solution becomes soiled very quickly and servicing of the solution must occur. The servicing of these units increased our disposal of waste oil and water from 34,710 gallons to 275,819 gallons. The new process also produced a lead-contaminated filter element that required special disposal operations. Even with the additional disposal issues, the payback of changing the process from trichloroethylene to water-based cleaning was under five years.

With all the new washers in operation for over two years, we have found a new area to tackle for minimization. We have began using ultrafiltration on all wastes from the washers. Two types of ultrafiltration are used to improve the process at this time. One is to filter a solution from a washer and then refill the system with that solution, adding more chemical to bring it into the proper concentration range for cleaning. In the second, when the solution becomes too soiled to properly be reused, we then dump it in our waste system and recycle it with a tighter membrane. This tighter filtered material is then stored in a tank for feeding back out to the plant to recharge first operation washers, use in floor scrubbers, and for mopping production floors. With these two methods in place we have been able to reduce our waste oil and water disposal to an estimated 90,000 gallons in 1997. We are working with Bowden Industries on a concept they have developed called the Split-Matic. This unit is an add-on device that is designed to split the oils and detergent apart by inducing heat. The water vapor from the process is then condensed and used in the washer as makeup water, while the detergent is reinstituted into the main wash tank to maintain the detergent concentration. We feel that we can continue to evaluate methods to minimize this waste stream until we reach 1992 levels of 34,715 gallons of an oil and water mix being generated. With the present system in place, the only emission from the facility is the exhaust of the evaporators that are used to split off the oils and detergents left in the ultra-filtered water. These evaporators are used to vaporize the balance of the water recycled into washers, scrubbers, and mop water used by plant personnel. Any oils or detergents skimmed off the evaporators is collected as the waste oil and water wastes we send for off-site disposal. The oils found in this waste stream are reclaimed at the disposal site and used in their processes of oil blending. The waters are then discharged to their local POTW for final disposal.

Further assistance in minimizing the waste oil and water disposal was found when we consolidated cleaning materials in the plant and found that changing all units to alkaline-based cleaners eliminates the lead contaminated filter and the need for disposing of it. With the consolidation of the cleaners we also find that our Ultra filter system is becoming more efficient.

The last project undertaken was to modify and upgrade our metal shaving handling equipment. The old equipment at the facility was limited by the types of shaving that could be processed through it. New automated spinners with chip-crushing systems as well as modified manual units allows us to centrifugally separate approximately 97% of all the coolants contained in the machined shavings. The new equipment allows us to spin 95% of all the shavings generated at the facility. Design changes are presently underway to allow the other 5% to be separated as well.

This process change allowed us decrease the purchase of cutting oils for the facility by 39% from 31,150 gallons per year to 19,150 gallons per year.

The next area being addressed under waste minimization is the DOO8 lead contaminated waste generated by plant cleanup activities. This is the only toxic waste generated at the facility through the normal production process. The waste is made up of residuals from daily plant cleanup from floor scrubbers, mops and buckets, and absorbent tubes used around old machine bases. This material contains small shavings of brass that, during the preparation process for TCLP testing, allows lead to leach out at approximately 5 PPM. This then determines the waste is hazardous, and it is handled as such. The process we are looking into to render this waste non-hazardous is a patented chemical reaction process designed by RMT of Madison, Wisconsin. RMT, a consulting firmed used by Dana Corporation, has instituted this process at other facilities with excellent results. We hope to complete the planning stages on the project by May 1, 1997. Implementation of the project should be completed by August 1, 1997.

If the project works out as predicted, the facility will eliminate all disposal of process generated hazardous waste and will potentially become a small quantity generator. This means that the facility's hazardous waste disposal will drop from 50 tons per year to less than 1 ton per year. This is a significant achievement for a facility that in 1989 was listed as a Treatment Storage and Disposal Site.

Waste minimization is a practice that can be implemented in any type of manufacturing environment. It takes commitment from all levels of the corporate structure, from the president to the shipping clerk. If it is not supported or embraced as a method to reduce costs, protect the worker and the community, it will not succeed. If it is implemented with an open mind and a committed spirit, success is inevitable.

# INDEX